# SELECTION
## The Mechanism of Evolution

GRAHAM BELL
Molson Professor of Genetics
and Director of the Redpath Museum, McGill University

**International Thomson Publishing**
Thomson Science
New York • Albany • Bonn • Boston • Cincinnati • Detroit • London • Madrid • Melbourne
Mexico City • Pacific Grove • Paris • San Francisco • Singapore • Tokyo • Toronto • Washington

Printed in the United States of America

For more information contact:

Chapman & Hall
115 Fifth Avenue
New York, NY 10003

Chapman & Hall
2-6 Boundary Row
London SE1 8HN
England

Thomas Nelson Australia
102 Dodds Street
South Melbourne, 3205
Victoria, Australia

Chapman & Hall GmbH
Postfach 100 263
D-69442 Weinheim
Germany

International Thomson Editores
Campos Eliseos 385, Piso 7
Col. Polanco
11560 Mexico D.F.
Mexico

International Thomson Publishing - Japan
Hirakawacho-cho Kyowa Building, 3F
1-2-1 Hirakawacho-cho
Chiyoda-ku, 102 Tokyo
Japan

International Thomson Publishing Asia
221 Henderson Road #05-10
Henderson Building
Singapore 0315



2 3 4 5 6 7 8 9 XXX 01 00 99 98 97

**Library of Congress Cataloging-in-Publication Data**

Bell, Graham, 1949 --
Selection : the mechanism of evolution / by Graham Bell.
p. cm.
Includes bibliographical references (p. )
ISBN 0-412-05521-X (hc : alk. paper)
1. Natural selection. 2. Evolution (Biology). I. Title.
QH375.B44 1996 95-17458
575.01'62--dc20 CIP

*Visit Chapman & Hall on the Internet http://www.chaphall.com/chaphall.html*

To order this or any other Chapman & Hall book, please contact **International Thomson Publishing, 7625 Empire Drive, Florence, KY 41042.** Phone (606) 525-6600 or 1-800-842-3636. Fax: (606) 525-7778. E-mail: order@chaphall.com.

For a complete listing of Chapman & Hall titles, send your request to **Chapman & Hall, Dept. BC, 115 Fifth Avenue, New York, NY 10003.**

**For Susan**

# Contents

## 3. Selection on Several Characters



### *3.A. Selection Acting on Different Components of Fitness*

## 5.B. *Selection Within a Single Diverse Population: Frequency-Dependent Selection*



## 5.C. *Several Populations: Kin Selection and Group Selection*

# Introduction

This book has been written to make a point and to fulfill a need. The point is that the importance and the distinctiveness of the process of selection have been undervalued by most biologists. There is, consequently, the need for a book that describes the principles of selection in a simple but reasonably comprehensive way.

## *Selection Is a Distinct Kind of Process*

Although we are now well into the second century of Darwinism, the theory that Darwin and Wallace announced in 1858 has not yet made much progress beyond a small coterie of professional biologists. The reason is that it is jarringly unfamiliar to our normal experience of how things come to be. Few of us would be able to design a light bulb or a lathe, still fewer the computer and its attendant software with which this sentence is being written. But we all have a clear idea of what is meant by "design", and we readily, too readily, transfer this notion to the natural world. A light bulb or a lathe are prefigured in the mind, and constructed according to a plan. It is entirely reasonable to assume that beetles and daisies must be constructed after the same fashion, especially because they are much more complicated than anything that human ingenuity has so far managed to devise. There is, however, a second route to complex organization, through the selection of random variants that propagate nearly exact copies of themselves. It is of very little consequence in our daily lives, because if is so much more laborious and expensive than deliberate design. However, it is another way of constructing things. Indeed, so far as I know, it is the only other way of constructing things that we have ever been able to imagine. Selection is not merely a theory antagonistic to some other possibility, in the same way that oxygen and phologiston or open and closed circulatory systems are antagonistic theories; it is one of the only two categories of theory competent to explain the living world. It is as important as all other scientific theories combined, because it offers a different *category* of explanation.

Of course, it might be wrong, and it could then be relegated to the footnotes of historically-minded textbooks of biology. There are, indeed, two influential schools committed to the footnote position. One consists of religious people who with different degrees of stridency assert on their own authority that the world was designed and created by an intelligence superior to ours. The other consists of a mixture of different kinds of people, mainly scientific, who think that life embodies a principle of inherent self-organization, with selection playing a minor role, if any. These two schools are mistaken, root and branch. Living complexity cannot be explained except through selection, and does not require any other category of explanation whatsoever. This is the proposition that this book sets out to assert, to elaborate, and to exemplify.

## *The Importance and Distinctive Character of Selection Are Widely Ignored*

It may seem a little strange to insist on this position so strongly. There are good reasons for doing so. The weaker reasons are supplied by the continual slight assault on Darwinism—the discovery every two or three years that the theory is outdated, logically invalid, overtaken by recent discoveries—underwritten by the popular press. These mosquitos must simply be swatted as they swarm. A much stronger reason is the near-invisibility of selection in the curriculum of scientific education.

This has the very serious consequence that the great majority of students in science enter their university programs with little if any training in selection. They are required to master calculus, inorganic chemistry and the cranial nerves of the dogfish; but beyond a vague and fleeting memory of peppered moths they have no acquaintance with the basic organizing principle of biology. I do not exaggerate; I have taught these students for twenty years. One might expect, for example, that in Britain, the cradle of evolutionary biology, natural selection might be accorded an honored and conspicuous position. But in the published curricula of the programs that students follow preparatory to university, it is no more prominent than aquaculture. In the Canadian system with which I am familiar, about half our introductory class has never been introduced to natural selection, or at least has retained no recollection of it.

The Evolution course is the core of any good university program in biology. Unfortunately, there is often only *one* course in evolution, which tends to attract all the bits and pieces for which there are no separate courses. This is reflected by the textbooks used in these courses, full of good stuff, but resembling a little those compendia of games that can be used to play everything from ludo to backgammon. The few lectures on selection at the start of the course are quickly forgotten in the rush to get through biogeography, classification, embryology, palaeontology and large parts of ecology and animal behavior before the end of term. Only a very strong-minded lecturer will insist, week after week, that the material on biogeography, or phylogenetics, or whatever is being taught that week, can be understood in terms of selection, and in no other way.

A further weakness is that when selection is described, it is often treated as one of several possible mechanisms of evolution. In the best textbook currently available, for example, random genetic change is treated at about the same length as selection, and it is quite common to see the neutral theory of molecular evolution being treated on a level with—that is, at the same length as—the theory of natural selection. This is surely unjustifiable. Evolution is the development and maintenance of complex organization that functions appropriately in given conditions. There are many forces that hinder evolution—mutation, sampling error, immigration, and so forth—but selection is the only process that *causes* evolution.

The treatment of selection in textbooks usually follows more or less the same course. There is an introductory section on genetics, sometimes even outlining the chemical structure of DNA, followed by an account of Hardy–Weinberg equilibrium, a page or two of population genetics, and of course the history of the peppered moth. The whole is illustrated by a picture of Darwin, looking stern, as well he might. This is a caricature, of course, but by no means an unrecognizable one. Thousands of students have left courses on evolution with the vague impression that selection is something to do with the Hardy–Weinberg law, and study for the examination by trying to remember which it is that adenine pairs with.

In short, current texts give too little space to selection, give an unbalanced treatment of other kinds of processes, and fail to explain clearly the distinctive character of the selection process. Nor is there any alternative book devoted to giving a reasonably comprehensive account of selection, to which the student can turn. Evolutionary biology is in the extraordinary position that the best general accounts of the central principle of the subject are given in popular essays and books by Richard Dawkins and Stephen Jay Gould. Their books are the recommended readings for the course that I teach. There are excellent accounts of particular aspects of selection by a variety of other people (John Maynard Smith, George Williams, and Mark Ridley are the other three names on the course list), but there is nowhere else that students can get a general account of selection. That is the main reason for writing this book.

## *The Content and Arrangement of the Book*

This book is about selection. It is not about any of a host of related topics. It is not primarily about evolution; that is to say, it does not mention most of the subjects that a book on evolution would deal with. In particular, it is not about adaptation. This is because I am more concerned to explain how selection can act, in general, than to describe how it has acted, in particular cases. Consequently, I shall refer only very briefly, or not at all, to many of the most familiar and important issues of evolutionary biology, which are discussed adequately in other books.

The usual approach to selection in modern accounts is through population genetics. I have not used this approach, for several reasons. The first is that it is difficult to give an adequate treatment of population genetics without a great deal of algebra, which I have found to be a very ineffective way of teaching the theory of selection. Several excellent algebraic treatments are already available; I have used only the simplest and most basic mathematical arguments. A more important reason is that the genetic details tend to overgrow the basic principles of selection involved. Almost all texts of population genetics deal with diploid, sexual, multicellular organisms whose genetics are rather complicated. These complications are quite unnecessary and, tend to obscure the underlying mechanism of selection. I have, therefore, taken the opposite approach of framing the argument in terms of asexual, haploid microbes, where only the simplest rules of inheritance are involved, and treating any other kind of organism as a special case. A final reason is that selection is a much broader topic than current population genetics. The great bulk of population genetics is concerned with a process that I shall refer to as sorting, the selection of one or a few types from some range of pre-existing variation. The most important consequences of selection, however, involve the appearance of new types over long periods of time. Population genetics, as currently understood, is therefore an inadequate foundation for a general account of selection.

**Selection Experiments.** The emphasis on the process of selection leads to an emphasis on the experimental investigation of evolution. There are many different kinds of experiment, and much of the experimentation in evolutionary biology uses concepts and techniques that are widely used in other fields. The unique character of selection, however, leads to a unique kind of experiment, in which populations evolve under known conditions in a controlled environment. The *selection experiment* is the hallmark of experimental evolutionary biology; it has no counterpart in other sciences. I have tried, whenever possible, to illustrate general points by means of appropriate selection experiments, and have neglected most other kinds of experiment. The reason is that it is only through selection experiments that the mechanism of evolution can be studied directly. There is a serious objection to such experiments: how do we know that the processes occurring under highly simplified conditions in the laboratory or on the experimental farm can account for the adaptations that we see in natural populations? Selection experiments may establish the possibility of a certain outcome of selection, but they can scarcely prove that any particular adaptation did in fact arise in the same way. The point is entirely valid, and the analysis of particular adaptations is the concern of comparative biology. The value of selection experiments is that they test the validity of broad general principles of selection that must be invoked to explain the adaptedness of natural populations. There is no other way of doing this. An holistic approach to evolution that refers despairingly to the almost unimaginable complexity of natural communities can easily slide into obscuranticism, and has, I feel, led to the neglect of selection experiments in most treatments of evolution; a glance at the contents of any current text on evolution will confirm this. I hope that this book helps to redress the balance.

**The Organization of the Book.** The book is constructed in an unusual manner, consisting of two parallel parts. The first is the plain text, an almost uninterrupted prose argument that can be read in isolation and will, I hope, be palatable to students, especially undergraduate students. The second part consists largely of additional material, figures, and references to the technical literature. I have made this division because I think that a reasonably extensive bibliography of experimental evolution will be useful to professional biologists, whereas the continual interruption of the text by citation (e.g., Smith, Jones, and Brown 1978) is often distracting.

The ideas that have informed my views on selection have been developed by theoreticians to whom I have given very short shrift. It is possible that the reader might think that many of the views advanced here are original. This is not the case. I have attempted to synthesize a vast body of theoretical literature that is scarcely referred to, either in the text or in the reference list. To all who know that their ideas, or the ideas of illustrious predecessors, are being advanced without adequate acknowledgment, I offer my most sincere apologies, with the excuse that the theoretical literature has been treated at length in other works and would inflate the present book beyond endurance. I have attempted to remedy this defect to some extent by supplying a General Bibliography of major works not referred to in the text, that will direct the reader to the theoretical literature.

The plain text is divided into sections by a series of declarative sentences, in a manner that is now fairly familiar. The object is to systematize the most important principles and results of the theory of selection, and to emphasize the fact that selection can be treated in the same mechanistic, reductionist way as immunology or organic chemistry. I hope that this arrangement will also make it easier for students to learn the subject. Wherever possible, I have tried to find experiments that substantiate the assertion made by the section heading; this has often not been possible, and I have then sometimes inserted comparative material, though as sparingly as possible. On a few occasions, the experimental evidence goes against the assertion, where I have felt that the assertion is so likely to be true that the experimental results should be laid aside. Naturally, all of the assertions are provisional and liable to be upset by future experiments, but I do not think that the framework of the argument will be jeopardized.

# *Acknowledgments*

I owe a large debt to my doctoral students, who have never stinted the time to train me in evolutionary biology: Austin Burt, Vassiliki Koufopanou, Mark Chandler and Clifford Zeyl. I am especially grateful to Xavier Reboud and Jack da Silva for their patient instruction. The other denizens of the lab while I was writing this book were Hans Koelewijn, Patrick de Laguérie, Nick Colegrave, Torsten Bernhardt, Doug Collins, Shaun Goho, Rees Kassen, Arne Mooers, Ian Rae, Jessica Rabe, Rima Hallik, Jamie Bacher, Stéphane Duboeuf, Lori Pilkonis, Kathy Tallon and Yannick Ducharme. They have been an unfailing source of inspiration.

# SELECTION
## The Mechanism of Evolution

# 1

# *Simple Selection*

The main purpose of evolutionary biology is to provide a rational explanation for the extraordinarily complex and intricate organization of living things. To *explain* means to identify a mechanism that causes evolution and to demonstrate the consequences of its operation. These consequences are then the general laws of evolution, of which any given system or organism is a particular outcome. The very complexity that stimulates investigation, however, long obscured the possibility of a mechanical interpretation, and the richness of the subject matter continues to encourage a piecemeal and anecdotal description of nature, and inhibits the development of a strong central research program in experimental evolution. Nevertheless, to explain complexity it is first necessary to study simplicity; although the *products* of evolution are exceedingly complicated, the *processes* of evolution can be seen most clearly and manipulated most easily when these complications are stripped away. One way of asking, what is the most fundamental principle of evolution? is instead to ask, what is the simplest system capable of evolving?

We can imagine organisms that lack fins or photosynthesis or any of a thousand other attributes, but we cannot imagine organisms that are unable to reproduce, because they would lack both ancestors and descendants. The simplest type of organism would be one that reproduced, but did nothing else. There is, of course, no cellular organism of this kind, because cell growth and division are supported by multifarious processes of metabolism. Underlying the reproduction of cells and individuals, however, is the replication of the nucleic acid molecules that encode this metabolism and ensure its hereditary transmission. The simplest evolving system is therefore a self-replicating nucleic acid freed from its dependence on cellular machinery.

# 1. *RNA viruses are the simplest self-replicators.*

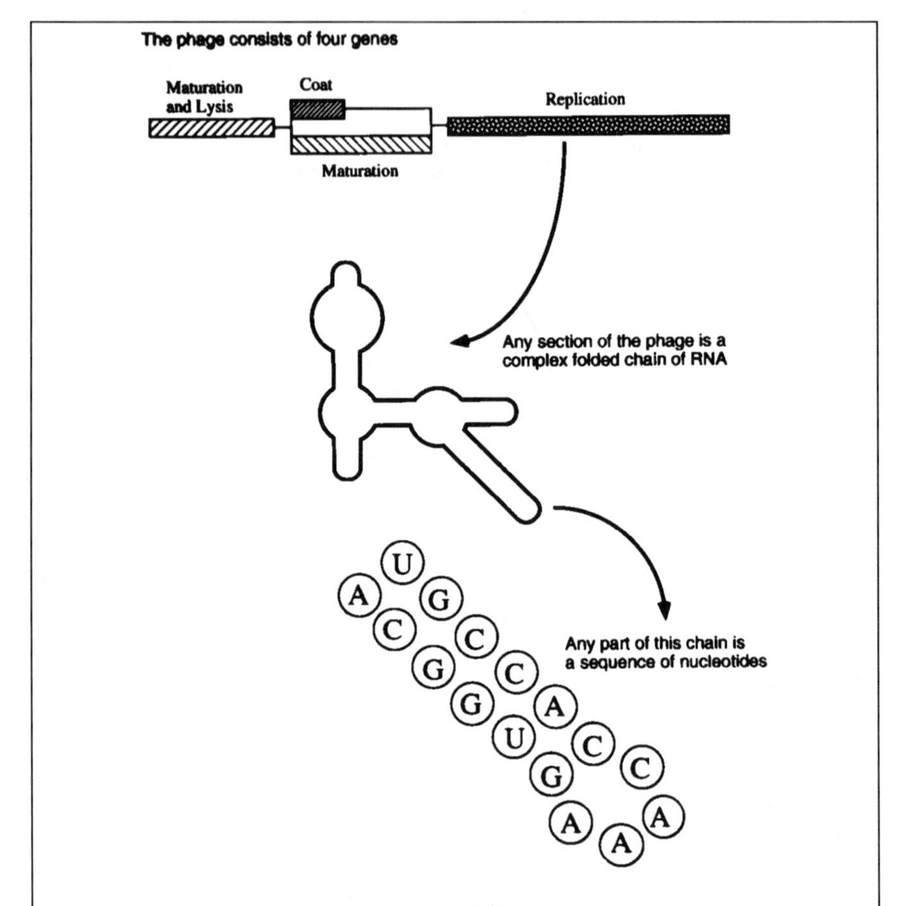


RNA phages are small organisms about 250 Å ($2.5 \times 10^{-8}$ m) in size. Q$\beta$ comprises just four genes: two encode proteins responsible for attachment, penetration, and lysis; one is a coat protein; and the fourth is the replicase. The second maturation gene is expressed by occasionally reading through the stop codon delimiting the coat protein gene. When magnified, the organism consists of a complex bundle of loops formed from a single strand of RNA. When magnified still further, the RNA strand is defined by a linear sequence of the four nucleotides Adenine, Uracil, Cytosine, and Guanine. This sequence completely determines the properties of the organism—including self-replication—in a particular environment, such as the interior of a bacterial cell. The four genes of the phage are embodied by a sequence of somewhat more than 4000 nucleotides.

Q$\beta$ is a virus that infects bacteria. It uses RNA, rather than DNA, as its genetic material. In cellular organisms, RNA is used only as messages to decipher the DNA code, and there is no machinery for replicating RNA. The Q$\beta$ RNA encodes a number of proteins, including one that specifically catalyzes the replication of Q$\beta$ RNA—the Q$\beta$ replicase. The host cell provides the rest of the apparatus for producing the protein, and a supply of raw material from which new Q$\beta$ genomes can be constructed. The virus can thus be thought of as a small wormlike creature with an unusually simple morphology that is completely specified by the sequence of nucleotides in a single RNA molecule. Simple though it is, it can be simplified further. The viral genome encodes several kinds of protein used to transmit itself from one host cell to another, but not while replicating within the host cell. Once inside the cell, all that is needed is the replicase, together with a supply of nucleotides got from the host. Because the replicase can be isolated and purified from infected bacteria, these simple requirements can be provided in a culture tube. A solution of replicase and nucleotides provides a chemically defined environment in which Q$\beta$ RNA will replicate itself as though it were inside a bacterial cell.

## 2. *Exponential growth can be maintained by serial transfer.*

When the culture medium is inoculated with a small population of Q$\beta$, each RNA molecule immediately proceeds to produce copies of itself; the copies themselves produce copies, and the copies of copies do likewise, until the population is hundreds or thousands of times as large as the original inoculum. This process, however, cannot continue for very long. Because each copy of one of the original molecules is itself copied, the number of molecules increases *exponentially*. The finite supply of nucleotides, the building- blocks of RNA molecules, soon begins to become exhausted, and after a time no further growth is possible. Growth can be renewed only when a fresh supply of raw materials is provided. This is most easily done by using a small amount of the grown culture to inoculate a fresh culture tube. When this too has become depleted, a small amount is withdrawn and used to inoculate a third tube, and so on, until it is the patience of the investigator that has become exhausted. This is the technique of *serial transfer*. It is a simple and effective way of keeping a population in a state of perpetual growth.

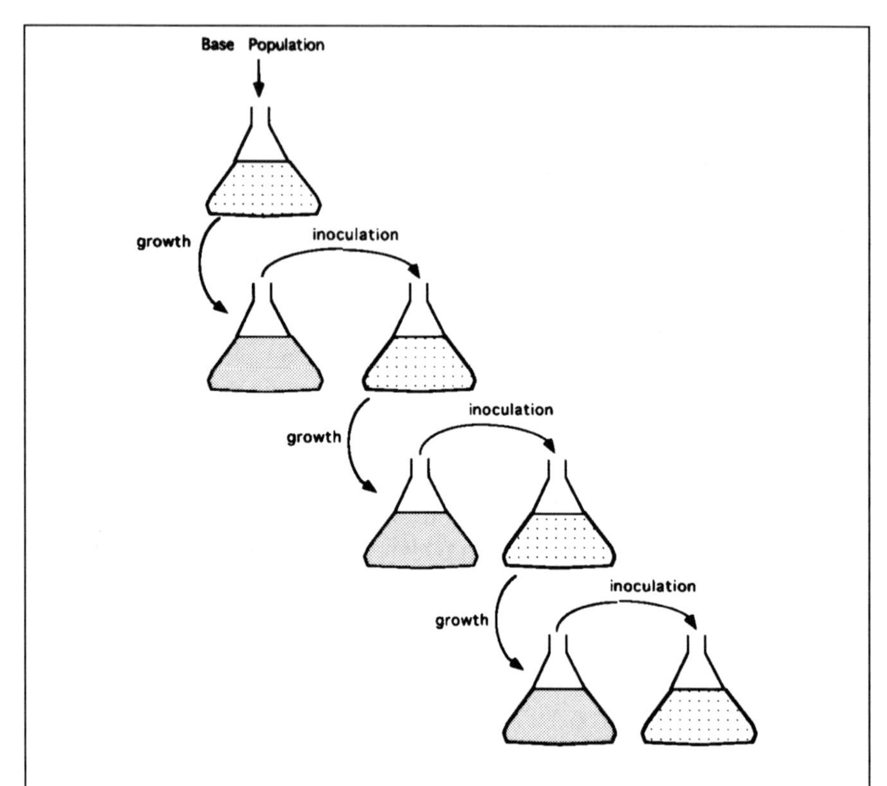


To grow an organism in batch culture, a number of individuals are inoculated into a volume of growth medium. They will continue to reproduce until the nutrients in the medium are exhausted and growth ceases. A sample of the population, however, can be withdrawn at any time and transferred to a fresh batch of medium. So long as this process is repeated, the population will continue to reproduce. The rate of growth of the population is set by the frequency of transfer and the volume of inoculum.

## 3. *Replication is always imprecise.*

As long as any raw materials remain, each RNA molecule produces copies of itself, through the mediation of the replicase. The replicase is not a perfect machine, however, and many of these copies will not be perfect simulacra of the parent molecule. In some cases, the wrong nucleotide will have been added at one or more positions along the growing RNA molecule; in others, extra nucleotides may have been inserted, or resident nucleotides deleted. More drastically, severely abnormal molecules may

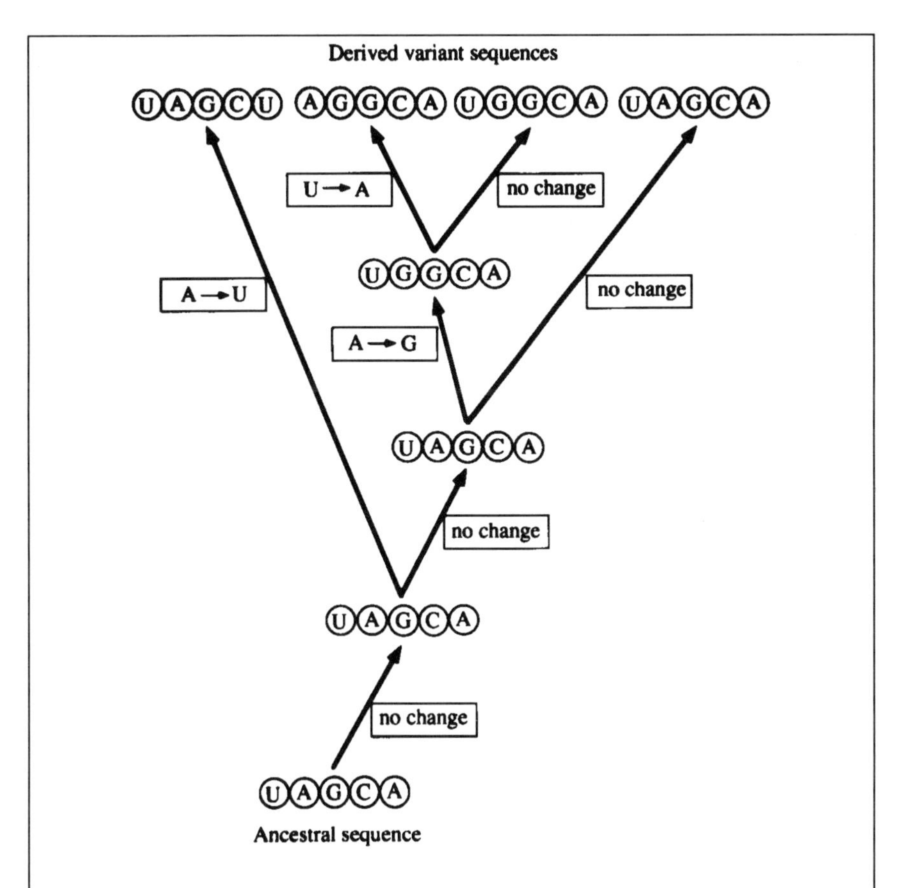


The relationship between ancestors and descendants can be represented as a family tree, also called a tree of descent or a phylogenetic tree. This diagram is the most direct representation of descent, as the copying of a sequence of nucleotides. The lines connect sequences that are copies, the arrow pointing from the ancestral to the descendant sequence. A line is thus a succession of copies. Any line, or any set of lines descending from the same ancestral sequence, is a lineage. When a sequence is copied incorrectly, the altered or mutated sequence is the ancestor of a distinct lineage. A descendant copy of this altered sequence, however, may itself be copied incorrectly, giving rise to a variant that differs in two positions from the original ancestral sequence. This alteration of alterations is responsible for the branching, treelike ramification of lineages through time.

be produced by the failure to copy large parts of the RNA sequence. Errors of this sort are quite often made during RNA replication, because the replicase has no ability to review its work and correct any errors that it has made. It has no ability to engineer molecules, either: errors are made at random, or, rather, without respect to their present or future utility. It is, therefore, not possible to create a large clone of RNA molecules, all of which are exactly alike; instead, a large population descending from a single founding molecule will consist of numerous variations on a theme, each of which is a sequence that differs to a greater or lesser extent from the original sequence. Populations of self-replicating RNA molecules are markedly diverse, because of the low fidelity of the RNA replicase, but the same principle applies to the replication of DNA, although this is a more rigorously controlled process, and therefore applies to clones of DNA-based organisms. Indeed, it is a very general principle that applies to any copying process. Information cannot be transmitted without loss: therefore, no message can be copied perfectly with certainty. Self-replication is always to some degree imprecise, and variation is a property of self-replicating systems that does not in itself require any special explanation.

Each variant sequence is itself copied, although of course imprecise copies are themselves copied imprecisely. A growing population of $Q\beta$ does not, therefore, consist merely of a diversity of sequences; it consists of a diversity of *lineages*, each propagating the altered sequence of its founder.

## 4. *Imprecise replication leads to differential growth.*

The technique of serial transfer is based on a very simple ecological principle: a population that grows exponentially rapidly exhausts the resources that support its growth. In diverse populations, this principle has an evolutionary corollary. Each variant lineage will tend to increase exponentially, but different lineages may increase at different rates. After any given interval of time, some lineages will have become more abundant than others, even if all were initially equal in numbers. Lineages that are increasing at different rates will therefore diverge in frequency, some becoming more common and others more rare. This can occur very quickly, because *different exponential processes diverge exponentially*. The diversity of the growing population is thus not merely a passive consequence of the infidelity of self-replication, but also reflects variation in the rates of self-replication among the variant lineages. The composition of the population will change through time as more rapidly replicating lineages become increasingly prevalent.

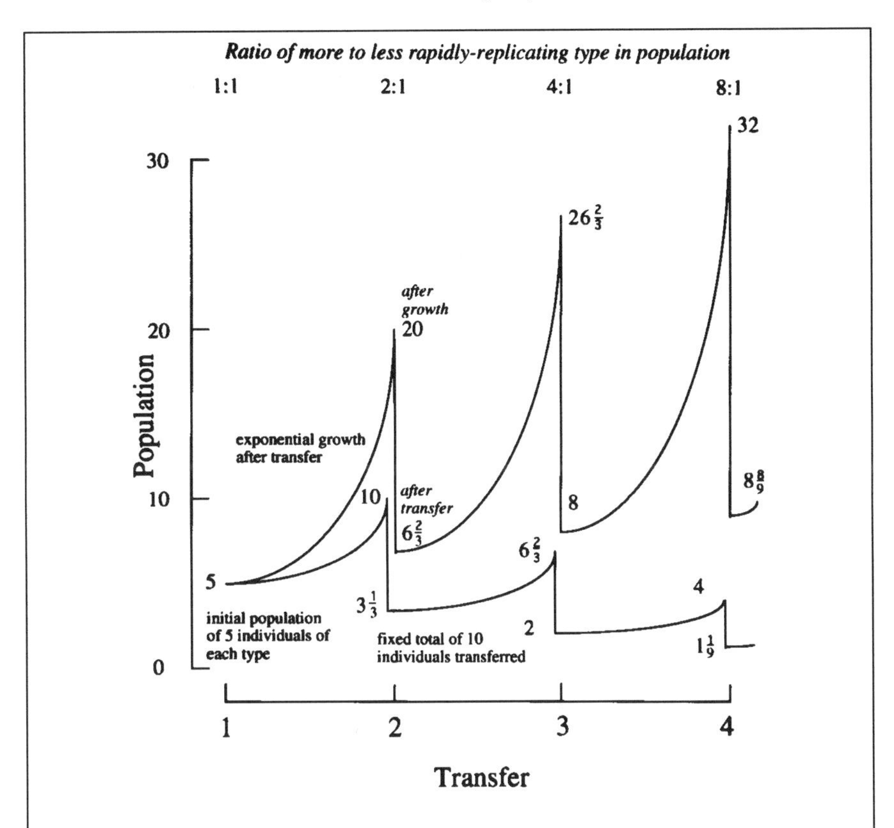


Two types of individual, one of which reproduces twice as fast as the other, are mixed in equal proportions, with five individuals of each type. After a period of growth, one type has increased by a factor of four, giving rise to 20 descendants; the other has doubled, giving rise to 10 descendants. The more rapidly reproducing type has thus increased in frequency from 1/2 to 2/3. After the period of growth, 10 random individuals are transferred to a fresh medium. Because the more rapidly reproducing type has a frequency of 2/3 among the individuals in the population after growth, it will have a frequency of 2/3 among the individuals chosen at random for the inoculum. Differential growth leads to a rapid divergence in frequencies as the more rapidly reproducing type spreads through the population.

## 5. *Selection acts directly on rates of replication.*

The serial transfer of self-replicating RNA molecules was the basis of one of the most remarkable experiments in modern biology. It was performed

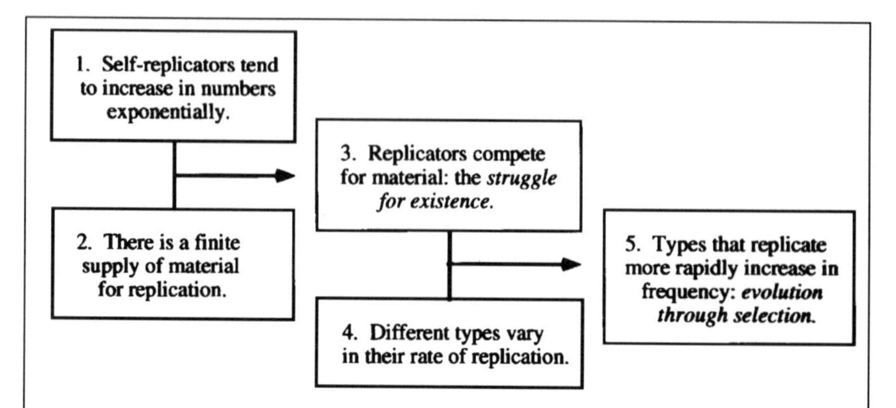


The simple Darwinian logic of evolution in a population of self-replicators.

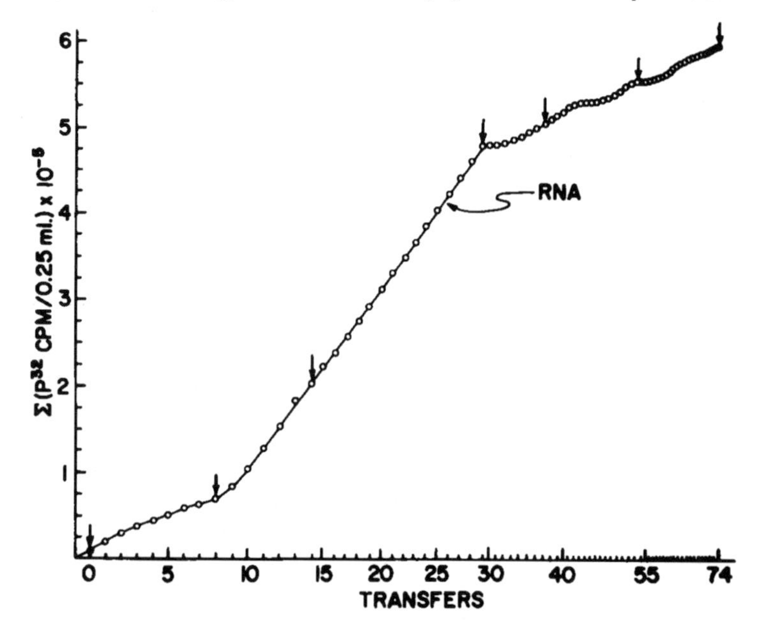


This figure represents the original $Q\beta$ experiment, from Mills et al. (1967). The $y$-axis is the cumulative incorporation of labeled radionucleotide into RNA and the $x$-axis the number of serial transfers. The cultures were initially allowed to grow for 20 minutes; this was reduced to 15 minutes after transfer 13, to 10 minutes after transfer 29, to 7 minutes after transfer 52, and thereafter to 5 minutes only. The changes in the length of the growth period, and the logarithmic scaling of the time axis, make it difficult to infer changes in the rate of increase of the population

(*Continues*)

(*Continued*)

of RNA molecules, but there is obviously a sharp advance between transfers 8 and 9. The early Q$\beta$ experiments were reviewed by Spiegelman (1971), who gives a wealth of background information about the system. There are subsequent reviews by Orgel (1979), Eigen (1983), and Biebricher (1983). Biebricher and Orgel (1973) reported similar experiments with the RNA polymerase of *E. coli*.

by a group of scientists in Sam Spiegelmann's laboratory at the University of Illinois in 1967. They designed their experiment around the following question:

> What will happen to the RNA molecules if the only demand made on them is the Biblical injunction, *Multiply*, with the biological proviso that they do so as rapidly as possible?

The elementary considerations that I have outlined are enough to provide an answer. The initial inoculum of viral RNA encounters a new and strange world, an unusually benign world in which it did not have to deal with the usual problems of parasitizing complex and hostile bacterial cells. Being provided with replicase and nucleotides, it begins to increase exponentially in numbers. This increase is soon checked by the finite supply of these resources. The tendency to increase in numbers while resources are in short supply creates competition, because not all can prosper: there will arise, as one might put it, a *struggle for existence*. But the growing population is necessarily diverse, and not all variants will be equally able to replicate themselves. Those that replicate more rapidly will increase in frequency, replacing their competitors. This process can be referred to as the *selection* of the more rapidly replicating types. Because each type tends to reproduce itself, selection will involve the replacement of some lineages by others or, in other words, will cause a permanent change in the genetic composition of the population, which constitutes *modification through descent*, or evolution. The experiment will result in the evolution of RNA molecules that are better *adapted* to the novel conditions of growth furnished by culture tubes: better able, that is, to replicate themselves at high rates in this novel environment.

This chain of reasoning was first forged by Darwin and Wallace more than a hundred years before the Q$\beta$ experiments. However, it predicts the leading result of these experiments very clearly: the emergence and establishment of variants with much greater rates of self-replication. This happens quite quickly. After about 70 transfers, amounting to about 300 rounds of RNA replication, the population consists of types able to replicate in their new environment about 15 times as fast as the original virus.

In the simplified world of the culture tube, many of the attributes that are essential to the success of Q$\beta$ as a parasite are no longer necessary. The virus no longer needs to be able to encode proteins for packaging its genome or for lysing the host cell, because the experimenter transfers it from tube to tube. It no longer even needs to encode a replicase, because that, too, is provided for it. All of the elaborate metabolic machinery deployed by living organisms has been eliminated, and we are left with their essence alone, a naked replicator with unlimited opportunity to replicate itself. It is not really very surprising that the experiment succeeds in illustrating the Darwinian thesis. The more important point is not that the rate of self-replication is an attribute that can be selected, but rather that *the rate of self-replication is the only attribute that can be selected directly*. There is no natural tendency that favors strength or beauty or cunning, or any other quality, as a quality in its own right. There *is* a natural tendency that favors high rates of self-replication, because more efficient replicators will tend to increase in frequency without the intervention of any external agency. So far as we know, this is the only natural tendency of this kind. The process of selection acts directly on rates of self-replication. It does not act directly on any other attribute or characteristic whatsoever. The direct response to selection is therefore always an increase in the rate of replication in given circumstances.

## 6. *Selection may act indirectly on other characters.*

The ability to grow, or to grow more efficiently, in a given environment cannot exist in itself without any external reference: it must be caused by morphological or physiological or behavioral attributes of some sort. These attributes can also be selected; indeed, we are usually far more concerned with them than we are with replication rate itself. They are, however, selected only indirectly; selected, that is, through their connection with altered rates of self-replication. Q$\beta$ has little in the way of morphology, physiology, and behavior, especially when cultured in tubes; being defined by a particular sequence of nucleotides, it is changes in this sequence that cause changes in rates of replication and that are selected indirectly as a consequence. One change that is easy to demonstrate, and whose effect on rates of replication is easy to appreciate, is that Q$\beta$ gets smaller. The intact virus with which the experiment was originally inoculated is a chain of 3300 nucleotides, encoding the proteins necessary for functioning as a transmissible intracellular parasite. After 70 transfers, the evolved variant, with its much greater rate of replication, was only 550 nucleotides in length. The reason is very simple: other things being equal, a smaller molecule will be replicated more rapidly than a larger one. In the benign environment of the culture tube, more than 80% of the viral genome is unnecessary, and variants that lack the unnecessary sequences are favored by selection by virtue of their greater rates of replication. What eventually remains is not a random fragment—random pieces of Q$\beta$ RNA are unable to replicate—but rather a minimal sequence that supports efficient replication. The experimenters, however, made no deliberate attempt to select small molecules. Rather, they contrived a situation in which novel types with higher rates of replication would evolve, and small size evolved as a side-effect of this procedure because of its correlation, in these circumstances, with greater rates of replication. All attributes of organisms evolve in this way, as an indirect consequence of selection for greater rates of increase. Their evolution thus represents an *indirect* or *correlated response to selection.*

## 7. *The indirect response to selection is often antagonistic.*

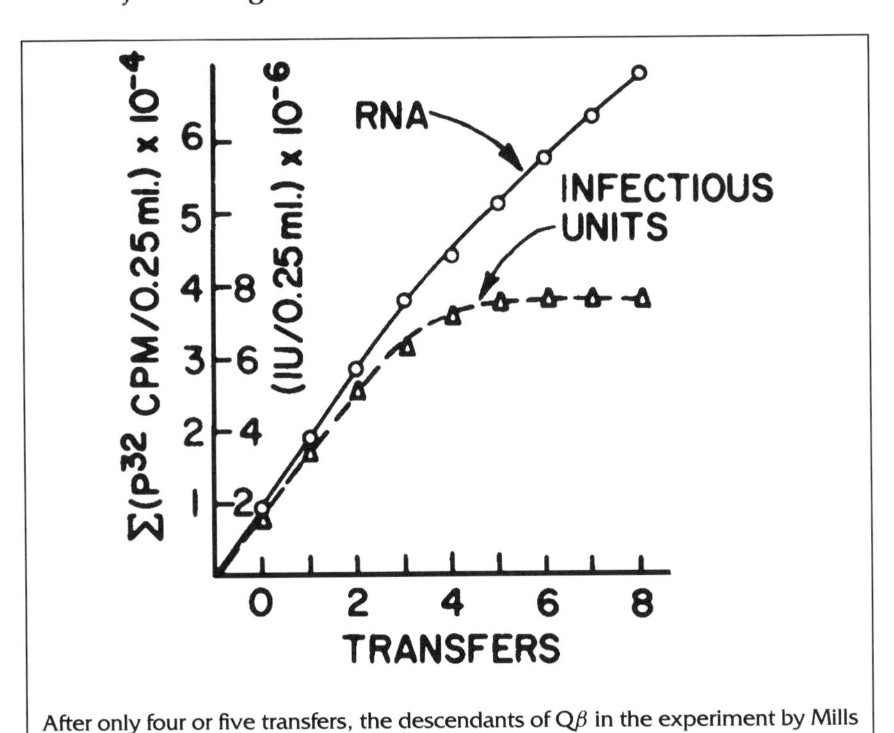


After only four or five transfers, the descendants of Q$\beta$ in the experiment by Mills et al. (1967) had completely lost their ability to infect *E. coli*.

In its natural state, Q$\beta$ cannot jettison 80% of its genome, because it would then be unable to infect bacteria. The intact virus is not very well adapted to growing in culture tubes; equally, the much shorter RNAs that evolve in culture tubes are not very well adapted to growing in bacteria. Indeed, the ability to infect bacteria is almost completely lost after four or five transfers. This is a type of correlated response: the loss of an ability that was formerly present, as the indirect consequence of selection for higher rates of increase in a novel environment. It can be distinguished as an *antagonistic* response, to recognize how selection that has enhanced performance in one environment has had the effect of reducing performance in another environment. Selection will cause progress, but it is a constrained and restricted sort of progress, with advance in one direction being associated with regress in others. Perfection is unattainable.

## 8. *Evolution typically involves a sequence of small alterations.*

After 70 or so transfers, Q$\beta$ has become fairly well adapted to the comfortable environment of the culture tube. What if we now make it uncomfortable? The difference between a bacterium and a culture tube is so great that it is difficult to interpret the evolutionary changes that take place, except in a very broad sense (the overall reduction in size, for example). Comfort is a general state of affairs. Once the population has approached an evolutionary equilibrium, however, with further change taking place only very slowly, it can be made uncomfortable in some specific way by introducing a defined source of stress. Any further evolution can then be attributed to that stress, and if the biochemistry and physiology of the stress and the response to it are known, then the evolutionary sequence of events that builds adaptation can be analyzed in great detail. RNA molecules can be made uncomfortable in a number of ways: one is to add to the culture medium a small amount of ethidium bromide, a substance that binds to RNA molecules and makes it difficult for them to replicate themselves. A very low concentration of ethidium bromide is at first sufficient to inhibit RNA replication almost completely. Within a dozen transfers—fewer than a hundred rounds of replication—however, the molecules have evolved a resistance to the drug, and by raising the dosage every dozen or so transfers, and thus maintaining a continual process of selection for resistance, it is possible to produce types able to grow at a concentration 20 times as high as that which was originally sufficient to suppress growth entirely.

The physiological basis of this adaptation is straightforward: the evolved molecules bind ethidium bromide less strongly. For this reason, their rate of replication exceeds that of the original type whenever the toxin is present in appreciable quantities. The reverse, however, is also true: in the original culture medium, with no ethidium bromide, the resistant variant grows less rapidly than the original type. This is another example of an antagonistic correlated response, showing how advance and regress may be coupled even when alternative environments differ only by a single defined chemical feature.

The simplicity of the system makes it possible to analyze the genetic basis of this physiological change. In one experiment, for example, resistance to ethidium bromide was caused by altering three nucleotides at different positions along the RNA molecule. Note that the genetic change is rather slight; from a structural point of view, the molecule has hardly changed at all. The functional change, on the other hand, is pronounced. Indeed, it is as pronounced as it can possibly be; the functional difference is life or death. There is no simple scaling between genotype and phenotype: very different genotypes may behave in very similar ways, whereas a trivial genetic change may have the most profound phenotypic consequences.

From a physiological or genetic point of view, this is as deep as any analysis can be pursued: nucleotide sequence is the ultimate character, from which all other attributes are eventually derived. From an evolutionary point of view, however, the fundamental issue is the manner in which these alterations occur in time. There are two possibilities. The first is that the original population was so diverse that it contains all possible sequences, including the sequence that happens to be successful when ethidium bromide is present in the growth medium. Evolution through selection is then simply a matter of *sorting* this initial variation, until the best-adapted sequence has replaced all others. The second possibility is that the original population includes only a small fraction, perhaps an extremely small fraction, of all possible sequences, so that the well adapted sequence that eventually predominates evolves through the *sequential* replacement of sequences by superior variants that have arisen during the course of the experiment. The evolution of resistance to ethidium bromide occurs sequentially, with the three altered nucleotides being substituted one at a time, so that the well adapted sequence that eventually evolves is built up in a step-like manner. It is easy to appreciate that this will almost always be the case. The number of possible combinations of nucleotides is so large that it is inconceivable that any but the tiniest fraction of them will be present in the original population. The evolution of any but the simplest modifications will generally involve the *sequential substitution* of several or many slight alterations, leading eventually to a state that was not originally present.

The variant with three alterations is resistant to ethidium bromide. Whether or not any or all of the three variants with a single alteration show any resistance has not been ascertained directly, but at least one of them must do so: otherwise, sequential change would be impossible. Similarly, once this variant has become established in the population, at least one of the two variants with two alterations must show increased resistance; and the evolved variant that is the final outcome of selection must be more resistant than its predecessors. Resistance can evolve only if the initial, susceptible genotype is connected with the evolved, resistant genotype through an unbroken series of intermediate states, each of which confers an advantage. Whether or not any particular attribute can evolve depends on the *connectance* of the system; on whether or not there is a route leading to a genotype specifying that attribute, along which a process of sequential substitution can lead the population, step by step, each successive alteration representing an improvement. The steps are, generally speaking, unit alterations in nucleotide sequence. This need not be taken quite literally. If two alterations are necessary to create a better adapted sequence, although neither confers any advantage by itself, the probability that both should occur in the same genome will be small, but such sequences may nevertheless arise quite often in large populations. Sequences with several predefined alterations, however, will be vanishingly rare, even in the largest populations. Evolution can leap only very small gaps.

The three alterations conferring resistance not only occur one after another, they also occur in the same sequence whenever the experiment is attempted. This seems likely to mean that only one of the alterations confers resistance when it occurs by itself; the other two increase resistance once the first has spread through the population. Their spread is therefore *contingent* on the prior establishment of the first alteration; until this has occurred, they do not cause increased adaptation by themselves. They do not, in other words, have independent effects on resistance to ethidium bromide; rather, their effect depends on the presence of another alteration. Whenever genetic changes interact in this way, the connectance of the system is reduced. It is not enough to show that a particular unit change increases adaptation, because it may do so only in a particular context, when other changes have already occurred. If this is the case, then the evolution of adaptation does not only require that the appropriate alterations should be substituted sequentially, but also that they should be substituted in a particular sequence, because only this sequence—or perhaps several sequences, out of the very large number of possibilities—involves a continuous increase in adaptedness.

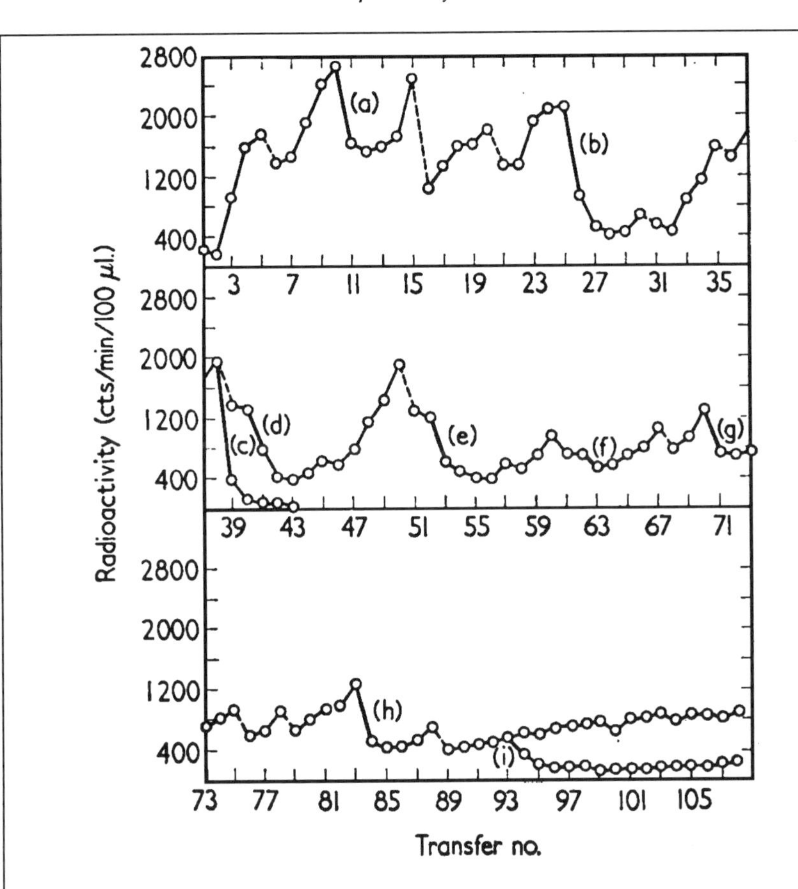


Saffhill et al. (1970) selected Q$\beta$ populations for resistance to ethidium bromide (Et Br). The base population was the V-2 variant, produced by selecting V-1, the variant evolving in the original experiment by Mills et al. (1967), for growth from extremely dilute inocula. This was grown in batch culture by serial transfer in the presence of a low concentration (2 $\mu$g/mL) of Et Br until, after 8 or 9 transfers, it grew about as fast in these conditions as the original V-2 did in the absence of Et Br. The concentration was then increased to 4 $\mu$g/mL, and by repeating this procedure a population capable of growing at 50 $\mu$g/mL Et Br was eventually selected, after 108 transfers. These increases in concentration of Et Br, each resulting in a transient decrease in replication, are indicated by letters a–h on the diagram.

(*Continues*)

*(Continued)*

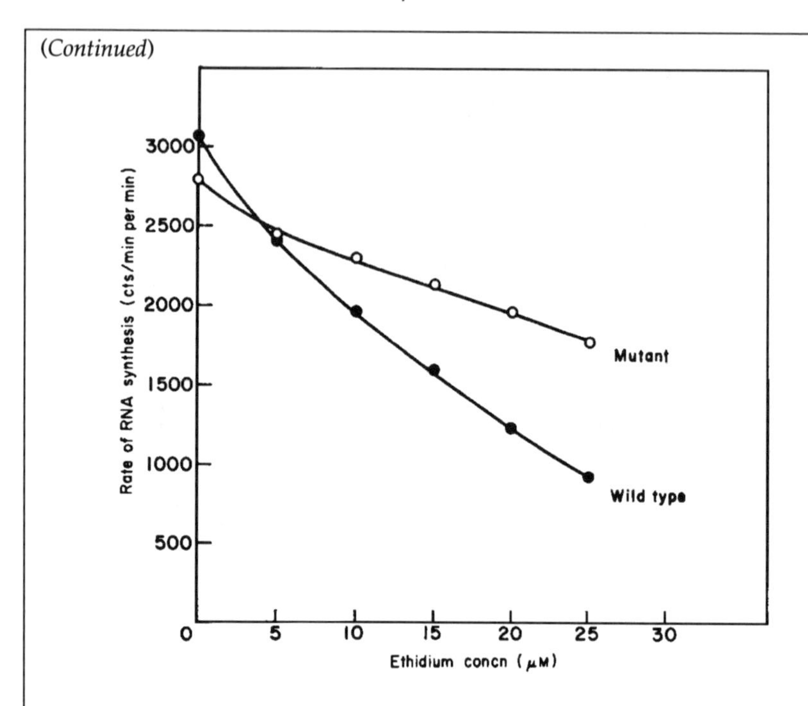


Kramer et al. (1974) selected for Et Br resistance in MDV-1, a 221-nucleotide molecule evolving from much shorter sequences. Very small inocula—to maximize the rate of selection—were grown in 15 $\mu$g/mL for 25 transfers. The evolved strain grew much better than MDV-1 when the concentration of Et Br exceeded about 2 $\mu$g/mL, but the original MDV-1 grows faster than the evolved strain when there is no Et Br in the medium.

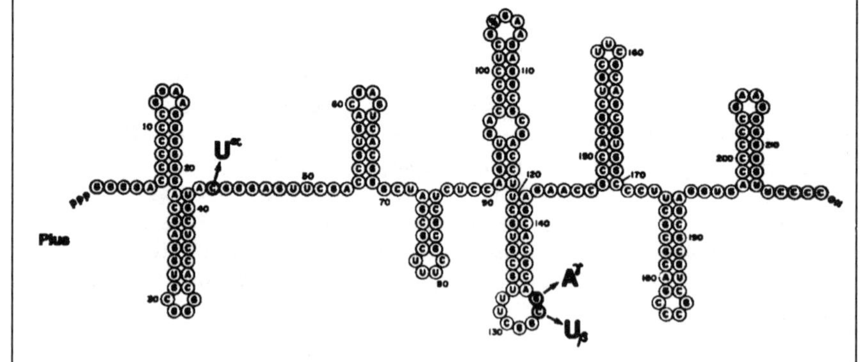


MDV-1 is short enough for the sequence changes underlying the evolution of Et Br resistance to be identified. Three alterations, without any change in total size,

*(Continues)*

*(Continued)*

reduce Et Br binding by about a third, giving the evolved strain its advantage over MDV-1. The figure shows the position of the three alterations, arbitrarily labeled $\alpha(C \rightarrow U)$, $\beta(C \rightarrow U)$, and $\gamma(G \rightarrow A)$, on the plus RNA strand.

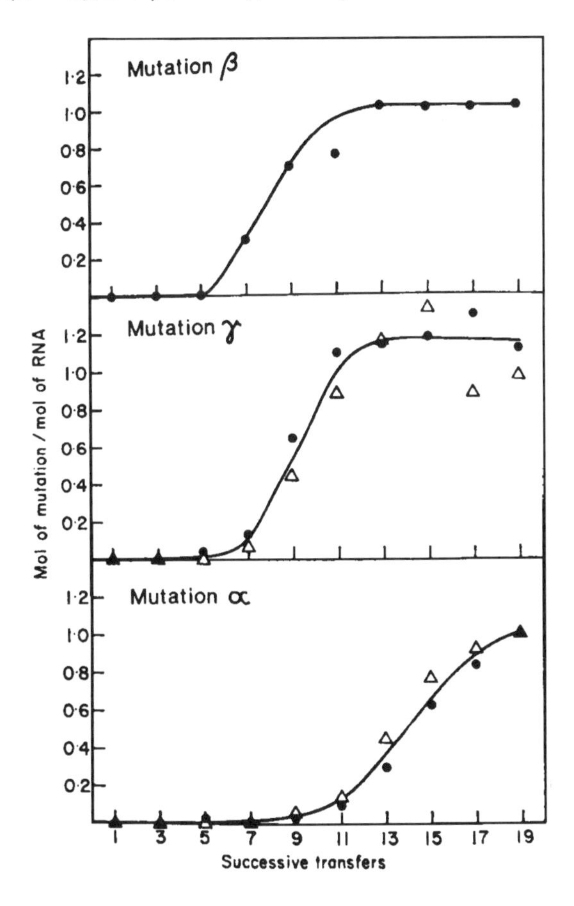


The sequential substitution of these mutations seems always to occur in the same order: $\beta \rightarrow \gamma \rightarrow \alpha$, so the population evolves from its original state to $\beta$, then to $\beta\gamma$, and finally to $\beta\gamma\alpha$. Evolution in these populations is presumably more complicated than this simple scheme, because the low fidelity of RNA replication will rapidly create a heterogeneous population containing several hundred variant sequences; the dynamics of the system are discussed in more detail by Davis (1991).

## 9. *The evolution of increased complexity is a contingent process.*

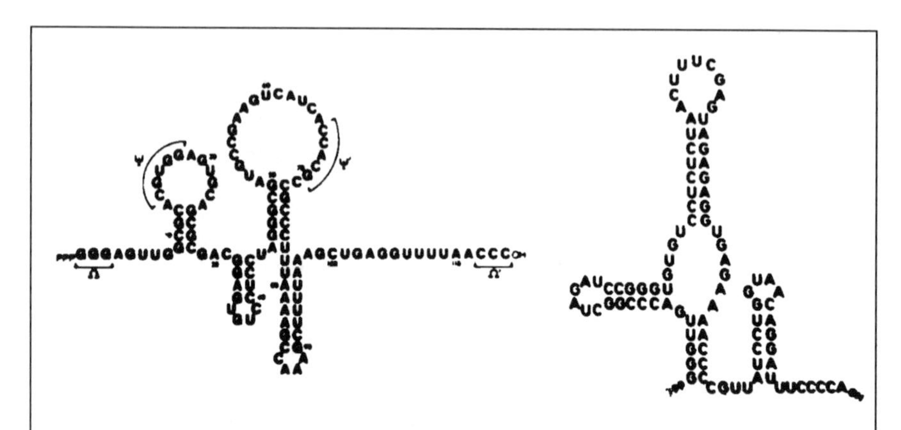

These are the nucleotide sequences and secondary structures of two of the self-replicators apparently evolving from very short sequences that arise spontaneously in solutions of nucleotides. The longer molecule is the "microvariant" reported by Mills et al. (1975); the shorter sequence, only 91 nucleotides in length, was found by Schaffner et al. (1977). Other sequences are described by Biebricher and Luce (1993). The evolution of these molecules de novo was first reported by Sumper and Luce (1975) and Biebricher et al. (1981a, 1981b), and has since been confirmed by Biebricher et al. (1993), although Hill and Blumenthal (1983) and Chetverin et al. (1991) have argued that they, or similar ancestral molecules, are present as contaminants in the original reaction mixture. Some remarkable experiments have been reported recently by Bauer et al. (1989) and McCaskill and Bauer (1993). They stabilized nucleotide triphosphate solutions in very long, narrow capillary tubes, and observed waves of evolving RNA spreading in both directions from nucleation sites representing a single initial template molecule of RNA. When no template was supplied, similar wave-fronts were detected, although they appeared only after a much longer initial lag.

It is not, perhaps, very surprising that the relatively complex intact virus evolves in a benign environment into a simpler structure. The objection may be raised, however, that the important point is not how complex things can become simpler, but how complexity can evolve in the first place. This is much more difficult to study. Nevertheless, it has been found that self-replicating RNA molecules will appear in the culture tubes, even if the cultures are not inoculated with $Q\beta$ RNA. It seems that they evolve from very short RNA sequences that form spontaneously in solutions of single nucleotides, or else are present in minute quantities as contaminants, and that have some very rudimentary ability to replicate themselves in the presence of the replicase. (These results are still somewhat controversial. Some people think that the self-replicating molecules found in the cultures have not evolved, but enter as contaminants. Here, I have accepted that any contaminants are much shorter than their eventual descendants.) They are themselves quite short, typically between 80 and 250 nucleotides in length, and, indeed, include the smallest and simplest self-replicators known: they resemble the defective viral genomes often found late in viral infections, being replicated by the replicase supplied by intact virus.

These spontaneously evolving RNAs are very variable; as a general rule, every experiment yields a different result, and there is little similarity between the sequences that evolve in replicate experiments. And yet all are efficient self-replicators, though some may be superior to others. I presume that in any given chemically defined environment there is some sequence superior to all others that would eliminate all its competitors if it ever evolved. There is, however, no sign that it ever does evolve. Selection does not seem to hack out a straight path to an optimal solution, giving the same result in every case. It seems instead to drive historically unique processes of evolutionary change, because of the contingent nature of sequential substitution in poorly connected systems.

Populations that are evolving in similar environments come to differ because evolution through selection has no foresight: it does not provide any way in which a population can map out the best route towards a predefined goal. There may be several, or many, states that represent a high degree of adaptation to particular circumstances, as there are many small RNAs that replicate well in the same standard conditions in culture tubes. A population of much less well adapted types cannot choose one of these states and proceed to evolve towards it. The route taken by the population will instead be contingent: it will depend on the fortuitous occurrence of variation on which selection can act, and the enormous range of possible variation means that different variants are likely to arise at different times in different populations. Each population is likely to evolve by a different route and arrive at a different destination. Evolution is modification through descent; selection can do no more than to build on the foundations laid by previous generations. The course of evolution may therefore be irreproducible, with populations diverging through time even though the environments in which they are being selected are the same. The diversity of self-reproducing RNAs evolving from simpler ancestors is an example of this pattern. All of them behave in much the same way, of course, being able to replicate fairly efficiently in culture tubes: it is the genetic basis of this adaptation that varies so widely.

A population may fail to become perfectly adapted because superior variants are poorly connected. The 550-nucleotide RNAs evolving by simplification from intact virus do not continue to evolve into the 100- or 200-nucleotide RNAs evolving from simpler predecessors, or at least do not do so in the short term, despite the fact that the shorter molecules have higher rates of replication. Evolution through selection may produce improvement, but it must produce continual improvement. Only those variants that possess an immediate advantage over their competitors can increase in frequency, and the route taken by the population must be traced out by a succession of variants, each superior to its predecessor. This blind and fumbling process may lead the population towards a state that cannot be improved upon by any feasible modification, even though superior designs exist. If progress can be made only by the demolition and wholesale reconstruction of the present design, then progress will simply not be made; the abrupt appearance of a novel creature superior to its ancestors is so unlikely that it can be neglected. Selection provides no blueprints, and no alternative engineer is available.

## 10. *Very improbable structures readily arise through the cumulation of small alterations.*

The process of sequential substitution is one of the most fundamental aspects of evolution, because it is responsible for the historical character of evolutionary change. It is also one of the least well-understood aspects of evolution. It has been thoroughly misunderstood even by well-educated scientists, with the amusing, if rather embarassing, consequence that every few years an eminent physicist or astronomer announces that they have discovered an elementary logical flaw in the theory of evolution through selection. The argument has often been framed in terms of the complexity of hemoglobin, the oxygen-carrying protein of vertebrate blood. A molecule of hemoglobin is a chain of 128 amino-acids, and functions properly only if the appropriate amino-acid is present at each place along the chain. If evolution depends on the selection of variants arising by chance, what is the probability that this sequence of amino-acids arose by chance, so that it could subsequently be selected? There are 20 different kinds of amino-acids, and therefore $20^{128}$, or about $10^{200}$, sequences 200 amino-acids in length. This is a very large number. If the surface of the Earth were filled a metre deep with protein molecules 128 amino-acids long; if all the molecules had different sequences; if every molecule changed its sequence at random every second, without the same sequence ever recurring; and if this process had been going on since the origin of the Earth, nearly five billion years ago—then it is still exceedingly improbable that the hemoglobin sequence would yet have been generated. Selection is therefore powerless, because the appropriate variation will never arise. Evolution (it is concluded) requires a guiding hand, supplied in some versions by the inheritance of acquired characters, and in others by divine intervention.

A good way of understanding the fallacy in the argument is through a word game in which one word must be changed into another by altering

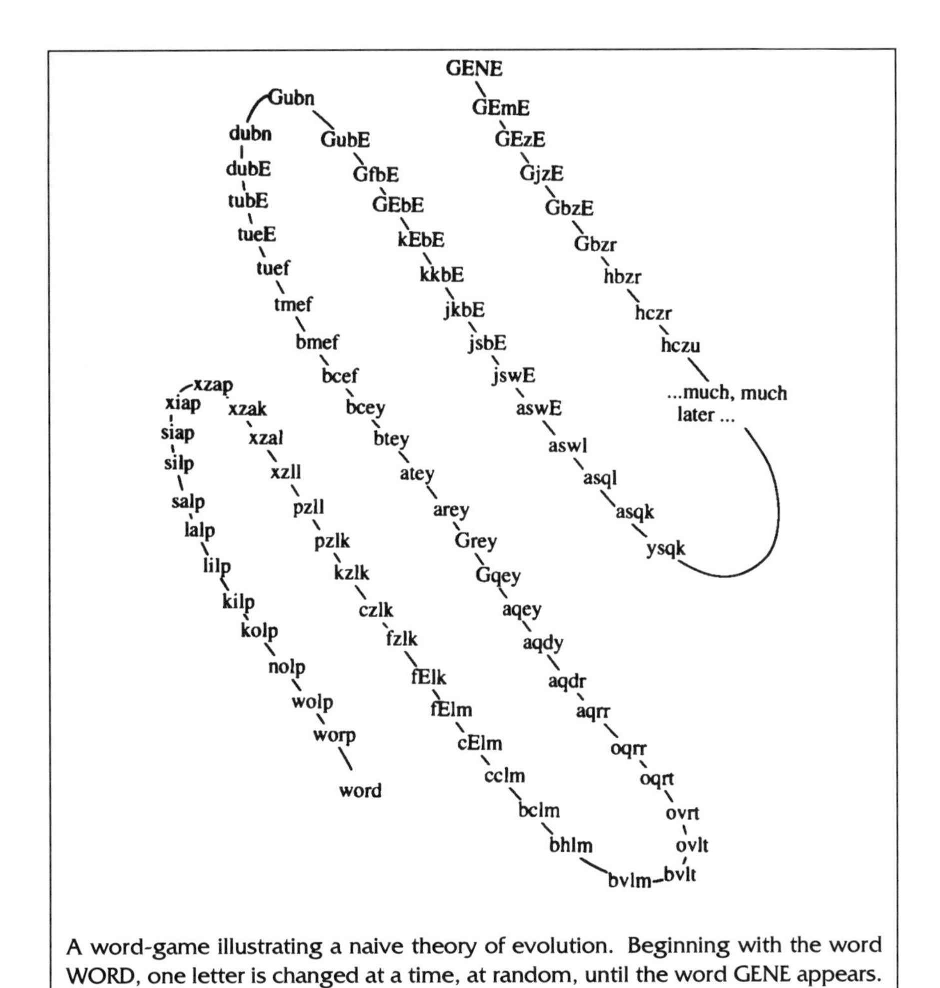


A word-game illustrating a naive theory of evolution. Beginning with the word WORD, one letter is changed at a time, at random, until the word GENE appears.

(*Continues*)

(*Continued*)

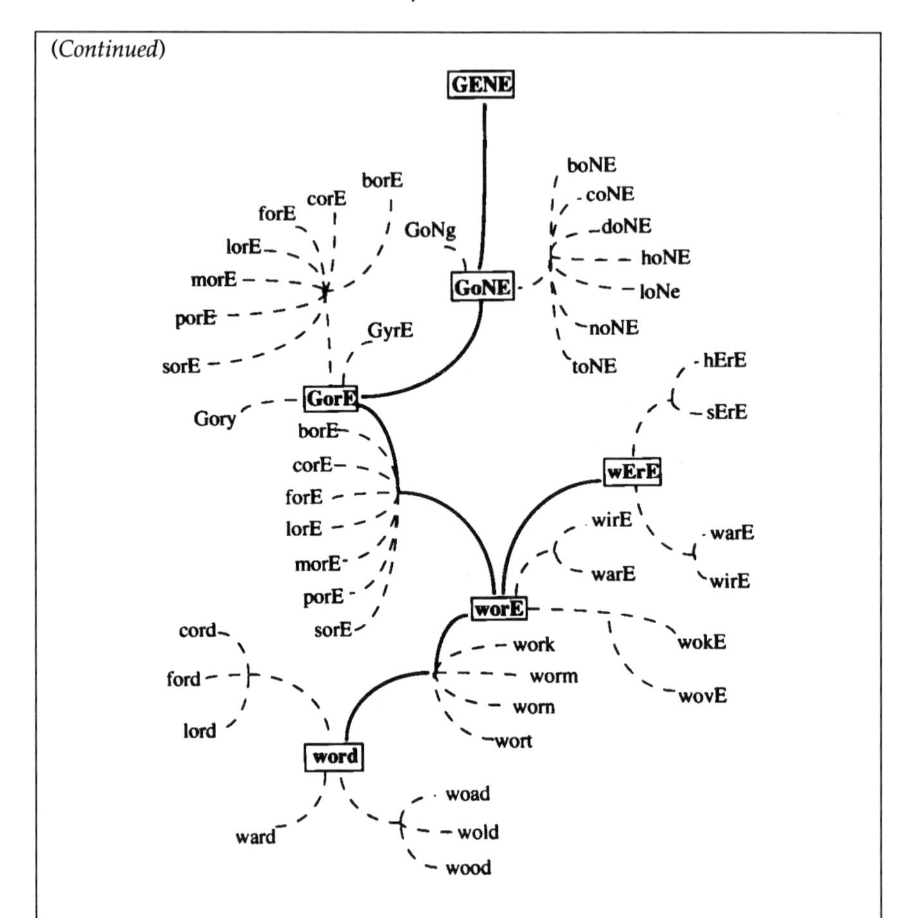


A word-game illustrating the Darwinian theory of evolution. Single-letter variants still arise at random, but only meaningful English words are viable sequences, and any variant that more closely resembles GENE is selected, rapidly replacing competing sequences. The solid line traces the ancestry of the successful lineage, which involves four successive substitutions; unsuccessful lineages are shown as broken lines.

one letter at a time. The argument is not rigorous, but it illuminates the main issue very clearly. We begin with a word, say, WORD, that must be transformed letter by letter into another word, say, GENE. The elegant analogy between words as linear sequences of letters, proteins as sequences of amino-acids, and nucleic acids as sequences of nucleotides, justifies regarding the game as an analogy of evolution. How long will the transformation take, given that one letter at a time is changed at random?

There are 26 different letters, and therefore $26^4$, or about $5 \times 10^5$, words four letters long. If we change one letter at random every second, we can expect with average luck to spend about 200,000 s, between two and three days, before getting GENE. The time that would be required to create a longer word, still less a short sentence, can be left to the imagination. It is inconceivable that any complex message can appear in a game of this sort. Evolution cannot work like this. It does not.

To show how evolution does work, we can replay the game with two additional rules. The first is that only alterations that give rise to meaningful English words are allowed. The second is that any variants that are more similar to GENE, in the sense of having the right letter in the right place, replace any that are less similar. These rules are analogous to assuming that different sequences of nucleotides in nucleic acids or of amino-acids in proteins have different properties: some are entirely non-functional, and cannot give rise to a lineage, whereas, among the functional variants some are superior to others and therefore replace them fairly quickly. The game now proceeds rather more briskly. There is only one variant that, by changing a single letter, increases the resemblance of WORD to GENE, which is WORE. This can be expected to arise by chance in 10–15 s, and will then replace all competing sequences. There are now two possible further improvements: WERE and GORE. WERE turns out to be a blind alley: no further improvement is possible from this position. (As I have pointed out, the evolution of an improved but imperfect state may not be unusual.) GORE, however, will be succeeded by GONE, which will in turn be replaced by GENE. This is a much more rapid process than waiting for the right variant to turn up fully formed: the transformation will be complete within a minute or two, rather than taking days. The difference between the two processes, which at first seems almost magical, is the difference between simple sorting and sequential substitution. A Micawberish strategy of merely waiting for something to turn up is just as poor as a theory of evolution as it is as a guide to conduct. It is the successive selection of slight improvements that makes it possible for apparently improbable transformations to occur so rapidly. Moreover, this process will still work fairly quickly even for rather complicated messages that would not turn up spontaneously if you were to wait around until Doomsday.

We do not, of course, have to rely on the analogy provided by word games; we can point directly to the results of experiments. Take the highly adapted 218-nucleotide RNA able to replicate in culture tubes containing ethidium bromide, for example. We do not know that it is perfectly adapted, but we do not know that this is the case for hemoglobin, either. There are $4^{218}$, or about $10^{128}$, different sequences of this length. How long would we have to wait around for the right one to turn up? A culture tube contains about $10^{16}$ molecules. If we use the whole of the world ocean, we can fill about $10^{22}$ culture tubes, containing in total $10^{38}$ molecules. By the same ingenious biochemistry as before, we change the sequence of every molecule every second, never allowing any sequence to be repeated. We shall have tried out all the possible sequences, and found our adapted RNA, after about $10^{128}/10^{38} = 10^{90}$ s, or about $10^{72}$ years. This is about $10^{62}$ times as long as the universe has existed. Because highly- adapted RNAs can, in fact, evolve within a few days or weeks, it is clear that their evolution is not a matter of waiting for something to turn up. No-one, however, has suggested that the normal rules of RNA replication are suspended for the benefit of these experiments, or that the experiments are the target of continual divine intervention. The paradox disappears when it is realized how evolution through selection proceeds in a stepwise fashion. Random alterations in the sequence of nucleotides occur from time to time, because the replicase is not a perfect copying machine. Most changes are deleterious, and quickly die out without giving rise to a lineage. Occasionally a change is advantageous, in the sense of permitting more rapid replication in the particular circumstances of an experiment. Any such variant quickly replaces the resident type. The whole population has now changed, and the altered sequence can serve as the basis for further improvement. In this way, it is possible very rapidly to evolve sequences which at first sight seem hopelessly improbable.

## *11. Competitors are an important part of the environment.*

When a culture tube is inoculated with RNA, all resources are at first present in excess, and growth is exponential, or nearly so. Under these conditions, variants with greater exponential rates of growth are selected. The outcome of competition among a number of variants could, therefore, be predicted by growing each variant for a short time in a separate tube: the variant which grew fastest would be the one which replaced all others in a mixed population. If growth is continued, however, before very long resources will begin to become scarce. In particular, there will no longer be enough replicase to go round. The nature of competition now changes; it is no longer purely reproductive, but operates more directly between molecules for access to resources. Variants with high rates of replication will, of course, continue to be selected; but they must now achieve these high rates of replication by obtaining an adequate supply of replicase, and thereby directly or indirectly preventing other molecules from doing so. In practice, this means that the most successful sequences will be those which are best at seizing scarce replicase molecules whenever they encounter them. Now, there is no reason to suppose which sequences which can replicate fast when replicase is abundant will also be best able to scavenge for replicase when it is scarce. In fact, new variants do appear and spread when selection is prolonged to the point where the availability of replicase becomes limiting. These variants perform poorly in the early stages of growth, and their superiority becomes evident only when molecules begin to compete amongst one another so that resources cease to become available (for a time) to others, and thereby deny them the opportunity to replicate. The crucial feature of the environment is no longer the physical conditions of growth—the presence or absence of ethidium bromide, for example—but rather the behavior of the other variants in the population. Absolute measures of performance, such as the exponential rate of increase in pure culture, may then be unreliable predictors of evolutionary success or failure. What matters is not so much the ability to do well, but rather the ability to do better.

I cannot resist stealing an anecdote from Matt Ridley to illustrate the crucial importance of relative fitness. A wise man lived with his disciple in a remote forest. One day when they were out gathering berries, the two were suddenly attacked by a ferocious bear. As they fled pell-mell through the trees the bear was clearly gaining, and the disciple panted despairingly to the sage, "I am afraid, sire, that it is no use: we cannot outrun the bear." But the sage only replied, "It is not necessary that I run faster than the bear. It will be enough if only I can run faster than *you*."

## 12. *Evolution through selection is a property of self-replicators.*

The very simple nature of the Qβ system, and in particular the tight coupling between nucleotide sequence and behavior, makes it possible to exemplify evolutionary principles in a particularly direct and straightforward manner. In other respects, choosing to open a book about selection with an account of degenerate viruses in culture tubes may seem rather eccentric. Evolution is the study of living complexity: are viruses (let alone RNA molecules a few hundred nucleotides in length) even alive? The question is important, not because it can be given any definite answer—'alive' is a category invented long before the nature of viruses was elucidated—but rather because it is not necessary to give it an answer at all. Whether or not they are regarded as living, viruses, and even their degenerate descendants in culture tubes, can evolve; and 'evolvable' is a more fundamental category than 'living'. Anything that replicates itself will do so imprecisely;

some of the variants that appear will have altered rates of replication; those with higher rates of replication will be selected. Evolution through selection is, therefore, a property of all self-replicating things.

## 13. *Selection can be used to engineer the structure of molecules.*

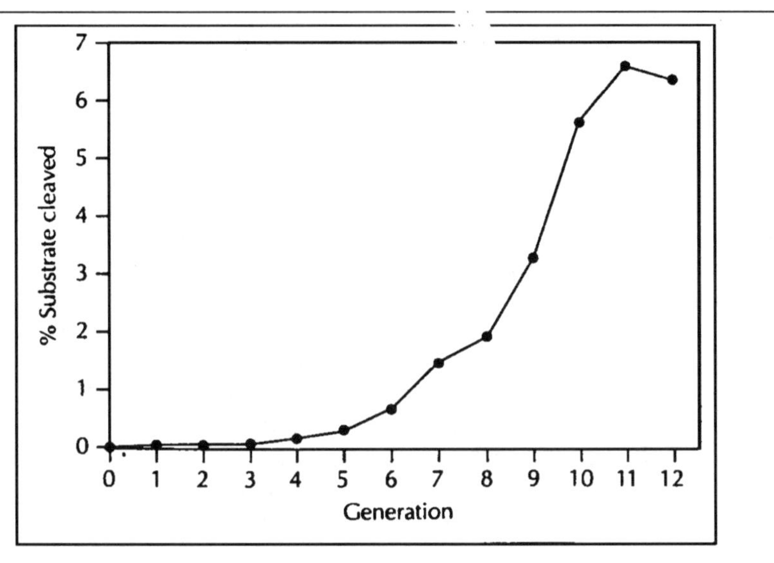


The system devised by Joyce uses a ribozyme from the ciliate *Tetrahymena*. This ribozyme binds a partially complementary substrate RNA molecule, cleaving it at a particular site and bonding part of the cleaved product to the ribozyme itself. It normally requires $Mg^{2+}$ or $Mn^{2+}$ to operate in this way. To evolve variants functional in the presence of $Ca^{2+}$ alone, it is necessary to amplify lineages that are more successful in catalysis under these conditions. Very briefly, a heterogeneous population of catalytic RNA molecules binds the substrate RNA. This complex is translated into cDNA by reverse transcriptase, using a primer corresponding to the bound substrate sequence. RNA is transcribed from this DNA in such a way that only the catalytic part of the molecule is synthesized, and then amplified by RNA polymerase in an error-prone fashion that recreates a large, highly variable population of catalytic RNA molecules. Because only those RNA molecules that bind the substrate are translated into DNA and subsequently amplified, the procedure is highly selective for successful catalytic activity. The results of engineering molecules in this way are described by Joyce (1992), Beaudry and Joyce (1992) and Lehman and Joyce (1993a, 1993b). The figure shows the increase in the

(*Continues*)

(*Continued*)

catalytic ability of the ribozyme in the presence of $Ca^{2+}$ alone as the result of 12 cycles of selective amplification.

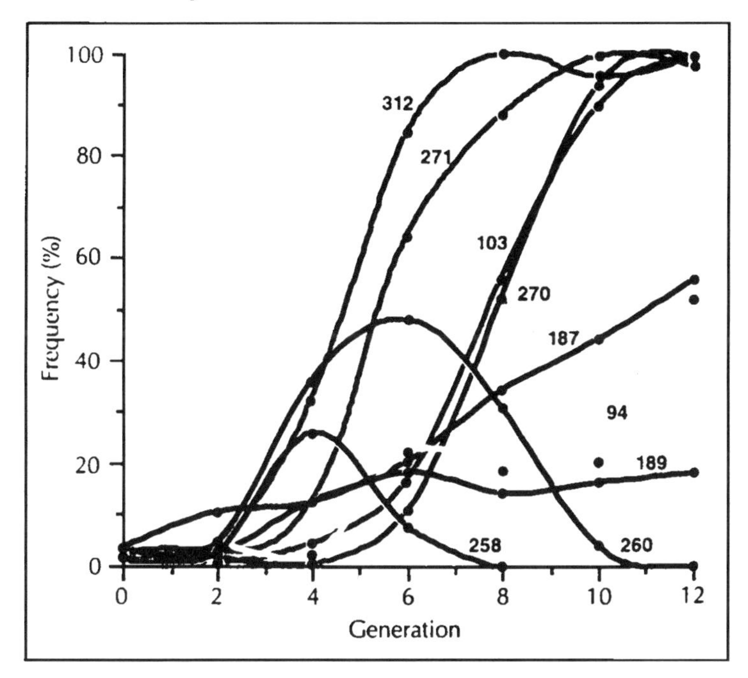


This advance was caused by a complex series of substitutions. The diagram shows how the frequency of some of the mutations that were detected changes through time; the numbers are their positions on the ribozyme molecule. Some spread through the population to become fixed by the end of the experiment; others increase in frequency during the early stages of selection, but then decline and are eventually lost. Not surprisingly, genetic variance increased rapidly in the first few cycles, but later fell as the population converged on a few highly superior variants that shared mutations at seven sites. Many of the alterations were highly epistatic, their effect depending on their genetic context; some were even mutually exclusive, being effective only if the other did not occur.

We are only now beginning to understand, and therefore to be able to exploit, the generality of this principle. RNA maintained in culture tubes is not, strictly speaking, an autonomous self-replicator, because the replicase, which it normally encodes, is supplied by the experimenter. RNA, like protein, however, is capable of acting as an enzyme; and, in particular, RNA enzymes, or ribozymes, can catalyze the breakage and reunion of other RNA sequences. It is perfectly conceivable that there exist RNA sequences that are able to catalyze their own synthesis without the need for a protein replicase. No such molecule has yet been identified (although it is only a question of time, I think, before it is), but the evolutionary potential of self-catalytic RNA has begun to be studied by people such as Gerald Joyce and his colleagues at the Scripps Institute in California. They have used an RNA isolated from the ciliate *Tetrahymena* that is able to cut a different substrate RNA molecule and attach part of the substrate to itself. This reaction normally requires the presence of $Mg^{2+}$ ions. How might it be possible to create a molecule that would carry out the same reaction when $Ca^{2+}$ but not $Mg^{2+}$ were present? It would be extremely difficult to *design* such a molecule; the relevant biochemistry is just not known in enough detail. However, it is possible to devise a system in which variants that can perform the reaction, however inefficiently, succeed in replicating, whereas those that cannot perform the reaction are unable to replicate. (The details of the amplification procedure are quite complicated, but are not important here. They involve several proteins acting as replicases of different kinds.) The system can then be left to design itself. The initial level of activity is very low. Variants soon arise, however, that are relatively successful—not very successful, but more successful than any others—and because these are replicated preferentially they spread through the population of RNA molecules. Some increase in frequency for a time, and later decline and disappear, supplanted by superior variants. Others serve as the basis for further change, and are retained. Some alterations are successful independently; others are effective only when combined with certain other alterations and, therefore, spread only when these have become established. This ribozyme system, in short, undergoes the same process of sequential substitution as evolving $Q\beta$ RNA. After only a dozen cycles of selection and amplification, a new RNA capable of cleaving the substrate RNA in the presence of $Ca^{2+}$ alone with quite high efficiency has evolved. This is an interesting evolutionary system in its own right; it is also a remarkable illustration of how evolution through selection can produce functional structure with an ease and rapidity that a conventional engineering approach could not hope to emulate.

## 14. *Self-replicating algorithms evolve in computers.*

Among the most illuminating and certainly the most bizarre of self-replicators are the digital organisms invented by Thomas Ray of the University of Delaware. They are algorithms: string of instructions written in the

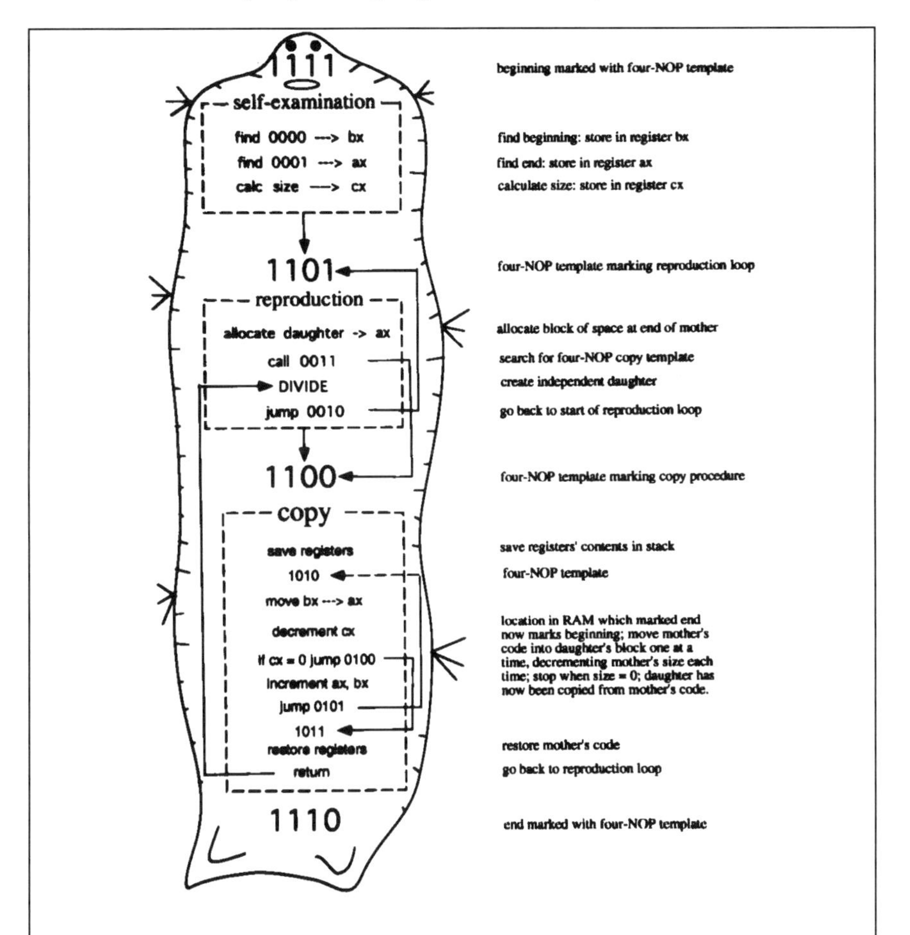


The Tierran organisms are still not well known. So far as I know, they have still not been described in any of the conventional journals, and so I shall give some details of their physiology from a manuscript kindly sent to me by Thomas Ray. They are CPUs of a virtual computer, giving them the status of digital individuals that can communicate with the physical environment provided by the CPU of the real computer. Each individual comprises two address registers, two numeric registers, a flag register, a stack pointer, a 10-word stack, and an instruction pointer that addresses the core memory of the real computer. Each such CPU continually performs a cycle of operations. It first fetches the machine-code instruction currently addressed by the instruction pointer into the virtual CPU. This is then decoded and executed. The instruction pointer is then moved to the next position in RAM, unless it has been manipulated directly by the previous instruction, and the instruction it finds there is imported.

*(Continues)*

*(Continued)*

The Tierran language is distinguished by two features that are designed to be analogous with the metabolic and genetic systems of conventional organisms. In the first place, it has a very small set of instructions: only 32 in total, that can be represented by 5 bits, so it is about the same size as the 26 instructions used to make proteins encoded by the six bits of a triplet of nucleotide bases. Unlike previous attempts to simulate living organisms, the language contains no integers, so that it cannot refer directly to a new address in memory: it must instead construct an integer by a series of instructions (such as, "change low-order bit and shift left") acting on a numeric register. Its second important feature is that the addresses of data or code are not given directly, but are instead specified by template. The instruction pointer is ordered to move by a sequence of four binary numbers (No Operation, or NOP, procedures), for example, 0010. The system then searches outward in both directions from the instruction in order to find the complementary sequence, in this case 1101. If it fails to do so within a certain period of time, an error condition is flagged, and the instruction is ignored. Otherwise, the instruction pointer moves to the end of the complementary sequence, and resumes execution. This is intended to simulate the interaction between protein molecules, or between enzyme and substrate, where there is no procedure for using spatial coordinates to bring the two into contact.

The operating system of Tierra defines the primary features of the growth, reproduction, metabolism, death, and mutation of its inhabitants. Each creature occupies a block of memory within the core memory of the real computer. This block defines the size of the creature—80 instructions, say—within which it has an exclusive privilege to write instructions. Other individuals, however, may read or execute instructions within this block. An individual may also write instructions into a second block, representing growth. At reproduction, the mother has finished copying her instructions into a block representing a daughter cell. The mother now loses write privileges on the space of the daughter cell she has created (but can then be allocated a second block of memory). These operations take place within a slice of CPU time allocated by the real computer to the vitual CPU of the Tierran individual. The individuals are not immortal: they enter life at the bottom of a queue of similar individuals, and move upwards in this queue every time they generate an error, provided that the individual ahead of them in the queue has committed fewer errors. When they reach the top of the queue, they are killed by de-allocating their space in memory. The errors themselves are generated by randomly flipping bits with some given probability, or by wrongly executing instructions. Because of this mutational input of slight novelties, the population quickly diverges into a bush of lineages, representing variant sequences whose fate can be followed in any degree of detail by a gene-bank manager.

With this introduction, the figure represents the basic Tierran self-replicator with which the real CPU is inoculated.

*(Continues)*

*(Continued)*

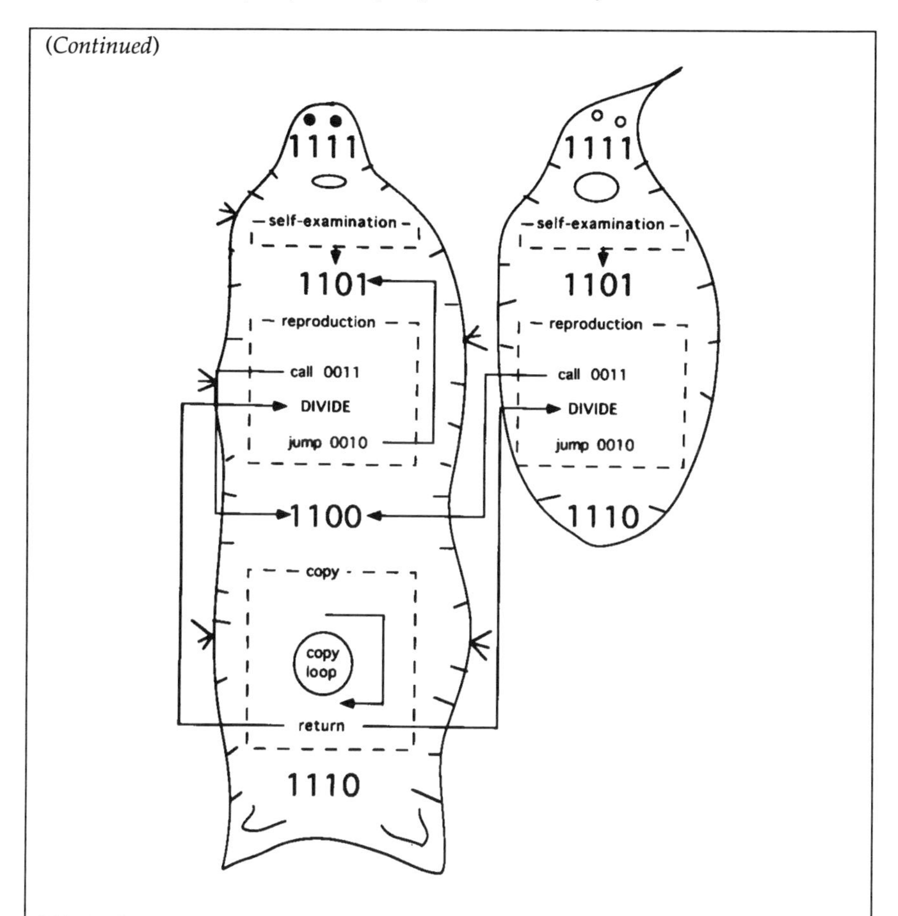


This is the parasitic algorithm that often appears in evolving populations. It is produced by a point mutation in the low-order bit of instruction 42 of the ancestor, NOP_0 to NOP_1. This changes the template for the copy procedure from 1100 to 1110, which is now taken to mark the hind end of the individual. The parasite thus lacks a copy procedure of its own, and cannot replicate in pure culture. If it is close enough to the copy procedure of a host, however, it can locate, read, and execute it. The host is not harmed directly, but the parasite, by virtue of its smaller size and shorter life cycle, is a superior competitor, provided that enough hosts remain in the population.

machine-code language of a computer. The basic algorithm, a list of 80 instructions, is designed to replicate itself—to place a copy of the same set of instructions into the memory of the computer. The process of replication is deliberately made somewhat imprecise, so that instructions are occasionally modified slightly as they are being copied. The algorithm is designed to replicate itself, and it is not designed to do anything else; that is, no inherent tendency to change in certain directions, or in any direction, is programmed into the list of instructions that defines each individual. The creatures proliferate within the central processing unit of a computer. The resources they require are memory space (for storing their code and that of their descendants) and computing time (for executing their code). It is not too fanciful to imagine them as algorithmic parasites, utilizing the computer's operating system to achieve their own self-replication, somewhat as Q$\beta$ uses the protein-synthesis machinery of its bacterial host. The space and time that has been allotted to them is sooner or later insufficient to allow all the copies of the original sequence to replicate themselves. Some must then be discarded, by erasing them from the memory of the computer; those whose code has suffered the most errors during copying are the most likely to be lost in this way. The population is at this point a clone, whose members reproduce and die in their allotted reserve within the computer. If they are allowed to do this for long enough, however, something happens: they begin to evolve.

The first sign that the system is something more than an eternally recurring set of computer instructions is that new creatures 80 instructions in length appear, usually differing from each other, and from their ancestor, by a minor change in a single instruction. In different runs, different variants arise and spread, often replacing the ancestral type. This was unexpected: the original sequence had been very carefully written as a minimal self-replicating algorithm, and yet several thousand different self-replicating variants were derived from it through slight modifications of one sort or another. Much more surprising, however, was the abrupt appearance of a completely different creature, in which a trivial change to a single instruction has a marked effect on its appearance, reducing its size from 80 to 45 instructions. This variant does not have the ability to replicate itself autonomously, because the code that directs the copying procedure has been lost. It is, however, able to use the copying procedure of the intact algorithm and is, therefore, maintained in the population provided that there are enough 80-instruction self-replicators to supply it with the missing instructions. It is, in short, a parasite that is able to utilize the replication machinery encoded by the intact host. The parallel with Q$\beta$ is astonishingly close: in both systems defective variants arising through slight modifications that cause the deletion of a large part of the genome parasitize the ancestral sequence by utilizing its replication machinery—a replicase in the one case, a copying procedure in the other.

This does not by any means exhaust the evolutionary possibilities of the system, whose full extent, indeed, remains to be explored. Host variants

79 instructions in length appear that are resistant to the parasite; they are outmaneuvered by modified parasites 51 instructions long. A peculiar creature that is 80 instructions long, but differs from the ancestral sequence by 20 or so alterations, has sometimes been observed; it is a hyperparasite able to force parasites to replace themselves with copies of itself. Even more remarkable are 61-instruction sequences that are unable to replicate when alone, but achieve a cooperative self-replication when associated in groups, occupying adjacent blocks of computer memory. They are eventually exploited by tiny creatures only 27 instructions long that insert themselves between these cooperating sequences and exploit the cooperative system of replication to their own advantage. This whole menagerie of different kinds of creature descends from the single self-replicating sequence with which the computer is originally inoculated. The process is completely transparent, because the complete sequence of any or all algorithms in the population can be ascertained at any time. Despite the complete lack of any specific directing principle, the sequential substitution of slight variants drives the evolution of a diversity of replicators of different kinds. It is perhaps the purest example of evolution through selection that can be imagined, and provides a striking demonstration that evolution through selection is a general property of self-replicators, irrespective of whether by any conventional standards they would be considered to be living.

## 15. *Evolution through selection is governed by a set of general principles.*

The experiments that I have described exemplify a series of generalizations about the behavior of systems of self-replicators.

- Heritable variation in the rate of replication causes evolution through selection.
- Heritable variation arises as random, or undirected, alterations of nucleotide sequence; it does not in itself direct the course of evolution.
- The rate of replication is the only attribute that is selected directly.
- Characters that cause changes in the rate of replication will be selected indirectly and may evolve as a consequence.
- Adaptation caused by selection in given conditions is likely to be associated with loss of adaptation in other conditions.

- Evolution proceeds through the sequential substitution of superior variants, not exclusively by sorting pre-existing variation.
- A given state can evolve from a prior state only if the two states are connected by a continuous series of slight modifications, each of which is individually advantageous.
- Selection causes the modification of prior states of organization, but cannot abruptly give rise to wholly novel states. The course of evolution is an historically unique, contingent process conditioned by the fortuitous occurrence of particular variants.
- Selection tends to improve performance in given conditions, but does not always, and may never, optimize performance.
- Because selection is caused by differences in rates of replication, the outcome of selection will often depend on what kinds of competitor are present, rather than on the physical conditions of growth.

The way that these general laws, and others yet to be discussed, mold the evolution of living organisms will be modulated by the developmental, physiological, genetic, and ecological circumstances in which they operate. That is what the rest of this book is about. It is important to realize, however, that they are not themselves developmental, physiological, genetic, or ecological laws. Nor can they be described in terms of or reduced to developmental, physiological, genetic, or ecological principles. They are a separate set of laws that defines a separate discipline, based on the single irreducible quality of self-replicators of any kind: evolution through selection.

# 2

# *Selection on a Single Character*

## *16. Phenotypic selection is caused by the selection of genes as replicators.*

Genes are replicators, and genes in organisms will evolve in the same way as RNA molecules in culture tubes, through selection for higher rates of replication. In this formal sense, indeed, they will evolve in exactly the same way: variants with higher rates of replication will replace those with lower, and the rate of replication is the only character on which selection will act directly.

In some cases this is brutally clear. Most genomes of cellular creatures contain a whole zoo of competing replicators. Transposons distribute copies of themselves throughout the genome, infecting new lineages through sexual fusion. Plasmids are separate from the rest of the genome and utilize the machinery it provides to replicate themselves. Supernumerary chromosomes in multicellular organisms are able to guide themselves into the germ-line. Elements such as these, which are discussed at more length in Part 4, are essentially infective, maintaining themselves by their ability to replicate, without contributing to the well being of the cell or the organism. Their structure is similar to that of normal genes, but their behavior is more like that of a virus.

Most genes are, of course, domesticated, and their transmission is governed by Mendelian rules. They are replicators, but they are neither *autonomous* nor *independent* replicators.

The machinery for gene replication is extremely complex. In a sense, it encompasses the whole life of the organism. Disease resistance, for example, is not conventionally associated with DNA replication; but if a rabbit dies from myxomatosis because it lacks the genes that confer resistance, then none of its genes can be replicated. Gene replication, then, depends on the genome as a whole; each of its component replicators is necessary to ensure its own replication, and none is sufficient to replicate alone.

Nor are genes replicated separately. They are associated as circular or linear arrays of hundreds or thousands of genes that are replicated and transmitted together at the same time. This has the obvious consequence that a gene cannot be replicated faster than its neighbors and, indeed, all the genes in a genome will in normal circumstances be replicated at the same rate.

## 17. *The replication of genes depends on the reproduction of organisms.*

The physiological and mechanical interdependence of genes means that the success of any particular gene depends on the success of the genome as a whole. The genome itself, however, cannot replicate as an isolated entity. It encodes the intricate array of structures that make up the individual organism, and its transmission is absolutely dependent on the ability of the organism to reproduce. The rate of replication of a gene, therefore, depends on the effect that it has on the rate of reproduction of individuals that bear it. Different variants of a gene may be replicated at different rates, not because as isolated systems they would be more or less efficient self-replicators, but because the individuals that bear them reproduce at different rates. This does not in any way challenge the basic principle that selection will act directly only on differences in the rates of replication of genes. These differences, however, can be expressed only through the differences they cause in the rates of reproduction of individual organisms.

## 18. *Characters evolve through their effect on reproduction.*

Individuals do not evolve; they develop, reproduce, and die. The characteristics of organisms, however, may change through time. A lineage may acquire the ability to live in fresh water rather than salt, to use ribulose rather than lactose as a carbon source, or to grow limbs rather than fins. These differences in character state are encoded by different variants of genes. The average state of a population will change through time when these variants change in frequency as a consequence of differing rates of replication, expressed through the different rates of reproduction of the individuals that bear them.

If each gene encoded a single character and each character were encoded by a single gene, then we could simply replace 'character' with 'gene', and 'character state' with 'gene variant'. In situations where this is possible, the process of evolution is relatively straightforward and easy to understand.In most cases, however, there are two major complications that make it necessary to reformulate the basic argument.

The first is that not all differences in character state among individuals are caused by differences in the genes that those individuals bear. If one plant is short and another tall, the two may bear different variants of a gene or genes influencing height; but equally, the short plant may have grown under crowded conditions with insufficient nutrients. Differences in phenotype do not necessarily imply differences in genotype.

The second consideration is that the relationship between genotype and phenotype may be very complicated. As a general rule, each gene encodes a single protein, and each protein is encoded by a single gene. Any given protein, however, may have manifold effects on the phenotype, and moreover, any given aspect of the phenotype may be affected by many different proteins. Thus, different variants of a gene are likely to cause variation in several characters, and the variation in any given character may be caused by variants of several genes.

In simple systems, it is enough to recognize that evolution through selection will occur as the consequence of differences in the rate of self-replication among variant gene sequences. In the more complex organisms with which we are usually concerned, we must retain this basic principle while reformulating it in terms of the evolution of characters and the selection of individuals. Populations of individuals have a tendency to increase exponentially in numbers, whereas there is a finite supply of the resources necessary for reproduction. As these resources become depleted, there will be a struggle for existence among individuals competing for opportunities to reproduce. Individuals will vary to a greater or lesser extent with respect to all characters because the process of hereditary transmission is not perfectly precise. They will, therefore, vary with respect to those characters that affect reproduction. Some individuals will possess character states that enable them to reproduce more successfully than average in the environment they occupy. The offspring of these individuals will be disproportionately represented in the next generation: this is the process of selection. If the successful individuals differed from the average of the population because they bore different variants of certain genes, these variants will have been transmitted to their offspring, who will then express the same aberrant character state as their parents. Because this character state is more frequent in the population than it was in the previous generation, the average of the population will have changed as a consequence of the increase in frequency of the variants responsible for encoding the altered character state. This permanent genetic change in the composition of the population is the process of evolution.

This is, perhaps, a deplorably long-winded way of presenting a very simple argument. I think it is important, however, to stress that the basic principle that evolution through selection acts directly only on rates of self-replication applies with the same force to genes in organisms as it does to RNA molecules in culture tubes. The complexity of the relationship between genotype and phenotype in cellular organisms makes it necessary to reformulate it in a more complicated way in order for it to be useful, but the powerful simplicity of the basic principle remains.

We can, in fact, recover some of this simplicity by taking a short cut. The reproductive success of an individual is often referred to as its fitness. A more precise definition of the fitness of a given type of individual, distinguished by a particular character state, is the rate of increase of a population consisting exclusively of that type. We can then encapsulate the whole argument by saying that *evolution through selection is the necessary consequence of heritable variation in fitness.*

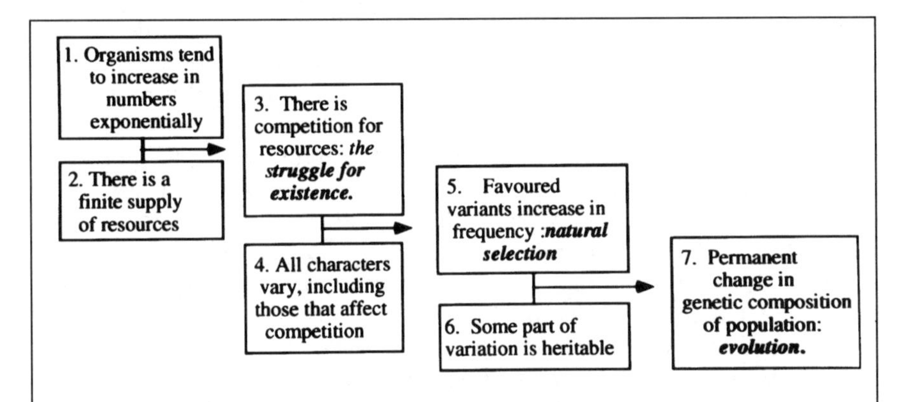


Character evolution, caused by the selection of individuals, requires a slight elaboration of the simple Darwinian logic sketched in Sec. 5.

The definition of fitness given here, as the rate of increase of a lineage in isolation, is close to a general concept of adaptedness. It has been adopted at this point because it provides a simple basis for explaining simple evolutionary processes. It does not apply when there are social or sexual interactions among different kinds of individuals (Parts 5 and 6). Competitive measures of fitness are introduced in Sec. 40. Concepts of fitness have been reviewed by Hedrick and Murray (1983) from an empirical point of view and by de Jong (1994) from a theoretical point of view.

# 2.A. Single Episode of Selection

## *19. The unit event of evolution is an episode of variation followed by an episode of selection.*

The history of a lineage can be represented as a phylogenetic tree linking ancestors and descendents. This may be evolution on the grand scale,

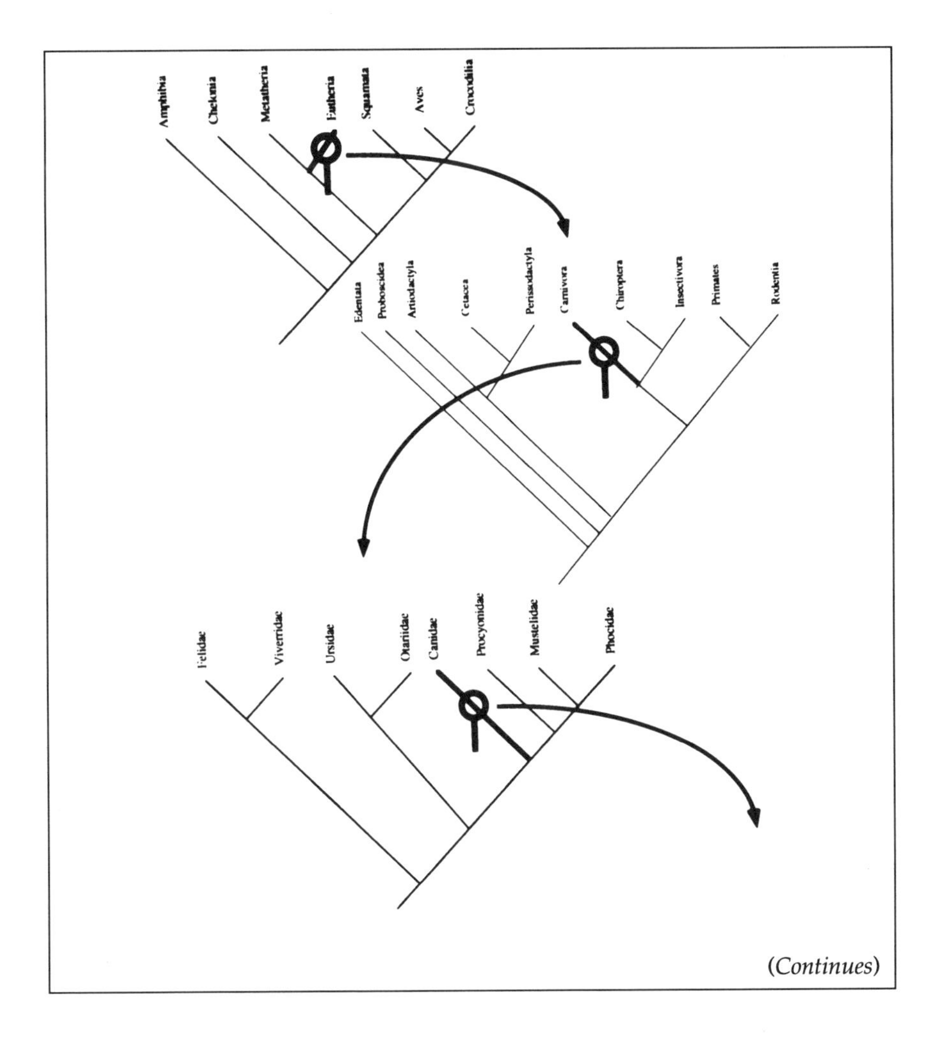

(Continues)

(*Continued*)

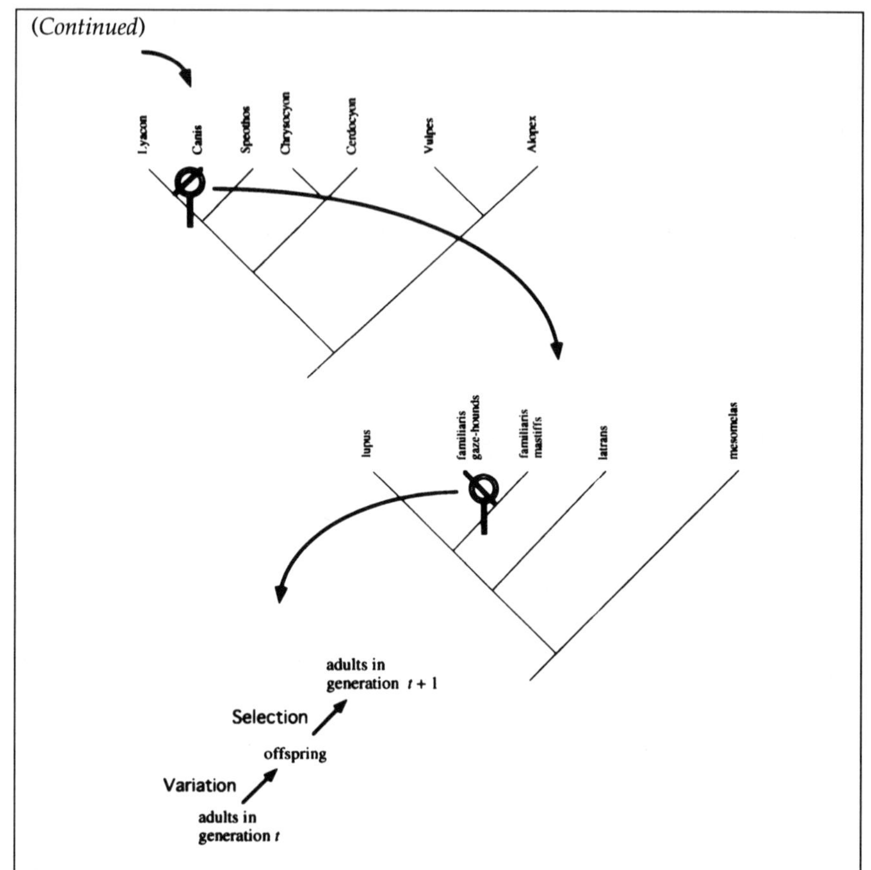


Phylogenetic trees can be drawn at different genetic scales. Each of the trees here is a magnified view of the branch shown as a thick line in the preceding tree. The first tree represents tetrapod vertebrates, where the earliest branching shown occurred about the middle of the Carboniferous (lower Pennsylvanian), about 300–320 million years ago. The second tree shows placental mammals, with most orders diverging in the Eocene or before, about 50–70 million years ago. The third is the family tree of one of these orders, the carnivores; the separation of canoid and feloid lineages is Eocene, but subsequent divergence was spread over about 25 million years, into the middle Oligocene. The fourth tree is for canids, the dogs, where the earliest branching shown occurred in the early Miocene, about 10 million years ago. The final tree shows the ancestry of dogs and wolves, *Canis*, which have diverged over the last 1 million years or less. Continuing to magnify branches in this way will lead eventually to a population of dogs at a scale of generations, in each of which there is an episode of variation followed by an episode of selection. This is the lowest genetic scale; it cannot be magnified any further.

depicting the ancestry of the Metazoa or the origins of land plants. If we examine any part of such a tree in more detail, however, we shall see the same treelike structure on a smaller scale. A tree representing the ancestry of vertebrates, for example, would have a time-scale of hundreds of millions of years, because the earliest vertebrates are jawless fishes from the upper Ordovician, some 400 million years ago. It would probably be drawn with a single line to represent the divergence of each major group of vertebrate, such as the branch corresponding to the subclass Eutheria, the placental mammals. If we look more closely at this line, however, we find that it is itself a tree, whose branches are orders of mammals—bats, whales,primates, and so forth. This tree would have a time-scale of tens of millions of years to accomodate the Cenozoic radiation of mammals. If we continued this process of magnification, we would eventually have reduced the time-scale to hundreds of years and the taxonomic scale to populations. At the limit, we would be looking at a single generation in a single population: a cohort of newborn offspring that grow up, reproduce, and die. These offspring are not exact replicas of their parents; setting the complexities of sexuality aside, the imprecision of reproduction will introduce alterations of various kinds, causing them to vary to some extent. This new variation will be sorted by selection, with some variants surviving and reproducing more successfully than others. This coupling of an episode of variation with an episode of selection during a single generation is the unit event in evolution; evolutionary history on the grand scale consists of a long series of such events occurring successively through geological time.

## 20. *Evolution is caused by a 'lack of fit' between population and environment.*

It is possible to imagine a population that is perfectly adapted, in which offspring are always exact replicas of their parents and that inhabits an environment that never changes. In these circumstances, evolution would come to a halt, and the population would propagate itself, generation after generation, without any kind of change. In practice, this is never the case.

Even if the environment were to remain constant, the population would change. Genotypes inevitably suffer changes in structure, especially when being replicated. Genetic alterations, or mutations, are of various kinds, from a change in a single nucleotide of DNA to the duplication, deletion, or rearrangement of large sections of chromosomes. Some mutations may have no effect on the phenotype, but many have appreciable phenotypic effects. Even if perfect adaptation were attainable, therefore, it could not be permanently maintained, because of this internal process of genetic change.

Even if mutation could be neglected, the environment would change. The physical environment is continually changing on all time-scales: from hour to hour, from day to night, from season to season, from year to year. Organisms that live on different time-scales—bacteria on a scale of hours, mammals on a scale of years—will be affected by different types of change, but none are insulated from change altogether. Perfect adaptation to any given environment could not be permanently maintained, because of this external process of environmental change.

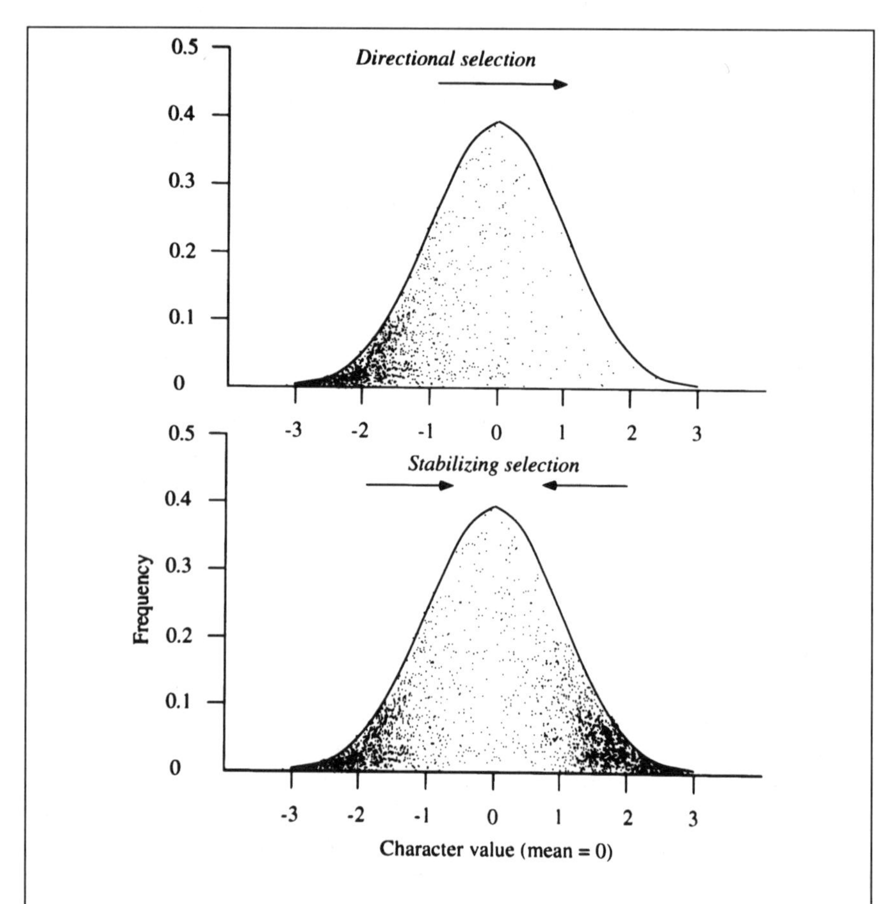


These diagrams are frequency distributions of character state: the $x$-axis is the character state, or score, in units of standard deviations in a Normal population

*(Continues)*

*(Continued)*

with mean zero (see Sec. 42); the $y$-axis is the frequency of individuals with a given score. Fitness is indicated by the depth of stippling: heavier stippling indicates a greater dosage of deleterious mutations or of genes that have become inferior because the environment has recently changed. The upper diagram illustrates directional selection: fitness increases from left to right, and directional selection will favor individuals with relatively high scores, towards the right-hand end of the distribution, thus increasing the mean score of the population. (The direction of selection I have illustrated here is immaterial; selection could just as easily act in the other direction.) The lower diagram illustrates stabilizing selection: individuals with nearly average scores have the highest fitness, with those at either end of the distribution having high dosages of deleterious mutations. Selection will thereby maintain the mean value of the character, while reducing its variance through the differential elimination of individuals with extreme phenotypes. Note that selection is always directional when the character concerned is fitness.

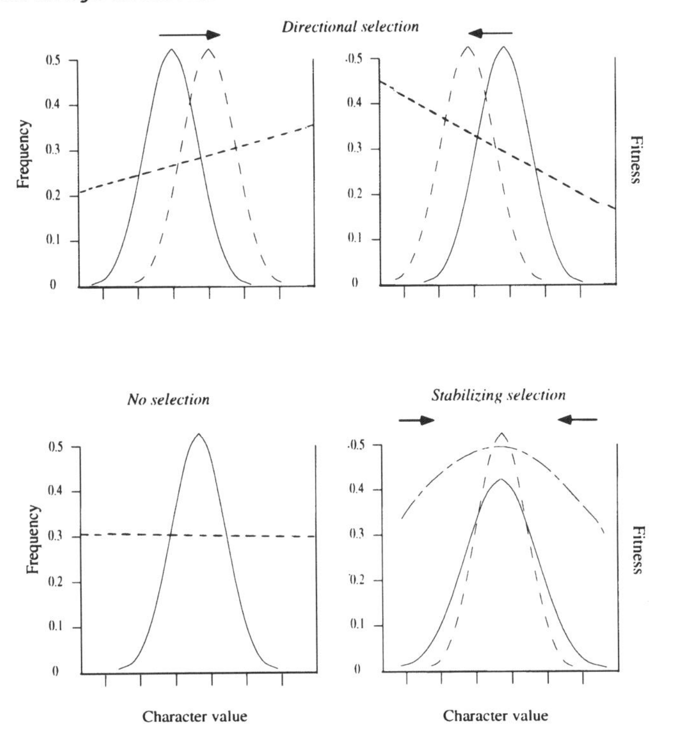


Another way of representing directional and stabilizing selection is through the effect of the fitness function (the graph of fitness on character value) on the population frequency distribution.

Adaptedness is not maintained by default; there is no passive process ensuring that a high level of functional organization will continue to be preserved as long as no specific catastrophe occurs. *Change is generally for the worse.* A well-adapted population is fitted to its environment, in the colloquial sense of being appropriately formed, like a key to a lock. If either the population or the environment change haphazardly, it is most unlikely that the fit between them will be improved. Both internal and external change, therefore, cause in every generation an increasing lack of fit between population and environment that if it were to continue unchecked would break down adaptedness completely. The maintenance of adaptation therefore requires the perpetual operation of selection, acting in every generation to restore the fit of population to environment.

Selection will act in two ways. Firstly, it will be a conservative force that eliminates new mutations so as to preserve the current adaptedness of the population. This is called *stabilizing* selection. Secondly, it will be a progressive force that favors new mutations when the environment changes. This is called *directional* selection. These categories are not exclusive. Both internal and external change will occur in every generation, eliciting both stabilizing and directional selection in different degrees on different characters.

## 21. *Mutation is not appropriately directed.*

The coupled process of variation and selection recurs in every generation. There are therefore two theories of evolution: the first is that it is directed by variation, and the second is that it is directed by selection.

The first is Lamarck's theory of the inheritance of acquired characteristics, published in the year of Darwin's birth. During their lifetimes, individuals perceive the state of their environment and respond appropriately to it. This may involve some trivial change, such as the thickening of part of the cuticle or the enlargement of a muscle, or a more profound reorganization, perhaps through the formation of new body cavities as the result of a change in the pattern of circulation of fluids. There is thus an inherently progressive principle in nature, causing a general increase in the level of organization. Selection may occur, but it plays a subordinate role; adaptation to novel environments is caused primarily by the spontaneous appearance of appropriately directed variation. We can translate this into modern terms by saying that the environment elicits favorable mutations.

So far as we know, this theory is wrong. It is not wrong as a matter of principle; it is an internally consistent and intellectually satisfying theory

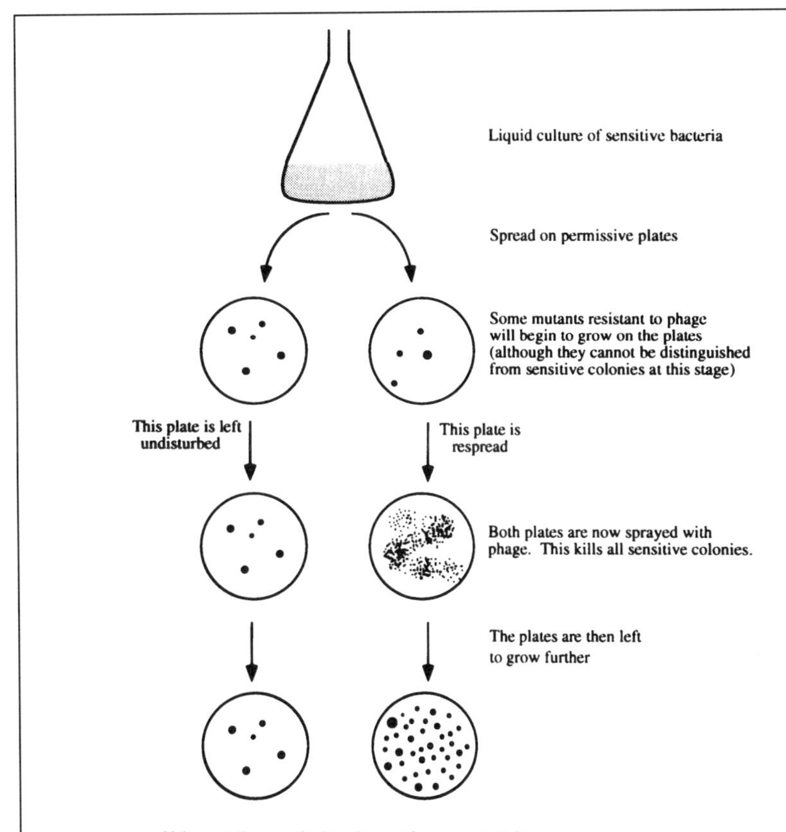


If the mutations conferring phage resistance occurred in liquid culture before exposure to phage, there will be many more colonies on the respread plates.

A number of experiments were conducted in the 1940s and 1950s to investigate the appearance of mutations in relation to the specific stress that they conferred resistance to. These experiments were responsible for the current concensus that mutation is not appropriately directed. The classical experiment was by Luria and Delbruck (1943). This figure shows a simpler experiment by Newcombe (1949). A strain sensitive to phage is grown up in the absence of phage, and samples from the culture are then spread on two agar plates. One plate is respread after a short time, so that the cells from each of the growing colonies are distributed all over the plate. The other plate is not disturbed. Both plates are then sprayed with phage. If a few mutant individuals resistant to phage have arisen in the liquid culture, before exposure to phage, then these will be represented by a few colonies on the undisturbed plates. Each of these colonies, however, would have contributed hundreds of isolated resistant cells on the respread plates, so that respreading

(*Continues*)

would cause a dramatic increase in the number of colonies developing on the infected plates. On the other hand, if mutations occur only on the plates, as the result of a period of exposure to phage, respreading will have little or no effect on the number of colonies. In fact, the respread plates bear many more colonies than the undisturbed plates, suggesting that the resistant mutants had arisen before they were exposed to phage.

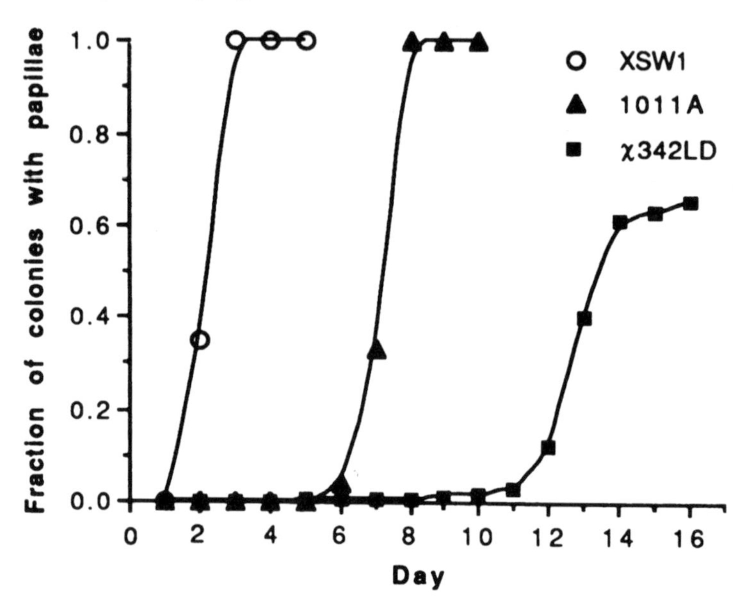


Cairns et al. (1988) pointed out that the conventional wisdom has not been rigorously tested. The environmental stress used in experiments like Newcombe's is extremely severe: cells that are susceptible to phage infection are killed immediately (Newcombe 1949). It is conceivable that a less severe stress that permitted unadapted cells to survive, and perhaps even reproduce slowly, for some time would induce appropriate mutations. This is how Cairns interpreted the evolution of lactose fermentation in populations starving in the presence of lactose; Hall (1990) reported a similar case involving the reversion of tryptophan auxotrophs. The publication of such results has led to a lively controversy: see the issue of Nature for 8.xii. 1988 for the immediate response to the Cairns et al. paper, and the paper by Lenski et al. (1989).

The figure illustrates a similar phenomenon described by Hall (1988). Utilization of salicin by *E. coli* is controlled by the beta-glucoside (*bgl*) operon, which consists of an upstream promoter *bglR* and a group of three structural genes, one of which, *bglF*, encodes a transport protein. The X342 strain used by Hall could not grow on salicin because it bore two mutations in *bgl:* a point mutation in *bglR* and the

(*Continues*)

*(Continued)*

insertion of the 1.4-kb element IS103 into *bglF*. The rate of reversion of the point mutation (in a strain not bearing the insertion) is about $5 \times 10^{-8}$, whereas the rate of excision of IS103 (in a strain not bearing the promotor mutation) is about $2 \times 10^{-10}$. The double mutant should therefore appear at frequencies of only about $10^{-17}$. When single mutants are cultured on salicin plates, they exhaust other sources of carbon in the medium when colonies have grown to about $10^9$ cells. These colonies cannot grow further; but after some days many of them show small papillae that represent clones within the colony that have recommenced growth. These are mutants: promotor revertants (of strain XSW1) appear after a lag of about three or four days, and IS103 excision revertants (of strain 1011A) appear after a week or so. The double mutant should never show the growth of papillae, since it will hardly ever occur in a colony of only $10^9$ cells. Nevertheless, papillae begin to form in X342 colonies after a lag of about two weeks. These double revertants are never observed in normal growth media. They appear to arise sequentially, rather than simultaneously. The scenario favored by Hall (1988) is that there is a burst of excision in starving cells after about 10 days of culture, creating a population of single-mutant cells large enough for a few promotor revertants to occur within a day or two. However, the excision by itself does nothing to increase the fitness of the cells; it seems to anticipate the alteration in the promoter that will enable them to grow on salicin. Shapiro (1984) previously described a similar case, in which the excision of phage Mu permits growth to occur on lactose-arabinose medium, but never occurs in growing cells.

These very interesting experiments have reawakened interest in the origin and very early history of the mutations that are subsequently selected in novel and stressful environments. I do not think that they will overturn the conventional scheme outlined in the main text (see Lenski and Mittler 1993). It seems that populations can turn over and undergo selection even in starvation regimes (Sec. 129). I am also impressed by the number of cases of apparently directed mutation that involve transposon excision. We tend to think in terms of the fitness of the cell, when perhaps we should also be thinking in terms of the fitness of autonomously replicating entities, such as transposons within the cell (Sec. 122). The genetic mechanism of apparently directed mutation has been investigated by Foster and Trimarchi (1994), Harris et al. (1994) and Rosenberg et al. (1994); it appears to involve simple single-base deletions within runs of a given base, presumably through polymerase errors. At the time of writing, the controversy is shifting so as to focus on the molecular genetics of starving cells.

of evolutionary change. It is wrong as a matter of fact. No mechanism that would act as a specific directing principle to produce appropriate genetic variation has ever been identified. During the early years of the study of bacterial genetics a number of experiments were devised to investigate this issue. They involved challenging bacteria with a new and hostile environment, and then tracing the origin of the adaptations that evolved.

In all cases, it was found that the mutations that conferred adaptation, and that increased in frequency through selection in the new environment, occurred before the environment changed. These results have been very widely accepted as showing that adaptation occurs through the selection of pre-existing variation and not through the elicitation of appropriate variation. Nevertheless, the conventional wisdom is challenged from time to time. This is usually the pastime of literary gents such as Samuel Butler or Arthur Koestler, who might hesitate to pronounce on the principles of inorganic chemistry or hydrodynamics, but regard evolution as a subject where everyone has the right to their own opinion. Together with the ha emoglobin paradox, this is the main source of the periodic announcement of the demise of Darwinian orthodoxy.

**Directed Mutation in Bacteria.** Not all challenges, however, can be shrugged off so casually. John Cairns and his colleagues at Harvard and Barry Hall of Rochester have recently published a series of experiments that they interpret as evidence for Lamarckian evolution in bacteria. For example, Cairns used a *lac*$^-$ strain unable to metabolize lactose because of a chain-terminating mutation in the gene encoding $\beta$-galactosidase, the enzyme that hydrolizes lactose. Cultures of this strain were grown in standard medium, which does not contain lactose, and then plated onto medium in which lactose was the sole source of carbon. The only colonies that could grow were those in which a mutation had restored $\beta$-galactosidase activity. These mutations might occur either during the preliminary growth of the cultures or after they have been replated. There is good evidence that mutations occur at both times. Mutations that restore *lac*$^+$ function, however, are detected after replating *only* if the medium contains lactose, suggesting that the presence of lactose somehow causes the appearance of mutations enabling it to be used. These might be quite non-specific; it is conceivable that mutation rates generally increase when bacteria are stressed and that the very small proportion of these that happen to restore the ability to hydrolyse lactose are then selected on lactose plates. This possibility, which although interesting is perfectly compatible with Darwinism, is discussed further in Sec. 116. It does not appear to explain Cairns' results, where the mutation to *lac*$^+$ seems to be specifically induced by lactose. Hall has reported a similar experiment, in which loss-of-function mutations in the genes responsible for tryptophan synthesis reverted to their normal functional form in cultures starved for tryptophan (so that acquiring the ability to make it is strongly selected) much more often than in cultures where tryptophan was supplied in abundance (so that the ability to make tryptophan was only weakly selected, if at all). The evidence again suggests that the stressful environment specifically induces a class of mutations that restore high fitness, rather than merely increasing the overall rate of mutation. These experiments continue to be controversial, and a number of complicated but conventional processes that might explain them have been suggested.

**The Selection of Undirected Variation.** The remaining possibility is Darwin's theory of evolution through selection. This does not require that mutations occur at random: on the contrary, it is well-known that some genes mutate more frequently than others, and that different sites within a gene may have different rates of mutation. Nor does it assert that mutations are not induced by the environment: it is equally well-known that many mutations are caused straightforwardly by physical agents, such as ionizing radiation. The crucial point is that mutations are not *appropriately* induced by environmental factors. Suppose that we expose bacterial cultures, or populations of any organism, either to ultraviolet light or to ethyl methyl sulphonate (EMS), a mutagenic chemical. Both are toxic, and inhibit growth; both are also powerful mutagens, and in the appropriate doses will cause a range of mutations. It will not be found, however, that the mutations induced by ultraviolet light are specifically resistant to ultraviolet light, and not to EMS; nor that the mutations induced by EMS are specifically resistant to EMS, and not to ultraviolet light. Adaptation to either agent, or to any less dramatic environmental challenges, is through the selection of *undirected* variation. The direction (or lack of direction) taken by evolution is determined by the action of selection in each generation, and not by any inherent directional property of variation itself.

## 22. *Most mutations have slightly deleterious effects.*

Another way of expressing the undirected nature of variation is to say that mutation does not generally increase fitness. This is not unexpected. The phenomenon that we are trying to explain is the complex and highly integrated architecture of living organisms. It is not very likely that changing a circuit at random in a pocket calculator will enhance its mathematical capabilities. It is, perhaps, even less likely that rearranging the skull bones of a fish or the amino-acid sequence of cytochrome *c* should improve their design. The great majority of mutations, introducing random changes into complicated machinery, will reduce fitness.

Most newborn individuals are capable of surviving if they are protected from disease and starvation. If the entire seed output of a plant or the entire egg output of a fish is collected, most seeds and most eggs can with sufficient care be raised to adulthood. Mutation may be inevitable; most mutations may inevitably be deleterious; nevertheless, most individuals that are born can live. Most mutations, therefore, are not very deleterious. The assault on adaptedness is not carried out primarily by a storm of mutations that kill or maim, but rather by a steady drizzle of mutations with slight or inappreciable effects on health and vigor.

## 23. *Adaptedness can be maintained only if mutation is rare.*

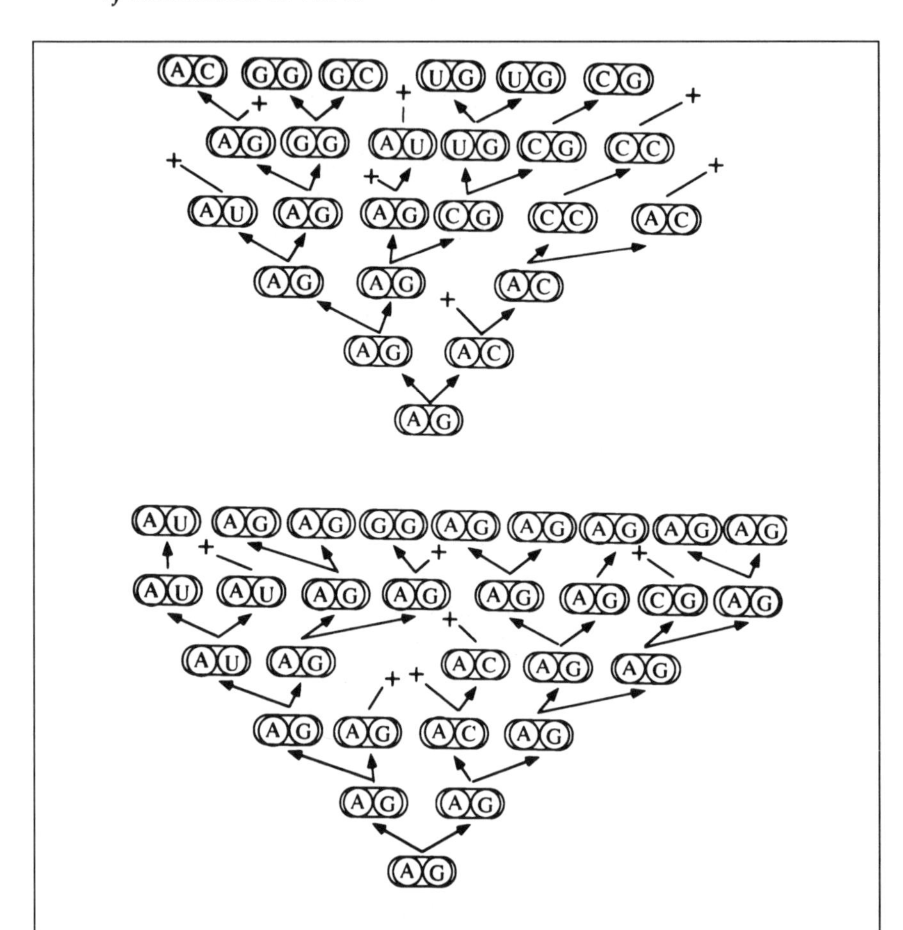


These two diagrams show the fate of a two-nucleotide replicator. In both cases, a parental AG sequence gives rise to a clone of descendants. In every generation, a replicator may die before replicating (marked with a +), survive but fail to replicate, replicate but then die, or replicate and survive. When it replicates, the sequence it transmits to its offspring may be altered. In the upper diagram, the error rate per nucleotide is greater than 50%, so that about 80% of the two-nucleotide sequences are replicated incorrectly. The AG sequence of the founding sequence is only half as likely to die as other sequences, but the infidelity of replication nevertheless prevents it from prevailing, the population becoming dominated by a cloud of mutant sequences. In the lower diagram, the error rate is less than 50% per nucleotide per replication, so that about 40% of sequences are transmitted without alteration. The greater fitness of the AG sequence now ensures that it will dominate the population.

The reduction in fitness that mutations cause may be individually slight, but collectively substantial. It is easy to appreciate that a large number of slightly deleterious mutations may have collectively the same effect as a single crippling mutation. Adaptedness is therefore endangered by the *accumulation* of mutations with slight effects. The greater the mutation rate, the faster this accumulation will proceed, eventually to the point where adaptedness can no longer be maintained.

**The Replication Limit.** This argument can be used to show that there is a limit to the rate of mutation that genomes can sustain. A replicator, such as a self-replicating RNA molecule or a DNA gene, is a sequence of nucleotides, and functions properly only if this sequence is preserved. Mutation causes changes in the sequence. Suppose that each molecule replicates $R$ times before being destroyed by some thermodynamic accident. On average, at least one of these $R$ new molecules must be a perfect copy of the parental sequence; if this is not so, then the number of copies of that sequence must inevitably decrease through time. Thus, if the precision of replication (the probability that no errors are made during replication) is $Q$, it must be the case that $Q > 1/R$ if adaptedness is to be stable. Suppose that the replicator is a sequence of $N$ nucleotides and that the error rate per nucleotide during replication is $u$. The probability that no errors are made is $Q = (1 - u)^N$, or about $e^{-Nu}$. The minimal requirement for stability is $Q > 1/2$, or roughly $u < 1/N$. Replicators that are long sequences of nucleotides can maintain themselves only if the mutation rate is sufficiently low.

The existence of this *replication limit* places severe constraints on the design of self-replicating systems. Nucleic acids can replicate themselves without enzymes, but the error rates for non-enzymic replication are about $10^{-1}$–$10^{-2}$, so that no very sophisticated replicator can be maintained. RNA replication mediated by specific replicases has error rates of about $10^{-3}$–$10^{-4}$, so that the largest single self-replicating RNA molecule is about $10^4$ nucleotides long. This is why the genomes of RNA viruses such as Q$\beta$ do not greatly exceed a length of about $10^4$ nucleotides. DNA replication has a much lower error rate, about $10^{-9}$–$10^{-10}$ per nucleotide per replication, allowing the genomes of DNA-based organisms to range up to about $10^9$ nucleotides. The genome of the bacterium *E. coli* comprises about $3.8 \times 10^6$ nucleotides; the human genome is about a thousand times larger, with about $3.5 \times 10^9$ nucleotides, approaching the upper limit for stable replication.

It is not, of course, merely a happy accident that mutation is sufficiently rare that large genomes and complex organisms can evolve: the rate of mutation is a character that varies among genomes, like all other characters, and selection will favor any reduction in this rate. The evolution of mutation rates is taken up further in Sec. 119.

## 24. *Mutation provides a constant input of variation in fitness on which selection acts.*

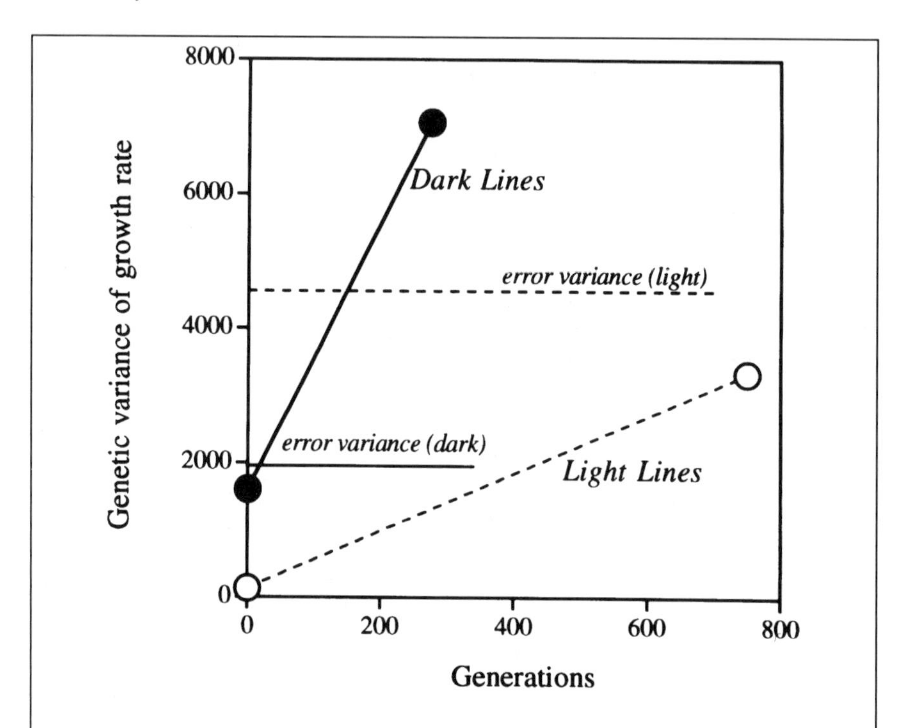


This diagram shows the outcome of culturing isolates of *Chlamydomonas* either in the light or in the dark for about a year. Growth is slower in the dark, so there are fewer generations in a given period of time. The measure of productivity is growth after ten days (more precise measures, the maximal rate of increase and the carrying capacity—see the introduction of Part 5A—gave similar results). The data are the estimates of overall genetic variance of growth rate within selection lines before and after selection. The error variances, representing uncontrolled microenvironmental differences between replicate cultures, are also shown; they did not change appreciably through time. The slopes of the graphs are estimates of the rate at which genetic variance in growth rate accumulates in the clonal populations, from Bell and Reboud (1996). This can be standardized by dividing by the error, or environmental, variance. The rate of input of new genetic variance can then be calculated as $(3330 - 131)/750 = 4.265/4655 \approx 1.0 \times 10^{-3}\sigma_e^2$ per generation in the Light lines, and $(7059-1605)/275 = 19.833/1988 \approx 1.0\times10^{-2}\sigma_e^2$ per generation in the Dark lines. The value of about $10^{-3}$ of the environmental variance per generation is comparable with other guesses and estimates (e.g. Clayton and Robertson, 1957) although it is not clear why microenvironmental

(*Continues*)

(*Continued*)
effects causing the variance of replicate cultures should provide a standard for comparing different systems.

Mutation is a stochastic phenomenon that occurs independently in different genomes. The average rate of mutation drives a process of accumulation, so that after a given period of time individuals bear some average load of mutations. But the number of possible mutations is exceedingly large, and the number of combinations of mutations is inconceivably larger; the accumulation of mutations is therefore a unique historical process, because the mutations that actually occur in any given lineage will be a very small random sample from a very large number of possibilities. Some lineages will by chance accumulate many more mutations than average, others many fewer. All will accumulate different combinations of mutations in different genes. If we isolate a single individual and then allow it to reproduce, the lineages that descend from it will come to differ in fitness as the result of acquiring different mutations. The stochastic nature of mutation and the very large number of possible mutations and combinations of mutations cause lineages to diverge, creating genetic variation in fitness.

**The Rate of Accumulation of Genetic Variance in Fitness.** Imagine a serial transfer experiment that is orginally inoculated with a single cell of some rapidly reproducing microbe. After a few generations this will have given rise to a large population, from which we can extract a number of cells and measure their fitness. They will probably all have nearly the same fitness, because there has not been enough time for very many mutations to have occurred. If we continue to transfer the population for hundreds or thousands of generations, lineages will diverge through the stochastic accumulation of mutations, and when we repeat the assay we shall expect to find substantial genetic variation in fitness among cells. At what rate will the variation of fitness caused by mutation increase through time? We can compare the genetic variation in fitness with the variation among replicate cultures of the same genotype; this variation is largely attributable to slight uncontrolled differences in the conditions of growth in different culture tubes and, thus, represents environmental variance. It has been suggested on theoretical grounds that new genetic variation in fitness should arise at a rate equivalent to about one-thousandth of the environmental variation per generation. I have done several serial transfer experiments with the unicellular green alga *Chlamydomonas* that show that this guess is about right: after about a thousand generations of culture in favorable conditions (for an alga growing photoautotrophically, this means adequate supplies of light, carbon dioxide, and mineral nutrients), the genetic variation of fitness has increased from zero to a level about equal to environmental variation.

This slight incessant input of mutational novelty is the original source of all subsequent evolutionary change. Because most mutations are deleterious, they will generally reduce the fit between the population and the environment, which must then be restored by selection. On the other hand, should the environment change, it is the only process by which new favorable variants can appear and become available to selection.

## 25. *Selection preserves adaptedness by preventing the spread of mildly deleterious mutations.*

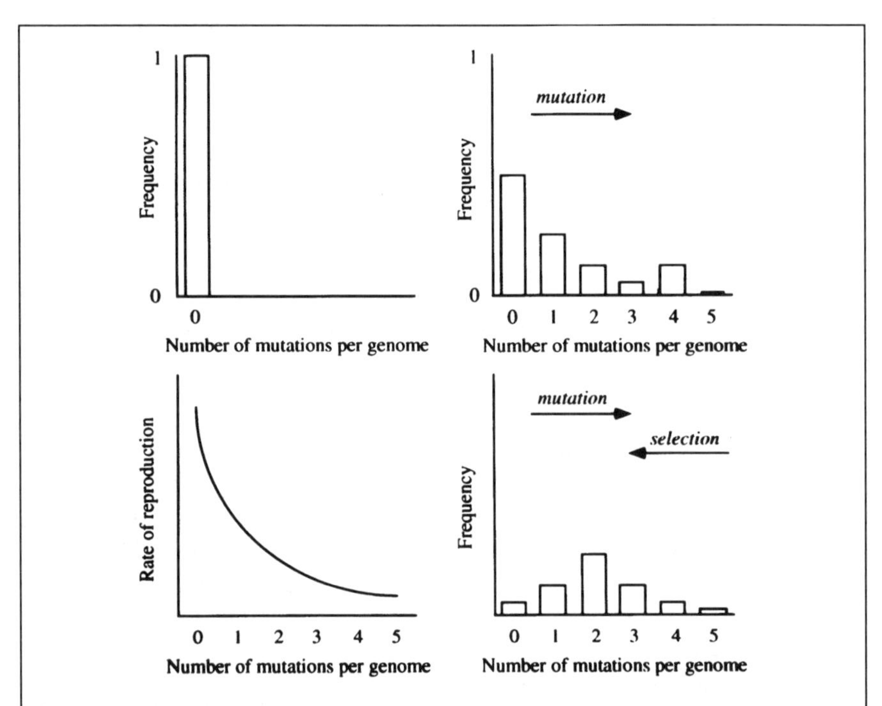


These four diagrams illustrate the argument of Sec. 25. A population that is initially perfectly adapted to its environment, with no individuals bearing any deleterious mutations, will inevitably accumulate mutations through time. Individuals with more mutations, however have lower fitness, and so the decrease of fitness through mutation is opposed by an increase through selection. The outcome is a frequency distribution of mutational load representing a balance between mutation and selection.

Imagine, again, a population that at some point becomes perfectly adapted to its environment. Every gene has attained the best possible state; no individual bears any deleterious mutations. This perfection cannot be maintained even for a single generation. Some progeny will receive one or more new mutations, so that the population includes, not only the class of individuals with no mutations, but also classes of individuals that have suffered different degrees of mutational damage. Each will transmit these mutations to its offspring, together with any new mutations that have occurred in its germline. The number of mutations borne by any lineage will tend inexorably to accumulate, eroding the adaptedness of the population. This process is countered by selection. Individuals which bear more mutations are less likely to reproduce successfully than those which bear fewer. The next generation will therefore be recruited predominantly from individuals in the current generation which bear fewer mutations than average, so that selection tends to reduce the average number of mutations per individual. In this way, the tendency for mutations to accumulate without limit is checked by the tendency for individuals which bear fewer mutations to produce more offspring. At some point the number of old mutations removed through selection will be equal to the number of new mutations appearing. The opposed tendencies of mutation and selection are then balanced: the population does not consist entirely of perfect, undamaged individuals (that is forbidden by mutation), nor of a few hulks riddled by mutations and scarcely able to survive (that is prevented by selection), but rather of a mixture of individuals, some with fewer mutations and some with more. This mixture represents an equilibrium state, a population in which the frequency distribution of mutations per genome is stable from generation to generation. In each new generation there is an episode of variation, in which new mutations arise, and an episode of selection, in which an equivalent number of mutations are removed through the death or sterility of individuals that bear them, restoring the fit between population and environment. At this point, the population is said to be in *mutation-selection equilibrium*.

**Mutation-Selection Equilibrium.** The point of balance is set by the relative rates of mutation and selection. Suppose that each generation begins with an episode of mutation. The mutation rate of a given gene is $u$. This means that the frequency of the mutant after mutation ($p'$) is related to its frequency before mutation ($p$) in this way: $p' = p(1 + u)$. There is then an episode of selection. The rate of selection acting on the mutation is $s$, in the sense that the frequency of the mutant after selection ($p''$) is related to its frequency after mutation but before selection ($p'$) in this way: $p'' = p'(1 - s)$. At equilibrium the rate of mutation $u$ is balanced by the rate of selection $s$, and the frequency of the mutant in the population is: $p = u/s$. In other words, selection cannot completely eliminate deleterious mutations from the population, but instead drives them down to a frequency $u/s$ that is sustained by recurrent mutation. The mutation rate is usually taken to be the collective rate of all changes in the structure of a gene, so that the selection rate is an average taken over all the different kinds of change, which may have quite different effects on fitness. Most genes are about $10^4$–$10^5$ nucleotides in length, so that an error rate of $10^{-10}$–$10^{-9}$ per nucleotide implies an overall error rate of roughly $10^{-5}$ per gene. The average rate of selection acting against mildly deleterious genes is commonly supposed to be about $10^{-2}$ (this owes more to the decimal system than to accurate measurement), so that mutant variants of a gene should occur in the population at a frequency of the order of $10^{-3}$.

This argument can be extended to the genome as a whole. If the combined rate of mutation at all gene loci is taken to be $U$ and the average rate of selection against mutants to be $S$, then the average number of mutations borne by each individual at mutation-selection equilibrium is $U/S$. The genome of a metazoan or a vascular plant typically comprises about $10^5$ genes and, therefore, typically bears about 100 slightly deleterious mutations, to which one new mutation is added at every replication.

These figures should be taken only as very crude rules of thumb. They serve to emphasize how the continuous operation of selection stems the continuous accumulation of mutational load.

## 26. *Characters other than fitness have intermediate optima.*

Selection acting on fitness is always directional, because variants with greater rates of replication always tend to increase in frequency. This is not true for the structure and biochemistry that underlies fitness. If any character is exaggerated too far in any direction, it has the effect of reducing fitness. This is because any character affects fitness in many ways; some effects increase fitness when the character is altered in one direction, others increase fitness when it is altered in another direction. If it is altered too far in any direction, the effects that reduce fitness come to outweigh those that increase fitness. A resistant spore produced when conditions are unfavorable for growth must persist for a considerable period of time, and then

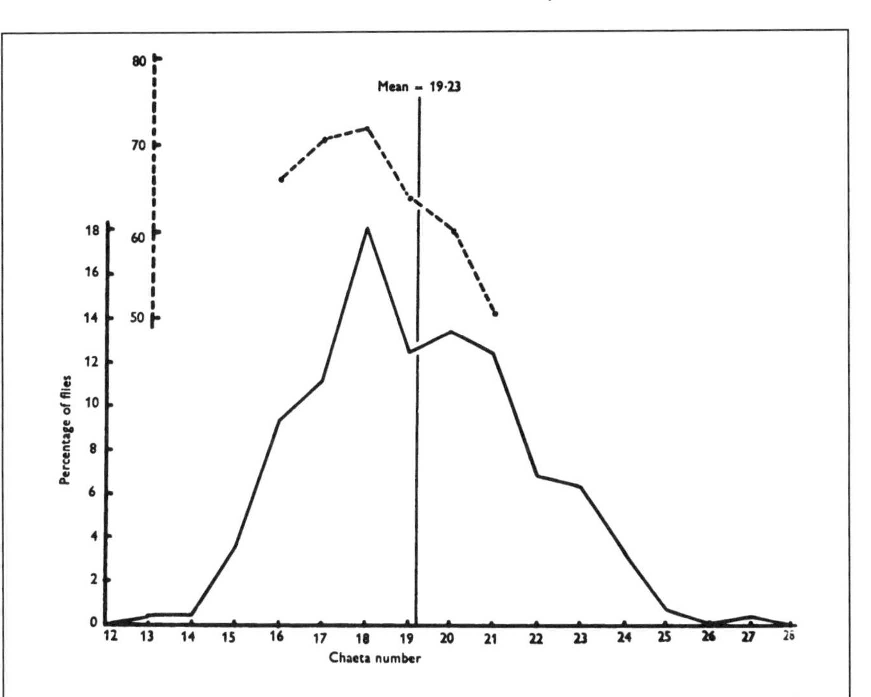


The usual example of a character in which fitness falls off on either side of the mean is human birth weight, as reported by Karn and Penrose (1951) and described briefly in the text. The figure shows a less familiar example of stabilizing selection, but involves a very familiar system in experimental studies of evolution: the number of sternopleural bristles in *Drosophila*. McGill and Mather (1972) crossed two wild-type strains, obtaining an $F_2$ in which bristle number was approximately Normally distributed with a mean of about 19 and a mode at 18 bristles. A pair of these flies with a given number of bristles and a pair of flies from the 6CL strain, which bear many more bristles, were permitted to mate freely, and their progeny were then reared together. The proportion of wild-type flies emerging from this $F_3$ was then used as a measure of the fitness of the test flies. When fitness was plotted as a function of bristle number, it was maximal at 18 bristles and decreased on either side of this value. This suggests that the distribution of bristle number in the population is maintained through stabilizing selection. Other experiments on stabilizing selection in *Drosophila* include Falconer (1957; abdominal bristles), Thoday (1959; sternopleural bristles), Mather and Cooke (1962; sternopleural bristles), Barnes (1968; sternopleural bristles), Gibson and Bradley (1974; sternopleural bristles), Prout (1962; development time), Scharloo Hoogmoed and ter Kuile (1967; length of wing vein), Bos and Scharloo (1973a, 1973b, 1974; thorax length), Tantawy and Tayel (1970; thorax length) and Grant and Mettler (1969; escape behavior).

(*Continues*)

(*Continued*)

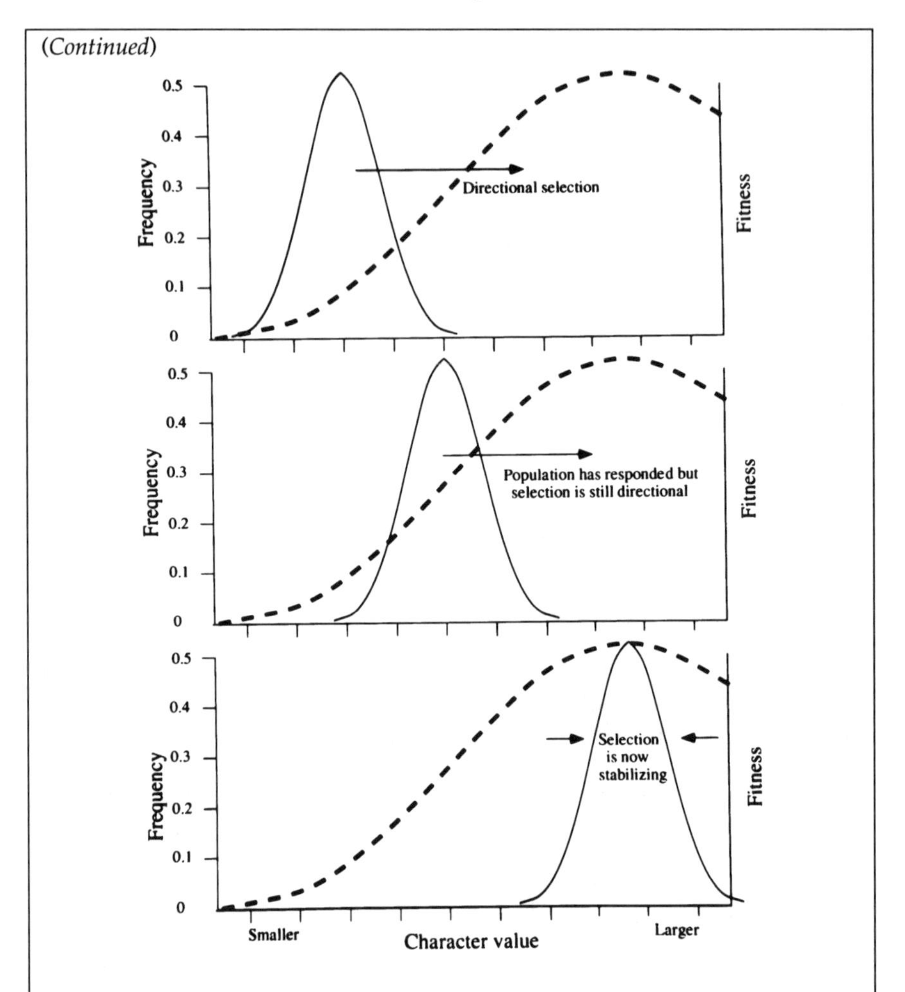


Because all characters will have intermediate optima, the fitness function will, in general, be a humped curve. We can then draw the general case of the various fitness functions discussed in Sec. 20. The effect of selection on a character depends on the relationship between the population frequency distribution and the fitness function. If the mean character value does not coincide with the maximum of fitness (for example, if the environment has recently changed), directional selection will drive the population so as to increase mean fitness. This process slows down and finally halts as the population mean approaches the maximum of fitness, at which point selection is stabilizing.

germinate when conditions improve. It must have a thick wall in order to resist desiccation, abrasion, and the attacks of fungi and grazing animals. The thicker its wall, the more likely it is to survive. It must also have a thin wall in order to be able to germinate easily. The thinner its wall, the more likely it is to germinate successfully. These effects are incompatible. A spore with a very thin wall will have nearly zero fitness, because it is almost certain to dry out or rot. A spore with a very thick wall also has nearly zero fitness, because it is almost certain to be unable to germinate. The optimal value for the thickness of the wall is, therefore, intermediate between these extremes.

**Stabilizing Selection on Birth Weight.** Human birth weight is a classical example of a character with an intermediate optimum. M. Karn and L.S. Penrose of London investigated the effect of the size of newborn babies on their survival in the first month of life, in a cohort of children born in Britain in the 1930s and 1940s. Their average birth weight was about 3.25 kg. Babies of average size had a nearly maximal survival rate of about 98%. Very small babies and very large babies were much more likely to die: mortality increased steeply as birth weight decreased, so that less than 30% of babies smaller than 1.8 kg survived, and increased rather less steeply as birth weight increased, with about 90% of babies 4.5 kg in weight surviving. The average mortality was about 5%, enough to cause appreciable selection against the extremes of birth weight.

Selection will seldom, if ever, continue to exaggerate any character in any direction without limit. The optimal value of a character, defined as the value that maximizes fitness, will be intermediate between possible extremes.

## 27. *Stabilizing selection reduces variation around the optimal value of a character.*

In a well adapted population, the average value of a character will be close to its optimal value. This does not mean that all individuals will express the optimal value of the character. If this were ever so, mutation would in the next generation give rise to individuals that deviated in either direction from the optimum. Mutation has a directional effect on fitness, reducing it in nearly all cases. It does not have a directional effect on characters: it may either increase or decrease the value of a character, reducing fitness as a consequence. The population will thereby come to display a distribution of character values, centred near the optimum. Human birth weight is a character for which the values of individuals are distributed around an average value close to the optimum.

In every generation mutation will cause an episode of variation, reducing the frequency of individuals with nearly optimal values for a given character and increasing the frequency of individuals with extreme values. Individuals with extreme values have low fitness; there will therefore be an episode of selection in which the frequency of individuals with extreme values is reduced, and the frequency of individuals with nearly optimal values increased. When mutation and selection have come into balance, the frequency distribution of the character in the population will represent an equilibrium: the input of new variation by mutation is canceled by the selective removal of individuals with extreme values. The population breathes, so to speak: in every generation, variation is first inflated by mutation and then deflated through selection.

We are accustomed to thinking of evolution as involving change—an increase in the adaptedness of the population. It is also important to appreciate that adaptedness is constantly being degraded by mutation, and cannot be maintained passively. However paradoxical it may seem, selection will normally act as a conservative force that resists evolutionary change. Stabilizing selection is therefore a very general and very important evolutionary process, despite the fact that it causes no change in the average state of the population. What it does do is change the amount of variation: we expect to find that variation declines during the lifespan of a single cohort of offspring because of the selective mortality of extreme individuals.

**The Shape of Snails.** An elegant example of stabilizing selection was described by A.P. di Cesnola in the early days of quantitative evolutionary biology, nearly a century ago. Snail shells grow by accretion, with new whorls being added to those formed previously. Every snail therefore carries around with it an account of its individual developmental history that can be read by sectioning the shell. Di Cesnola found that the variation of shell shape among young snails was greater than the variation among the first-formed whorls of the shells of older snails. This strongly suggests that individuals with aberrant shells are less likely to survive.

**The Skeletons of Newt Larvae.** As a graduate student, I studied the morphology of a cohort of newt larvae in a pond near Oxford. By measuring a variety of skeletal characters from the time when the larvae hatched in early spring to the time when most of them left the water in midsummer,

I was able to show that there was a general and continuous decline in morphological variation during this period. This did not happen when the larvae were carefully reared in the laboratory, supplied with adequate food and protected from predators, and so it was probably caused in the field by stabilizing selection.

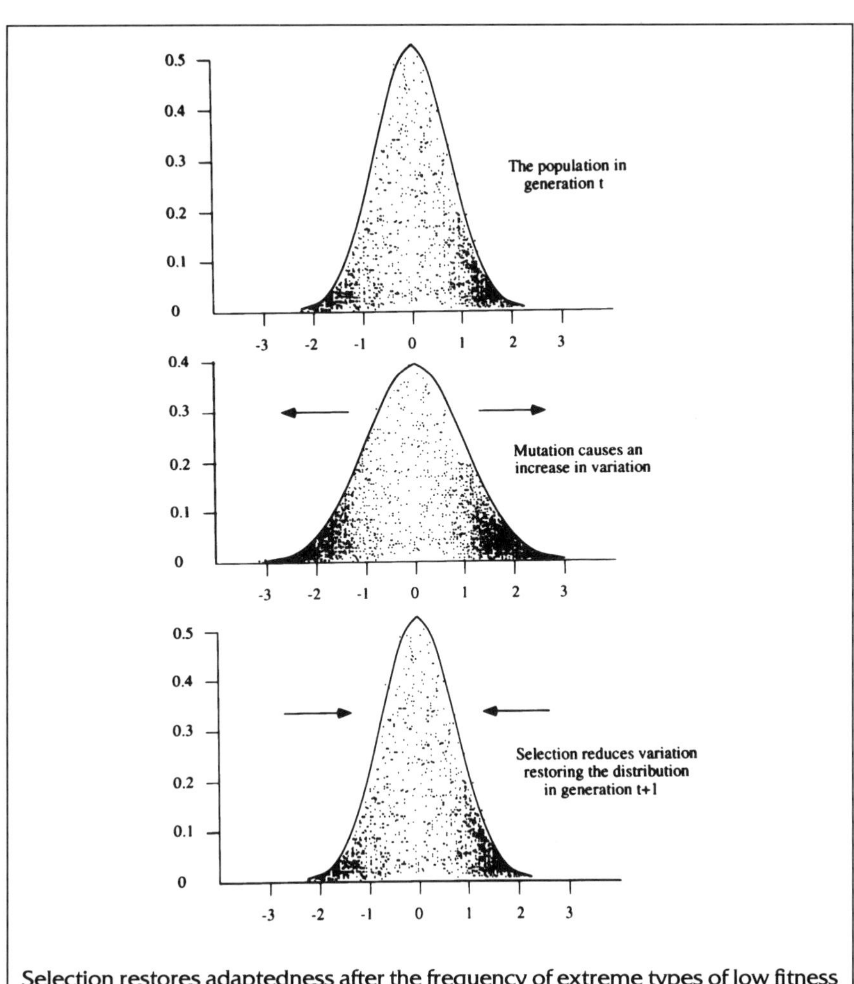


Selection restores adaptedness after the frequency of extreme types of low fitness has been increased by mutation.

(*Continues*)

(*Continued*)

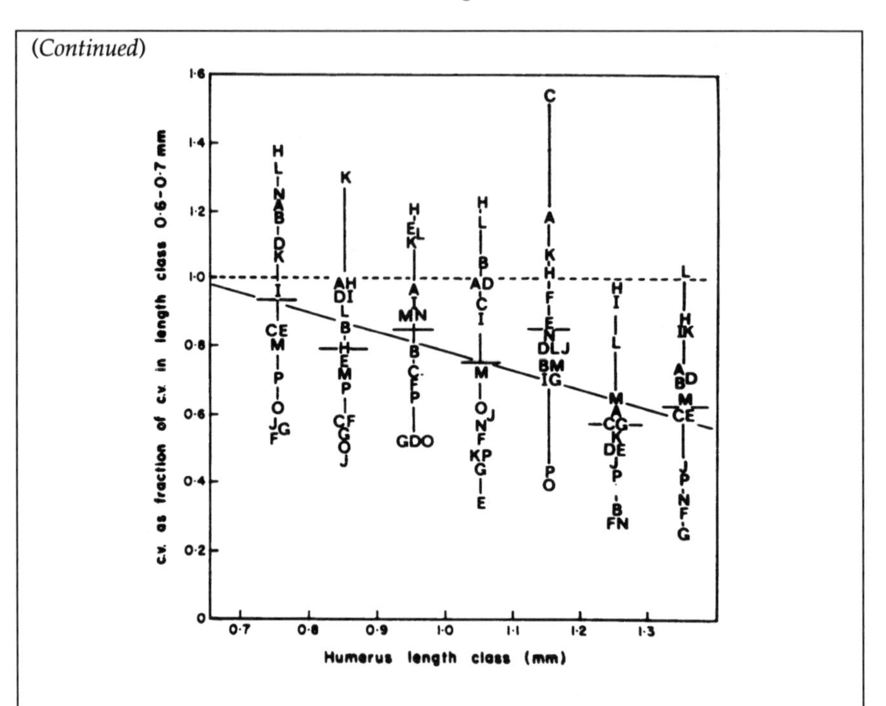


This figure shows the decline of morphological variation with age in a cohort of newt larvae, from Bell (1978). The $y$-axis is the coefficient of variation, a measure of variation scaled to the mean, relative to the variation in the youngest individuals; the $x$-axis is humerus length, which is used as a surrogate for age. Each letter represents 1 of 16 different characters, measured by clearing the soft tissues and staining the skeleton with alizarin. A decline in the variability of size or shape during the lifespan of a cohort caused by a single episode of stabilizing selection has been reported from sparrows (Bumpus 1899; also see O'Donald 1973, Lande & Arnold 1983), ground finches (B.R Grant 1985, P.R. Grant et al. 1976), ducks (Rendel 1943), snakes (Inger 1942), lizards (Fox 1975), sticklebacks (Gross 1978, Hagen & Gilbertson 1973), land snails (di Cesnola 1907), whelks (Berry & Crothers 1970), bivalves (Palenzona et al. 1971), crabs (Weldon 1901), moths (Crampton 1904), beetles (Mason 1964, Scheiring 1977) and corals (Potts 1984). Most of the experimental work with *Drosophila* (Sec. 26) describes the effect of artificial stabilizing selection on phenotypic and genetic variation.

## 28. *Bacterial screens and crop trials are simple examples of directional selection.*

When a population is well adapted to an environment that has remained constant for a long period of time, almost all mutations are deleterious, and selection is stabilizing. If the environment changes, it is likely that for some loci the most common variant is no longer the best version of the gene. Some mutations which would previously have been deleterious actually improve performance in the altered conditions. These mutations will tend to increase in frequency through selection of the individuals that bear them. In this case, selection, instead of preserving the characteristics of the population, tends instead to cause a change in the average values of some characters. The lack of fit between population and environment caused by environmental change is thus restored by directional selection.

The mutant screens that are carried out on bacteria and other microbes illustrate directional selection at its simplest. The object of a screen is often to isolate extreme variants that are able to grow in a novel and hostile environment, such as that created by adding an antibiotic to the growth medium. A large population is cultured in a permissive environment, often in the presence of a mutagen that will increase the extent of genetic variation. Samples are then spread onto plates, using growth medium containing the antibiotic. The great majority of cells will die; a very few cells will survive and grow to form colonies that can be picked off and used to found resistant strains. It requires a little more ingenuity to select for defective strains that grow slowly. There is a repertoire of methods for isolating mutants with specific lesions. For example, nitrate in the medium must be reduced before it can be incorporated into protein by a chain of reactions, the first of which is its reduction to nitrite by a nitrate reductase. Strains that lack nitrate reductase activity can be identified by plating samples onto medium containing sodium chlorate. This is harmless; but it is reduced by nitrate reductase to sodium chlorite, which is highly toxic. Cells with a functional nitrate reductase system are thus killed by their own metabolic competence, and only a few loss-of-function mutants survive. A more general technique is to grow a culture in minimal medium with an antibiotic, such as penicillin, that kills only growing cells. The survivors, replated onto complete medium without the antibiotic, will be auxotrophic mutants of various kinds that can grow only when the medium is supplemented with the substances they are unable to synthesize. In short, a bacterial screen is run by constructing an environment, or a series of environments, in which the frequency of the desired type will be enormously magnified through selection.

A crop trial furnishes another familiar example of a single episode of directional selection. The goal of most trials is to identify plants in which some desirable characteristic, such as the weight of cotton bolls or the sugar content of cane juice, is most highly developed. A large and diverse set of entries is created, usually by crossing parents in which this characteristic is already well-developed , and allowed to grow. The boll weight or sugar content of each individual is then measured, and the most extreme individuals chosen to found lines for commercial release or further study.

## 29. *Artificial selection allows characters to be selected directly.*

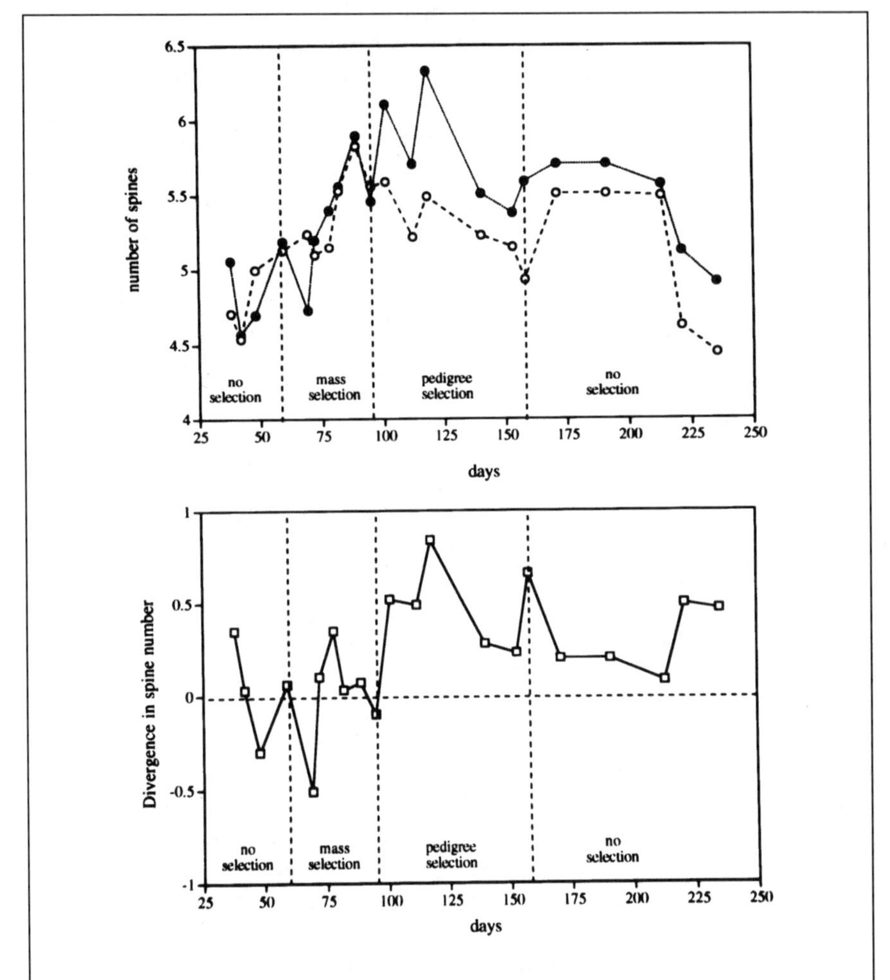


There were a number of attempts to select within asexual clones in the first two decades following the rediscovery of Mendel's work. Jennings (1916) selected for the number of spines on the test of the rhizopod *Difflugia* and succeeded

(*Continues*)

in producing a diversity of lineages with more or fewer spines from a single clonal stock. The upper diagram shows the divergence of upward and downward selection lines in his family 326; the lower diagram is the difference between the two lines. The experiment fell into four periods: an initial period during which no selection was applied, a period of mass selection for the most extreme individuals, a period of pedigree selection for the parents of the most extreme individuals, and a final period of relaxed selection. Mass selection had no discernible effect, probably because it was practiced only briefly on small populations of a relatively long-lived organism—*Difflugia*, though unicellular, can live for weeks or even months. Pedigree selection was immediately effective, perhaps capturing a new genotype very quickly, for there was little sign of a continued response. The difference between the lines was maintained, perhaps at a decreasing level, when selection was relaxed.

Middleton (1915) likewise obtained a response to selection for higher fission rate in clones of the ciliate *Stylonichia*. However, most experiments ended in failure: Hanel (1908) and Lashley (1916) attempted to select for increased tentacle number in *Hydra*; Jennings (1908) for size in *Paramecium*; Agar (1913) for the rate of reproduction in the cladoceran *Simocephalus*; Ewing (1914) for rate of reproduction in *Aphis* (also see Agar 1914); Surface and Pearl (1915) for yield in oats; Mendiola (1919) for the shape and rate of growth of *Lemna*. These results confirmed the pure-line theory of Johannsen (1903; English account by Johannsen 1911; see Jennings 1910 for a review of selection experiments in relation to the theory). In sexual populations, on the other hand, selection was generally effective. The work of the de Vilmorin family on carrots, and later sugar beet, dates back to 1840. Very few other cases, however, are mentioned by de Varigny, in a book published in 1892. Systematic investigations began shortly after this date, in the United States, where experiments with corn (notably the Illinois corn-oil experiment, first described by Hopkins 1899, see section 62; Smith 1908 and Pearl & Surface 1910 also conducted experiments with maize) and fowl (Pearl 1917) were set up at agricultural research stations in the closing years of the 19th century. The comparable tradition in university laboratories was founded by selection for coloration in rats by Castle and Phillips (1914) and for bristle number and eye facet number in *Drosophila* by Macdowell (1919) and Zeleny (1922; also see Zeleny & Mattoon 1915, May 1917 and Zeleny 1921). There is a good contemporary review by Pearl (1917), who cites a number of other early references and emphasizes the lack of any evidence for the cumulative effect of selection on slight novel variations. He makes the very perceptive point that selection is most effective when applied to the germ rather than to the soma, so that it is more effective to select individuals according to the qualities of their offspring rather than their own qualities; Jenning's successful experiment with *Difflugia* (Jennings 1916) used pedigree selection of this kind. All of this work involved artificial selection, and contemporary experiments on natural selection are exceedingly scarce. Fernandus Payne (1912) cultured *Drosophila ampelophila* in the dark for 69 generations in order to see whether it would evolve the characteristic adaptations of cave animals; he found no response in either morphological or behavioral characters.

Bacterial screens and crop trials both exemplify the unit event of evolution, in which an episode of variation is followed by an episode of selection, but in one respect they are quite different. The mutant screen is an example of *natural selection*, even though it is a deliberately contrived situation. The population is exposed to a novel environment in which some variants are able to replicate and others not; those which are able to replicate most successfully are the founders of the next generation. The process is a faithful if rather dramatic representation of organic evolution. In crop trials, on the other hand, the rate of replication is scarcely ever the character sought by the experimenter: lines with high rates of replication are weeds, and do not need to be encouraged. Rather, the experimenter applies selection by deliberately choosing the most desirable individuals, which in other respects may be very feeble. This is *artificial selection*. In natural selection, nothing does the selecting; the fittest individuals select themselves, so to speak, by virtue of their greater rates of replication. In artificial selection, a human agency is responsible for life and death, almost regardless of the ability of individuals to grow and reproduce. The distinction is not, to be sure, a very profound one. If it appears to be so, it is only because of our habit of thinking of ourselves as being apart from nature. We can easily conflate the two processes by observing that the individuals which are successful in a crop trial are indeed those with the greatest rates of replication, in the special environment furnished by the presence, not of an antibiotic, but of an agronomist. Nevertheless, the distinction between natural and artificial selection is a useful one, because it is easy to apply selection deliberately in ways that would be difficult to simulate by a process of natural selection in which populations reseed themselves. In particular, it is easy to ensure that only those individuals with a character value exceeding a given threshold succeed in breeding. Character evolution is often studied more easily and more effectively through artificial selection than through natural selection.

**Early Selection Experiments.** The earliest selection experiments are pre-Darwinian. Phillippe-André de Vilmorin planted seeds of the wild carrot in 1833, selecting the plants with larger and thicker roots. He continued this process for seven generations, until 1839, by which time his stocks were producing a large proportion of roots comparable in size to those of cultivated varieties. No doubt many other such experiments went unrecorded, as farmers and stockbreeders following the example of Robert Bakewell attempted, perhaps in a less conscious and systematic fashion, to improve the breeds of corn and cattle. One might expect that the announcement of the theory of natural selection in 1858, and the controversy that followed the publication of the *Origin of Species* in 1859, would have stimulated a flood of experimental work; Darwin, after all, had begun his exposition of natural selection with a description of artificial selection. However, this was not the case. Indeed, I am not aware of a single selection experiment, deliberately contrived to investigate the action of selection, being performed between the publication of the *Origin of Species* and the close of the century, the years when the evolution controversy was at its height. Because of this remarkable circumstance, the intellectual tradition in evolutionary biology was early established as descriptive, comparative, and interpretive, rather than analytical, experimental, and demonstrative. It was not until the first decade of the twentieth century that selection experiments began to appear in the literature; but they were stimulated, not by Darwinism, but by the newly rediscovered principles of Mendelism, and they were performed, not in Britain, but in the United States.

The genetical principle at issue was the stability of pure lines. Johannsen, in his oft-cited experiment of 1902, had demonstrated the uniformity of inbred homozygous lines of beans by failing to cause any response through selection, but for 10 or 15 years this remained a controversial result because of the brevity of the experiment. A number of authors tried the effect of selecting for many generations in asexual populations of creatures as diverse as ciliates, hydras, cladocerans, aphids, lentils, and duckweed. In almost all cases, selection was entirely ineffective, and Johannsen's result became established as one of the main proofs of Mendelism. There were some disquieting indications of quite contrary results from bacteria; but bacteria were known to be such fickle and fluctuating creatures that these could safely be ignored. At the same time, it became clear that when selection was practised on sexual populations it was almost always effective. The first experiments involved crop plants and chickens, but in the second decade of the century the foundations of modern experimental evolution began to be laid by people such as W.E. Castle of the Carnegie Institute and Edwin Macdowell at the Cold Spring Harbor station, who began to report the results of work on laboratory populations of rats and *Drosophila*. The ease of selecting in sexual populations, and the lack of progress in asexual populations, was generally interpreted as a *falsification* of Darwinism. The Mendelian elements were fixed and finite in number; they could be reassorted and recombined, but not altered, destroyed, or created. Consequently, selection would sort any variation initially present in the population, but could never produce new kinds of organism. With hindsight, it is not difficult to understand the situation. The attempts to select within clones involved populations that were small, and characters (such as the number of tentacles in *Hydra*) that were sensitive to environmental or developmental variation. It is not surprising that they failed, and it is a pity that the bacterial results were not taken more seriously. Their failure was a decisive influence on the history of evolutionary biology for the next 50 years. It led to the neglect of experimental work in favor of an increasingly elaborate theory of sorting in sexual populations. This did not begin to be remedied until the *Drosophila* program, heralded by L'Héritier and Tessier in the 1930s, began to gather pace in the 1950s, with experiments on bacteria following the invention of the chemostat at about the same time. The basic principles of population genetics 40 years later have been illuminated by a considerable body of experimental work; but the lasting scar left by the failure to develop experimental programs in the early years of Darwinism has been, as we shall see, the continued obscurity of the effects of selection in the long term.

## 30. Selection is a commonplace process in natural populations.

The efficacy of selection in certain contrived situations has never been in dispute. It has often been argued, however, that this does not license extending what we know from the laboratory or the experimental farm to natural populations. Experiments might tell us little about nature. Selection might rarely occur in nature, or, if it did occur, might have slight, inappreciable effects; the routine occurrence of powerful selection capable of causing rapid genetic change was felt by many biologists, 30 or 40 years ago, to be out of the question. This position was buttressed by the formidable masonry of mathematical population genetics. It was undermined by the efforts of a group of British naturalists—E.B. Ford, Arthur Cain, Philip Shepherd and others—in the 1950s and 1960s, and demolished by their innumerable students, collaborators and colleagues. Their efforts have given us most of the classical accounts of selection in natural populations.

The effect of a single episode of natural selection is measured by the change in frequency of alternative types during the lifespan of a single cohort. In principle, any change in frequency should be measured between equivalent stages of life, for example, between egg and egg. In practice, mortality is much easier to measure than fecundity, and most studies are incomplete, measuring the change in frequency attributable to differential survival over part of the life cycle. This can be done with different degrees of sophistication. The simplest approach is to measure the frequency of variants at different stages of life in unmanipulated populations. My study of stabilizing selection in newt larvae is an example of this kind. The main objection to it is that alternative explanations, for example, the tendency of individuals to resemble one another more nearly with advancing age for purely developmental reasons, must be eliminated by further experiments. A more direct and more satisfactory experimental approach is to construct an artificial cohort, by raising individuals in the laboratory and releasing them into the natural population. If very many individuals are released relative to the number in the population, this is a *perturbation experiment*: the frequency of variants in the population has been changed, and we expect selection to restore the *status quo ante* . The objection to this design is that it may change the absolute population density of the organism and thereby change the nature of the selection previously acting on it. If the number of releases is small compared with the number in the population, then we can measure changes in the frequency of variants among the released individuals by marking them in some way so as to make them easily recognizable. This is a *mark-release-recapture experiment*. The principle objection to it is that the marks themselves may be a confounding source of selection. This objection can usually be discounted by careful experimental design; the mark–release–recapture experiment is the most elegant and satisfactory method of detecting a single episode of selection in natural populations.

There are, of course, few places in Britain that are untouched by human activity, to the extent that sites in the Canadian forest or the Pacific abyss are untouched. Natural means little more than "outdoors". Indeed, some of the clearest examples of natural selection have been provided by the vast unplanned experiment of the Industrial Revolution.

**Industrial Melanism.** About the middle of the nineteenth century, black varieties of several moths, formerly rarities, began to spread to a remarkable extent in the industrial regions of midland and northern England. The canonical example is the peppered moth, *Biston betularia*, in which a marked increase in melanic pigmentation is caused by a single, almost completely dominant, mutation. The spread of the melanic type followed the appearance of clouds of coal smoke above the newly industrialized towns, smoke that was washed down by the rain to blacken walls and tree-trunks with soot. It is easy with hindsight to appreciate that the pepper-and-salt markings of the type originally common blended with the lichen-covered tree-trunks of the unpolluted countryside, whereas the black wings and abdomen of the melanic variant were inconspicuous when the moth was resting on the soot-encrusted bark of urban trees. Visual predators such as birds would detect the pepper-and-salt variety more easily against a blackened background, creating selection that favored the melanics and caused their spread. Bernard Kettlewell of Oxford showed that birds preyed on moths resting on tree-trunks, taking the more conspicuous kind first. Their effect on the frequency of melanics during a single episode of selection was demonstrated by following the fate of a relatively small number of moths, melanic and pepper-and salt, released into natural populations. In the countryside, where lichens still grew thickly on tree-trunks, the pepper-and-salt variant increased in frequency over the course of a few days; in polluted areas, it was the melanics that increased. A single episode of selection, therefore, showed that melanic variants tended to spread in the novel environment furnished by sooty towns.

**The Patterns of Snail Shells.** The ideal witnesses of selection for crypsis, or concealing coloration and pattern, would be the birds themselves. Another Oxford project pressed them into service. The snail *Cepaea nemoralis*, common in limestone regions of Europe, has a very variable shell; the background color ranges from pale yellow through pink to dark brown, and it may bear a number of dark bands running around the whorls. There are an enormous number of possible variants, which might be (and were) dismissed as being of no functional importance. They might equally be regarded as characters that conceal the snail more or less effectively against visual predators in different kinds of habitat. Thrushes eat snails. They cannot swallow whole any but the smallest; larger ones they carry to a nearby stone, hammering them until the shell is broken and the snail can be extracted. The same stone is used over and over again, so that the preferred anvil is surrounded by the broken remains of the thrushes' prey. Biologists are more effective predators than snails, and can probably find almost all the adult snails in a small area, including those that have been missed by the thrushes. This gave Arthur Cain and Philip Sheppard the opportunity to compare the snails that had been detected and eaten by the thrushes in a small area of woodland with the remnant population that had escaped predation. In early April, when the floor of the wood was covered by sodden dark leaves, snails with yellow shells were disproportionately frequent on the anvils; as the season advanced and the understorey greened, the preference of the thrushes changed, and yellow shells became less frequent on the anvils than in the population at large. The obvious conclusion, that the thrushes, visual predators, were choosing the more conspicuous snails, was checked experimentally by releasing marked snails in the wood. Those with yellow shells were more at risk earlier in the season, and survived better later in the season, confirming that selection favored the more cryptic individuals.

**The Beaks of Darwin's Finches.** At the other end of the world, Peter and Rosemary Grant, working from McGill and Princeton, had taken up the task of explaining the diversity of form, first noticed by Darwin and later studied by David Lack, among the finch-like birds of the Galapagos Islands. Birds with different kinds of beak feed more efficiently on different kinds of fruit. *Geospiza magnirostris* has a large beak and feeds on large, hard-shelled fruits; *Geospiza fortis* has a smaller beak and prefers smaller, softer fruit. The difference in beak size, like all differences between species, is a matter of degree; individuals of *G. fortis* vary in beak size, and this variation is inherited. In 1976 there was a severe and prolonged drought that greatly decreased the amount of food available to the finches. This drought affected some plants more than others; those with large, thick-walled fruits survived better, so the composition of the assemblage of fruits available to the finches changed. The decline in the overall availability of food caused a great deal of mortality; the change in the frequency of different kinds of fruits ensured that this mortality was selective. The increased proportion of thick-walled fruits meant that *G. magnirostris* could live where *G. fortis* might perish; but it also favored individuals of *G. fortis* with stouter beaks than average. As the result of selection caused by the drought, the mean beak size of *G. fortis* increased by about 4% in a single generation. In 1982 the opposite happened: heavy rainfall was followed by the lush growth of plants with small, soft fruits, and selection acting contrarily caused beak size to decrease by about 2.5%.

Peppered moths, snails, and Darwin's finches are good examples of natural selection; but they are only three examples in a very long list. John Endler of Santa Barbara has collated hundreds of examples of selection in natural populations, together with an account of the various methods that have been used to demonstrate selection in different circumstances. His work has brought to an end the period in which the ubiquity of natural selection could legitimately be questioned. It is evidently a commonplace process that can readily be discovered in almost any population. The strength of selection is more surprising: there are many cases in which the rate of selection causes substantial changes in the frequency of different types within a single generation. Such cases are, no doubt, disproportionately represented in the literature; given the brevity of graduate programs and research grants, most people prefer to work on systems where positive results may confidently be anticipated. It is, nonetheless, remarkable that despite the difficulties of fieldwork a single episode of selection has been found on so many occasions to alter the composition of natural populations. Selection in nature is sometimes very powerful and is, therefore, capable of causing rapid change in phenotype.

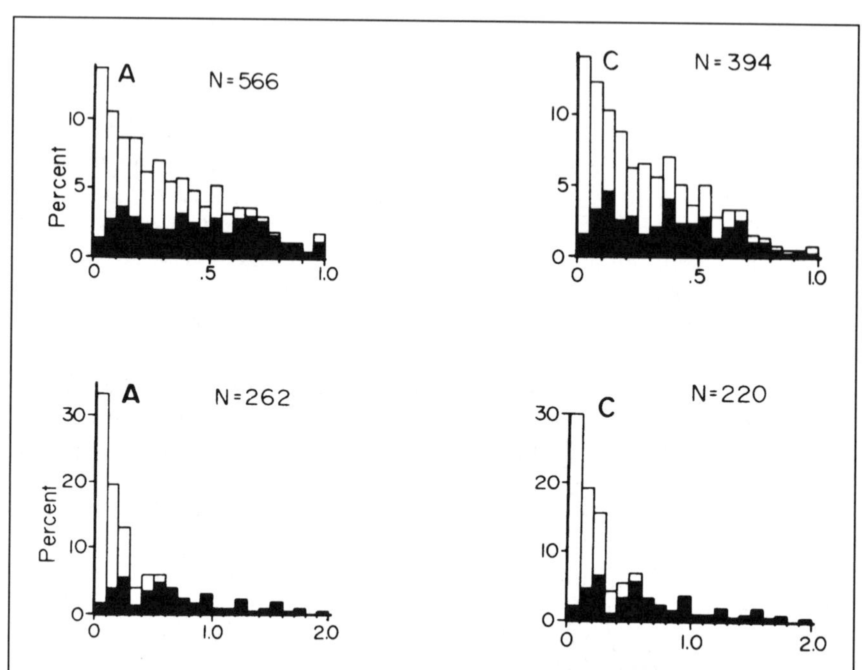


Accounts of selection in natural populations have been thoroughly reviewed by Endler (1986), whose excellent book makes it unnecessary for me to give an extensive list of examples here. These figures summarize a large number of field studies, as frequency distributions of estimates of rates of selection. The upper pair of figures refer to selection coefficients for discrete polymorphic traits, such as color. In each case, the selection coefficient expresses the relative fitness of two (or more) types, from the numbers of each present before and after selection. Thus, if the numbers of two types are $m_1$ and $m_2$ before selection and $n_1$ and $n_2$ after selection, then their fitnesses are $w_1 = n_1/m_1$ and $w_2 = n_2/m_2$. Supposing the fitness of the first type $w_1$ to be the greater, we can set it equal to unity and then calculate the relative fitness of the second type as $w_2/w_1$. The selection coefficient is then $s = 1 - w_2/w_1$. Larger values of s indicate stronger selection, with a maximum value of unity. The lower pair of figures refers to selection differentials for continuous traits, such as size. They are calculated as the standardized difference between the population before selection and the population after selection. Thus, if the mean value before selection is $z_1$ and the population standard deviation is $\sigma_p$ whereas after selection the mean value is $z_2$, the selection differential is $(z_2 - z_1)/\sigma_p$. Larger values indicate stronger selection, without any well-defined limit. In both cases, the left-hand figure refers to studies of undisturbed populations, and the right-hand figure to populations that have been experimentally manipulated or are known to have experienced some unusual stress. The figures

(*Continues*)

probably do not represent unbiased estimates of the distribution of rates of selection, because situations in which selection seems likely to be detectable are more likely to be studied and reported, but they are the most complete information currently available. They establish the main point that selection is usually weak, but can on occasions be very strong.

Industrial melanism in insects, the color and banding polymorphisms of Cepaea, and the morphology of Darwin's finches all have large literatures. There are extensive reviews of industrial melanism by Kettlewell (1973); of *Cepaea* by Jones et al. (1977); and of Darwin's finches by Lack (1947) and by Grant and Grant (1986).

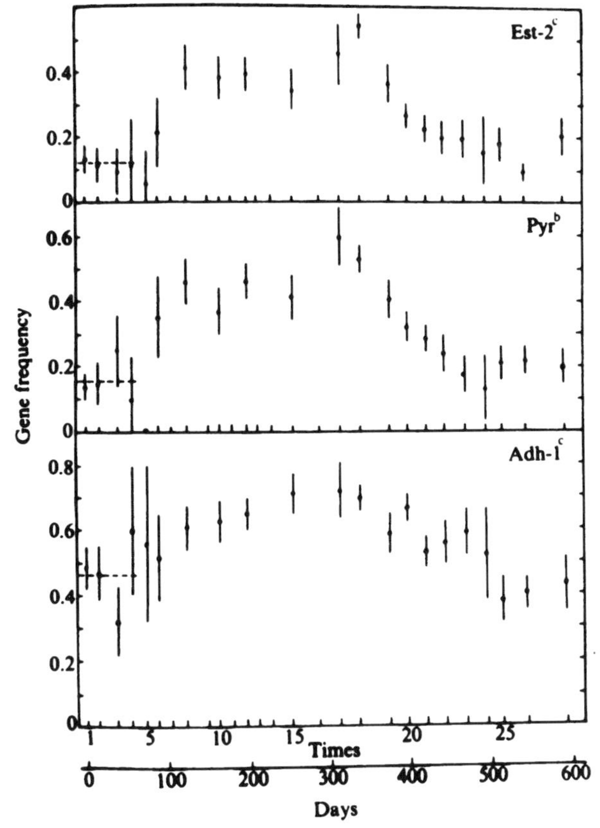


A more rigorously experimental approach to selection in natural populations involves the deliberate perturbation of gene frequencies. If selection is primarily responsible for maintaining the genetic composition of natural populations, then gene frequencies should tend to return toward their original values when the perturbation is discontinued. The diagram shows the effect of perturbing natural

(*Continues*)

*(Continued)*

populations of *Drosophila buzzatii*, from Barker and East (1980). Isofemale lines made homozygous for alleles of esterase (*Est-2*), pyranosidase (*Pyr*), and alcohol dehydrogenase (*Adh-1*) loci were used to increase the frequency of a given allele at each locus in an isolated population of flies, either by allowing adult flies to escape from population cages, or by inoculating cacti, the food plant of this species, with eggs and larvae. This caused sustained and substantial shifts in gene frequency over a period of about 250 days. When the introductions were discontinued, gene frequencies returned to their original values within about 300 days. This was not caused by immigration, because the three loci responded at different rates; the nearest patch of cactus was in any case some 3 km from the study site, and immigration would be much too infrequent to cause the observed shifts. The re-establishment of the original genetic composition of the population was therefore caused by natural selection acting on the enzyme loci or on loci linked to them. Comparable experiments, with less positive results, were carried out by Jones & Parkin (1977) and Halkka et al. (1975).

## 2.B. Selection of Pre-existing Variation

### 31. *The unit process of evolution is the substitution of a superior variant.*

In a more or less unchanging environment, stabilizing selection will preserve the adaptedness and the existing genetic composition of the population through time. If the environment fluctuates without any consistent trend, each episode of selection will be directional, but the genetic changes that occur in each generation will not cumulate through time, the effects of one episode being canceled by the next. There may, however, be a consistent secular trend persisting over many generations, or there may be an abrupt long-lasting shift in the state of the environment, or migrants may enter and permanently occupy an environment different from that of their parents. In these circumstances, successive episodes of selection will have the same direction: the same set of variants will be selected, generation after generation. The variation that was present in the population before the change will be sorted by selection until, if the altered state of the environment lasts long enough, the type that was originally common has been largely supplanted by another variant. The simplest case of this process of genetic replacement is the substitution of one allelic variant of a gene for another. (It is unfortunate that the term "substitution" is used to mean either the replacement of one nucleotide by another in a DNA molecule

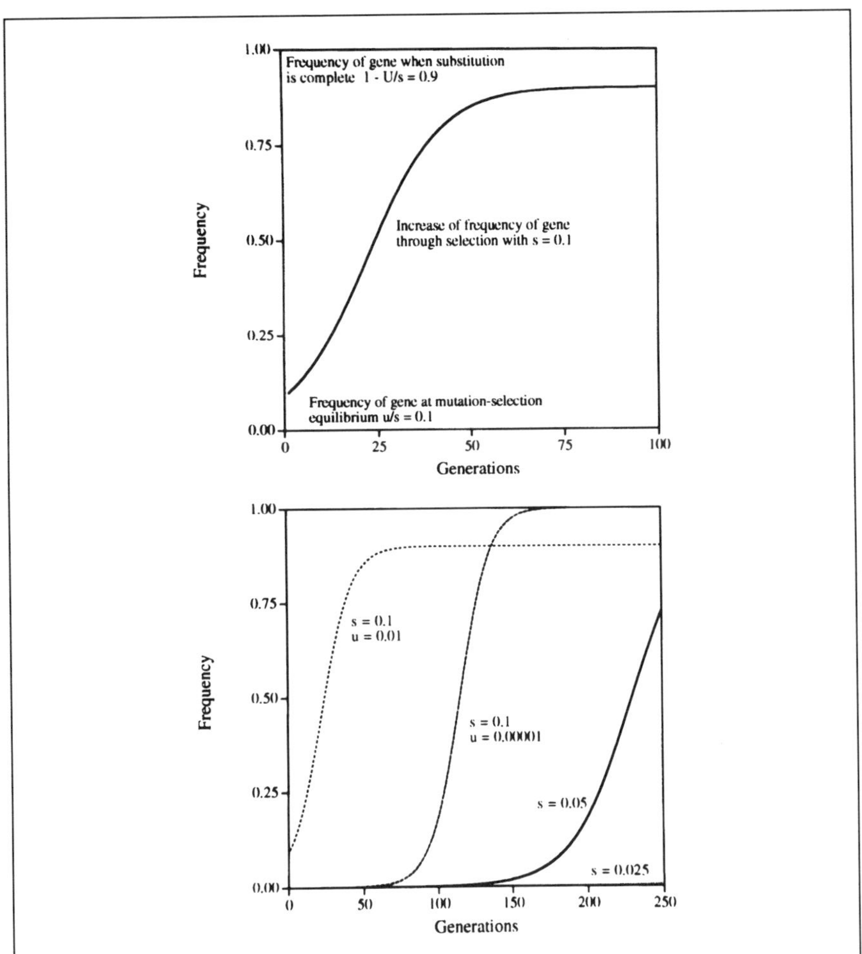


These graphs illustrate the time course of the substitution of a favorable allele in a haploid population. The upper graph shows the (unrealistic) situation in which $s = 0.1$ and $u = U = 0.01$. From its frequency at mutation-selection equilibrium of $u/s = 0.1$, the gene spreads rapidly through the population, attaining its final frequency of $1 - U/s = 0.9$ within 100 generations. In more realistic situations, the rates of selection and mutation will be much lower. The lower graph shows the substitution of a gene with $u = U = 10^{-5}$. When $s = 0.1$, the gene, although increasing exponentially in frequency, remains rare for a considerable period of time until its frequency becomes substantial, when it spreads very rapidly through the population. With weaker selection the gene remains rare for a longer period of time; when $s = 0.025$ its increase is scarcely apparent (on an arithmetic scale)

(*Continues*)

after 250 generations. The dynamics of gene frequency in diploid populations are complicated by dominance. In the simplest case the heterozygote is intermediate in fitness between the two homozygotes, and the rate of increase of the favored allele is just half that in the haploid case. Recessive alleles occur mostly in heterozygotes while they are rare, and being only weakly expressed are only weakly selected, remaining rare for a long period of time even when the selection coefficient $s$ is large. These complications are discussed further in Sec. 47.

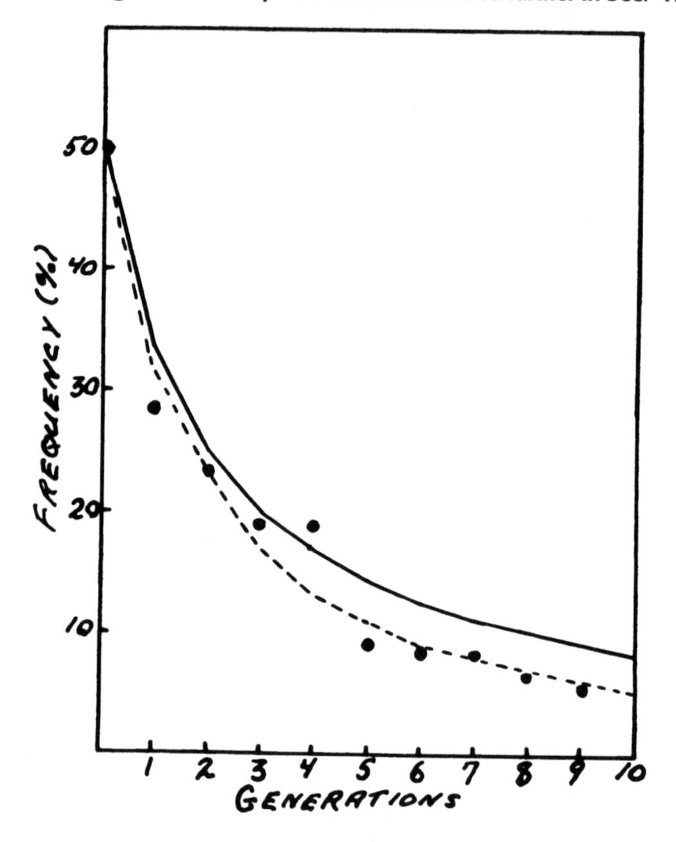


It is surprisingly difficult to find good examples of the substitution of a single allele through a predetermined rate of selection in the laboratory, presumably because the experiment is too trivial to be worth reporting. This figure shows the elimination of a recessive lethal gene in a cage population of *Drosophila*, from Wallace (1963a). The rate of elimination is slightly faster than that expected from theory, presumably because the allele is slightly deleterious in heterozygotes. In a haploid population, a lethal would, of course, be eliminated in a single episode of selection.

(*Continues*)

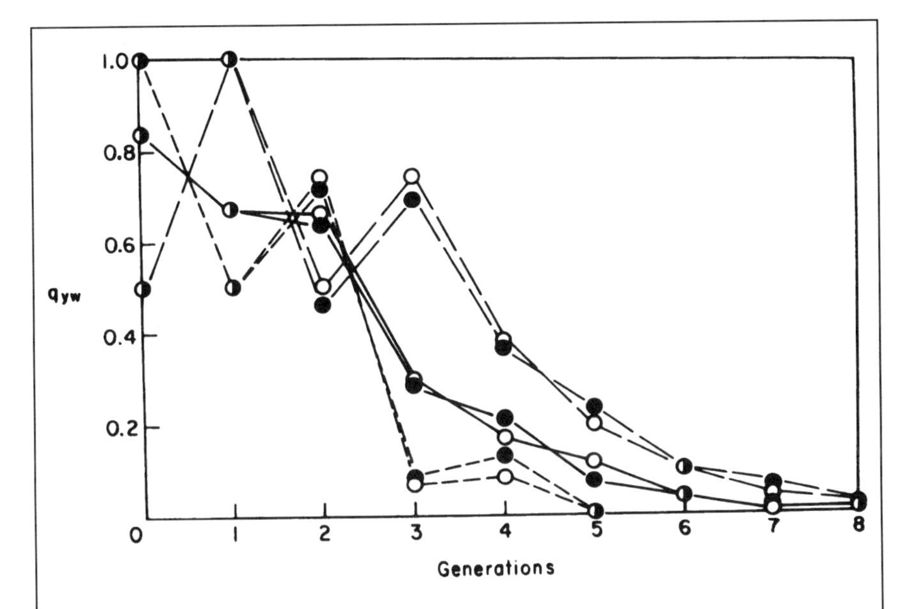


The elementary theory of sorting continues to give adequate predictions of the dynamics of gene frequency even in more complicated situations. In this example, Hedrick (1976; also see Hedrick & Murray 1983) followed the frequency of chromosomes bearing the X-linked mutations yellow *y* and white *w* in an experimental population of *Drosophila*. The dotted lines represent the frequency in females, the broken lines the frequency in males, and the solid lines the mean frequency. The solid circles are observed values. The *y w* chromosome is deleterious, primarily because it reduces male mating success and is, therefore, driven down in frequency. The experiment, however, was set up with unequal frequencies in the two sexes: all males (with XY genotype) bore only the *y w* X-chromosome, but all females (with XX genotype) were heterozygous. The frequency of the *y w* chromosome decreases through time, but its frequency in either sex fluctuates from generation to generation, these fluctuations taking the form of a damped oscillation. This is because males get their X chromosome from their mothers, so that the frequency of an X-linked gene among males in one generation must be equal to its frequency among females in the previous generation; females, on the other hand, receive an X chromosome from each of their parents, and the frequency of an X-linked gene among females will, therefore, be the mean of its frequencies in males and females in the previous generation. These arguments can be expressed mathematically as a set of equations that describe how gensotype frequency in males and females is expected to change from generation to generation. The fluctuations in genotype frequency can then be predicted from the results of an independent evaluation of the effect

(*Continues*)

(*Continued*)

of the $y\ w$ chromosome on fitness, and the predicted frequencies are shown as open circles. There is clearly a good agreement between observation and prediction.

or the replacement of one allele by another in a population. I hope that context will make it clear which meaning is intended.)

In the changed environment, the fitness of a certain allele is much greater than it was before, and the allele spreads in the population through selection. Before the change, it was a deleterious allele with a frequency $m = u/s$, determined by the balance between mutation and countervailing selection. The rate $u$ is essentially the frequency with which the common type gives rise by mutation to this particular variant. After the change, it is favorably selected and will continue to increase in frequency until the allele that was common before the change has been reduced to the frequency that is just maintained by recurrent mutation. At this point, the combined frequency of all the variant forms of the gene is $M = U/S$, say, where $U$ is the frequency with which the type now common gives rise by mutation to any variant; $S$ is the average rate of selection acting against these variants. The allele favored by the change has therefore risen in frequency from $m$ to $1 - M$. In the simplest case, there would be only two alleles, favored in different environments, and the allele favored in the new environment would increase in frequency from $m$ to $1 - m$. The time taken for this substitution is roughly $t = 2(1/s)(-\log_e m)$, provided that selection is fairly weak. It is, therefore, inversely proportional to the rate of selection, as we would expect: substitution occurs more rapidly when the difference in fitness between the two alleles is greater. If $s = 10^{-2}$ and $u = 10^{-5}$, then $m = 10^{-3}$ and, thus, $t = 1380$ generations. The time-scale of gene substitution is likely to be typically a few hundred or a few thousand generations.

The consistent sorting of pre-existing variation for a few hundred or a few thousand generations causes a permanent genetic change in the population, which serves as a new basis for further evolution. *Allelic substitution is thus the unit process of evolutionary change.*

## 32. *Allelic substitution can occur very rapidly when selection is intense.*

The rate of sorting depends on the rate of selection, and in novel and stressful environments where types vary widely in fitness the composition of a population may change radically within a hundred generations. Rapid evolutionary change has often been witnessed in environments that have been disturbed by human activities, whether deliberately or not.

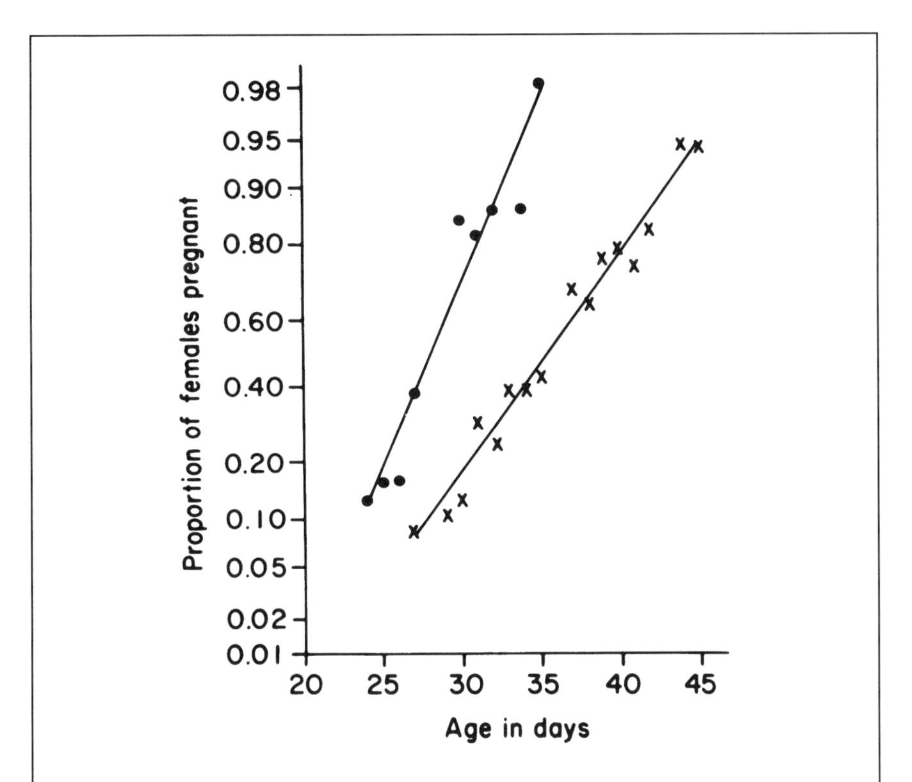


This figure shows the maturity schedule of females selected by serial transfer in the laboratory for high rates of increase (circles), compared to that of females newly isolated from the same natural population (crosses), from Doyle and Hunte (1981a; also see Doyle and Hunte 1981b). Note the much shorter cycle of the selected females, evolving within 20–30 generations. For a general discussion of the evolutionary effects of domestication, see Kohane and Parsons (1989). The recent evolution of antibiotic resistance in bacteria is discussed by Neu (1992). The experiment on mass selection in maize described in the text is by Genter (1976).

(*Continues*)

(*Continued*)

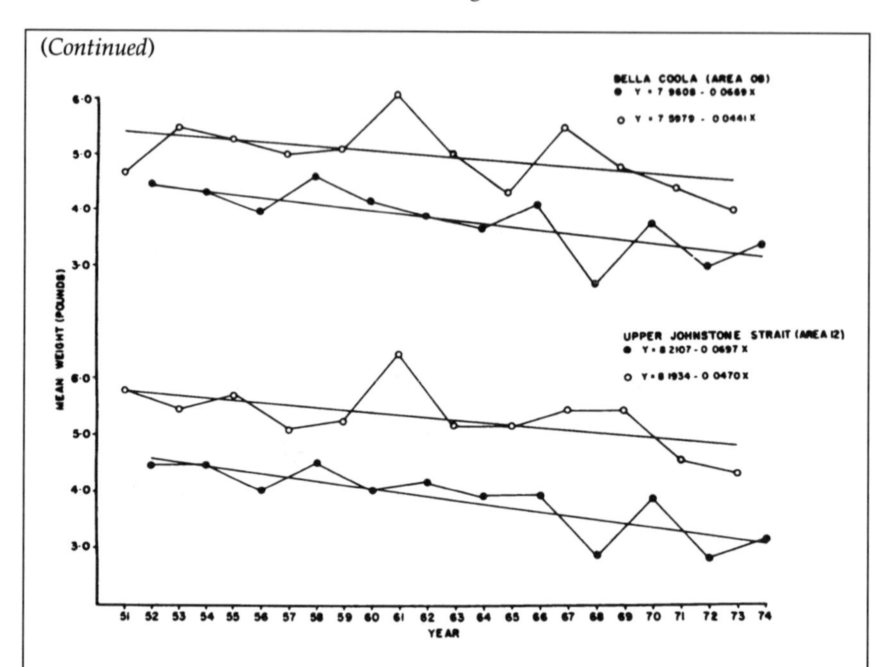


This figure shows the decline in mean whole weight of net-caught pink salmon *Oncorhynchus gorbuscha* between 1951 and 1974 at two stations on the coast of British Columbia (Ricker 1981). The adults return after their second year at sea; consequently, the odd-year and even-year stocks are separate and are somewhat different in size. Both respond to selection, although the response is somewhat steeper in the even-year stock (solid circles). Seine catches decrease even more steeply in size, perhaps because the cultural response of the fishermen (in decreasing the mesh size of their nets) lags behind the genetic response they elicit in the fish population. Changes in growth-rate caused by the selective effects of harvesting have also been documented in lake whitefish *Coregonus* by Handford et al. (1977) and in cod *Gadus* by Borisov (1978); changes in exploited populations of trout *Salmo trutta* are analyzed theoretically by Favro et al. (1979). The effect of selective fishing on the life history is discussed further in Sec. 96.

**Conscious Selection: Crop Plants.** The rapid modification that can be caused through the deliberate generation and sorting of variation has been responsible for producing modern varieties of crop plants and livestock, and the wide differences between modern strains of corn, cattle, or dogs and their wild progenitors are familiar to everybody. From the innumerable examples available, I have chosen a sorting experiment on maize by C.F. Genter of Blacksburg, Virginia. The object of the experiment was to broaden the genetic basis of standard Corn Belt cultivars by developing entirely unrelated strains with which they could be crossed. The initial material for the experiment was 25 Mexican races, including indigenous, pre-Columbian and prehistoric stocks, that varied widely but were generally unsuited to cultivation in Virginia because of low yield, late maturity, high ears, and susceptibility to corn smut. The base population comprised material from over 200 crosses among these races. This was then sorted for 10 generations; in every generation, 10,000 seeds were planted at Blacksburg, and about 600 ears selected at the end of the growing season from erect, healthy, productive plants. Because the base population was poorly adapted to the climate and soil of Virginia, 10 generations of selection caused rapid and substantial change. The selected population was much more productive, with yield increasing from 1.2 to 3.2 tons per hectare. It comprised shorter plants (232 cm vs 289 cm) that bore ears lower down (at a height of 115 cm vs 179 cm). The development time had decreased, from 90 to 78 days at silk, and the proportion of diseased plants dropped from 44% to 19%.

**Unconscious Selection: Domestication.** It is perhaps not surprising that the deliberate selection of a few percent of the population in each generation should cause such dramatic changes. Most of the modification of a domestic breed early in its history, however, is probably caused by natural selection in the novel conditions afforded by the domestic environment of the farm, the garden or the laboratory. A few organisms, for example, have been domesticated in the laboratory because of their suitability for scientific studies: the gut bacterium *E. coli*, yeast, the green alga *Chlamydomonas*, the ciliate *Tetrahymena*, the fly *Drosophila*, the nematode *Caenorhabditis*, the crucifer *Arabidopsis*, mice, and a few others. Isolating new strains of such organisms from the wild involves sorting a few unusually sturdy and productive organisms from the myriads of less suitable creatures that do not flourish in laboratories. Even when a range of isolates has been obtained, many are relatively feeble and likely to die out, so that it is often several generations before the collection settles down to laboratory life. The process of settling down is, of course, a process of selection for genotypes adapted to the unusual features of the laboratory, such as a continuous and plentiful supply of nutrients.

Domestication in the less familiar context of aquaculture has been described by Roger Doyle and Wayne Hunte at Dalhousie. They collected 100 pairs of the amphipod *Gammarus* from an estuary in Nova Scotia and then propagated their descendants in mass culture for about 25 generations. They were not deliberately selected, but the cultures were provided with an excess of food and thinned at intervals, so as to keep them in a state of continuous growth. At the end of the experiment, the domesticated animals were compared with a fresh sample of *Gammarus* from the same locality. The outcome of selection in the lush conditions of the laboratory was a near-doubling of the rate of growth: the exponential rate of increase $r$ increased from 0.18 to 0.31 per generation. This was caused primarily by a reduction in the age at maturity, but the domesticated animals also had better survival and greater fecundity.

**Unintentional Selection: Fisheries.** Fisheries often cause considerable mortality, relative to other causes of death, in the populations that they harvest, as the collapse of most of the commercially important stocks during the last few decades demonstrates only too vividly. This mortality is often selective, because the fishing gear tends to catch some types of individual more readily than others. The population will respond to this process of selection, although the gear is not designed intentionally to cause selection, and indeed the response is usually undesirable from a commercial point of view.

Changes in the size of Pacific salmon, *Oncorhynchus*, caught off the coast of British Columbia as they return from their feeding grounds in the ocean to the streams where they spawn, have been described in detail by W.E. Ricker of the Pacific Biological Station at Nanaimo. The fish are caught mainly with seines, with gill-nets, and by trolling. A seine is a close-meshed net supported by a cork-line that is drawn around a group of fish in shallow water; even the smallest fish cannot pass through the meshes, so seines probably catch fish with the same efficiency, irrespective of their size. Seine-caught fish are thus taken to be a random sample of the population, and the selectivity of other kinds of gear is estimated by comparing the fish they take with the seine-caught sample. (There is an interesting problem here: if the selective effect of a particular procedure is to be established by comparing the selected sample with a random sample from the population, how can we be sure that the methods used to capture the supposedly random sample are not themselves selective? The short answer is that we cannot be sure, and a selection differential that is estimated in this way always represents the difference between two sampling techniques, both of which may be selective.) Gill-nets are passively suspended in the water, and capture fish which are too large to pass through entirely, but which are prevented by their gill-covers from withdrawing their head. The selectivity of gill-nets depends on their mesh size: a large mesh allows small fish to pass through, while retaining larger individuals, and therefore selects for small size. A very small-mesh net would select for large size, because large individuals would be unable to insert their heads; but in practice gill-nets with such small mesh are never used. The most profitable mesh size depends in part on how the fishery is conducted. In the salmon fishery, payment was originally by the piece; this encouraged fishermen to use small-mesh nets which are rather unselective with respect to size, in order to capture the greatest possible number of individuals. In 1945, the method of payment was changed to payment by weight. This made it more profitable to use large-mesh nets that maximize the total weight of fish caught, by increasing the take of large individuals. For the same reason, the troll fishery specialized in capturing, or retaining, the larger fish. The selection against large size caused by this change in economic policy had two sorts of effect on the salmon populations. Most of the species of salmon—chum, sockeye, chinook and coho—return from the sea at different ages and, therefore, at different sizes, and selection for smaller fish will cause selection for earlier maturity, as well as for smaller size at any given age. The situation is simpler for pink salmon, which always return after two years at sea. The mean size of pinks decreased between 1945 and 1975 in almost every fishing area along the coast. The rate of decrease has averaged about 20–30 g per year, so that over 30 years the average size of the fish caught decreased from about 2.3 kg to about 1.7 kg, a very considerable change, from either a biological or a commercial point of view. This change was almost certainly caused by the selectivity of the fishing gear, rather than by any trend in environmental factors, such as sea temperature or salinity. For example, the rate of decrease in size has been greater in fishing areas where a larger proportion of the population is captured.

**Unintentional Selection: Resistance to Pesticides and Antibiotics.** The outcome of selection caused by human activity is sometimes not merely unintended, but—from our point of view—highly undesirable, because it tends to frustrate our efforts. A classical example, with enormous social and economic importance, is the evolution of pesticide resistance in insect populations. It is generally caused by simple changes involving one gene, or a few genes. The first case was recorded in 1908. By 1948 there were some 14 species resistant to one chemical or another; after the widespread use of pesticides in the next two decades over 200 species had evolved resistance, and the number continues to increase.

A similar sequence of events followed the introduction of antibiotics to control bacterial infections. The $\beta$-lactams, such as penicillins and cephalosporins, kill bacteria by inhibiting cell wall synthesis. They were introduced into general practice in the early 1940s, and were at first dramatically successful in suppressing infections by *Staphylococcus*, *Streptococcus*, and other bacteria. The first cases of resistance, however, began to be reported only a few years after their introduction, and most populations throughout the world are now resistant to doses hundreds or thousands times larger than those that were once effective. There are various sources of resistance. Penicillin normally acts by binding to the proteins that link peptidoglycans in the cell wall; the wall is then weakened, and ruptures under the osmotic pressure of the cytoplasm. Changes in the structure of these cell wall proteins reduce their tendency to bind penicillin and enable the cell to survive. Some bacteria are resistant because they produce a $\beta$-lactamase that cleaves the antibiotic and renders it harmless. More simply, some strains are just less permeable to antibiotics, as Gram-negative bacteria are generally less sensitive than Gram-positive bacteria. Unfortunately, the response to the appearance of strains resistant to currently prescribed doses is very often to increase the dose until the infection is just brought under control; this is an efficient procedure for selecting increased levels of resistance. The various sources of resistance are sometimes encoded by chromosomal genes; in other cases, however, they are borne by plasmids or transposons, and a gene selected in one lineage of bacteria can quickly be transferred to a large proportion of the bacterial community. Despite the ingenuity of pharmacologists in devising new kinds of antibiotic, every hospital in the world now harbors bacterial strains resistant to a range of antibiotics, and whenever a new antibiotic is released the evolution and spread of resistance follows, sometimes with bewildering rapidity.

**Accidental Selection: Industrial Melanism.** Melanic variants of the peppered moth were first recorded in 1849. In 1875 they were still listed in catalogs as curiosities and were presumably still rather infrequent. By the mid-1880s, however, collectors were noticing that in some areas they were more common than the original pepper-and-salt type, and by 1898 they had reached a frequency of 98% in the Manchester–Liverpool area. Thus, within 50 generations—the moth has a single generation per year—the melanic phenotype had increased in frequency from about 1% to about 95%. Since in this case melanism is caused by a single dominant allele, the frequency of the melanic allele must have increased more than a hundredfold during this period, from about 0.005 to about 0.775. This implies intense selection, with $s = 0.2$ or so.

## 33. *The rate of evolution is limited by the "cost of selection".*

Rapid evolution can be observed easily enough by stressing natural populations, and happens spontaneously from time to time after droughts or floods or fires. Catastrophes will often create high rates of selection that will cause rapid allelic substitution. Although such events are very instructive, because they allow us to study evolution on a human time-scale, they are surely exceptional. No population can sustain powerful selection acting simultaneously on many characters. If a bacterial population is exposed to a novel antibiotic, 99.999% of cells may be killed, but the tiny fraction of resistant cells will build the population up again. If the same population were exposed simultaneously to two novel antibiotics, only a tiny fraction of a tiny fraction would bear mutations giving them resistance to both, and very few cells indeed, if any, would survive. It would probably be impossible to adapt to three new antibiotics presented simultaneously: there would be no triply-resistant mutants already present in the population and no opportunity to build up the triply-resistant genotype in a stepwise fashion over a number of generations. Instead of adapting, the population would become extinct. There is therefore a limit to the amount of evolution that can occur in a given period of time because there is a limit to the rate of selection that a population can sustain. J.B.S. Haldane expressed this principle by referring to a *cost of natural selection*, by which he meant the excess mortality or sterility needed to drive evolutionary change.

This term is misleading. Selection is not in itself costly. It does not damage the population any more to have 99% of its members killed selectively than it does to have them killed at random. It would be more precise to say that the *opportunity for selection* is limited by the capacity of the population to regenerate itself. Though imprecise, however, the term is probably too familiar to discard, and the principle that it embodies is an important constraint on the rate and pattern of evolutionary change.

While a newly favored allele is spreading after a change in the environment, some proportion of the population must be eliminated in every generation in order for selection to continue. At first, when the newly favored allele is still rare, a fraction $s$ of individuals bearing the common gene must be eliminated, because their fitness is $1 - s$. When the newly favored allele has become very common, selective mortality will be negligible, because it now affects only a very few individuals. One might guess that on average the proportion of the population that has to be eliminated in every generation during the substitution of the allele is about $s/2$. The process is complete in $t$ generations, so the total selective elimination required by the gene substitution is $st/2$. From the equation describing the length of time required for the substitution, this is equivalent to $-\log_e m$. The cost of selection is therefore independent of the rate of selection. Very weak selection acting over a very long period of time, or stronger selection over a shorter period of time, will involve similar total costs per allele substitution. (This is approximately correct only when selection is not too strong; if $s = 0.1$ or more, the cost is greater than these expressions suggest.) The cost is directly proportional to the initial frequency of the allele, because most of the cost is incurred early in the process of substitution, while the newly favored allele is rare. Alleles that are only mildly deleterious and that are, therefore, maintained at fairly high frequencies at mutation-selection equilibrium are more likely to be substituted when the environment changes than are severely deleterious and, therefore, extremely rare mutations. Even so, the cost is substantial. As a rough rule of thumb, the number of genetic deaths required by the substitution of a single mildly beneficial allele is about 10 times the number of reproducing individuals in the population at any one time. These genetic deaths represent individuals that die or fail to reproduce because of their genotype at this locus; most death or sterility is likely to be accidental, or caused by other genetic effects, and the actual death-rate is certain to be much greater then this figure suggests. A population of 1000 individuals, exposed to a novel environment, can proceed to substitute alleles at a rate equivalent to one new substitution every generation at the cost of 10,000 genetic deaths per generation. This is not impossible, because the selective elimination could take place among young offspring. It does, however, severely restrict the rate of evolution in slow-growing organisms that produce few offspring. Even in fecund and fast-growing organisms, excess production will never be sufficient to permit selection to act strongly on several independent characters at the same time.

## 34. *Selection in natural populations is usually weak.*

The cost of selection, or limited opportunity for selection, leads to a very simple conclusion. A specific stress, eliciting a specific response, may generate intense selection leading to rapid genetic change; but most evolution must involve weak selection. Because the rate of evolution depends on the rate of selection, most evolution must be slow.

## 35. *Weak selection is easily capable of driving observed rates of allele substitution.*

This is a very simple point, that has plagued evolutionary biologysince the first announcement of the theory of selection. If selection is weak and evolution slow, does even the geological scale supply enough time to accommodate the diversification of living organisms? Suppose that 10% of excess production were eliminated selectively; then if each allele substitution requires selective deaths equal to 10 times the number of adults in the population, the rate of substitution driven by selection can be about $10^{-2}$ per generation. The actual rate of substitution can be estimated from differences among modern organisms in the amino-acid sequence of proteins, or the nucleotide sequence of genes, provided that the phylogeny of these organisms is known. This rate differs widely among proteins, but a typical value (for ha emoglobin, say) is about $10^{-9}$ per site per year, or about $10^{-7}$ or $10^{-6}$ per protein per generation. The imprecision of this estimate is beside the point; weak selection is capable of driving evolution at such rates at many loci simultaneously. The adequacy of selection for driving observed rates of evolution is not in dispute; clearly, selection is capable of causing substantial change on time-scales that are nearly instantaneous on geological scales. A more interesting issue, indeed, is why evolution proceeds so much more slowly than it might.

## 36. *Only mutations of small effect are likely to be beneficial.*

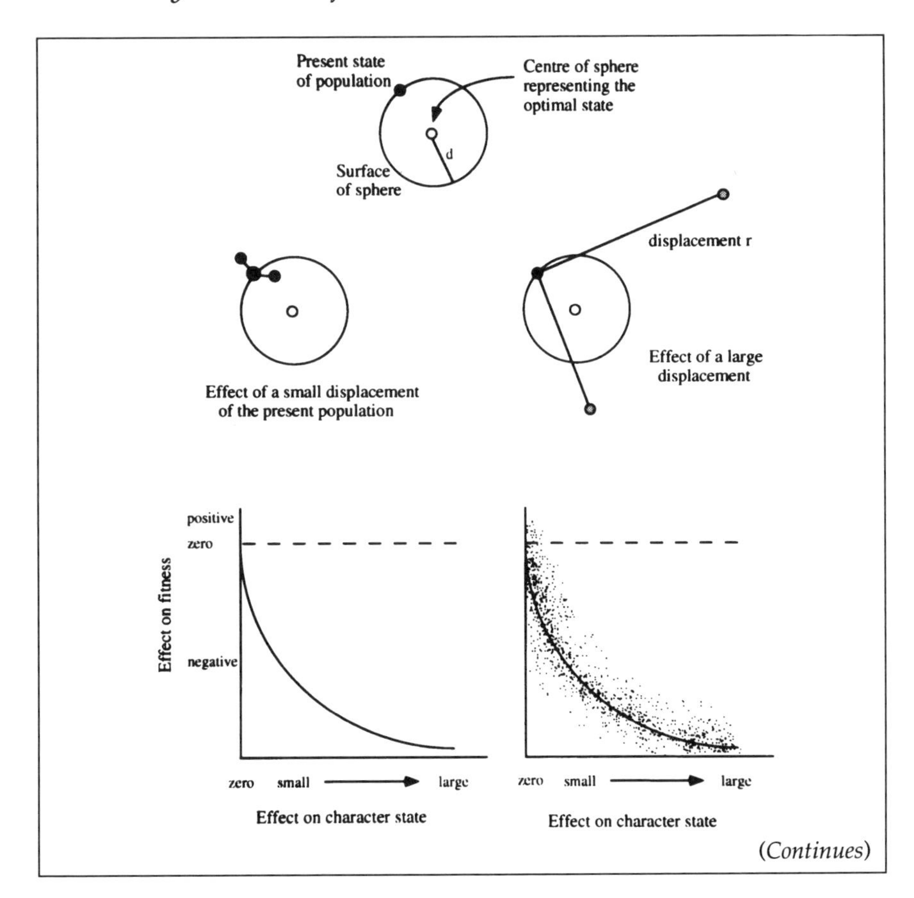


(*Continues*)

*(Continued)*

Fisher (1930) interpreted the effect of a change in some structure or character on fitness in the following way. Let us represent the current state of the population, or mean value of the character, as a point in space (solid circle in the upper drawings); some other point represents the optimal state that maximizes fitness (open circle). The adaptedness of the population is increased if these two points move closer together. The space in which the points fall has many dimensions, corresponding to the multifarious effects of any character on fitness, but the essentials of the situation can be represented in three dimensions. The current population then lies on the surface of a sphere of diameter $d$ whose center is the optimal state; this sphere encloses all the points that constitute an increase in adaptedness, and the current lack of adaptedness is expressed by $d$. If the current state is displaced by mutation a distance $r$ in any direction (stippled circle), adaptedness is increased if this displacement carries it within the sphere, but is worsened if it is moved outside the sphere. If $r$ is small relative to $d$, the chances of it being moved inside or outside the sphere are about the same; in fact, as $r$ becomes very small, they approach equality. On the other hand, if $r$ is large relative to $d$, then the character state will almost certainly move outside the sphere, regardless of the direction of change. In fact, for $r < d$ the probability of improvement is $\frac{1}{2}(1 - r/d)$. (If $r > d$ this probability is, of course, zero.) Thus, as $r$ increases relative to $d$, the probability that a change will increase adaptedness becomes steadily smaller, as indicated in the lower left drawing. However, not all mutations that cause an equivalent change in character state will have precisely the same effect on fitness; they will rather show some distribution of effects around an average value. When the expected effect of a small change in character state is nearly zero, some mutations of this sort will actually improve fitness; however, this is very unlikely to be the case for mutations of large effect. This is illustrated by the lower right drawing, in which the density of stippling represents the occurrence of mutations with a given effect on fitness, distributed around the expected value. For some experimental results, see Gregory (1965).

Evolution proceeds in the main by small genetic changes, the substitution of one variant of a gene for another. This is because large beneficial changes are improbable; it is simply not very likely that all the changes necessary to restore adaptation when the environment changes should occur simultaneously. It does not by itself imply that evolution generally involves small phenotypic changes. There is no regular or proportional scaling between genotype and phenotype: a trivial genetic change can have a profound effect on the phenotype. This is most obvious at the level of fitness. Switching a single nucleotide, altering a single amino-acid in a single protein, may confer resistance to an antibiotic and there by make the difference between life and death. There could scarcely be a larger phenotypic difference. The same is true for any other character, which might be scarcely modified or completely disrupted by a single mutation. It is perfectly conceivable, then, that evolution might involve abrupt shifts in phenotype even though the underlying genetic change were as smooth and continuous as possible. However, this is not very likely to be true.

Mutations of whatever sort that have a large effect on character state are almost certain to be severely deleterious. The argument is merely a reasonable extension of the one already applied to mutations in general; if altering a single circuit at random in a calculator is unlikely to improve it, rewiring it completely will be disastrous. A random rerouting of all the ducts and vessels in the vertebrate body would equally produce abortion, not evolution. At the other extreme, an imperceptible change will have no effect on fitness, and some very slight rearrangement is likely to be only slightly deleterious; altering the precise pattern of blood-vessels on the surface of the retina is unlikely to disrupt vision very markedly. In general, therefore, the degree of alteration of a character will be related to its effect on fitness: the larger the change in character state, the greater will be the reduction of fitness.

The scaling of character state and fitness, however, is in turn far from uniform. Minor changes in characters may have unexpectedly deleterious consequences; slight variation in the dimensions of the mitral valve or the epiglottis can threaten life. It is likewise true that slight variations may actually be beneficial: running the blood-vessels behind the retina, for example, or routing the spermatic cords directly from the testes to the ureter, are minor changes that would probably be improvements. Wholesale rearrangement of the anatomy, on the other hand, might kill or merely cripple, but it will certainly not result in a better design. Only minor changes in characters, then, have any substantial chance of turning out to be improvements in any circumstances.

Putting these arguments together, it will generally be the case that *increased adaptedness evolves through minor genetic changes that cause small shifts in phenotype*. This is not a point of dogma that admits no exceptions: some events in the history of life, such as the origin of eukaryotes, probably did occur rather abruptly as the result of an improbable combination of circumstances. Such events may have been very important. But the generality of evolutionary change occurs through the selection of slight variants with individually small effects on phenotype. This is the *gradualist* point of view.

## 37. *Evolution through weak selection is hindered by sampling error.*

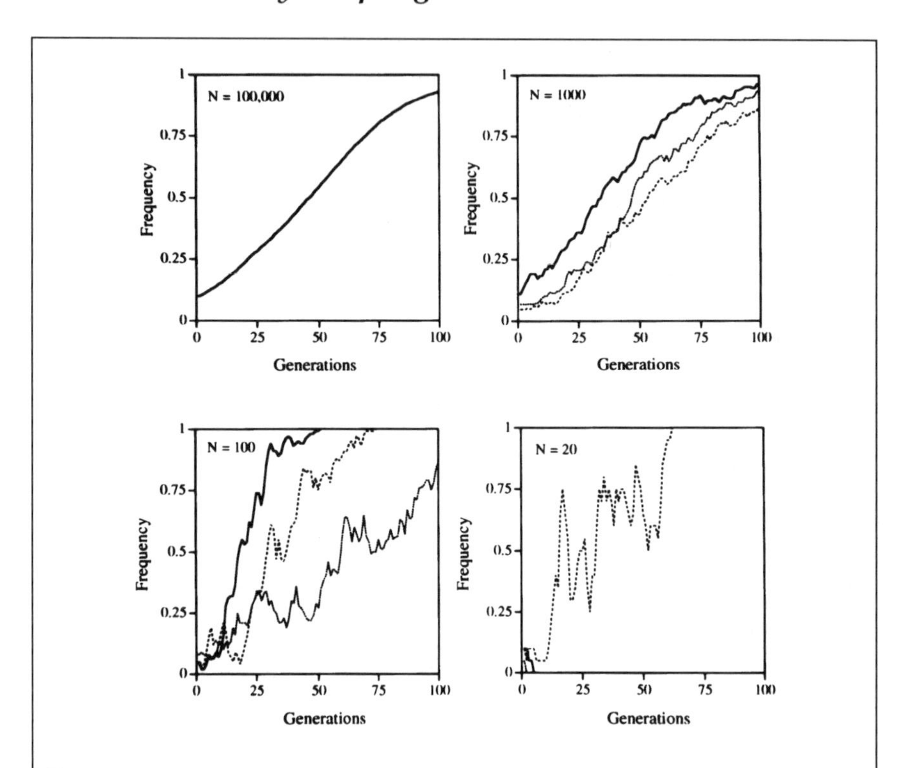


These diagrams show how the frequency of a gene with a selective advantage of 5% spreads through haploid populations of different size when introduced at a frequency of 0.1. In large populations of $N = 100,000$ or more, it increases in frequency smoothly through time, and all replicate populations behave in the same way. In smaller populations of $N = 1000$, its increase in frequency is noticeably somewhat erratic, and replicate populations follow somewhat different courses, primarily because they diverge slightly in the early generations of selection, when there are only a hundred or so copies of the gene present. Nevertheless, the replicate populations follow similar courses, and the gene will approach fixation at about the same time in each of them. In small populations of $N = 100$ or so, only 10 copies of the gene are present initially. Gene frequency fluctuates widely through time, and replicate populations diverge markedly. All three populations shown here eventually become fixed, but they do so at different times, and the time that will be taken by a particular population cannot be confidently predicted. When population size approaches $N = 1/s = 20$, only two copies of the gene

(*Continues*)

are initially present, and whether the gene spreads at all is largely a matter of chance. In this case, it quickly became extinct in two populations; in the third, there were broad fluctuations in frequency before it was abruptly fixed after about 60 generations. It cannot be predicted with confidence how long it will take for the gene to be fixed, or even whether it will be fixed at all.

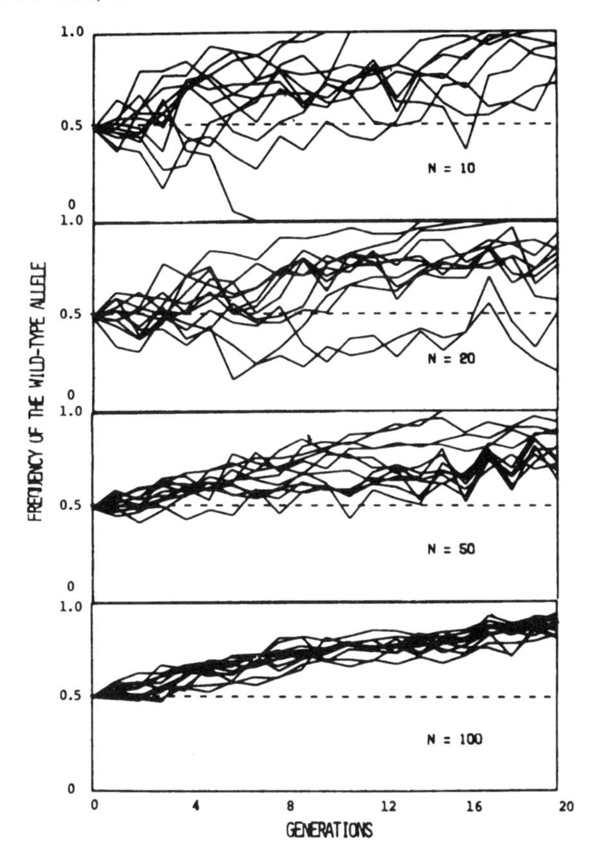


These graphs show the behavior of gene frequency in experimental populations of the flour beetle *Tribolium* studied by Rich et al. (1979). A mutation for body color, black, and the wild-type allele are initially equally frequent. The organism is diploid, but heterozygotes can be distinguished, and so in subsequent generations the frequency of black can be estimated directly by counting. Wild-type individuals are fitter, and in populations of $N = 100$ (since the organism is diploid, with 100 individuals there are 200 copies of the gene) the wild-type allele increases in frequency fairly regularly and predictably. In small populations of $N = 10$ the increase is much more erratic, the behavior of a given population cannot be

(*Continues*)

(*Continued*)

predicted in detail, and in one case the allele, despite its greater fitness, is lost from the population.

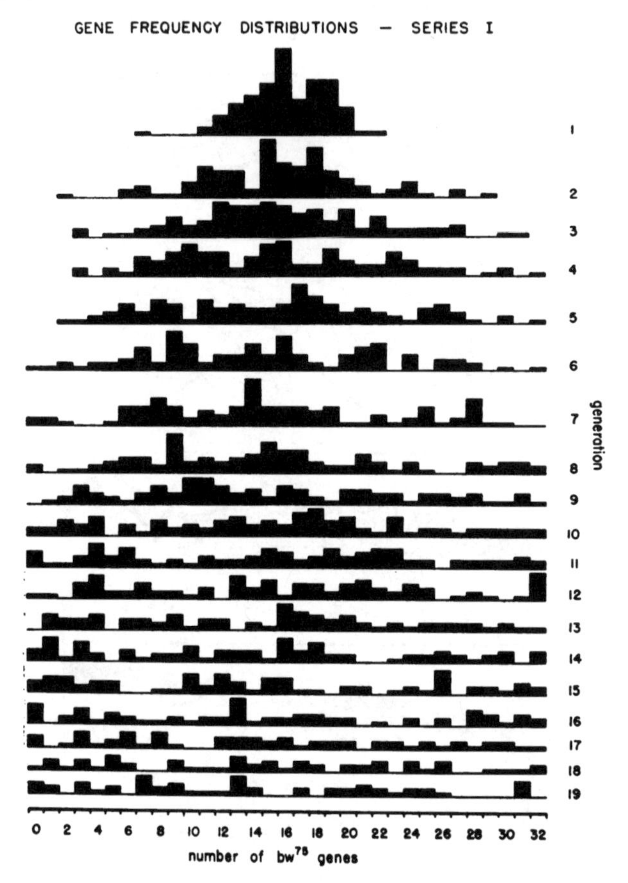


Thus, the frequency of alleles that are equivalent in fitness will diverge widely in replicate populations, and eventually all populations will become fixed for one or the other allele, without it being possible to predict the fate of any given population. This figure shows the frequency of two alleles at the brown *bw* locus in 105 experimental populations of *Drosophila* with $N = 16$ (Buri 1956). The two were initially equally frequent, with 16 copies of each. Over the course of 19 generations, gene frequencies diverge; one after another, the populations become fixed for either allele (and having become fixed, are then omitted from these frequency distributions), and if the experiment were continued for long enough all would become fixed, about half for one allele and half for the other.

In every generation, better-adapted individuals will be more likely to survive and reproduce. This is only a tendency, however, not a strict deterministic rule. A snail living in a English hedgerow is less likely to be eaten by a bird if its shell is striped than if it is plain. But it is not very likely to reproduce in any case. It may be eaten by a shrew, or die from heatstroke or starvation; it may even be eaten by a bird after all. *Selection is a process of sampling.* The variation of characters among individuals ensures that the sample that reproduces is a biased sample of the population as a whole, but its composition cannot be precisely specified in advance. In artificial selection, we can indeed specify that all individuals with certain character states, but no others, will contribute to the next generation. But there is nobody actually responsible for selecting snails at the bottom of hedgerows, and no individual, no matter how well endowed, has any guarantee of success, but only a greater or lesser chance. Richard Lewontin once prefaced a lecture on this theme with a quotation from Ecclesiastes: the race is not always to the swift, nor the battle to the strong; but time and chance happen to them all.

The nature of selection as sampling implies that evolution is a stochastic process that is subject to *sampling error*. The composition of a population at any point in time will have been determined by three factors. One is historical, the composition of the previous generation from which it descends. The second is selection, which tends to cause an increase in the frequency of some kinds of individual and a decrease in others. The third is chance. The actual composition of the population will inevitably differ from what we expect on the basis of descent and selection, because the life of each individual is an historically unique succession of events whose eventual outcome is influenced by a multitude of factors. The next generation is formed in a stochastic, or probabilistic, fashion from the success or failure of many such lives. We may be able to predict its average properties with some assurance, but its composition will fluctuate to a greater or lesser extent through time in ways that we cannot predict or account for.

If it were possible to distinguish two variants which were in all other respects functionally equivalent, and which in all circumstances had equal fitness, then neither would tend to change in frequency under selection; but they would nevertheless fluctuate in frequency from generation to generation through sampling error. Sampling error is in this case the only process causing the composition of the population to change through time. It will continue to operate if the two variants, rather than being strictly equivalent, express an extremely small difference in fitness. In a perfect world, any difference in fitness, however small, would drive a gradual but perfectly regular directional change in the composition of the population. It is easy, however, to appreciate that if the difference in fitness be sufficiently small, the change in composition caused by selection may be smaller than the change caused by sampling error. The *expected* outcome of selection is unaffected; but in any particular case, this expectation is unlikely to be realized. If sampling error could be neglected, then two populations identically constituted will change in precisely the same way when selection is applied, however slight the selection might be. But this is not the case. If selection is very slight, its effect may be small relative to sampling error, and replicate populations will then come to differ, despite being selected in the same way. Sampling error, acting at the level of the population, is analogous to mutation acting at the level of the individual. It is impossible, in either case, to transmit information with perfect fidelity: mutation makes it impossible for a nucleotide sequence to be transmitted perfectly, whereas sampling error makes it equally impossible for a population distribution to be transmitted perfectly. Both are undirected processes of variation that reduce the effectiveness of selection and therefore directly retard or inhibit increase in adaptedness.

**Genetic Drift in Replicate Populations of *Drosophila*.** I have described how deleterious alleles will be retained at low frequencies in the population as the result of a balance between mutation and selection. If such alleles are only very slightly deleterious, their frequencies will also be affected by sampling error and will fluctuate from generation to generation. Experiments with *Drosophila* have confirmed that some alleles will by chance be lost entirely, whereas others will by chance increase to substantial frequency and may even occasionally become fixed at a frequency of 100%. In different populations, *genetic drift* will occur independently, with certain alleles spreading in some populations and quite different alleles in others. Sampling error therefore leads to undirected variation and divergence.

**Neutral Evolution Through Genetic Drift.** This process has been elaborated as a distinct theory of non-Darwinian evolution: almost all changes in the nucleotide sequence of genes or the amino-acid sequence of proteins have an inappreciable effect on function, so that the variation within populations and the variation among species are equally the consequences of genetic drift. This is the *neutral theory of evolution*, developed principally by Motoo Kimura of Tokyo, but elaborated by a host of mathematical population geneticists. Richard Lewontin has given a brilliant account of the struggle between selectionist and neutralist interpretations of variation, published at the time when this struggle was the main preoccupation of population geneticists. The struggle, indeed, still continues, somewhat muted, and all the texts of evolutionary biology that I know of devote large sections to a description of the neutral theory. I shall pass it by, only remarking that it is not really a theory of evolution at all. Sampling error may be very important, as the next two sections will discuss, but it does not produce evolution; it prevents evolution. Mutation and sampling error both reduce the effectiveness of weak selection by introducing random changes into genes or populations. They act directly to retard adaptedness. The occurrence of deleterious genes within a population is attributable to mutation; part of the divergence of gene sequence between populations is attributable to genetic drift. But we do not speak of a mutational theory of evolution. An important indirect effect of mutation is to cause a small proportion of changes that although deleterious in many circumstances are beneficial in others; an important indirect effect of genetic drift may be that it retains within the population a diversity of alleles that although neutral in most circumstances are beneficial in others. This issue is discussed further in Secs. 40 and 47. We still do not speak of a mutational theory of evolution. Evolution is caused by selection. Mutation and sampling error are important processes that act indirectly to reduce the effectiveness of selection and indirectly to fuel selection. They are not themselves theories of evolution capable of explaining the organization and history of living organisms.

## 38. *Selection is less effective in small populations.*

The effect of sampling error will be most pronounced in small populations. Rolling dice provides a familiar analogy. If six fair dice are rolled, we expect one to show a six. We are not astonished, however, if none do; the probability that this will happen is $(5/6)^6$, or about one in three. If 6000 fair dice are rolled, we expect 1000 to show a six, but are not astonished if 990 or 1010 do so. If none do, we would dispute that they are fair, because the probability of this happening by chance is $(5/6)^{6000}$, a probability too small for my calculator to deal with or my credulity to accept. In large populations, large fluctuations caused by sampling error are very unlikely to happen, and selection proceeds nearly deterministically. In small populations, the frequency of an allele under weak selection will fluctuate appreciably because of sampling error. A slightly beneficial allele may be lost; a slightly deleterious allele may be fixed. Sampling error thus reduces the effectiveness of both directional and stabilizing selection, hindering adaptation to a new environment and the maintenance of adaptedness in an unchanging environment.

Roughly speaking, sampling error will be a serious hindrance to selection if $s < 1/N$, where $N$ is the number of individuals in the population. For most purposes, the effects of sampling error are negligible in populations which always consist of a few thousand individuals or more. It cannot be neglected in populations of a few hundred individuals or less. The expected outcome of selection is the same in populations of any size, but the *reproducibility* of evolution is less in small populations. Small populations with the same initial composition will diverge through time until their genetic composition becomes widely different, despite being exposed to the same agents of selection acting in the same manner. The sorting of pre-existing variation by selection would have caused them all to change at the same rate in the same direction; thus, genetic drift has hindered or prevented at least some of them from attaining the degree of adaptedness that we would expect, if selection were the only force operating.

**The Founder Effect.** Large populations are much less sensitive to sampling error, so long as they remain large. But all populations fluctuate in numbers to some extent, and many are reduced to relatively few individuals from time to time by brief but severe environmental deterioration. Chance will play a large part in determining the composition of such residual populations and may again cause inappropriate shifts in gene frequency. In the case of a population permanently inhabiting a given area, this is called *intermittent drift*; when a few individuals from a large population colonize a new area, it is called the *founder effect*. The two are essentially equivalent.

## 39. *New beneficial mutations are often lost by chance.*

There is one special case in which the population is always small. This is when a beneficial mutation first begins to spread in a population. The number of individuals bearing the mutation is initially very small, and the dynamics of the gene are at this point strongly influenced by sampling error. It is therefore often the case that a mutation fails to become established in a population, even though it increases fitness. The probability that a newly arisen beneficial mutation will eventually become fixed depends on the balance between the directional process of selection and the nondirectional fluctuations caused by sampling error: the greater the rate of selection, the greater the probability that the mutation will continue to increase in frequency. In simple models, indeed, the probability that a novel beneficial mutation will become fixed is $2s$. A mildly beneficial mutation with $s = 0.01$ will spread quite rapidly once it has become established; but it is unlikely ever to become established at all.

The stochastic fate of novel mutations does not cause any qualitative change in the process of adaptation. Novel mutations are a random sample of possible mutations; a second round of randomization caused by sampling error does not randomize the sampling of mutations any more completely. However, there is a quantitative effect. A mildly beneficial mutation may have to occur dozens or hundreds of times before it evades sampling error and becomes sufficiently numerous to spread in a nearly deterministic fashion. Sampling error thereby slows down adaptation to new environments.

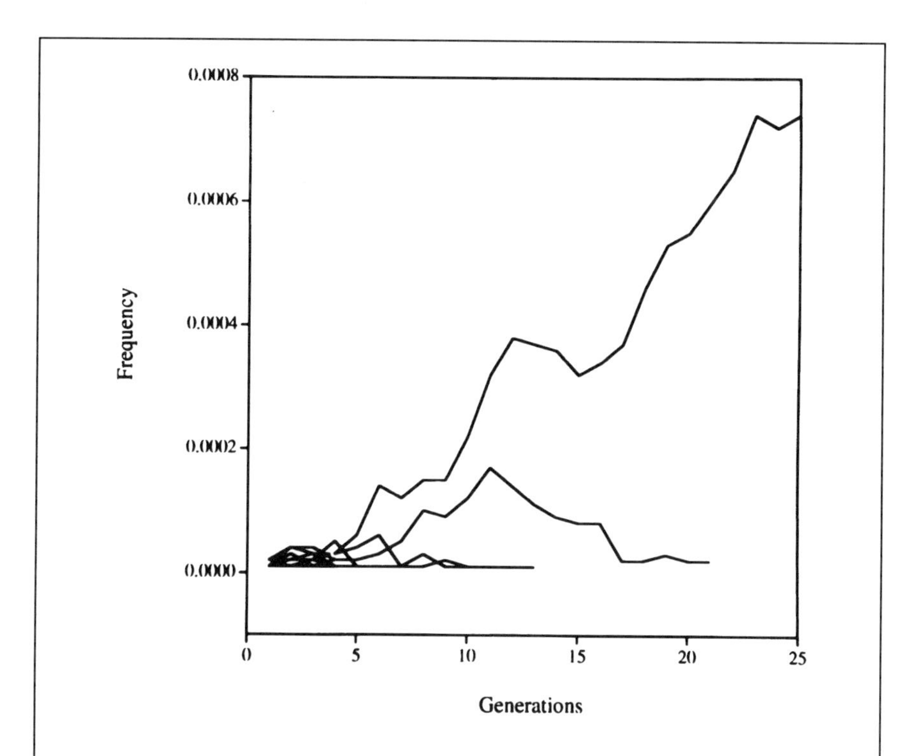


The lines in this diagram are 20 simulations of the initial history of a rare beneficial mutation. The mutation conferred an advantage of 5% and should spread through a population of $N = 100{,}000$ individuals as shown in Sec. 37. In fact, it will often fail to spread at all. The probability that the mutant is not represented in the second generation is approximately $(1 - 1/N)^N = e^{-1} = 0.37$, and so a large fraction of new beneficial mutations are never propagated. In these simulations, the mutation exceeded a frequency of $10^{-4}$ and persisted for more than 10 generations on only 2 occasions out of 20; in one case it subsequently declined, becoming extinct in the 20th generation, whereas in the remaining case alone it continued to increase, eventually becoming fixed after about 200 generations.

(*Continues*)

(*Continued*)

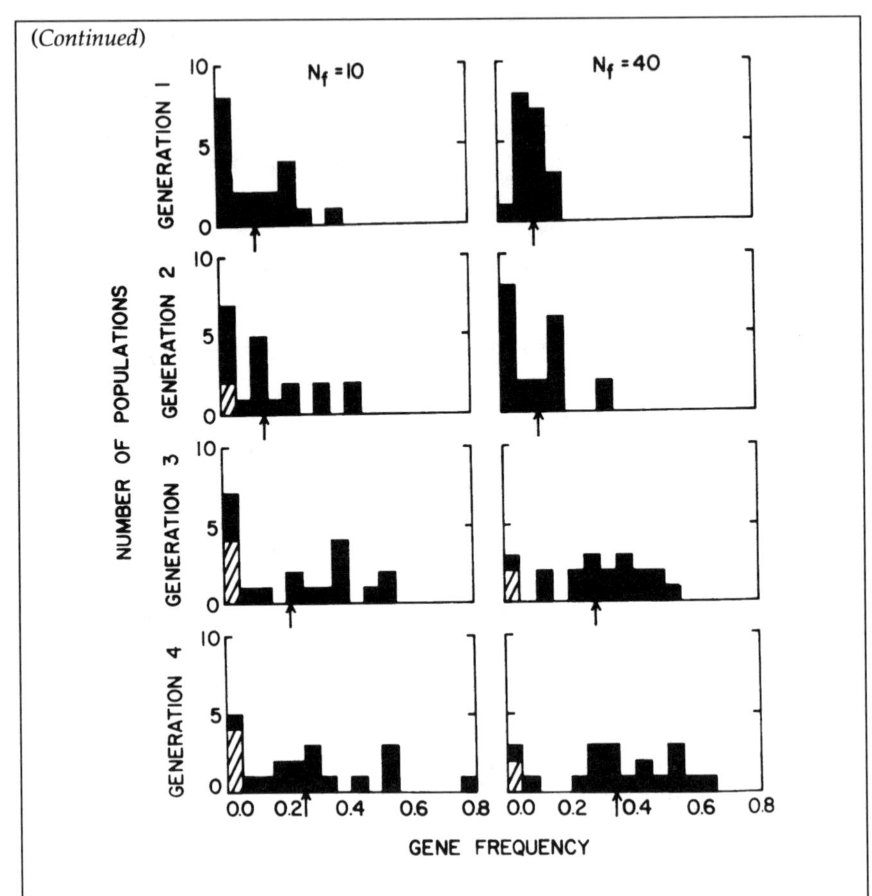


These diagrams show the frequency distributions of allele frequencies through four generations of selection in Hedrick's experiment (Hedrick 1980). The hatched portion of bars represent fixed populations.

The magnitude of the fluctuations caused by sampling error, and thus the probability that a novel mutation will by chance become extinct, does not depend on its frequency in the population, but rather on the number of individuals bearing it. Once several hundred individuals bearing the mutation are present in the population, its dynamics are nearly deterministic. In a large population, genes will escape from sampling error and begin to spread deterministically at very low frequencies; in small populations, their dynamics may be largely stochastic even when they are very common.

**The Establishment of New Mutations in *Drosophila* Populations.** There have been few attempts to study the fate of new mutations experimentally, no doubt because of the difficulty of following events that are taking place at very low frequency. Philip Hedrick at the University of Kansas set up replicate populations comprising 10 or 40 individuals of a white *w* strain of *Drosophila*. He then introduced the wild-type eye-color allele at a frequency of 0.1, so that it was represented by a single individual in populations of 10 flies, and by 4 individuals in the larger populations. This allele tends to spread rapidly through the populations, and reached an average frequency between 0.3 and 0.4 after only four generations. Replicate populations of either size diverge, however, so that after four generations it had reached a frequency as high as 0.8 in some populations—but had been lost entirely from others. It was among the smaller populations that gene frequency varied more widely, and the introduced wild-type allele became extinct more often, as expected. There is clearly a need for more extensive experiments on larger populations with more realistic mutation rates.

## 40. *Weak selection is readily detected by selection experiments.*

There is an important practical objection to gradualism: if selection is usually so weak, how can we study it? or even detect it? Imagine being given two strains of microbe and asked to determine whether they differ slightly but consistently in fitness. Their fitness can be estimated as a rate of increase in pure culture. Whether or not the two strains differ, however, replicate measurements of the same strain will certainly differ because of the slight differences between culture tubes, even under laboratory conditions. This makes it difficult to identify any real difference between the two strains if this difference is small in comparison with the difference between replicate cultures of the same strain. If the ratio of the real variation between cultures of the two strains to the total variation among cultures is $h^2$, then the number of cultures of each strain that we shall have to measure in order to establish the difference between them beyond reasonable doubt is, roughly speaking, proportional to $1/h^2$. It might easily be that the genetic variation of fitness amounts to only 1% of the environmental variation. Then, to be 95% sure of establishing that the 1% difference that we observe between the average fitnesses of the two strains would not arise by chance alone in more than 5% of experiments, we would have to measure fitness in more than 1000 cultures of each strain. This is a very onerous procedure, and to study the properties of a large number of strains would be a very laborious undertaking.

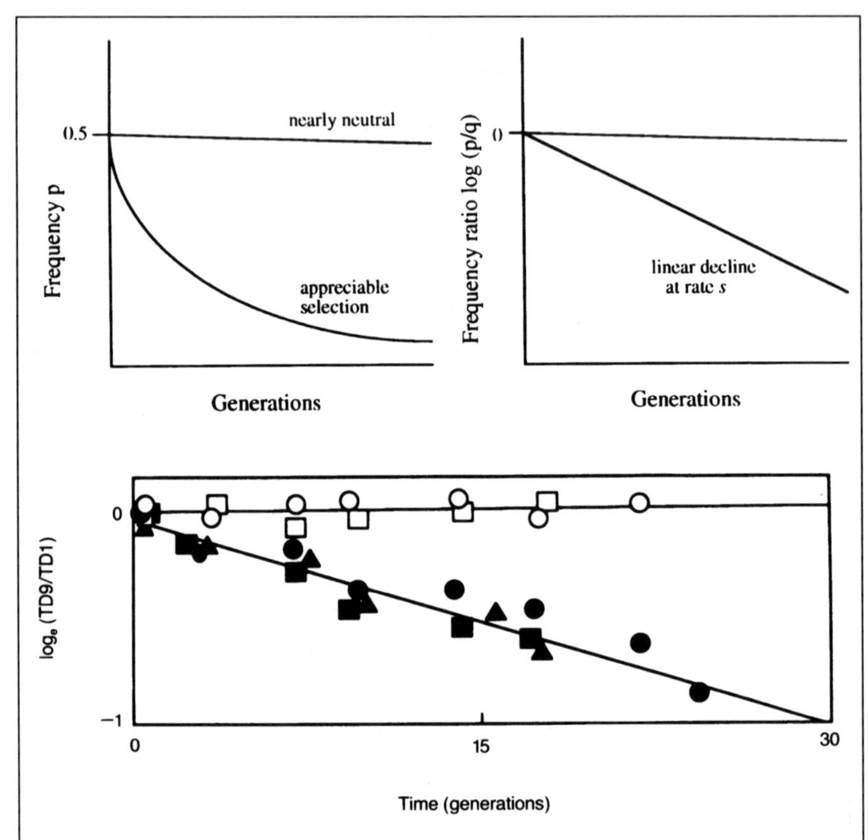


When two strains are mixed together and allowed to propagate for a number of generations, their frequencies will change through selection. The upper left drawing shows a situation where the two are initially equally frequent, with $p = q = \frac{1}{2}$. If they are nearly neutral in the prevailing conditions of culture, they will change only slightly, and perhaps imperceptibly, in frequency, but if one is appreciably less fit its frequency will decline rapidly. The upper right drawing shows that $\log_e(p/q)$, where $p$ is the frequency of the less fit type, declines linearly through time. The rate of decline is equal to the rate of selection (a consequence of the fact that exponential processes diverge exponentially: Sec. 4). Thus, if the initial frequencies of the two strains are $p_0$ and $q_0$, and their frequencies at a subsequent time $t$ (in units of generations) are $p_t$ and $q_t$, then, as long as the culture continues to grow exponentially, it can be shown that $\log_e(p_t/q_t) = \log_e(p_0/q_0) + st$. The selection coefficient $s$ can then be estimated as the slope of the linear regression of $\log_e(p_t/q_t)$ on time $t$. The lower figure is an example of this analysis from Dykhuizen and Dean (1990), using two strains of

*(Continues)*

*E. coli* cultured in a chemostat. The open symbols represent the frequency ratio in replicate experiments in glucose-limited chemostats. The slope of the pooled data is +0.0045, with standard error 0.0078; thus, selection is so weak that it cannot be detected reliably. The filled circles represent similar experiments in lactose-limited chemostats. Here, the slope is −0.0334, with standard error 0.0023; in this environment, therefore, selection involving a difference in fitness of about 3% is readily detected. Selection coefficients as small as $s = 0.01$ can routinely be estimated by this procedure, and the actual limit of detection in longer-term experiments is substantially lower.

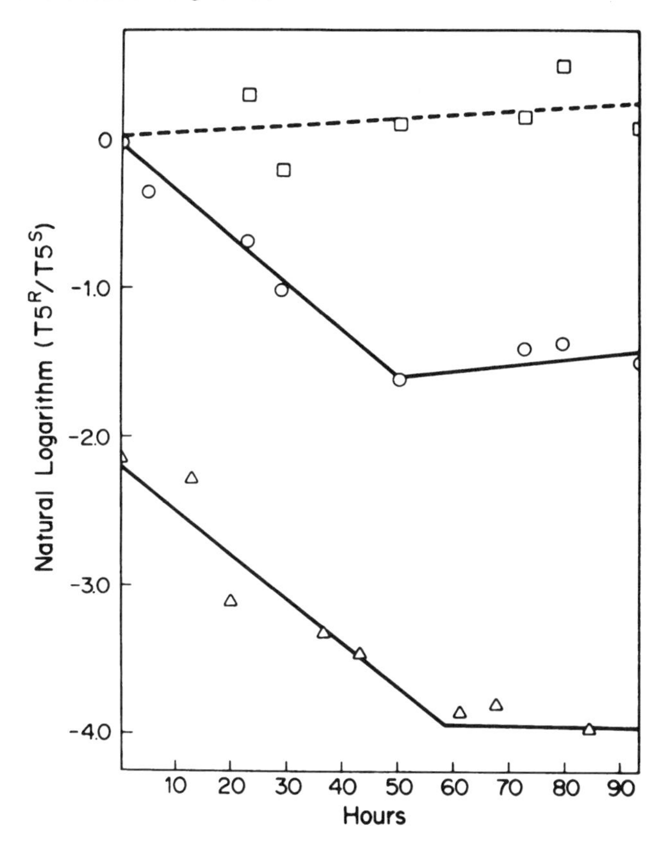


This diagram shows selection at the *gnd* locus in *edd*$^-$ strains, from the experiment by Dykhuizen and Hartl (1980; see also Hartl and Dykhuizen 1981) described in the text. The y-axis is the frequency of F2, marked with T5. The broken line and squares show the outcome of growth in a ribose-limited chemostat, where there is little or no change in frequency. The solid lines show that F2 declines in

(*Continues*)

*(Continued)*

frequency when competing against S8 in gluconate-limited chemostats, whether the strains have previously been selected in gluconate (circles) or not (triangles). In an $edd^+$ background no appreciable change in frequency occurs. There are comparable studies of phosphoglucose isomerase alleles by Dykhuizen and Hartl (1983) and of lactose alleles by Dean (1989).

The large literature on selection at the alcohol dehydrogenase locus in *Drosophila* has been reviewed by van Delden (1982) and Chambers (1988). There are also many comparable studies of other systems: for example, amylase (e.g. Powell and Andjelkovic 1983), esterase (e.g. Árnason 1991), and mitochondrial genotypes (e.g. MacRae and Anderson 1988).

**Measuring Fitness Through the Outcome of Competition.** This difficulty can be circumvented by using a more evolutionary approach to the problem. We first mix the two strains in equal proportions, and spread a sample of the mixture onto plates. The initial frequencies of the two strains can be estimated by counting the number of colonies of each type, provided, of course, that they have been made distinguishable. This can be done, for example, by incorporating into one of the strains a gene that is not expressed during growth in culture tubes, but that causes a visible phenotypic change when the cells are growing on plates using a different culture medium. A second sample is then used to inoculate a culture tube, and the mixed culture propagated by serial transfer as a selection line for a number of generations. This culture is then again plated out and the frequency of each strain estimated by counting colonies. If the two types differ in fitness by 1% ($s = 0.01$), selection will cause the frequency of the fitter strain to increase from 0.5 to 0.55 in about 40 generations. This difference in frequency can be measured easily and precisely by plating out the culture and counting 1000 or so colonies. Replicate selection lines must be run to eliminate the possibility that gene frequencies are changing at random, sometimes one strain increasing and sometimes the other, but the precision with which the frequencies can be estimated in each case means that a few (about 10) replicate lines will be adequate. Weak selection is, therefore, readily detected by allowing selection to sort a mixed population so that a small, almost imperceptible difference in fitness gives rise to a substantial and easily estimated difference in frequency.

The technique of serial transfer has one important drawback for estimating fitnesses: that the conditions of growth are continually changing as the population increases and resources are depleted. Rates of increase are therefore changing too. It is possible to devise an apparatus in which the conditions of growth remain constant by continually draining the culture at the same rate that fresh growth medium is admitted. This apparatus is a *chemostat*. The continuous culture of microbes in chemostats is the most sensitive procedure for estimating relative fitness.

This procedure has three important drawbacks. One is purely technical: how can one be sure that the genetic markers used to discriminate between the two strains are not themselves under selection, confounding the difference that is being studied? This is ruled out, or controlled statistically, by arranging that one strain bears the marker in some replicate lines, and the other strain in other lines. A more troublesome objection is that 40 generations can be run through in a few days when working with microbes, but would take a year or more with an organism like *Drosophila* and most of a career with mice or maize. This is a powerful incentive to use microbes to investigate fundamental evolutionary principles. A related, and perhaps even more troublesome, objection is that running selection experiments (especially with microbes) is much more difficult in the field than in the laboratory. This is an equally powerful incentive to use laboratory systems. To scientists in other disciplines, it may not seem very revolutionary to choose to work on simple and convenient systems under controlled conditions, but population biologists have always resisted doing so. It is true that there are many interesting evolutionary problems, for example, those involving the development of multicellular organisms, or the relative importance in nature of physical and biotic agents of selection, that cannot be studied in laboratory cultures of microbes. Nevertheless, they supply the most elegant way, and often the only feasible way, of investigating the general principles of gradual evolutionary change.

**The Alcohol Dehydrogenase Polymorphism of *Drosophila*.** The conflict between neutralist and selectionist interpretations of allelic diversity in natural populations (Sec. 37) led to a large literature concerned with the differences in fitness, if any, among alleles. These differences turned out to be unexpectedly difficult to characterize. The most direct approach is to measure the viability or fecundity of strains bearing different alleles. This is a crude procedure, however, that is incapable of resolving small differences in fitness, as I have explained earlier. Moreover, any differences that were detected might be attributable, not to the locus being studied, but to genes at linked loci. Finally, the failure to find any substantial differences in fitness would be inconclusive, because they might be expressed only in some other, untried, conditions of culture.

These difficulties can be illustrated by work on the alcohol dehydrogenase (*Adh*) locus of *Drosophila*. This encodes an enzyme catalyzing the oxidation of alcohols to aldehydes or ketones. The locus has two common alleles, named fast $F$ and slow $S$ because of their respective mobilities on electrophoretic gels, that differ by a single amino-acid substitution. Most natural populations contain both alleles, and the object of most studies of the population genetics of the locus has been to explain why both persist, rather than one or the other becoming fixed. In the absence of alcohol, laboratory populations that start with different allele frequencies tend to converge on the same composition, usually close to that of the base population from which the lines were derived. This suggests that this composition represents an equilibrium that is restored through selection when it is experimentally perturbed. It is possible, however, that the $F$ and $S$ alleles are themselves nearly neutral, but happen to be associated with different alleles at a nearby locus that is strongly selected. When ethanol is added to the food medium at concentrations of 5–15%, there is a clear and consistent difference between the two forms: the $F$ enzyme has the higher activity in vitro, and $FF$ homozygotes increase in frequency toward fixation in the presence of ethanol. The details of the *Adh* system are, of course, more complicated than this simple account suggests; in particular, there are several other alleles, besides numerous minor variants of the two common alleles, whose behavior is not as straightforward. Nevertheless, the clear functional link between enzyme activity and changes in gene frequency is convincing evidence that the $F$ and $S$ alleles differ in fitness, at least in some environments, and it would be surprising if selection did not determine how they are maintained in natural populations.

**Selection of 6-Phosphogluconate Dehydrogenase Allozymes in *E. coli*.** More sophisticated experiments are possible in microbes. Daniel Dykhuizen and Daniel Hartl, working at Purdue, attempted to detect fitness differences among alleles in chemostat populations of *E. coli*. To understand their experiment, and to appreciate the elegant and powerful work that can be done in such systems, it is necessary to give a few biochemical details. The locus they studied was *gnd*, which encodes 6-phosphogluconate dehydrogenase (6PGD), an enzyme involved in the pathway by which glucose-6-phosphate is converted to a pentose phosphate. It is quite diverse, with 15 alleles being detected in about 100 natural isolates; Dykhuizen and Hartl studied four of these, with the unmemorable names of S4, S8, F2 and $W^+$. The other locus involved in the experiments was *edd*, which encodes phosphogluconate dehydratase; this enzyme leads to an alternative pathway for the metabolism of 6-phosphogluconate, making it possible to bypass

6PGD. They also used—merely for purposes of estimation—genes conferring resistance ($T5^R$) or susceptibility ($T5^S$) to phage T5. To construct the initial strains, the appropriate combination of genes was inserted into the genome of a standard laboratory strain by viral transduction, so that they were as nearly identical as possible except for carefully specified differences in *gnd, edd*, and T5. The strains bearing different *gnd* alleles could then be put into competition with one another in order to estimate the strength of selection. Each experiment involved four trials. Two were the experimental lines, reciprocally marked with T5. Because *E. coli* is completely asexual (in the conditions of this experiment), any two loci in the same genome are completely linked; hence, changes in frequency at *gnd* could be followed (much more conveniently) by changes in the frequency of T5 alleles, measured by exposing the cultures to phage. The other two lines were controls, incorporating a point mutation $gnd^-$ that abolished 6PGD synthesis. There was strong selection against $gnd^-$ on gluconate medium, supplying a positive control ensuring that selection would be detected if it occurred. On the other hand, $gnd^+$ and $gnd^-$ alleles are nearly neutral in a ribose-succinate medium that contains no gluconate. The growth of the experimental lines on the ribose-succinate medium thus gives a negative control showing that any difference betwen the functional *gnd* alleles when growing on gluconate is indeed attributable to differences in 6PGD activity.

The general result of these experiments was that the four *gnd* alleles had indistinguishable effects on fitness when tested in an $edd^+$ background; the sensitivity of the assay is such that the result implies that any selection that occurs is very weak, with $s < 0.01$ per generation. There were, however, two complications. The first is that the S8 $T5^S$ strain was consistently selected over the S8 $T5^R$ strain in gluconate, although not in ribose-succinate. There is, therefore, some functional interaction between 6PGD and phage resistance, and the dynamics of selection of *gnd* alleles will depend to some extent on their genetic background. A more serious complication is that quite different results were obtained when the experiments were run with an $edd^-$ background. The details are complicated, because $edd^-$ strains do not grow well on gluconate, so that before the experiment is completed new *gnd* alleles have arisen by mutation. It was clearly established, however, that S4 is consistently superior to F2. It seems that $edd^+$ genotypes can compensate for any small differences among *gnd* alleles in their ability to process gluconate, but these differences are expressed in $edd^-$ genotypes, where no compensation is possible.

Now, if we ask whether or not the *gnd* alleles are neutral, it will be difficult to give a straightforward answer. In most of the conditions where they were tested, they have very similar fitnesses, but differences appeared when they were tested on a particular genetic background (*edd*$^-$) in a particular environment (gluconate growth medium). I think it is very likely that most loci will display this sort of conditional neutrality. Selection among alternative alleles is weak or non-existent across a broad range of conditions, but may be appreciable when the same alleles are expressed in a different genome or in a different environment.

## 41. *The response to selection can be predicted from first principles.*

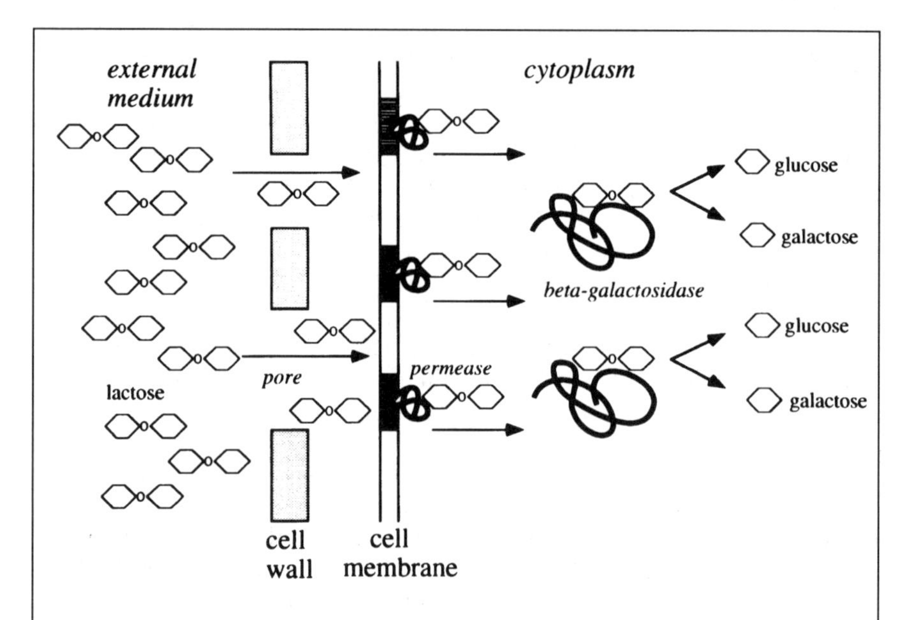


Lactose present in the medium diffuses passively through pores in the cell wall into the periplastic space. It then passes across the cell membrane into the cytoplasm by an active process requiring the enzyme *lac* permease. In the cytoplasm, it is hydrolyzed by a $\beta$-galactosidase, yielding glucose and galactose, which undergo further metabolic transformations. The work on lactose metabolism summarized here has been published by Dean (1989), Dean et al. (1986) and Dykhuizen et al. (1987), and reviewed by Dykhuizen and Dean (1990).

(*Continues*)

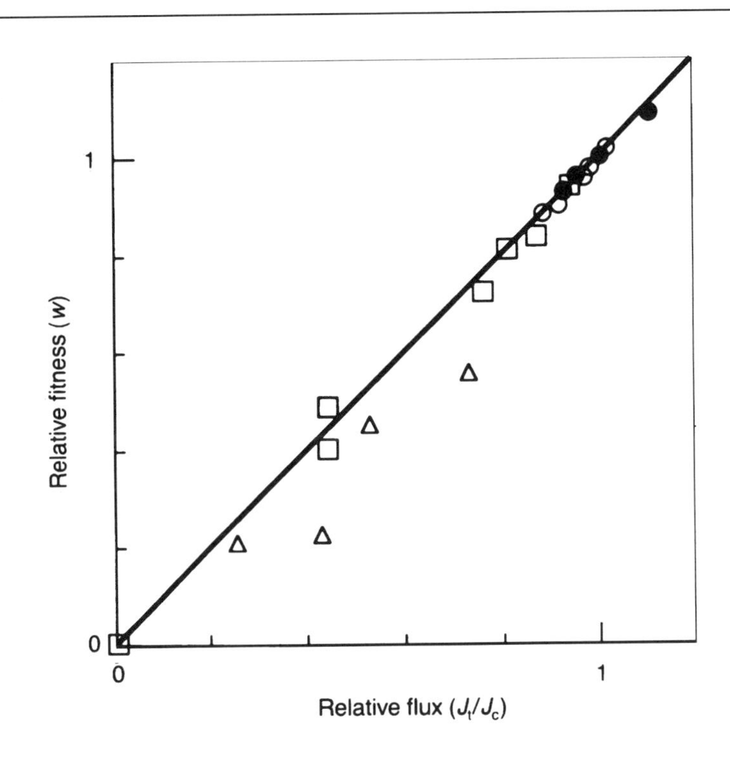


When lactose is the limiting resource, the fitness of a given type depends only on the rate at which it generates glucose and galactose from environmental lactose. Thus, relative fitness is an increasing linear function of the flux of lactose through the pore–permease–$\beta$-galactosidase system. The diagram here shows the fitness of various types relative to the standard K12 strain of *E. coli*; these types, which are mostly deficient in lactose metabolism, include natural isolates, strains bearing mutations in the $\beta$-galactosidase structural gene, strains with a novel $\beta$-galactosidase (see Sec. 75), and strains with regulatory mutations.

(*Continues*)

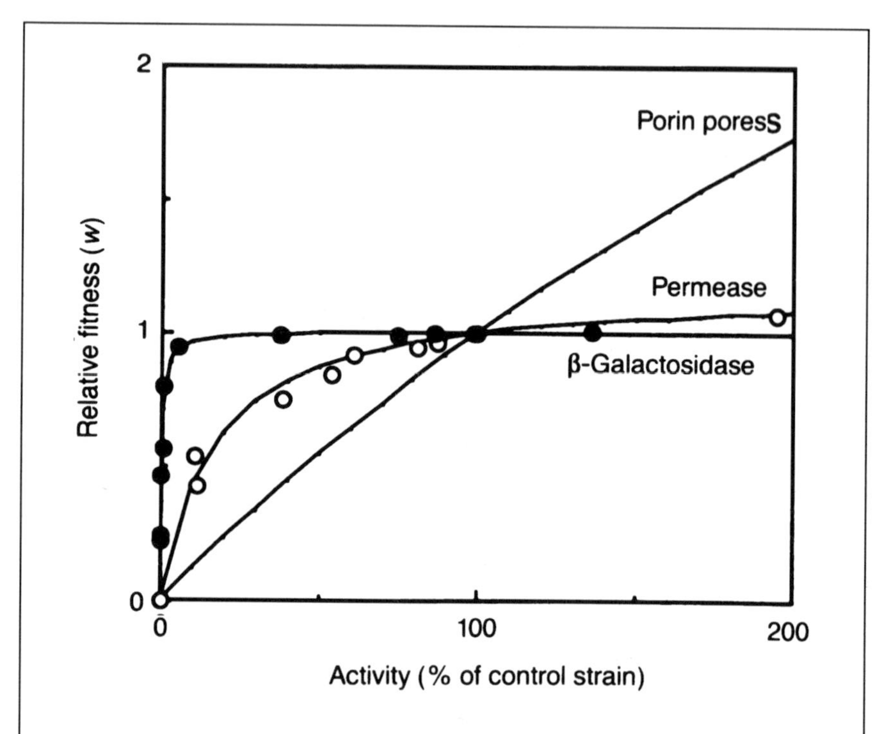


The rate $V$ of an enzyme-catalyzed reaction can be described in terms of the substrate concentration $S$ by the Michaelis–Menton equation, $V = V_{max}S/(S + K_s)$, where $V_{max}$ is the maximal rate at high substrate concentrations and $K_s$ is the half-saturation constant, i.e., the substrate concentration at which the rate of the reaction is half its maximal value $V_{max}$. In metabolic pathways, a series of reversible reactions are linked together by a series of enzymes with different Michaelis–Menton dynamics; the overall flux of material through the pathway depends on all these enzymes, and moreover the flux at each step depends on the flux at every other step in the pathway. The lactose system is relatively simple, because it is terminated by the irreversible hydrolysis of lactose by $\beta$-galactosidase. If the concentrations of substrates are low (less than their $K_s$) the system is unsaturated, and the flux $J$ will be approximately

$$J = S/[(1/E_1) + (1/E_2K_{1,2}) + (1/E_3K_{1,3})]$$

The subscripts 1, 2, and 3 refer to the three components of the system: the cell-wall pores, *lac* permease, and $\beta$-galactosidase. $E$ is a measure of enzyme activity $V_{max}/K_s$; for the pores, this is equivalent to their diffusion constant. $K$ is the thermodynamic equilibrium constant for the reaction (often symbolized $K_{eq}$, and not to be confused with $K_s$). Writing out the equation in this way makes it clear how a change in any component of the system will affect flux through the

(*Continues*)

*(Continued)*
pathway as a whole, and will thereby affect fitness. All the parameters involved can be measured independently and then used to predict the outcome of selection in a lactose-limited culture. The figure shows the predicted effects (curves) of alterations in the three components of the lactose system, relative to a control strain. Mutants unable to utilize lactose through a failure of any of the three components have zero fitness, there being no alternative carbon source; mutants that are equivalent to the reference strain have a relative fitness of unity. In general, relative fitness increases in an asymptotic fashion as the activity $E$ of a given component increases. Note that fitness is much more sensitive to changes in the activity of some components than to comparable changes in others. A very modest level of $\beta$-galactosidase activity yields nearly wild-type performance, and further increases in activity give little increase in fitness. Minor changes in the permease are more likely to cause appreciable changes in fitness, and the system is most sensitive to pore diameter. The plotted points are actual relative fitnesses, measured in competition trials, of a series of mutant alleles of *lac* permease and $\beta$-galactosidase (no information is available for pore mutants). The outcome of selection seems to be adequately predictable from the biochemical properties of the enzymes concerned.

Because the physical conditions of growth in the chemostat remain constant, there is a simple relationship between phenotype and fitness. If we know enough physiology and biochemistry, therefore, we should be able to predict how the population will evolve, at least in the short term. Some very elegant experiments of this sort have been done by Daniel Dykhuizen, Anthony Dean and their colleagues at Stony Brook, using chemostat populations of the bacterium *E. coli*.

**The Lactose-Limited Chemostat.** If all resources, except one, are supplied in excess, then the rate of growth will be proportional to the rate at which the limiting resource is supplied. Dykhuizen and Dean used growth medium in which the limiting resource was carbon, supplied as lactose. The rate of increase of a culture is then proportional to the rate of supply of lactose. When the lactose concentration and the rate of inflow of the growth medium are held constant, therefore, the fitness of any given strain will depend only on the rate at which the cells can acquire and utilize lactose. This is easy to measure in cultures of any given strain, and we predict that strains with a greater rate of lactose metabolism will increase in frequency when competing with other strains in the chemostat. We can, therefore, use a purely biochemical measurement to make an evolutionary prediction. It might be objected that strains that differ in lactose metabolism might differ in many other ways too, so that the prediction might often be right by chance. This objection is readily circumvented in *E. coli*, where it is possible to construct strains that differ at the *lac* operon (that encodes the enzymes responsible for processing lactose) but that are identical in all other respects. The prediction is then verified: in lactose-limited conditions, variants with a greater rate of utilization of lactose replace their competitors. Moreover, this is quantitatively as well as qualitatively true: the greater the difference between two strains in their rate of lactose utilization, the greater their difference in fitness, and the more rapidly one replaces the other. The difference in rates of lactose utilization, measured in pure cultures of two or more strains, can thus be used to predict the direction and the rate of allelic substitution when the strains are cultured together. The difference in the rate of lactose utilization that can be detected in this way are quite small: strains that differ in fitness by only 0.5% diverge reliably in frequency, allowing very small differences in fitness to be studied. The chemostat is a more precise instrument than batch culture for studies of this sort, although, as one might expect, it is more difficult to set up and maintain.

Furthermore, it is possible to predict the type of genetic change that is likely to be favored by selection, from a knowledge of the biochemistry of lactose utilization. Lactose diffuses through pores in the cell wall, is actively transported across the cell membrane by a permease, and once in the cytoplasm is hydrolized by $\beta$-galactosidase to form glucose and galactose, which then enter central metabolism. The permease and the $\beta$-galactosidase are both encoded by genes of the *lac* operon. Changes in pore diameter or in the activity of either of the two enzymes will cause changes in the rate of lactose utilization and, therefore, in fitness. Different types of change, however, have quantitatively different consequences because of the dynamics of lactose metabolism. These can be inferred from purely physiological and biochemical arguments. The rate of lactose utilization at low lactose concentrations is extremely sensitive to the diameter of the cell-wall pores, less sensitive to the permease-mediated transport across the cell membrane, and rather insensitive to alterations in the activity of $\beta$-galactosidase. The genetics of pore size is not well characterized, but the fitnesses of the permease and the $\beta$-galactosidase are very closely in accord with theoretical expectation. We can therefore argue directly from the biochemical properties of variant enzymes to predict whether and how fast one mutant allele will be substituted for another under selection in an experimental population. In environments that are more complicated than lactose-limited chemostats, the biochemical arguments would be correspondingly more complicated, but these simple experiments provide the assurance that we can, in principle, provide a complete accounting of the functional basis of gene substitution in terms of the chemical properties of the variant enzymes.

## 42. *The sorting limit in asexual populations is the limit of extant variation.*

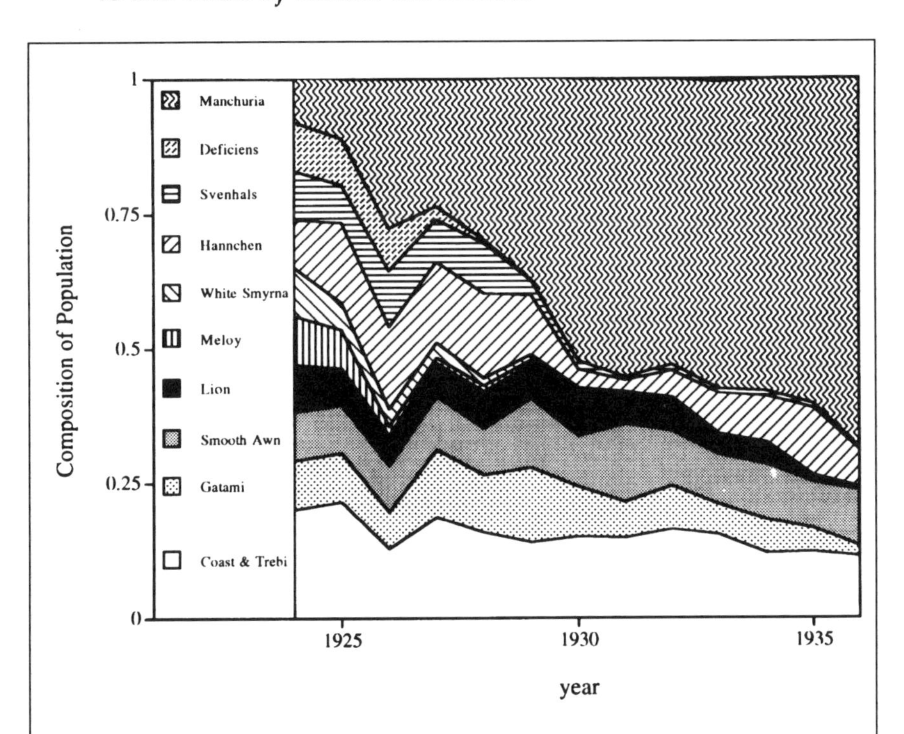


This diagram shows the change in composition of the mixture of barley varieties set up by Harlan and Martini (1938) over 12 years at Ithaca, New York. Coast and Trebi could not be reliably distinguished and have been pooled. In this case, varieties such as Meloy, White Smyrna, and Deficiens are eliminated quite rapidly, whereas others, such as Hannchen and Smooth Awn, persist for much longer, but Manchuria quickly becomes the most frequent type, and by the end of the experiment comprises two-thirds of the population. Results at other localities, and selection in populations with an indefinite composition, set up from crosses, are discussed in Sec. 103. For a comparable experiment, see Blijenburg and Sneep (1975).

(*Continues*)

(*Continued*)

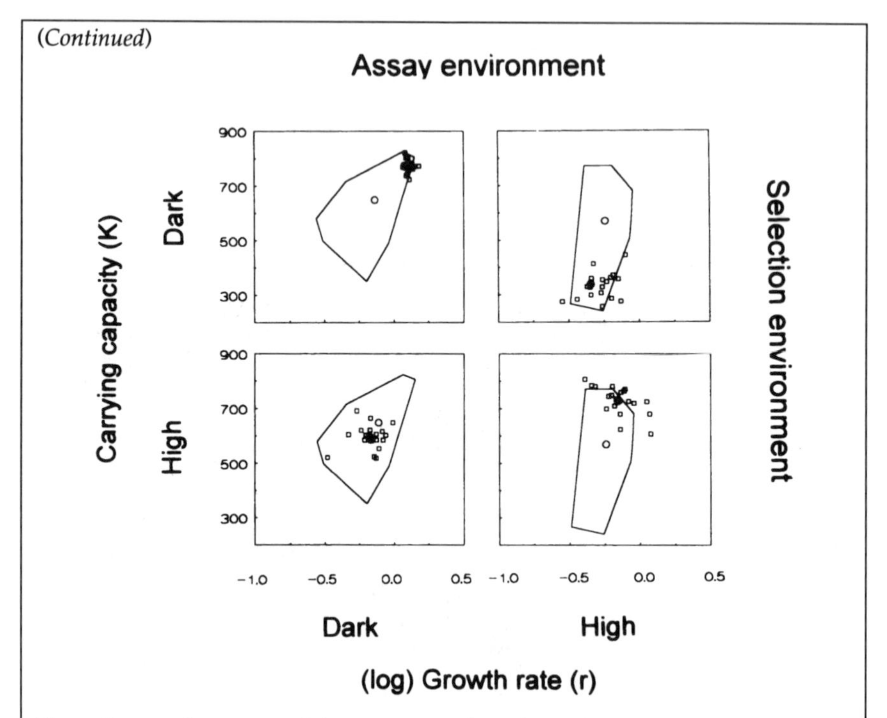


These figures show part of the outcome of the Farnham sorting experiment described in the text, from Koelewijn, de Laguérie, and Bell (unpublished). The envelopes show the joint distribution of the limiting rate of increase $r_{\max}$ and the carrying capacity $K$ (see the introduction to Part 5.A) in the set of spores isolated from the field, the open circle marking the bivariate mean. The open squares are genotypes reisolated after about 70 generations of selection in the light in minimal medium, or in the dark in medium supplemented with acetate. The mean of the selected population is marked with a solid circle. In the environment of selection the sorting limit for both measures of fitness is quickly approached (and even somewhat transgressed). Note that selection in one environment often has little effect on the mean or variance of fitness in other environments (as in the case of the population selected in the light and tested in the dark), showing that sorting can cause specific adaptation to particular conditions of growth (Sec. 103). Indeed, selection in one environment may actually cause a deterioration of performance in other environments (as in the case of the population selected in the dark and tested in the light); see Sec. 110.

Using isogenic lines of bacteria, genetically identical except for a single carefully specified difference, is a very powerful way of isolating unit processes in evolution. It does not, of course, correspond to the situation in natural populations, where a great diversity of genotypes may contend together at any given time. The principle, however, remains the same. We can phrase the problem in several different ways, corresponding to different possible experiments. We might begin with a well-adapted population, in which variation around a nearly optimal type is maintained by mutation. We might instead begin with an arbitrary collection of types, perhaps representing new isolates from natural populations. In either case, we expose a diverse population to a new environment.

**Variation.** Selection will proceed to increase adaptedness by sorting the variation initially present in the population. To describe this process more precisely, we need to define parameters that express the quantity of variation. The simplest is the observed range of individuals, the difference between the most extreme individuals in the population. For many purposes, this is not a very satisfactory parameter, because it varies with the number of individuals sampled: the larger the sample, the more likely it is to include rare or extreme types of individual. The parameter that is usually used to express the quantity of variation is the variance, which is the mean squared deviation of observations from the mean. The variance of a character $z$ can be symbolized $\mathrm{Var}(z)$. The square root of the variance is called the standard deviation. Roughly speaking, about two-thirds of the population will differ from the population mean by less than one standard deviation; about 95% by less than two standard deviations; and about 99% by less than three standard deviations. Estimates of variance and standard deviation are independent of sample size, but for many purposes require that the population is approximately Normally distributed, that is, that the frequency distribution of the character being analyzed is a bell-shaped curve. A more detailed account of the variance can be obtained from the introductory sections of any text on statistics.

**The Sorting Process.** Those types which happen to be better adapted to the new environment will tend to increase in frequency. As selection sorts the initial population, favoring types with greater fitness, the mean fitness of the population as a whole must increase continuously. Thus, a type may at first have a fitness greater than the average of the population and will increase in frequency; but after inferior types have been eliminated, its fitness may then be less than the population mean, and after its initial rise it will decrease in frequency. In this way, all but one of the genotypes initially present will eventually decline to extinction, often after a shorter or longer period of increase. The population will eventually come to consist almost entirely of one of the types, initially rare, that have the greatest fitness in the changed conditions. The very fittest type might not be fixed, because it might be lost by chance while it is still very rare, but, at any rate, the mean fitness of the population when selection is complete cannot exceed the fitness of the fittest type present before selection. The end-point of the sorting process is thus set by the *range* of types present in the initial population.

The range of a population depends on its size, although the range does not increase proportionately with population number. Suppose that the population is a mixture of distinct strains. The actual population can be thought of as a sample from an infinite population in which all possible strains occur at different frequencies. The probability that a sample of $N$ individuals does not include a given strain whose frequency is $p$ is $(1 - p)^N$, or about $e^{-Np}$ if the strain is fairly rare. The probability that a strain does not occur is rather large if the population is small, and falls as the population increases in number. A strain that arises by mutation at low frequency is therefore unlikely to be present in a small population, but almost certain to occur in a very large population. A similar principle holds when individuals are Normally distributed with respect to a character such as fitness. The ratio of the range to the standard deviation is about 3 for a population of 10 individuals; about 5 for 100 individuals; about 6.5 for 1000 individuals, about 7.75 for 10,000 individuals; and so forth. Sorting is therefore more effective in large populations, which will become, in the short term, more highly adapted to a novel environment.

**Sorting of a Barley Mixture.** H.V. Harlan and M.L. Martini, two agronomists in the U.S. Department of Agriculture, set up several large-scale, long-term crop trials in the 1920s. Some of them are still running. The simplest of them is an excellent illustration of sorting. They mixed 11 easily distinguished barley varieties in equal amounts and sowed plots at several localities in the northern and western U.S.A. Barley is almost entirely self-fertilized, and the varieties can be relied upon to remain distinct. Every year between 1924 and 1936, the plots were harvested, using routine agricultural methods, and some of the seed saved to be sown in the following season. A second sample of seed was sent to a central station, where it was planted out in spaced rows so that the varietal composition of the seed population could be scored. These stands of barley thus resembled self-propagating populations whose genetic structure could be observed directly. At every station, there were ill-adapted varieties that soon disappeared from the population. Others persisted throughout the experiment, but in most cases one variety would increase rapidly in frequency, so that after 10 or 12 generations it made up the bulk of the population, and would doubtless have eventually spread to fixation had the experiment lasted longer.

**The Farnham Sorting Experiment.** A great variety of motile green eukaryotic microbes can be isolated from soil samples. The isolates that

Hans Koelewijn, Patrick de Laguérie, and I got from the corner of a wheatfield near the village of Farnham, in southern Quebec, showed widely differing abilities to grow in different conditions: in dark or light, at high or low nutrient concentrations, in liquid medium, or on agar plates. When we cultured a mixture of these isolates in given conditions for a few growth cycles, amounting to 70 generations or so, this variation was greatly reduced, and most of the genotypes extracted from the cultures at the end of the experiment were equivalent to the fittest genotypes of the initial mixture. This was particularly clear for the more variable characteristics, such as the microbe's ability to grow heterotrophically in the dark on medium supplemented with an organic carbon source; 70 generations of sorting apparently reduced the diverse original population, comprising more than sixty genotypes, to a single clone.

## 43. *The rate of sorting is proportional to the amount of variation in fitness.*

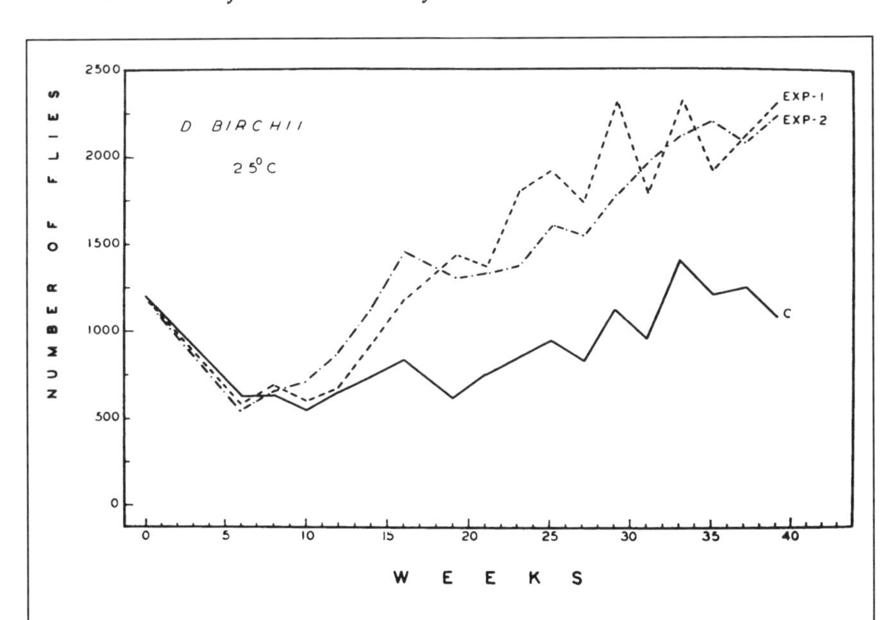


This diagram shows the size of two irradiated populations of *Drosophila birchii* cultured at 25 C, compared with that of an unirradiated control line, from Ayala (1966). Similar results were observed with *Drosophila serrata* at 25 C and 19 C, but

(*Continues*)

(*Continued*)
not with *D. birchii* at 19 C, where there was no difference between the irradiated lines and the control. Similar experiments have been carried out by Dobzhansky and Spassky (1947), Wallace (1957, 1958, 1959), Carson (1964), and Crenshaw (1965).

The rate at which sorting occurs depends on the magnitude of the variation among individuals in the initial population and on the fidelity with which these differences are transmitted to descendents. If there is initially a large difference in fitness between alternative types, the rate of selection will be greater. If the differences among individuals are predominantly inherited, selection will be more efficacious in causing evolutionary change. The rate of sorting does not depend on the range, which may be strongly influenced by the fortuitous occurrence of a single extreme individual, but rather on the variance of the population. The greater the variance of fitness, the greater the change in mean fitness that will be caused by a single episode of selection. This increase will be expressed in the following generation only if the differences in fitness among individuals are inherited; that is, the effect of selection on mean fitness in successive generations depends on the genetic variance of fitness. The greater the genetic variance of fitness, the more rapidly adaptation will occur. The Farnham sorting experiment provides an example of this.

In the simple situations that we are presently considering, we can phrase this conclusion even more tersely: *the rate of evolution is equal to the genetic variance of fitness*. If we write the mean fitness of the population at a given time as $w$, and the standardized genetic variance of fitness as $V_w^2$, then:

$$(1/w)\,\mathrm{d}w = V_w^2$$

The standardized genetic variance of fitness is the genetic variance divided by the square of the mean, $V_w^2 = \mathrm{Var}(w)/w^2$. This conclusion was first reached by Sir Ronald Fisher, who called it "The Fundamental Theorem of Natural Selection." It has been extensively debated by population geneticists ever since. The reason for this is partly that it seems tractable to algebraic analysis, without the need to understand the ecological details of particular situations, and partly that it seems to attach an arrow to evolutionary change, pointing in the direction of ever-increasing adaptedness. It is very generally applicable, being an algebraic identity that relates the unweighted mean of a set of numbers (mean fitness before selection) to their weighted mean, the weights being the numbers themselves (mean fitness after selection). It is the most succinct way of summarizing the dynamics of sorting in a population, relating the change in the mean to the quantity of variation.

It is worth emphasizing that it is only the genetic variance of fitness that underlies the response to selection. If all variation in fitness is caused by environmental factors, then mean fitness will remain the same from generation to generation, regardless of the operation of selection within each generation. Moreover, the environmental variation itself will remain constant, being restored anew in every generation. This is not true of the genetic variance. Under selection, less fit types are replaced by more fit types; the less fit types thereby decline in frequency and are eventually lost from the population. Eventually, the population, however diverse it may have been originally, becomes genetically uniform when the fittest type has become fixed. At this point, the genetic variance of fitness has been exhausted, and no further response to selection is possible. We might liken genetic variance to a fuel that drives selection and that is used up in the process. This implies that the rate of sorting will slow down as genetic variance is depleted, so that the short-term response to selection is a decelerated rise towards the limit set by the range of the initial population.

**The Effect of Increasing the Mutation Rate on the Response to Natural Selection.** It has often been argued that populations might adapt more quickly to novel environments and would reach a higher level of adaptedness in the short term, if the amount of variation they exhibit were increased by exposure to a mutagenic agent such as ionizing radiation. Francisco Ayala, then at The Rockefeller University, used the productivity of outbred *Drosophila* populations maintained in vials as a measure of their mean fitness or adaptedness. The flies had been maintained in the laboratory for two years before the experiment started, but under less crowded conditions in mass culture. Control lines were simply extracted from the mass cultures and transferred to fresh vials at intervals. The experimental lines were exposed to 2000–4000 R of X-rays at the beginning of the experiment. The immediate effect of the treatment was often a decrease in productivity, but in three cases out of four the irradiated lines thereafter increased faster than the controls, both in productivity and in population number and biomass. After about 40 generations, most of the irradiated populations produced about 50% more individuals than the control lines. In this case, the increased variation caused by radiation-induced mutations seems to have enhanced the response to selection. Not all experiments have given the same result. Hampton Carson, for example, reported a similar study in which he continued to irradiate the selection lines during the course of the experiment, administering a dose of 1000 R for over two years, resulting in a total dose of 65,000 R. He failed to observe any consistent or sustained increase in productivity. In this case, it would seem that any beneficial mutations were likely to be associated deleterious mutations because of the very high overall mutation rate continually induced by the treatment, so that selection was ineffective.

## *44. The rate of evolution of a character is proportional to its genetic covariance with fitness.*

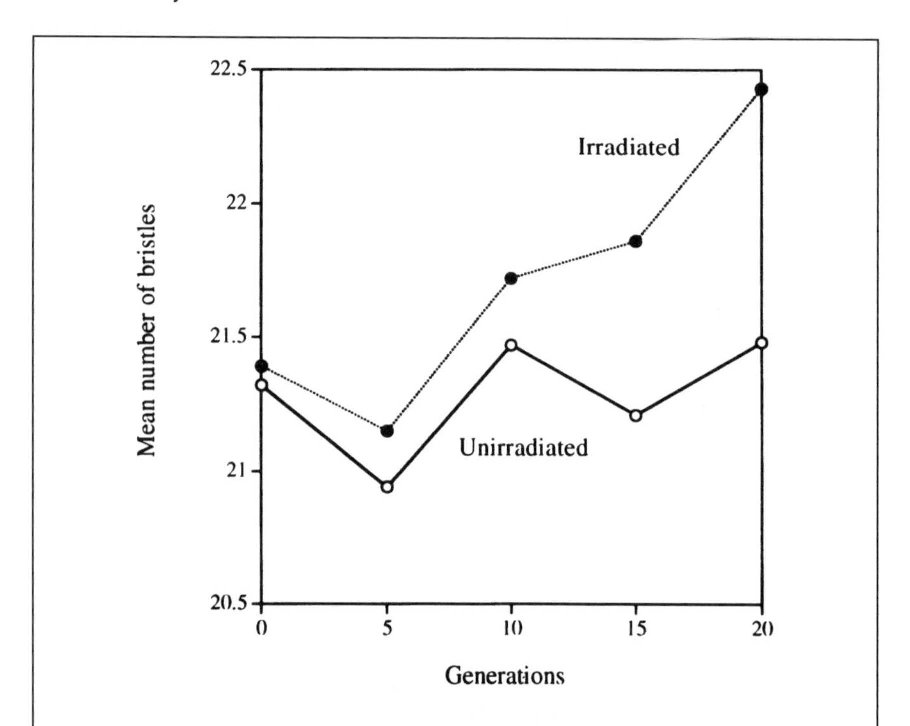


In the experiment reported by Hollingdale and Barker (1971), lines that received 1000 R of X-rays each generation responded more rapidly to artificial selection for bristle number; this graph shows the average values for five irradiated selection lines and three unirradiated controls subject to the same selection procedure, from Table 4 of the original paper. Similar experiments reporting the effect of irradiation on the response to artificial selection have been published by Scossiroli (1954), Harrison (1954), Clayton and Robertson (1955), R.E. Scossiroli and S. Scossiroli (1959), Gardner (1961), and Kitagawa (1967). The topic has been reviewed by Gottschalk and Wolff (1983) and Gründl and Dempfle (1990).

The evolution of fitness through sorting in asexual populations is easy to understand. We are, however, more likely to be interested in the beak shape of finches, the coloration of moths, or the lactose metabolism of bacteria. Selection cannot act directly on characters such as these; it acts only indirectly, through the effects that they have on fitness. Suppose that we plot the fitness of individuals expressing a given character as a function of character value. If the plot is flat, with zero slope, individuals have the same fitness, regardless of the value of the character. In this case, selection will have no effect on the mean value of the character. On the other hand, if individuals expressing different values of the character have different fitnesses, selection will tend to cause a change in the mean value of the character. In the chemostat experiments that I have described, it was possible to define the relationship between the value of a phenotypic character—the rate of lactose utilization—and fitness, and, therefore, to predict how selection would drive phenotypic change. More generally, we can define $b_{wz}$ as the slope of the graph of fitness $w$ on character value $z$, in which case the fractional change in the mean value of a character during a single episode of selection, from generation to generation, will be

$$(1/z)\,\mathrm{d}z = b_{wz} V_z^2$$

where $V_z^2$ is the standardized genetic variance of $z$. This is a straightforward extension of Fisher's fundamental theorem. The steeper the slope $b$, the greater the effect that variation in character value has on fitness and the faster the mean of the character will change under selection. The change caused by selection is established permanently in the population only to the extent that the character is inherited. If the character is fitness itself, then, of course, $b = 1$, and the relationship becomes the simpler one described before.

It is possible to express this relationship more simply. The extent to which a character varies is called its variation, and is appropriately measured by its variance. The extent to which two characters vary together, so that change in one is associated with change in the other, is called their covariation, and is appropriately measured by an analogous quantity called their covariance. In general, when a character varies, some part of that variation will be genetic. Selection acting on that character will then cause a change in its mean value through time. Fitness is the only character that is directly selected in this way. Similarly, when two characters covary, some part of that covariation is likely to be genetic. This means that individuals that possess a certain combination of the two characters are likely to produce offspring that express a similar combination. Selection on one character will then cause a change in the mean value of the other character. Thus, when there is genetic covariation between a given character and fitness, selection acting directly on fitness will cause an increase in mean fitness and will also indirectly cause a change in the mean value of the other character:

$$(1/z)\,dz = \mathrm{Cov}(z, w)$$

where $\mathrm{Cov}(z, w)$ is the standardized genetic covariance of the character with fitness. The rate of change of a character is equal to its genetic covariance with fitness. The mean value of any character in a population therefore changes in a predictable way, depending on the effect that the character has on fitness.

**The Effect of Increasing the Mutation Rate on the Response to Artificial Selection.** When the effect of a character on fitness is given, its response to selection will depend on its genetic variance. The rate of response to artificial selection might then be increased by raising the mutation rate. Barbara Hollingdale and J.S.F. Barker of Sydney administered 1000 R of X-rays per generation to lines of *Drosophila* selected for increased bristle number. The base population was an inbred stock expected to be largely homozygous and genetically uniform, and phenotypic variation in unirradiated control lines remained low throughout the 20 generations of the experiment. They did, however, respond slightly to selection, increasing by about 0.3 bristles from the original value of about 21 bristles. Irradiation caused a steady increase in phenotypic variance, presumably by inducing mutations in genes affecting bristle number, and the mean value of the irradiated selection lines increased by about 1.1 bristles. This is roughly equivalent to 0.5 phenotypic standard deviations, so although irradiation had a detectable effect on the response to selection, this effect was rather small, amounting to little more than 0.025 phenotypic standard deviations per generation.

## 45. *In sexual populations, genes that are transmitted independently can be selected independently.*

The process of sorting as I have described it so far involves the selection of certain types of individual, and thereby the proliferation of certain lineages, in clonal populations. The single clone with greatest fitness eventually replaces all others, which are maintained only by mutation. I have described how different *lac* alleles flourish or dwindle in the bacterial populations of chemostats. This description, however, is valid only because we are able to construct clones of *E. coli* that differ only at the *lac* operon. If we merely took *lac* variants at random from unrelated bacterial strains, they

(*Continues*)

would probably differ at hundreds of loci. For example, two strains with different *lac* alleles might also differ at the *trp* operon, which is responsible for encoding the ability to synthesize the amino-acid tryptophan. In minimal medium, a defective *trp* allele is lethal, and the clone that bears it will be eliminated, even if it has the superior *lac* allele. *The genome as a whole proliferates as the result of selection in asexual populations.*

**Hitch-hiking of Linked Genes.** In the example I have given, a superior allele fails to spread in a population because the genome in which it occurs happens also to include a deleterious mutation at another locus. The

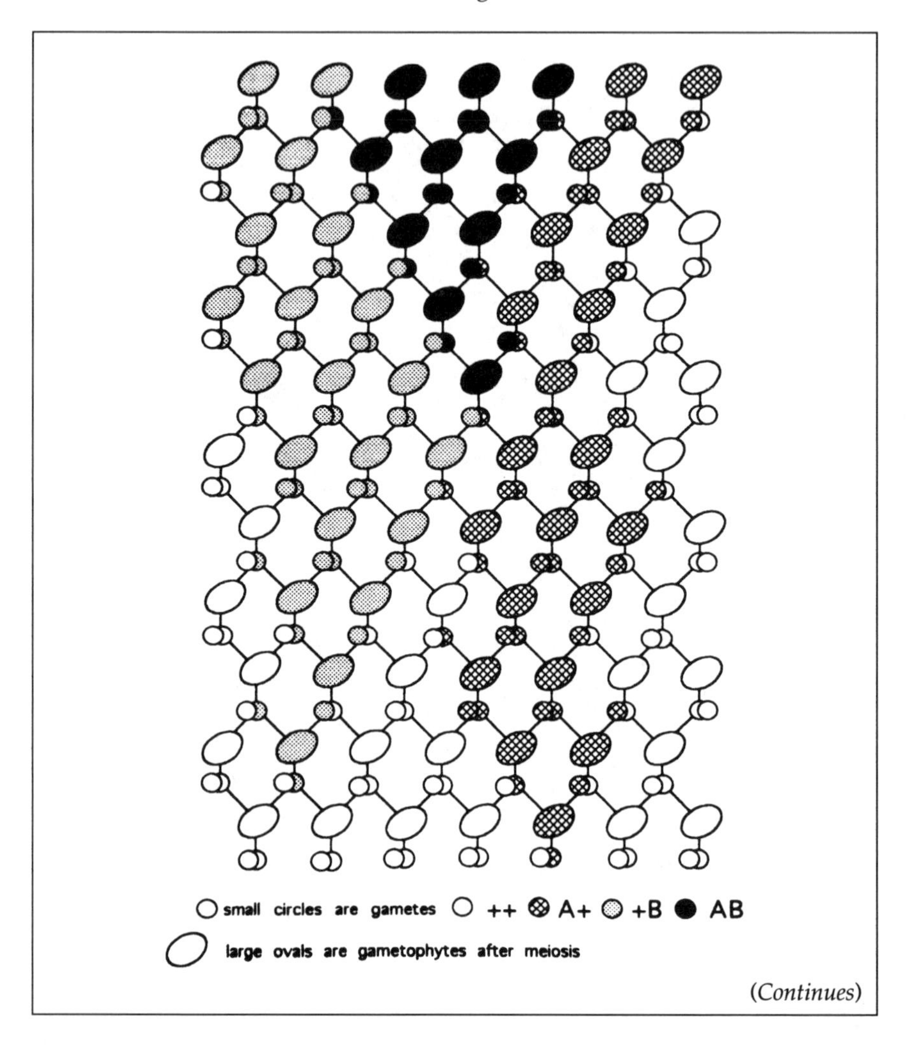


(*Continues*)

converse may also occur. If a small amount of tryptophan is added to the medium, a mutant defective in *trp* will be able to grow, although it will grow more slowly than types able to synthesize their own tryptophan, and will tend to be eliminated through selection from mixtures. Suppose, however,that, as before, one clone bears the defective *trp* allele and a superior *lac* allele (it may be designated $trp^-$ $lac^+$), whereas the other bears a functional *trp* allele but an inferior *lac* allele ($trp^+$ $lac^-$). If tryptophan is present in the growth medium, it is perfectly possible that the $trp^-$ $lac^+$

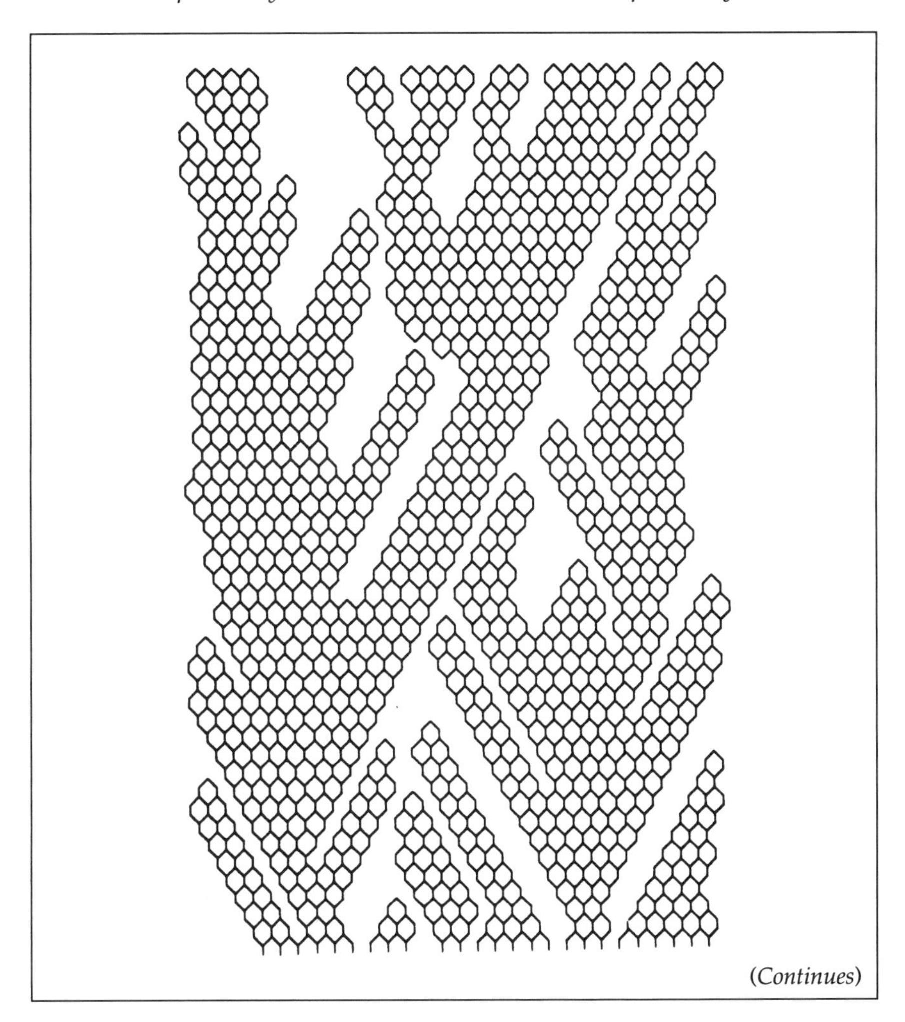

(*Continues*)

clone will become fixed, because its more efficient use of lactose gives it an overall competitive advantage. The *trp*$^-$ gene has thereby become fixed in the population, despite being an inferior variant. It has become fixed because it happens to be associated with a variant at another locus, whose superiority overrides its own inferiority. This process has been called *hitch-hiking*.

**Independent Selection of Recombining Genes.** Selection has somewhat different effects in sexual populations, because sexual lineages are not

(*Continued*)

These pictures represent successive generations in asexual and sexual populations, with each row representing a generation. Two life stages are shown, the young individuals as small circles and the mature adults as larger ovals. There are 11 successive generations shown up the page. The creatures inhabit a flat world, in which each interacts only with its neighbor on either side.

In the asexual population, the transverse lines running from the large ovals to the small circles indicate the production of spores; each individual may produce up to two spores. The vertical line running from each small circle to a large oval indicates development. The *branching system* of lines connecting individuals in different generations is thus a phylogeny. Two favorable mutations at different loci occur early in the history of the population, *stipple* and *hatch*. Both are superior to the wild-type genotype and have replaced their wild-type neighbors by the 5th generation. When they in turn come into contact and compete, *stipple* is superior to *hatch*, and increases in frequency, until by the 10th generation only a single hatch individual remains.

A similar chain of events has different consequences in the sexual population. Here, the large ovals are haploid gametophytes that produce haploid gametes (transverse lines to small circles) by mitosis. The gametes produced by neighboring individuals fuse to form a zygote that undergoes meiosis, with one meiotic product developing into the adult gametophyte (vertical line connecting pair of fusing circles with large oval). The two mutant types replace their wild-type neighbors, as before. (The process is slower, because I have assumed that individuals beyond the periphery of the picture are wild-type, and fertilize their neighbors with wild-type gametes.) Note that the phylogenetic diagram for the sexual population is not a branching system, but a *network*. When hatch and stipple neighbors mate, some of their offspring will be *hatch stipple* recombinants. This genotype, bearing both beneficial mutations, is superior to either single mutant, so the *hatch stipple* genotype increases in frequency at the expense of its neighbors and will eventually become fixed. Beneficial mutations that have arisen independently in different lines of descent must compete in an asexual population, but in sexual populations both can be combined into the same lineage.

At a larger scale, the phylogeny of a group of sexual organisms is a combination of branching systems, consisting of taxa that do not interbreed (like the lines in the diagrams in Sec. 19), and networks, consisting of the interbreeding lineages that make up a sexual taxon. In this diagram, one set of sexual lineages, comprising a single monophyletic group, is spreading at the expense of another group in a manner similar to the spread of *stipple* at the expense of *hatch* in the first diagram.

independent. There are two essential components of the sexual process: the combination of genes from different lineages through gamete fusion, followed by the recombination of genes during meiosis. In asexual populations, lineages form a strictly branching tree; in sexual populations, lineages form a network. An important consequence of sex is that genomes are not perpetuated as units, because in every sexual generation the genescon-stituting the genome of one individual are recombined with those of its sexual partner. There are, therefore, no permanent associations between genes at different loci. Loci that are close together on the same chromosome are unlikely to be separated by crossing-over in any given generation, but, otherwise, genes in sexual populations are inherited nearly independently of one another. Thus, a beneficial *lac*$^+$ allele would not be permanently associated with a *trp*$^-$ allele, because mating between *lac*$^+$*trp*$^-$ and a *lac*$^-$*trp*$^+$ individuals would produce some *lac*$^+$*trp*$^+$ progeny by recombination. The superior alleles at both loci can be fixed in sexual populations, because genes that are transmitted independently can be selected independently.

## 46. *Selection for genes with independent effects is more effective when they are transmitted independently.*

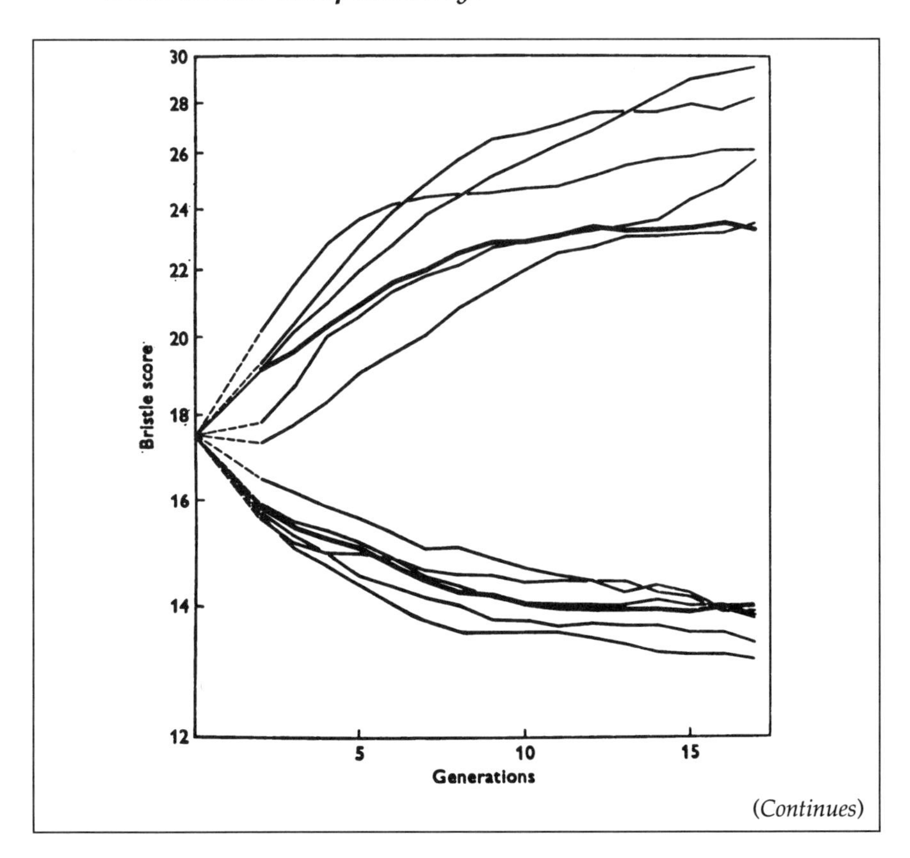


(*Continues*)

(*Continued*)

These are the results of an experiment by McPhee and Robertson (1970). The lines are five-generation moving averages of mean bristle number in upward and downward selection lines. The lines with normal levels of recombination are shown individually as thin lines. The thick lines represent the average of all of the selection lines in which recombination was suppressed by inversions. These results should be compared with those obtained by Thompson (1977), who failed to observe any consistent effect of suppressing recombination on the response to selection.

The linkage of genes in asexual lineages hampers selection, because the fate of an allele depends to some extent on the company it keeps. A mutation may on average tend to increase fitness and yet fail to spread, because it happens by chance to arise in an inferior genome. More generally, the fate of a gene in an asexual lineage will depend, not only on its intrinsic properties, but also on the way it interacts with the rest of the genome. The intrinsic properties of a gene are those which it displays independently of genes at other loci with which it happens to be associated. The overall effects of a gene can be decomposed into two components: those which are expressed independently of other genes and those which are modulated by alleles at other loci. Imagine a motile green alga in which genes at different loci control whether or not it is phototactic (swims towards light) and whether or not it is photosynthetic (uses light to fix carbon). Is the allele that directs phototaxis beneficial or not? The question cannot be answered by a simple yes or no. It may be advantageous, if the cell can utilize the light it has found; but motility is only an expensive burden if the cell would be equally successful in the dark. The fitness of an allele affecting phototaxis, therefore, will depend on the state of loci that affect photosynthesis. This is a simple example of *epistasis*, the interaction of the effects of different genes. Epistasis would make no difference to the outcome of selection if the population were infinitely large, since in that case all possible combinations of genes would be present, and selection would sort out the best possible combination. In a finite population, only a sample, typically an extremely small sample, of all possible combinations of genes is present. There might, for example, be only non-phototactic photosynthetic strains and phototactic non-photosynthetic strains present. In an asexual population, one or the other would win, depending on the balance of advantage in given conditions of growth: in the light, the non-phototactic photosynthetic strains might win, and in the dark the phototactic non-photosynthetic strains might have the edge. Recombination produces strains that are superior either in the light (phototactic and photosynthetic) or in the dark (non-phototactic and non-photosynthetic) because it allows the independent effects of each locus to be expressed. Sex thereby permits selection to cause a greater degree of adaptedness, by permitting the independent effects of genes to be selected independently.

The converse is also true. In sexual populations, it is *only* the independent effects of genes that are sorted by selection. If adaptedness requires a particular combination of genes, then recombination may build up that combination, but it will destroy it just as efficiently, so long as other genotypes are present in the population. If the best combination already exists, then it will be selected more rapidly in an asexual population, because its increase in frequency under selection will not be opposed by its destruction through recombination.

**The Effect of Recombination on the Response to Selection.** Recombination is suppressed in regions of the genome that are heterozygous for an inversion. This makes it possible to construct strains of *Drosophila* in which recombination is greatly reduced in females. (It is a peculiarity of certain Diptera, including *Drosophila*, that no recombination occurs during spermatogenesis in the male.) The response of these strains to artificial selection can then be compared with the response of comparable strains with normal levels of recombination. The result will depend to some extent on population size. In the first few generations, selection will for the most part sort the variation that is already present, and recombination will have little effect; the larger the population and the more variation it contains, the longer this phase will be. Variation is subsequently generated to an increasing extent by recombination, so that the response to selection in lines where recombination is suppressed should then be slower. C.P. McPhee and Alan Robertson of Edinburgh selected upwards and downwards on bristle number in *Drosophila*. In one set of lines, wild-type males were mated in every generation with females heterozygous for inversions on the two largest autosomes, causing a substantial reduction in crossing-over. In a second set of lines selected in parallel, males carrying the inversions were mated with wild-type females, generating normal levels of crossing-over. Directional selection on bristle number was effective and followed the expected course: the two sets of lines initially responded at about the same rate, but after about 10 generations the response was markedly less in the lines where recombination was suppressed. Vinton Thompson of Chicago, however, failed to replicate this result. He selected for phototaxis, using a similar scheme to create high-recombination and low-recombination lines, but could not confirm that recombination had any detectable effect on the response to selection after about 20 generations. It is possible that the number of flies maintained in the selection lines contributed to this disparity.

McPhee and Robertson selected only 10 flies in each generation in each line, whereas Thompson selected about 50; it is possible that the initial phase of sorting pre-existing variation was prolonged in Thompson's experiment, so that the subsequent phase of selection on variation generated mainly by recombination was not witnessed.

## 47. *Selection in the diploid phase is complicated by allelic interaction.*

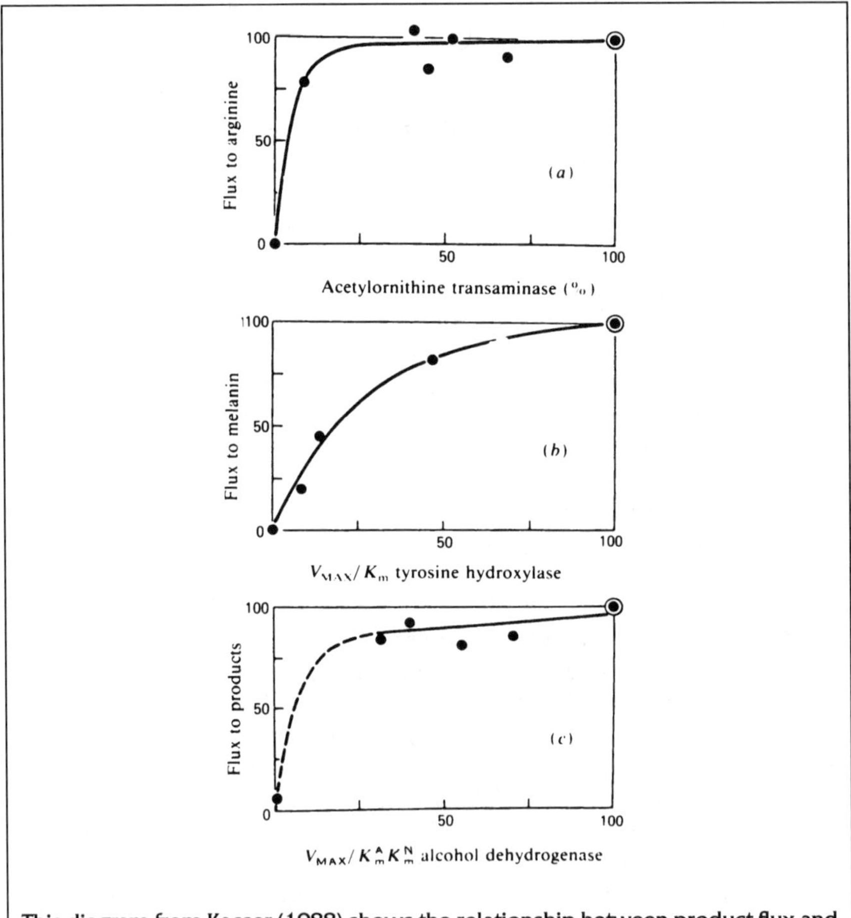


This diagram from Kacser (1988) shows the relationship between product flux and enzyme activity. These convex curves imply that loss of function will usually be recessive.

*(Continues)*

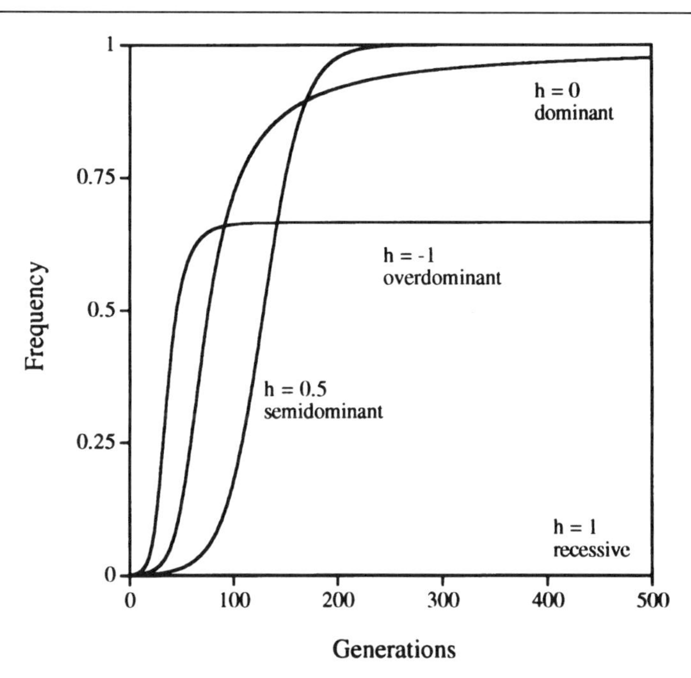


This diagram shows how dominance affects the spread of a rare beneficial mutation in a sexual diploid population. The fitnesses of the three genotypes at a locus with two alleles can be represented as: $w_{AA} = 1$; $w_{Aa} = 1 - hs$; $w_{aa} = 1 - s$. The degree of dominance is expressed by the value of $h$. When $h = 0$, Aa heterozygotes have the same fitness as AA homozygotes, i.e., the A allele is dominant. Conversely, when $h = 1$ the A allele is recessive. A value of $h = 0.5$ represents the situation when the A and a alleles combine additively, so that the heterozygote is intermediate between the two homozygotes. Values of $h$ that fall outside the range 0, 1 represent a genetic interaction that causes the heterozygotes to lie outside the range of the homozygotes; if $h < 0$ the heterozygote is fitter than either homozygote. The curves show the spread of the A allele from an initial frequency of $10^{-5}$ caused by selection with $s = 0.05$. Any new beneficial mutation occurs almost exclusively in heterozygotes so long as it is very rare. If it is recessive, selection is very weak, and its frequency increases only very slowly. If it is dominant, it is expressed in every individual it occurs in and is selected as rapidly as it would be in a haploid population; when it becomes very frequent, however, selection does not eliminate the inferior allele as effectively (because at this stage it occurs mostly in heterozygotes, where it is not expressed), and the rate of spread of the superior allele falls off. A semidominant allele with $h = 0.5$ is selected symmetrically, either allele being equally exposed to selection in homozygotes and heterozygotes, and a beneficial mutation is fixed quite

*(Continues)*

*(Continued)*

rapidly. An overdominant allele with $h < 0$ cannot spread beyond a certain frequency that represents a stable equilibrium. This is because the genotype with the greatest fitness, the heterozygote, cannot produce exclusively heterozygous progeny. This equilibrium frequency is, in general, $(w_{\mathrm{aa}} - w_{\mathrm{Aa}})/(w_{\mathrm{AA}} - 2w_{\mathrm{Aa}} + w_{aa})$, or for the model used here $(h-1)/(2h-1)$, which is equal to $\frac{2}{3}$ for the case of $h = -1$ illustrated in the diagram. If $h > 1$ the heterozygote is inferior to either homozygote and there is an unstable equilibrium, the A allele will become fixed or lost, depending on its initial frequency, and even if $w_{\mathrm{AA}} > w_{\mathrm{aa}}$ it will not spread when rare.

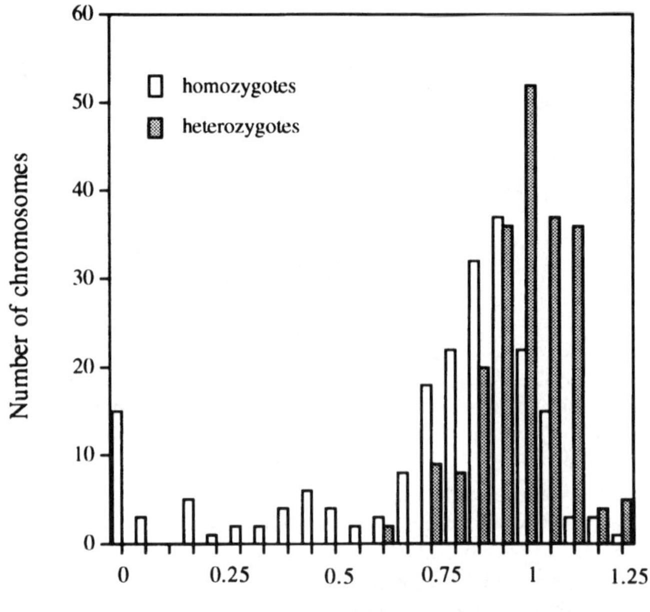


This diagram shows the distribution of viabilities among genotypes homozygous or heterozygous for wild-type chromosomes, using data on about 200 second chromosomes extracted from a South American population of *Drosophila pseudoobscura* by Dobzhansky et al. (1963). For the extraction procedure and the method of calculating viabilities, see Lewontin (1974), who gives an excellent account of the context in which these and similar results have been interpreted.

For studies of the dominance of new mutations induced by irradiation, see Wallace (1958, 1963b) and Falk (1961).

Sex necessarily involves an alternation between haploid (gamete) and diploid (zygote) stages in the life cycle. I have assumed so far that growth and reproduction are confined to the haploid stage, which is the case in most algae and fungi. In many organisms, however, it is the diploid stage that grows and reproduces. Since most plants and animals have life cycles of this sort, a great deal of evolutionary genetics has been developed for the special case of sexual diploids. Although this is understandable, it has had the unfortunate effect of introducing complications into the very simple arguments that apply to asexual haploid organisms, thereby making it difficult for them to be presented clearly. The most serious complication concerns the interaction of allelic genes, recognized by Mendel and known as *dominance*. This has two consequences for how selection acts.

**Selection on Dominant and Recessive Genes.** Completely dominant alleles, which direct the same phenotype regardless of their allelic partner, can be treated essentially as if they were haploid. Recessive alleles are quite different. When they are rare, they will occur almost exclusively in heterozygotes and will therefore not be expressed at all. The initial stages of the spread of a beneficial recessive allele, or the final stages of the elimination of a deleterious recessive allele, will occur very slowly indeed, depending as they do on the very rare exposure of homozygotes to selection. Intermediate degrees of dominance produce intermediate effects, the partial shielding of the more recessive allele slowing down the response to selection. Thus, the general effect of allelic interaction in diploids is to hinder selection and retard evolution.

**The Expression of Novel Mutations.** There is a very simple reason for expecting most loss-of-function mutations that may be severely deleterious as homozygotes to have little effect on fitness in heterozygous combination with a normal functional allele. The rate of the reaction catalyzed by an enzyme will rise asymptotically towards a maximum value, so that it will not be increased much by additional synthesis of the enzyme beyond a certain level. Conversely, if the current rate of synthesis of the enzyme is sufficient to drive the reaction at somewhere near its maximal rate, a somewhat lower rate of synthesis will not decrease the rate of reaction very much. A heterozygote bearing a defective allele may synthesize the enzyme it encodes at little more than half the normal rate, but the reaction it catalyzes may still proceed at a nearly normal rate.

One way of estimating the degree of dominance of severely defective genes that are lethal as homozygotes is to extract them from heterozygotes in samples from natural populations and then test them in heterozygous combination with a normal gene. It has only recently become possible to extract single genes, and most of the information that we have comes from classical experiments with *Drosophila* in which individuals can be made homozygous for a particular chromosome through ingenious inbreeding procedures. Once a lethal chromosome has been identified, through the failure of homozygotes to appear in the progeny of a cross, it can be combined with a normal chromosome and its fitness can be measured. It is inevitably the fitness of the chromosome as a whole, rather than a single gene, that is scored in experiments of this kind, but in most cases the assumption that lethality is caused by a single gene of large effect is probably justified. The general result of experiments of this kind has been that heterozygotes for a lethal allele suffer only a minor handicap, amounting to a loss of 1–5% in viability. This is not unexpected. Natural populations will contain a biased sample of deleterious mutations: those that have the smallest effect in heterozygotes will accumulate at higher frequencies and will be more likely to be sampled and tested.

**Selection for Heterozygotes.** Allelic interaction may be so pronounced that heterozygotes lie outside the range of homozygotes. A familiar example is provided by alleles specifying different sequences for ha emoglobin A in human populations. One allele causes a fatal anemia when homozygous; the other is associated with normal erythrocytes that are, unfortunately, susceptible to infection by malarial parasites. The heterozygote is slightly anemic, but is also resistant to malaria. Which allele is superior, given that malaria is endemic? As in the case of phototaxis and photosynthesis in algae, there is no simple answer. It is the combination of the two alleles, in the heterozygote, that has the greatest fitness. It is easy to see that in an extreme case, in which both alleles were lethal as homozygotes, only the heterozygote would survive. But this contradicts the argument that has been built up so far, that selection will consistently reduce genetic variation and cause the fixation of the fittest type, with any residual variation being caused by mutation. Selection for heterozygotes will actually maintain genetic variation in sexual diploids, because they will continually generate homozygous progeny. This will be the case whenever the heterozygote is fitter than either homozygote. Allelic interaction, like mutation, may thus hamper adaptation and maintain a range of maladapted genotypes, although only in the special case of sexual diploid organisms.

There is abundant evidence that heterozygotes are generally fitter than homozygotes. The simplest experiment is simply to obtain inbred families from an outbred population; it is a very general result that these are less viable and less fecund than normal outbred families. This might well be caused by the homozygosity of a few severely deleterious genes, however, without implying that genes with normal levels of activity generally confer higher fitness when they are matched with a somewhat different allele. A more sophisticated approach is to extract chromosomes from natural populations and compare the viability or fecundity of individuals that are homozygous or heterozygous for these chromosomes. Several such experiments, pioneered by Thedosius Dobzhansky in the 1960s, have given a consistent account of the difference between chromosomal homozygotes and heterozygotes. The distribution of viability or fecundity among homozygotes is bimodal: there is a lesser mode at zero, probably reflecting the segregation of a few recessive lethals, with the rest of the genotypes broadly distributed around a larger mode representing more or less normal performance. The heterozygote distribution is unimodal; it shows less variation than the homozygotes, and the mean of the heterozygotes exceeds the mean of the normal homozygotes by about 10%. Heterozygotes are thus substantially more fit than homozygotesat the level of whole chromosomes. It is not clear, however, whether this difference is created by the slight superiority of heterozygotes at many loci, or by a few genes that are substantially inferior as homozygotes, and extending the result to single genes is not easy. Bruce Wallace created homozygous stocks of *Drosophila* and then induced mutations with low doses of X rays. By crossing irradiated males with their non-irradiated sisters, he obtained flies that were heterozygous at only a few loci, rather than for a whole chromosome. These could then be compared with the progeny of crosses between non-irradiated partners, which should be homozygous at all loci, barring a very few spontaneous mutations. An alternative approach is to use spontaneous mutations that have been accumulated on inversions. Experiments of this sort have given rather ambiguous results. There is a general tendency for individuals that are heterozygous for an irradiated chromosome to be fitter, but the effect is so small that despite the vast size of the experiments (Wallace scored over two million flies) both its existence and its interpretation remain doubtful .

Whether or not heterosis (the greater fitness of heterozygotes) is a frequent or general phenomenon has been highly controversial because of its relevance to contrasting theories of the genetic structure of populations. According to one school of thought, natural populations are homozygous except for rare deleterious mutations maintained at mutation-selection balance and even rarer beneficial mutations destined to spread rapidly through the population. The opposed point of view is that natural populations are exceedingly diverse, with different alleles being maintained in the population by some form of balancing selection. This balancing selection was almost universally taken to be caused by heterosis, presumably because almost all the population geneticists involved worked with, or thought in terms of, diploid organisms. It proved to be impracticable to distinguish between these two views by the obvious procedure of measuring fitnesses directly, as I indicated earlier. The issue was not resolved by the discovery in the late 1960s and 1970s that many, perhaps most, enzymes have several variant forms. It was quickly argued that these variants were essentially equivalent in fitness, fluctuating at intermediate frequencies through sampling error (see Sec. 37). It was largely resolved by intensive DNA sequencing in the 1980s that showed that the third base of the coding triplet is much more variable than the first two. The genetic code is partly redundant, so that changes in the third base often do not change the amino-acid that the triplet encodes. Other non-coding sequences (such as introns) are likewise exceptionally variable. The difference between coding and non-coding sequences implies equally that the bulk of variation, in non-coding regions, is neutral and that the remaining variation, in coding regions, is constrained by selection. Both sides may therefore claim a victory. In retrospect, the conflict was not as important as it appeared to be at the time. The genetic structure of populations is important because it determines how, and whether, they will respond to selection. But we know from innumerable selection experiments that populations will usually respond to natural or artificial selection, and this simple fact seems to provide direct and incontrovertible evidence that the amount of genetic variation they contain is generally adequate to support quite rapid evolutionary change.

## 48. *Genetic combination and recombination can be selected.*

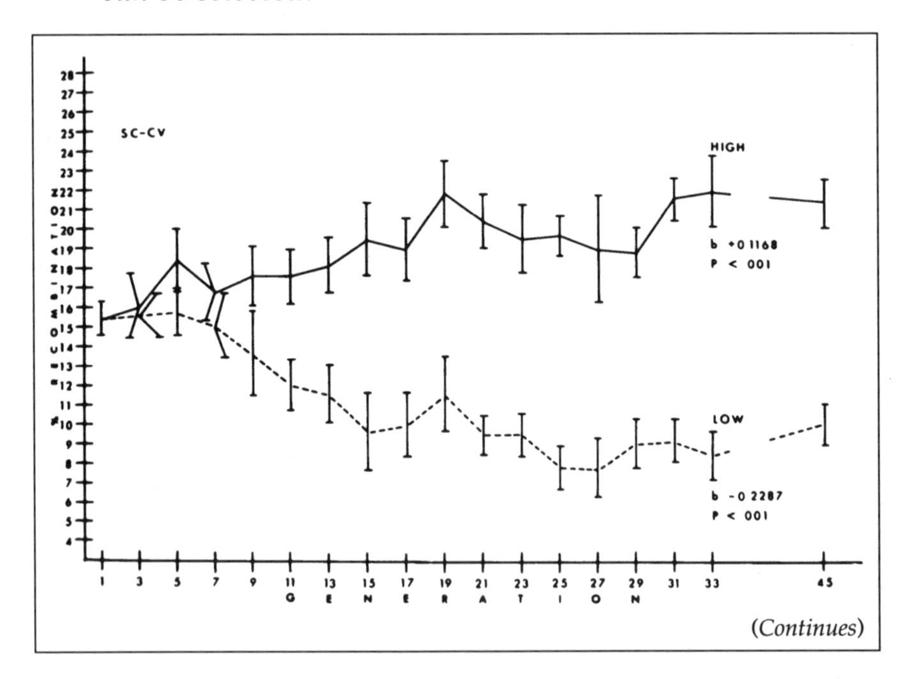


(*Continues*)

*(Continued)*
This figure shows the response to selection for recombination between two X-linked genes in *Drosophila*, from Chinnici (1971). There are other reports of artificial selection for recombination in *Drosophila* by Detlefson and Roberts (1921), Parsons (1958), Acton (1961), Mukherjee (1961), Kidwell (1972a, 1972b), Abdullah and Charlesworth (1974), and B. Charlesworth and D. Charlesworth (1985). Some experiments on other species have been described by Calef (1957; *Neurospora*), Allard (1963; lima bean, *Phaseolus*), Dewees (1970; flour beetle, *Tribolium*), and Turner (1979), Ebinuma and Yoshitake (1981) and Ebinuma (1987; silkworm, *Bombyx*).
The study of epistasis in phage described in the text is by Malmberg (1977). Ford (1940) described the modification of dominance in *Abraxis*.

Thus, the effect of selection depends on how genes interact when they are associated and on whether they continue to be associated when they are transmitted. Genetic interaction and linkage are not fixed and unalterable properties of genes; rather, they are themselves characters that will vary and that can therefore be selected. Selection may thus modify the context in which selection occurs.

## Selection for Recombination

A number of genes in various organisms are known to affect the rate of recombination, either in some defined region between two loci, or across the genome as a whole. These genes are identified by making a set of crosses and then looking for families with unusually high or low numbers of recombinant offspring. Even if a large number of offspring can be scored from each family, this is a very coarse procedure that will detect only genes with extreme effects, usually those that abolish recombination altogether. Such genes are often loss-of-function mutations that tell us very little about how rates of recombination evolve. It is impracticable to set up a screen that would detect small differences in the proportion of recombinants, however, because these differences will be obscured by sampling error unless the number of offspring scored from each family is extremely large. The point at issue is whether recombination rates can evolve through the selection of individuals whose differences are attributable to genes of slight effect at many loci. Both the evolutionary question and the genetic question are best answered by a selection experiment. If recombination rates cannot be altered through selection, there is no heritable variation for the character in the experimental population. If the rates shift abruptly under selection, rapidly attaining a new value and thereafter failing to respond to further selection, recombination is controlled by one or a few genes with large effects on the character, of the sort that would be detected by conventional screens. If recombination rates change rather smoothly and continuously over time, the selection line is probably accumulating alleles at different loci with individually small effects on recombination, of the sort that would be missed by a single-generation screen.

**Artificial Selection for Recombination in *Drosophila*.** Most experiments of this kind have used *Drosophila*, although there are a few involving other organisms. They involve crossing individuals bearing different mutant genes with easily visible effects on the phenotype. The mutations are at loci on the same chromosome, so that the combinations of characters seen in the progeny are generated by crossing-over rather than merely by the random assortment of chromosomes. There are two ways in which the combinations of characters affected by these marker genes can be used to select for higher or lower rates of recombination. The first is through straightforward individual selection: individuals bearing recombinant genotypes are identified and used to propagate a high-recombination line. To propagate a corresponding low-recombination line, one would deliberately select non-recombinant individuals that retained the parental combination of genes. There is one difficulty with this otherwise simple procedure: if a gene in one of the parents changes the rate of recombination between the marker loci, it may also affect recombination between itself and the marker loci. The recombinant offspring may not bear the gene responsible for their condition. This sort of problem is discussed further in Sec. 87. An alternative procedure is to scrutinize each family separately and select individuals from those families with the highest or lowest rates of recombination. Family selection is described in Sec. 115. This procedure will be more likely to trap genes responsible for altered rates of recombination, but it is also much more laborious.

Margaret Kidwell, working at Brown University, used family selection to alter recombination between Glued *Gl* and Stubble *Sb*, two genes on the third chromosome. The mutant alleles at both loci are dominant, and their phenotypes are easily scored. In the base population used to initiate the experiment, an average of about 15% of the progeny were recombinant. This rate was estimated by crossing *Gl Sb* / + + females, where the + indicates the wild-type allele, to wild-type males and counting the number of *Gl* + / + + and + *Sb* / + + progeny produced. Attempts to increase recombination beyond 30% seemed to be unsuccessful because the fertility of the selected flies declined markedly once recombination exceeded about 20%. Thus, the tendency of artificial selection to increase recombination was countered by natural selection for lower rates of recombination. The nature of limits to the response to selection is considered later, in Secs. 50 and 62.

Brian and Deborah Charlesworth carried out a similar experiment at Sussex, but performed individual rather than family selection. In the upward selection line, *Gl* + / + *Sb* females were mated with wild-type males, and the recombinant *Gl Sb* / + + female progeny selected. These were crossed again with wild-type males in the next generation. This generated *Gl* + / + + females, which could then be crossed with + *Sb* / + + males to regenerate *Gl* + / + *Sb* females, and the cycle was repeated. Downward selection was simpler, with *Gl* + / + *Sb* males and females being mated and the non-recombinant *Gl* + / + *Sb* progeny selected. This procedure was almost entirely ineffective. Only 1 of 16 lines showed any appreciable response to selection, an increase to about 22% recombination. This was presumably because, as anticipated, individual selection is less effective than family selection. One must be cautious, then, in extrapolating from the outcome of artificial selection to evolution in natural populations: natural selection, after all, will usually act through the phenotypes of recombinant individuals rather than through the average recombination of families. More fundamentally, such experiments demonstrate that one cannot predict with certainty the outcome of selection, even when it is applied by a rigorous protocol on experimental populations in a controlled environment. Replicate selection lines may show quantitatively different behavior, some responding whereas others remain unchanged, at least in the short term.

Joseph Chinnici of the University of Maryland used a combination of individual selection and family selection to increase and decrease recombination between two loci on the X chromosome. Both upward and downward lines responded smoothly and more or less continuously, so that after about 30 generations of selection recombination had fallen from about 15% to about 8% in the downward line and had increased to about 22% in the upward line. Recombination in neighboring regions of the chromosome had not changed, so that crossing-over had been specifically altered in the selected region rather than being redistributed to or from nearby regions. Thus, the general conclusion from theseexperiments is that recombination can be modified through selection and that by inference the linkage between elements in the genome represents an evolved character rather than an arbitrary restraint on selection.

## *The Evolution of Epistasis*

Genes at different loci may interact so that the phenotype of an individual cannot be accurately predicted from knowing the state of either locus separately, but depends instead on the combination of genes at both. These combinations of genes can be selected, just as individual genes can be selected, with the most fit combinations tending to increase in frequency. This selection, however, is likely to be ineffective in sexual populations, where the genes are often separated by recombination. Epistasis may occur nonetheless, because of biochemical or physiological constraints on gene expression; but there will be little tendency for the interactions among genes to be strengthened by selection when combinations of genes cannot be inherited. In asexual organisms, on the other hand, the genome is transmitted as a unit, so that the best-adapted combinations of genes can be selected, with little regard to the properties of these genes as independent entities. We expect, therefore, that epistasis will be much more marked in asexual than in sexual populations.

**Selection for Epistasis in Phage.** This theory was tested in an ingenious experiment by Russell Malmberg of Madison, using phage T4, a viral parasite of *E. coli*. Phage do not, of course, have any process resembling the sexuality of eukaryotes; however, when two different viral clones infect the same bacterial cell, many of the progeny phage will be recombinants, with a mixture of parental genes. Malmberg controlled the amount of recombination occurring in a culture simply by dilution. A fixed quantity of phage was added to a large or a small volume of medium containing a fixed density of bacteria. In the larger volume, the ratio of phage to bacteria is low, and almost all bacteria will be infected by a single phage; the phage population that grows up inside the cell will be a clone, and there will be no recombination. In the smaller volume, the ratio of phage to bacteria is higher, and any given bacterium is quite likely to be infected more or less simultaneously by two or more phage; this phage population is mixed and will produce recombinant offspring. The phage were grown by serial transfer in batch culture for about 20 growth cycles, in medium containing proflavin, which is toxic to phage because it interferes with DNA replication. The phage, of course, evolve resistance to proflavin during the course of the experiment. This resistance, however, may be conferred by alterations of several phage genes. The point of the experiment was to see whether the evolved resistance could be adequately explained by the independent effects of each new mutation or whether it was dependent on particular combinations of genes that were much less effective when expressed separately. The genome of the evolved phage was broken into eight separate segments, any one of which might have borne a mutation conferring partial resistance to proflavin. Each segment was then inserted separately into an unevolved phage, and this construct tested for proflavin resistance. If the modified genes act independently, the resistance of the intact evolved phage should be equal to the sum of the partial resistances of the lines that were each transformed with a different segment of the evolved genome. The less dilute treatment, with a high ratio of phage to bacteria and thus a high rate of recombination, showed this additive relationship among the modified phage genes. In the more dilute treatment, however, the resistance of the intact evolved phage substantially exceeded the resistance that would be expected simply by adding up the independent effects of parts of the genome. Epistatic interactions among genes will thus tend to evolve when there is little recombination, so that combinations of genes are transmitted intact and selected as units.

## *The Evolution of Dominance*

The general tendency for deleterious mutations to be nearly recessive can be accounted for, to some extent, in purely physiological terms. The degree of dominance that is exhibited by an allele, however, can also be thought of as an aspect of phenotype that is liable to be selected and may thus evolve. There are two ways in which this might occur.

Firstly, I have already mentioned that lethal genes isolated from natural populations are expected to be nearly completely recessive, because those allelic lethals with the least heterozygous effect will necessarily become the more abundant through selection. A similar argument applies to loci that display heterosis. If one homozygote is the fittest genotype in a given environment, it will become fixed in the population; if a heterozygote is the fittest genotype, then the locus will continue to segregate for both alleles. Consequently, the number of heterotic loci will tend to increase through time and might constitute a large fraction of loci that are observed to be variable; even a modest input of mutations with heterotic effects will eventually lead to a large number of segregating heterotic loci.

A second possibility is that the degree of dominance of a specific allele might be modified by genes at other loci. Selection would then favor modifiers that reduced the level of expression of deleterious mutations and enhanced the expression of beneficial mutations. This theory was the subject of a celebrated controversy between Sir Ronald Fisher (supporting it) and Sewall Wright (opposing it) in the early years of population genetics. The main reason for scepticism is the weakness of selection acting on the modifiers; because the mutant heterozygotes are rare, selection on the modifiers will be of the order of the mutation rate. Moreover, it seems reasonable to suppose that any such modifier would also affect the expression of the common homozygote; and the homozygote is so much more abundant that this effect would swamp any effect that it had on the heterozygote. The controversy has had a curious history. Fisher felt the point to be so important that he devoted the third chapter of his classic book to it; but since the 1930s it has attracted so little interest that most current texts do not refer to it at all.

**Pigmentation in the Currant Moth.** The currant moth *Abraxis grossulariata* normally has pale white or cream wings with some black markings. The variety *lutea* has yellow-brown wings, the difference being caused by a single gene. The heterozygotes are more or less intermediate. E.B. Ford of Oxford, the founder of ecological genetics, crossed heterozygotes and selected from their progeny those individuals, judged to be heterozygous, that were most extreme, with darker wings in the upward selection line and paler wings in the downward selection line. After three generations, the gene was nearly dominant in the upward selection line and nearly recessive in the downward selection line. It is, thus, rather easy to modify the dominance of a mutation through artificial selection, although the selection that is imposed in this way is, of course, much more powerful than the selection that would act on the modifier of dominance of a rare mutation in the wild.

## 49. *The short-term response to selection of sexual populations can be predicted from their genetic structure.*

The genetic basis of sorting in asexual populations is straightforward, and it is the physiology and biochemistry of the characters under selection that receives most attention. In sexual diploid populations, on the other hand, the genetics of sorting are more complicated, and the aim of most experiments is to understand the genetics while ignoring the function of the characters under selection. For this purpose, any character can be chosen on the grounds of convenience and equated with fitness by making it the criterion for choosing the parents of the next generation. This settles the question of function, at least in the short term: the experimenter defines the function. We can now ask whether the evolution of the population through artificial selection can be predicted from knowing how the character is inherited; whether the genetic dynamics can be predicted from the genetic statics.

**Variation, Selection and Response.** The variation of individuals around the mean of the population may be caused by genetic or environmental factors. The two can be distinguished by raising families in isolation and measuring the degree of resemblance among relatives. Roughly speaking, the variation among families is genetic, and the variation within families is environmental. The genetic variation is itself compound: some part is attributable to variation of the independent effects of genes on the character, another part to variation of the effects of genes in combination. The independent effects are usually defined as *additive*, in the sense that the independent effects of genes sum to the character value; thus, if one gene causes the character value to differ from the population mean by an amount $x_1$ and another gene causes it to differ by an amount $x_2$, then the total difference in individuals bearing both genes is $x_1 + x_2$. The interaction between genes is measured as the deviation from this strictly additive relationship, and it can in turn be decomposed into a component arising from

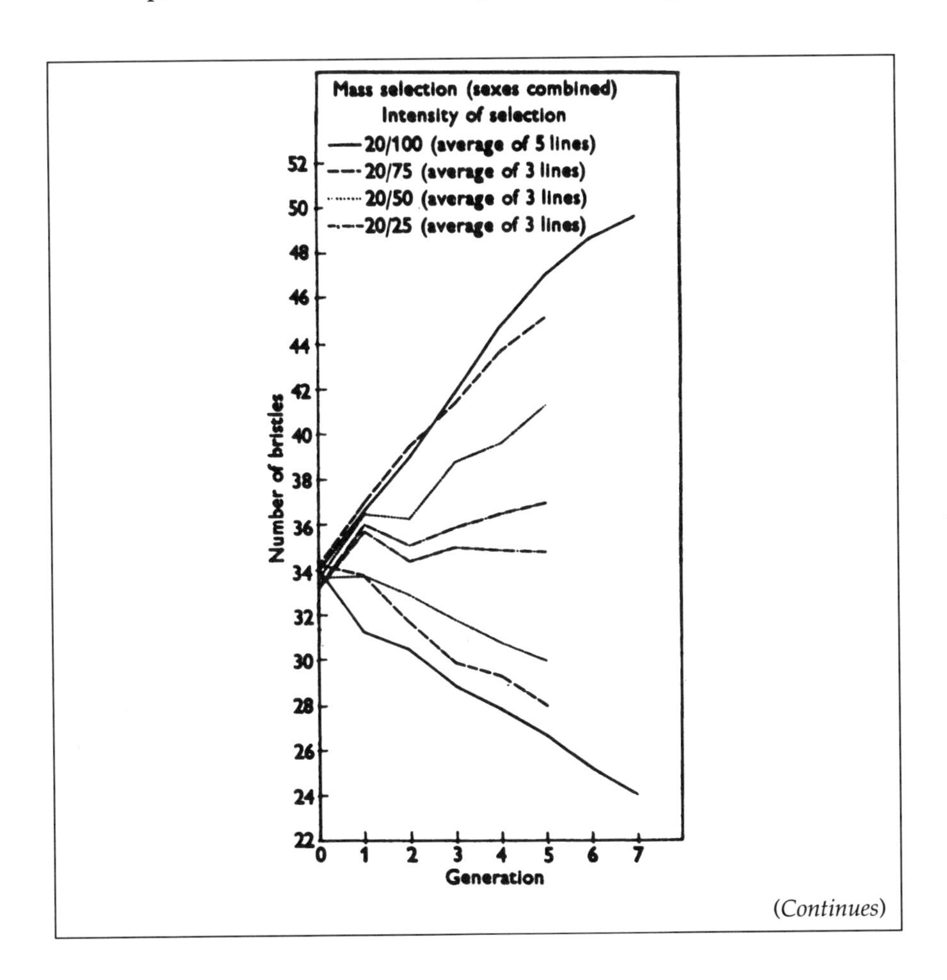


(*Continues*)

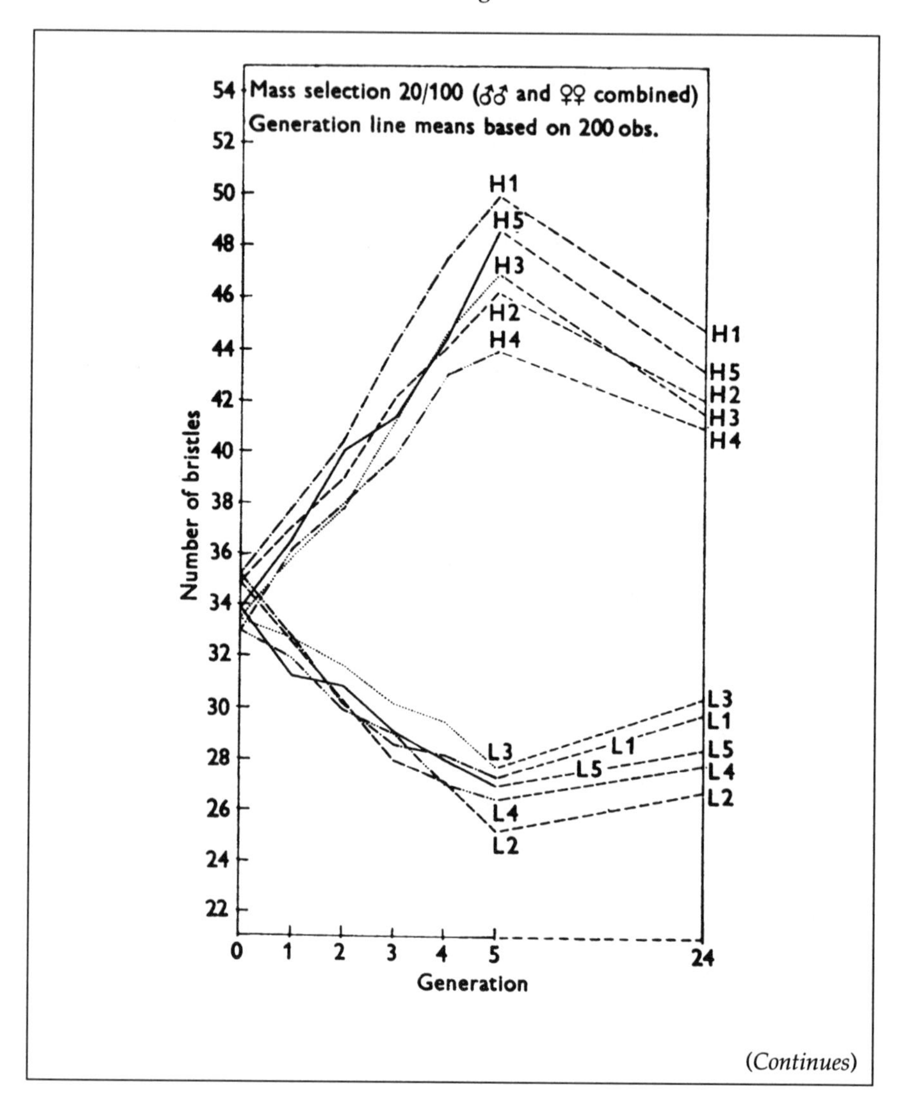
Mass selection 20/100 (♂♂ and ♀♀ combined)
Generation line means based on 200 obs.
Number of bristles
54
52
50
48
46
44
42
40
38
36
34
32
30
28
26
24
22
Generation
0
1
2
3
4
5
24
H1
H5
H3
H2
H4
H1
H5
H2
H3
H4
L3
L1
L5
L4
L2
L3
L1
L5
L4
L2

(Continues)

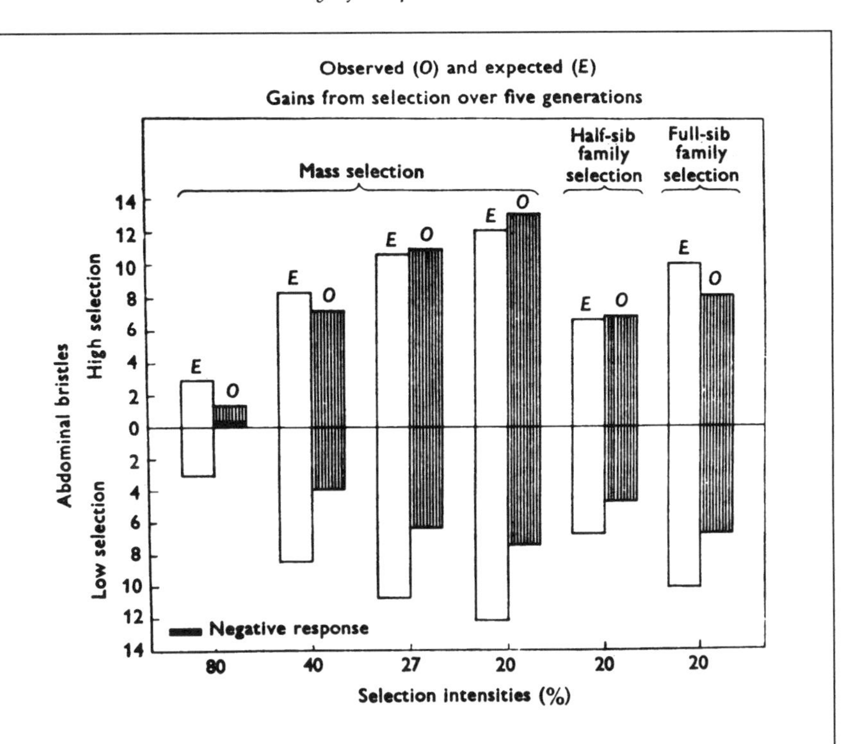


These three figures show some of the results of the experiment by Clayton et al. (1957). The first shows the response at different rates of selection. With intense selection (20 flies selected from a sample of 100), there is a steep and regular response in the direction of selection; when selection is much weaker (20 flies selected from a sample of 25) there is little or no consistent response in the short term. The lines drawn in this diagram are mean values for 3–5 replicate selection lines in each treatment. The second diagram shows the responses of the replicate selection lines separately, for the highest rate of selection. Note the divergence of the lines. The broken lines, after generation 5, show the outcome of 19 generations in which selection was relaxed, the flies being chosen at random in each generation. These relaxed lines tend to return towards the phenotype of the base population, although rather slowly. This shows that the lines still contain genetic variation for bristle number (not surprising, after only five generations of selection), and that natural selection in the culture vials is antagonistic to artificial selection. Note that the ranking of the lines established after five generations of selection tends to be maintained during the 19 subsequent generations. The third diagram compares the response to selection after five generations with the response predicted from estimates of the quantity of additive genetic variance for bristle number present in the base population. It shows different rates of

(*Continues*)

mass selection and also the outcome of selecting on family means rather than on individual values (see Sec. 115). Similar work has been done with *Drosophila* by Martin and Bell (1960), Sheldon (1963a,b), and Latter (1964); with *Tribolium* by Enfield et al. (1966); with fowl by Lerner and Hazel (1947); with rats by Chung and Chapman (1958); and for swine by Hetzer (1954). A wide range of *Drosophila* experiments is reviewed by Mather (1983).

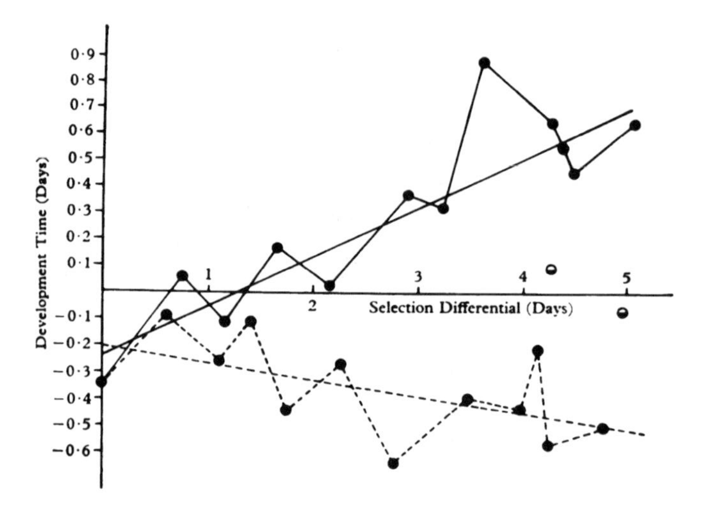


The response to bristle number is asymmetrical, being greater in the upward than in the downward selection lines. This shows the asymmetrical response to selection for development time in *Drosophila* from Clarke et al. (1961). Note that the $x$ axis is scaled in units of selection differential, rather than in generations. In each generation, the selection differential is the difference between the mean value of the character before selection and the mean after selection. Cumulating this difference over all preceding generations gives the cumulative selection differential in a given generation. The regression of response on cumulative selection differential is the "realized heritability" that in simple mass selection experiments should be equal to the proportion of additive genetic variance in the base population. A general interpretation of asymmetric responses to selection is that they are caused by an underlying genetic asymmetry that can be represented as a nonlinear regression of offspring on parent. Prior natural selection will tend to have elevated the frequency of genes associated with higher fitness. The offspring–parent regression should therefore be shallower for the range of character values associated with higher fitness under natural selection and steeper for character values associated with lower fitness. The response will then be

(*Continues*)

greater for artificial selection applied in the direction of greater heritability, that is, for character values associated with lower fitness under natural selection. In this case, more rapid development will lead to earlier reproduction and thus greater fitness (Sec. 98); consequently, lines should respond more quickly to selection for slower development than to selection for faster development, which is what is observed. This interpretation of asymmetric responses has been reviewed critically by Frankham (1990), who lists about 30 studies of bidirectional selection in various organisms on characters which contribute directly to fitness. These include female fecundity (Falconer 1955, 1971, Narain et al. 1962, Richardson et al. 1968, Land & Falconer 1969, Lambio 1981; de la Fuente & San Primitivo 1985); male mating (Manning 1961, 1963; Siegel 1965; Tindell & Arze 1965; Kessler 1969; Sherwin 1975; Spuhler et al. 1978); size (Prevosti 1967, Baptist and Robertson 1976) and rate of development (Marien 1958, Hunter 1959, Clarke et al. 1961, Moriwaki & Fuyama 1963, Dawson 1965, Sang & Clayton 1957, Bakker 1969, Englert & Bell 1970, Smith & Bohren 1974, Drickamer 1981, Soliman 1982, Hudak & Gromko 1989). Frankham concludes that, on balance, the experimental evidence favors the interpretation of asymmetric responses to selection given here.

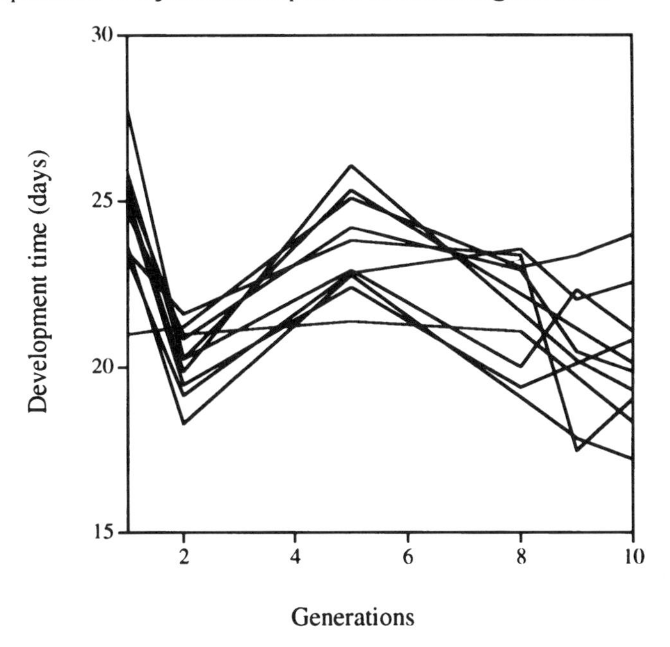


This shows the response of 10 replicate selection lines in another experiment using selection on development time in *Drosophila*, by Marien (1958). The trend

(*Continues*)

downward is caused by selection; the large average fluctuations are caused by environmental differences between generations; the divergence of the lines during the first 10 generations of selection is caused by sampling error. Frankham et al. (1968a) also give detailed information on the behavior of replicate lines.

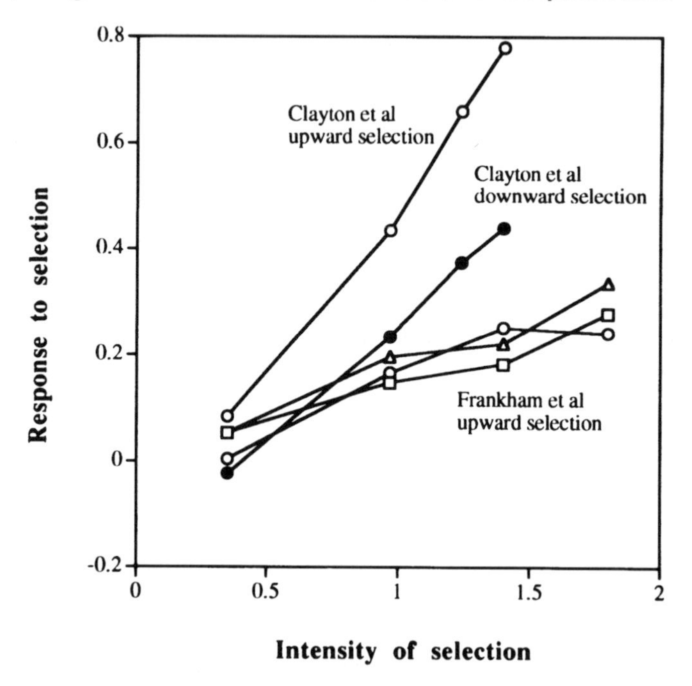


This graph shows the effect of increasing intensity of selection on the short-term response to selection, using data from two studies of bristle number in *Drosophila* by Clayton et al. (1957) and Frankham et al. (1968a). The response to selection is scaled in units of phenotypic standard deviations in the base population. The intensity of selection is the selection differential, again in units of phenotypic standard deviation.

It is obvious that greater short-term gains will be made by more intense selection, but this may be at the expense of longer term progress, because very intense selection is likely, by chance, to fix some undesirable genes among the few individuals selected. Theory developed by Robertson (1960) suggests that the eventual response will be maximized by selecting half the population in each generation. This has been broadly confirmed by experiments with *Tribolium* (Ruano et al. 1975) and *Drosophila* (Frankham 1977).

(*Continues*)

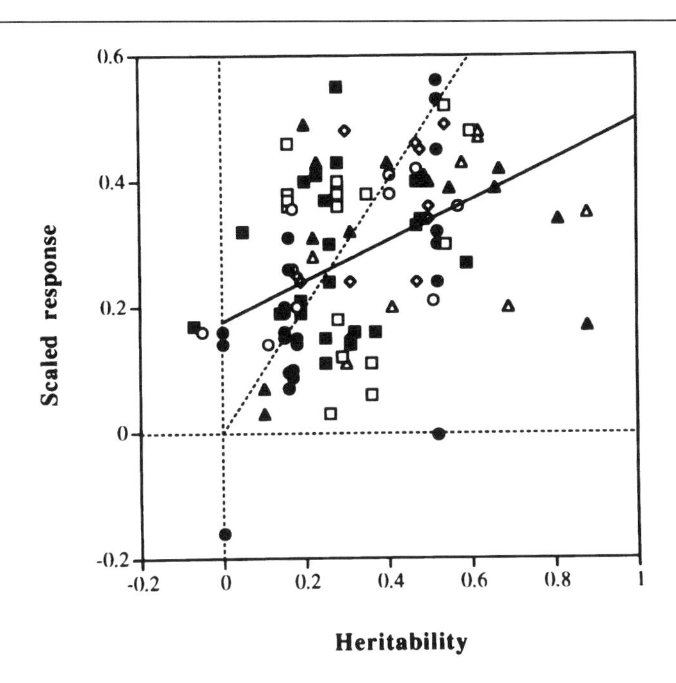


The expected response to selection is $R = Sh^2$, where $S$ is the selection differential and $h^2$ the heritability; thus, the scaled response $R/S$ (estimated from the selection experiment) should be predicted by $h^2$ (estimated in the base population by breeding trials). This figure shows the relationship between the scaled response and the heritability for about 100 comparisons, based on the valuable review by Sheridan (1988). The studies included in the survey are as follows:

*Drosophila* (solid circles): Clayton et al. (1957), Sheldon (1963a, 1963b), Frankham et al. (1968a,b), Sheridan and Barker (1974a).

*Tribolium* (open circles): Wilson (1974), Meyer and Enfield (1975).

Mouse *Mus* (solid square): Sutherland et al. (1970), Wilson et al. (1971), Falconer (1973), Hurnik et al. (1973), Cheung & Parker (1974), Frahm and Brown (1975), Eisen (1978), Nagai et al. (1978), de la Fuente and San Primitivo (1985); Sanders (1981, rat).

Quail *Coturnix* (open square): Godfrey (1968), Marks and Lepore (1968), Macha and Becker (1976), Lambio (1981), MacNeil et al. (1984).

Fowl *Gallus* (solid triangle): Manson (1973), Nordskog et al. (1974), Pym and Nicholls (1979), Garwood et al. (1980).

Swine *Sus* (open triangle): Hetzer and Harvey (1967), Bernard and Fahmy (1970), Rahnefeld and Garnett (1976), Fredeen and Mikami (1986a, 1986b).

Sheep *Ovis* (diamond): from review by McGuirk et al. (1986).

Other references that I have not used are given by Sheridan (1988). I used

(*Continues*)

(*Continued*)
studies in which the response was measured over the first 12 generations or fewer of selection and gave preference those in which the base population was a large mass-mated population that had been raised for some generations in the laboratory. I chose regression estimates of heritability, when available, in the order son–sire, offspring–sire, and offspring–midparent; and preferred estimates from the base population rather than the selection lines. Consequently, the estimates I used differ somewhat from those cited by Sheridon (1988). I took averages over replicate selection lines, but separated different characters, sexes, directions, and rates of selection reported in the same study. The observed response is assigned a negative value when it is opposite in sign to the direction in which selection is applied. Negative heritabilities refer to variance-component estimates. The least-squares regression equation is $y = 0.32x + 0.18$, with $r^2 = 0.20$, based on 106 observations.

dominance, the interaction between allelic genes, and a component arising from epistasis, the interaction among genes at different loci. These different genetic effects can be estimated from the resemblance among different kinds of relatives. The response to selection in sexual populations, however, depends primarily on the independent effects of genes, and therefore on the *additive genetic variance.* This can be estimated directly from the resemblance between parents and offspring, since this necessarily represents the differences among individuals that are transmitted from generation to generation, and that can therefore be selected so as to cause a permanent genetic change in the population. When several pairs of adults are mated and their offspring reared, the average value of a brood of offspring can be plotted against the average value of their two parents; the slope of this graph is the fraction of the total variance among offspring attributable to additive genetic variance. This quantity is called the *heritability* and is symbolized as $h^2$. We can modify the fundamental theorem to state that the rate of change in fitness in a sexual population will be proportional to the additive genetic variance of fitness. In artificial selection experiments, we can use this result directly to predict character evolution, because the character that we are selecting deliberately is a surrogate for fitness.

Selection that is applied to an experimental populations, whether of flies or of crop plants, by choosing individuals on the basis of the character values they express is called *mass selection,* as distinct from various types of family selection. This may be done in one of two ways. The simpler procedure is to choose a certain fraction of the population, decided in advance, made up of those individuals with the most extreme values of the character. The alternative is to choose as parents individuals in which the character being selected exceeds a certain value, and rejecting all the others. This is called *threshold selection* or *truncation selection*. In the context of natural selection, the relative fitness of distinct types is expressed by the *selection coefficient*, $s$, which is the difference in fitness between the types; for example, it is common practice to set the fitness of one type equal to unity, $w = 1$, and express the fitness of an alternative type as $w = 1 - s$ (see Section 31, for example). In experimental situations, however, we are more likely to be interested in continuously varying characters, such as seed yield, fleece quality, or fitness. In this case the experimenter creates two types, the selected and the rejected individuals, according to some phenotypic criterion that imposes stronger or weaker selection. This criterion can be expressed as the *selection differential*, which I shall symbolize as $D$: the selection differential is the difference between the mean of the population before selection and the mean of the selected individuals. The selection differential is clearly related to the fraction of the population that is accepted: the greater the selection differential, the smaller the fraction that are chosen. In truncation selection, setting the threshold for acceptance further from the population mean increases the selection differential. To compare the response of different characters, it is often convenient to use a common scale for the selection differential. This is usually supplied by the phenotypic standard deviation of the base population, $\sigma_p$: the standardized *intensity of selection* is thus $i = D/\sigma_p$, or in other words a selection differential in units of phenotypic standard deviations.

If all the differences among individuals in the population were heritable, then the response to selection $R$ would be equal to the selection differential $D$. Because only a fraction $h^2$ of the phenotypic variation will be heritable, the response to selection will be the corresponding fraction of the selection differential: $R = h^2 D$ or, equivalently, $R/\sigma_p = h^2 i$. This modification of the fundamental theorem is the basic predictive equation for the change in population mean value under artificial selection. The selection differential is under the direct control of the experimenter, who may choose to select the top 1% or the top 90% of the population, applying powerful selection in the former case and weak selection in the latter. The heritability is a property of the base population and is not under the direct control of the experimenter, although the quantity of genetic variation initially available for selection can be specified to some extent through an appropriate choice of the number and relatedness of individuals constituting the base population. The theory of sorting in a sexual population can therefore be tested by using estimates of the heritability, derived from breeding trials, to predict the change in population mean, estimated from selection experiments.

**Selection for Bristle Number in *Drosophila*.** Alan Robertson and his colleagues at Edinburgh carried out an experiment specifically designed to test the ability of this elementary theory to predict the response to selection in terms of the heritable variation of a character in the base population. The character they used was the number of bristles on the abdomen of *Drosophila*. The bristles are sensory chaetae borne on the thorax and abdomen of the fly. Their main virtue for experimental work is that are easy to count, and they have been the subject of dozens, if not hundreds, of selection experiments. Their base population was a large outbred population maintained in a cage in the laboratory for many generations. The mean bristle number was about 30 or 40 (it differs between males and females), with a standard deviation of 3 or 4, and showed no tendency to change through time. They first estimated the heritability of bristle number in this base population, using a variety of techniques that consistently gave values of about 0.5. They then established upward and downward selection lines, in each of which a different proportion of individuals was selected. In each line, 40 flies—20 males and 20 females—were selected, but the 20 of each sex were chosen as the most extreme individuals among 25, 50, 75, or 100 scored, in order of increasing intensity of selection, in different lines. The difference among lines in the rate of selection gave a range of predictions of the expected response. The lines did all respond to selection within five generations, with mean bristle number increasing or decreasing; this was not unexpected, given the ease of scoring the character and its rather high heritability. Moreover, they responded in the expected way, with more intense selection causing a proportionately greater response. Altering the intensity of selection, then, causes a regular and predictable change in the response of the population.

There were, however, two unexpected trends. The first was the *asymmetry* of the response: the upward selection lines behaved in almost precisely the way anticipated by theory, except at the lowest intensity of selection, but the downward lines showed consistently less response than expected. Indeed, in most of the downward lines the response was only about half the expected response; at the lowest selection intensity, there was no response at all. Asymmetrical responses, which are often met in experiments involving divergent selection lines, can be produced either by the genetic composition of the base population, or by the nature of gene action. If genes which tend to increase the value of a character are less frequent than genes which tend to decrease it, downward selection will be the more effective. This effect is often observed when the character selected has a strong influence on fitness. Genes that increase fitness, having been selected in the past, are likely to be near fixation, and upward selection is likely to be ineffective: it is more difficult to evolve a better fly. In artificial selection experiments where the character under selection is rather remote from fitness in normal circumstances, an asymmetrical response to selection may be caused by an asymmetrical relationship between genotype and phenotype. Several loci may be segregating for genes that increase or decrease character value, so that individuals vary in their dosage of high-value genes. If these loci interact in such a way that the effect of a given gene substitution decreases with dosage, there will be a process of diminishing returns: a gene that increases character value will have a large phenotypic effect and will thus be selected effectively when low-value genes are present at other loci, but its phenotypic effect, and thus the response to selection, will be much less when there are many other high-value genes in the genome. The same, of course, applies to low-value genes. This will necessarily be the case when there is a natural limit to variation in one direction: for example, it is impossible to select flies with a negative number of bristles, whereas there is no clear upper bound to bristle number. More generally, there may be no scale on which equivalent genetic changes cause equal phenotypic effects in different directions. In diploid organisms, there is a second source of genetic interaction that may cause an asymmetrical response to selection. If dominance happens to be directional, such that genes increasing the value of the character tend to be dominant, then downward selection will be more effective, and vice versa. It is not known why the response to selection was asymmetrical in this experiment; in other experiments, using different strains for the base population, downward selection on bristle number has been more effective.

The second trend was the *variability* of the response: lines that were selected identically responded at different rates. The lines all responded in the same direction, and the differences in response are not very great. Nevertheless, they are consistent, with any given line tending to remain above or below average in successive generations. This is caused by sampling error. The lines are set up as rather small random samples from the base population, and differ somewhat in their initial composition. This difference will subsequently tend to increase, as the lines are propagated, as biased samples, from generation to generation. If the experimenter could scrutinize the genotypes of the flies directly, choosing only those individuals bearing the appropriate genes, these errors would not exist, or would be very slight. The experimenter must instead select phenotypes, however, and will often choose flies which express extreme phenotypes because they have obtained somewhat more or less food than average, which have developed from somewhat larger or smaller eggs, or which vary for any of a multitude of reasons attributable to the unique circumstances of their individual development. Consequently, any two selected samples will be somewhat different genetically, and because only genetic differences are transmitted independent selection lines will tend to diverge. A single selection line thus represents a unique historical process that cannot be precisely repeated. Unreplicated selection experiments are therefore difficult to interpret: in this case, the predictions of the theory were borne out on average, but observations, however detailed, of a single, unreplicated selection line would have been much less informative. As selection proceeds, the supply of genes causing variation in the direction of selection is exhausted more quickly in the lines that respond more quickly, and replicate lines should then begin to converge on the phenotype that represents the limit to selection; whether this is actually the case is discussed later in Secs. 50 and 62.

**The Effect of the Amount of Variation in the Base Population.** When the composition of the base population is given, the response will depend on the rate of selection; conversely, if the rate of selection is given, the response will depend on the quantity of variation available for selection. The quantity of useful variation that can be winnowed by selection in the short term will depend on two factors. The first is the genetic diversity of the population, as in the case in asexual populations, the response to selection will be greater when there are more distinct kinds of individual initially present. The second is the rate at which new diversity is made available to selection by recombination in a sexual population.

Genetic diversity is manipulated most simply by maintaining selection lines with different numbers of individuals. R. Frankham and his colleagues in J.S.F. Barker's laboratory in Sydney set up an experiment in which they began with a large outbred cage population and selected at different rates on abdominal bristle number, as in the experiment just discussed. In contrast to the previous experiment, however, they ran lines of different size at each selection intensity, made up in each generation of 10, 20, or 40 pairs of flies. As expected, the response during the first 10 generations of selection increased with the rate of selection; however, the response was also proportional to population size. For any given rate of selection, the larger populations tended to respond more rapidly, presumably because a larger sample is likely to include a greater diversity of genotypes. This effect can be large enough to overcome a substantial difference in selection intensity. For example, after 10 generations, the response of lines with 40 pairs, selected by choosing the top 40% of the population, was as great as that of lines with only 10 pairs, selected more intensely by choosing the top 20% of the population. Moreover, population size influenced the repeatability of the response. Replicate lines, at a given population size and selection intensity, diverged through time, as in Robertson's experiment. The larger lines diverged less than the smaller lines, however, because sampling error will be smaller in larger populations. Thus, larger populations give both a greater and a more reliable response to selection in the short term.

A range of characters that vary in different degrees and that are selected at different intensities can be compared by scaling the response to selection on the selection differential to form the realized heritability $R/D$. The simple theory I have just outlined suggests that this should be equal to the heritability $h^2$ as estimated by a breeding trial in the base population. This is a matter of some importance to animal and plant breeders, who can manipulate the relative magnitude of response by adjusting selection intensities or (within limits) population sizes, but who would like to be able to predict the absolute value of the advance they are likely to make. In practice, such predictions are not very successful. There is a tendency for the scaled response to increase with heritability, but in many cases the response has been much greater or much less than was anticipated. This is presumably because the intensity of selection and the size of the selected population can be specified precisely, whereas the estimate of the heritability is subject to a considerable sampling error.

In outbred populations where mating is more or less random, the selection of genes that increase (or decrease) a character such as bristle number, will not be appreciably hindered by their linkage with genes at other loci that have the opposite effect, and the population will immediately respond smoothly to selection in either direction. If variation is constrained by linkage, however, there may be little or no response to selection for a few generations, until these constraints are loosened by recombination. For example, suppose that the base population is set up by crossing two unrelated, highly inbred lines. The $F_1$ progeny are, of course, genetically uniform. The $F_2$ are highly diverse, but retain parental combinations of genes. The early reponse to selection thus depends on the segregation of large blocks of linked genes, or whole chromosomes, and is likely to be slow. It will accelerate only when recombination breaks up these blocks, allowing genes to be selected independently. The same principle applies during selection. If lines are deliberately inbred by mating relatives, the number of different combinations of genes is reduced, and the response to selection will be retarded.

In these *Drosophila* experiments, the observed response to selection usually corresponds quite closely to the value predicted from basic theory. However, such predictions are rather limited. They are made on the basis of an estimate of heritability in the base population, and are therefore, strictly speaking, valid for only the first generation of selection; when this has been completed, selection will not only have changed the mean, but will also have reduced the additive genetic variance, and the heritability must be estimated anew. In practice, the heritability is unlikely to change substantially in the first five or six generations of selection, so that the original value can be used as the basis for predicting short-term sorting on time-scales of 10 generations or less. On longer time-scales the theory breaks down, as we shall see.

## 50. *The sorting limit in sexual populations is the limit of potential variation.*

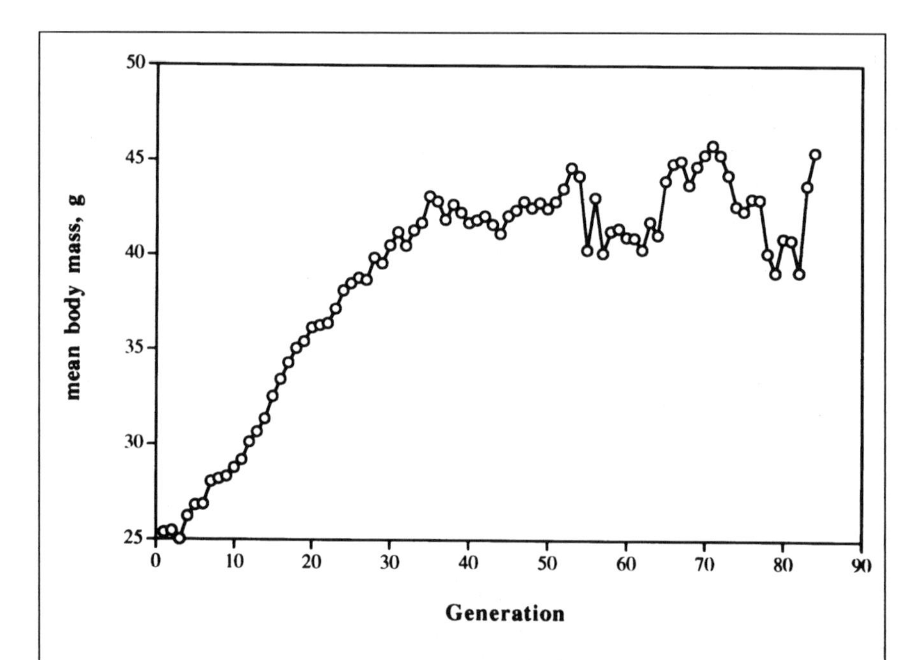


This graph shows the attempt by H.D. Goodale to breed mice as large as rats. Selection began in 1930 and was continued for over 30 years, spanning over 80 generations of mice. The base population comprised 8 males and 28 females, obtained from a commercial breeder. The mice initially averaged about 25 g, with a phenotypic standard deviation of about 2.5 g. After 35 generations of selection with an average intensity of about $i = 1$, mean weight had risen to about 43 g, for a total advance of about 7 phenotypic standard deviations. This represented a limit: in 50 subsequent generations, mean weight fluctuated erratically without any definite trend, despite the continued application of selection of about the same intensity as before. This graph was drawn from the data presented by Wilson et al. (1971). Long-term selection for body size in mice has been reviewed by Eisen (1980).

(*Continues*)

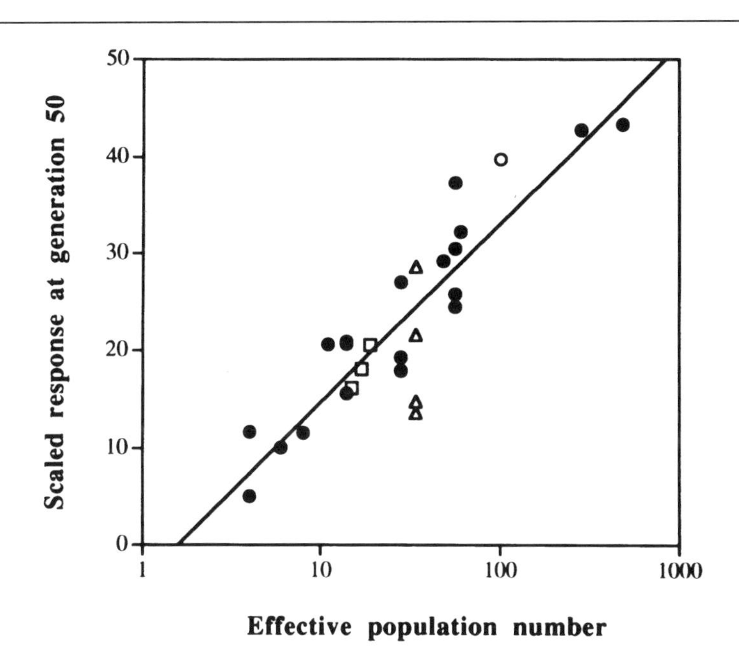


The graph shows the relationship between the limit to selection in sexual populations and their effective population number. (Effective population number is a concept developed in population genetics to accomodate the fact that the dynamics of neutral genes may depend on sex ratio, variance in offspring production, and several other factors, as well as on census number. Populations with the same effective size decrease in heterozygosity at the same rate. The difference between the effective number and the census number is of little practical importance except in extreme situations.) The data are taken from 10 studies, summarized and analyzed by Weber and Diggins (1990). The limit is taken to be the total response at generation 50, $R_{50}$, although some lines were still responding at this time, and the figures are thus underestimates. It is scaled by the response in the first generation of selection, $R_1$; this is a measure of the quantity of selectable variation in the base population, and is equivalent to scaling by the initial quantity of additive genetic variance. The plotted points are $R_{50}/R_1$. The effective population size is calculated by the original authors, or recalculated by Weber and Diggins, using standard approximations for *Drosophila* populations. The studies epitomized here are as follows:

*Drosophila* (solid circle): Mather and Harrison (1949), Robertson and Reeve (1952), Rasmuson (1956), Jones et al. (1968), Yoo (1980a), Weber (1990), Weber and Diggins (1990).

(*Continues*)

*Tribolium* (open circle): Enfield (1982).
Mouse *Mus* (square): Roberts (1966a, 1966b).
Maize *Zea* (triangle): Dudley (1977).

Goodale's experiment could not be included on this graph because there is no precise information on the numbers of individuals in each generation, which varied in different generations between about 100 and about 1000. With $R_{50} = 18$ g and $R_1 = 0.34$ g, however, (based on response in the first 10 generations), $R_{50}/R_1$ is about 50, which is consistent with an average population size of a few hundred individuals.

The regression equation is $x$ (scaled response) $= 18.4\ y$ (effective population number) $-3.7$, with $r^2 = 0.79$. Further analysis of these data shows that total response is positively correlated with initial response, i.e., with the selectable variance originally present ($r^2 = 0.84$, or $r^2 = 0.53$ if the outlying study, by Enfield, is excluded), but weakly and *negatively* correlated with selection intensity ($r^2 = 0.24$).

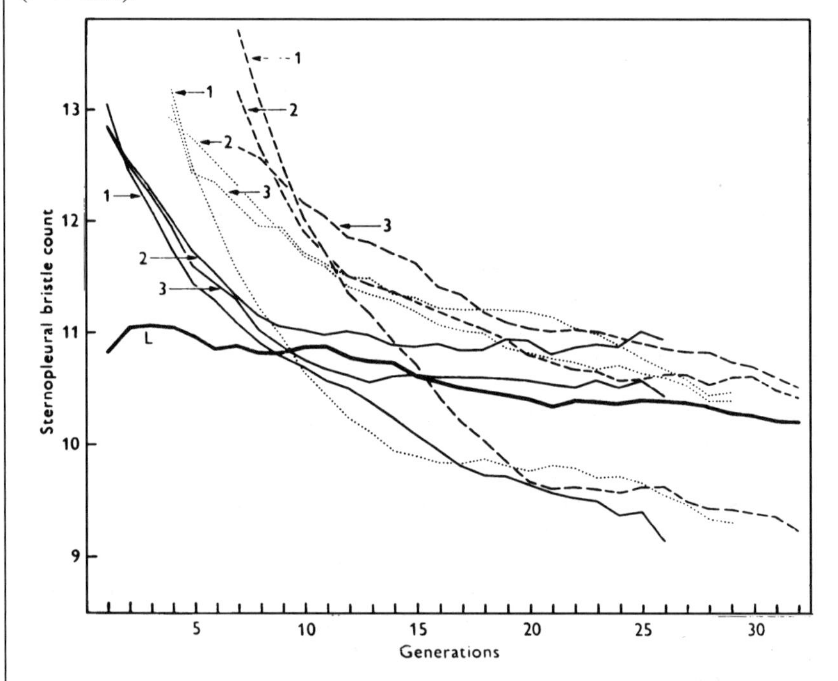


Osman et al. (1968) attempted to transgress the limit apparently approached after 25 generations of downward selection on bristle number that reduced the

*(Continues)*

population mean from 17.5 to 12.4. Crossing two replicate selection lines gave flies with a mean bristle number close to the average of the two lines, and a further 30 generations of selection yielded a very modest response that brought mean bristle number down to 11.0. There was thereafter scarcely any response to either upward or downward selection. At this point the selection line was crossed to the base population, and downward selection applied again. The experiment was quite complicated, with treatments involving the preselection of the base population, a pause of a few generations before beginning to select, and different intensities of selection. The figure shows the behavior of lines where the base population was selected before crossing and selection was weak; the thick line is the original selection line that shows only a very slow response to 30 generations of further selection. The most important feature of this result is simply that some lines that have been crossed back into the base population transgress the limit approached by the original selection line, although many others do not. Some other treatments caused more transgression, others less. Comparable experiments with mice have been reported by Falconer and King (1953) and Roberts (1967); both were successful. The procedure is related to the reciprocal recurrent selection often practised by agronomists and stockbreeders (Sec. 66). Two other procedures that have been suggested are indirect selection (Sec. 83) and selection in an environment where new genetic variation is expressed by the plateaued line (see Sec. 103).

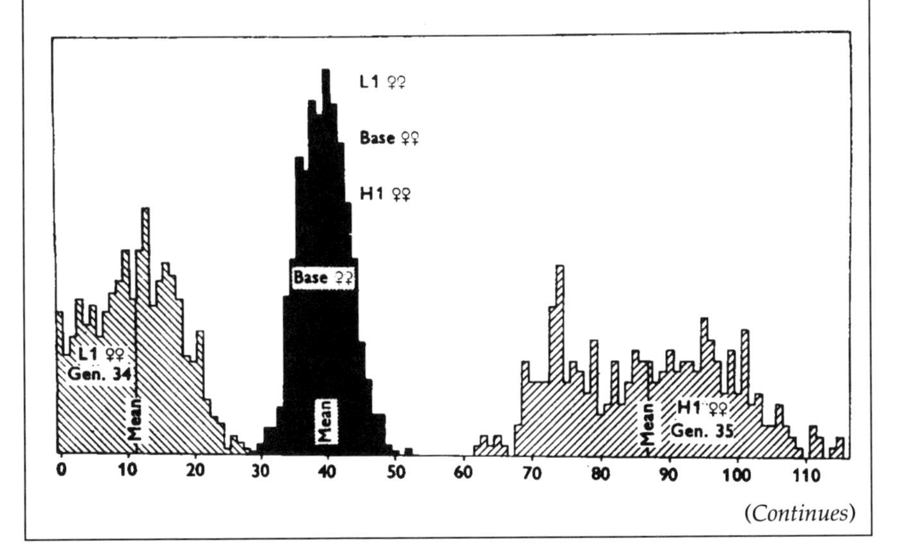


(*Continues*)

(*Continued*)

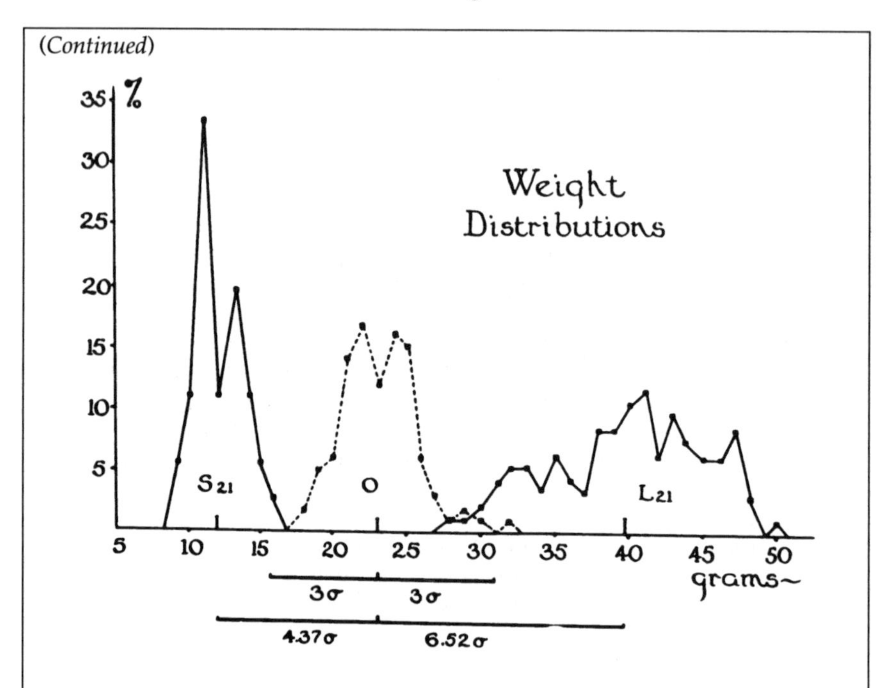


After 20 or 30 generations of selection, divergent selection lines may no longer overlap with themselves or with the original population. These examples are from *Drosophila* after 35 generations of selection for bristle number (Clayton & Robertson 1957) and from mice after 21 generations of selection for body weight (MacArthur 1949).

In an asexual culture of bacteria inoculated with $lac^{+}trp^{-}$ and $lac^{-}trp^{+}$ strains, the limit of sorting would be $lac^{+}trp^{-}$, whereas in a corresponding sexual culture, the limit would be $lac^{+}trp^{+}$. A greater degree of adaptedness can evolve in the sexual culture because recombination allows the two genes to be selected independently. The population will eventually be dominated by a genotype that is more highly adapted than either of the initial genotypes. This genotype was not originally present; but neither is it entirely new. Rather, it has been constructed by recombination from elements of the genotypes originally present.

This principle is quite general. Suppose that selection were acting on some quantitative character such as body size. This might be affected by very many genes, but let us consider just 12, for simplicity, each of which has two alleles. One allele causes an increase in size, and is designated +; the other allele, written as -, causes a decrease in size. A sequence of plus and minus alleles defines a genotype, for example, - - + - + + + - - - + -. We can imagine that increased size is being favored by selection, and that the fitness of an individual is the count of + alleles in its genotype, in this case, five. The population initially comprises a range of genotypes, such as

$$\begin{array}{cccccccccccc}
- & + & - & - & - & + & - & - & - & + & - & + \\
+ & + & - & - & - & - & - & + & - & - & + & - \\
- & + & - & - & + & - & + & - & + & - & - & + \\
+ & + & - & + & - & - & + & + & + & - & + & - \\
- & - & + & - & + & - & - & - & + & + & - & -
\end{array}$$

and so forth. The fittest genotype in this set is the second from last, with a score of seven. This would be the limit of sorting in an asexual population; this genotype would become fixed despite bearing the deleterious - allele at many loci. If the population is sexual, it can transcend this limit. Imagine a mating between the second-from-last genotype and the last. This would produce recombinant progeny with a range of phenotypes. Some would by chance inherit mostly - alleles, and would have very low fitness (the worst is - - - - - - - - + - - -, score 1). Others, however, would inherit mostly + alleles, and would have very high fitness (the best is + + + + + - + + + + + -, score 10). Recombination has the effect of making all of the allelic diversity present in the initial population available to selection, which then drives the population beyond the original limits of variation. If the individual with a score of 10 mated with the first genotype on the preceding list, some of its progeny would have + alleles at all loci and would have reached the theoretical limit of sorting in sexual populations.

The maximum response of sexual populations to selection depends on the number of segregating loci that affect fitness, or the character under selection, and on the number of individuals present. If there are only four loci, say, with - and + alleles equally frequent at each, then the extreme + + + + genotype will arise at a frequency of 1/16 and will be present even in small populations. If there are 400 loci, on the other hand, extreme genotypes are extremely rare. With very many loci contributing to a character, the effect of each must necessarily be very small. We have seen that the frequencies of alleles with small effects on fitness are strongly influenced by sampling error, so that some of these loci will become fixed by chance for the - allele, preventing the theoretical limit of sorting from being attained. The full theory of sorting limits is, therefore, quite complicated. Roughly speaking, if the population is very large relative to the number of segregating loci, the limit is set by the number of loci; when the number of loci is very large, the limit is proportional to population size. In most artificial selection experiments the number of individuals selected is quite small—20 pairs or so would be a typical value for *Drosophila* experiments—and the limit to selection will therefore be sensitive to increases in population number. So long as the number of loci affecting the character greatly exceeds the number of individuals selected $N$, each individual added to the selection line will contribute a more or less constant number of new alleles. The response at the limit $R_L$ will then be related to the response in the first generation of selection $R_1$, as $R_L = NR_1$. As the population becomes large, however, there will be a process of diminishing returns, with additional individuals contributing fewer and fewer new alleles. The limit would then tend toward an asymptote in large populations and, in practice, the limit seems to be proportional to the logarithm of population number.

**The Effect of Recombination on Short-Term Limits to Selection.** The effect of recombination on the sorting limit was shown clearly by McPhee and Robertson's experiment on selection for bristle number in *Drosophila* (Sec. 46). After 15 or 20 generations of selection, the lines were approaching their short-term limit. At this point, progress in the low-recombination lines was only about 70% of that in the high-recombination lines for upward selection, and about 80% for downward selection.

**Breaking the Limit.** Once a population has reached a plateau, it is natural to ask how it can be made to progress further. The most reasonable suggestion is to cross it with a replicate selection line, then select for valuable combinations of genes from both lines. This approach assumes that evolution is generally contingent, with adaptation having a different genetic basis in lines selected in the same way. Although it can be tested by experiment, it is often of little use to practical breeders, who may possess only one highly superior breed or cultivar. In that case, it is possible to cross the selection line back into the base population, reasoning that some favorable genes in the base population will have been lost by chance during selection. Both techniques have been tried, with variable success. More important than the fact that they do not always work is the fact that they sometimes do work, providing a good demonstration of chance and contingency, and showing how these principles can be put to practical use.

**The Limits to Selection in Sexual Populations.** The important qualitative conclusion to be drawn from these arguments is that when selection is applied to sexual populations, it may produce a response that takes the population far beyond the limit of variation that is originally expressed. This principle has been clearly demonstrated by the classic experiment involving artificial selection on bristle number in *Drosophila* conducted by Robertson's group at Edinburgh, the initial phase of which I described earlier. In the base population, the mean bristle number in female flies was about 39, with a standard deviation of about 3.5; virtually all the flies in this population would bear between 25 and 55 bristles. After 35 generations of upward selection, mean bristle number had increased to about 87; in the downward selection lines, mean bristle number had decreased to about 11. After 30 or 40 generations of selection the population mean can shift by more than 10 phenotypic standard deviations—equivalent to about 20 genotypic standard deviations—from its original value; no population, however large, would initially contain individuals so extreme. The sorting of recombinants produces surprisingly rapid and extensive modification of quantitative characters.

## 2.C. Continued Selection

The initial effect of selection is to sort the variation expressed in the base population—for example, the variation attributable to major genes, or to blocks of linked genes—acting in more or less the same way in sexual and in asexual populations. This process is well understood, and its outcome can be predicted reliably by simple theories. In sexual populations, it is followed by a phase in which selection sorts the variation that is newly expressed in every generation as the result of recombination. This phase lasts for about 50 generations in artificial selection experiments; it is not as well understood, and so it is treated largely through generalizations, rather than by predictive theory. The third phase of selection is the history of the population after most of the pre-existing variation, actual or potential, has been expressed and sorted; this is very poorly understood, and the paucity of long-term experiments makes even generalization hazardous. These three phases are not, of course, sharply distinguishable in most cases. The beginning of the second phase in sexual populations can sometimes be recognized as an acceleration of the response to selection in experiments where variation in the base population is constrained by linkage. It comes to an end, in sexual or asexual populations, as the population approaches a limit to the response that can be achieved in the short term, either because selectable genetic variation has been exhausted or because other, less easily foreseeable, factors have come into play. To this point, the outcome of selection has been dominated by the variation present in the base population; after this point, any continued response must depend on novel variation that arises during the course of selection. The nature of the evolutionary process shifts on long time-scales from being rather regular, predictable, and repeatable, to becoming increasingly contingent and historical. The very different principles involved in the continued response to selection are the subject of the next sequence of sections.

## 51. *Adaptedness may be lost through continued internal or external deterioration.*

Selection is acting continually to restore the lack of fit between population and environment caused by mutation and environmental change, but it is not necessarily successful in doing so. Selection is not a very efficient way of producing or maintaining adaptedness—not nearly as efficient as deliberate design—and instead of becoming adapted, populations very often become extinct.

## 52. *Deleterious mutations may continue to accumulate in small asexual populations.*

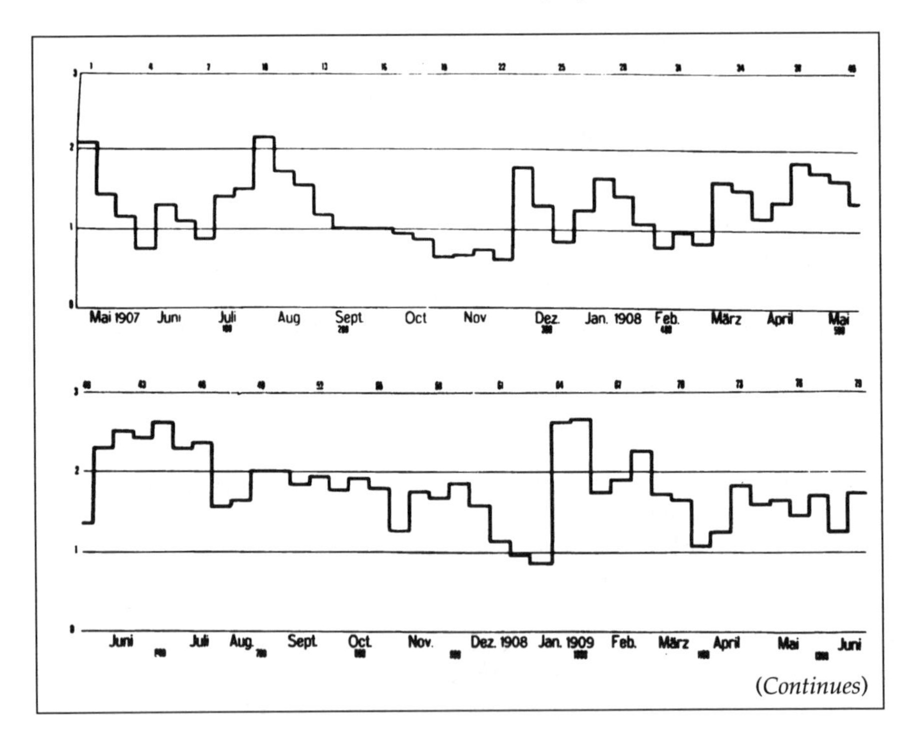


(*Continues*)

The history of isolate asexual cultures of ciliates can be followed by plotting the number of fissions occurring in successive intervals of time as a histogram. This example is taken from an experiment by Woodruff (1911). It shows the mean fission rate of an isolate culture of *Paramecium* at 10-day intervals for the first 1200 generations of the experiment. There is, in this, case no perceptible decline in fission rate through time, and these lines were, in fact, eventually kept for over 10,000 generations (Woodruff 1926). They did, however, periodically go through a cryptic sexual process involving a sort of internal self-fertilization.

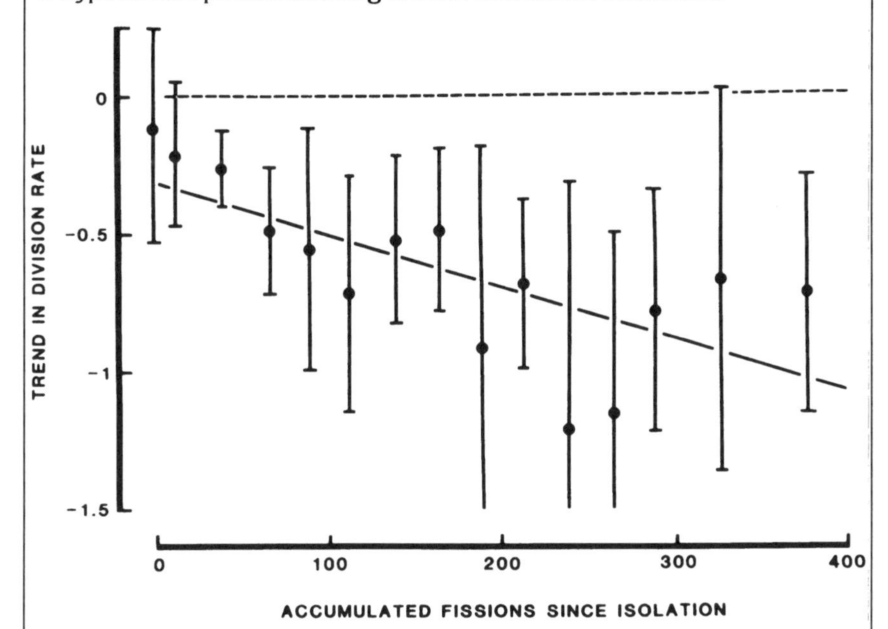


I have reviewed 26 experiments of this kind, involving about 80 lines maintained for up to 4000 generations (Bell 1988). These are the longest running of all selection experiments (though they are now being surpassed by some experiments with bacteria), the record being held by L.L. Woodruff (1926), who propagated lines of *Paramecium* for 19 years, representing about 11,000 generations. This graph summarizes the trend in fission rate in all lines for which information is available during the first 400 generations of culture, at intervals of roughly 25 generations. Note that the trend is always negative, showing that in each interval of time the rate of reproduction is tending to fall; moreover, it becomes increasingly negative through time. These small asexual populations thus become increasingly enfeebled, until eventually they become extinct.

(*Continues*)

(*Continued*)

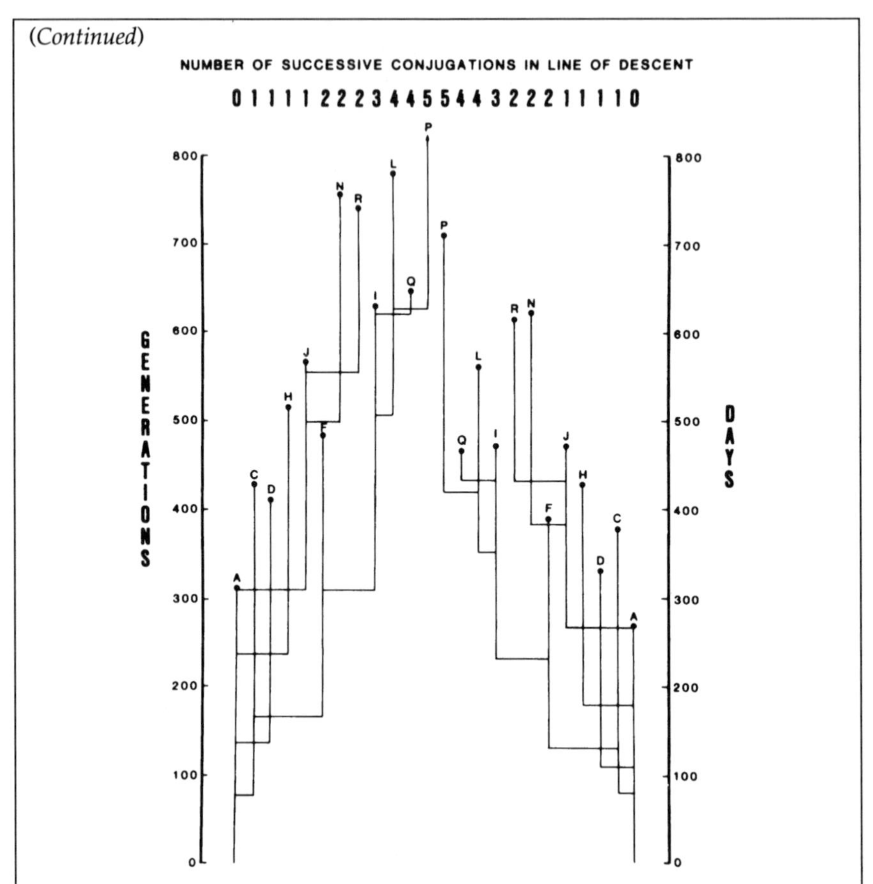


The senescent decline of these cultures can be arrested if they are allowed to produce sexual progeny occasionally. In this diagram, which summarizes experiments on *Uroleptus* by Calkins (1919, 1920), the vertical lines indicate the lifespan of an asexual culture, the solid circle marking the point at which it becomes extinct. The horizontal lines indicate the sexual origin of a new culture. Time is given in generations on the left and in calendar time on the right. Although each asexual culture declines quite rapidly, becoming extinct within a few hundred generations, the sexual lineage as a whole can be propagated indefinitely.

For estimates of mutation rates based on the accumulation of mutations in inversions, see Mukai et al. (1972).

Mutation does not inevitably cause a progressive collapse of adaptation, because individuals with more deleterious mutations are less likely to reproduce than those with fewer. When mutation and selection are balanced, there is a range of individuals in the population, some with very few mutations, some with very many, and most with some intermediate number. This distribution represents a stable equilibrium that is propagated from generation to generation. It may be destabilized, however, if the population is small, when the randomizing effect of sampling error is combined with that of mutation.

**Muller's Ratchet.** A population can be divided into classes of individuals bearing different numbers of deleterious mutations. Each class has a certain frequency, and once the population has reached equilibrium these frequencies do not change consistently through time. However, they may, and in fact they must, fluctuate to some extent through sampling error. Even if the population is rather large, a class that occurs at low frequency may comprise very few individuals, and the number of individuals in this class will fluctuate wildly from one generation to the next. It is usually the extreme classes that will be the least frequent, including the class of individuals that bear no deleterious mutations. This class is, therefore, very vulnerable to sampling error, and very likely sooner or later to disappear entirely. Once lost, it cannot be regained. The least damaged individuals in the population, following the loss of the completely undamaged individuals, are those which bear a single mutation. In principle, this class could give rise to undamaged individuals by back-mutation; but the alteration of a non-functional gene to produce a functional one is much less likely than loss of function, so rates of back-mutation are very low, and long before the class of undamaged individuals would have been restored, the class with a single mutation has itself disappeared. The geneticist H.J. Muller was the first to realize that asexual populations are subject to a ratchetted process, in that the number of mutations borne by the least heavily loaded class of individuals can never decrease. If this class is sufficiently small, as it is likely to be in small populations, the combination of mutation and sampling error will cause a cumulative and irreversible loss of adaptedness.

**Deterioration of Asexual Isolate Cultures of Ciliates.** This process is very neatly demonstrated by a series of experiments, performed by several biologists between about 1880 and 1920, that were intended to test the idea that sex halted a process of racial senescence and rejuvenated the lineage. Most of the experiments used ciliates, which reproduce quite rapidly but are big enough to manipulate easily. A single individual is isolated on a microscope slide and left for a day or so, until it has divided; the two individuals produced by fission may in turn have divided before the slide is inspected again, and so a small population of between two and eight individuals will be found. One of these is then chosen at random and moved to a new slide, the rest being discarded. This technique of *isolate culture* is the limiting case of serial transfer, a single individual being transferred in each line at the end of the growth cycle. In most cases, it was impossible to propagate isolate cultures indefinitely. Over the course of a few hundred generations, the rate of fission declined, and the line eventually became extinct. The failure of isolate lines to maintain their adaptedness is almost certainly caused by the accumulation of slightly deleterious mutations through sampling error. When the line is propagated by choosing a single individual, it is very likely that the individual with fewest mutations will be discarded. In each cycle, the individual that is chosen is likely to bear more mutations than the founder, and the load of mutations must tend to increase through time, degrading adaptedness, eventually to the point where individuals are no longer viable and the line becomes extinct.

This process is much less important in large populations, such as the usual batch cultures of microbes that number millions of individuals, because the class of individuals with fewest mutations is so large that stochastic fluctuations are unlikely to eliminate it completely. It is also much less important in sexual populations. The reason for this is that recombination creates variation in the number of mutations per genome; a mating between two individuals each bearing a single mutation at different loci will produce some progeny bearing neither and others that bear both. Mating and recombination thus bring mutations that have arisen independently in different lines of descent into the same lineage, which is then likely to be eliminated by selection, restoring the frequency of undamaged individuals in the population. In this sense, sex rejuvenates populations by forestalling the ratchet-like accumulation of deleterious mutations, thereby permitting adaptedness to be maintained indefinitely. This is a special case of the more general principle that selection acts more effectively on the independent effects of genes when the genes are transmitted independently.

**Deterioration of Non-recombining Regions of the Genome.** Although sexual genomes are generally less vulnerable to mutation accumulation, their integrity depends on recombination, and regions that do not recombine will be subject to the same irreversible degradation as asexual genomes. It has been suggested that this process explains the partial loss of function by non-recombining elements such as Y chromosomes. This principle has been put to use by T. Mukai and his colleagues in experiments with *Drosophila*. Regions of the genome that have become inverted do not cross-over with the corresponding uninverted region of the homologous chromosome. In the absence of recombination, recessive deleterious mutations tend to accumulate in these inverted regions. The rate of accumulation of genetic variance caused by mutation can then be measured by crossing the experimental flies, creating progeny homozygous for the inversion in which the mutations are expressed.

## 53. *The population is a small sample of potential variation.*

Any population consists of a finite number of types that represents only a very small fraction of possible combinations of genes. When the environment changes, some possible types will be fitter than others, and these will tend to increase in frequency, provided that they occur in the population. Sorting will increase adaptedness, until the sorting limit is reached. If the environment changes radically, however, those types that would be well-adapted to the changed conditions of life may simply not exist in the population; the new adaptation required to cope with environmental change may exceed the sorting limit. Selection is then powerless to maintain adaptedness. It might be that even poorly-adapted individuals can continue to survive and reproduce, so that the population is able to replace itself and persist. If the rate of replacement is inadequate, with adults giving rise on average to fewer than one descendant (two descendants, in an outcrossed sexual population) in the next generation, the population will dwindle and eventually disappear. Or the change may be so drastic that the population is wiped out immediately; when a lake dries up, the fish will not evolve into amphibians, and most have no means of surviving even short periods in dry mud. This is the extreme case of the cost of selection or of limited opportunity for selection; when the relevant variation is not available during the life-span of the population, the result of environmental change is not adaptation, but extinction.

## 54. *Adaptation may be inaccessible because intermediate types are inferior.*

It is an obvious and familiar idea that environmental change may be lethal. It is not quite as obvious that populations may fail to become adapted to changed conditions, even when they are able to persist indefinitely. This follows from the general principle that evolutionary change generally occurs in small increments, with each stage representing an advance on the previous one. Organisms are never perfect, because it is always possible to imagine some capability that they lack and that would be beneficial, provided that it could be acquired without cost. However, organisms are also not perfectible; however desirable some new feature might be, it might never evolve. No doubt trees would benefit from being able to move around in search of light or water or nutrients, but the gradual transformation of a lineage of trees into creatures with muscle and nerve is most unlikely ever to occur. It might be perfectly feasible to design such creatures—perhaps we shall do that eventually—but evolution is not a designed process, and the requirement that it proceed through a series of intermediates, each being slightly superior to its predecessor, places severe limits on the types of creature that can exist. This requirement is often referred to as a *developmental constraint*, meaning that abrupt changes in the development of organisms that give rise instantaneously to a superior type of fundamentally different construction are so improbable that for all practical purposes they can be neglected.

**The Protein Space.** Consider first the constraints on the evolution of a single gene. The clearest representation of this situation is the concept of the protein space, developed by John Maynard Smith of Sussex. A typical protein consists of a sequence of, say, 100 amino-acids of 20 different kinds. All $20^{100}$ possible sequences can be represented as a 100-dimensional graph with sides of length 20. At any given time, the population occupies a small

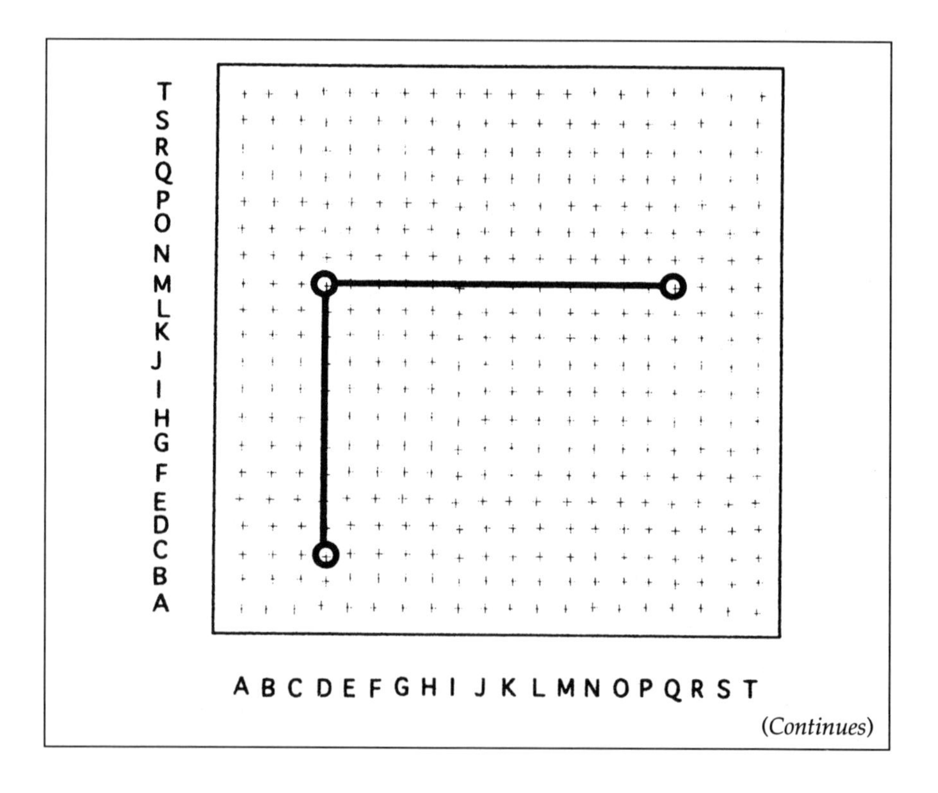


(Continues)

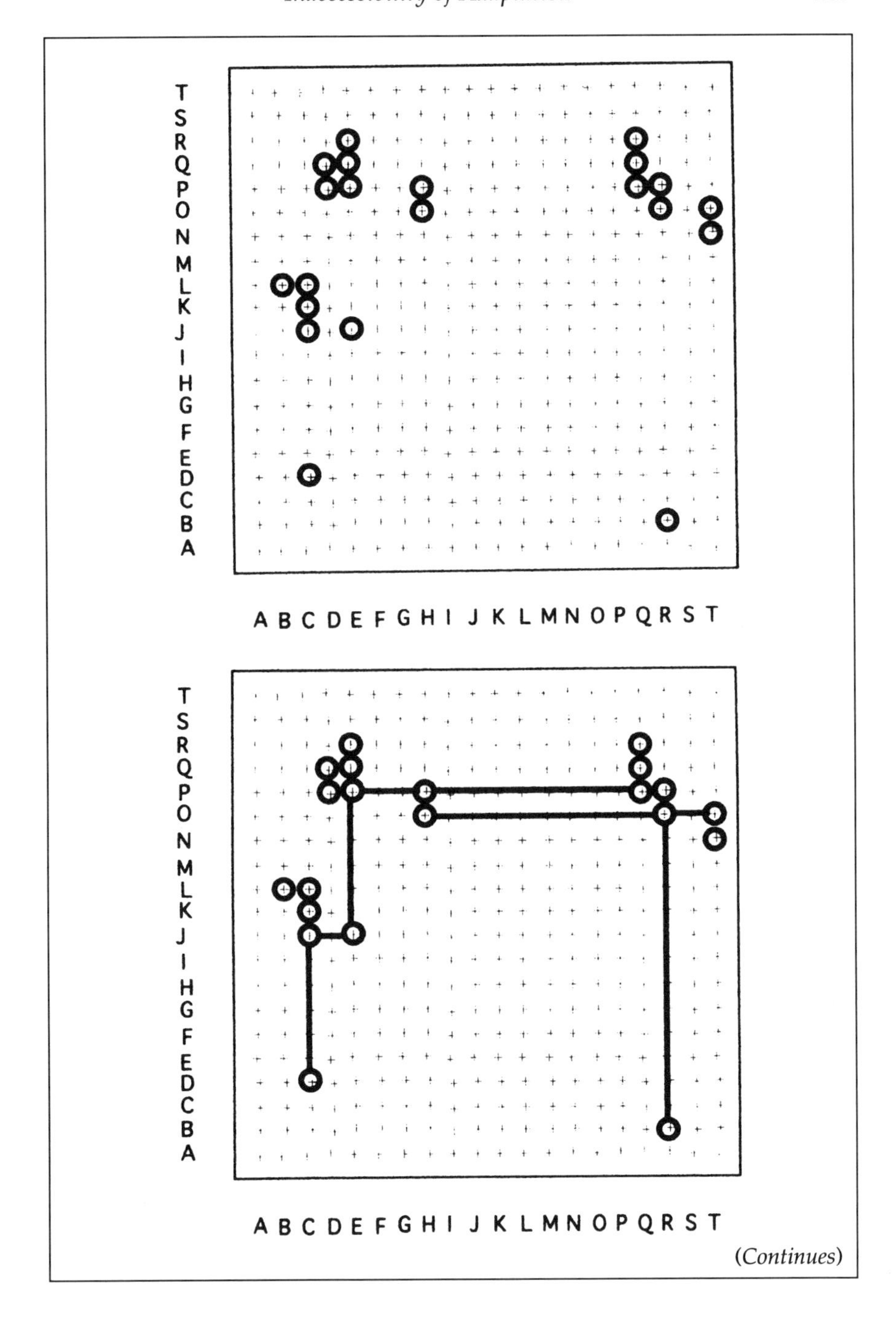
T
S
R
Q
P
O
N
M
L
K
J
I
H
G
F
E
D
C
B
A
A B C D E F G H I J K L M N O P Q R S T
T
S
R
Q
P
O
N
M
L
K
J
I
H
G
F
E
D
C
B
A
A B C D E F G H I J K L M N O P Q R S T

(Continues)

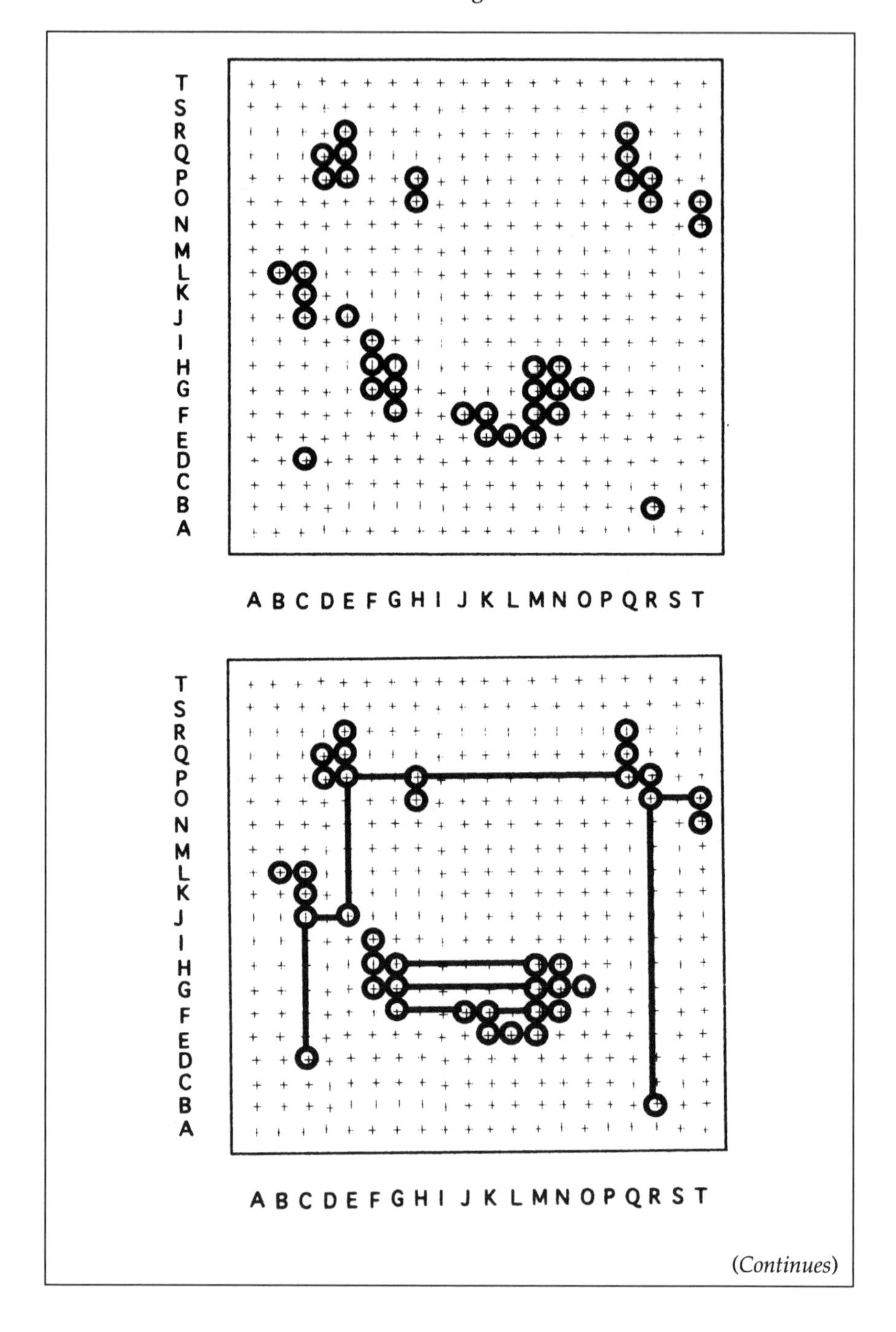
T
S
R
Q
P
O
N
M
L
K
J
I
H
G
F
E
D
C
B
A
A B C D E F G H I J K L M N O P Q R S T
T
S
R
Q
P
O
N
M
L
K
J
I
H
G
F
E
D
C
B
A
A B C D E F G H I J K L M N O P Q R S T

*(Continues)*

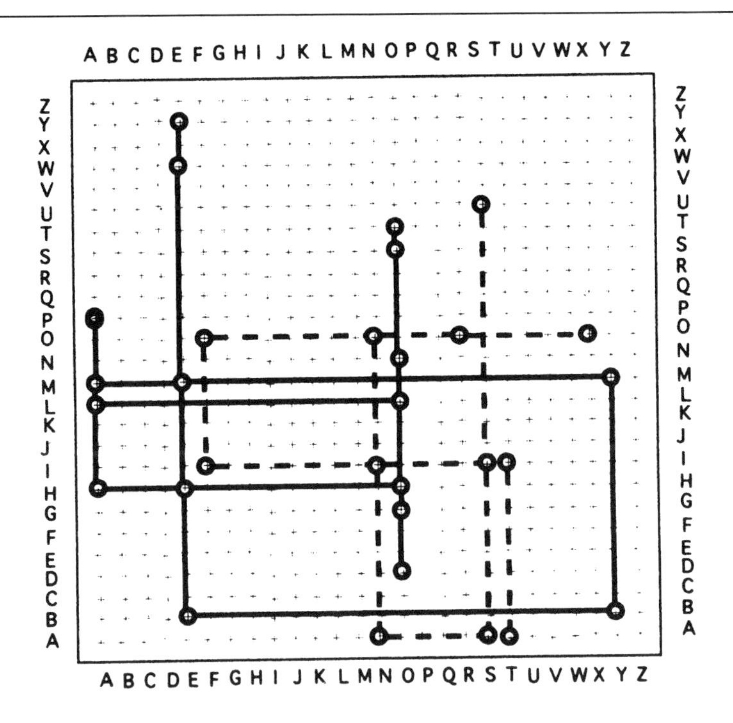


The first diagram illustrates the simplest version of the protein space, the $20 \times 20$ dipeptide matrix. I have taken the idea from an essay by John Maynard Smith (1970); it does not seem to have been developed much beyond its original brief description (see Kauffman, 1993). If only single substitutions are allowed, the possible evolutionary transitions are rook's moves such as the CD–MD–MQ transitions shown here in the first diagram. The second figure shows one possible set of functional dipeptides. Because they can all be connected by rook's moves (third figure), any one dipeptide can evolve into any other by successive single substitutions. Although the next set looks superficially similar, it turns out on inspection to consist of two clusters of dipeptides (fifth figure). Any dipeptide can evolve into any other within the same cluster, but cannot evolve into any dipeptide in the other cluster. The opportunity for and limits to selection are thus constrained by the inaccessibility of some functional states. The lower figure shows the analogous $26 \times 26$ matrix of all meaningful two-letter words $yx$ in English. There are two clusters, one comprising the consonant-vowel sequences such as DO and BY, the other the vowel-consonant sequences such as IN and AT. The two clusters are separate because there are no meaningful vowel-vowel or consonant-consonant sequences. The reader may like to confirm that the matrices for all three-letter words $y{*}x$ and $yx{*}$, where $*$ is any letter, are fully connected.

(*Continues*)

*(Continued)*
I do not know whether the three-dimensional space $zyx$ is fully connected, but, as Maynard Smith points out, spaces of four dimensions or more are not.

region of this figure, corresponding to the prevailing sequence or sequences and their mutational variants. Stabilizing selection restricts it to as small a region as possible by trimming random deleterious mutations that would otherwise cause it to expand indefinitely. When the environment changes, another region corresponds to a new optimal sequence, and directional selection tends to chivvy the population through the 100-dimensional protein space by causing sequences that increase fitness through possessing a different amino-acid at a given position to increase in frequency. Eventually the optimal sequence is attained and is subsequently maintained by stabilizing selection. This process will work only so long as the former sequence and the newly optimal sequence are connected by single amino-acid substitutions, each superior to the last. Evolution then proceeds by a series of rook's moves from one square to another of a chessboard with 100 dimensions. But this may not be possible. Selection may cause the spread of a molecule that cannot be improved by any single change, even though greatly superior molecules exist in distant regions of the protein space. The stringency of the model can be relaxed somewhat; we can allow changes that have no effect on function, provided that they are not actually deleterious, and a superior molecule with two or three alterations may arise occasionally in large populations. But it is perfectly conceivable that the population will be stranded in a region where the only possibility of improvement requires making alterations at 20 or 30 positions simultaneously. It will then remain stranded; evolution through selection has no means of leaping so large a gap. The degree to which adaptation can be improved through selection thus depends on the *connectance* of the protein space: on how often changing a single amino-acid at a given position results in a molecule that is potentially superior if the environment should change. The protein space must be reasonably well connected (otherwise adaptive evolution would rarely occur), presumably because most minor changes in sequence have very little effect on function. Connectance, however, has never been measured, and we have no precise idea of how such constraints affect the outcome of evolution. Modern techniques of molecular genetics would make it possible to begin to explore this important issue.

**The Adaptive Landscape.** There may also be constraints on the evolution of genotypes as combinations of several or many genes. These are expressed by the well-known analogy of the *adaptive landscape*, an attempt by Sewall Wright of Chicago to illustrate how genetic interactions mold the outcome of selection. Imagine two loci, each with a series of alleles. These can form the axes of a graph, so that any point on the plane is occupied by the two-locus genotype representing a combination of alleles at the two loci. The third dimension of the graph is the fitness of this genotype. The surface of the graph then resembles a landscape whose elevation differs from point to point, higher elevations corresponding to genotypes of greater fitness. At any given time, a population that is dominated by a single genotype, or a small set of similar genotypes, occupies a certain place on this landscape. The selection of genotypes with greater fitness causes the population to move about the landscape in such a way that its mean fitness continually increases; that is, it always moves uphill. Suppose that the alleles at either locus are arranged in order of increasing fitness. This will be possible only if they act independently, so that the fitness of an allele at one locus can be defined without reference to the allele present at the other locus. A combination of two low-fitness alleles thus gives a low-fitness genotype, and a combination of two high-fitness alleles gives a high-fitness genotype. The landscape then resembles a ramp, with its elevation increasing continually in one direction. Selection pushes the population up the ramp until the fittest allele has been fixed at both loci. If there are epistatic interactions among loci, however, it will not be possible to arrange the alleles at either locus in a series of increasing fitness, because the fitness of an allele at one locus cannot be defined without reference to the state of the other locus. The topography of the adaptive landscape is now much more complicated than a simple ramp and will, instead, resemble a hilly terrain of peaks and valleys. A population at any particular point will be pushed upward by selection until it reaches the top of a peak; the population will then cease to evolve, because movement in any direction is downhill, and is therefore prevented by selection against genotypes of lower fitness. This peak, however, may not be the highest point in the landscape. There may well be higher peaks elsewhere, but they cannot be reached through the stepwise selection of unit genetic changes, because they are separated by valleys of low fitness. The simplest case would be if the two loci each had two alleles A, a and B, b. If AB and ab are fitter than Ab and aB, then a population fixed for AB will remain indefinitely in this state even if the genotype ab has the greater fitness, because evolution from AB to ab through sequential substitution would involve passage through the valleys represented by the low-fitness genotypes Ab or aB. Populations that are placed initially at different points on the landscape might therefore move towards different peaks, and end with differing degrees of adaptedness; moreover, even if they began at the same point, any minor shift in location caused by sampling error, or by the chance occurrence of one of several possible advantageous mutations rather than another, might cause them to move towards different peaks. Thus, populations are not necessarily perfectly adapted to their environment. Of course, the landscape is not static; envionmental change corresponds to a change in the elevation of peaks and valleys, and what was formerly a mountain of high fitness may subsequently become a very modest hill. The population that occupied the mountain-top, however, is trapped on the hill, so long as it is completely surrounded by areas of lower relief. Selection is then powerless to increase adaptedness in novel environments if fitness cannot be increased through sequential substitution, despite the relatively low fitness of the current population.

The protein space and the adaptive landscape are useful devices for beginning to think about how genetic interactions contribute to the distinctively historical character of evolutionary change. These are topics that will be taken up at greater length in subsequent sections.

## 55. *Stasis is likely to be a frequent outcome of selection.*

When the population is sufficiently well-adapted to maintain itself, it may be unable to advance further even though superior genotypes exist in some inaccessible region of protein space or the adaptive landscape. Stabilizing selection will then conserve the present state of adaptation, even though it may be far from ideal. If the environment were to remain constant for long periods of time, it would not be surprising to observe little evolutionary change. Even if the environment fluctuates through time, however, it does not necessarily follow that organisms will respond. Paradoxically, the larger the change in the environment, the less likely it is to lead to a response; a sufficiently large change is likely to lead either to extinction or to the continued maintenance of a suboptimal phenotype. It has been strongly emphasized by Stephen Jay Gould of Harvard that lineages often persist unchanged (in fossilizable hard parts, that is) for immense periods of time. *Character stasis* may indicate that the relevant features of the environment remain unchanged, so that stabilizing selection maintains a highly adapted state; but it may equally indicate that the environment often changes so greatly that all minor variants have reduced fitness, and stabilizing selection therefore maintains a poorly adapted state. It is not known which of these alternatives applies more generally, but the requirement that evolution through selection must involve an unbroken sequence of slightly superior variants implies that many organisms will be poorly adapted because of the inaccessibility of superior variants. It should therefore be possible in some cases to increase adaptedness substantially by techniques such as saturation mutagenesis. The failure of such an attempt would be inconclusive (the organism chosen for study might be well adapted), but a single success would be very persuasive evidence for the argument that I have developed here. So far as I know, no experiments of this kind have yet been attempted, although mild irradiation of plateaued populations has not been successful.

## 56. *Selection may be effective only when the environment changes gradually.*

The population is most likely to respond to directional selection when the environment changes gradually, so that superior variants are accessible and the cost of selection is tolerable. Gradual environmental change will then lead to gradual evolutionary change. This is easy to appreciate when the environmental change is one-dimensional. $Q\beta$ evolved resistance to high levels of ethidium bromide because the concentration of the toxin was increased gradually during the course of the experiment; the original sequences cannot replicate and, therefore, cannot maintain a population at the concentration of ethidium bromide that the final evolved sequences are able to tolerate. It is true that flies abruptly exposed to high concentrations of novel insecticides, or bacteria exposed to high concentrations of new antibiotics, often evolve resistance quite quickly. This resistance, however, usually involves a single genetic change, often a loss-of-function mutation, such as defective ribosome assembly, and the resistant genotype is thus likely not only to be readily accessible but actually to lie within the sorting limit of the unadapted population. It would be interesting to confirm experimentally that natural selection is more effective when a novel stress is applied gradually than when it is applied abruptly, providing that the stress is sufficiently severe. I do not think that such experiments have been done; there are, of course, many experiments in which a stress is applied gradually, and many in which it is applied suddenly, but few if any in which both treatments are used.

## 57. *On long time-scales, repeated sorting results in cumulation.*

Evolution on short time-scales occurs by simple sorting, because the variation on which selection acts is dominated by the variation initially present in the population, and the input of new variation can be ignored. On longer time-scales this balance shifts, as novel mutation becomes an increasingly important source of variation. Over the course of thousands of generations the balance reverses, and the variation initially present, however great, is of little consequence relative to the new variation arising during evolution.

In any given generation, of course, selection sorts the variation that happens to be present. When selection is continued for long periods of time, however, this sorting process has in every generation a new point of departure, as novel mutations modify the genotypes that have been established in the population as the result of previous generations of sorting. Long-term evolution is therefore a process of *cumulation*. There is an important difference between sorting and cumulation. Simple sorting is a repeatable process whose outcome can be predicted when the environment can be controlled and the composition of the initial population specified. This is not necessarily the case for cumulative change. We cannot, in general, predict which mutations will occur or in what order they will occur. If we establish replicate selection lines, maintained in identical environments, they will each experience a different sequence of beneficial mutations and will therefore follow a different pathway to adaptation. The opportunistic nature of selection means that we cannot, in general, predict the route of cumulative change; we may not even be able to predict its eventual outcome. We must be prepared to interpret evolution as a unique sequence of historical change, rather than as a regular and repeatable process of development.

## 58. *Continued change occurs through successive substitution.*

A simple example of cumulative change would be the history of a bacterial population exposed to novel conditions of growth in a chemostat. Ideally, the chemostat would be engineered so that directional selection acted exclusively on a single enzyme encoded by a single gene. We can imagine that the population, though initially poorly adapted, is able to replace itself in the chemostat and that it becomes well adapted once different amino-acids have been substituted at several positions along the enzyme molecule. The appropriate variants will arise by mutation, most of them involving a single amino-acid alteration. One of these variants will spread through the population, until the clone in which the mutation happened to occur has eliminated all others, barring a low frequency of mutants. If the population is sufficiently large that all single variants occur with appreciable frequency, the variant that spreads first is likelyto be that with the greatest fitness. Once this clone has become established, a second beneficial mutation will sooner or later occur in a member of the clone and will, in turn, spread to fixation. This process will be repeated, until all the mutations that confer high fitness in the novel environment have one by one become established in the population. The history of the chemostat thus includes longer or shorter periods of stasis, interrupted by episodes of selection in which a newly-arisen beneficial mutation spreads through the population. This is the process of *periodic selection*. Whether the population is static most of the time depends on the number of novel beneficial mutations arising every generation which will be determined by the population size $N$ and the mutation rate $u$. If $Nu$ is large, then new beneficial mutations are likely to arise and spread in every generation, and the population will be

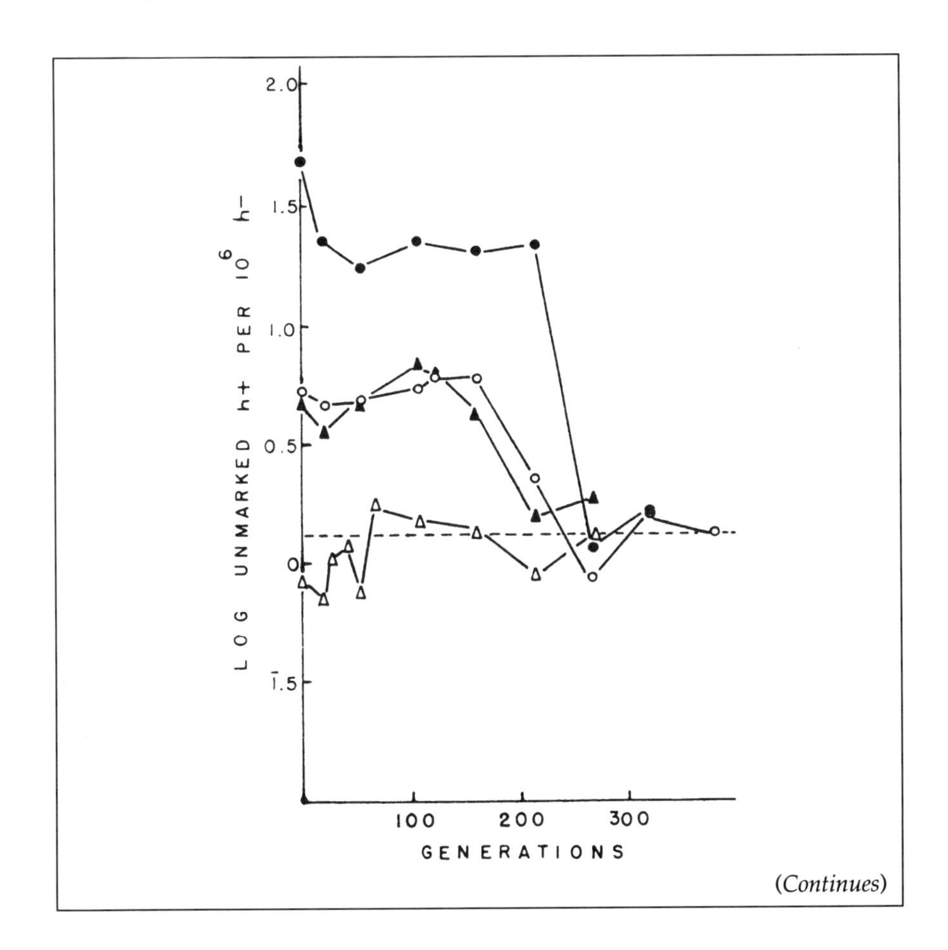


(*Continues*)

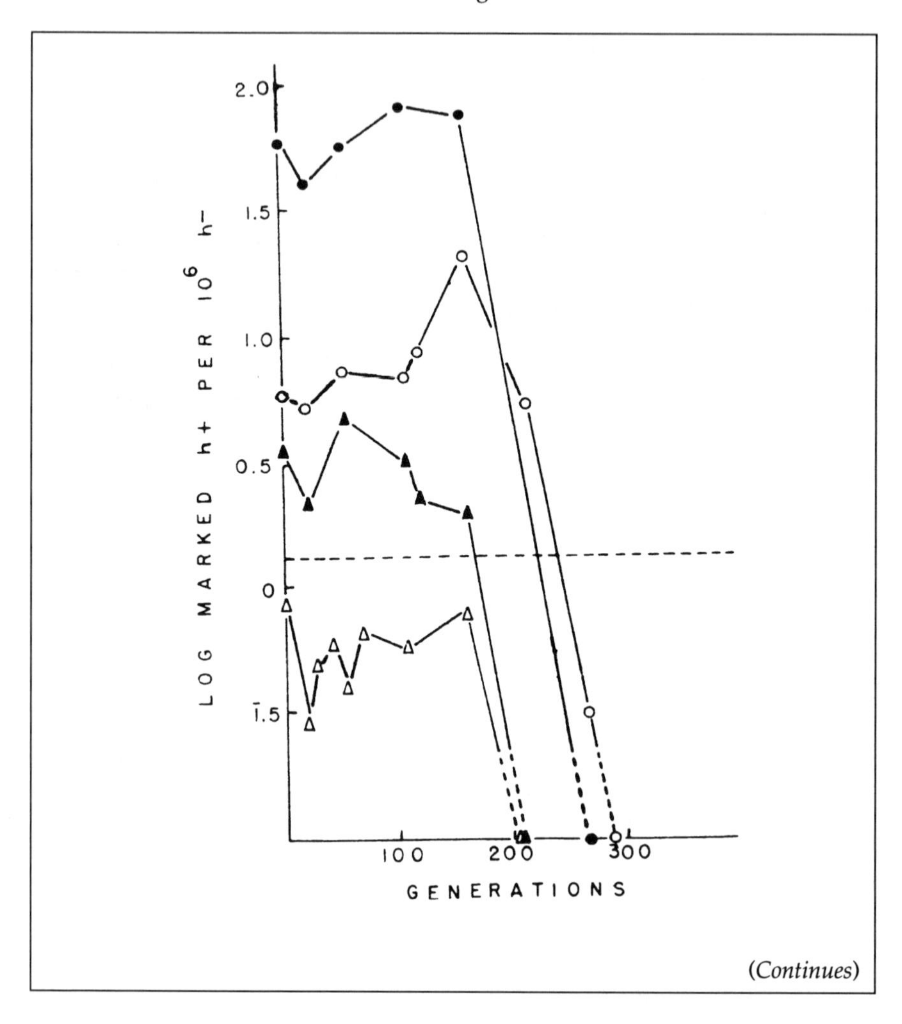
2.0
1.5
1.0
0.5
0
$\bar{1}$.5
LOG MARKED h+ PER $10^6$ h−
100
200
300
GENERATIONS

*(Continues)*

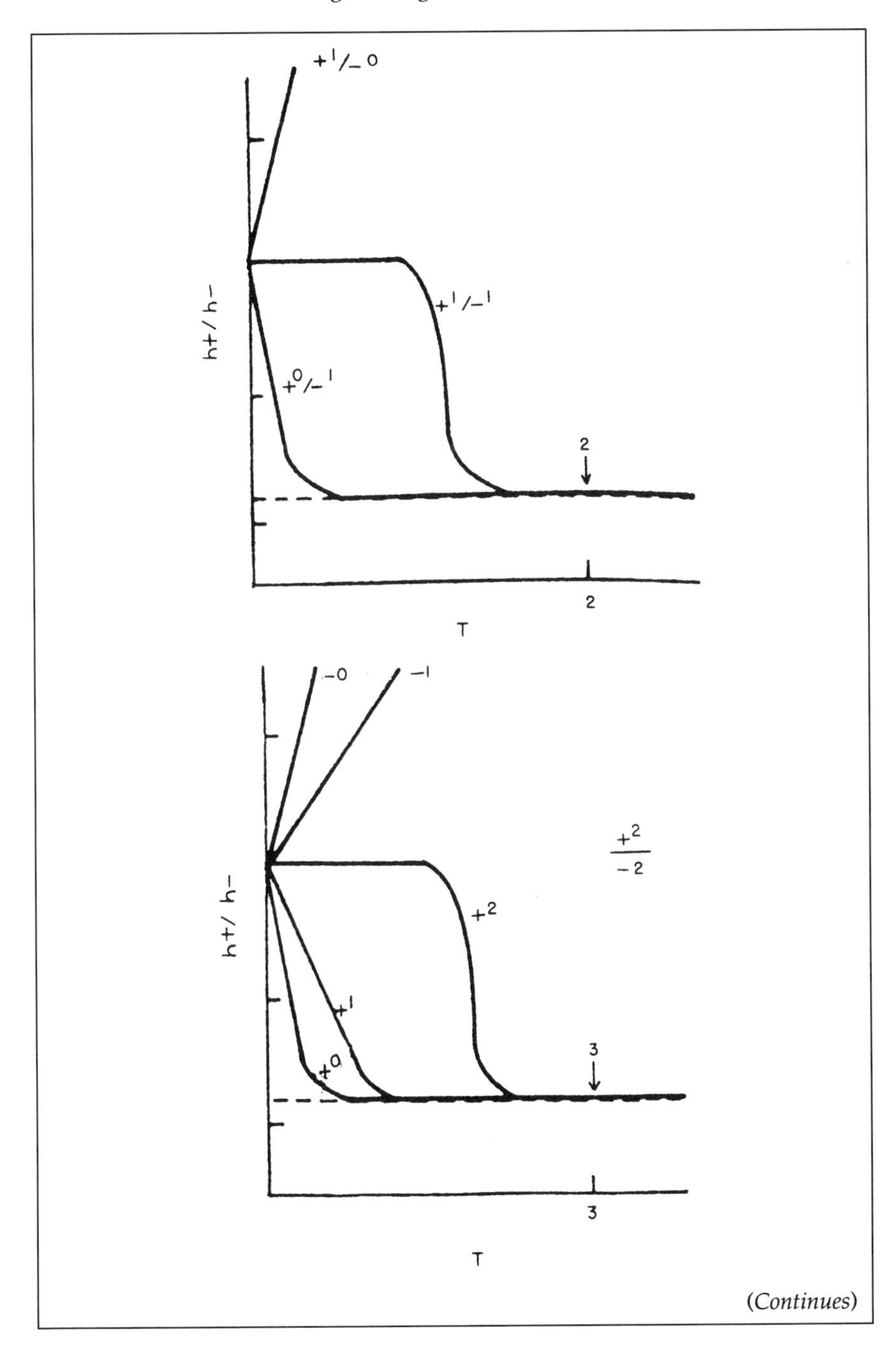


(*Continues*)

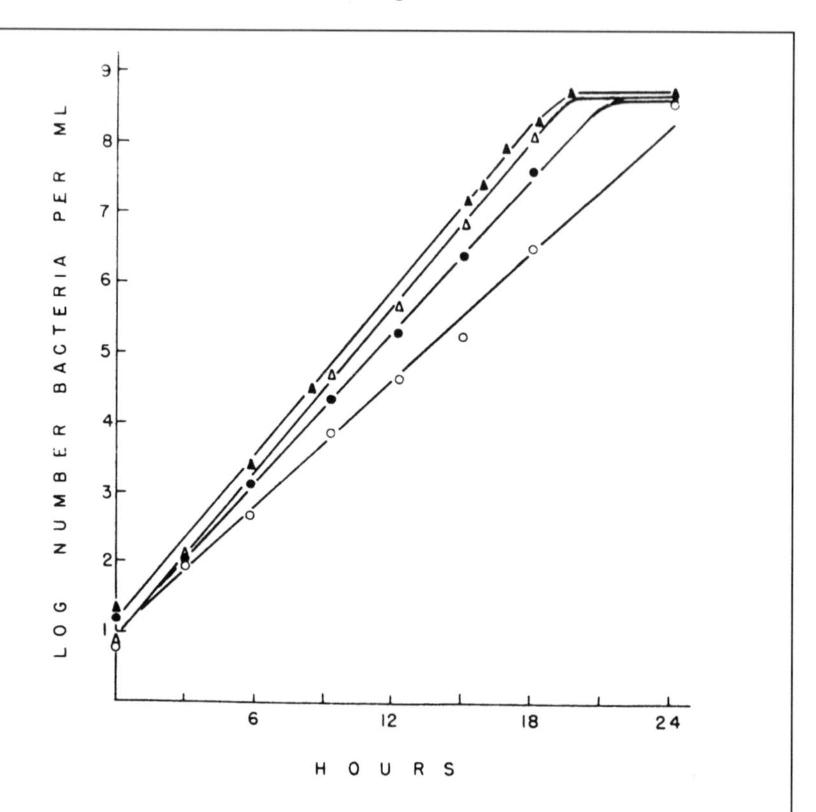


The earliest work on the evolutionary dynamics of bacteria in chemostat or batch culture was by Novick and Szilard (1950, 1951) and Atwood et al. (1951a, 1951b). These figures are taken from Atwood et al. (1951b). They measured the ratio of histidine prototrophs $his^+$ and auxotrophs $his^-$ in medium supplemented with histidine, where $his^+$ and $his^-$ can be treated as selectively neutral markers. The $his^+$ cells arising by mutation in an $his^-$ population increase to a frequency of about $1.3 \times 10^{-6}$, which is then maintained, with some fluctuations, for hundreds or thousands of generations. This equilibrium value is not explained by the ratio of forward ($his^- \to his^+, 2.7 \times 10^{-8}$) to backward ($his^+ \to his^-, 1.2 \times 10^{-6}$) mutation rates, which would predict a frequency at equilibrium of about $2 \times 10^{-2}$. The $his^+$ cells must therefore be selected against. If they are deliberately introduced at higher frequencies, they remain abundant for a time (usually about 200 generations), before abruptly shifting downwards to the equilibrium frequency (first figure). This is caused by the passage of a novel beneficial mutation that fixes the $his^-$ genotype in which it had arisen. It arises in an $his^-$ rather than in an $his^+$ genotype simply because the $his^-$ cells are vastly the more abundant. If $his^+$ cells bearing a second distinguishable marker are used, the marked strain is completely lost during the passage of such beneficial mutations (second figure); the $his^+$ cells observed in the earlier experiment must therefore be recent mutations, not survivors

*(Continues)*

from the original inoculum. By isolating cells from different times after inoculation, it can be shown that there is a sequential increase in fitness associated with the periodic selection of novel mutants. The third figure is their diagrammatic representation of the outcome of competition between the original *his*$^+$ cells ($+^0$) or those isolated subsequently ($+^1$) and the corresponding *his*$^-$ population ($-^0$, $-^1$). The original *his*$^+$ are rapidly eliminated, without any lag, by the evolved *his*$^-$; conversely, the evolved *his*$^+$ (descending from mutants *his*$^-$ $\rightarrow$ *his*$^+$ in the *his*$^-$ strain that acquired the novel beneficial mutation) quickly spread in the unevolved *his*$^-$ population. Thus, there has been a permanent increase in the adaptedness of the population. If the evolved *his*$^+$ and *his*$^-$ are mixed, with *his*$^+$ being the rarer, *his*$^+$ maintains the same frequency for about two hundred generations before abruptly declining to its characteristic equilibrium frequency. This marks the passage of a second beneficial mutation. If we now make pairwise mixtures of *his*$^+$ and *his*$^-$ cells from different time intervals, as in the fourth figure, we shall find that an *his*$^+$ strain spreads when it has evolved after the *his*$^-$ population into which it is introduced at low frequency and is eliminated when it evolved before; if it is contemporaneous with the *his*$^-$ population, then it remains at the same frequency until the next beneficial mutation arises, almost certainly in an *his*$^-$ cell. This demonstrates the cumulation of adaptive modification through periodic selection. When cells are isolated after successive episodes of selection, the more recent isolates have the higher growth rates (fifth figure). Thus, the sequential substitution of successively better-adapted genotypes can be followed through the abrupt shifts in frequency of neutral markers, and confirmed by extracting and assaying lineages from the culture when these events have been detected.

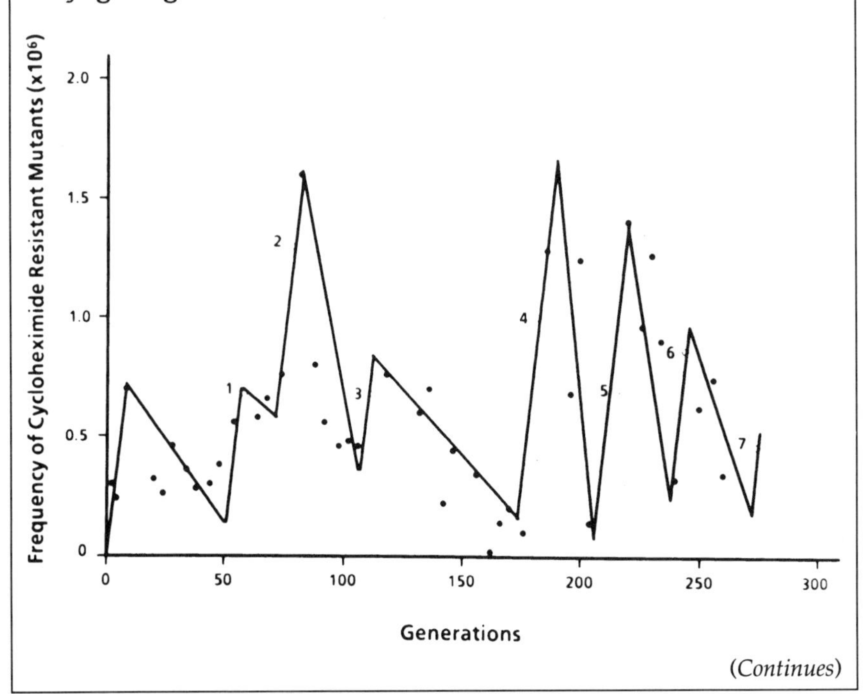


(*Continues*)

(*Continued*)

Adams et al. (1985) have described periodic selection in a population of eukaryotic microbes. They grew a standard strain of budding yeast in a glucose-limited chemostat for 270 generations, measuring at regular intervals the frequency of cells resistant to cycloheximide. Because such mutations have little effect on fitness when cycloheximide is not present, they increase in frequency through mutation. They are still very rare, however, when a mutation that increases the adaptedness of a lineage to chemostat conditions arises, and that mutation therefore almost certainly arises in a cycloheximide-sensitive cell. As the mutation passes to fixation, it drives down the frequency of cycloheximide resistance, because the genes conferring the resistance are linked to an inferior allele at the locus where the mutation conferring increased fitness has arisen. Fluctuations in the frequency of cycloheximide resistance thus mark the passage of seven successive substitutions (sixth figure, above). The nature of adaptive change can then be studied by withdrawing samples from the chemostat after each episode of selection. The major adaptation evolving in this system was, not surprisingly, an increased rate of uptake of glucose.

The simple picture of periodic selection given here is almost certainly inadequate. In particular, the rate of turnover observed in chemostat experiments (a substitution every 40 generations or so in the yeast experiment) seems too rapid to be consistent with the assumption that the population is genetically uniform (or very nearly so) during the intervals between substitutions. Adams and Oeller (1986) have suggested that chemostat populations are usually so large that most possible single mutations, including beneficial mutations, will be present at any given time. The evolving population may then contain hundreds of superior types, all of which are increasing in frequency at different rates. As the most strongly selected types come to predominate, culture conditions will change, altering the selection coefficients associated with all the other types. This may cause a continual flux in the composition of the population that cannot be adequately interpreted through the fluctuation in the frequency of neutral markers, since all lineages are roughly coextensive. It seems likely that future studies of evolution in chemostats will uncover a much richer and more dynamic system than has hitherto been envisaged. This topic is discussed further in Secs. 64–67 and 144–145, and has been reviewed by Dykhuizen (1990).

undergoing the replacement of one clone by another most of the time. If $Nu$ is small, then new beneficial mutations will become established at appreciable frequency only rarely, and the population will be static most of the time. In chemostats, $Nu$ will usually be large simply because $N$ is so large. The unit process involved is the substitution of a single allelic variant; the entire process of adaptation comprises the *successive substitution of a series of beneficial alleles*. Contrary to what I suggested in the previous paragraph, this might be, in principle, a rather regular and repeatable process. There will be a more or less predictable succession of clones, those bearing mutations that confer the greatest increase in fitness being selected first; and the end-point is reached when all the beneficial alterations have become established in the population. Replicate lines may differ in the details of their dynamics—sampling error, for example, may cause some variation among lines in the order in which alleles are substituted—but the course and outcome of evolution can be predicted rather precisely. I do not assert that continued evolution, even in chemostats, is usually like this, but, rather, use this simple imaginary experiment as the basis for discussing what actually happens in experimental populations once the horizon of simple sorting has been passed.

## 59. *A single clone of microbes will readily respond to natural selection.*

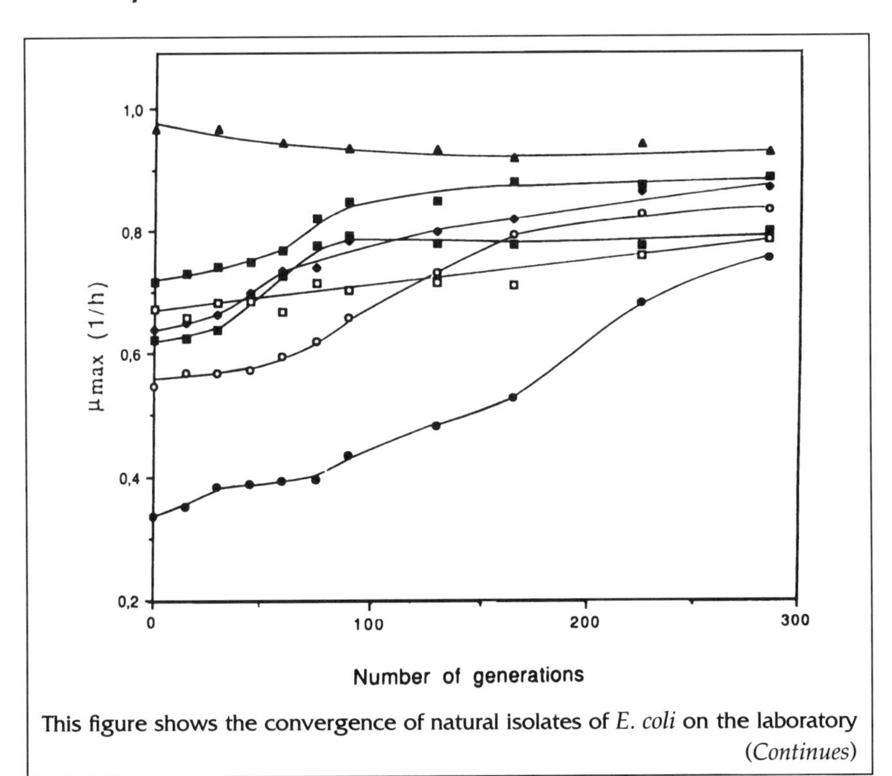


This figure shows the convergence of natural isolates of *E. coli* on the laboratory

*(Continues)*

phenotype of rapid growth during 280 generations of culture, from Mikkola and Kurland (1992). For comparable results with *Drosophila*, see Ayala (1968) and Marinkovic (1968).

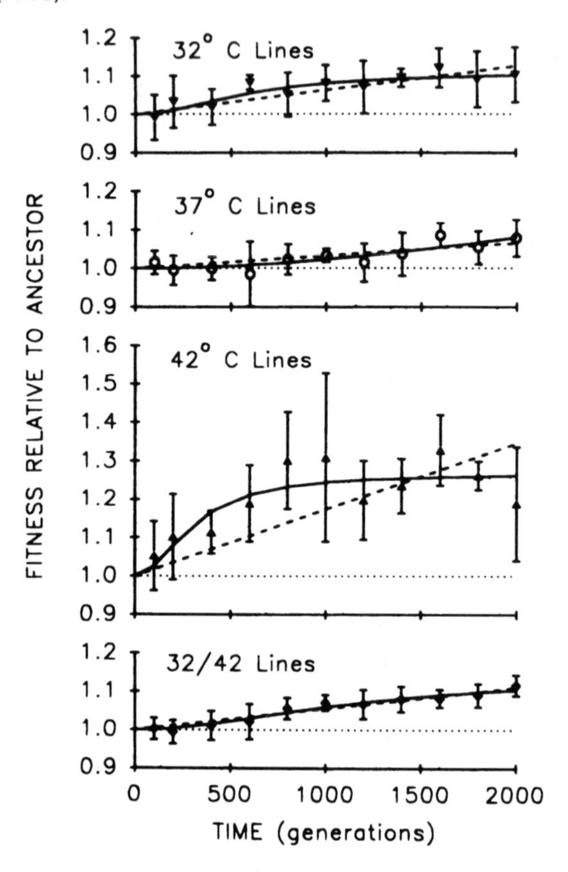


This figure shows the response of laboratory-adapted populations of *E. coli* to culture at different temperatures, from the experiment by Bennett et al. (1992; see also Bennett et al. 1990). Clones were isolated from the selection lines at intervals and tested against the ancestral strain; the plotted points are fitness relative to this ancestor, with 95% confidence limits of the mean of six replicates. The solid line is the fit to a sigmoidal curve; the broken lines are the average of the linear regressions for the six replicate lines.

*(Continues)*

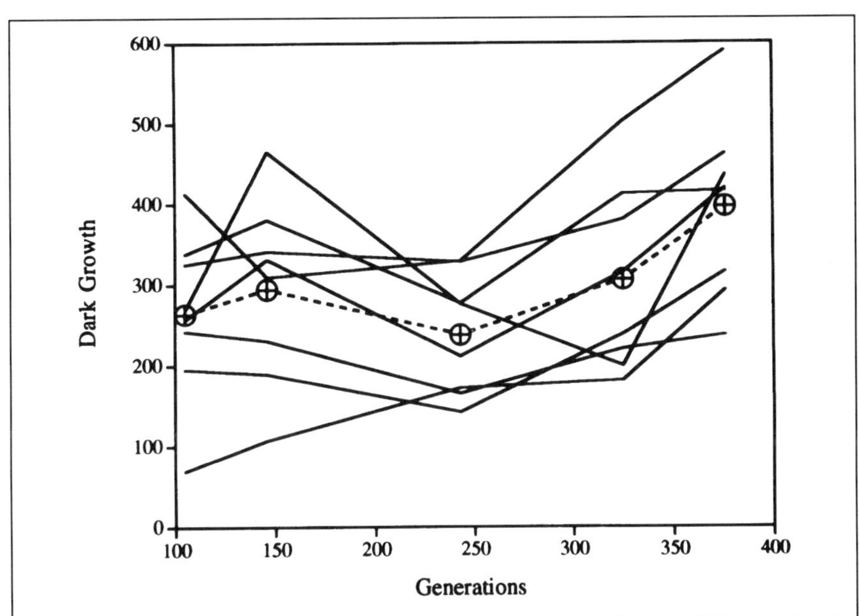


This diagram shows the amount of growth in the dark of eight different clonal lines of *Chlamydomonas* during nearly 400 generations of selection. The broken line joining the crosses within circles shows the mean value of the lines. Spores were isolated from the selection lines and grown as pure cultures, so growth is here a measure of general adaptedness to the dark environment.

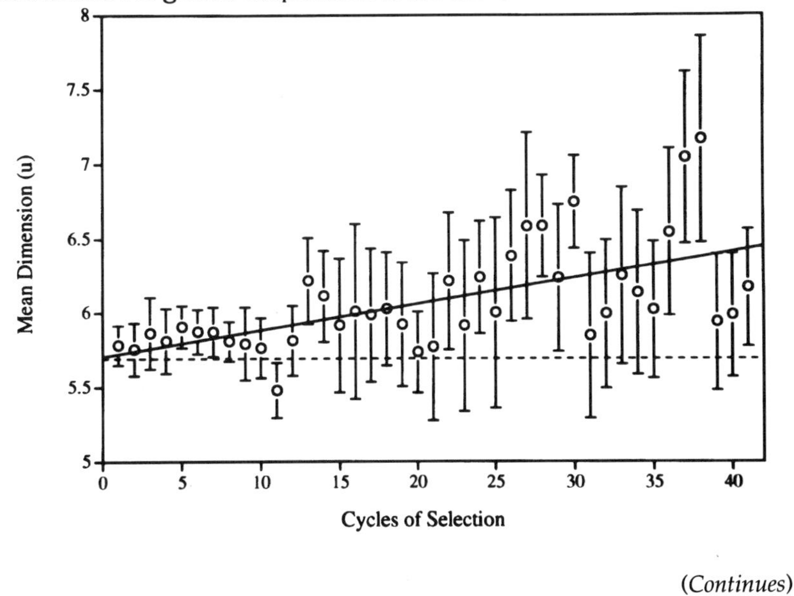


(*Continues*)

(*Continued*)
Size increased slowly through 40 cycles of artificial selection for larger size, imposed by filtration, within clones of *Chlamydomonas*. The points plotted here are the means of four replicates of each of four founding spores; the bars mark one standard deviation among lines each side of the mean. The equation of the regression line is: $y = 5.7 + 0.018x$; the broken line represents no response to selection.

Because the response to selection is proportional to the quantity of genetic variance, it is natural to assume that evolution can proceed at an observable rate only in highly variable populations and that selecting within a clone would be a waste of time. This is not correct. If a population is large enough, the flux of novel mutations will be large enough for selection to drive evolutionary change fast enough for it to be detected, measured, and analyzed. Moreover, to use a single clone as the founder of a selection line necessarily implies that any change that is observed will have been caused by a cumulative process of some kind, since no appreciable variation was initially available for sorting.

**The Domestication of Microbes.** It is not usually necessary to concoct very exotic environments in the laboratory in order to study adaptation; for most microbes, the laboratory environment itself is exotic enough. The natural environment of the bacterium *E. coli* consists of short periods in a mammalian gut interrupted by long periods in the outside world, short periods of plenty punctuated by long periods of penury, a way of life dominated by continual struggle against a host of enemies and competitors when conditions of growth are favorable, and by the necessity of resisting starvation and physical stress when they are not. A warm flask supplied with glucose and salts is as foreign to *E. coli* as a culture tube supplied with nucleotides and replicase is to $Q\beta$. Riitta Mikkola and C.G. Kurland have described how natural isolates of *E. coli* rapidly converge on the laboratory phenotype of high rates of growth in rich medium. Most isolates grew much more slowly than laboratory strains, with average doubling times of about 70 min. After 280 generations of growth in glucose-limited chemostats, the isolates had converged on rates of growth typical of laboratory strains, apparently as the result of increases in ribosome efficiency.

**Adaptation to High and Low Temperature in Bacterial Populations.** Albert Bennett and Richard Lenski, working at Irvine, California, selected *E. coli* for growth at high (42 C) and low (32 C) temperatures. A single clone was first cultured at normal temperature (37 C) for 2000 generations, so that it was thoroughly domesticated and so that any subsequent adaptation would be likely to be specific to the altered temperature, rather than to general laboratory conditions of growth. A single clone was then extracted from this line and divided into replicate lines that were then cultured for a further 2000 generations at 32 C, 37 C, or 42 C. Adaptation was measured at intervals during the experiment by competing these selection lines against their common ancestor. As expected, there was little further improvement at 37 C, but both the high-temperature and the low-temperature lines became substantially better adapted. By the end of the experiment, the low-temperature line was more than 10% fitter than its ancestor, an average rate of increase in fitness of about $6 \times 10^{-5}$ per generation; the high-temperature line responded more steeply, ending up about 20% fitter, equivalent to an increase in fitness of about $17 \times 10^{-5}$ per generation. As will be seen later, these rates of increase in fitness are consistent with the rate at which variance in fitness arises through mutation.

**Adaptation to Growth in the Dark by *Chlamydomonas*.** In my laboratory, we cultured lines of the unicellular green alga *Chlamydomonas* in the dark for several hundred generations, assaying the response to selection by measuring the rate of growth in pure culture of clones extracted from the selection lines. *Chlamydomonas* can grow heterotrophically in the dark by using acetate as a carbon and energy source; some strains grow well in the dark, whereas others are almost unable to grow at all. The latter improved during the course of the experiment, until after about 600 generations they were well adapted to heterotrophic growth. This is a much more dramatic change than was seen in the *E. coli* experiments, because it changed an almost obligately photoautotrophic creature into one that could also live heterotrophically, a substantial new capacity. As in the *E. coli* experiments, however, it was possible to show (by estimating the variance within lines) that genetic variance in fitness was arising fast enough to explain the observed response to selection.

**Selection for Large Size in *Chlamydomonas*.** In a very different kind of experiment, we selected for large size in *Chlamydomonas* by passing cultures through a fine filter, discarding the cells that passed through and using those retained by the filter to inoculate fresh growth medium. This procedure is surprisingly ineffective, perhaps because the filters are not very uniform, and it imposes only very weak selection differentials of about 0.1 phenotypic standard deviations per generation. After 40 cycles of selection, beginning with clonal cultures, cell diameter had increased on average by about 1 phenotypic standard deviation. This is consistent with the cumulation of mutational effects through time, given that these contribute about a quarter of the overall variance in size.

The input of new beneficial mutations alone, then, is sufficient for selection to increase adaptedness in large asexual populations. Moreover, the cumulative increase in adaptedness seems to be an orderly process that obeys rules similar to those of sorting, with adaptation increasing at a rate determined by the quantity of genetic variance of fitness. Different lines may start out from different positions, or respond for a time at different rates, but all eventually become adapted. There is a general tendency for *phenotypic convergence*; lines may at first vary widely in their ability to grow in some novel environment, but after a few hundred generations of selection all can grow reasonably well and in this sense resemble one another more closely. There are two respects in which cumulation may differ from sorting. The first is that there is no clear limit to the cumulation of change, in the sense that the actual or potential genetic variation within a population defines the sorting limit. The second is that the underlying genetic basis of phenotypic convergence is not clear. When a defined set of strains is sorted, the fittest will replace the others, and the process is predictable in terms of both phenotype and genotype. When strains replace one another successively, the result will be predictable only if there is a limited set of possible mutations, all of which will eventually be tested.

## 60. *Phosphate utilization in experimental populations of yeast evolves by successive substitution.*

The simplest way in which adaptedness can increase through selection is by a succession of point mutations, each of which alters a single amino-acid in a given protein. The evolution of resistance to ethidium bromide by evolved Q$\beta$ RNA provides an example of this process. I do not know of any strictly comparable examples, involving a defined succession of alterations in the nucleotide sequence of a single gene in response to a single environmental stress, in experiments with cellular organisms. This may be due in large part to the greater difficulty of working with more complex organisms, but their greater complexity may itself be a factor. When several or many genes affect the expression of a character, periodic selection may involve a succession of alterations at different loci. The evolution of phosphate utilization in chemostat populations of yeast is a good example of this process.

Yeast obtains a usable supply of phosphate by hydrolyzing phosphate esters to release orthophosphate that is then transported into the cell by a permease. The hydrolysis is mediated by an acid phosphatase located in the cell wall. P.E. Hansche and J.C. Francis set up chemostats so that phosphate utilization was limiting to growth. The only source of phosphate was $\beta$-glycerophosphate, which is hydrolyzed by the phosphatase to inorganic phosphate. The optimal pH for this reaction is 4.2; by maintaining the culture medium at pH 6, phosphatase activity is made limiting to growth, and thereby exposed to intense selection. The experiment began with a single clone of brewer's yeast, *Saccharomyces cerevisiae,* with a population size of about $10^9$ individuals. It grew slowly for about 180 generations, when there was a sudden increase in population density. This was attributable to more efficient assimilation of orthophosphate, presumably as the result of

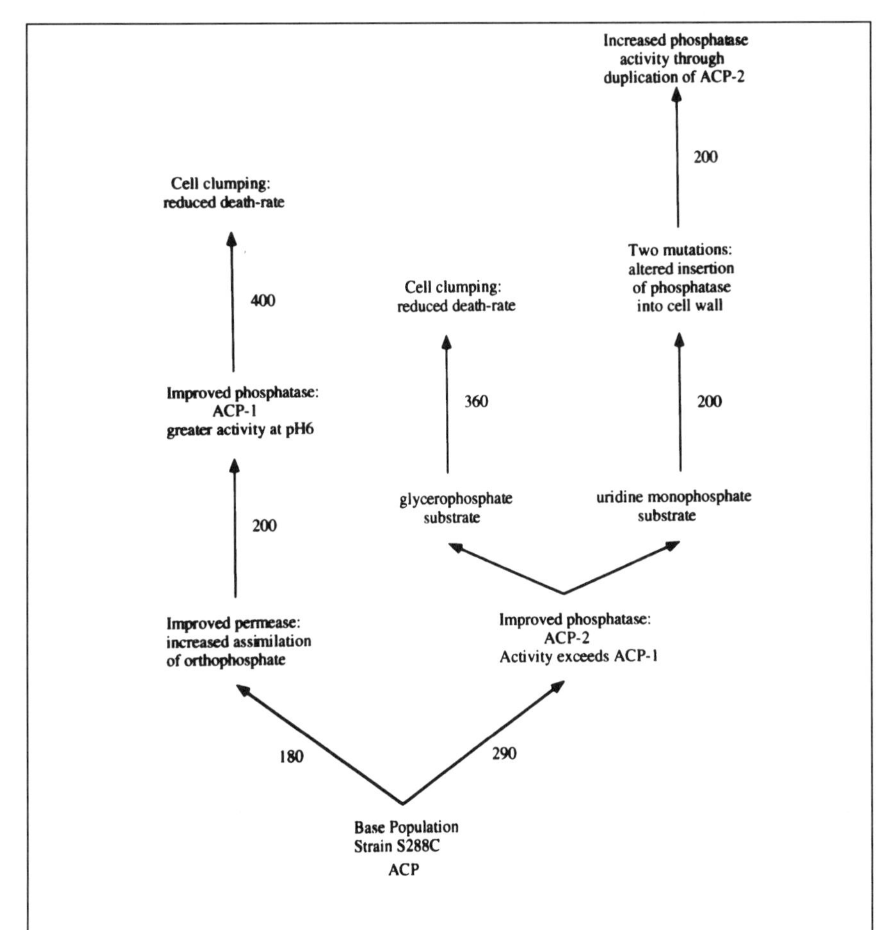


This diagram summarizes the course of events during the evolution of phosphate metabolism in the yeast populations studied by Francis and Hansche (1972, 1973) and Hansche (1975). The numbers by the lines are the number of generations between events; see also Sec. 67.

an improved permease. A second mutant appeared after about 400 generations. This was an altered phosphatase, with greater specific activity and a higher pH optimum than the original enzyme. A third mutant, appearing after about 800 generations, caused the cells to clump, and spread because clumped cells tend to settle and are thereby less likely to be washed out of the chemostat. Thus, the population had become well adapted to utilizing organic phosphate under chemostat conditions within 1000 generations, as the result of a sequence of three independent genetic changes, affecting in turn the assimilation of inorganic phosphate, the hydrolysis of organic phosphate, and the retention of cells within the culture vessel.

## 61. *Continued selection causes adaptation to novel environments.*

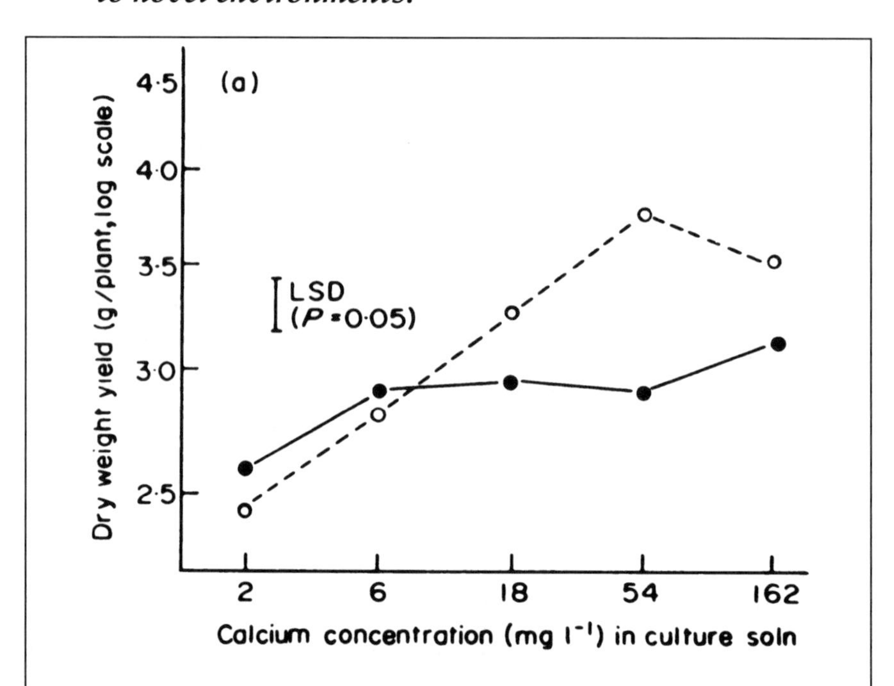


Davies and Snaydon (1973a) compared the growth of plants from limed and unlimed plots of the Rothamstead Park Grass Experiment. Plants were collected

*(Continues)*

(*Continued*)

from eight plots, four of which had been limed at four-year intervals since 1903, the other four being unlimed. These were grown in potting soil and then broken up into ramets and transferred to sand culture at five levels of application of calcium (as calcium chloride). The diagram shows the response of *Anthoxanthum odoratum*, as dry weight yield. Plants from the limed plots (broken line) have clearly evolved a greater capacity to respond to high levels of soil calcium. The same authors have also described the response to aluminium (Davies & Snaydon 1973b) and phosphorus (Davies & Snaydon 1973c); see also Snaydon (1970) and Snaydon and Davies (1972).

The evolution of tolerance to heavy metals is described in more detail in Sec. 111.

**The Rothamsted Park Grass Experiment.** The nearest approach to a chemostat experiment out-of-doors is the Park Grass Experiment, set up at Rothamsted Experimental Station in the middle of the last century and maintained ever since. It consists of a number of small adjacent plots that have been receiving different nutrient treatments for 100 years. The aim of the experiment was to find out the best procedure for cultivating crops of hay; I do not know whether it has succeeded in this, but it has certainly provided an excellent example of short-term adaptation to novel environments with defined characteristics. The basic observation is as follows. Many plots have received high inputs of specific nutrients, such as phosphate. Naturally, the plants in these plots grow taller and lusher than plants in neighboring plots which, although treated similarly in other respects, have received much lower phosphate inputs. This is an environmental effect, caused mainly by the difference in soil phosphate concentrations between the plots. However, there are also genetic differences. This can be demonstrated by extracting clones from the plots (easy to do, because most grasses proliferate vegetatively and can be broken up into individual ramets that can be planted separately) and growing them under standard conditions in the greenhouse. The plants extracted from the high-phosphate plots are much more responsive to phosphate addition than are those from low-phosphate plots; that is, when high levels of phosphate are applied to plants in the greenhouse, these high levels can be utilized for growth much more effectively by the plants from the high-phosphate plots. During the few hundred ramet generations since the beginning of the experiment, the grass population in these plots has adapted to a novel nutrient environment.

**Evolution of Tolerance to Heavy Metals.** More intense stress leads to more dramatic adaptation. The metal ores found in the igneous rocks of North Wales were being mined when the Romans came to Britain, but the pace of exploitation increased sharply when the Industrial Revolution created new markets for copper, lead, and zinc. The deposits were relatively small and had been worked out by the 1870s or so, leaving a fenced patch of rubble and spoil surrounding each old mineshaft. This patch was sterile; nothing grew on it, because high concentrations of heavy metals were leached from the low-grade ore that had been discarded there into the soil, creating a bare circle in the surrounding sheep pasture. After a while, a few plants from the pasture succeeded in growing on the old mine area, which eventually became tufted with grasses such as *Agrostis* and *Anthoxanthum*. By taking the progeny (either ramets or seeds) of plants growing on pasture or mine, A.D. Bradshaw and his colleagues at Liverpool showed that the mine plants had evolved a heritable resistance to high levels of heavy metals and were able to grow fairly well at concentrations that were lethal to most pasture plants. Resistance is not attributable to a single gene, like melanism in *Biston* or most cases of insecticide resistance, and adaptation no doubt involves changes at many loci. Nevertheless, it can occur quite quickly: it has been little more than a century since the mines were abandoned, but resistant populations can appear on spoil-heaps within 50 years or so. It has even been reported that linear populations of plants resistant to zinc, a metre or so in width, appear beneath galvanized-iron fences within a decade of their construction.

## 62. *The limits to continued artificial selection are not well defined.*

**Laboratory Selection Experiments.** Natural selection can be applied to asexual populations of microbes for hundreds or even thousands of generations in the laboratory. It is seldom, if ever, possible to study sexual populations at such length, because the sexual cycle is much more time-consuming than the asexual cycle; the complete sequence from gamete production or induction, through sporulation, to the eventual release of new vegetative spores takes several days in organisms such as yeast or *Chlamydomonas*, which go through an asexual cycle of reproduction in a few hours. Using artificial selection rather than natural selection also slows down the experiment, simply because of the time taken by the experimenter to carry out the selection. A number of ingenious devices have been invented to speed up selection experiments, usually by automating the process of selection. One biologist, for example, lowered a glue-covered screen onto a platform where anaesthetized flies had been strewn; the flies with the longest wings stuck to the screen first and could thus be selected readily in large numbers. Another biologist, applying selection for tolerance to ethanol fumes, invented the "inebriometer". Nevertheless, the time and labor involved in applying artificial selection to organisms with life-spans of several days or weeks means that few such experiments extend beyond 20 or 30 generations. In a few cases, a deliberate attempt has been made to continue selection for as long as possible. These heroic experiments, which have extended observations of artificial selection out to a hundred generations, have involved selecting for body weight in mice, for bristle number in *Drosophila*, and for pupal weight in *Tribolium* .

**Body Weight in Mice.** Goodale's attempt to select rat-sized mice has already been described in Sec. 50. The response ceased after about 35 generations, but the experiment was continued for over 30 years, until 84 generations had been completed. During the last 50 generations there was little or no consistent increase in size, but quite marked fluctuations.

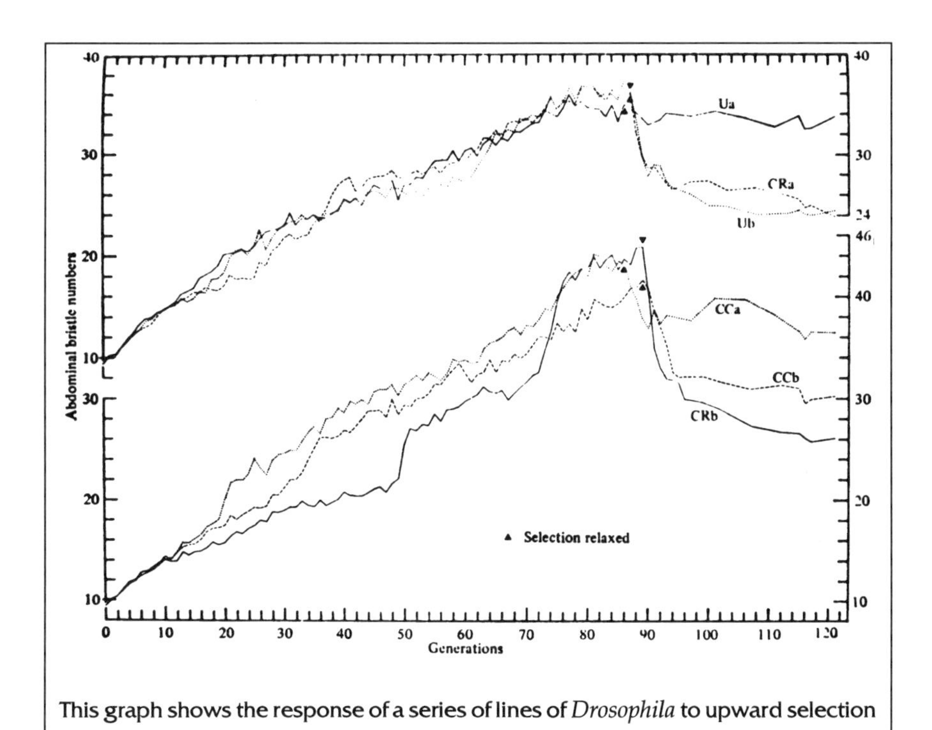


This graph shows the response of a series of lines of *Drosophila* to upward selection

(*Continues*)

on abdominal bristle number over about 90 generations, from Yoo (1980a). Selection was relaxed at the end of this period, causing a partial regression toward the base population; see Sec. 63. Other *Drosophila* experiments that involve selection for more than 50 generations include Buzzati-Traverso (1955, fitness components), Jones et al. (1968, abdominal bristles), Weber (1990, wing length), and Weber and Diggins (1990, ethanol knock-down). The longest *Drosophila* experiment on record seems to be that reported by Ricker and Hirsch (1985, 1988), in which they selected for geotaxis over about 600 generations, although selection was applied only intermittently.

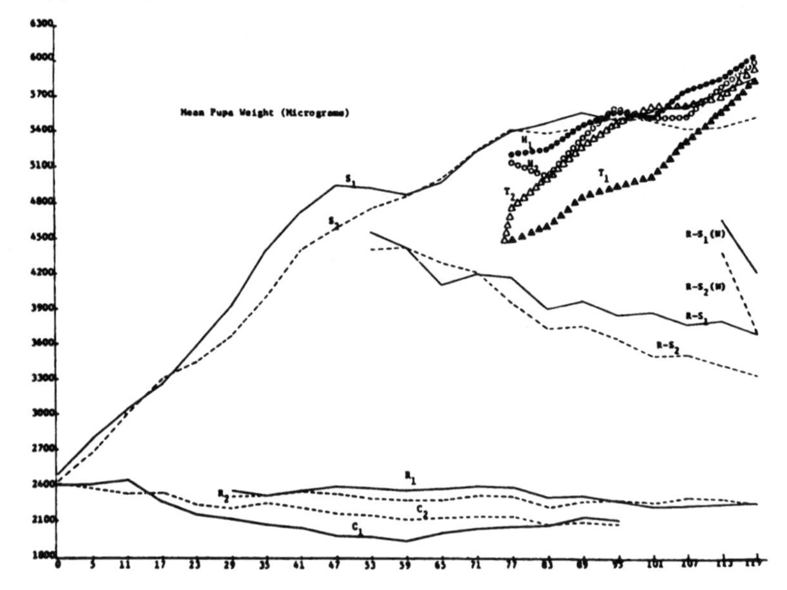


This graph shows the response of two replicate lines of *Tribolium* to upward selection on pupal weight over nearly 120 generations, from Enfield (1982). The

(*Continues*)

**Bristle Number in *Drosophila*.** B.H. Yoo of Sydney selected several lines of Drosophila for increased abdominal bristle number for nearly 90 generations. The most surprising feature of the experiment was that most of the lines were still responding when the experiment was ended, though some seemed to have reached a plateau after 75 or 80 generations. The pattern of response, however, was quite variable. Some lines showed a rather smooth increase in bristle number, but in others bristle number seemed to reach a limit that was maintained for 10 or 20 generations before abruptly resuming their response to selection.

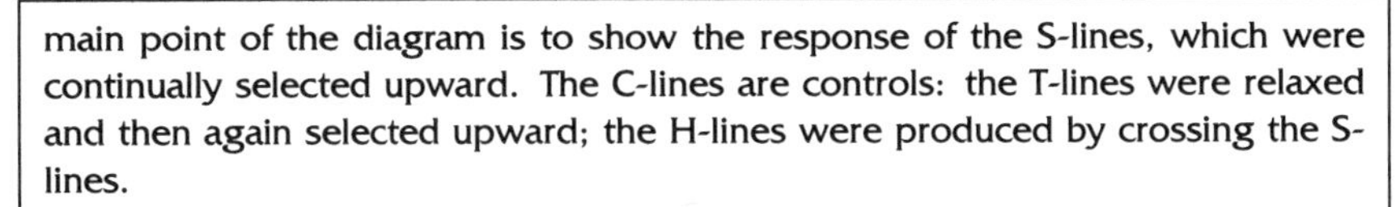
main point of the diagram is to show the response of the S-lines, which were continually selected upward. The C-lines are controls: the T-lines were relaxed and then again selected upward; the H-lines were produced by crossing the S-lines.

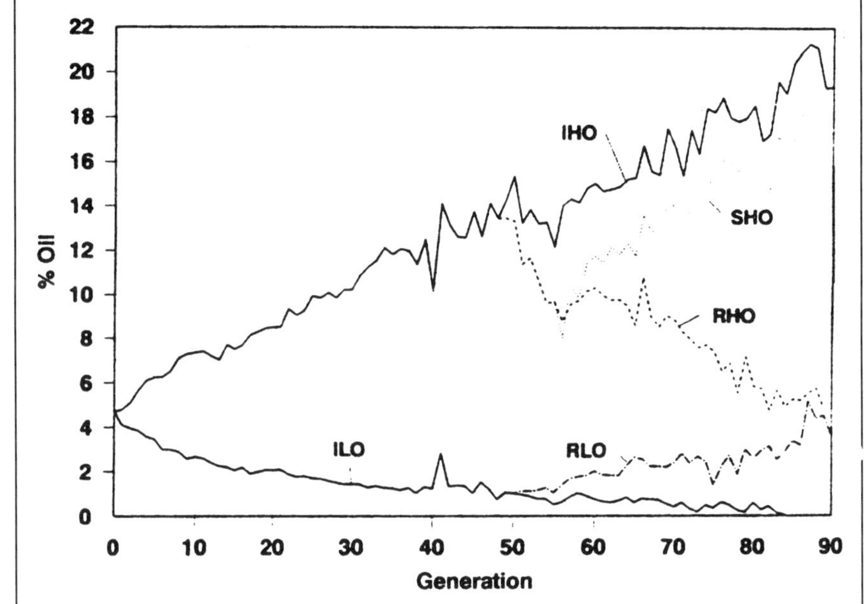


The most recent report of the Illinois Corn-Oil Experiment is by Dudley and Lambert (1992), who show the effects of continued upward and downward selection

*(Continues)*

**Pupal Weight in *Tribolium*.** F.D. Enfield of the University of Minnesota selected increased pupal weight in flour beetles for 120 generations. Two replicate lines continued to respond throughout the experiment, although the response was slow from about generation 75 onward. Again, the pattern of response was not the same for the two lines; in particular, one appeared to reach a plateau for several generations at about generation 50, but subsequently resumed its advance.

**The Illinois Corn-Oil Experiment.** Flour beetles and fruitflies have generation times of weeks, so that selection experiments lasting for a hundred generations represent an enormous commitment of time and labor. The generation time of mice is months, so that Goodale's experiment required the whole of a research lifetime to complete; a nnual plants represent the over about 90 generations, as well as the outcome of relaxed selection. Previous reports were published by Hopkins (1899), Smith (1908), Winter (1929), Woodworth and Jugenheimer (1948), Woodworth et al. (1952), Dudley et al. (1974) and Dudley (1977).

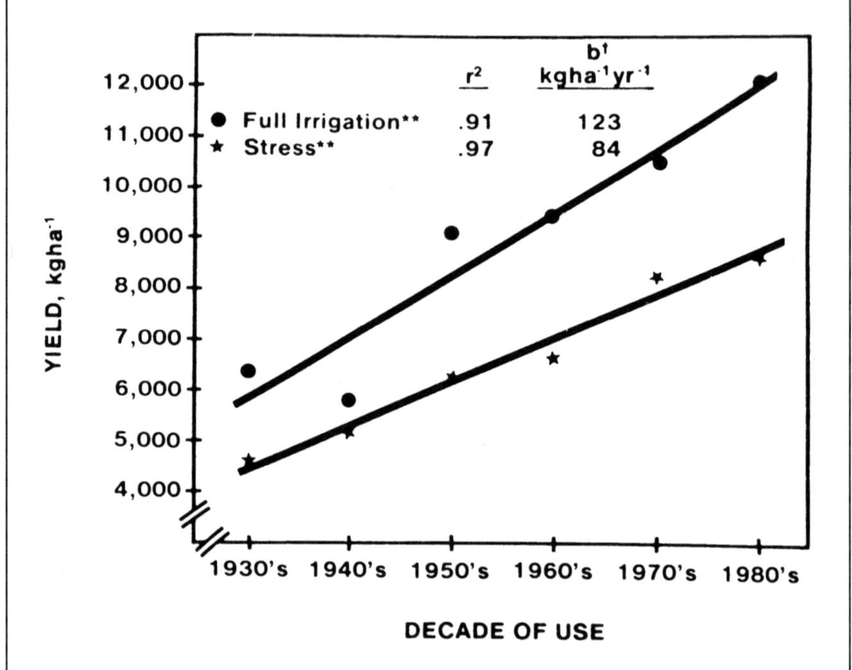


Growing stored seed from cultivars developed at different times in the past in a common garden illustrates the effect of selection in improving crop plants over the last 50 years. This figure, from Castleberry et al. (1984), shows how maize yields have been improved. Note that the improvement is more marked in irrigated than in unirrigated plots; it is also more marked in fertilized than in unfertilized plots. Comparable data for cotton production has been published by Meredith and Culp (1979), Bridge et al. (1971) and Bridge and Meredith (1983); for other differences between obsolete and modern cultivars, see Wells and Meredith (1984a, 1984b, 1984c).

*(Continues)*

(Continued)

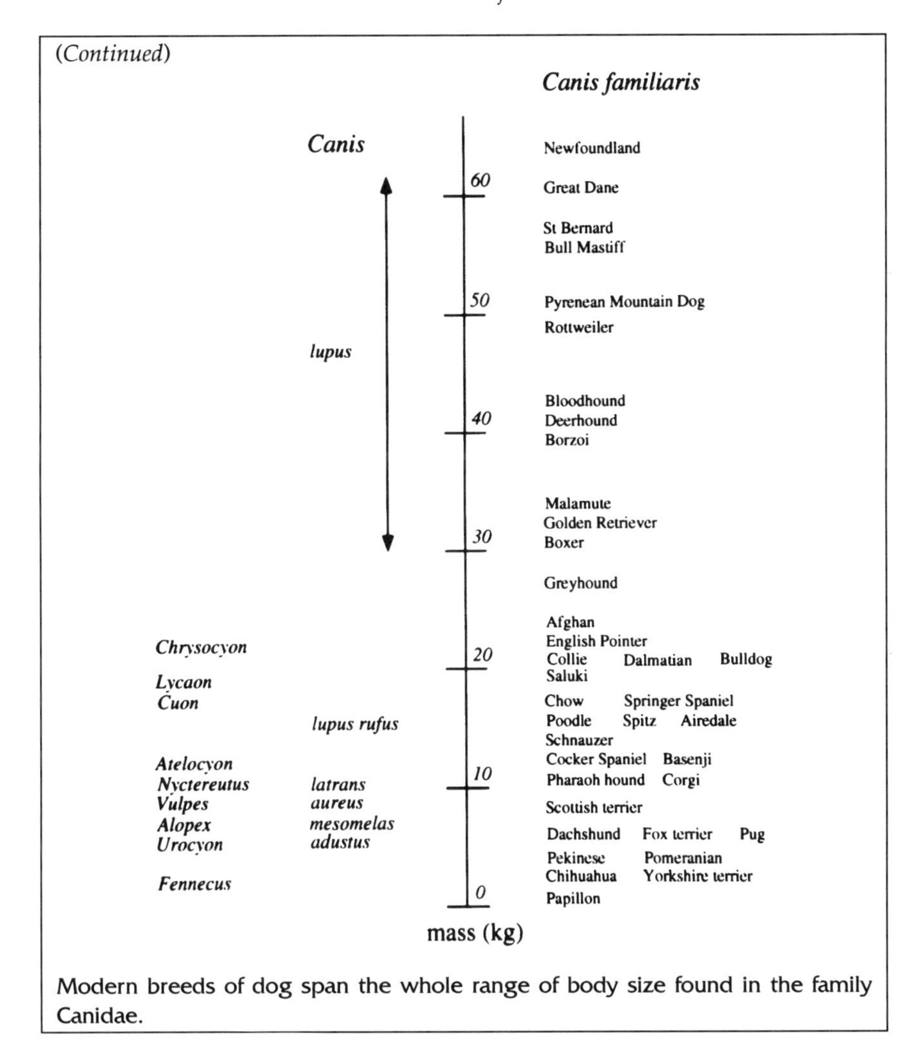


Modern breeds of dog span the whole range of body size found in the family Canidae.

limit of hundred-generation experiments that must span the lifetimes of a succession of experimenters, few of whom will see the outcome of the experiment they are conducting. The Corn-Oil Experiment that was begun in 1896 and is still being continued at the University of Illinois is probably unique. Corn seeds have been selected for higher and lower oil content, beginning from a base population with an oil content of about 5%. After 87 generations, seeds from the low line had substantially less than 1% oil, and selection was discontinued because of the technical difficulty of measuring such low oil contents. Oil content in the high line has increased more or less linearly to about 20%, and when last reported, after 90 generations of selection, the response showed no sign of slackening. After about 40 generations, however, the fluctuations in oil content from generation to generation became more marked.

The limit to selection that is often reached after 20 or 30 generations may then be transgressed if selection is continued for 100 generations. Further response is often much slower than the initial response, as would be expected if it is fueled largely by new mutation. The response, however, varies from experiment to experiment. In Goodale's mice, response ceased altogether after 35 generations and was never resumed; in Enfield's flour beetles, the lines were still responding slowly after 120 generations; in the Corn-Oil experiment, there has been little diminution of response in the upward line after 90 generations of selection. Continued selection may thus drive populations far beyond the limits of variation, actual or potential, of the base population. In almost all cases, however, the response, whatever its overall value, has become markedly more irregular during the later generations of selection. This suggests the appearance of new factors not anticipated by the elementary theory; this possibility is further considered in the next section.

**Selection for Yield in Crop Plants.** A more serious and extensive version of the Park Grass Experiment has been conducted over the last few thousand years by millions of farmers throughout the world and more systematically over the last hundred years by European and North American agronomists using modern theories of genetics and evolution. Crop yields have certainly increased many-fold since 1900, but agricultural practices have changed greatly over the same period, and it is not immediately clear whether improvements in yield should be credited to the selection of superior varieties or to greater inputs of fertilizer and pesticide. Fortunately, seed of commercial varieties has been kept in storage, so that long-obsolescent cultivars of corn or cotton can now be woken from their long sleep, and planted out in company with the modern varieties that have replaced them. These trials have shown that about half the increase in crop yield since about 1920 has been caused by selection. The difference between modern and obsolete varieties is especially marked when both are grown under modern agronomic conditions, so, as in the Park Grass Experiment, part of the improvement can be attributed to selection for increased responsiveness to high nutrient input.

**The Evolution of Dogs.** This point is evident to anyone who has pondered the variety of dogs to be seen every day in streets or fields. All modern breeds descend from the wolf, a large and rathervariable canid that struck up an acquaintance with human bands toward the end of the last glacial advance, some 10,000 years ago. They were long selected, no doubt largely unconciously, for several useful characteristics, particularly for hunting and guarding, and have more recently been deliberately selected and inbred to create the vast range of modern breeds. In body size alone, the variation among domestic dogs not only exceeds the variation among wolves, but actually exceeds the variation among all members of the family Canidae: wolves, jackals, wild dogs, and foxes. The extent of behavioral modification has been even more extraordinary. The capture of prey by wild canids may involve a whole sequence of behaviors: flushing; tracking by scent; detection by sight; pursuit, either individually or in a pack; herding; crippling; killing; and finally carrying the prey back to be eaten. Domestic dogs have been selected to excel at one of these tasks, while often suppressing all others. Spaniels will flush game from low undergrowth; bloodhounds follow a scent trail; gazehounds, such as salukis, have exceptional visual acuity, whereas pointers will actually detect prey and then freeze without proceeding to pursue it; foxhounds and beagles hunt in packs; sheepdogs will herd flocks of sheep (and almost anything else) without attacking them; bulldogs were bred to grip their prey and hang on; mastiffs are fighting and killing dogs; retrievers will fetch dead or wounded prey without (in theory) damaging it; every component of the strategy of wild canids for capturing prey is represented by a specialist breed. Moreover, different breeds tackle different prey: deerhounds and otterhounds for large errant animals pursued in the open, the whole tribe of terriers for small game hiding in burrows and crevices. The exuberant diversity of dogs is a striking testimonial to the power of selection to direct adaptive change far beyond the limits of the original population within a few hundred generations.

## 63. *New and unexpected constraints may set a limit to continued selection.*

**Variation Within Selection Lines.** Crop plants and dogs represent huge experiments in which the diversity of the material seems to nullify any particular limit to modification. This is not necessarily the case in simpler situations when a single character is selected in a single lineage or restricted set of lineages. In the Illinois Corn-Oil Experiment, no limit to upward selection is yet apparent. Selection for bristle number in *Drosophila*, on the other hand, reached a plateau after about 35 generations of selection in the experiment by George Clayton and Alan Robertson described

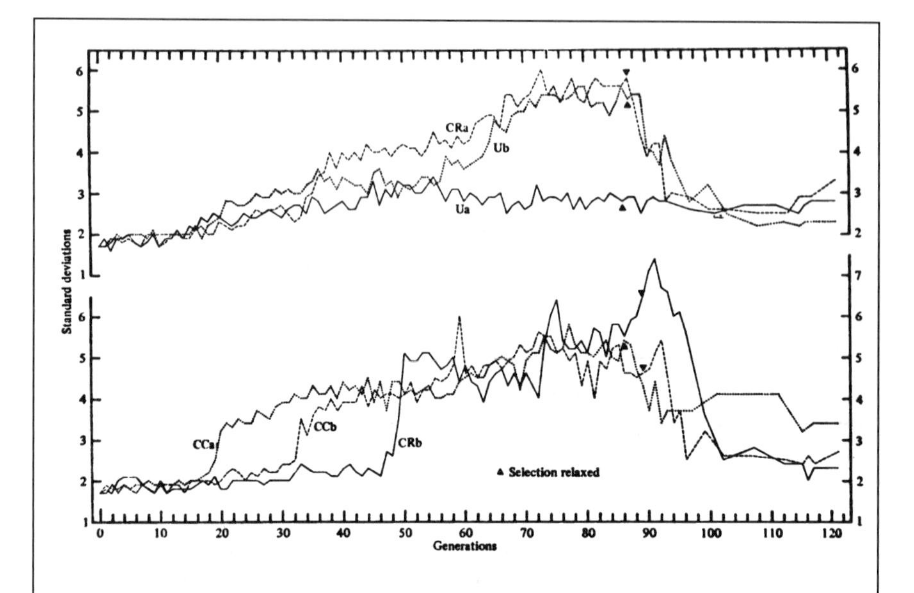


The quantity of additive or selectable genetic variance remaining after a few dozen generations of selection can be estimated by breeding trials or reverse or relaxed selection.

1. Experiments in which genetic variance was nearly exhausted: Brown and Bell (1961), Roberts (1966b), Scowcroft (1968).
2. Experiments in which a substantially diminished quantity of genetic variance remained: Clayton and Robertson (1955), Yamada et al. (1958), Latter (1966).
3. Experiments in which genetic variance was not substantially diminished, or even increased: Lerner and Dempster (1951). Falconer (1953), Reeve and Robertson (1953), Dickerson (1955), Roberts (1966a), Gall and Kyle (1968), Wilson et al. (1971), Eisen (1972).

The accumulation of recessive lethals in selection lines has been documented by Yoo (1980b), Garcia-Dorado and Lopez-Fanjul (1983), and, especially, Frankham (1982). Frankham suggests that the gene responsible for the response observed by Clayton and Robertson (1955) was the mutation *bobbed* at the ribosomal-RNA locus on the X chromosome. In some cases (e.g., Yoo 1980c), however, genes of small effect are involved; it is not definitely ascertained whether these were present at low frequency in the base population, or arose by mutation during the course of the experiment. The evolution of reduced viability or fecundity in selection lines has been observed on many occasions, e.g., Mather and Harrison (1949), Fowler and Edwards (1960), Latter and Robertson (1962), Dawson (1965), Gall (1971), etc., in addition to the studies cited earlier; see Sec. 89. *(Continues)*

(*Continued*)

The effect of stabilizing selection on genetic and environmental variance has been investigated by Prout (1962), Scharloo (1964), Scharloo Hoogmoed and Ter Kuile (1967) and Gibson and Bradley (1974); the long-term experiment on pupal weight in *Tribolium* was reported by Kaufman et al. (1977).

The figure shows the increase of phenotypic variance in Yoo's selection lines (Yoo 1980b). Variation increased under directional selection (until about generation 90), then decreased when selection was relaxed (generation 90 onward, after arrowheads on the graph). In some lines there was as rather gradual increase in variation, whereas in others there were abrupt increases to a new level that was afterwards maintained until selection was relaxed. These abrupt increases in variation mark the passage of recessive lethal genes with opposed effects on fitness and character value.

in Sec. 49. An enormous shift in character value, amounting to some 20 phenotypic standard deviations in the upward selection lines, had been achieved by then, but further selection was ineffective. The obvious interpretation, suggested by the simple theory that I have already described, is that the decelerated rise of bristle number towards a plateau represents the gradual exposure of genetic variation through recombination and its fixation through selection, until no further genetic variation is available in the short term and the response to selection ceases. Nevertheless, there was still plenty of phenotypic variation present in the population. Where did it come from? It did not seem to represent any of the sources of variation that were identified in the base population: it was not developmental variation, because asymmetry had not increased, and it was not environmental variation, because conditions of culture were more or less the same throughout the experiment. It did not seem to represent selectable genetic variation, because no further response to upward selection could be obtained; but there was an immediate and rather rapid response to back-selection, when flies with fewer bristles were selected in lines that had previously been under upward selection. In several lines, there seemed to be a gradual reduction in variation, as expected, for about 20 generations, and then a sudden increase in variation that after another dozen generations could be selected downwards but not upwards. This is quite a common observation in selection experiments that last longer than 20 or 30 generations; in some cases the additive genetic variance seems to be exhausted, or has at least been markedly reduced, but they are outnumbered by experiments in which genetic variance remains high or even increases. These experiments show that the simple theory of sorting is inadequate to describe the evolution of populations after a few dozen generations. It breaks down at this time-scale because it does not take into account the generation through mutation of a quantity of new variation and of qualitatively new kinds of variants.

**The Evolution of Constraints.** In most organisms, the result of Clayton and Robertson's experiment would simply have remained deeply mysterious; fortunately, the sophisticated chromosomal genetic techniques that had been worked out for *Drosophila* made it possible to find out what was going on. A mutation had arisen that increased bristle number in heterozygotes, but was lethal when homozygous. Artificial selection favored the mutant form, because the flies chosen by the experimenters were likely to be heterozygotes; but when these mated, a quarter of their progeny died, so that natural selection favored the normal gene. There was no further response to artificial selection because it was opposed by countervailing natural selection: any advance created by the choice of flies with more bristles by the experimenters was annulled because, among this selected sample, flies with fewer bristles (normal homozygotes) produced more progeny than flies with more bristles (mutant heterozygotes). No survey could have detected this limit to selection in the base population. It arose during the course of selection, as an unexpected consequence of the selection regime itself.

Similar recessive lethal mutations have appeared in several other cases, and seem to be a routine feature of long-term selection experiments. It is possible that mutation is biased, so that mutations that enhance expression of the selected character necessarily reduce fitness, for physiological reasons that differ from character to character. There is no reason, however, to suppose that this must be so. Imagine a population undergoing directional artificial selection, in which mutations of all kinds are constantly arising. Some will enhance the expression of the character and also increase viability or fecundity; these will rapidly be fixed. Others will hamper expression of the character, while at the same time reducing fitness; these will rapidly be lost, or will rather fail to spread. Mutations that have opposed effects on the character being selected and on fitness will be selected more slowly, and will therefore tend to accumulate in selection lines. In diploid organisms such as *Drosophila*, such genes may even be stably maintained in the population if their effect on viability is recessive but their effect on the selected character is partly dominant, because the heterozygote then has a greater fitness than either homozygote. The reason that experimental populations tend towards a limit at which response ceases is, then, not that genetic variance is exhausted, but that natural selection comes to act antagonistically to artificial selection. This argument is not limited to lethals, or other genes of large effect. There are many instances of lines that cease to respond to selection but are not segregating for recessive lethals; these lines have presumably accumulated many mutations of lesser effect, all tending to reduce viability while enhancing the character under selection. The very general tendency for continued artificial selection to depress viability or fecundity is discussed further in Sec. 89.

**The Limits to Stabilizing Selection.** Artificial stabilizing selection, practised by removing extreme individuals, should tend to reduce phenotypic and genetic variance and will reach a limit when the genetic variance has been exhausted. The phenotypic variance should then be equal to the environmental variance of the base population. Franklin Enfield's group at the University of Minnesota investigated the limits to stabilizing selection by selecting pupae closest to the median weight within half-sib families of *Tribolium*. The experiment continued for 95 generations and is the only long-term experiment on stabilizing selection that I have been able to find. Genetic variance did decrease, but not dramatically, and the heritability of the character fell only slightly, from about 0.25 to about 0.2. As with directional selection, this may have been caused by the accumulation of genes with antagonistic effects on pupal weight and fitness, although there is no direct evidence for this. It should be noted that stabilizing selection does not cause the general reduction in fitness associated with directional selection: viability and fecundity were both greater in the selection lines than in the control (randomly selected) lines. However, another complication in such experiments is that the environmental variance may itself decline, because selection will favor genotypes that are insensitive to fluctuations in conditions of culture. The evolution of the environmental variance is considered in Part 3.B.

**The Amelioration of Antagonistic Effects.** The evolution of reduced fitness in lines selected for up to about 100 generations is by now well established. Its most characteristic manifestation is the tendency to evolve back towards the character state of the base population when selection is relaxed. In one experiment, however, the opposite effect was observed. This happened in lines of *Drosophila* selected for positive geotaxis that were then selected in the reverse direction, for negative geotaxis, for several generations before selection was relaxed. The response to relaxed selection was increased positive geotaxis, suggesting that the extreme behavior evolved through artificial selection was now favored by natural selection. The observation is unique, perhaps because the experiment is unique: Jeffry Ricker and Jerry Hirsch of the University of Illinois had been selecting these lines for 25 years, or about 500 generations. It is possible, then, that the harmful side-effects of mutations enhancing a selected character may themselves be modified and eventually ameliorated by selection extending over hundreds of generations.

## 64. *The cumulative response to continued selection may generally be nonlinear.*

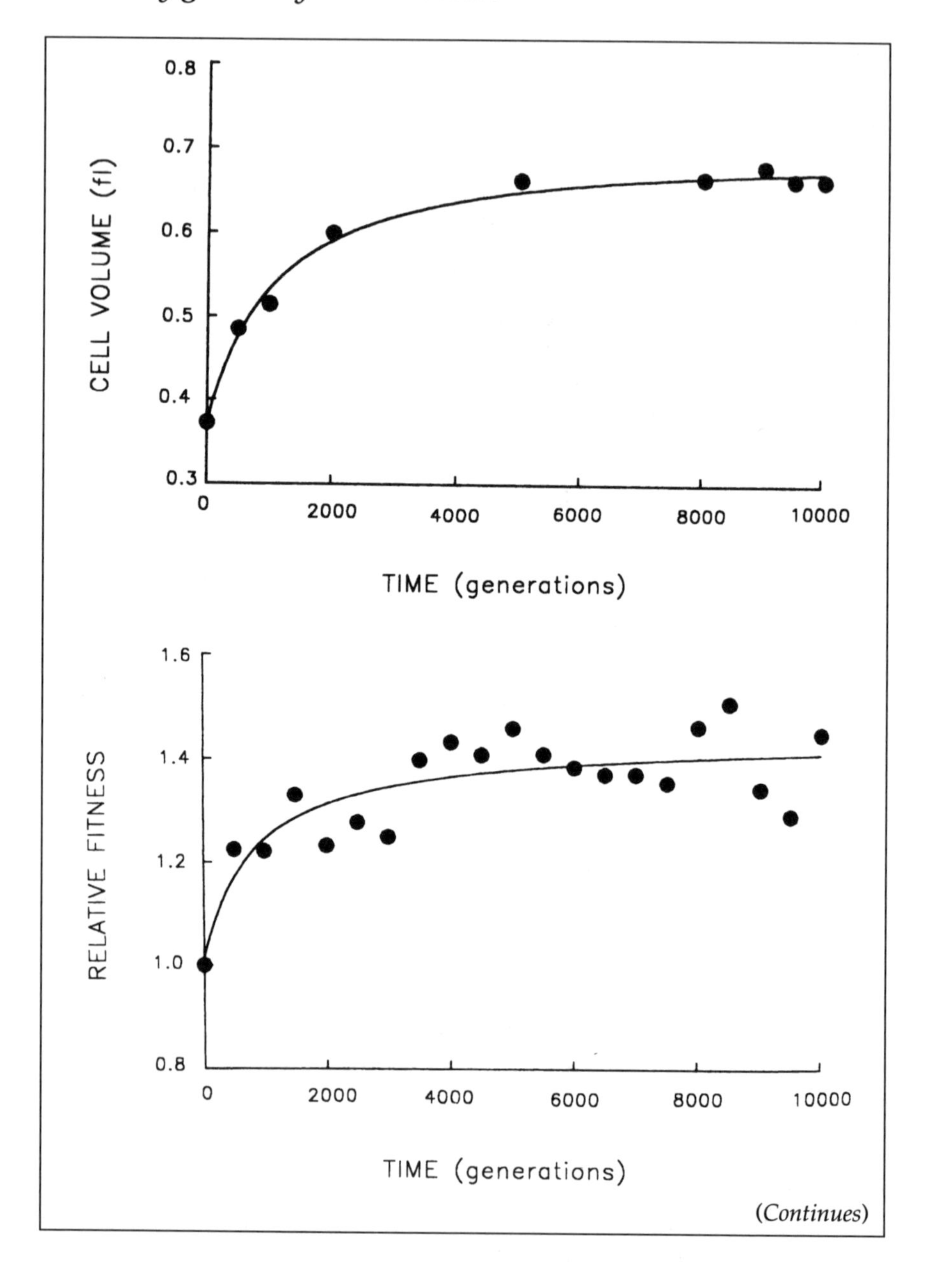


(Continues)

(*Continued*)

These figures show the response of cell size and fitness to natural selection in populations of *E. coli* maintained in glucose-limited chemostats for 10,000 generations (from Lenski & Travisano 1994). The smooth curve fitted here presumably reflects the underlying discontinuous change caused by a parade of substitutions at progressively longer intervals and, indeed, the authors interpret it in this way; see also Dykhuizen & Hartl (1983).

The response to artificial selection may continue for 100 generations or more, either because rare beneficial alleles present in the base population continue to increase in frequency or because new variation arises by mutation. Nevertheless, the rate of response almost invariably declines through time, either because genetic variation is eroded or, more probably, because of opposed natural selection. Such experiments, however, do not take us very far into the realm where substantial quantities of new genetic variation are being created and exposed to selection and, as I have pointed out, it is unlikely that they will ever be prolonged much beyond the hundred-generation boundary. To investigate the pattern of change over longer time-periods, we must turn to experiments using natural selection in microbial systems.

**Ten Thousand Generations of *E. coli*.** The simplest realization of long-continued change would be a long, regular succession of substitutions each involving genes of slight and more or less equal effect, driving a roughly linear increase of adaptedness through time, without any definite limit. This has sometimes been observed, as in the Corn-Oil Experiment, but in general it does not seem to be a good description of natural selection on time-scales of between 100 and 1000 generations. In chemostat experiments with bacteria, the interval between successive substitutions typically lengthens with time or, equivalently, the rate of increase in fitness declines through time. For example, Daniel Dykhuizen and Daniel Hartl, then working at Purdue University, grew *E. coli* in glucose-limited chemostats for 500 h (equivalent to 200 generations for the more rapidly cycling lines). Such populations evolve primarily by increasing the rate of glucose assimilation. Most of the observed response, however, occurred in the first 200 h, with the rate of assimilation changing much more slowly thereafter.

In many cases, this pattern might be caused by the rapid sorting of variants that accumulated while the strains were being stored prior to the experiment. Richard Lenski of Michigan State and his colleagues have studied the evolution of fitness in bacterial populations founded by newly isolated single clones and then maintained by serial transfer for thousands of generations. In these experiments, fitness is measured as competitive ability, the change in frequency of an evolved strain mixed with some test strain, usually the founder. Fitness increased steadily for 2000 generations, at which point the evolved population was nearly 1.4 times as fit as the founder. This advance must have been caused by selection acting on the small quantity of genetic variance in fitness introduced in every generation by novel mutation. From the fundamental theorem that the rate of increase of fitness is equal to the genetic variance of fitness, we can calculate the fitness variance that is required to support the observed change; this turns out to be $2.7 \times 10^{-4}$. This can be compared with estimates obtained by measuring the fitness of replicate cells isolated from the same selection line; the median value of these estimates over all selection lines was $2.5 \times 10^{-4}$. Therefore, it seems likely that most, if not all, of the observed changes in fitness are caused by the selection of novel mutational variation. This experiment has now been extended to 10,000 generations. Both fitness and cell volume followed the same sort of evolutionary trajectory: rapid change during the first 2000 generations of selection, with little further change during the last 5000 generations.

The simplest interpretation of this result is that continued selection follows a rule analogous to that of short-term sorting. In the short term, response decays and eventually ceases as the variation present in the base population is used up. In the longer term, there may be a finite, and rather small, set of *accessible* mutational changes (Sec. 54) that increase adaptedness. Those with larger effects are substituted first; those with lesser effects are substituted more and more slowly and, as their supply fails, at longer and longer intervals. This hypothesis is especially attractive because the large populations and long periods of time involved in such experiments imply that almost all single point mutations appear during the course of the experiments. It is almost certainly wrong, however, because it implies that replicate populations, although they may diverge transiently, will eventually converge on the same phenotype and the same genotype, which as we shall see is not the case.

## 65. *The variance among replicate selection lines increases with time.*

If a large clonal or inbred population is exposed to a new environment or deliberately selected for a certain character, its response is predictable, insofar as fitness will increase under natural selection and the mean character value will shift in the desired direction under artificial selection. If two replicate selection lines are set up at the same time from the same clone,

Lines derived from different populations responded at different rates to selection for resistance to DDT in *Aedes* (Inwang et al. 1967) and for sternopleural bristle number in *Drosophila* (Lopez-Fanjul & Hill 1973). In both cases, however, the same genes seemed to be responding to selection. The response to selection for ethanol tolerance in *Drosophila* (David et al. 1977) also varied among populations. Cohan (1984) selected for expression of an interrupted wing vein in *Drosophila* and found that although there were substantial differences between lines, these seemed to be no greater between lines from different populations than between lines from the same population. The initial difference between the populations, however, was rather small.

These graphs show the results of the experiment in which larger *Chlamydomonas* were selected by filtration (Bell 1996b) (see Sec. 65). Variance increased among cells within lines, showing that genetic variation tended to accumulate in the initially uniform cultures. The rate of increase in genetic variance was in this case

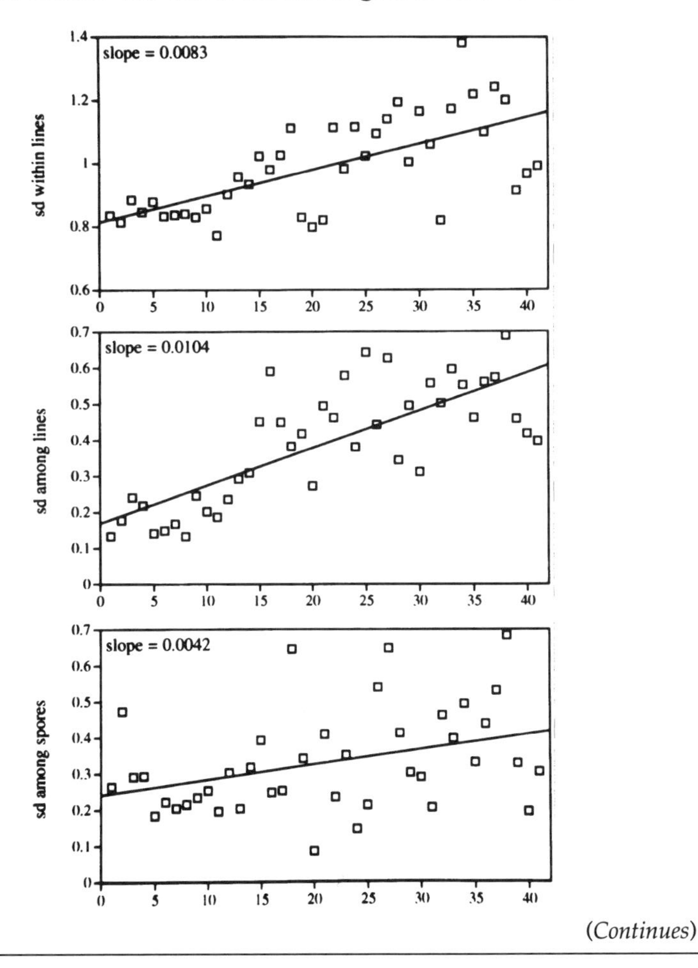


*(Continues)*

about $2 \times 10^{-3}$ Var(spores within lines) per generation, where Var(spores within lines) is the environmental variance expressed by the initial cultures. The cells were not selected every generation, but only every growth cycle of about 13 generations. The variance among replicate lines founded by the same spore increased through time, as these replicate lines responded to selection at different rates; the rate of increase, expressed in terms of the initial variance among lines, was about $23 \times 10^{-3}$ Var(lines) per generation, where Var(lines) is here the initial variance among lines. Independently of this divergence, the sets of replicate lines established from different spores also tended to diverge through time, showing some tendency for the rate of reponse to selection to vary among base populations. The rate of divergence among sets of replicate lines was about $3.6 \times 10^{-3}$ Var(sets of lines) per generation, where the initial variance among sets of lines is partly genetic and partly environmental. Subsequent genetic analysis showed that offspring from crosses between lines tended to be intermediate between their parents; this implies that genetic variance for size was largely additive and that given time the lines might have converged on a similar phenotype.

These figures show the continued divergence of *E. coli* populations propagated for 10,000 generations by serial transfer in glucose-limited medium (Lenski & Travisano 1994). The variance among replicate lines, founded from the same clone, of both cell size and fitness increased through the experiment. It seems that diversification is asymptotic, with variance among lines increasing more rapidly in early generations; the parallel with adaptive radiations is suggestive.

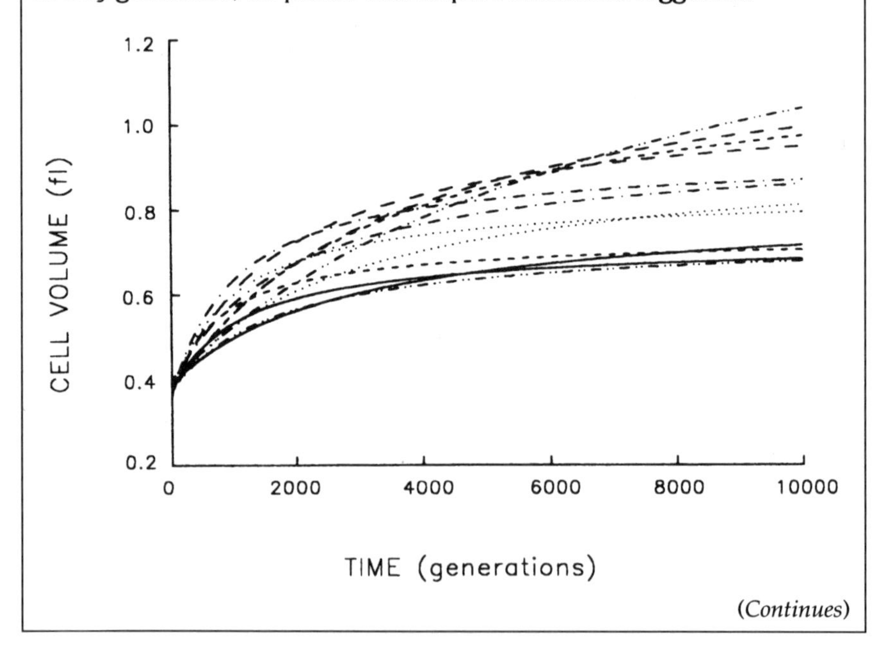


(*Continues*)

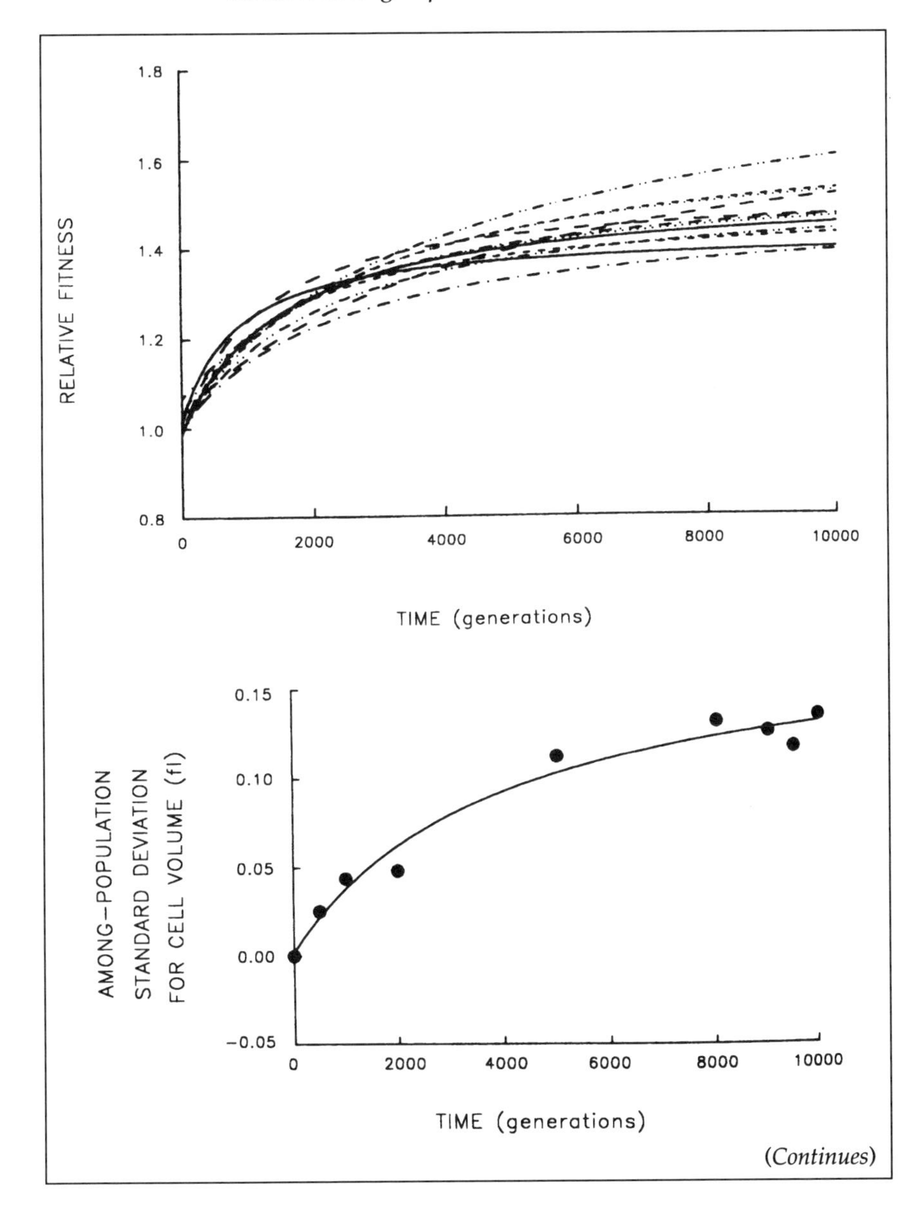
1.8
1.6
1.4
1.2
1.0
0.8
RELATIVE FITNESS
0
2000
4000
6000
8000
10000
TIME (generations)
0.15
0.10
0.05
0.00
−0.05
AMONG-POPULATION
STANDARD DEVIATION
FOR CELL VOLUME (fl)
0
2000
4000
6000
8000
10000
TIME (generations)

*(Continues)*

(*Continued*)

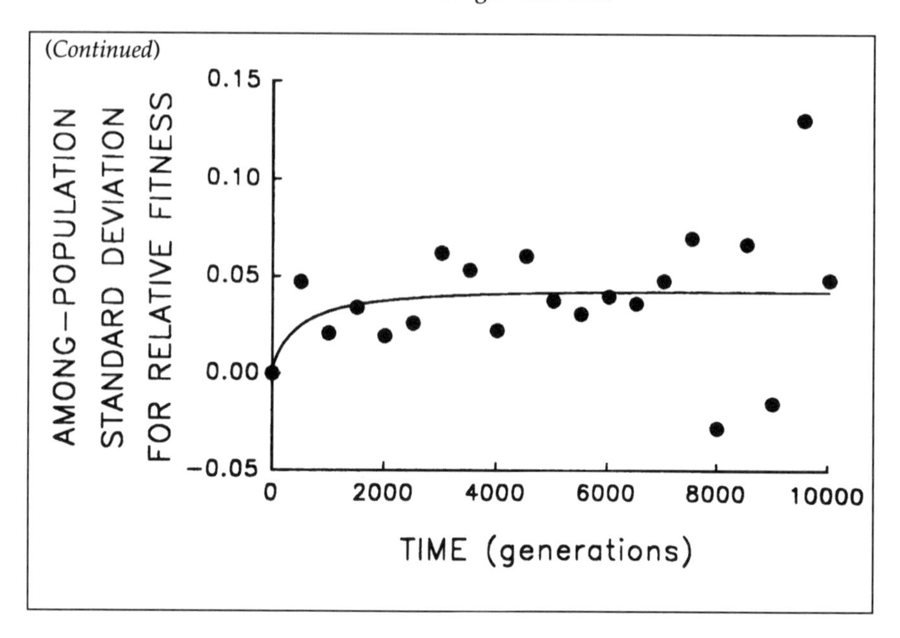


then both will respond in the same direction, but they may not respond at the same rate. Because evolutionary change is cumulative, with each generation modifying whatever modifications have occurred previously, the differences among replicate lines will also tend to cumulate through time. The variance of fitness or character value among replicate lines will tend to increase; or, equivalently, the covariance of the means of lines in successive generations will increase. This applies regardless of the rate of selection and will hold for strong selection, for weak selection, or for no selection; genetic drift alone will cause the variance among lines to increase through time. The null model for a population on which no selection is acting is not a stochastic fluctuation around a fixed mean, but rather a random walk in which the mean itself changes stochastically through time.

Two principles that I have described may seem to be contradictory: a principle of cumulative change requires that selection lines starting from identical base populations will diverge through time, whereas a principle of phenotypic convergence requires that selection lines starting from different base populations will become more similar through time. Both principles are required for a complete description of the history of replicate selection lines. They will at first diverge, because they will evolve at different rates; but all evolve towards the same endpoint, and after a certain time they will begin to converge, until at last all are identical. The variance among lines, therefore, is initially zero, will increase for some time, but will then decrease, until it is again zero. The rate of this process depends on the rate of selection: if selection is strong, phenotypic convergence will be rather rapid whereas for strictly neutral genes there will be no tendency for convergence, and lines will continue to diverge indefinitely. This is the classical account of continued selection.

**Convergence of Lines with Different Origins.** Similar sources of selection often lead to different kinds of adaptation; for example, resistance to insecticides or to heavy metals often involves quite different physiological mechanisms. There is then convergence at the level of fitness, but not at the level of physiology. Similar sources of selection need not be identical, however, and it is possible that different kinds of adaptation are required in environments that although contaminated with the same toxin differ in water stress, or in the abundance of competitors, or in some other way. We must therefore turn to laboratory experiments to investigate the convergence of different lineages selected identically. Lines that are founded from different strains of the same species, typically isolated from distant localities, usually evolve at different rates. Whether or not this implies that they are evolving in different ways can be investigated by crossing the base populations and selecting the hybrids; if the base populations had the potential to evolve differently, then the hybrids, combining both sources of variation, should respond more rapidly or more effectively to selection. Selection on bristle number in *Drosophila* has shown that although two strains may respond to selection at different rates, the population formed by crossing them does not respond more rapidly than its parents. It seems likely, therefore, that the two base populations bore the same genes affecting bristle number, but at different frequencies. Such lines would converge on the same state if selection were continued for sufficiently long. On the other hand, when two strains of *Drosophila* were selected for greater tolerance of ethanol, the lines derived from one strain repeatedly evolved tolerance through elevated alcohol dehydrogenase activity, whereas the second strain evolved tolerance through some other unidentified mechanism. Whether such lines would ever converge is uncertain. Thus, identical selection of different strains of the same species may lead to convergence at three levels. Firstly, lines may respond in the same direction, for example, evolving larger bristle number or greater ethanol tolerance. This is almost invariably the case. Secondly, lines may respond at the same rate. This is rare; most experiments, as expected, show that selection lines initially tend to diverge. Thirdly, lines may respond in the same way, all evolving the same kind of adaptation and thus converging after a prolonged period of selection. It is not surprising that this should often happen; the more interesting cases are those in which the same process of selection led to different outcomes, showing that initial differences in the genetic composition of the population predicate different evolutionary pathways.

**Divergence of Lines with Identical Origins.** Experiments in which the same source of selection leads to different outcomes suggest that adaptation is not always accessible; if it were, populations would converge on the same state, regardless of their initial state. The same principle should apply to experiments in which the same selection is applied repeatedly to the same base population, because novel mutations occurring independently in the selection lines will create differences among them. In the *Chlamydomonas* filtration experiment (Sec. 65), for example, the variance among replicate selection lines increased over the 40 cycles of selection. The objection could be raised, however, that none of the experiments that demonstrate a variable outcome to similar processes of selection have been continued for sufficiently long to exclude the possibility that all lines will eventually converge on the same state. This objection has now been met by the results of Lenski's 10,000 generations of selection on *E. coli* in glucose-limited chemostats. The variance of fitness among lines continued to increase (though at a decreasing rate) during the experiment, with no hint of any eventual convergence. This result suggests that the classical model of continued selection is inadequate, and that new historical principles must be invoked to explain the course of evolution in the longer term. These principles are the subject of the next two sections.

## 66. *The genetic basis of adaptation may differ among replicate selection lines.*

**Historicity and Contingent Change.** Replicate selection lines nearly always respond in the same direction, in the sense that fitness increases through natural selection, or the mean of a character shifts appropriately through artificial selection. Nevertheless, they may repond at different rates, and if they continue to show different levels of fitness, as seems to be the case in Lenski's experiments, we may infer that different genotypes have evolved in different lines, despite the fact that the same source of selection was acting in each. This conclusion is even more secure when it can be shown that adaptation involves different physiological, morphological, or behavioral changes. The converse is not true: phenotypic convergence, for fitness or for character state, need not imply genotypic identity. The adaptedness of replicate lines after a period of selection in a novel environment may be substantially the same, in the sense that all have undergone similar phenotypic changes leading to similar mean fitness, but this does not mean that the same genotype has prevailed in every line. It is conceivable that a different genotype becomes fixed in every line, but that all these genotypes express nearly the same phenotype. The reason for thinking that this may often be the case involves three principles, or generalizations, described in previous sections. Firstly, the population may at any one time represent only a small sample of possible combinations of genes. Secondly, the effects of a gene may depend on the state of other loci elsewhere in the genome, so that any combination of genes has effects that cannot be predicted precisely from knowing the independent effects of each gene. Finally, evolution is a process of cumulation in which each episode of selection modifies the prior composition of the population. Replicate selection lines may initially diverge because a different beneficial mutation arises and becomes fixed in each. To the extent that the effects of a gene depend on the state of other loci, the properties of new mutations arising subsequently in each line will depend on which mutation happens to have been substituted first in that line. The initial divergence of lines will thereby predispose selection in each to favor different mutations in the next period of time. The differences among lines will continue to cumulate, with each line evolving high fitness, or a similar character state, but with a unique genetic basis. This is the principle of *contingency*: lines diverge genetically because of the contingent nature of cumulative change under continued selection.

This is an important principle, because it colors our whole approach to evolution. If evolution is a regular and predictable process, then the history of life on Earth was more or less preordained from the beginning, and mankind, or something very like mankind, is the inevitable outcome of three billion years of continued selection. If evolution is a contingent process, Earth history is only one of an uncountable number of possible outcomes, and we are the unrepeatable product of a unique experiment. What applies to us applies, of course, with equal force to the crab under the rock or the rose in the hedge. Which of these two extreme views is more nearly correct cannot be established simply by documenting the course of evolution to date; the question can be decided only by comparing the history of life on different planets or, more conveniently, by analyzing the genetic basis of adaptation in replicate selection lines.

Contingency is difficult to grasp, because it is at odds with our normal perception that natural laws produce orderly and predictable change. I therefore suggest a couple of analogies, which although imperfect, may serve to clarify what I mean.

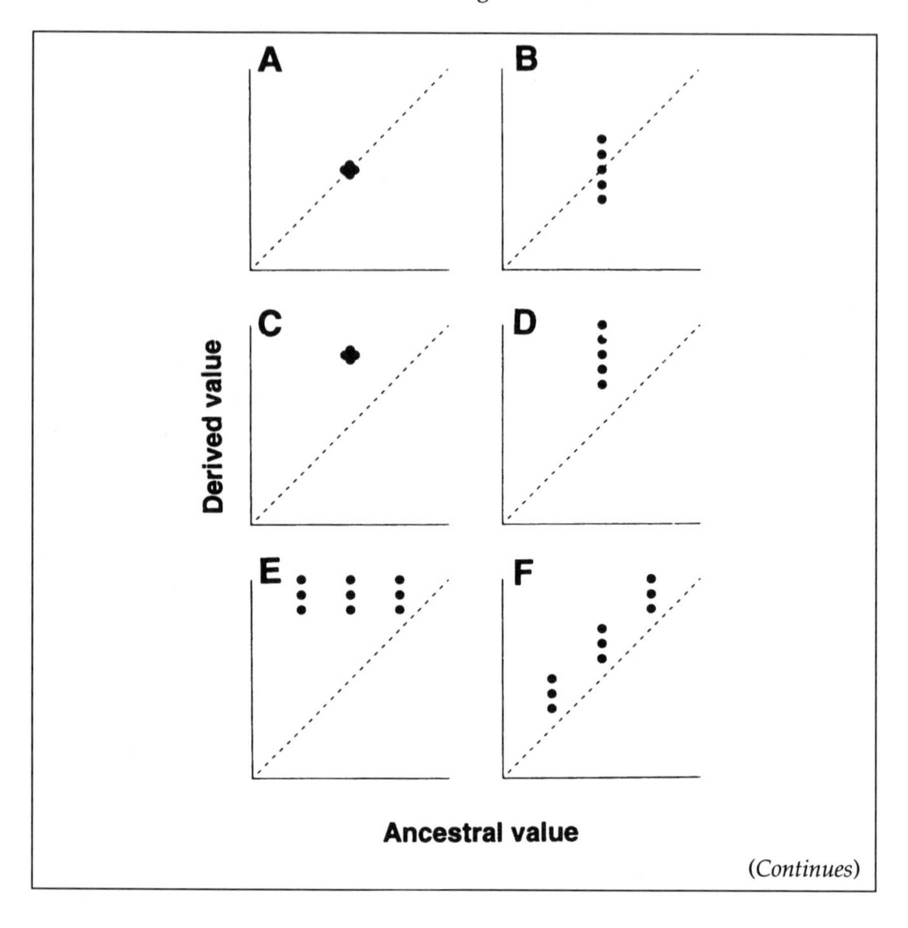


(*Continues*)

**A Word Game.** The first analogy is a word game similar to that involving stepwise changes in the sequence of letters in a word that I used to explain the resolution of the hemoglobin paradox. In this case we begin with a given word and assume that any change that reduces the length of the word, by deleting a single letter, will be favored, as long as the shorter sequence is itself an English word. This is analogous, if you like, to selection for shorter, more rapidly replicating sequences in Q$\beta$ populations living in culture tubes. In some cases, it will be found that the sequence of changes is strictly preordained, and the final result completely predictable. YEAST evolves in this way towards the shortest possible sequence of a single letter (YEAST–EAST–EAT–AT–A). Other starting points yield a number of lineages, but eventually converge on the same sequence. CREAM

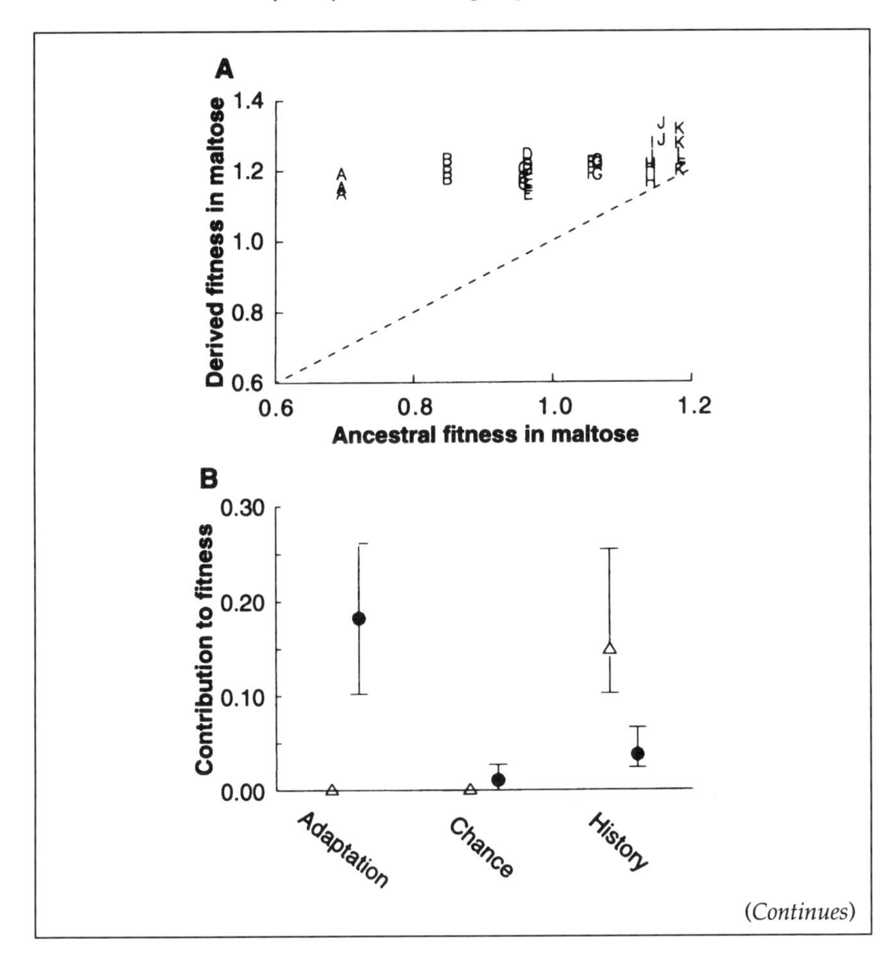


(*Continues*)

is an example, in which every divergent pathway leads eventually to the minimal sequence: CREAM–CRAM–RAM–AM–A, or CREAM–CRAM–CAM–AM–A, or CREAM–REAM–RAM–AM–A. Other words, however, diverge to different end- points, and if letters were deleted at random the outcome of continued change would not be predictable. SOLID is a word of this sort, because the two lineages SOLID–SOLD–OLD and SOLID–SLID–LID (–ID–I, if "id" is permissible) never converge. This is an example of contingent change.

**Evolution Is a River, Not a Road.** The second analogy contrasts the course of a river with that of a road. The initial course of a river, close

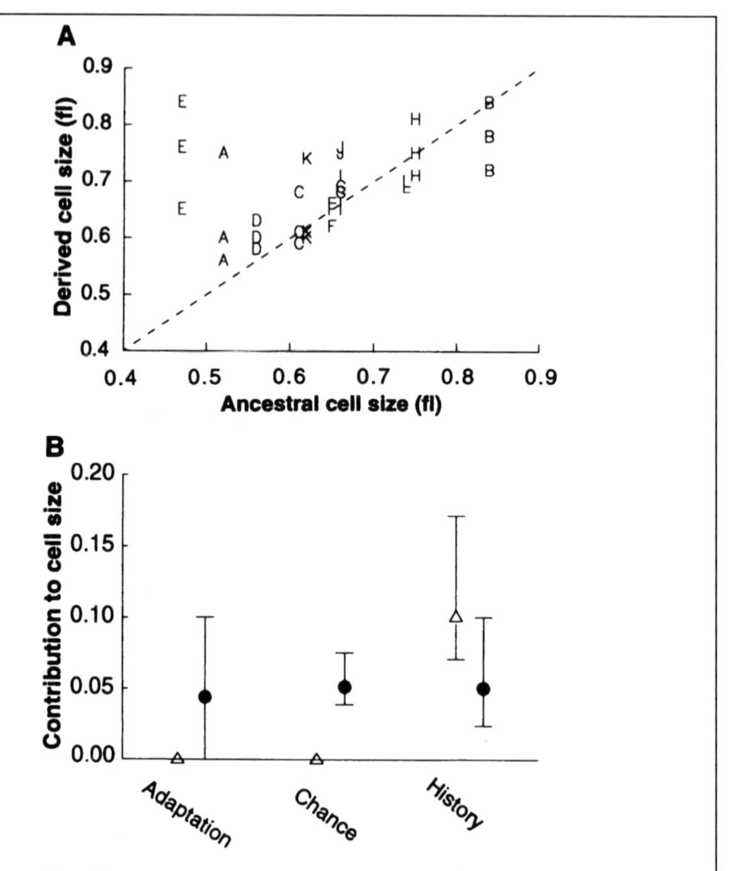


The long-term *E. coli* experiments run in Richard Lenski's laboratory offer an opportunity to distinguish the contributions of adaptation, chance and history to evolutionary change; these figures are from Travisano et al. (1995). The upper figure is the conceptual basis of their interpretation:

A. With no initial variation, there will be no evolutionary change.

B. Divergence with no change in the mean represents the effect of chance alone. This is measured by the variance among replicate selection lines.

C. A shift in the mean, with no divergence, represents the effect of adaptation alone. This is equivalent to the main effect of treatment.

D. The joint effect of chance and adaptation.

E. An initial difference among populations is eliminated by adaptation, with or without divergence caused by chance.

F. The initial historical difference is maintained, with effects of adaptation and chance superimposed. The historical effect is measured by the variance among lines descending from different base populations.

(*Continues*)

The lower diagrams show the outcome of selection for 1000 generations in maltose-limited chemostats, using lines that had previously been selected for 2000 generations in glucose-limited chemostats. These lines differed in their ability to utilize maltose. The upper diagram of each pair shows the ancestral and derived populations, as in the schematic given. The lower pair show estimates of the effects as variance components. Fitness in maltose, measured through the change in frequency of the derived line when competing against its ancestor, follows pattern E: the increase in fitness is caused by adaptation, with chance and history making little if any contribution. Cell size more nearly resembles pattern F: although there may be some consistent change, the role of adaptation is slight, whereas the effects of chance and history are substantial. This is as expected, with fitness being strongly convergent whereas the characters underlying fitness are influenced by the initial and subsequent states of the populations.

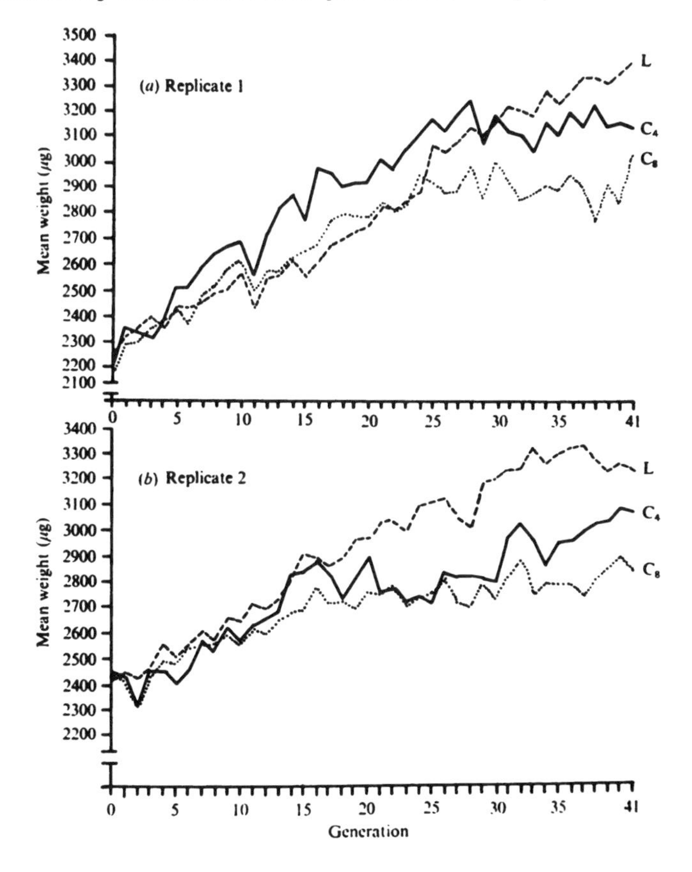


This figure shows the results of an experiment by Katz and Enfield (1977), who selected upward for pupa weight in *Tribolium*. It contrasts the response to selection over 40 generations of a single large population (L), propagated by 24 males and

*(Continues)*

(*Continued*)

48 females in each generation, with that of a set of six small populations, each comprising 4 males and 8 females. At intervals of four ($C_4$) or eight ($C_8$) generations, the best two lines were selected and crossed, the set of six lines being reconstituted from their progeny. In this case, population subdivision seems to retard progress under selection. A similar experiment, with a similar outcome, was reported by Goodwill (1974). Madalena and Robertson (1975) and Rathie and Nicholas (1980) tried similar but more complex schemes when selecting for bristle number in *Drosophila*, without finding any clear-cut benefits in subdividing the population. It is possible that crossing the selected populations breaks up any epistatic combinations of genes, thereby frustrating the process that the experiments are designed to exploit. Experiments in which the more responsive lines are expanded, rather than crossed, have been performed on *Tribolium* by Wade's group in Chicago; they are described in Sec. 148. Katz and Young (1975) and Banks (1986) selected for body weight in *Drosophila* while allowing some restricted movement of individuals among lines; in these experiments, subdivided populations showed an appreciably greater response to selection. The subject has been reviewed by Barker (1988). A similar procedure, often used by animal and plant breeders, is *reciprocal recurrent selection*. Two different source populations are selected and crossed at intervals, the best *parents* being chosen to propagate the lines. The object is to build up superior epistatic combinations of genes. There are experimental studies of the efficacy of reciprocal recurrent selection in *Drosophila* by Bell et al. (1952), Rasmuson (1956), and Kojima and Kelleher (1963); in *Tribolium* by Bell and Moore (1972); and in fowl by Saadeh et al. (1968) and Calhoon and Bohren (1974).

to its source, is strongly influenced by very slight variations in relief; a minor bump or slope predisposes its course in one direction or another, and each slight shift of course shuts off a host of future possibilities. A difference of a few feet one way or another near the source may mean that hundreds of miles downstream it flows on one side of a mountain range rather than the other. The flow of a river is therefore an historical process, in which events at any point influence its whole future course. Two rivers that arise in the same meadow may yet follow very different courses, diverging further and further, until they eventually debouch into different oceans. The patterns of relief that cause their divergence may be slight, but their effects are cumulative and irreversible. By contrast, a road follows a course that has been engineered to be optimal. If by chance it is thrown a little out of its way—to avoid a church, for example, or to bypass a small lake—then having avoided the obstacle it will bend back again towards the best path. The course of a road is not an historical process, because deviations from this course have no effect on its future direction or its eventual destination. It is rather goal-directed , so that roads built by different people at different times from similar starting points will follow similar courses; the Roman road and the modern highway take nearly the same route. The principle of contingency amounts to saying that evolution is a river, not a road.

**Selection Among Selection Lines.** These highly abstract ideas have important practical consequences. Breeds of livestock are maintained as a series of herds, with a certain degree of migration, physical or genetic, among the herds. How should the herds be managed so as to maximize response to selection for some desirable character? I have until now assumed mass selection in a single homogeneous population, a single large random-mating herd, so to speak. Suppose that epistasis produces a rather hilly adaptive landscape (Sec. 54), however, through which any given selection line takes a unique and essentially unpredictable path. A single selection line, however large, follows a single path that may lead to a peak of only modest elevation. A set of selection lines, comprising the same total number of individuals, will set off in different directions and climb different peaks. Perhaps these peaks will be on average lower than that climbed by the single line because of the smaller number of individuals, and thus the smaller quantity of variation, in each line. The tallest peak climbed by one of these lines, however, is likely to surpass the peak climbed by the single large line. The limit of response for a structured population, consisting of several or many herds, may thus exceed the limit for a single large herd under mass selection.

This argument suggests two ways in which population structure might be manipulated so as to take advantage of multiple-peak epistasis. The first is to select at intervals among lines that have been propagated independently for a number of generations, retaining those lines that have shown the greatest response and discarding the others. The set of independent lines can then be reconstituted either by expanding and subdividing each of the selected lines, or by crossing them and re-extracting lines from the progeny. The procedure of selecting and crossing the most responsive lines has been tried on a number of occasions with *Drosophila* and *Tribolium*, but such experiments have generally failed to demonstrate any clear advantage in subdividing the population and, indeed, in most cases a single large population showed the greater response. An alternative procedure is to allow a certain low level of migration in each generation, so that lines of greater mean fitness are able to infect, so to speak, lines of lower fitness, causing all lines to converge on the highest available peak. Some experiments of this kind have shown that the limit to selection in structured populations exceeds that in single large populations. There are, however, as yet too few experiments of this kind for any generalization to be secure.

## 67. *The contingent nature of evolution can be investigated through the behavior of replicate selection lines.*

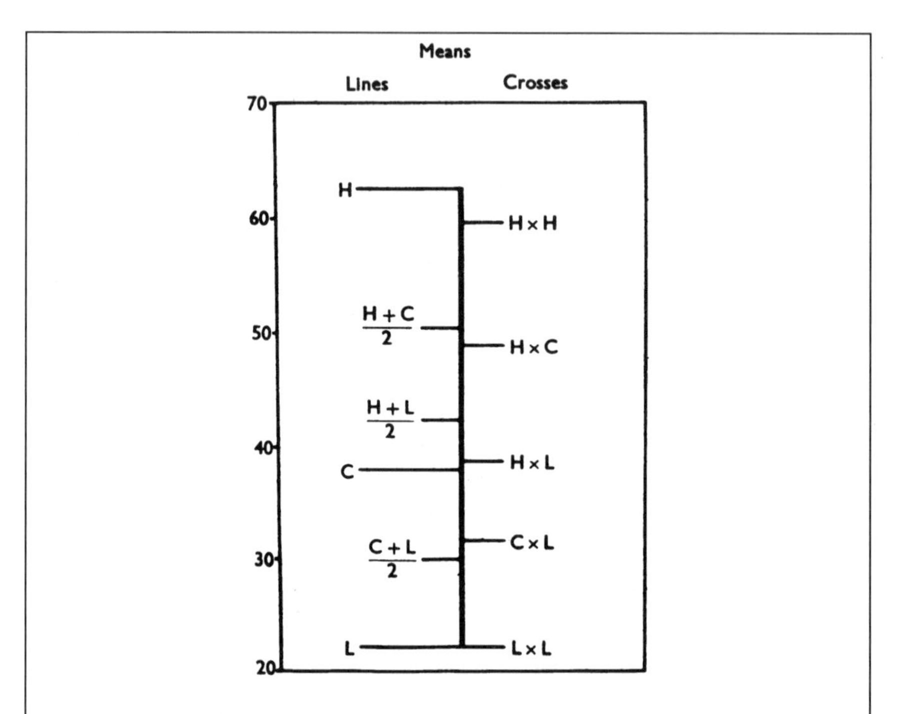


Crossing lines selected for high and low bristle number in the experiment by Clayton Morris and Robertson (1957) gave offspring intermediate between their parents. Similar results were reported by Milkman (1964, 1979) for the expression of crossveinless, and by Cohan et al. (1989) for ethanol resistance. Some degree of hybrid breakdown, however, was found after selection for DDT resistance (King 1955), for expression of *cubitus interruptus* (Cohan 1984) and for pupal weight (Enfield 1977). These are all *Drosophila* experiments, except that Enfield worked with *Tribolium*. For direct determination of the contingent response to selection of the yeast alkaline phosphatase system, see Sec. 60.

The genetic basis of adaptive change can be studied in several ways, depending on the organism and the character under selection. If the selection lines differ phenotypically when tested in a standard environment, then it can, of course, be concluded that they are genetically different. This difference may be transient, however, reflecting different rates of evolution in lines that will eventually converge to the same point. We wish to know if selection lines become permanently different, whether or not they express similar phenotypes.

**Fitting Models to the Observed Response.** The behavior of the long-term *E. coli* lines reported by Lenski was analyzed by fitting mathematical models of contingent and non-contingent evolution to the observed patterns of change. This gave no support for contingency: the data could be adequately explained by the successive substitution of the same set of three or four mutations with independent effects in each line, with the rate and order of substitution varying among lines. This is not a very convincing refutation of contingency, since modeling observed patterns of change in mean fitness can test only phenotypic convergence. The genetic basis of increased fitness in this experiment remains unknown.

**Testing Replicate Lines in Exotic Environments.** An alternative approach is to test the selection lines in different environments. They may all have the same phenotype in the environment of selection; but any genetic differences between them may become apparent when they are cultured in other environments. This is essentially what was done with the six lines that were extracted from the long-term *E. coli* experiment, where they had been cultured at 37 C and then cultured for 400 generations at 42 C. When these lines were tested at the end of the experiment, five had evolved higher rates of growth both at 42 C and at 37 C, whereas the sixth grew faster at 42 C but not at 37 C. Thus, growth at 37 C shows that the genetic basis of adaptation to high temperature was not the same in all lines. The drawback of this technique is that negative results are inconclusive, because it is not known how to specify the type of environment that would reveal differences between the lines, if any exist.

**Crossing Replicate Lines.** One procedure is to cross the selection lines and compare the progeny with their parents. If the genes that have been substituted in different lines have independent effects on fitness, the fitness or character value of an individual will be the sum of the effects of the genes it has received from its two sexual parents. The average value of the progeny should therefore be equal to the average value of the two parents. On the other hand, suppose that in each selection line adaptation is based on a unique combination of genes. In the progeny produced by a sexual cross, the well-adapted combinations of genes borne by the parents will be broken up by recombination. The progeny will be, on average, inferior to the parents. This is a general, easy, and sensitive test for contingency, although it requires using a sexual eukaryotic microbe and cannot be applied to bacteria. It also assumes that the genes responsible for adaptation are nuclear, with Mendelian transmission. When flies from

Clayton and Robertson's bristle-number experiment (Secs. 49 and 50) were crossed after about 10 generations of selection, the average bristle number of progeny was about the same as that of the parents. When *Chlamydomonas* lines selected for dark growth or for size were crossed, the results were very variable, perhaps because of the effects of the chloroplast or mitochondrial genomes, but there was no consistent tendency for offspring to be inferior in either case. Frederick Cohan and his colleagues of Wesleyan University obtained lines of *Drosophila* from different natural populations, selected for resistance to ethanol vapor. The lines responded differently, but there was no breakdown of adaptation when they were crossed. The difference between the lines probably arose, not through epistasis, but simply because the founding populations contained different alleles affecting the character that combined more or less additively. So far, crossing replicate selection lines has given little evidence that evolution is generally contingent, under artificial or natural selection, in the short term or in the long term.

**Direct Determination of Genetic Change.** A final procedure is to study genetic change directly by sequencing the genes or proteins involved. This is an appealing but restrictive approach that supplies an unequivocal answer, but requires an experimental design that defines the target of selection. It cannot be used routinely to analyze how adaptedness increases in novel environments. It has, however, been used with brilliant success to uncover the genetic basis of the evolution of improved or new metabolic pathways in microbes. The acid phosphatase system of yeast, described earlier to illustrate successive substitution, provides an example. In the experiment that I summarized, selection acted to change first orthophosphate uptake, then glycerophosphate hydrolysis, and finally cell flocculation. In a second experiment, using the same strain as a base population, events took a different course. The first change affected the phosphatase, but the mutant enzyme was different from (and more active than) that evolving in the first experiment. A second change again affected cell retention within the chemostat. A final experiment used the strain with the altered phosphatase that had evolved in the second experiment as a base population. In this case, however, the limiting substrate was uridine 5′-monophosphate, which is less easily hydrolyzed than glycerophosphate, in order to impose more stringent selection on the phosphatase, which is a rather non-specific enzyme. Within 200 generations, two mutations with approximately independent effects had become established, both affecting the insertion of the enzyme into the cell wall. After about 400 generations, a further change that increased phosphatase activity had occurred. This modification was not an alteration in enzyme structure, however, but a duplication of the phosphatase locus that increased gene expression. All three experiments, then, resulted in an increase in adaptedness to the novel phosphate-limited chemostat environment, but all three took different routes and gave different genetic outcomes. There are many similar experiments with bacterial systems; but in many cases these experiments involve changes that lead, not merely to improved performance, but to entirely new metabolic capacities and will be described in the following sections. In general, detailed studies of the precise genetic changes occurring in replicate selection lines give strong support to the principle of contingency, in contrast to the experiments using crossing as a means of detecting differences among lines that I have described.

## 2.D. The Evolution of Novelty

In the previous section, I have emphasized the constraints that must often prevent selection from producing adaptation through the cumulation of slight modifications. If the environment changes abruptly, the new optimal phenotype may not be accessible because there is no route from the prevailing phenotype to the new optimum through a continuous series of slight advantageous intermediates, whereas the sudden appearance of a well-adapted type is extremely improbable because the population is only a small sample of potential variation, and the random alteration of complex, integrated systems is almost certain to be severely deleterious. The result is more likely to be extinction than adaptation. Nevertheless, the world is full of complicated creatures following recondite ways of life, to which they are clearly well-adapted. This has always been felt to be a powerful objection to natural selection as a general mechanism for evolution. The most dogged advocates of religious or transcendental theories of nature have now ceased to deny the operation of selection and have turned instead to deny its scope. Selection may modify the bands on snail shells, but it is quite unjustifiable (the argument runs) to infer that a simple extension of this process will produce a new kind of snail. Selection may maintain the type, or cause slight changes within the type, but can never produce a new type. This is a generalization of the hemoglobin paradox: how can we explain organized complexity without recourse to design?

### *68. Very simple and very complex structures are connected by a series of intermediate forms.*

The classical example of a complex structure, whose evolution through the selection of a long series of intermediate forms is difficult to conceive, is the vertebrate eye. We are visual animals, and the structure of our primary sense organ has always impressed philosophers and biologists as demonstrating perfection in organic design, an optical instrument on which human ingenuity could scarcely improve. The basic elements of this structure are familiar to everybody. Light enters the eye through a transparent cornea and is focused on the retina by a crystalline lens, whose position can be adjusted by delicate muscles. The retina itself contains cells whose light-sensitive ends are directed inward and which are supplied by nerves that trail over its inner surface before diving through the retina via the fovea and connecting with the brain as the optic nerve. In such a complex and highly integrated structure, any random change is most unlikely to result in improvement, any more than a camera could be improved by unskilful tinkering with a wrench. How, then, can it have evolved through a continuous series of slightly superior variants? Given that the correct integration and nice adjustment of all its components is necessary for it to function properly, what use is half an eye?

My favorite answer to this query was given by Richard Dawkins: half an eye is about half as good as a whole eye, and even a little bit of an eye may be better than none at all. In *Chlamydomonas*, the eye, if it can be called that, is no more than a cluster of pigment granules on the surface of the chloroplast. It is more remarkable for what it does not contain than for what it does and for what it cannot do than for what it can. It has no cornea, no iris, no lens, no retina, and no nerve. It cannot form an image, nor even detect movement. In fact, there is only one thing that it can do: it can tell whether the light is on or off. This may not seem very impressive to a mammal, but for a motile green alga that feeds on light it is a very important thing to be able to do. It enables the organism to swim toward the light; when you cannot eat other creatures, and could not avoid your predators even if you could see them, this is as useful a task as any eye could accomplish. Even the simplest eye, then, may serve useful ends, depending on the way of life of the organism that possesses it.

It may still be doubted that from so simple a beginning so complex an organ could have evolved by a process of gradual modification. We do not have to rely on imagination, however, to show that the transition from the simplest to the most complex can be traced through a series of intermediate forms. We can instead point to these series among extant organisms such as molluscs. I have abstracted the following series of examples from Libbie Hyman's classic text on invertebrate zoology.

Some molluscs have no trace of eyes. Monoplacophorans are bizarre segmented molluscs whose only living representative closely resembles its early Palaeozoic ancestors. It lives in the lightless abyssal depths of the ocean, where an eye would serve no function and has presumably never evolved. Aplacophorans, which are sluggish wormlike molluscs living on the sea bed, are also primitively blind. Scaphopods are the tusk-shaped shelled molluscs that burrow in marine sands and sediments; since they burrow head-down they need no eyes, but they probably evolved from eyed ancestors, so that their lack of eyes is derived. This is certainly true for blind molluscs whose nearest relatives are sighted, but which themselves live in environments where vision is not required. *Janthina* is a good example; it is a pelagic marine snail, that is buoyed up by a cluster of bubbles and drifts with wind and tide, eating jellyfish whenever it bumps into them. Although it lives in the open, it has no ability either to pursue its prey or to flee from its predators and, therefore, also has no need to see them.

There are several opisthobranch molluscs that also eat jellyfish but that are able to swim clumsily by undulating their whole bodies and these often have very simple eyes. Perhaps the simplest is found in the nudibranch *Phylliroe*: merely a patch of pigment granules underlying an unmodified epidermis, and closely appressed to the pedal ganglion. It is no more complicated than the eyespot of *Chlamydomonas*, and its usefulness is likewise restricted to telling the difference between light and dark. In other nudibranchs and pteropods the eye is a distinct vesicle embedded in the dermis, but is otherwise no more complex or versatile. In such cases the extreme simplicity of the eye is primitive, but simple eyes, like complete blindness, may also evolve through degeneration. Freshwater snails that live in caves, for example, often have eyes that are no more than pigmented vesicles sunk in the dermis, but these are clearly derived from the much more complex eyes of their ancestors in streams and lakes.

In some pteropods, a simple patch of pigment, or a pigment-bearing vesicle, is protected by a modified epidermal layer that represents the first approach to a cornea. *Styliola* has an eye of this kind, in which the pigmented vesicle is supplied with a single nerve cell. At this point, then, we have an eye with a very limited functional range but with three components: a device for collecting light, or at least permitting it to enter; a device for absorbing light; and a device for transmitting the information. In *Styliola*, these components are extremely simple. More sophisticated eyes evolve by one of two routes: as an innervated patch in a pit or as an innervated vesicle sunk in the skin.

The eyes of limpets are open cups, consisting of a sheet of epidermal cells whose distal ends are pigmented and supplied with an optic nerve. They are infolded to form a depression with a broad opening. Bivalves have eyes of this sort, often arranged in long rows so that they are able to detect the movement of a shadow overhead, triggering the closure of the valves. In other archaeogastropods the opening is narrower, and the internal space may be partly filled with a translucent gelatinous material. The pigmented zone of the epidermal cells begins to resemble a retina. The most advanced eyes of this kind are found in nautiloids, which are active pelagic carnivores able to detect and pursue prey. Their eyes have no lens, but the opening of the optic cup has become so restricted that the eye can function like a pinhole camera, capable of forming a distinct image.

In many prosobranchs and opisthobranchs the eye is a multicellular vesicle embedded in the skin in which cells in different regions become specialized for different functions. The interior is filled with fluid and may contain a hyaline body that represents a crude lens, capable of collecting light although not able to form an image. The cells on the outer surface of the vesicle are thinner, to admit light; those on the inner surface are heavily pigmented and constitute a retina. The eyes of polyplacophorans (chitons) and pulmonates (land snails) are basically of this type, although pulmonate eyes are more highly organized than the others. Heteropods have quite complicated eyes, in which light enters through a transparent cornea, is focused by a nearly spherical lens, and is then absorbed by a complex, layered retina. These are naked pelagic molluscs that pursue active prey. The most advanced eyes belong to another group of active pelagic molluscs, the cephalopods. Their eyes are in many respects astonishingly similar to those of vertebrates. They include the full range of structures: a transparent cornea, an iris diaphragm, a lens that can be adjusted by ciliary muscles, and a retina in which light-sensitive cells with their pigmented ends directed outwards are innervated distally by nerve fibers that are collected into an optic nerve leading to a large brain.

The cephalopod eye illustrates the same problem as the vertebrate eye: how could so complex a structure have evolved gradually through a series of intermediate types? In molluscs, however, the mystery is resolved because a complete series of intermediates exists, from the simplest pigment patch to the perfected optical instrument. The degree of sophistication of the eye is related to the way of life of the organism. Simple eyes are found in forms that have little need for any but the most limited vision; complex image-forming eyes are borne by active pelagic carnivores. There is no need to suppose that the most complex and highly integrated structures evolve by any process other than the cumulation of adaptive change through continued selection.

Indeed, one can go further than this. The vertebrate eye has several features that can scarcely be explained, other than as the contingent outcome of cumulative modification. The light-sensitive ends of the retinal cells are directed *inward*, away from the source of light. It is therefore the inner surface of the retina that must be innervated, creating a tangle of neurons that straggles across the retina, not only interfering with light reception, but also making it necessary to bore a hole through the retina in order to get to the brain, thereby creating a permanent blind spot in the field of vision. This can scarcely be defended as good design. It is not made necessary by some unanticipated quirk of optical design in living organisms because cephalopods, such as squid, with a life-style similar to that of pelagic fish have eyes of comparable acuity but more logical design. It cannot even be argued that it is made necessary because eyes must, for some unexplained reason, be designed differently in vertebrates and molluscs, because some molluscs, such as onchidiacean opisthobranchs, have eyes with the same inverted design as that of vertebrate eyes. The detailed structure of the vertebrate eye does not make sense in terms of an optical instrument engineered for optimal performance. It makes sense only as the outcome of a long process of cumulative change, during which selection continually acts to modify pre-existing structures.

The comparative biology of the eye shows that we do not need any principle other than selection to account for the evolution of novelty. Nevertheless, any comparative analysis can provide only circumstantial evidence, however compelling. When we find a trout in the milk, we may suspect; but to be quite sure, we must watch the milkman by the river. To uncover the mechanisms by which novelty emerges from continued selection, we must turn to experimental studies.

## 69. *Bacterial metabolism is a classical example of a complex, integrated system.*

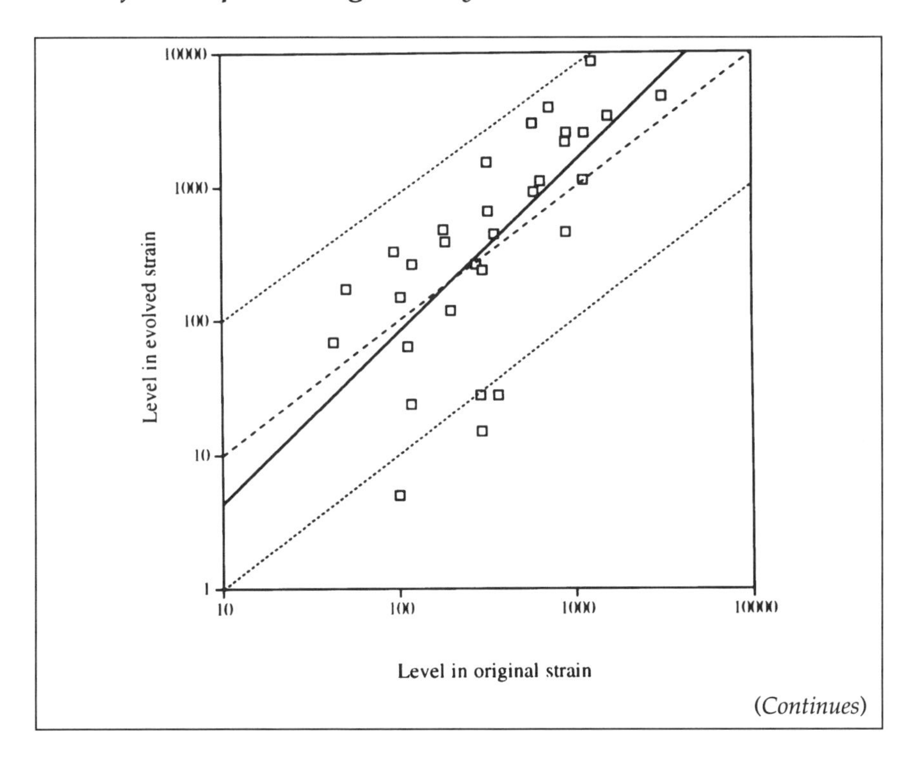


(*Continues*)

*(Continued)*

Bacteria selected in chemostats undergo some predictable changes, evolving more efficient lactose uptake, say, when lactose is supplied in the medium. The complex and highly integrated nature of bacterial metabolic systems, however, means that changes in any one component may cause shifts in many more. Kurlandzka et al. (1991) isolated *E. coli* from a glucose-limited chemostat after about 800 generations of selection and then attempted to identify as many proteins as possible, by two-dimensional gel electrophoresis, in the ancestral and the derived lines. The position and intensity of a spot on the gel can be used to attempt to identify the protein concerned and measure its level of expression. About 300 spots could be examined in this way, about 40 of which could be confidently identified with a known protein. The scatter diagram shows the level of these proteins in the ancestor and in one of the evolved strains. The broken line is the line of equality; points should fall close to this line unless the level of a protein has changed in response to selection. The solid line is the regression, with $r^2 = 0.5$; obviously, there is considerable scatter around this line, representing deviations of the selected line from its ancestor with respect to many proteins. The dotted lines represent tenfold differences in level, to indicate the magnitude of the differences. Furthermore, three proteins that could be detected in the ancestor but not in the descendant have been omitted. Most of this divergence was probably caused by regulatory changes. There were also two cases, however, in which a shift in the position of a spot showed that a change in the sequence of a structural gene had occurred. In comparable experiments, adaptation to the chemostat environment would be caused by about 10 successive substitutions; this study shows that the number of loci whose level of expression is altered is at least an order of magnitude greater.

One of the most remarkable achievements of modern biology has been to chart the routes by which substances are built up and broken down inside the cell. The web of chemical reactions driven by enzymes that constitutes metabolism is more important, more complex, and more highly integrated than any morphological structure, such as the eye, which is by comparison a crude and trivial device. Two features of the enzymes of intermediary metabolism are especially important. The first is *specificity*. There are many different kinds of enzyme (a bacterial cell contains at least 1000 kinds) because each kind is specialized for a particular reaction. Despite this division of labor, many enzymes catalyze a fairly broad range of reactions, and few, if any, are absolutely specific. The specificity of an enzyme is completely determined by its structure—ultimately by its amino-acid sequence—and any structural change is likely to cause a functional change by altering its ability to catalyze some or all of the reactions within its range. The second feature is *inducibility*. Some enzymes are needed all of the time, but most are not. The *lac* system is required only when lactose is present, and it is very wasteful to express it otherwise. Indeed, it would be impossible to express all enzymes all of the time, except at rates too low to be useful. Most catabolic enzymes are therefore expressed only at very low levels unless their substrate (or a derivative or analogue of their substrate) is present, in which case this substance induces a much higher level of expression. An enzyme is useful only when it is regulated properly, because a breakdown of regulation is likely to cause either wasteful overproduction when there are no appropriate substrates or lack of production when substrates are present; in either case, the cell is likely to starve. The properties of a particular kind of enzyme, therefore, will be altered by changes in its structure or by changes in its regulation. Moreover, each enzyme is expressed coordinately with several or many others. A given substrate is often processed by a group of enzymes involved, for example, in both its transport into the cell and its initial processing; such groups of enzymes are often encoded by adjacent genes under the control of a common regulator. The complete metabolism of the substrate usually involves a long sequence of reactions catalyzed by different enzymes, linked functionally into metabolic pathways. The whole of this intricate system must function properly if the cell is to survive.

The problem of how the eye evolves can be restated with even more force in the case of intermediary metabolism. Where do all these different enzymes come from? How does a new enzyme evolve? How does it become appropriately regulated? How are new metabolic pathways established? We could proceed as before, adopting a comparative approach to describe the diversity of enzyme structure and the existence of incomplete and intermediate versions of metabolic pathways; there is certainly no shortage of comparative material. However, although metabolism is more complicated than eyes, it is also a great deal more convenient to work with. The genetics and physiology of bacterial cells are known in great detail, and bacteria are easy to grow and manipulate. We can therefore adopt an experimental approach, ask whether new metabolic abilities can evolve in the laboratory, and, if so, investigate the mechanism involved.

## 70. *Novel metabolic abilities can evolve through exaptation following deregulation and amplification.*

There are several excellent reviews of the experimental evolution of metabolic pathways in bacteria. The articles by Lin et al. (1976) and Clarke (1983) are especially approachable. There are more extensive and technical reviews by

(*Continues*)

(*Continued*)
Clarke (1978) and Mortlock (1980), who cite a large number of studies that I have not described. Reviews of the evolution of pentitols (Mortlock 1984), ribitol dehydrogenase (Hartley 1984), the fucose pathway (Lin & Wu 1984), a novel $\beta$-galactosidase (Hall 1984), amidases (Clarke 1984) and the alcohol dehydrogenase of yeast (Wills 1984) are collected together in the book edited by Mortlock (1984).
The more general ideas introduced in Secs. 71 and 72 have modest literatures; for gene duplication in experimental systems, a good source is Ornston and Yeh (1979; see also Rigby et al. 1974), and for retrograde evolution Horowitz (1945, 1965).

It is easy to challenge bacteria with an exotic substrate that they cannot use because no enzyme they possess is adapted to deal with it. If this substrate is the only source of some essential resource, then they will probably die. If they continue to grow, it can only be because some enzyme, normally serving a quite different function, can be pressed into service to deal with the new substance, however inefficiently. Once any degree of function can be expressed, the enzyme will tend to be modified by selection until it becomes much more efficient in its new role. The modification of a structure originally adapted for one function to serve another dissimilar function is called *exaptation*. The term was introduced by Stephen Jay Gould and Elizabeth Vrba to replace the older term preadaptation that has the same meaning but conveys an inappropriate sense that structures can evolve so as to meet some future need.

**The Panda's Thumb.** Gould's classic example of exaptation is morphological: the panda's thumb. Like us, pandas have an opposable digit, which they use for stripping the leaves from stems of bamboo. In primates that do not use the front feet for locomotion, the opposable digit evolves as a metacarpal, the thumb. Pandas, however, have remained plantigrade on all four feet, which makes it difficult to use a metacarpal in this way; we do not have opposable big toes. The panda's thumb has therefore evolved in a quite different way, from the scapholunate and proximal carpal that normally form part of the articulation of the radius with the palm of the hand. The structure that represents the functional thumb was thus never a digit at all, and in its original form it served functions that have little in common with those of a digit. Its use as a thumb was not anticipated; it just happened to have some rudimentary capacity to function in a thumb-like manner that could be improved through selection.

**The Growth of Bacteria on Novel Substrates.** Examples of exaptation are commonplace in morphology, but the same principle applies to biochemistry. When bacteria are exposed to an environment so novel that no pre-existing system is adapted to deal with it, they can adapt only through the modification of enzymes that originally evolved in response to quite different selective regimes. However, this will rarely solve the problem. Some enzyme may possess minimal activity toward an exotic substance, but it would be a striking coincidence if it were regulated so as to be induced by the same substance. Even if such an enzyme exists, therefore, it will not allow growth on the novel substrate because it will not be expressed. And even if it were expressed, its activity is likely to be too low to support growth, because enzyme specificity is an evolved character, and no enzyme is likely to be specifically active toward a substrate that the population has not encountered before. A population that is forced to depend on a novel substrate therefore encounters three formidable difficulties: no enzyme has the appropriate structure; no enzyme is appropriately regulated; and weakly-active enzymes, if expressed at all, are expressed too feebly to be effective.

Let us suppose that some enzyme exists that is able to metabolize the novel growth substrate slowly and inefficiently. Given the limited specificity of enzymes, this is not unreasonable; it will not be true for all substrates or for all bacterial strains, but it will often be true. The enzyme is not induced by the substrate, and before an appropriate regulatory system could evolve the population would be extinct. It is quite likely, however, that constitutive mutants, in which transcription occurs whether or not the substrate of the gene product is present, will be fairly frequent. For example, mutations that alter a regulatory locus so that it is no longer able to produce the protein that normally represses transcription of the structural loci are simply loss-of-function mutations, deleterious in normal circumstances, that are likely to occur quite often and that will normally be present in large populations. The gene is now expressed; but the enzyme is still so inefficient that growth is at best very slow; it could be increased if enzyme structure were modified through selection, but this requires gain-of-function mutations that are almost certain to be exceedingly rare. It is much more likely that selection will favor making more enzyme, rather than making a better enzyme. This could be achieved by modifications to the promotor that controls the rate of transcription through RNA polymerase; it is much simpler, however, simply to duplicate the gene. Gene duplication occurs spontaneously at a fairly high rate, and once two copies are present, cells with many copies will arise through unequal crossing-over or some equivalent process. These are normally unstable, but can be retained in the population and, perhaps, eventually be stabilized, given sufficiently strong selection. This will enable cells to grow successfully, and selection can now begin to modify enzyme structure, and eventually the system of gene regulation, to form a new, exapted metabolic pathway.

There are four stages in this process of exaptation. The first is simply incomplete specificity, so that some pre-existing enzyme is able to process the new substrate, however inefficiently. The second is deregulation, so that the enzyme is expressed constitutively. The third is duplication and further amplification, so that the total activity of the inefficient enzyme is adequate for growth. The final stage is the normal process of continued selection, causing cumulative improvement in the structural gene and its regulatory system.

## 71. *Duplication followed by divergence leads to increased metabolic versatility.*

Gene duplication can be selected directly because it will usually permit higher rates of transcription. Once established for this reason, however, it creates an opportunity. If both copies of the gene are modified through selection to deal with the new substrate, the population will be poorly adapted if the environment should return to its previous state, and the whole process of modification will have to start again. If only one copy of the gene is modified, however, then the organism will have the ability to process either substrate. Gene duplication can thereby enable organisms to become adapted to a greater range of environmental conditions. Of course, there is no way that the population can foresee the return of previous conditions and mold itself appropriately. Rather, we expect that independent copies of the gene will respond to the selection caused by the presence of a new substrate at different rates and will therefore diverge in structure. In some lineages, this divergence will by chance be more marked than in others. In some, one copy will become well adapted to the new conditions of growth, whereas the other will have scarcely changed. If the previous conditions now return, such lineages will be at a marked advantage, relative to lineages in which both copies of the gene have responded so quickly to selection that neither is capable of processing their previous substrate efficiently. Provided that the environment changes often enough, gene duplication will thereby enable selection to favor metabolic versatility.

## 72. *Intermediary metabolism is thought to have evolved in a retrograde fashion.*

A theory of this sort can be extended to provide, not merely an explanation of some particular episode in bacterial ecology, but a general explanation of cellular metabolism. This was first done by N.H. Horowitz 50 years ago. The earliest self-replicators would have been able to obtain resources such as ribose or nucleotide triphosphates directly from the primitive ocean. Once fairly efficient replicators had evolved, however, they would quickly increase in numbers and by doing so would deplete the supply of dissolved nutrients. There would then be very strong selection for a creature with metabolism, even of the simplest sort; any creature that encoded a protein that acted as a very inefficient enzyme, converting some previously unexploited substance into the desired resource. This would represent the simplest possible metabolic pathway, with one step, a single enzyme that modifies a single substrate so as to produce energy or a substance useful to the cell. The success of such creatures, however, would in turn deplete the substrate. As the substrate became limiting to growth, variants able to scavenge low concentrations of the substrate would spread; some of these variants would have two copies of the gene, thereby producing greater quantities of the enzyme. If the environment contained substances chemically similar to the substrate, selection would favor any type able to convert one of these substances into the substrate. This might be achieved by an enzyme similar to the one that the organism already possessed, given that the two substances involved were similar. Variants would arise in which one copy of the duplicated gene encoded a protein able to catalyze the reaction, however inefficiently, and selection would subsequently enhance its ability to do so. The organism would now have a metabolic pathway with two substrates and two enzymes, whose end product would be the resource that the cell required. The repetition of this process would eventually lead to the construction of complex metabolic pathways in which the initial substrate and the final product were separated by a long series of steps, in each of which a specific enzyme caused a simple modification of an intermediate metabolite.

This theory was originally advanced to explain the clustering of related genes coordinated for the same function (for example, the uptake and hydrolysis of lactose) under a common regulatory system. It can be applied very broadly to gene evolution, however, and perhaps its most important feature is to show how the different kinds of gene that comprise the genome have a branching phylogeny similar in structure to the phylogeny that describes the descent of different kinds of organism. Its great merit is that it invokes processes such as gene duplication and functional divergence that can readily be studied in experimental populations of bacteria.

## 73. *Utilization of the fucose pathway for propanediol metabolism is an example of exaptation.*

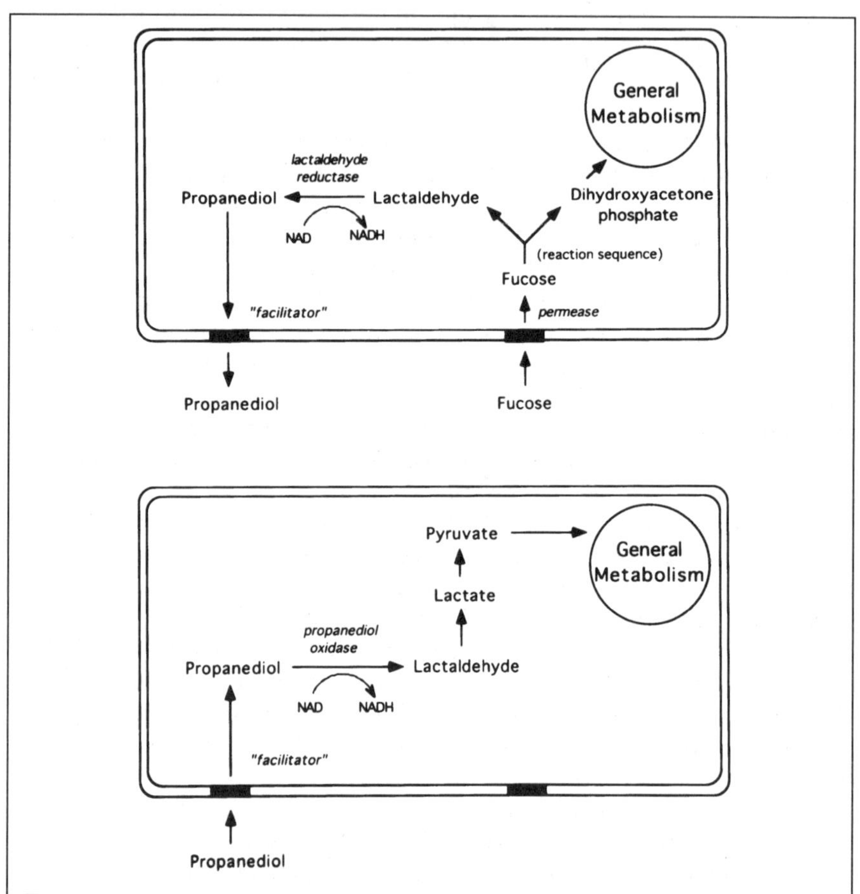


This sketch shows the main features of the evolution of propanediol utilization in *E. coli*, from Lin and Wu (1984). In normal anaerobic metabolism, the fermentation of fucose produces lactaldehyde, which is reduced to propanediol and excreted from the cell. Selection on propanediol yields types in which the NAD-linked oxidoreductase that normally reduces lactaldehyde functions as a dehydrogenase that oxidizes propanediol; the lactaldehyde formed is further oxidized to lactate, a normal substrate for growth.

*E. coli* cannot normally use propanediol as a food source. E.C.C. Lin of Harvard and T.T. Wu of Northwestern University, however, have shown that it is possible to select variants that have evolved the ability to grow on propanediol as the only source of carbon and energy by exaptative modification of the pathway that is normally used to metabolize fucose.

In normal *E. coli* cells growing anaerobically, the hexose L-fucose is eventually split to yield L-lactaldehyde and dihydroxyacetone phosphate. The lactaldehyde is then reduced to L-propanediol by an NAD-linked enzyme and excreted from the cell via a "facilitator" that promotes movement in either direction across the cell membrane. (Discarding half of the carbon skeleton of the original substrate in this way improves the efficiency with which the dihydroxyacetone phosphate can be used as a carbon source.) When propanediol is supplied at high concentration as the only source of carbon and energy, some variants are able to utilize what was previously a waste product as a food. The crucial genetic change involved is the modification of the enzyme that normally reduces lactaldehyde to propanediol under anaerobic conditions into a new oxidoreductase that oxidizes propanediol to lactaldehyde under aerobic conditions. Once this has been achieved, the lactaldehyde is readily oxidized to lactate and enters general metabolism as a source of carbon and energy. Adaptation to a novel environment in which the original strain cannot live thus involves modifying the final enzyme in an anaerobic pathway so as to function as the first enzyme in an aerobic pathway.

## 74. *The catabolism of exotic five-carbon sugars demonstrates the importance of deregulation and duplication.*

The importance of the initial breakdown of gene regulation to permit selection to modify inefficient enzymes has been very clearly established by Robert Mortlock of Cornell and his collaborators. They have studied the utilization of exotic five-carbon sugars by *Klebsiella* (*Aerobacter*) *aerogenes* in great detail. Some of these sugars, such as D-ribose or D-xylose, are relatively common in nature, and it would not be surprising to find that many bacterial strains have appropriately-regulated pathways for their catabolism. Others, however, such as D-arabinose and xylitol, are rare in nature and cannot be utilized by the great majority of bacteria. Mortlock's strategy was use a strain of *Klebsiella* that was able to metabolize these exotic sugars, after a lag indicating that selection was first necessary, to investigate how this ability evolves.

**D-arabinose.** For D-arabinose to enter the main pathway of pentose metabolism, it is necessary to convert it to D-ribulose by an isomerase. The base populations do not possess any such enzyme, but L-fucose isomerase has a low activity for D-arabinose. This enzyme is induced by its usual substrate, L-fucose, not by D-arabinose. Cultures are therefore unable to grow when D-arabinose is the sole carbon source present. Mutants that are able to grow turn out to be constitutive for L-fucose isomerase, producing it whether or not its substrate is present. This is normally wasteful and would be selected against, but when D-arabinose is the only carbon source available, constitutive expression is favored because it permits the deregulated cells to grow. The structural enzyme itself is unchanged. Further selection yields types with greater rates of growth on D-arabinose. These produce an altered isomerase. Thus, the course of events was first a change in the regulatory gene and then modification of the structural gene. Once viable cultures have been established, it is possible to select for inducibility by the appropriate substrate (D-ribulose in this case), restoring appropriate regulation.

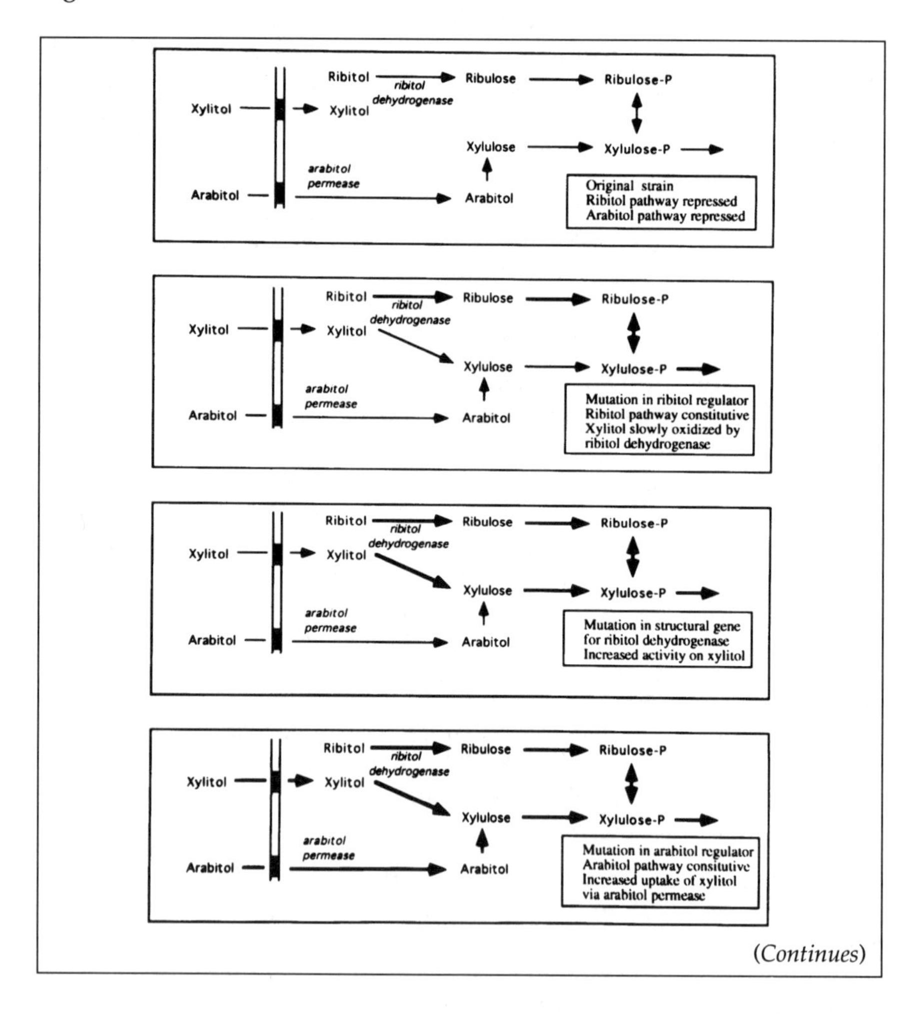


(*Continues*)

These two figures show different aspects of the experimental evolution of xylitol metabolism in *Klebsiella* (*Aerobacter*). The first describes how a xylitol pathway evolves by recruiting elements of two other pathways. Thin lines represent very low rates of reaction in repressed systems; thicker lines represent higher rates of reaction in induced or constitutive systems. The vertical bar is the cell membrane. The evolutionary sequence was worked out by T.T. Wu, Robert Mortlock, and their colleagues (Lerner et al. 1964, Wu et al. 1968, Mortlock & Wood 1964, Mortlock et al. 1965, Wu 1976, Lin et al. 1976).

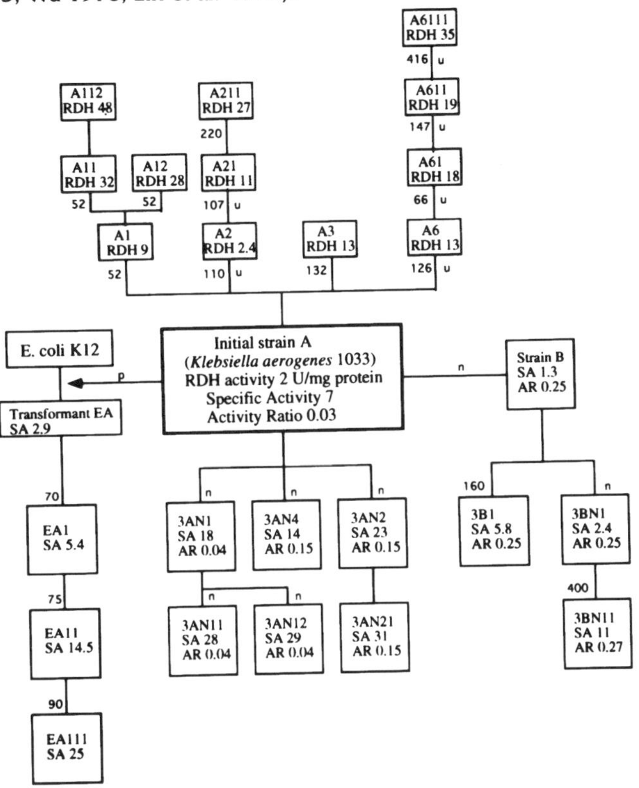


The second figure shows the genealogy of strains selected for growth on xylitol in several experiments. The number of generations and the use of various genetic manipulations are shown alongside the branches: *u* signifies ultraviolet mutagenesis, *n* chemical mutagenesis with nitrosoguanidine, and *p* phage transduction of the *Klebsiella* ribitol and arabitol operons into *E. coli*. The ribulose dehydrogenase activity per unit total cell protein (RDH), the specific activity of the enzyme toward xylitol (SA), and the activity ratio (AR, activity on xylitol as a fraction of activity on ribulose) of the evolved strains were measured in different experiments. The

(*Continues*)

(*Continued*)

uppermost tree shows the results of isolating strains from a xylitol-limited chemostat, after a drop in effluent xylitol concentration had signaled the passage of an improved mutant. These strains, evolving spontaneously or with mild ultraviolet mutagenesis, produce large quantities of ribitol dehydrogenase that increase from 2% of total cell soluble protein to 30% or more. In some cases (such as strain A1), this increase in production seemed to be caused by more efficient transcription, but in most cases (such as strain A11) it was caused by gene duplication. The enzyme itself seemed to be altered little, if at all, having nearly the same $K_m$ in all strains. Chemical mutagenesis resulted in strains with altered enzymes, as shown in the two genealogies on the lower right. These may have undergone several simultaneous changes in amino-acid sequence; however, culturing such mutants often leads to the spontaneous appearance of strains such as 3B1 with even more active enzymes. The tree on the lower left shows the result of selection in *E. coli* to which the genes responsible for xylitol metabolism in *Klebsiella* have been transferred. The EA strains evolve an increased rate of production of ribitol dehydrogenase, presumably largely again through gene duplication; other experiments, not shown here, demonstrate the evolution of altered enzymes with greater activity toward ribitol. The diagram as a whole is a highly condensed summary of work by Brian Hartley and his students and colleagues, reviewed by Hartley (1984); for the original papers, see Burleigh et al. (1974), Rigby et al. (1974), Hartley et al. (1976) Inderlied and Mortlock (1977), and Thompson and Krawiec (1983).

Such evolved strains catabolize D-arabinose by constructing a new metabolic pathway from elements of two others. The first enzyme in the pathway is an isomerase borrowed from the L-fucose pathway that converts D-arabinose to D-ribulose. The second is a kinase from the pentose pathway that phosphorylates D-ribulose, so that it subsequently follows the normal course of pentose metabolism. It is in this way possible to acquire a new metabolic capacity by modifying pre-existing metabolic machinery.

**Xylitol.** Like D-arabinose, xylitol is not normally metabolized, and wild-type strains do not possess any enzyme specifically adapted to deal with it. The ribitol dehydrogenase that converts ribitol to D-ribulose in the pentose pathway also has weak activity toward xylitol, however, converting it into D-xylulose. This can be phosphorylated by kinases of the pentose pathway and proceeds through normal pentose metabolism. Mutants that are constitutive for ribitol dehydrogenase are therefore able to grow on xylitol.

In chemostat culture, the rate of xylitol metabolism by constitutive strains often increases by a factor of four or so within fifty or a hundred generations, and then after a time increases again, until specific activity may be twenty times as high as in the parental strain. The enzyme itself is structurally unchanged. The first advance seems to involve promotor mutants; the second is attributable to an increase in the number of copies of the gene. The result is massive overproduction of an unaltered ribitol dehydrogenase that in evolved strains may have increased from less than 1% to nearly 20% of total cell protein. Under permissive conditions this is of course highly disadvantageous, and ribitol dehydrogenase activity falls rapidly, because supernumerary copies of a gene tend to be unstably inherited and are quickly segregated out when they are no longer actively maintained by selection. The ability to grow on the novel substrate is, then, solely the consequence of changes in the regulation of gene activity: constitutive expression, promotor activity, and the number of copies of the structural gene.

**Regulatory Changes in the L-Fucose System.** The exaptation of the fucose system for the utilization of propanediol can provide a basis for further modifications that extend the metabolic range of the strains. Most of these modifications involved changes in gene regulation.

In some cases, little modification is required. If the membrane facilitator, the oxidoreductase, and the lactaldehyde dehydrogenase are all expressed constitutively, ethylene glycol can be used as a substrate. It is converted first to glycoaldehyde and then to glycolate, which enters the glyoxalate pathway.

Xylitol can also be used, provided that it can enter the cell. Propanediol-using strains that are constitutive for the xylose pathway can use xylitol as the sole carbon and energy source. Xylitol is brought into the cell by the xylose permease and there converted to D-xylose because the novel oxidoreductase can act as a xylitol dehydrogenase. The xylose is then processed in the normal way.

Variants that grow on D-arabitol arise after mutagenesis from propanediol-utilizing strains, but not from normal strains. In such variants, the propanediol membrane facilitator is constitutively expressed, and admits D-arabitol into the cell. Once inside, however, it can be used only if a third change occurs to convert it into a metabolic intermediate. This involves the constitutive expression of a dehydrogenase that has a high affinity for galactose but is also capable of converting D-arabitol into D-xylulose, an intermediate in the xylose pathway. In this case, therefore, the use of a novel substrate depends on the prior acquisition of an unrelated mutation, permitting growth on propanediol and then on regulatory changes in two other systems that by switching them on permanently allow their specialized enzymes to process, rather inefficiently, the new substance.

The initial activity of ribitol dehydrogenase toward xylitol is rather low and, under the appropriate conditions, can be limiting to growth. It is therefore possible to select for mutants with greater specific activity that have structurally modified dehydrogenases. At least one such mutant, when cultured at very low concentrations of xylitol, evolved a higher growth rate because it was able to acquire xylitol at a greater rate from the medium. This was because the ribitol transport system, normally induced by ribitol, had become constitutively expressed, and xylitol uptake increased as a consequence. In this case, then, we can trace the evolution of high growth rates on an exotic carbon source through the constitutive expression of a catabolic enzyme, the capture of a pre-existing metabolic pathway, the structural modification of the enzyme, and the constitutive expression of a transport system.

B.S. Hartley of Imperial College carried these experiments further by growing *Klebsiella* in xylitol chemostats. Cultures that were initiated either with strains that carried several copies of the ribitol dehydrogenase gene, or with strains bearing the modified gene, were usually taken over by mutants with even greater production of the same enzyme. When the cultures were exposed to rather severe mutagenesis, however, new strains with altered enzymes appeared. Most of these alterations appeared to involve single amino-acid substitutions that accumulated in a stepwise manner. Many different point mutations gave rise to strains with increased specific activity for xylitol, leading to many different evolutionary outcomes. Similar results were obtained when the experiments were repeated in a different context by transferring the ribitol operon into *E. coli*: a series of genetic takeovers causing a stepwise increase in the specific activity of the enzyme for xylitol.

## 75. *A new β-galactosidase may represent a case of duplication followed by divergence.*

The usual approach to experimental evolution, exemplified by the experiments that I have just described, is to challenge the population with a novel environment and then to study its response. The alternative is to change the genotype rather than the environment and see how organisms compensate for the loss of an ability they normally possess. If they can respond at all, it must be by exaptation. The most fruitful system for this type of experiment has been the *lac* operon of *E. coli*, where the *lac*Z (*β*-galactosidase) structural gene can be completely deleted, while leaving the other elements of the system (the permease, for example) intact. A culture of *lac*Z-deleted cells is streaked onto plates containing the normal nutrient broth supplemented with lactose. The broth is the primary source of nutrients and allows colonies to develop; as itbecomes depleted, any variant that can utilize lactose would have a large advantage. When the structural gene has been completely deleted, the only way of recovering the ability to ferment lactose is to use or modify the product of some other gene. Rather surprisingly, this turns out to be possible, and white papillae appearing on the surface of the colonies herald the spread of variants with a new $\beta$-galactosidase. It has been discovered subsequently that wild-type cells synthesize an enzyme that can hydrolyze $\beta$-galactosides, although too slowly to support growth in the absence of other substrates, and it is this enzyme that is responsible for the growth on depleted plates. A single mutation alters the gene responsible, *ebg*$A^0$ (where *ebg* is "evolved beta-galactosidase"), into a much more efficient form *ebg*A that will support growth on lactose. The enzyme is quite dissimilar to the $\beta$-galactosidase encoded by the *lac* operon, and the gene itself is situated about as far away from the *lac* operon as it is possible to get. The main function of the original enzyme is unknown. The fact that the *ebg* system is induced by lactose suggests that *ebg*$A^0$ may represent a duplicate copy of *lac*Z that has diverged through mutation from its parental sequence, to the point where it is now normally functionless. It is also possible, however, that it is an unrelated gene whose function remains to be discovered.

## 76. *New amidases evolve through a sequence of structural and regulatory changes.*

One of the classical studies in experimental evolution is the selection of new amidases in *Pseudomonas* by Patricia Clarke of University College London. The wild-type amidase hydrolyzes the two- and three-carbon amides acetamide and propionamide, yielding ammonia as a nitrogen source and acetate as a source of carbon and energy. Both acetamide and propionamide are good substrates and good inducers. However, substrate and inducer specificities are distinct. Other amides may be substrates but not inducers, or inducers but not substrates, or neither inducers nor substrates. If they possess any activity, either as substrates or inducers, it is generally much lower than that of acetamide and propionamide, and base populations are unable to grow, or grow only very slowly, when amides with four or more carbon atoms are the sole carbon source.

The simplest amide that is not normally used for growth is the four-carbon butyramide, which is hydrolyzed only about 2% as fast as acetamide. It is easy to isolate mutants that overproduce the amidase and might therefore be able to grow on butyramide despite their inefficient utilization of the substrate. Some are regulatory mutants that express the

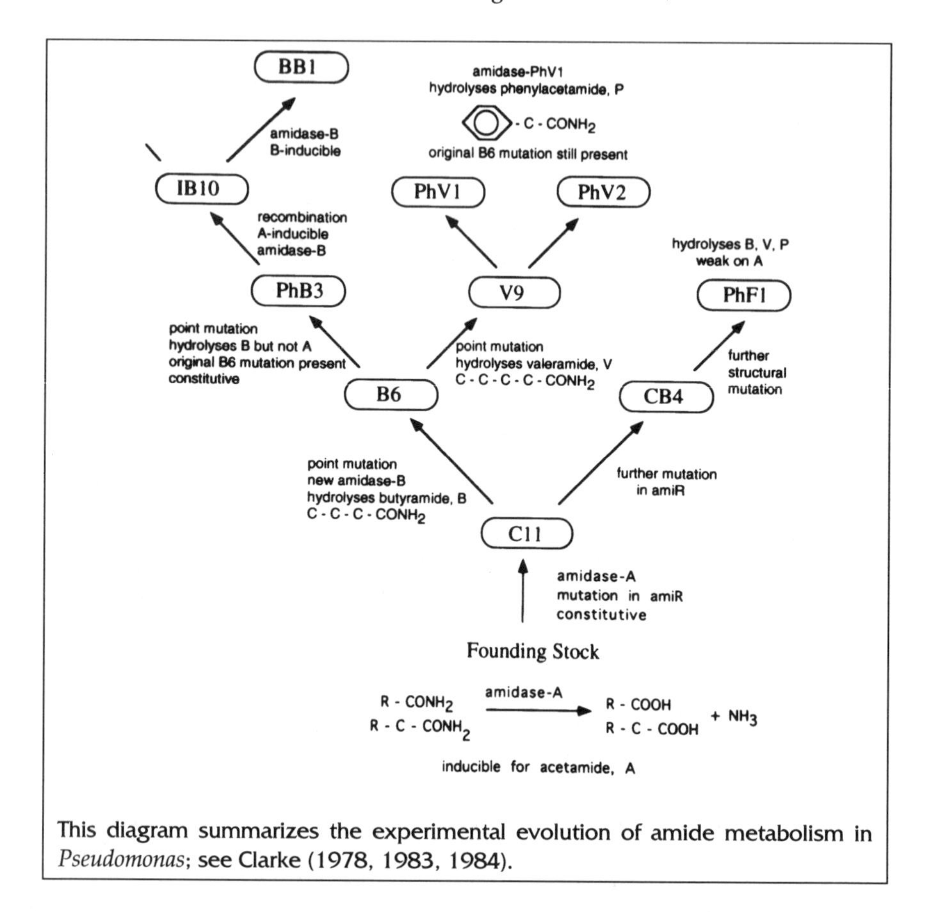


This diagram summarizes the experimental evolution of amide metabolism in *Pseudomonas*; see Clarke (1978, 1983, 1984).

amidase constitutively, as in the pentitol system studied by Mortlock. Others appear to represent mutations in the promoter region of the amidase structural gene that cause increased rates of transcription. Neither necessarily permits rapid growth, however, because butyramide, far from being an inducer, actually represses amidase synthesis. Effective adaptation requires one of two further changes. The first is another change in gene regulation, giving rise to the so-called CB strains, that causes a much higher rate of production of amidase. The second is the production of an altered amidase with higher activity toward butyramide. These B mutants, because they are able to hydrolyze butyramide efficiently, remove it from the medium and thus prevent it from repressing amidase synthesis. One such mutant, B6, was studied in detail. It produced an amidase-B that differed from the original amidase-A in a single amino-acid residue, the replacement of serine by phenylalanine in the seventh position from the N-terminus of the protein. This was presumably in turn caused by a single nucleotide change (UCU to UUU, or UCC to UUC) in the appropriate codon of the amidase gene. This is a good example of the lack of proportionality between genetic and phenotypic change. The genetic change is minimal; the phenotypic result is life or death.

The B6 strain was then selected on growth media containing more complex amides. A second mutation in the structural gene permitted growth on the five-carbon amide valeramide. These V mutants could in turn be used to select PhV mutants able to grow on phenylacetamide, which contains an aromatic ring, when this is supplied as a nitrogen source. These mutants have three alterations in the amidase. Phenylacetamide is neither a substrate nor an inducer for the original amidase system, so that by this point a genuinely new metabolic capacity had evolved. It had evolved through a cumulative process of successive substitution: the original B6 mutation is still present in the V strains, and both B and V mutations are present in the PhV strains, showing unequivocally how new capacities evolve through the stepwise modification of prior states.

This is not, however, the only way in which the ability to metabolize phenylacetamide can evolve. Another class of PhB mutants arose directly from the B6 strain by a second change in the amidase structural gene. These are able to metabolize either phenylacetamide or butyramide. They have acquired one capacity, however, at the expense of losing another: they are now unable to grow on the original substrate, acetamide. They are still constitutive; however, it is possible to select a strain that is induced by acetamide (although it cannot utilize it) by recombination with the basal stocks. A further regulatory mutation then produces a strain that is induced by butyramide. The result is a strain that both hydrolyzes butyramide efficiently and is appropriately induced by it.

Finally, a quite different amidase was selected directly from the CB regulatory mutants. It was able to hydrolyze butyramide, valeramide, and phenylacetamide, but could not grow on acetamide. The capacity to utilize phenylacetamide, therefore, can arise in a variety of ways, involving different changes in regulatory and structural genes.

## 77. *The contingent evolution of novelty implies parsimony.*

The phylogeny of phenylacetamide utilization in these experimental populations of *Pseudomonas* is an excellent example of contingent evolution.

Two lineages that are selected to grow on phenylacetamide may both become adapted, but will do so in different ways, according, in part, to the circumstances of selection and, in part, to chance. After selection, their adaptive phenotype (growth on phenylacetamide) tells us nothing about their relationship to one another, or to any other strain. Their descent is recorded in their genotype and may be discerned, perhaps rather dimly, in non-adaptive aspects of phenotype, such as whether they are still induced by acetamide. Because the genotypic outcome of selection will be unique, we can be confident that two lineages with very similar genotypes are closely related and, in broad terms, that genotypic similarity implies recent common ancestry. This may not apply to phenotypes because phenotypic convergence will occur very frequently, although we can apply it with caution to non-adaptive aspects of the phenotype that are interpreted as remnants of previous adaptations.

If two identical populations adapted to one environment are independently selected in a novel environment, they will become adapted to it, but will tend to diverge genotypically; if both are then placed back into their original environment, both will re-acquire their original adaptedness, but will diverge genotypically even further, as the result of being different to start with. Evolution is perfectly reversible at the level of adaptive phenotype, but is likely to be irreversible at the genotypic level. Again, this irreversibility may be displayed, although imperfectly, by characters that reflect past rather than current adaptedness. The experiments on the fucose pathway happen to provide an example: propanediol-utilizing variants that have lost the ability to grow on fucose can be back-selected successfully, but the genetic basis of their reacquired ability is quite different to that of the original base population.

If the outcome of cumulative change is likely to be unique and essentially irreversible, then we can infer that the same state is unlikely to evolve independently on different occasions. Evolution is therefore, in general, *parsimonious*, in the sense that the most likely way in which a particular combination of character state changes evolved from some other state is that involving the fewest sequential substitutions, rather than a longer one involving several reversals and subsequent reacquisitions.

Because genotypic change is cumulative, the genotypes of the strains could be used to work out their phylogenetic relationships, even if all records of the experiment were lost. Each new change that becomes fixed in the population will be transmitted to all subsequent lineages, although not appearing in any previous lineages. Each change thus represents a derived character that is shared by the descendants of the strain in which it first occurred. Because such *shared derived characters* cumulate through time, they will express the sequence in which changes occur. They can therefore be used to identify the order in which lineages arise, which is the branching structure of the phylogeny.

The principle of parsimony and the use of shared derived characters form the basis of the *cladistic* method of estimating phylogenies, which was introduced by Willi Hennig in the 1960s. The application of cladistic methods to phenotypic data is often contentious, mainly because it is doubtful whether characters are currently adaptive and therefore likely to show phenotypic convergence. These difficulties are beyond the scope of this book. Nevertheless, the cladistic interpretation of phylogeny is generally consistent with the contingent nature of evolutionary change, and therefore with the theory of continued selection that I have advanced here.

## 78. *Duplication followed by divergence of cells and tissues may give rise to novel kinds of organisms.*

When a gene is duplicated and subsequently modified, its product may be sufficiently improved that the original function of the duplication, to increase the rate of expression of an inferior enzyme, is no longer required. One copy of the gene is now redundant and can be modified for another purpose. The cooperative division of labor among replicated entities is a broad principle of organic construction that applies at all levels of organization.

A unicell can evolve straightforwardly into an organism containing $2^n$ cells through a mutation, or mutations, that direct each cell to undergo $n$ mitoses to produce a cluster of small cells that adhere to one another. If the unicell in question is *Chlamydomonas* and $n = 2$, the result is the small colonial form *Gonium*. The advantage of living as a small cluster of identical cells is unknown; perhaps it reduces the risk of being eaten by filter feeders. Notwithstanding, multicellular organisms have the opportunity of evolving cells specialized for different functions, just as duplicated genes can be modified to produce different enzymes. The simplest differentiated relatives of *Chlamydomonas* are *Pleodorina* and *Volvox*, forms with $n = 7–10$ and two cell types, small flagellated somatic cells and much larger unflagellated germ cells. The primary division of labor is thus between soma and germ. This has two functions. The first is simply to keep the germ cells in the water column without requiring that they should themselves be flagellated. The second, which is likely to be more generally applicable, is that the somatic cells act as a source and the germ cells as a sink of elaborated products (such as starch in these green algae). By translocating the products of photosynthesis to germ cells, somatic cells minimize end-product inhibition and thus increase the rate of growth of the cell assemblage as a whole. Vassiliki Koufopanou and I disrupted *Volvox* to obtain germ cells that can be grown independently in nutrient medium; we showed that these isolated germ cells grow more slowly than those of intact colonies, demonstrating the effect of the soma. A series of *Volvox* mutants in which this division of labor is disturbed has been studied by David Kirk of St. Louis. It includes both extremes, types that produce only somatic cells and types that produce only germ cells, and by combining them with mutants that affect the colony matrix, in which the cells are embedded, it is even possible to obtain strains that live and reproduce as unicells. The series of simple and differentiated colonial forms in the Volvocales makes it clear that they have evolved through duplication followed by divergence.

The same principle can be applied at the level of large aggregates of cells. Creatures such as arthropods and annelids are constructed on a segmental plan, with repeated blocks of tissue becoming variously modified for feeding, locomotion, or reproduction. Variants with suppressed or altered segment identity are well known in *Drosophila*, where the genes responsible are now being identified and characterized.

Thus, both multicellular and metameric organization represent a cooperative division of labor that arises through duplication followed by divergence. The availability of variants in which this division of labor is modified offers an obvious opportunity for evolutionary experiments, especially in simple creatures such as the Volvocales, but so far as I know none have yet been attempted.

## 79. *Fusion is another route to novelty.*

There are two routes to the production of novel kinds of organisms. The duplication of a pre-existing structure followed by the divergence of the copies may be the more frequent, but the alternative is the fusion of unrelated organisms to form a new compound organism. This is well known on a small scale, through phenomena such as sex and plasmid transfer. It also occurs on a much larger scale: plant-mycorrhizal assemblages and lichens are intimate partnerships that create compound organisms with new properties, and eukaryotes themselves evolved as bacterial communities. However, fusion itself is not enough; it usually leads to parasitism rather than to a continued and eventually obligate mutualism. In order to create a new kind of creature, we must have *fusion followed by domestication*. The circumstances in which fusion leads to cooperation rather than antagonism are discussed in Part 5.D.

## *80. Evolution is not necessarily progressive.*

Because the subject of this section is the evolution of novelty, it may have left the impression that evolution is inherently progressive, in the sense of leading naturally to new kinds of creature with extended and broadened abilities. This interpretation, with its heavy freight of social and political corollaries, has bedevilled the study of evolution since its inception. Some have taken for granted that evolution implies progress; some have emphatically denied that the concept of progress has any place in evolutionary biology. In fact, evolution is progressive in some senses, but not in others.

The plainest meaning of progress is increase in fitness or adaptedness. Evolution is very generally progressive in this sense. Organisms are very evolvable, so to speak: placed in a new environment, or deliberately chosen for certain characteristics, selection will always tend to modify the population appropriately and very often succeeds in doing so. There are a few recorded instances to the contrary: *Drosophila* does not seem to respond to selection for an altered primary sex ratio, for example. No doubt these are under-recorded, because there is inevitably a tendency to publish the positive results and neglect the negative. The documentation of selection is now so extensive, however, that there is no reasonable doubt that the elementary conclusion of the Fundamental Theorem is very often borne out in practice.

Progress, however, is often taken to mean an advance in some subjective measure, such as complexity of organization. This usage is especially prevalent in the popular literature of the subject. Evolution does not, in general, cause progress in this sense. Selection will favor attributes such as beauty, strength, or wit only to the extent that they are associated with increased fitness, and this is not by any means necessarily the case. Organisms may readily be selected to become smaller, simpler, or less aesthetically appealing: internal parasites, for example, are often considered to be degenerate, relative to their freeliving counterparts or ancestors, although it is more appropriate, instead, to view them as being differently specialized. There are, to be sure, circumstances in which progress, in this sense, is almost certain to occur. Our most remote ancestors, three billion years ago, were necessarily small and simple in structure, and because this provides more opportunity to become larger and more complex than to become yet smaller and simpler, an average increase in size and complexity was virtually inevitable. A more interesting argument is Darwin's original suggestion that most evolutionary change occurs in response to competing organisms and therefore involves ever more sophisticated structures and behaviors. This interpretation of progress, then, maywell be true in particular cases, and is fallacious only when it is asserted, or implied, that change of this sort will occur independently of its effect on current fitness.

The final meaning of progress, which has now disappeared from the technical literature but can still be found in some popular accounts of evolution, is change that is directed toward some preordained goal. The goal in question is almost always ourselves. According to this interpretation, the evolutionary history of organisms is analogous to the development of an individual, unrolling according to a fixed plan towards a specific destination. Evolution is never progressive in this sense. The most distinctive contribution of Darwinism to the theory of evolution is that selection acts only through current differences in fitness and has no foresight whatsoever. Evolution may have a direction, but it never has a destination—unless, indeed, that destination is chosen by a human experimenter. The history of life on Earth is a single, unreplicated selection experiment whose outcome to this date could not have been foreseen and whose future course is equally unknowable.

# 3

# *Selection on Several Characters*

## *81. Selection directed towards any given character is likely to cause changes in other characters.*

The evolution of a single phenotypic character is determined by its genetic covariance with fitness. Fitness can be regarded as a privileged character in that it alone is directly selected. Thus,analyzing how a single character changes under selection involves considering two characters—the character itself, and fitness—and the relationship between them. It should be clear from the previous section, however (and it is in any case a rather obvious thought), that evolution is not simply a succession of changes in single phenotypic characters. Adaptation generally involves coordinated changes in several or many characters, and to understand how adaptation evolves we must proceed to investigate how selection acts on several characters simultaneously.

The basic principle is simply an extension of the argument for single characters: *characters will evolve together if they are correlated genetically with one another and either or both are correlated genetically with fitness.* A genetic correlation between characters can arise in three ways.

**Linkage.** In any finite population it is unlikely that genes will be associated entirely at random, if only because new mutations will arise by chance in some lineages and not in others. In asexual populations, such associations are permanent, and the characters that are expressed by the genes concerned will be genetically correlated because the genes themselves are transmitted together. Hitch-hiking (Sec. 45) is an example of genetic correlation arising through linkage, but the phenomenon is quite general and applies irrespective of the effects on fitness of the genes concerned.

**Epistasis.** When different characters are encoded by different genes, they may become correlated through selection, either in sexual or in asexual populations, if their effects on fitness are not independent. This is because selection will cause some combinations of genes to increase in frequency and others to decrease. Suppose that there are two loci, A and B, each with two alleles. It happens that the genotypes A1B1 and A2B2 are more fit than the combinations A1B2 and A2B1. Selection will tend to increase the frequency of A1B1 and A2B2, creating a genetic correlation between the character states associated with the alleles at each locus. We might imagine, for example, that A controls the flower production of apple trees, with A1 producing many flowers and A2 few, whereas B controls fruit production, with B1 producing many fruit and B2 few. The genotype A1B1 is very fruitful in the short term, whereas A2B2 can invest the energy that it has not expended in reproduction on root growth, and thereby ensure that it will survive to reproduce in the following year; both genotypes may have high fitness. A1B2, on the other hand, wastefully produces many flowers that it cannot use to produce fruit, and A2B1 might mobilize resources for fruit production that cannot be achieved because of a paucity of flowers. The A and B loci will then evolve together if selection favors the combinations A1B1 and A2B2, creating a genetic correlation between flowering and fruiting.

**Pleiotropy.** Conversely, a single gene may affect several characters, which in turn affect fitness in different ways. One allele of a gene affecting fruit production, for example, might direct the production of large apples, another of small apples. If the total mass of fruit produced is unaffected, then a tree producing large apples must produce few, whereas a tree producing small apples can produce many. The two characters, fruit size and fruit number, will therefore be correlated genetically because they are expressed differently by alleles of the same gene.

Characters that are genetically correlated through linkage, epistasis or pleiotropy will tend to evolve together. To describe the situation, it is convenient to recognize three types of character. One is fitness, that responds directly to selection. The second is the character that is the phenotypic criterion of fitness, having the effect of increasing adaptedness in a new environment; this responds indirectly to selection. The third is a character or characters that are not themselves responsible for differences in adaptedness, but that change because they are genetically correlated with the character that is. It is this third category of characters that represents the new element in the situation. The change through time of such characters can be termed a *correlated response to selection*. In practice, the distinction among the three categories is not as clear as I have indicated. In the case of artificial selection, fitness is conflated with the character that is the target of selection, but there is a clear distinction between the direct and correlated responses. In the case of natural selection, the distinction between fitness and other characters is easily drawn, but often no one character can be designated as the target of selection, and there are instead a number of correlated characters, all contributing to fitness in some degree. Nevertheless, it may be useful to recognize three categories of character in order to emphasize the logical transition from self-replicators, through a character encoded by a gene and expressed by an organism, to a complex phenotype comprising many characters.

**Correlated Response to Selection for Size.** Individuals that differ in overall body size inevitably differ in the size of some, or many, of their body parts. A very simple example of a correlated response is thus the increase in the size of a particular body part as the consequence of selection for overall size. William Atchley and his colleagues at Madison conducted an unusually thorough experiment of this sort by measuring a series of skull characters in lines of rats that had been selected for about 20 generations for increased or decreased body weight. There was a pronounced, although asymmetrical, direct response, the upward lines increasing in weight by about 3.5 standard deviations, relative to randomly selected controls, whereas the downward lines decreased by about 1.3 standard deviations. This caused a correlated response of most of the skull characters. Not surprisingly, the most conspicuous feature of the correlated response was a change in the value of each skull measurement corresponding to the change in overall size: all of the characters increased in the upward lines and decreased in the downward lines. As is usually the case, the correlated response is smaller than the direct response because of the imperfect genetic correlation between the skull measurements and body weight. Although the correlated response was dominated by a general increase or decrease in character value, it also involved some changes in shape. In the upward selection lines, for example, the zygomatic (cheek bone) length increased, but did not increase in proportion to the increase in total skull length. These rats thus evolved skulls whose middle region (corresponding roughly to the tooth row) was short relative to either the snout or the brain case. Straightforward selection for size may thus have subtle and unexpected effects on the conformation of body parts.

**Correlated Response to Selection for Stress Resistance.** The occurrence of a correlated response can often be understood qualitatively because the primary and secondary characters are both particular cases of a more general category of character. A simple example is the evolution of resistance to a particular stress, such as starvation, high temperature, or the presence of a toxin. A part of the response to selection may be specific to the stress applied, but another part may be attributable to a more general response, such as lowered activity or metabolic rate. This general response is likely to give some protection against a broad range of stresses, and thus resistance to other sources of stress will evolve as a correlated response. Two Australian biologists, Ary Hoffman and Peter Parsons, selected *Drosophila* for resistance to desiccation by a very simple method that entailed placing vials of flies in a desiccating chamber, waiting until most have died, and then breeding from the survivors. There is a strong response to selection:

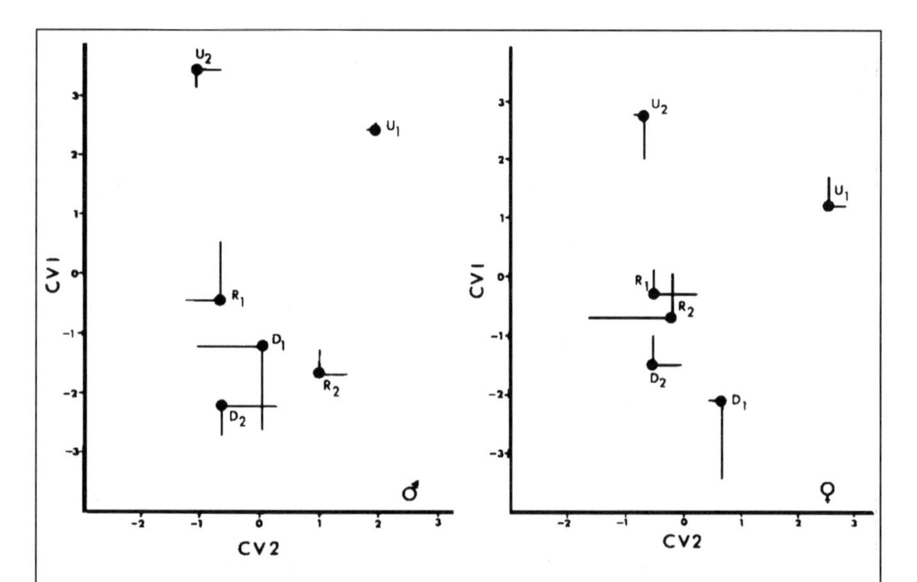


Atchley et al. (1982) expressed the correlated response of skull characters to selection for body size in terms of canonical vectors which are linear combinations of character values weighted in such a way as to maximize the difference between the groups being compared. This is a four-dimensional plot of the first four canonical vectors. The first two (CV1 and CV2) are the axes of the graph; the third is represented by the vertical line at each point (positive deviations above, negative below), and the fourth by the horizontal line (positive deviations to the right, negative to the left). Each point is the mean for a selection line: the two U lines were selected upward, the two D lines downward, and the two R lines randomly. CV1 is essentially a measure of overall size, and accounts for about half the total phenotypic variance; note that this accounts for the major part of the correlated response, at least in the upward lines. Males are shown in the left-hand figure, females to the right. The work on desiccation resistance in *Drosophila* cited in the text is by Hoffman and Parsons (1989).

within three generations, the proportion of females surviving 24 h in the desiccator has risen from 10–25% in the control lines to about 85% in the selection lines. The physiological basis of the response is a decrease in water loss through the spiracles, which can be viewed as an appropriate and rather specific adaptation to desiccation. Decreased spiracle activity, however, seems to be in turn the consequence of a lower metabolic rate: less demand for oxygen leads to decreased spiracle activity, which reduces water loss. It seems likely, therefore, that the response to selection was achieved simply by reducing the resting metabolic rate. Now, this is a rather generalized kind of change that might be effective against many different sources of stress. When the selection lines are tested, they are found to be resistant to temperature shock, ethanol and acetic acid fumes, and gamma radiation. Resistance to these factors is a correlated response rooted in the rather non-specific physiological mechanism evolving as an adaptation to desiccation.

## 82. *The correlated response can be predicted from the genetic structure of the base population.*

The response to selection of a single character is, in principle, predictable from the manner of its inheritance in the base population. A response can occur only if there is genetic variance for the character, and the rate of response is governed by the magnitude of genetic covariance between character value and fitness. Let us call this character the primary character. Similar principles apply to the correlated response of a secondary character. A correlated response can occur only if there is genetic variance for the secondary character and the rate of correlated response is governed by the magnitude of genetic covariance between the secondary and primary characters.

**Correlated Response and Genetic Covariance.** It may not be surprising to learn that selecting flies with longer wings yields lines that not only have longer wings but also have a longer thorax, and vice versa; selection for longer wings is likely to favor larger flies that also have a longer thorax. It is more interesting that the correlated response of thorax length to selection for wing length (or vice versa) can be predicted in the base population before selection is applied. I have described in Sec. 44 how the regression of character value on fitness $w$ determines the response of a single character, which in this case we may take to be the primary character, $z$, to selection. Some part of this response will appear as a correlated response of the secondary character $x$. The change in $x$ caused by unit change in $z$ is given by the regression $b_{xz}$ of $x$ on $z$. The correlated response is therefore predicted by

$$(1/x)\mathrm{d}x = b_{xz} b_{wz} V_z^2$$

The variances and covariances that comprise the right-hand side of this equation are standardized genetic variances and genetic covariances; in sexual organisms like *Drosophila* they are additive genetic variances and covariances. Perhaps a more convenient way to express this argument is to say that the direct response to selection $R$ will be depreciated by the genetic regression of $x$ on $z$, such that the correlated response is equal to $\mathbf{R} = r_G(\sigma_x/\sigma_z)R$. The genetic correlation $r_G$ and the genetic standard deviations $\sigma_x$ and $\sigma_z$ can be estimated in the base population, and the equation then used to predict the change in $x$.

**The Correlated Response to Selection on Bristle Number.** This procedure is subject to the same constraints as predicting the outcome of selection on a single character, where the quantity of genetic variance that determines the rate of response to selection itself changes as the result of selection. Where more than one character is concerned the difficulty is greater, because both genetic variances and genetic covariances may change because of selection. The predictions are therefore unlikely to be useful for more than a very few generations. Moreover, even in the short term the correlated response will be predicted less precisely than the direct response because of the additional error involved in estimating genetic covariances. A.K. Sheridan and J.S.F. Barker of Sydney selected for bristle number on two segments in *Drosophila*, measuring the correlated response of coxal bristle number to selection for sternopleural bristles, and vice versa. The predictions were reasonably successful, except that they consistently overstated the response that was actually achieved. Thus, the predicted response of coxals to selection for sternopleurals, over the first 10 generations, was 2.6 bristles, whereas the actual response was only 2.3 bristles; the disparity was greater in the converse experiment, where the correlated response of sternopleurals to selection for coxals was only 0.9 bristles, against a prediction of 1.8 bristles. Another way of expressing this result is to say that measuring the correlated response to selection gives a lower estimate of genetic correlation than measuring the resemblance among relatives in the base population: in this experiment, the realized genetic correlation was only about 0.24, against an estimate of 0.48 in the base population.

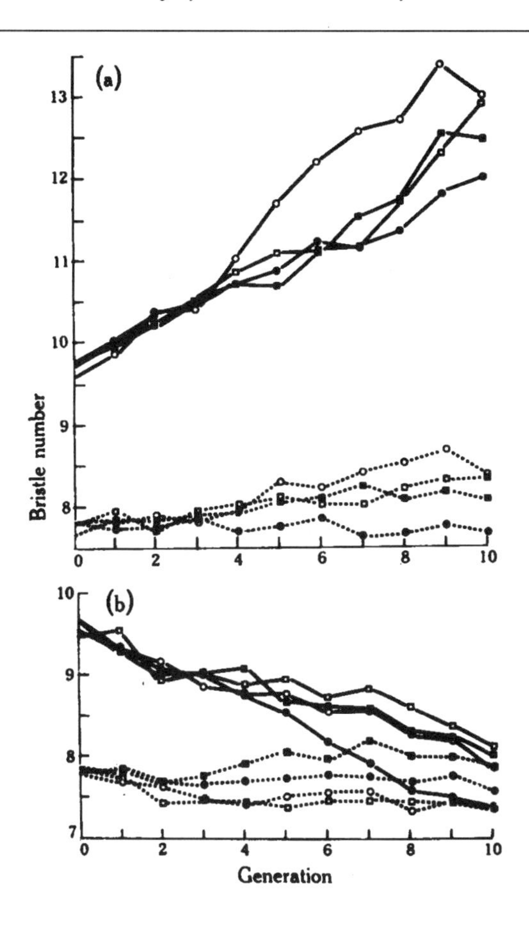


This figure shows the correlated response of coxal bristle number (broken lines) to upward and downward selection for sternopleural bristle number, from Sheridan and Barker (1974a). The four plots in each case are four replicate lines. Correlated responses were also scored in the Clayton–Robertson experiment (Secs. 49 and 50), but all that could be said was that the correlation between sternopleurals and abdominals was very slight, and in the short term the sternopleurals scarcely responded to selection for abdominals (Clayton, Knight, Morris and Robertson 1957). In the longer term (of 20–30 generations) there was a pronounced correlated response, as the result of new covariance appearing during the course of the experiment; see Sec. 86. Reeve and Robertson (1953) found that the correlated response of wing length to selection for thorax length, and vice versa, was adequately predicted by their genetic correlation, perhaps because this was

(*Continues*)

(*Continued*)

very high. There are detailed studies of the correlated response to selection for wing length in milkweed bugs, *Oncopeltus* (Palmer & Dingle 1986, Dingle et al. 1988). The extensive literature on correlated responses in livestock and crop breeding is mostly concerned with components of yield and is referred to briefly in subsequent sections; for an example of correlated responses to selection for a morphological character (lint yield in cotton), see Miller and Rawlings (1967).

## 83. *Indirect selection can be used if the correlated response exceeds the direct response.*

Although it may seem paradoxical, the correlated response of the secondary character may exceed the response of the primary character. This may happen if there is greater genetic variance for the secondary character than for the primary character, and the two are highly correlated genetically. This has occasionally been put to use in agronomy. Selecting directly on the primary character might be inefficient if most of its variance is environmental. In this case it is much easier to select individuals that are genetically superior with respect to the secondary character, using them to get leverage, so to speak, on the primary character. This technique, which has been called *indirect selection*, is especially useful when the real object of selection is very difficult or very expensive to measure precisely, so that genetic variance would be obscured by measurement error, whereas an easily measured surrogate character is available. Its relevance to natural populations is that selection may often cause unexpectedly large responses by characters that seem to have little to do with adaptation.

## 84. *Gradual evolution requires dissociability.*

The primary constraint on the evolution of single characters is the lack of available genetic variation. The corresponding constraint on the evolution of adaptation involving several characters is the presence of fixed genetic covariance. When the genetic correlation between two (or more) characters cannot be modified in the short term, adaptation is hindered in two ways.

Firstly, it is impossible to modify a particular character through selection without causing irrelevant and perhaps inappropriate modifications in other characters at the same time. If you wished to improve flight ability in *Drosophila*, for example, you might attempt to do so by increasing wing loading by selecting for longer wings. This would not succeed, however, if flies with longer wings are simply bigger flies that also have larger bodies. The genetic correlation with body size, assumed to be unalterable, would prevent wing length from being selected independently, so that it would be impossible to evolve a strain of flies that had longer wings but were unchanged in other respects.

Secondly, although particular combinations of character states may currently be favored, different combinations may be superior if the environment changes. If wing length and body mass are positively correlated then all individuals are able to fly about equally well. It is conceivable, however, that in some circumstances selection will favor specialists: highly mobile individuals with long wings and small bodies that can disperse to find distant sites suitable for growth, and sedentary individuals with short wings and large bodies that can exploit these sites when they have been found. In this population there would be a negative correlation between wing length and body size; so the existing correlation, again assumed to be unalterable, makes this outcome impossible.

New adaptations can evolve only if characters can be to some extent selected independently, either to exaggerate a particular character or to create new combinations of character state. This will not be possible if genetic correlation is complete, so that a given value of one character is always inherited together with the same value of some other character, or if an incomplete correlation cannot be modified, so that individuals that vary in a certain direction with respect to one character always vary in the same direction with respect to another. Characters must therefore be *dissociable*: that is to say, it must be possible to break down the constraints imposed by existing patterns of genetic covariance, permitting characters to be selected independently or permitting new patterns of correlation to evolve. I have already invoked a similar principle in discussing the difference between continued selection in sexual and asexual populations (Sec. 46): selection for genes with independent effects is more effective when they are transmitted independently. The converse remains true: combinations of genes are selected more effectively when they are transmitted together. We might, therefore, recognize two opposed constraints. Adaptation to novel environments is hindered by a complete or unalterable genetic correlation between characters; the maintenance of adaptation in a constant environment, on the other hand, is hindered if the existing pattern of genetic correlation that preserves superior combinations of genes is continually being disrupted.

## 85. *Index selection is used to maximize the response of a character in the presence of genetic covariance.*

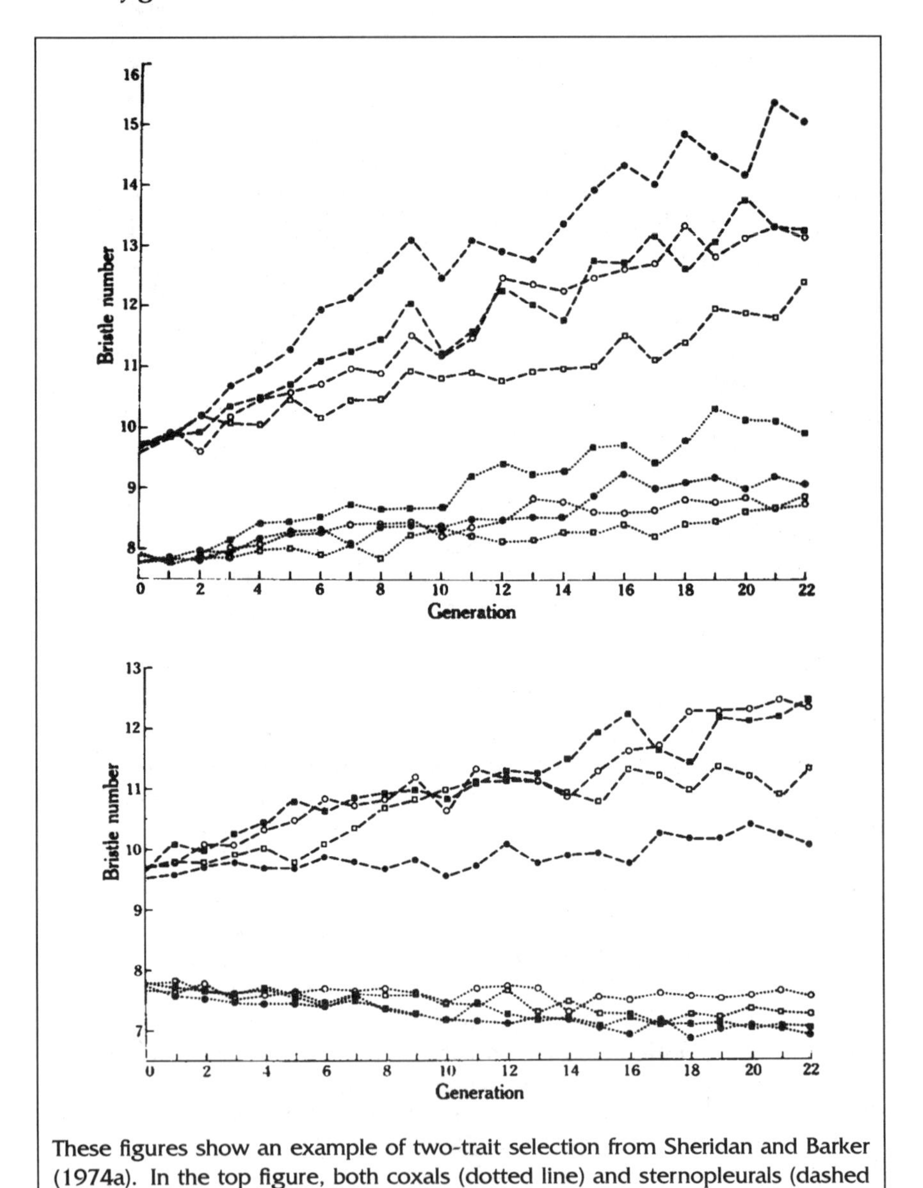


These figures show an example of two-trait selection from Sheridan and Barker (1974a). In the top figure, both coxals (dotted line) and sternopleurals (dashed

*(Continues)*

(*Continued*)

line) were selected upward, there being four replicate lines. In the bottom figure, coxals are selected downward and sternopleurals upward. In the latter case, it is possible to make progress in both directions simultaneously despite the fact that the two traits are positively correlated in the base population. The progress is markedly less than when both are selected in the same direction, because selection is hindered by the positive correlation between them. Note the divergence of the lines, especially with respect to sternopleural bristle number.

There is a large literature on index selection in agronomy and livestock breeding. Selection for economically desirable characters often involves one of two difficulties. The first is that the experimenter may want to select for one character while holding the value of another character constant, despite the fact that the two are correlated (restricted index selection). The second is that selection may oppose the existing correlation (antagonistic index selection); for example, the breeder might like to increase both egg size and egg number in fowl, but the two characters are negatively correlated. For reviews of experiments with laboratory models, see Campo and Velasco (1989) on *Tribolium* and Eisen (1992) on mice.

Genetic correlations often represent severe constraints in breeding programs designed to improve livestock or crops. The objective of such a program might be to increase the value of a certain character, say, the milk yield of cattle or the milling quality of wheat, that is known to be correlated with a number of other characters. These correlations are taken to be fixed constraints that will not change appreciably during the program. One can imagine two extremes in approaching this problem. One would be to select directly on the character of interest, ignoring the correlated characters; the other would be to select indirectly on one or more of the correlated characters, ignoring the character of interest. The most efficient procedure is to combine these two, by selecting simultaneously on all characters, in such a way as to maximize the response of the character of interest. *Index selection* involves constructing a composite character whose value is a linear function of the values of the character of interest and all the characters correlated with it,

$$I = b_1 P_1 + b_2 P_2 + b_3 P_3 + \cdots$$

The index value $I$ is the sum of the values of a series of phenotypic characters $P$, each weighted appropriately by a coefficient $b$. The weights are regression coefficients: each represents the genetic effect of a given character on the character of interest, when the values of all other characters are held constant. The index is calculated for each individual in a trial, and the individuals with the highest index values are chosen for breeding. Index procedures of this sort are very effective because they can incorporate much more information into the selection process than simply choosing individuals that have extreme values of a single character can. They are very laborious, however, if large numbers of individuals or varieties are being screened and are used most often when individuals are relatively few or costly, as in livestock breeding.

A second, and perhaps more important, reason to apply index selection is that it may be impracticable to select on the character of interest, so that selection must instead be applied to other characters that are thought to be correlated with it. The objective of most animal- and plant-breeding programs is to increase economic value; but to select directly on economic value, by releasing hundreds of trial varieties for commercial use and later assessing their performance, is unlikely to be feasible. It is possible, however, to infer that the economic value of a barley cultivar, for example, will be affected by grain size, heading date, mildew resistance, and a large number of other characters, all of which can be combined together into a single index whose value can be used as a criterion for selection. The weights are in this case guesses about the contribution of each character to the unknown economic value of the crop.

There is a clear analogy between this second method of index selection and selection in natural populations. Under natural selection, fitness replaces economic value as the quantity to be maximized, and the weights are the partial regressions of the genetic effects of each character on fitness.

**Two-Trait Selection.** The simplest form of index selection is simply selection at nearly equal rates on two characters simultaneously. For example, to select for coxal and sternopleural bristle number at the same time, one can choose the 11 pairs of the first 25 to emerge that have the most sternopleural bristles, and then choose the 5 pairs from these 11 that have the most coxals. One can then predict the expected response of either character as a complicated function of the direct and correlated responses that would be obtained if either were selected alone. This was exactly the procedure followed in the two-trait part of the Sheridan–Barker experiment described in Sec. 82. The predictions were, on average, borne out reasonably well, at least insofar as the sternopleurals responded much more than the coxals, although the responses were consistently less then expected. There was a very marked discrepancy between replicate lines that seems to be a general feature of experiments involving the response of two or more characters to selection. Because the replicate lines are founded from different small samples from some common stock, this may be caused in part simply by the somewhat different composition of the base populations. It is also conceivable, however, that the large number of different character combinations that can arise and be selected during the course of these experiments exaggerates the contingency of the response.

## *86. Genetic covariance may change through selection or recombination.*

During a crop trial, it is reasonable to neglect the appearance of new variation by mutation and to regard selection as a single episode of sorting constrained by the quantity of variation originally present. It is equally justifiable to take the existing pattern of correlation among characters as a fixed constraint that must be taken into account by a given protocol for index selection. In the longer term, in experimental or in natural situations, this is not necessarily the case; just as new variation can appear in the longer term so, too, can old patterns of correlation break down and new patterns emerge.

**Linkage.** The historical association between genes maintained by linkage is dissolved in sexual organisms by recombination. Recombination alone cannot create pattern; it always tends to reduce the magnitude of genetic correlation, whether positive or negative, towards zero. However tightly linked two loci might be, recombination reduces the correlation between them in every generation. The correlation between freely recombining loci, for example those on different chromosomes, is reduced in every generation to a value half-way between its value in the previous generation and zero, and therefore decays rapidly to insignificant values. More generally, the correlation moves toward zero at a rate equal to the frequency of recombination between the loci. Genetic correlations that are created exclusively through linkage are therefore rather evanescent, although they may represent important constraints in asexual organisms or for tightly linked loci in sexual organisms.

**Epistasis.** The corollary is that sets of genes that have high fitness in combination with one another are difficult to maintain in sexual populations. Epistatic combinations are attacked by recombination in every generation, but restored by selection. This offers the opportunity for selection to establish new combinations when the environment changes, although selection must be very powerful indeed to offset the randomizing effect of free recombination. In artificial selection, of course, it is possible to select directly for desired combinations of genes or characters.

**Pleiotropy.** The manifold effects of a single gene are unaffected by recombination. If they reflect physiological limitations, for example in the total quantity of resources that can be acquired or the rate at which they can be utilized, they are likely to be present permanently, either in sexual or in asexual populations. Genetic correlation attributable to pleiotropy will change only when the frequencies of genes with different pleiotropic effects change. It must be borne in mind, however, that genetic covariance, like genetic variance, is not a fixed and immutable property of a population, but also depends on the environment in which the population lives. Pleiotropic effects that reflect the total quantity of resources available to each individual, for example, will be expressed quite differently in productive and unproductive environments.

## 87. *Recombination may itself evolve as a correlated response to selection.*

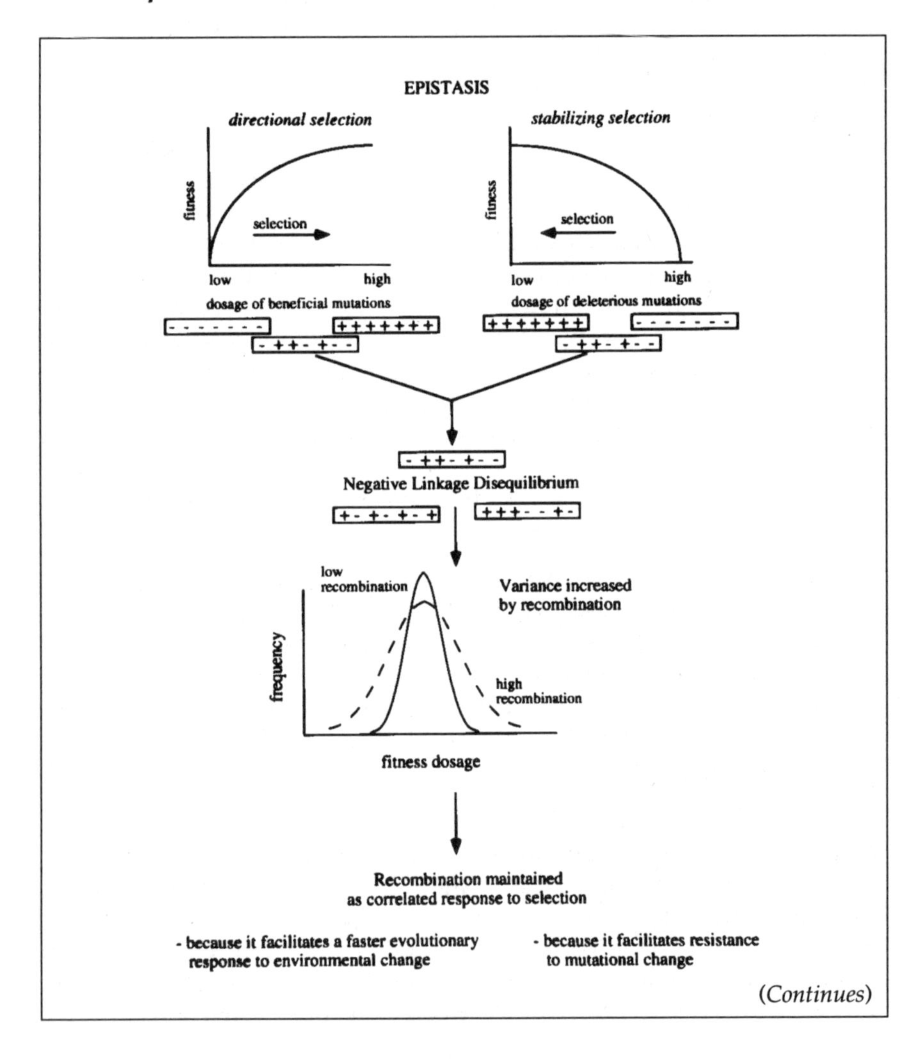


(*Continues*)

*(Continued)*

This sketch epitomizes the rather complex argument in the text describing how recombination might be selected through its effect on the rate of evolution. Stabilizing selection, as used in the drawing, means selection against new mutations.

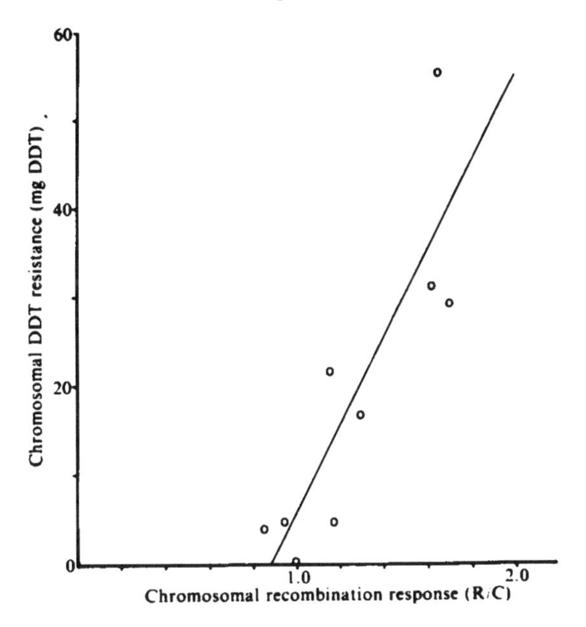


This figure shows how the response to directional selection for DDT resistance is related to recombination in chromosomes extracted from *Drosophila* selection lines, from Flexon and Rodell (1982). Resistance is measured as the median dose of DDT causing knock-down of the daughters of selected females; recombination was measured from progeny frequencies of loosely-linked markers. The parallel work on temperature resistance is by Gorodetskii et al. (1990).

Recombination will affect the direct response to selection because it will reduce interference caused by the correlated response at linked loci. Recombination, however, is not a fixed constraint: the rate of recombination is itself a character that can evolve through selection. In the laboratory, rates of recombination can be selected directly by choosing families with exceptional frequencies of recombinants (Sec. 48). Under natural selection, recombination must evolve as a correlated response to selection for other characters. It will do so if alleles causing higher or lower rates of recombination are selected along with the genotypes they create, a process of hitch-hiking (Sec. 45).

**Second-Order Selection of Recombination.** The correlated response of recombination to selection on characters that are encoded by several or many genes seems an appealing general approach to explaining how recombination evolves, but it involves a serious difficulty. If recombination is occurring at all, then any gene that influences the rate of recombination is likely to become separated from the genotypes that it creates. A gene that influences crossing-over on another chromosome, for example, will be inherited independently of the recombinant and non-recombinant chromosomes whose formation it directs. The situation will not be much different if it influences crossing-over at the other end of the chromosome where it is itself located. The selection of genes that modify the action of other genes is second-order and is always weak when the modifier locus is likely to be separated from the loci it modifies by recombination. The difficulty is obviously more acute for genes that increase rates of recombination, and are therefore more liable to distance themselves from their effects, than for genes that decrease rates of recombination and that could then be transmitted faithfully as part of the congealed chromosome they have created.

## *Selection for Low Recombination Through Stabilizing Selection*

**Recombination and Genetic Variance.** The key to understanding how recombination responds to selection is its effect on genetic variation. Let us use a simple model, such as that of Sec. 50, with a string of + and - signs representing a series of loci bearing alleles that increase or decrease the value of some character. The genetic composition of the population, in terms of the frequencies of individuals or chromosomes with different sequences of + and - alleles, depends on how the alleles at different loci are correlated. If they are positively correlated, the population consists mainly of extreme types of individuals bearing + + + + + + and - - - - - - chromosomes. This is called positive linkage disequilibrium. If the loci are negatively correlated, the population consists mostly of intermediate kinds of individuals bearing (say) + - + - + - kinds of chromosome, and is said to be in negative linkage disequilibrium. A population in which the loci are uncorrelated is in linkage equilibrium, and consists of a mixture of extreme and intermediate individuals whose chromosomes are random combinations of + and - alleles. Recombination is a randomizing process and, therefore, always tends to reduce any correlation among loci towards zero; a population that is currently in disequilibrium, positive or negative, tends to be restored to linkage equilibrium by recombination. The effect of recombination on genetic variation depends on the sign of linkage disequilibrium. With positive linkage disequilibrium, matings between the extreme + + + + + + and - - - - - - types will yield mostly intermediate offspring with + - + - + - chromosomes. The effect of recombination is thus to reduce variation. If there is negative linkage disequilibrium, on the other hand, matings between + - + - + - parents will yield some extreme progeny; in this case, the effect of recombination is to increase variation. Recombination ceases to have any effect on variation once the population has reached linkage equilibrium.

**Negative Linkage Disequilibrium is Generated by Stabilizing Selection.** Stabilizing selection (Secs. 26–27) favors individuals with intermediate phenotypes. These will have + - + - + - kinds of chromosomes; the loci will be negatively correlated because if a genotype bears a + allele at one locus, its fitness is likely to be enhanced if it has a - allele at some other locus. The population is thus in negative linkage disequilibrium.

**Selection for Reduced Recombination.** Recombination will therefore tend to increase variation by generating more extreme combinations of + and - alleles. Families with high rates of recombination will produce a greater range of offspring, with a greater proportion of offspring with extreme phenotypes and a smaller proportion with intermediate phenotypes. But under stabilizing selection, individuals with extreme phenotypes have lower fitness. Stabilizing selection therefore favors families in which the rate of recombination is relatively low, thereby increasing the frequency of genes that reduce rates of recombination. Low rates of recombination will thereby evolve as a correlated response to stabilizing selection on quantitative characters.

## *Selection for High Recombination through Directional Selection*

**Directional Selection with Epistasis.** The situation is somewhat more complicated in the case of directional selection, where there is no general rule for the relationship between recombination and variation. It depends on how the loci interact in determining the phenotype. Let us suppose that the population is currently dominated by - alleles, and that the environment changes so as to favor types with a larger number of + alleles, so that selection will move the population from types such as - - + - - - or - - - - + - towards types like + + - + + + or + + + + - +. In the simplest case, the loci act independently, in the sense that the fitness of any combination of alleles can be calculated by multiplying together their separate effects on fitness. Because the effects of the loci are independent, selection will not generate any correlation among them, and the population will remain in linkage equilibrium as it evolves. Recombination will have no effect on variation and will, therefore, have no effect on the progress of directional selection. On the other hand, suppose that genes interact in such a way that a few + alleles cause a large increase in fitness, relative to having none, whereas the addition of more + alleles causes only a small further increase in fitness. This seems likely to be generally true. Thus, the fitness of a genotype increases less steeply than multiplicatively as the number of + alleles it bears increases, rising asymptotically as it approaches the composition of the types favored in the new environment. This kind of epistasis will lead to negative linkage disequilibrium, because types with intermediate numbers of + alleles will have much higher fitness than those with few + alleles and will, therefore, increase swiftly in frequency; whereas extreme types with many + alleles are not much fitter than the intermediate types and will spread more slowly. Recombination will therefore tend to increase variation, and families with greater rates of recombination will produce a greater proportion of extreme offspring. Under directional selection, one category of extreme offspring has high fitness and will increase in frequency. These types are more likely to be generated in families with high rates of recombination, so that as they spread, genes for high rates of recombination will be carried along with them. Increased rates of recombination will thus evolve as a correlated response to directional selection.

Once the optimal phenotype has evolved, selection ceases to be directional and becomes stabilizing, preserving individuals with nearly average phenotypes. Increased recombination will evolve only as long as directional selection continues; once it ceases, selection will cause recombination rates to drop. Substantial rates of recombination can be maintained only if directional selection continues indefinitely, and because selection never favors the indefinite exaggeration of any character except fitness, this will occur only if fitness is continually eroded, either by environmental change or by mutation.

**Selection for DDT Resistance in *Drosophila*.** Philip Flexon and Charles Rodell investigated the response of recombination to selection for DDT resistance in *Drosophila*. They set up three selection lines by placing strips of filter paper soaked in a solution of DDT on the surface of the food medium, gradually increasing the concentration of DDT throughout the experiment. After about 20 generations of selection, DDT resistance had increased markedly, the flies being able to tolerate DDT at more than 10 times the concentration that was initially fatal. The rates of recombination between marker loci on the three major autosomes were then measured. On one chromosome, they were no different from rates in wild-type flies or unselected control lines. On the other two, however, they had increased substantially. Moreover, the two chromosomes where recombination had increased were the two that were primarily responsible for the increased resistance to DDT.

**Adaptation to Fluctuating Temperature in *Drosophila*.** *Drosophila* populations are usually maintained in the laboratory at 25 C. V.P. Gorodetskii and his colleagues at the Institute of Ecological Genetics in Moldavia subjected selection lines to diurnal fluctuations of increasing intensity, with the temperature eventually falling to 11 C and rising to 33 C every day. There was the expected direct response to selection: the selection lines evolved greater viability and fecundity after being cultured in these conditions for about 50 generations. There was also a correlated response of recombination, which increased substantially in several regions of the autosomes, and perhaps also in a restricted region of the X chromosome. In some regions of the autosomes, recombination rates were twice as great in the selection lines as in the controls.

**The Maintenance of Recombination by Deleterious Mutations.** These experiments show that a transient increase in recombination may evolve as a correlated response to directional selection in a novel environment. It is not obvious why selection should continually change direction so as to maintain substantial rates of recombination, but later in the text I shall argue that this is the general outcome of the coevolution of antagonistic organisms, such as hosts and parasites. Even if external conditions are constant, however, the continual slight degradation of adaptedness by mutation will cause continual weak directional selection for individuals bearing fewer deleterious mutations. This process will also (and for the same reason) favor recombination, provided that the mutations are epistatic. Suppose that fitness is plotted as a function of the number of deleterious mutations borne by a genome. It seems likely that one or two mutations might have rather little effect, but that beyond a certain load any further mutations would sharply reduce fitness, the last straw breaking the camel's back. Fitness would then decline more and more steeply as load increases. This is the same pattern of epistasis as before, with deleterious mutations taking the place of genes with low fitness in a changed environment. In the most extreme case, mutations would be harmless up to a certain threshold dosage, but lethal if this threshold were exceeded, even by a single mutation. An asexual lineage in which recombination did not occur would accumulate mutations until this threshold were reached, individuals dying if they sustained any more. All surviving individuals would at this point bear exactly the same number of mutations. If recombination occurred, however, some of the recombinant progeny would bear more than the threshold number of mutations and some fewer. Those that bore more would die. The mean fitness of the *surviving* progeny would therefore exceed the mean fitness of the progeny from a non-recombining lineage. The process is essentially the same as before, with high rates of recombination being selected because they increase the variability of the lineage in which they occur, thereby enhancing its response to selection.

**The Effect of Sex on Fitness in *Chlamydomonas*.** Jack da Silva and I maintained clones of *Chlamydomonas* for about 1000 generations. The clones were paired, each pair being founded by two sibs from the same cross, one of each gender, or mating type. At the end of the culture period all of the lines displayed substantial genetic variation in growth rate and were presumably close to mutation-selection equilibrium. We then crossed the sister lines. We expected that the lines would be in negative linkage disequilibrium because of selection against recurrent deleterious mutations with epistatic effects, so that recombination would increase the genetic variance of growth rate among the sexual progeny. Culturing these progeny as a mixed population would cause the selection of types with higher rates of growth, so the mean fitness of the population would rise. This effect would be transient if the sexual episode were not repeated, and mean growth rate would eventually fall, as new mutations again accumulated, to the level of comparable unmated lines. These expectations were not realized. Sex increased variation in some lines, but not in others. Sexual progeny suffered a slight decrease in mean fitness that was restored to its previous level by a few dozen generations of selection, but at no point consistently exceeded it. There is as yet no experimental evidence for the evolution of recombination as a correlated response to selection against deleterious mutations.

## 88. *The correlated response to selection may involve shifts in character correlations and the divergence of replicate lines.*

The elementary theory of the correlated response to selection assumes that genetic covariances remain constant, in the same way that the elementary theory of the direct response assumes that genetic variances remain constant. This assumption has been examined in a variety of different kinds of experiment.

**Continued Artificial Selection for a Single Character.** If two traits are correlated in transmission but selected independently, then selection on either will tend to reduce the genetic correlation between them. This is because when both traits are close to their optimal states, artificial selection on one will move the other away from its optimum; this will be opposed by natural selection favoring individuals in which the trait is less modified, reducing the correlation between the primary and secondary traits. If the two traits are selected coordinately, then a change in the value of either, caused by selection, will modify the optimal value of the other. Whether this will result in any alteration of the way in which they are related to one another depends on whether the present genetic regression will cause the secondary character to change in such a way as to maintain its optimal value, given the change in the primary character.

An experiment to investigate the effect of directional selection on the structure of genetic correlations among characters was conducted in Linda Partridge's laboratory in Edinburgh, using lines of *Drosophila* that had been selected upwards and downwards for thorax length for about 20 generations. In general, correlations with other morphometric traits related to overall size, such as wing length or tibia length, remained more or less unaffected by selection, presumably because the optimal values of these characters changed in proportion to the direct response of thorax length to selection. However, the correlation of these morphometric characters with bristle number shifted substantially: in both control (randomly selected) and downward selection lines, the low negative correlation expressed by the the base population changed sign, becoming much larger in magnitude and positive. Thus, when a character is selected, its correlation with some other characters may remain unaffected, whereas its relationship with others may be dissolved and recast in new form.

**Simultaneous Selection for Two Characters.** Two-trait selection or, more generally, index selection, is in principle a powerful way of altering genetic correlations. If two characters are selected in the same direction, genes that cause them both to vary in that direction will be fixed, whereas those that cause them both to vary in the opposite direction will be lost; after a sufficiently long period of time, therefore, most of the genes still segregating will affect the two characters in opposite senses, causing a negative genetic correlation between them. Conversely, selection in opposite directions will tend to create positive genetic correlations. Progress through simultaneous selection on two or more characters will thus eventually be hindered by antagonistic genetic correlations arising as the result of selection. Arguments of this sort have been most extensively developed for characters that affect components of fitness under natural selection, such as fecundity and longevity, and are discussed at length in Part 3.A. In the Sheridan–Barker experiment described in Secs. 82 and 85, the expected result was not obtained: the genetic correlation between coxal and sternopleural bristles actually tended to increase when both were selected in the same direction. There was so much variation among replicate lines, however, that it might be more accurate to say merely that genetic correlations are often severely disturbed by two-trait selection, often in unexpected ways.

**The Evolution of Composite Cross II.** The largest single sorting experiment ever run is the Composite Cross II, set up by H.V. Harlan and M.L. Martini at Davis, California, in 1928. The base population comprised equal quantities of seed from each of the 378 possible crosses between 28 barley cultivars taken from all of the regions of the world where barley is grown. Every year since, it has been allowed to set seed naturally and has then been harvested without deliberate selection, and replanted in the following year. The very detailed studies of this population that have been carried out since it was set up nearly 70 years ago have shown that all the 20 or so Mendelian loci identified through their clear-cut effect on a morphological or biochemical character have pleiotropic effects on other characters. When an allele has the effect of increasing plant survival or seed weight per spike, it usually tends to increase in frequency, often very rapidly. Such changes in gene frequency alter genetic correlations; but more importantly, the direction of pleiotropic effects on these characters changes from year to year. These changes were often associated with marked environmental fluctuations, especially in the availability of water, showing how genetic correlations caused by pleiotropy are sensitive to environmental fluctuation. Moreover, the genetic structure of the population, in terms of the prevalent combinations of genes, also shifted substantially between years, again apparently in response to year-to-year environmental changes. The pattern of genetic correlation among characters has therefore been in continual flux throughout the history of the experiment. The overall trend has been a gradual change from several groups of a few rather highly correlated characters to fewer groups of more characters that are less highly correlated. This may be attributable to the gradual breakdown of linkage by a low average rate of recombination (barley is about 99% self-fertilized). The membership of the sets of correlated characters also changed through time, however, and this is presumably due to selection, varying in direction from year to year. The general picture that emerges from this experiment is that the pattern of correlation among characters in a large population has an almost protean capacity to be molded and recast by natural selection.

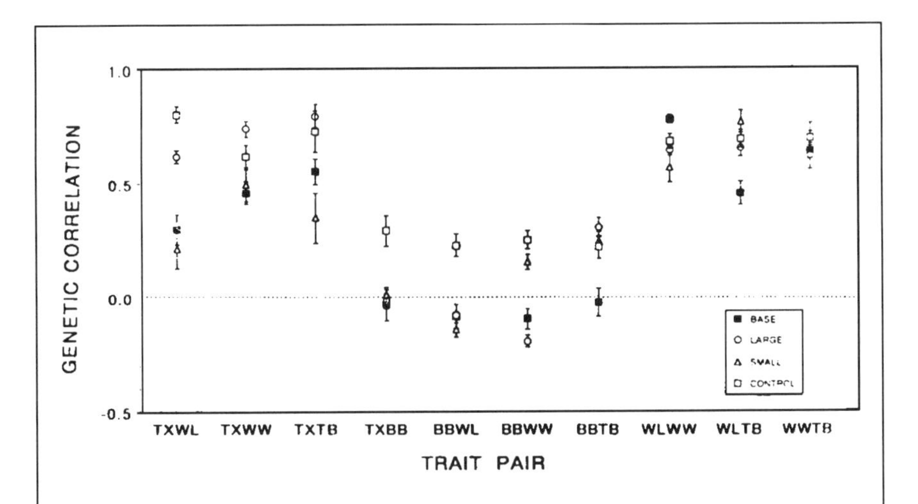


This figure shows the genetic correlations among pairs of characters in lines of *Drosophila* selected for thorax length, relative to the correlations in the base population, from Wilkinson et al. (1990). The characters are thorax length (TX), wing length (WL), wing width (WW), tibia length (TL) and sternopleural bristle number (BB). The solid symbol represents the base population in each case, the open symbols the selection lines (circle, upward; triangle, downward; square, random). Note the shift to positive correlation between body size and bristle number in at least some selection lines. Substantial changes in genetic correlations as the consequence of directional selection for a single character have also been reported by Clayton, Knight, Morris and Robertson (1957, bristles in *Drosophila*), Bell and McNary (1963, pupal weight in *Tribolium*), Bell and Burris (1973, larval and pupal weight in *Tribolium*), and Berger (1977, pupal weight and brood size in *Tribolium*).

For the effects of two-trait selection on the genetic correlation, see Sen and Robertson (1964), Bell and Burris (1973), and Sheridan and Barker (1974b).

The latest and most extensive report on Composite Cross II is by Allard et al. (1992).

# 3.A. Selection Acting on Different Components of Fitness

## 89. *Continued selection on one character causes a general regression of others.*

The way in which characters will change as a correlated response to selection is usually difficult to predict in detail for more than a few generations into the future. Nevertheless, it is possible to predict a general direction of change: *the advance made by the primary character will cause other characters to regress.*

The notion that by improving one character we make others worse implies that "worse" can be given an objective meaning. This meaning is that any change in the value of a character can be expressed in terms of its partial effect on fitness. Mean fitness increases as the consequence of selection on the primary character; the correlated response of other characters is such that were these characters to change independently in the same way fitness would be reduced. In this sense, the correlated response will generally be *antagonistic* to the direct response.

**Evolutionary Consequences of Domestication.** The most familiar example of antagonistic selection is the regress of unselected characters in crops. Plants such as potatoes or sugar cane are valued for their vegetative yield, and the same parts that provide the yield also usually provide the stock: potato tubers or short lengths of cane. Although breeding from seed is sometimes employed to generate new varieties, propagation in such crops is usually vegetative, and there is very little selection to maintain sexual function. Consequently, flowering and seed viability are often impaired. The effects of long-continued vegetative propagation on sexual competence are discussed further in Sec. 162.

**The Hierarchy of Characters.** All characters affect fitness in some way that could be described as a chain of physiological and ecological causes and effects. A particular mutation causes the synthesis of an altered protein, that modifies the development of pigmentation so that the shell bears five bands rather than three, making a snail living in short turf more conspicuous to visual predators, and thereby reducing its chances of surviving to reproduce and, hence, its fitness. It is often useful to think of the phenotype as being organized in this hierarchical way, as a series of levels intervening between the mutation itself and its eventual consequence in terms of fitness. These levels are successively genetic, physiological, developmental, and ecological. When a thrush kills the snail, this act is the manifestation of events at all of these levels and determines, in this particular case, their effect on fitness. More precisely, the summation of many such acts determines the fitness of the allele in question, which will become more or less frequent depending on the rate of increase of the lineage that bears it. Individuals do not have a rate of increase; rather, each contributes through reproduction to the rate of increase of a lineage. The final phenotypic manifestation of a given genetic change is therefore its effect on the reproduction of an individual.

The hierarchy of characters underlying fitness is analogous (no more than that) to the hierarchy of trophic levels that directs the flow of energy through an ecological community: green algae fix atmospheric carbon dioxide, are eaten by rotifers and copepods that are themselves eaten by fish larvae and ctenophores, and so forth up to the pelagic fishes and cephalopods that constitute the highest level of consumers. The one should be taken no more literally or seriously than the other. A fish occupies one trophic level as a larva, another as fry, yet another as an adult and may at each stage occupy different levels in different communities. In the same way, a character may in some circumstances stand remote from any substantial effects on fitness, yet in others be the primary determinant of failure or success. I do not mean to imply that physiologcal characteristics can be arranged in any fixed order that applies generally to all conditions of growth, but only that in given circumstances it will be possible to recognize a succession of events, from the gene sequence to the encoded protein and then through many strata of phenotypic effects, to the eventual outcome in terms of survival and reproduction.

The schedule of reproduction during the life of an individual constitutes its *life history*. The life history of organisms such as insects or annual plants is particularly simple: each surviving individual reproduces at a certain age and then dies. Its reproduction is therefore determined by whether or not it survives to reproduce, by the age at which it reproduces, and by the quantity and quality of offspring that it produces. These characters have directional effects on fitness: other things being equal, an allele that causes individuals to survive well, reach maturity early, and produce numerous and well-provisioned offspring will spread, because the lineage that bears it will proliferate rapidly. Such characters represent the topmost level in the hierarchy of phenotypic effects and can therefore be termed *components of fitness*. The principle of antagonism can be defined unequivocally in terms of components of fitness, because each component, considered separately, has a directional effect on fitness: if any one component of fitness advances, all others will regress.

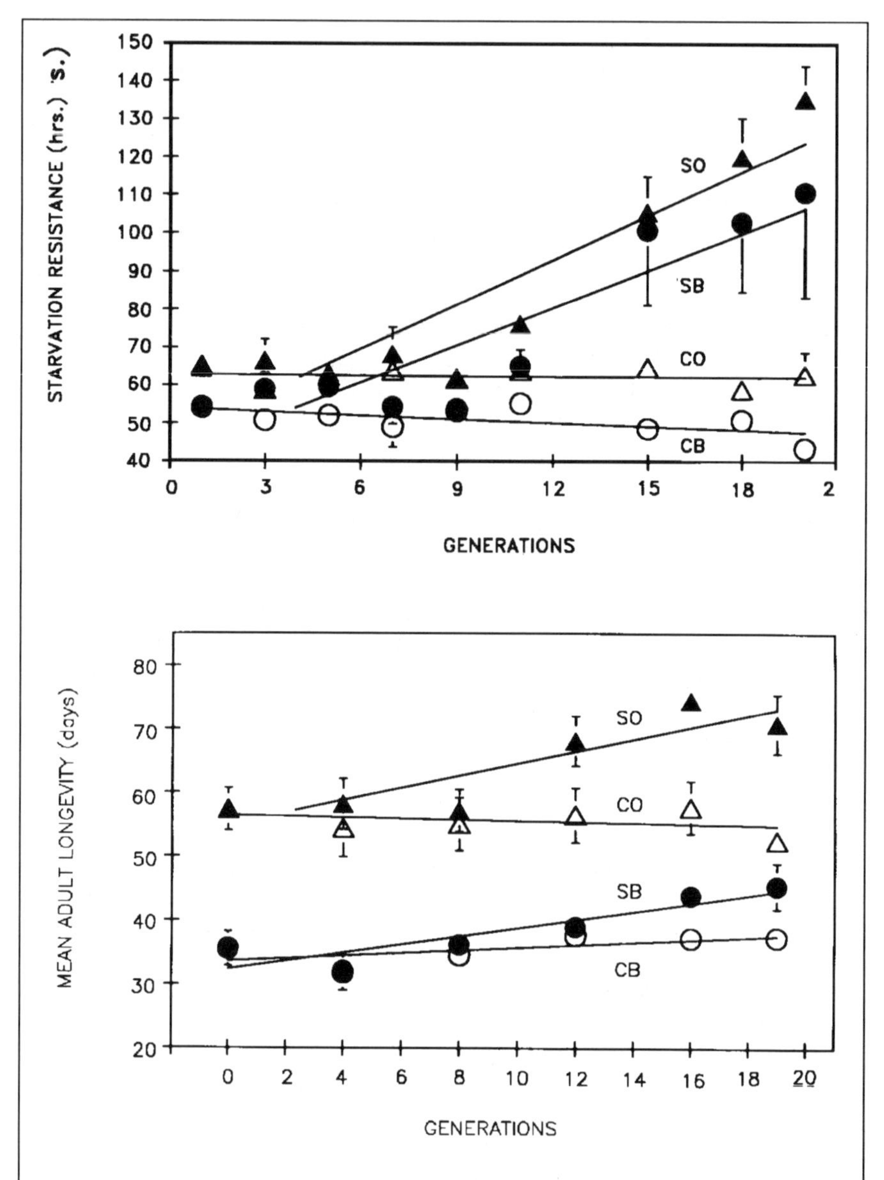


The upper figure shows the direct response to selection for starvation resistance in *Drosophila*. S and C denote selection and control lines; the O lines had previously been selected for increased longevity, the B lines being maintained on a normal culture schedule. The lower figure shows the correlated response of longevity,

(*Continues*)

(*Continued*)
the lines selected for increased stress resistance evolving greater longevity. These figures are from an experiment by Rose et al. (1992); see also Service et al. (1985), Service (1987), and Chippindale et al. (1993). For the correlated response of stress resistance to selection for increased longevity, see Graves et al. (1992). The background to these experiments, and others concerned with the evolution of longevity, is given by Rose (1991).

**The Correlated Response to Selection for Increased Longevity.** A very instructive series of experiments on life histories has been put in train by Michael Rose at Dalhousie and, subsequently, at Irvine. Their relevance to general theory will be described in Sec. 99, but their most fundamental outcome has been that it is possible to evolve lines of *Drosophila* with greater longevity. This is in itself of great interest because it demonstrates that components of fitness can be altered through selection and, therefore, that it is reasonable to interpret life histories as adaptations to particular circumstances. The hierarchical view of the phenotype that I have advanced, however, implies that any increase in longevity is underlaid by particular physiological changes that would evolve as the correlated response to selection aimed directly at longevity. This is readily demonstrated by assays of physiological characteristics in the selection lines. These have shown that the long-lived lines are characterized by a rather generalized increase in stress resistance. In particular, the storage of reserve substances, such as lipids and glycogen, is enhanced by selection for increased longevity.

**Increased Longevity as a Correlated Response.** The converse should also be true. If increased longevity evolves as the consequence of physiological mechanisms that increase resistance to stress, then selection for increased stress resistance should lead to increased longevity. This as also been tested experimentally. Lines selected for increased resistance to starvation or desiccation evolve greater longevity as a correlated response. The hierarchy of physiological characters that determine the life span can thus be identified and analyzed through selection experiments in a much more powerful and convincing way than has hitherto been possible.

## *90. Components of fitness are antagonistic because resources are finite.*

A general principle of antagonism is almost self-evident. If there were, in general, no deleterious consequences of selecting upward on a component of fitness, then it would be possible to select each component, simultaneously or sequentially, so as to produce an organism that could live forever and that as soon as it was born would begin producing enormous numbers of large offspring as fast as resources became available. This is made impossible by the primary constraint responsible for selection: that the world is finite. Each individual can gather only a limited quantity of resources and can, therefore, accomplish only a limited total quantity of reproduction. Insofar as different components of fitness impose separate demands on a common pool of resources, they must necessarily be antagonistic.

The simplest case of competing demands made by components of fitness is the partition of resources among propagules. Other things being equal, producing more offspring will increase fitness. But producing larger offspring will also increase fitness, because larger offspring are more likely to survive. Given a fixed total quantity of resources, producing more offspring must mean that they will be smaller, whereas producing larger offspring involves producing fewer.

**The Cost of Defense in a Water Flea.** There are two morphological types of the small planktonic water flea *Bosmina*: one is a long-faced type that develops a larger body, thicker carapace, and longer spines, and is less heavily preyed on by copepods. Development of the young, however, takes place in a dorsal brood pouch with a strictly limited volume. A short-faced type, therefore, although relatively defenseless against copepods, bears more young. W. Charles Kerfoot of Dartmouth College inoculated screened bottles hung in a lake with samples of both types and recorded changes in frequency for a week or two, amounting to between two and four generations. In these conditions, the short-faced types rapidly increase in frequency as long as resources are abundant, being protected from predation. In the lake, the long-faced type predominates in the open water, where predatory copepods are an important source of mortality, whereas the short-faced form is more abundant in the littoral, where the copepods are removed by planktivorous fish.

**Components of Yield.** This is a familiar problem in agronomy. In plants such as cereals that are grown for grain, the total yield of an individual $Y$ is determined by the number of infructescences per plant $I$, the number of fruits per infructescence $F$, the number of seeds per fruit $S$, and the mass per seed $M$,

$$Y = I \times F \times S \times M$$

$I$, $F$, $S$, and $M$ are *components of yield*. To what extent is it possible to increase total yield by selecting for increased values of a single component of yield?

## 91. *The correlation between fitness components depends on the balance between variance in allocation and variance in productivity.*

The components of yield represent successive partitions of total yield into different levels of structure. Because total reproduction is fixed for any given plant, this partition is necessarily antagonistic, with an increase in one component implying that other components must decrease. We therefore expect that components of yield will be negatively correlated. If a given genotype expresses an increase in one component, it will express a decrease in other components; there will be a negative *genetic* correlation among components of yield. Selection for advance in one component will thereby cause an antagonistic correlated response, reducing the value of other components and will produce little if any increase in total yield per plant.

This commonsense argument is often justified. Characters such as seed number and seed mass in cereals or pulses are often negatively correlated. Moreover, the correlation can be altered by manipulating the total quantity of resources available to the plant: when the plant is stressed by low nutrient availability, the conflict between components of yield for resources is exacerbated, and the correlations among them become more highly negative. The same phenomenon occurs in other organisms: increasing the brood size of altricial birds by adding eggs to the nest, for example, causes a reduction in mean weight at fledging. Nevertheless, components of yield do not necessarily, or invariably, show negative correlation. The reason for this can be understood through the "house–car paradox". Household income is limited, and what is spent on one commodity is not available for purchasing another. We might therefore expect, following the same line of reasoning as before, that people living in grand mansions would drive small, cheap cars, whereas large and expensive models would be found outside modest tenements in less desirable areas of town. This is notoriously untrue. The reason is that although the income of any given household is limited this income varies widely among households, and a rich family can afford to buy both a large house and an expensive car, whereas a poor family can afford neither. The negative correlation that would be created by variation in the allocation of resources is over ridden by the positive correlation created by variation in the total quantity of resources.

The total quantity of resources available for reproduction may vary either because individuals have different opportunities or because individuals have different capacities. A population dispersed in a heterogeneous environment will tend to display positive correlations between components of yield because some individuals will be growing in favorable sites, whereas others will have fallen on stony ground. This environmental variance can be eliminated experimentally by replication and randomization, but will always appear among natural populations of plants growing in situ. Even in a perfectly uniform environment, however, there may be genetic variance in the ability to gather resources. This will be particularly troublesome in two circumstances involving poorly adapted populations in novel environments. The first is when individuals or families newly collected from the field are tested in the laboratory. Some genotypes will by chance be well adapted to laboratory conditions of growth and will express high values of all components of yield; others will be poorly adapted and in all respects inferior. The second circumstance is when new lines are produced by mutation or inbreeding, so that some are likely to express unconditionally deleterious alleles. In either case, the result will be positive genetic covariance among components of yield.

The genetic covariance among components of yield, and hence the direction of the correlated response to selection, will thus depend on the ratio of two variances: the variance of total resource-gathering ability, or total yield, and the variance of allocation to different components of yield. When the variance of total yield is high, covariances will tend to be positive; this is likely to be the case in novel or heterogeneous environments. It is only when most of the variation of yield components among individuals is attributable to variance in patterns of allocation, rather than to variance in total yield, that components will be negatively correlated, directing an antagonistic correlated response to selection for any one of them.

**Comparative Biology of Yield Components in *Plantago*.** The relationship between components of yield in natural populations of plants is illustrated by Richard Primack's study of plantains, *Plantago*, a very abundant and widely distributed herb, with over 200 species growing in different habitats throughout the world. This offers the opportunity to look first at variation among species. All the components of yield vary among species, although $I$ and $F$ vary much more than $S$ and $M$, which is the usual case in plants. Thus, if all four components were independent, it is $I$ and $F$ that would be most important in regulating total yield. They, however, are not independent: $I$ and $F$ are strongly correlated, as are $S$ and $M$. Indeed, the regressions of $F$ on $I$, and of $M$ on $S$, have slopes that are close to $-1$, indicating a nearly perfect degree of compensation, with a unit increase in one character being accompanied by a unit decrease in the other. Thus, the total number of fruits per plant (fruit number, $I \times F$)

and the total mass of seeds per fruit (fruit mass, $S \times M$) seem to represent more or less fixed quantities that do not vary much among species and that show very clear negative correlation between their components. The relationship between fruit number and fruit mass is more complicated. Both are good components of yield in that both are correlated with yield $Y$. Fruit number, however, is the better predictor of yield because it is more variable than fruit mass. This is partly because annual and perennial species tend to produce different numbers of fruits: annuals have more fruits per plant than perennials. The correlation between fruit number and fruit mass is negative, as expected, although not very strongly so. It is the consequence of two different relationships. The first is that seed number varies according to life history: perennials bear more seeds per fruit than annuals and have fewer fruits per plant. Species with similar life histories show no pronounced correlation between seed number and fruit number. The second is that seed mass varies independently of life history: annuals and perennials produce seeds of about the same size, but among species with similar life histories those with fewer fruits produce heavier seeds. The overall negative correlation between fruit mass and fruit number is thus a consequence of negative correlations between life histories through seed number, and within life histories through seed mass.

The pattern of negative correlations among yield components among species is consistent with a basic principle of antagonism. However, Primack found that correlations among individuals of the same species were usually positive. The reason for this is that there is less genetic variation within species than among species, so that a greater proportion of the variation within species is environmental. Natural populations growing in a heterogeneous environment will vary in total reproduction, and the positive covariance contributed by the environment will obscure any negative covariance among genotypes. This can be demonstrated by raising plants from a single species in the greenhouse. When fruit number and fruit mass of individuals growing in the field and in the greenhouse are plotted on the same graph, they are strongly positively correlated, because the greenhouse plants, supplied with abundant light, water, and nutrients, are simply larger and more vigorous in all respects than the individuals from the field.

In short, the antagonism between components of yield is clearly seen when diverse genotypes that have long evolved in a given environment are compared. When similar genotypes are grown in heterogeneous or novel environments, antagonism can no longer be detected, and the correlations among yield components are predominantly positive.

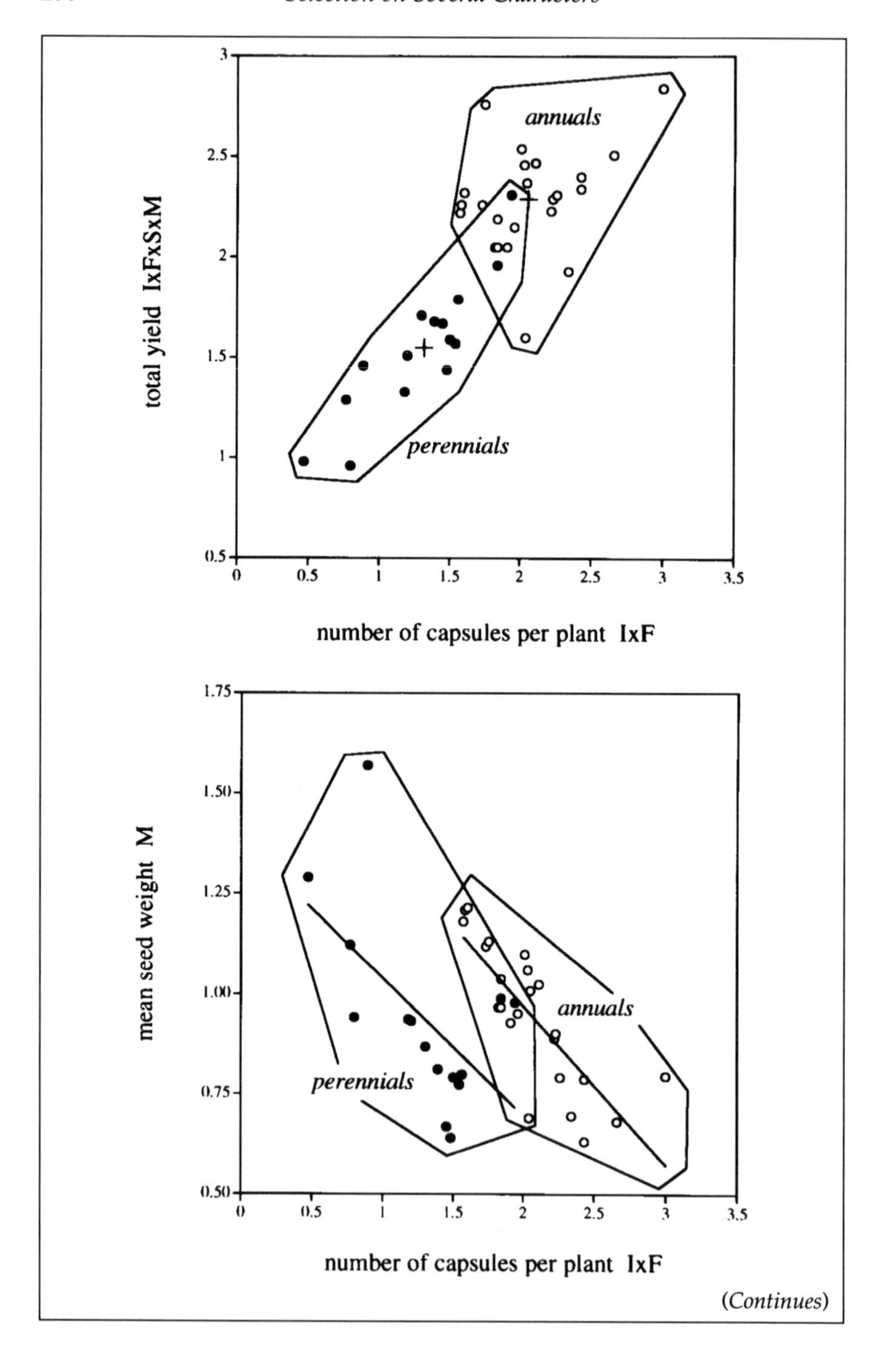

annuals
perennials
total yield IxFxSxM
number of capsules per plant IxF
3
2.5
2
1.5
1
0.5
0
0.5
1
1.5
2
2.5
3
3.5
annuals
perennials
mean seed weight M
number of capsules per plant IxF
1.75
1.50
1.25
1.00
0.75
0.50


(Continues)

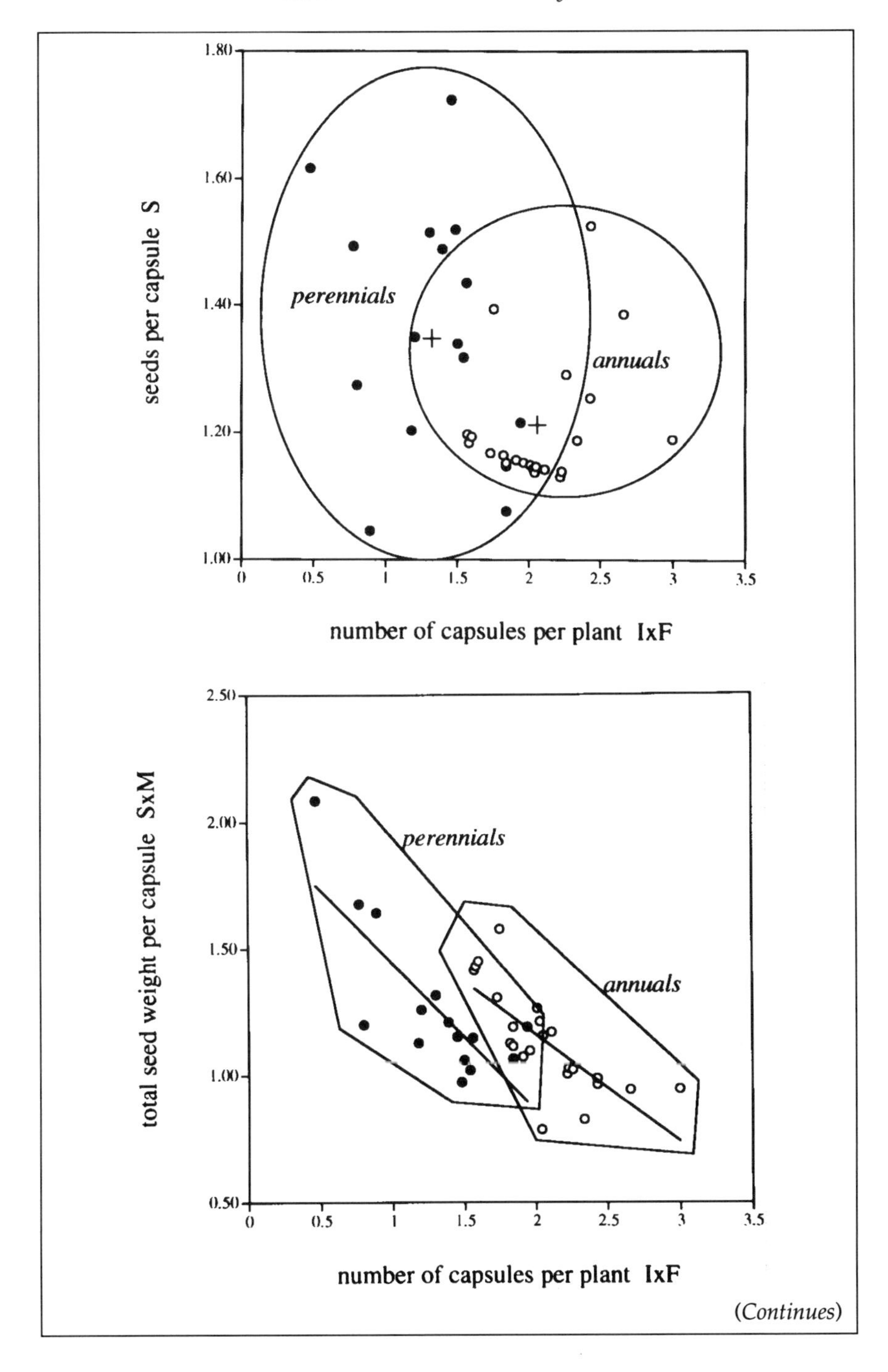
1.80
1.60
1.40
1.20
1.00
seeds per capsule S
perennials
annuals
0
0.5
1
1.5
2
2.5
3
3.5
number of capsules per plant IxF
2.50
2.00
1.50
1.00
0.50
total seed weight per capsule SxM
perennials
annuals
number of capsules per plant IxF

(Continues)

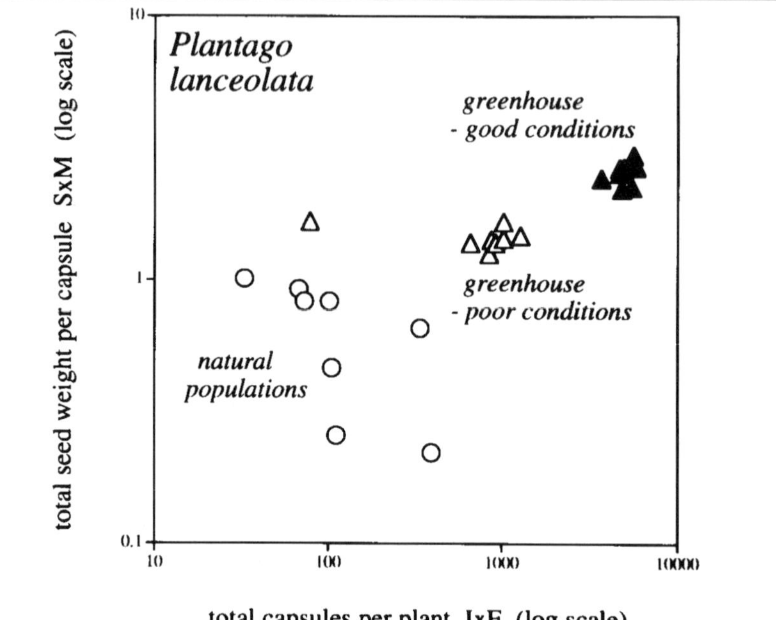


These figures are drawn from Primack's survey of reproduction in *Plantago*, used in the text to illustrate the antagonism of components of yield (Primack 1978, 1979; Primack & Antonovics 1981, 1982). All of these figures, except the last, are interspecific comparisons: each point is a different species of *Plantago*. They focus on the number of capsules per plant, $I \times F$; a similar analysis could be conducted for other components of yield. (There are two or three technical points. Primack obtained data from natural populations and from herbarium specimens; I have used both, so that some species are in fact represented by two points. In the first four figures, the number of capsules is in fact standardized so that it refers to a unit area of leaf, rather than to a whole plant, because overall plant size varies among species. Units of mass are milligrams.)

a) Capsule number is a good component of yield, being strongly correlated with total yield. This is, of course, a part–whole correlation. Note the difference between annual and perennial species, the perennials having fewer capsules and lower total yield; the ecological differences between annuals and perennials are discussed further in Sec. 93. The correlation has two components: the difference between the mean values of the two groups of annual and perennial species and the correlation among species within each group.

b) Although annuals and perennials bear different numbers of capsules, their seeds are about the same size. (Most annuals produce heavier seeds than most perennials. A few perennials with very heavy seeds, however, increase the average of the group as a whole.) Within either group, however, there is a pronounced negative correlation between seed weight and capsule number.

(*Continues*)

(*Continued*)

c) There is no tendency for either annual or perennial species to form fewer seeds per capsule if they produce more capsules. However, annuals on average produce fewer seeds than perennials. The annuals in this diagram that produce rather few seeds per capsule are the same as those in the previous diagram that produce rather heavy seeds.

d) Thus, there is a negative correlation of seed weight with seed number among species with a given life style, but no variation between life styles; whereas seed number per capsule varies between annuals and perennials, but not among species within either group. The combination of these two effects causes a marked antagonism between the two components into which total seed yield can be partitioned that is attributable in part to the difference between annuals and perennials and in part to the differences among species in either group. The overall regression slope is about $-0.5$, so about half of any increase in capsule production is offset by a decrease in capsule weight.

e) This correlation is less marked among populations of a single species and disappears entirely when plants from these populations are brought into the greenhouse. With more favorable conditions of growth, both capsule number and capsule weight increase greatly, creating a positive environmental correlation.

## 92. *Selection produces negative correlations between fitness components.*

A straightforward genetic approach to the evolution of components of yield, by measuring correlations among genotypes in the laboratory or greenhouse, is therefore likely to fail. In order to display the underlying antagonism of such characters,it is necessary to use an evolutionary approach, by showing that negative genetic correlations will evolve during the course of a selection experiment.

If we extract families from a natural population and raise them in the laboratory, it is quite likely that characters such as fruit number and fruit mass will show little correlation. However, suppose that we breed these families together, as members of the same experimental population, for many generations under laboratory conditions. Some genotypes will happen to be well adapted to this novel environment and will produce large numbers of heavy fruits: these will increase in frequency through selection and will quickly become fixed. Others will be very poorly adapted, producing a scanty crop of fruits with few seeds; these will decrease in frequency through selection and will rapidly be eliminated. Genotypes that produce many fruits each with few seeds, or that produce few but heavy fruits, will be selected less rigorously and will therefore persist in the population for longer. After some time, therefore, most of the genetic variation of components of yield will be contributed by such genotypes, in which the components are expressed antagonistically. This is a special case of two-trait selection, the argument being the same as that discussed in Sec. 88, except that it involves natural rather than artificial selection.

Antagonism may arise in this way as the outcome of a simple sorting process. However, it may be hindered by genetic constraints. Genes that by themselves have unequivocally favorable effects may be associated in the base population with inferior genes at other loci and cannot be fixed until this initial linkage disequilibrium has been broken down by recombination. If there are epistatic effects on yield, the most productive combinations of genes may initially be lacking and cannot be selected until they are created by recombination. Thus, sorting may, at first, produce positive correlations among yield components, by enhancing genetic integration. Nevertheless, as favorable combinations of genes spread and inferior combinations decline, continued selection will at last result in the emergence of antagonism.

The force of this argument is that yield components will be consistently antagonistic only in well adapted populations that are close to equilibrium under selection. It enables us to postulate that the evolution of yield is constrained by the antagonism of its components, despite the positive correlations that are often observed in novel or heterogeneous environments.

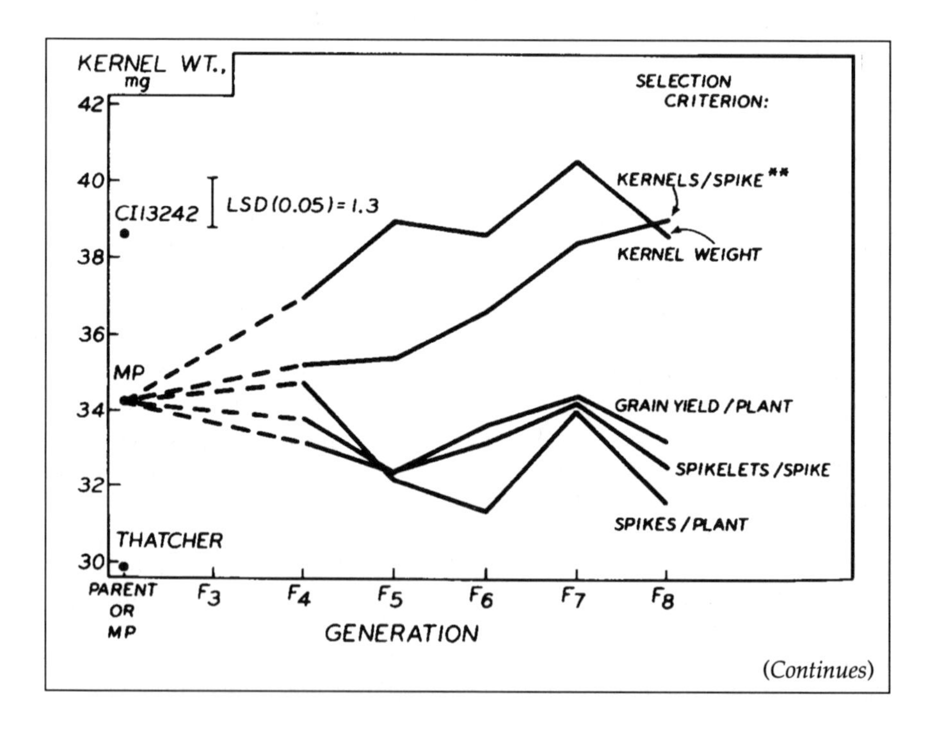


(*Continues*)

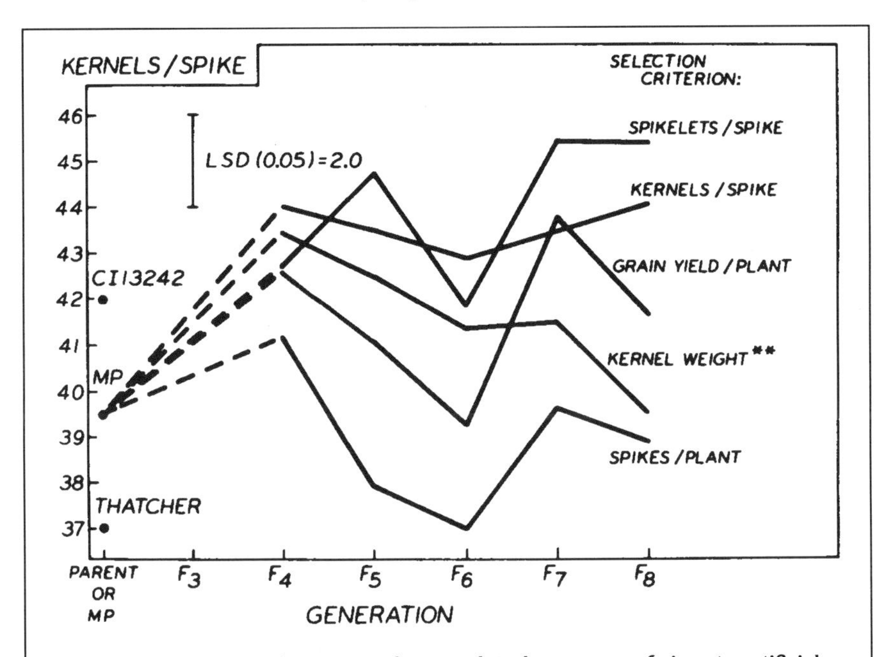


The classic experimental paper on the correlated response of vigor to artificial selection is by Latter and Robertson (1962). (I am using "vigor" in the sense of fitness under natural selection, as distinct from the fitness entailed by artificial selection.) They used lines propagated by only 10 pairs per generation, so that some loss of vigor was caused by inbreeding depression; after 25 generations, the vigor of randomly selected control lines, measured as competitive ability, had fallen to about 60% that of the base population. The vigor of lines selected for bristle number, however, was only 20–40% that of the base population, showing a substantial additional effect of artificial selection on vigor. It has been suggested, on the basis of experience in the poultry industry, that this decline in vigor could be halted by a special type of two-trait selection, selecting simultaneously for the desired phenotype and for high vigor. In practice, this would amount to culling weak, sickly, or infertile individuals from the selection lines. This idea was tested by Frankham et al. (1988), who selected *Drosophila* for resistance to ethanol vapor. There was a straightforward direct response, with 25 generations of selection doubling the inebriation time of the flies, but vigor, again measured as competitive ability, was halved as a consequence. In comparable lines that were selected simultaneously for vigor, by discarding the 20% of females who produced the fewest offspring, there was no detectable loss of vigor. From the explanation that I have given in the text, this result would follow from the antagonistic effects of pleiotropic genes, the culling having the effect of removing genotypes with the most strongly antagonistic effects. This can be expected to maintain vigor, although at the expense of retarding the response of the selected character. This was not the case: the culled

*(Continues)*

(*Continued*)

lines responded as fast as the others. Frankham et al. (1988) explain their result as the outcome of a non-linear heritability of vigor, the heritability being greater for lower vigor, because most deleterious alleles are recessive.

The figures shows the correlated response to selection for components of yield in wheat reported by McNeal et al. (1978). The base population was the $F_2$ of a cross between the two cultivars Thatcher and CI 13242. All four components of individual yield were assayed from generations 4 through 8: I (spikes/plant), F (as spikelets/spike), S (kernels/spike) and M (kernel weight). The two diagrams show the response of kernel weight and kernels/spike. Note that selection for kernel weight causes a direct response while also causing a negative correlated response of kernels/spike. Other responses, however, are inconsistent, perhaps because the experiment was so brief.

**Antagonism of Natural and Artificial Selection.** This antagonism will set a limit to selection: yield cannot increase without limit because at equilibrium the advance under selection of any one component of yield is matched by the regress of others. I have already described a related phenomenon that limits the response to artificial selection in Sec. 50: after the sorting limit of the population has been reached, the only variation available for selection is supplied by alleles that increase character value but decrease viability. In the Clayton–Robertson experiment, for example, response to selection was halted by the appearance of genes that increased bristle number in single dose, but were lethal when homozygous. In cases like this, continued selection creates an antagonism between artificial and natural selection that can be regarded as being analogous to components of fitness, the one supplied specifically by the experimenter and the other arising from the general conditions of culture. This is a great hindrance to the improvement of crops or livestock through selection. For example, the genetic correlation between growth rate and egg-laying rate in fowls is generally positive, although rather low. The selection of broilers with greatly increased growth rate might therefore be expected to pay a double dividend, with the selection lines having somewhat increased fecundity. In practice, however, broiler selection programs soon encounter, not an improvement, but a collapse of egg production. The pattern of genetic correlation in the base population is thus a poor guide to evolutionary change beyond the first few generations of selection.

**Selection for Yield Components in Wheat.** Despite the plenitude of studies of the phenotypic and genetic correlations among components of yield in crop plants, few selection experiments have been reported. F.H. McNeal and his colleagues of Davis, California, selected yield components of wheat for 8 generations. Selection for mean kernel weight was only marginally effective, but caused a decrease in the number of kernels per spike, as expected. Selection for the number of kernels per spike, on the other hand, was completely ineffective, and actually caused an increase in mean kernel weight. In neither case was there any change in total grain yield per plant, so that selecting a given component of yield would not be likely to be an effective way of increasing yield itself. This experiment, however, was set up by selecting a very small number of plants from the $F_2$ of a cross between two inbred lines, and was probably not continued for long enough for antagonistic correlated responses to appear. There is clearly a need for further experimentation to investigate the effect of selection on components of yield.

## 93. *Optimal patterns of reproduction evolve as compromises.*

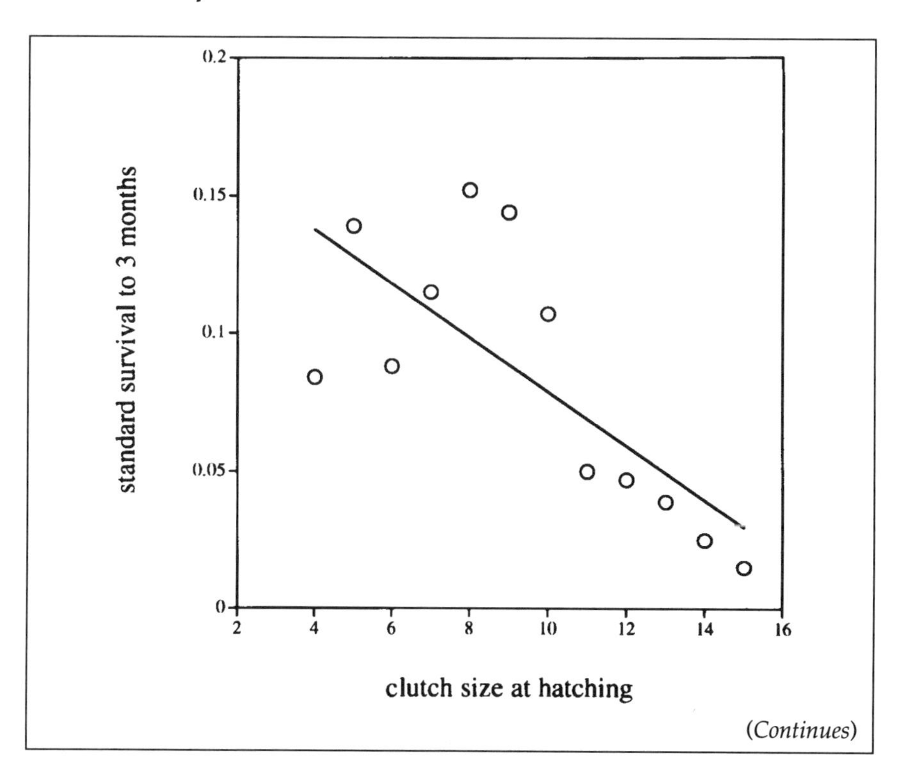


*(Continues)*

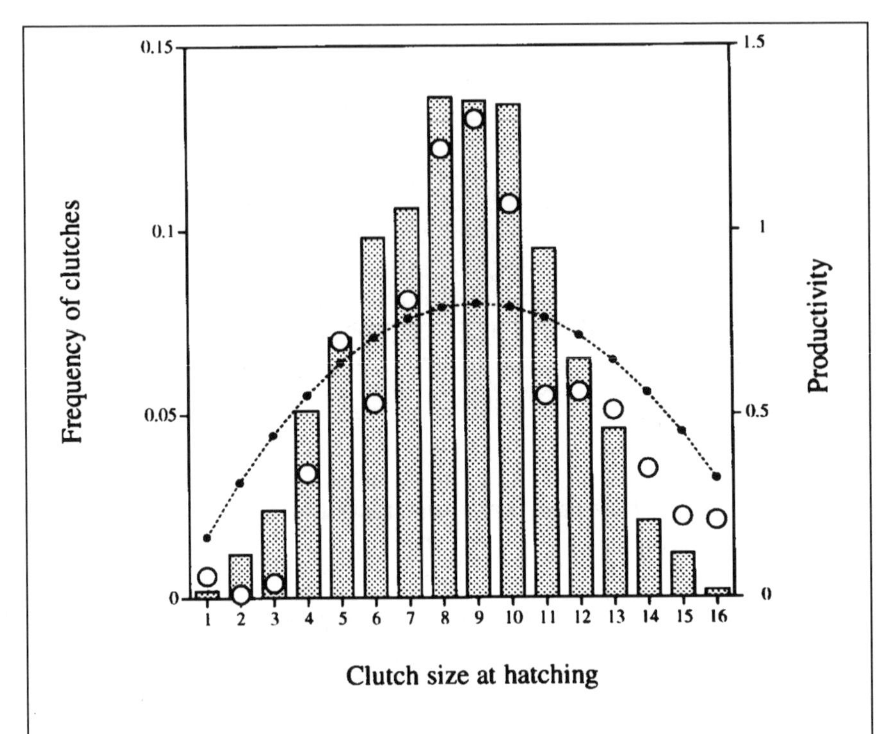


One of the most famous hypotheses of evolutionary ecology is David Lack's contention that the mean clutch size in a population of birds should be its optimal value under natural selection, that is, the clutch size that maximizes the production of independent young. Here, I have used this hypothesis to predict the optimal value of clutch size. The data on clutch size and the survival of young birds in the Wytham population of great tits is taken from Lack (1966, Table 7); a great deal more work has been done at Wytham since then, but there is a certain appeal in using Lack's original data to test his own hypothesis. The data set includes the number of clutches of given size and the number of young birds recovered three months or more after leaving the nest from clutches of given size, for 17 consecutive years. The mean clutch size in the whole sample of 1098 nests was about 8.5 eggs, although some birds produced clutches of as many as 15 or 16 eggs. Both sampling effort and the overall survival of nestlings and fledglings varied from year to year, so I have calculated a standard value of survival by multiplying the observed survival of clutches of given size in a given year by a quantity $(P_{\text{mean}}/P_{\text{year}})$, where $P_{\text{year}}$ is the mean survival rate over all clutches in a given year and $P_{\text{mean}}$ is the mean survival rate over all years. Other ways of dealing with year-to-year variation are possible, of course. The upper figure shows that this standard survival rate decreases with clutch size $B$; the regression is survival rate $= C - sB$, with $C = 0.177$ and $s = 0.0098$. (Very small clutches of 1–3 eggs,

(*Continues*)

*(Continued)*

and very large clutches of 16 eggs, do not fit this relationship, but were found only rarely.) Thus, both small and large clutches are relatively unproductive, and an intermediate clutch size may be optimal. More precisely, the total productivity of a brood, that is, the total number of surviving chicks, is the product of clutch size and survival rate, or $B \times (C - sB)$. The optimal clutch size is that which maximizes this quantity: by simple calculus, this is $B = C/2s$. The lower figure shows the relationship between the observed and predicted values of clutch size. The histogram bars are observed values, a nearly Normal frequency distribution with mean 8.48 and standard deviation 1.75. The large open circles show the actual productivity of clutches, in terms of survival to three months of age. This curve is also unimodal and approximately Normal, with a mode at 9 eggs. The small filled circles give the productivity as calculated from the survival regression, as $B(C - sB)$, yielding an optimal value of $C/2s = 9.03$ eggs. A much more extensive analysis, using data through 1982, has been published by Boyce and Perrins (1987); they calculate the optimal clutch to be 9.01. A straightforward optimization approach seems to give a satisfactory quantitative account of adaptation for characters such as clutch size.

The *Ambystoma* experiment was published by Semlitsch and Wilbur (1989).

Because selection will not favor the indefinite exaggeration of any given component of yield, the physiological and developmental machinery that underlies its expression will not continue to be enhanced beyond a certain point. The antagonism of fitness components prevents organisms from being perfectible. Instead of being perfectly adapted in all respects, organisms will instead evolve as *compromises*, expressing intermediate values of characters such as fruit size or fruit number. Nevertheless, selection will always favor an increase in total yield (or, more precisely, fitness). The combination of values of yield components that emerges through continued selection is therefore an *optimal* compromise, in the sense that it maximizes total yield.

The concept of optimality is very powerful because it enables us to predict how patterns of reproduction should evolve. One should always bear in mind that optimal design is not perfect design; optimality does not imply freedom from constraint, but on the contrary, implies that adaptation is molded by constraint. Nor does any particular pattern of reproduction represent a universal optimum, because the relationship between components of yield will depend on the way of life that an organism follows and the environment in which it lives. This is, indeed, the basis of the claim that patterns of reproduction are predictable. Finally, the concept of optimality applies only to components of yield (or fitness), and not to the multitude of morphological and physiological mechanisms that contribute to these components. There is nothing optimal about the panda's thumb or the xylitol metabolism of *Klebsiella*; these evolve in a contingent fashion through the cumulative modification of pre-existing adaptations and their present state reflects their previous history. The evolution of fruit size or seed number, by contrast, will depend on circumstance rather than ancestry, and at equilibrium the history of such characters will be largely effaced.

**Annual and Perennial Life Histories in *Plantago*.** The pattern of reproduction should therefore vary in a predictable fashion among organisms with qualitatively different ways of life, even if they are closely related species sharing a recent common ancestor. Annual and perennial species of *Plantago* allocate resources to seeds in different ways: annuals produce a smaller number of larger seeds, whereas perennials produce a larger number of smaller seeds. Why should they differ in this way? Perennials occupy their local site by growth and proliferation over several years, and a seed, even a large and well-provisioned seed, has little chance of succeeding in competition with a well-established vegetative plant. The seeds of perennials will thus be adapted primarily to seeking out new sites. Because sites that are favorable to growth but currently unoccupied are likely to be rare, only individuals that produce a large number of seeds are likely to besuccessful; but because their total reproductive capital is limited, the seeds they produce must be small. Annuals have no vegetative propagation and must recolonize the local site each year by reseeding. Larger seeds with more stored reserves and thus faster initial growth are likely to succeed in competition with smaller, less well-endowed seeds; but plants that produce larger seeds must produce fewer. The contrasted patterns of allocation that are found in annual and perennial species of *Plantago* show how the common constraint imposed by a limit on total reproductive output is modulated by the ecological context in which it is expressed.

**Optimal Clutch Size in the Great Tit.** It is, in principle, possible to predict the quantitative allocation of resources between components of yield if the relevant constraints are known. The great tit, a passerine bird of deciduous woodland, has been studied in great detail by David Lack and his successors at the Edward Grey Instute at Oxford. It can lay a clutch of up to about 20 eggs, but most clutches are much smaller: the mean is about 8.5 and the mode between 8 and 10. Why do most birds lay many fewer eggs than they are capable of doing? Parents which lay very small clutches of three or four eggs obviously have low fitness, because they can produce no more than three or four fledglings. Less obviously, parents which produce very large clutches of a dozen eggs or more also have low fitness; they are unable to provide enough food for so large a brood, and many of their offspring are small, fledge in poor condition, and die in their first winter. The number of surviving offspring is therefore maximized by an intermediate clutch size that represents an optimal compromise between antagonistic components of yield.

**Selection for a Major Life-History Dichotomy.** Many species of salamander can reproduce in one of two forms, as a metamorphosed adult or as a neotenic individual that retains many features of larval morphology, such as external gills. The advantage of neoteny is that growth rates are greater in water than on land; the disadvantages include the risk that the pond will dry out. One expects to see neoteny occurring only in permanent bodies of water, whereas populations inhabiting temporary pools should always metamorphose. Raymond Semlitsch and Henry Wilbur, of Duke University, established lines of *Ambystoma* from temporary and permanent ponds (the frequency of neoteny varied in the expected way) and selected them for four years in outdoor artificial ponds whose water level could be controlled. The outcome was an increase in the frequency of neotenic individuals in permanent ponds (from 0.50 to 0.63) and a corresponding increase of metamorphosing individuals in ponds that dried out every year (from 0.70 to 0.82), when tested in a common-garden experiment.

## 94. *The schedule of reproduction evolves through the antagonism of prospective components of fitness.*

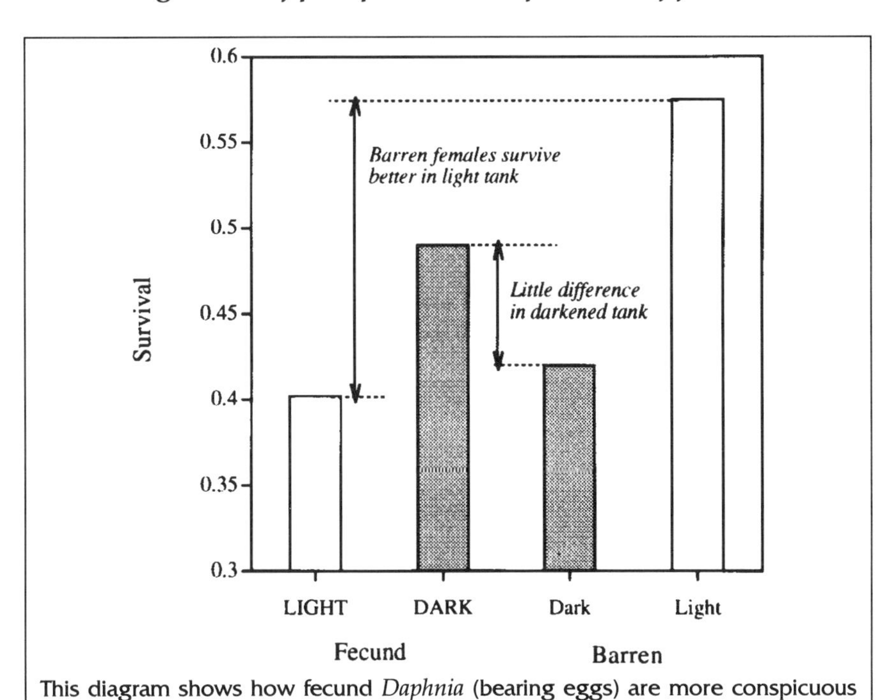


This diagram shows how fecund *Daphnia* (bearing eggs) are more conspicuous and thus more likely to be eaten by guppies than barren females, provided the tank is lighted so that the difference between the two is apparent.

(*Continues*)

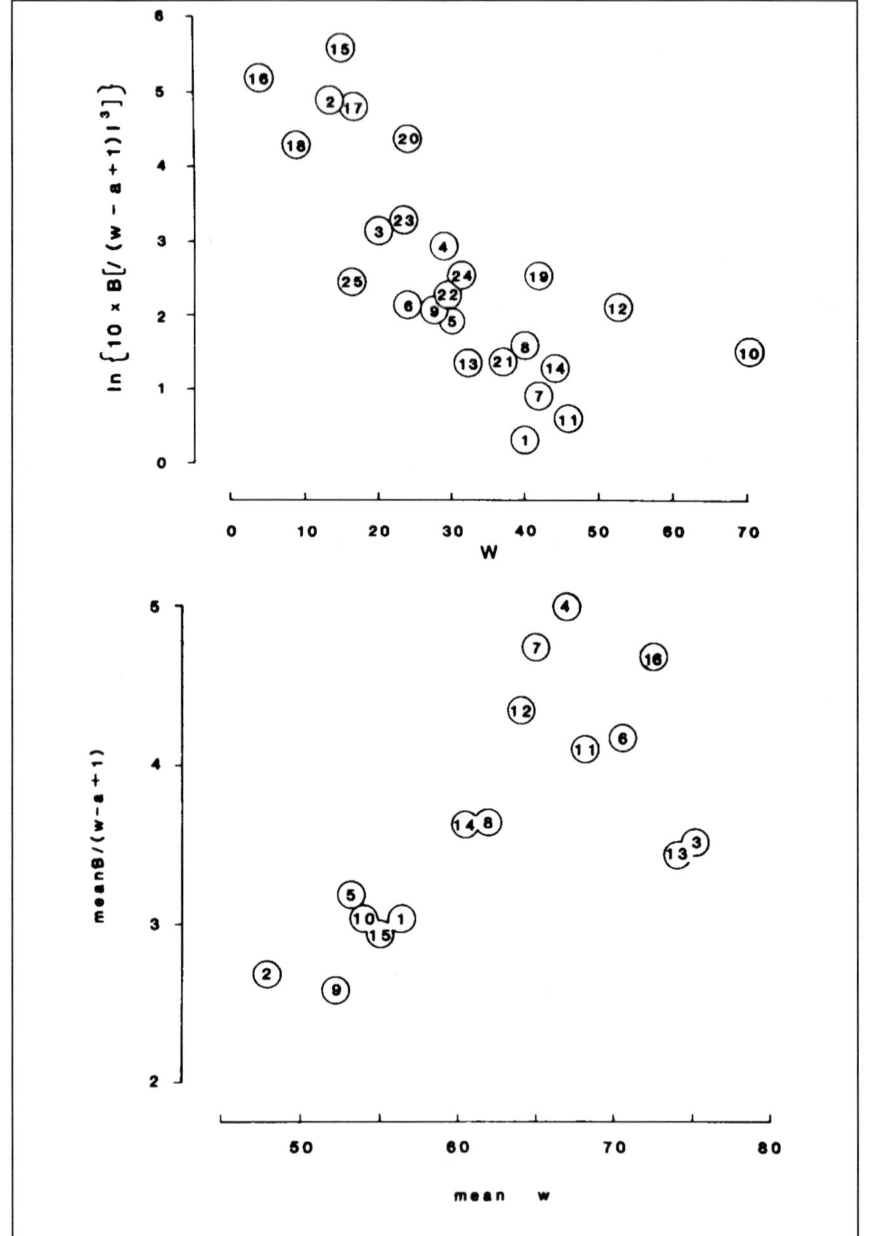


In these diagrams (Bell 1984) the $x$-axis is longevity and the $y$-axis is the rate of egg production during adult life. These two prospective components of fitness

(*Continues*)

(*Continued*)

are negatively related among species (upper diagram), but positively correlated among clones recently collected from natural populations within a species (*Daphnia pulex*) (lower diagram).

I have described in the preceding section how the compromise between seed size and seed number is adjusted differently in annual and perennial species of *Plantago*. This does not explain why some species should be annual and others perennial. The difference between the two life styles is that annuals die immediately after reproducing for the first time, whereas individuals of perennial species often survive to reproduce in the following season. Reproduction requires resources,that are exported as seeds or chicks; but survival equally requires resources that must be retained in order to grow, to accumulate storage tissue, to maintain defences against predators and pathogens, and so forth. In discussing components of yield, I have emphasized the antagonism of *current* components of reproduction, arising from a limited pool of available resources. These resources, however, may more generally be spent on present reproduction or hoarded in order to support survival and reproduction in the future. To the extent that individuals can garner a finite supply of resources at any one time, these resources must be allocated between present and future reproduction, and what is spent in the present will not be available in the future. There may then be antagonism between *prospective* components of fitness: if present reproduction is increased, the prospect of future reproduction may be threatened. This concept has been called the *cost of reproduction*. An increase in the current level of reproduction may represent an additional metabolic drain that reduces the quantity of resources available for reproduction later in life: this is the "fecundity cost". It may also decrease the likelihood of surviving to reproduce at all, either because individuals whose stored reserves have been depleted are more vulnerable to stress, or because activities associated with reproduction, such as courtship, may be inherently risky: this is the "survival cost".

**Riskiness of Reproduction in *Daphnia*.** It is fairly easy to set up experiments that identify and estimate a particular cost of reproduction. Water fleas (*Daphnia*) bear their asexual progeny in a dorsal brood pouch, releasing them as miniature versions of their mother. The brood pouch is transparent, and it is easy to distinguish gravid from barren individuals by sight. My student Vassiliki Koufopanou mixed gravid with barren individuals in equal proportions and found that small fish, such as guppies, ate the gravid individuals first. The obvious interpretation that the gravid individuals are more conspicuous to a predator hunting by sight was confirmed by repeating the experiment with a dark background, against which the gravid individuals are no more conspicuous than the barren, or with a predator that does not hunt by sight, such as *Hydra*; annulling the visual cue abolishes the differential in mortality. Moreover, by counting the number of offspring borne by gravid individuals before and after predation, it could be shown that increased fecundity entailed an increased risk. The experiment thus simulates a single episode of selection in which any increase in present fecundity reduced the chance of surviving to reproduce in the future.

**Positive Correlation of Fitness Components in *Daphnia*.** Studying the cost of reproduction by measuring phenotypic or genetic correlations between prospective components of fitness in experimental populations encounters the same difficulties as a similar approach to components of yield. Indeed, they are more severe. The longevity of different species of *Daphnia* and related genera is negatively correlated with the rate at which they produce offspring, as we might expect. However, if we raise clones in the laboratory using newly captured individuals from natural populations, the correlation between longevity and reproduction is usually positive. Part of this effect is straightforwardly environmental; individuals from the same clone also show a positive correlation between longevity and reproduction, no doubt because some culture vessels receive more food than others, or are subtly different in some other way. Another part is because some clones flourish better in laboratory culture than others. This pattern of positive correlation has been observed by myself and several other biologists; it applies to many organisms other than water fleas; it also applies to other prospective components of fitness, such as the number of offspring produced in different instars during adult life. Various statistical devices for manipulating the data so as to reveal an underlying pattern of negative genetic correlation have been suggested, but none are very convincing, and the only satisfactory way of demonstrating the fundamental antagonism among components of fitness is, as before, the selection experiment.

## 95. *Optimal schedules of reproduction evolve as compromises.*

The concept of the cost of reproduction can be used to explain why some organisms survive reproduction whereas others do not. The form of the relationship between the intensity of reproduction and the subsequent rate of survival will depend on circumstances. In the first place, reproduction may have little effect on survival up to a certain point, above which even a small increment in reproduction causes a substantial increase in mortality. This might be the case, for example, if reproduction simply created a drain on stored reserves: a modest drain could easily be tolerated and perhaps soon made good, but drawing down reserves beyond a certain point might weaken the individual so much that it would be unable to accumulate fresh reserves or, in the meantime, to resist predators and pathogens. In the extreme case, reproduction might have no effect on subsequent survival up to some threshold value, beyond which any additional reproduction is lethal. The optimal schedule of reproduction would then be to reproduce at the threshold value, the maximum value possible that would ensure that survival was not impaired, relative to individuals that did not reproduce at all. This is a perennial life history. The converse possibility is that even the most modest quantity of reproduction is virtually lethal, so that any individual that reproduces at all will almost certainly die afterwards, without the opportunity of reproducing for a second time. The optimal schedule of reproduction in these circumstances is to reproduce at the greatest possible intensity once reproduction becomes possible at all, because producing as many offspring as possible carries no greater penalty than producing only a few. This is an annual life history.

***Plantago* Again.** Annual and perennial species of *Plantago* not only allocate resources to seeds in different ways; they also allocate a different fraction of total resources. Primack's data shows that annuals allocate a much greater share of resources to reproduction and a correspondingly lesser fraction to vegetative tissue. Because they die after reproducing for the first time, annuals have no need to invest resources for maintenance and survival; perennials, on the other hand, must store reserves to tide them over from one growing season to the next, and these stored reserves cannot be used to make seeds.

**Spawning Migration of Salmonids.** Pacific salmon (*Oncorhynchus*) migrate as smolts from the streams in which they were born to the depths of the North Pacific Ocean, 2000 km or more away. Having fattened in the pastures of the sea, they then swim all the way back to their natal stream to spawn. This was a very hazardous journey even before the first gill-net was set on the Fraser River; few would survive it, and the probability of surviving a second round trip to the Pacific and back would be very low indeed. An adult salmon that has reached the spawning stream has therefore nothing to gain by restraint and should commit all its reserves to reproduction, rather than holding some back in a futile attempt to survive to breed for a second time. The suicidal reproduction of these fish is the consequence of a survival cost that makes any attempt to reproduce at all nearly certain to be fatal.

This interpretation is borne out by the life histories of other kinds of salmon and trout. Atlantic salmon (*Salmo*) undertake a similar but somewhat less arduous migration to the north-west Atlantic; they spawn with great vigor and many die, but a few survive to repeat the voyage and spawn for a second time. Sea trout move into coastal waters and back, without venturing into the open sea, and routinely survive for two or more spawning seasons. Exclusively freshwater salmonids, such as lake trout, and many related fish, such as charrs and graylings, make at most very limited journeys along the lakeshore to find their spawning grounds. Post-reproductive survival in this group of fish thus appears to be governed by a prospective cost, the risk entailed by the spawning migration.

The complete collapse of somatic maintenance after reproducing is an extreme case. More generally, current reproduction may be somewhat risky or may reduce the reserves available for future growth and reproduction, without leading inevitably to death. The timing and intensity of reproduction will evolve through selection towards an optimal schedule that will differ among species or populations, depending on how it is constrained by the current and prospective costs of reproduction in different ecological contexts.

## 96. *Age-specific selection causes changes in the schedule of reproduction.*

In many cases, it is possible either to identify a particularly important cost of reproduction in a natural population or to impose a specified cost of reproduction on an experimental population. The result is a shift in the schedule of reproduction that is predictable from the age-specific incidence of selection.

**Size-Selective Predators of Guppies.** The life history of guppies (*Poecilia*) in the mountain streams of Trinidad has been studied by David Reznick and John Endler. Guppies are small live-bearing fish that produce a succession of broods through their adult life. They are eaten by a variety of other fish, including *Crenicichla*, which is relatively large and preys mostly on adults, and *Rivulus*, a smaller fish that takes juvenile guppies. Young guppies are thus more at risk from *Rivulus*, and older, mature guppies from *Crenicichla*. When *Crenicichla* is the more abundant predator, adult survival is reduced and selection should favor reproducing intensely early in life. This is borne out by comparing populations that experience different patterns of predation: in streams where *Crenicichla* is abundant, guppies mature earlier, producing a greater number of smaller offspring at shorter intervals, whereas in streams where *Rivulus* is abundant, guppies tend to mature later in life and produce broods of fewer, larger offspring at longer intervals. Moreover, by raising families in the laboratory, it can be shown that this variation is inherited.

To confirm that predation is primarily responsible for causing the observed variation in life histories, guppies were introduced from a stream that contained only *Crenicichla* into a stream that contained only *Rivulus*. Within 11 years, or roughly 50 generations, the introduced population matured later in life and produced fewer and larger offspring in each brood,

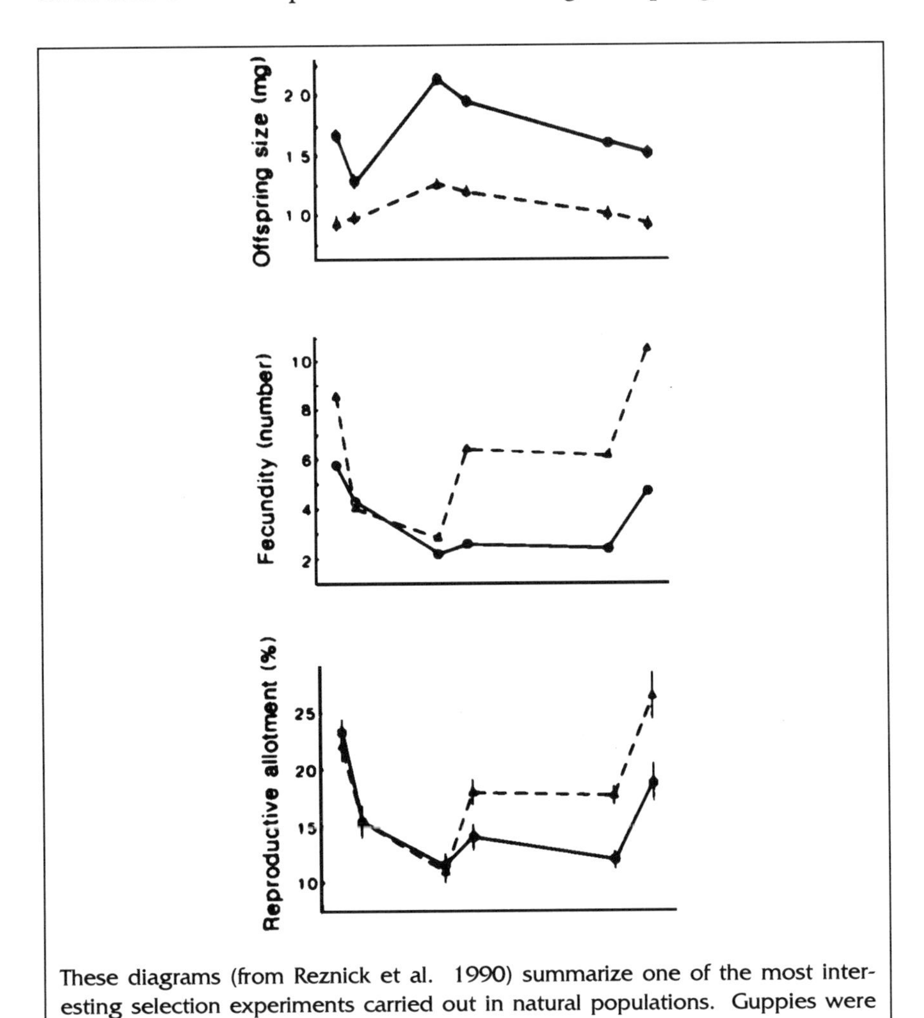


These diagrams (from Reznick et al. 1990) summarize one of the most interesting selection experiments carried out in natural populations. Guppies were

(*Continues*)

transplanted from a stream containing *Crenicichla* (a predator of adults) to a stream containing *Rivulus* (a predator of juveniles). Within 11 years, or 30–60 generations, the transplanted population had evolved smaller broods of larger offspring, making a smaller allocation per brood. The solid lines denote the introduction site and broken lines the native site (base population). The experiment was unreplicated. See, also, Reznick (1982), Reznick and Endler (1982), and Reznick and Bryga (1987).

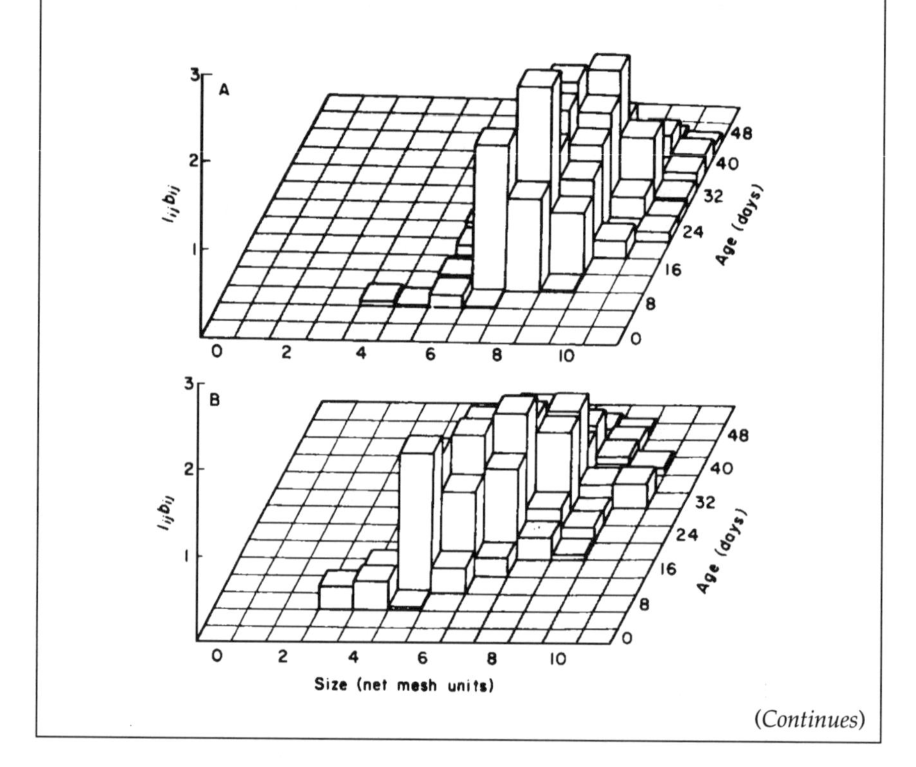


(*Continues*)

relative to the founding population. These differences were also heritable. The schedule of reproduction in these fish thus responds appropriately to selection acting in the short term.

**Size-Selective Culling of *Daphnia*.** Michael Edley and Richard Law of York performed a very similar experiment in more convenient circumstances by passing experimental populations of *Daphnia* through a series of sieves. By removing small individuals from some lines and large individuals from others they themselves acted as size-specific predators. The base population was a heterogeneous mixture of clones from different localities,

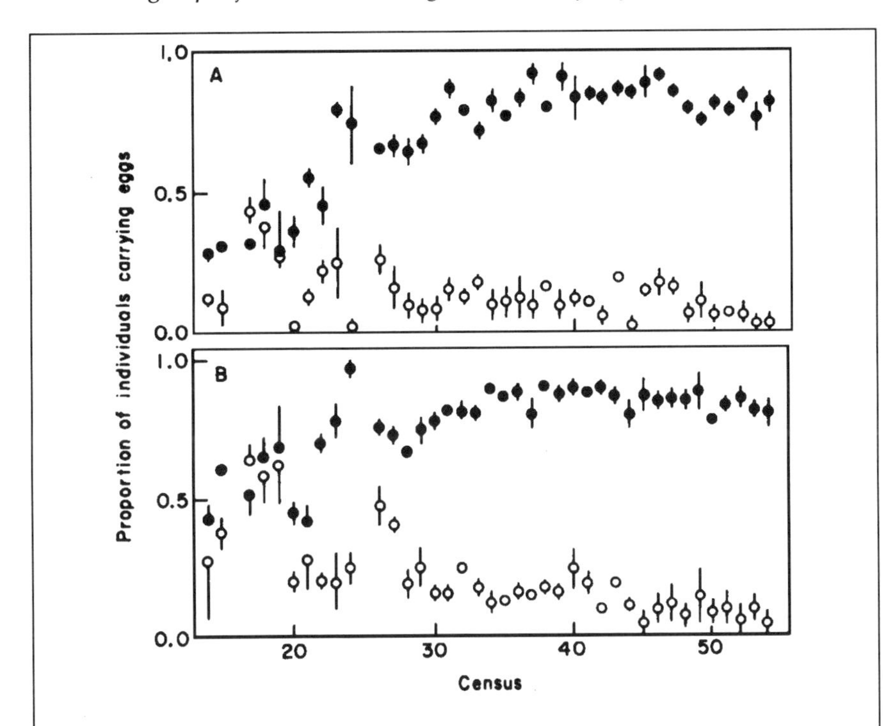


These three-dimensional diagrams from Edley and Law (1988) show the reproduction of populations of *Daphnia* selected for size. The two horizontal axes are age $i$ and size $j$, so that the two-dimensional projection of the data is a growth curve. The stippled area represents the size classes removed by sieving, so that selection favors large individuals in the upper diagram and small individuals in the lower diagram. The vertical axis is the expected reproduction of individuals of size $i$ and age $j$, as the product of survival $l_{ij}$ and fecundity $b_{ij}$. Note how the schedule of reproduction is molded by selection so that vulnerable classes contribute little to overall reproduction. This is partly because of the direct effect of

(*Continues*)

and lines of several hundred individuals were selected through about 40 four-day cycles of growth and culling, representing about a dozen generations of asexual reproduction. The outcome was no doubt attributable primarily to sorting the variation originally present. Lines from which small individuals were culled evolved more rapid growth because genotypes that escape more rapidly from the vulnerable size classes will increase in frequency. Individuals from lines in which large individuals were culled

*(Continued)*

culling on survival, and partly because of evolved changes in life history. The lower figure shows the proportion of individuals in the smallest fertile age classes that were carrying eggs during the experiment, the open circles representing samples from populations selected for large size (small individuals removed) and the filled circles populations selected for small size (large individuals removed). The census interval was four days; after about 80 days the two selection treatments diverge, removal of the larger individuals causing an increase in fecundity of the smaller individuals, and vice versa. The two diagrams refer to two successive adult instars. Selection also changed rates of growth, the removal of small individuals causing an increased rate of growth and thus an earlier entry into the fertile size classes.

grew more slowly, but reproduced at smaller sizes, because selection would favor genotypes that could reproduce before being sieved out from the population.

**Unconscious Selection in Fisheries.** By far the largest experiments imposing selection for altered schedules of reproduction through size-selective culling are those conducted by commercial fisheries. In many species, larger individuals have greater market value, and fishing methods are often designed to capture them preferentially. Indeed, some types of gear, such as gill-nets, act as filters that retain large fish while allowing smaller ones to pass through. This bias is often reinforced by regulations that protect smaller size classes or younger age classes. The same bias is evident in sport fisheries, where anglers usually seek to capture the largest individuals. This direct effect on size has already been mentioned in Sec. 32. The indirect consequence, as in the *Daphnia* experiments, will be to select for individuals that grow slowly and reproduce early in life. While working for the government of Alberta, Paul Handford, Tom Reimchen, and I found that gill-nets were selective, not only for size, but also for body form: they capture individuals that are relatively shorter and thicker than those caught by trawls, which because of their fine-meshed cod-end are much less selective. In lakes where gill-nets were set, whitefish (*Coregonus*) tended to become longer and thinner at given age, and reproduced earlier in life and at smaller sizes. Such evolutionary trends driven by size-selective predation are economically important, because they upset the steady-state assumptions on which fishery management is based. Heritable shifts in growth and size at reproduction are irreversible in the short term, and mean that populations are more likely to become extinct as fishing pressure shifts toward smaller and less valuable individuals, and less likely to recover if fishing is suspended. Despite the very widespread occurrence of such trends, evolutionary considerations are given little or no weight in the management of fisheries.

## 97. *Selection for early reproduction reduces vigor later in life.*

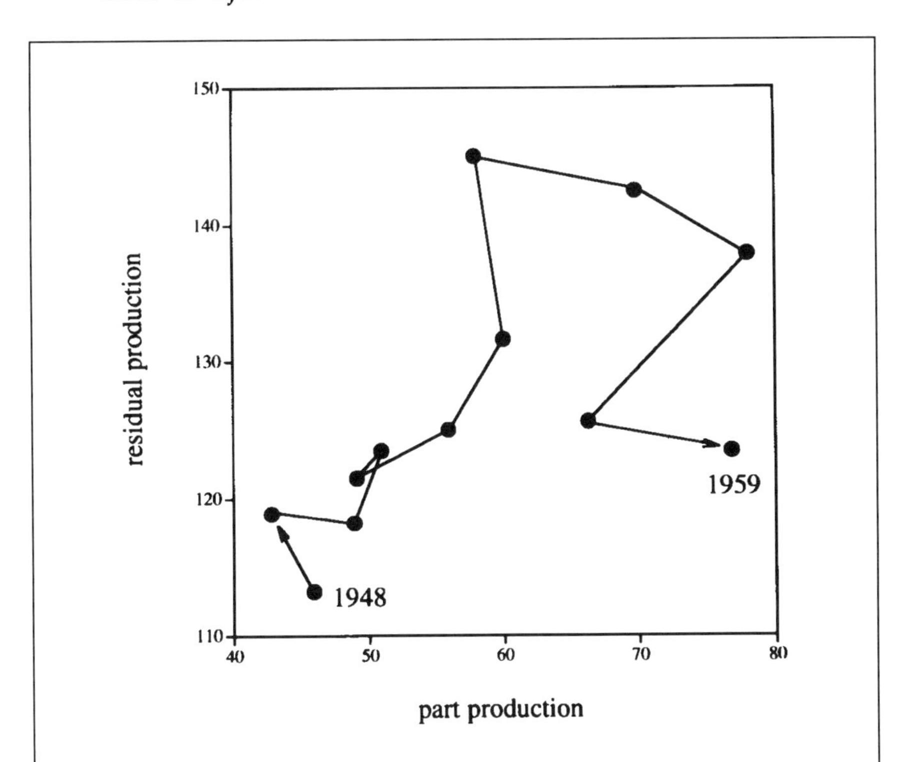


Selection for egg production in fowl during the first six months of life at first causes an increase in egg production during the remainder of life, but this is eventually succeeded by a negative correlated response (data from Morris 1963). The genetic correlation between part record and residual record was positive in the base population. See, also, Wallinga and Bakker (1978) and Mertz (1975).

A culling program that preferentially removes larger and older individuals selects for increased reproduction among smaller and younger individuals. This may in itself be undesirable from an economic point of view. However, it carries an additional penalty: the cost of earlier reproduction is likely to be a reduced survival or reproductionlater in life.

**Part-Record Selection in Domestic Fowl.** The theory that selection for any given component of fitness is likely to evoke an antagonistic correlated response at subsequent ages has an important economic consequence. Sustained reproductive output is an important character in livestock, such as domestic fowl kept for egg production. It is slow and inconvenient to select for total lifetime productivity in birds that may live for several years; a much more effective method would be to select for egg production early in life, provided that early vigor is a reliable guide to lifetime egg output. This is called *part-record selection*. It is known, however, that the correlation between first-year output and output in subsequent years falls steadily throughout life, so that egg production by birds five or six years of age is almost independent of their output in the first year of life; moreover, longevity is negatively correlated with early reproduction. This suggests that selection for early vigor will expose and exacerbate the antagonism of fitness components expressed at different ages. An experiment by J.A. Morris, of the CSIRO in Australia, seems to confirm this fear. He selected egg production during the first six months of life for 12 years in a closed flock of White Leghorns. There was a steep and continued response that amounted to a gain of about 35 eggs (from about 45 to about 80) by the end of the experiment. This was achieved largely through a decrease of nearly a month in the age at which the first egg was laid and, to a lesser extent, through an increase in the rate of lay thereafter. However, egg size decreased by about 5%. Residual egg production, in this case the number of eggs laid between 6 and 18 months of age, at first increased along with early production. This was consistent with the high positive genetic correlation between early and residual production in the base population. During the last five years of the experiment, however, residual production fell steeply, whereas early production continued to increase; during this period there was a gain of 20 eggs in early production, but a correlated loss of 19 eggs in residual production. At this point, part-record selection had ceased to be effective, with early gains being annulled by later losses. It had first caused an overall increase in productivity, offset by a decrease in egg weight as a correlated response to selection for egg number. It later failed to generate any overall gain in egg number, because of a decrease in egg number later in life as a correlated response to selection for early egg output.

**Selection for Size of First Litter in Mice.** J.H. Wallinga and H. Bakker selected a laboratory population of mice for the number of offspring in the first litter for 25 generations. They obtained a direct response of about six offspring. This effect extended to subsequent litters, although by the fourth or fifth litter the difference between selected and unselected lines was very small, and older animals in the unselected line may have produced larger litters. Selecting for the size of the first litter involves selecting for mice that survive to reproduce for the first time, and early survival was greater in the selected line than in the unselected control line. Animals in the selected line, however, survived much more poorly after their fourth litter. In this case, the antagonistic response to selection for early vigor seems to involve reduced survival among older individuals.

**Selection for Early Reproduction in *Tribolium*.** Similar results have been obtained in experimental populations of beetle and flies. Robert Sokal of Harvard and Daniel Mertz of Chicago selected populations of the flour beetle *Tribolium* for increased reproduction early in life by simply discarding individuals after a few days of egg laying. In Sokal's experiment, longevity was reduced in the selected lines. Mertz did not observe any effect of selection on longevity, but fecundity later in life was less in lines selected for early reproduction. The correlated response to selection for early reproduction may thus involve either a decrease in subsequent survival or a decrease in subsequent fecundity.

The general implication of experiments such as these is that selection for age-specific components of fitness may be no more effective than selection for components of yield. In the first few generations of selection overall fitness may advance, but if selection is continued, an antagonistic correlated response is induced, and although the character under selection may continue to advance, this advance is more or less completely countered by the regress of other fitness components.

## 98. *Selection generally favors early vigor because the force of natural selection weakens with age.*

Deliberate selection for increased reproduction early in life has the effect of decreasing subsequent survival or fecundity. This is a special case of the more general principle that correlated responses will be antagonistic. There is a second general principle guiding selection acting on schedules of reproduction that operates independently of the antagonism between prospective components of fitness. It leads to the conclusion that selection will generally favor more intense reproduction in early life, even when no deliberate age-specific selection is applied. The reason for this can readily be appreciated through an analogy with insurance and investment.

Suppose that you wish to purchase an annuity that becomes payable when you reach a certain age. This future income has a present value that depends on the probability that you will live to enjoy it: obviously, it might be very valuable if you are confident of living well beyond the redemption date, but almost worthless if you are likely to die beforehand. This value is reflected by the premium that you pay to the insurance company, which represents a bet that you will survive for long enough to profit from the transaction. At any given age, the amount of the premium that you must pay in order to purchase a given annuity depends on the insurance company's estimate of your probability of survival. The premium will be lower when your chances of survival are lower because the present value of the annuity is less. The present value of a future return on investment thus declines with the probability that you will survive to collect it. Thus, the later in life an annuity commences, the less valuable it is at present.

Alternatively, suppose that you wish to invest a sum of money at compound interest in order to realize the greatest possible capital gain at some fixed date in the future. It is assumed that you will still be alive on this date. Clearly, it is better to invest the money sooner than later because the sooner the investment is made the sooner the interest that it earns will itself begin to earn interest. This effect will be larger when the rate of interest is greater. Thus, the later in life an investment is made, the less valuable it will be.

Putting these two arguments together, we can see that postponing an investment will reduce its value in proportion to the rate of mortality and the rate of interest. Because populations grow at compound interest, there is a direct analogy between investment and reproduction. Postponing reproduction will reduce the number of your descendants alive at any given date in the future, because it postpones the time at which your offspring will themselves begin reproducing and makes it less likely that you will survive to reproduce at all. This principle was first expressed precisely by W.D. Hamilton of Oxford and University College London. Fitness has been defined as a rate of increase, which is wholly determined by the rates of survival and fecundity throughout life. A genotype whose rate of increase exceeds that of the population as a whole will tend to spread. Any age-specific change in survival or fecundity will therefore be selected according to its effect on the rate of increase of the population, given that the schedule of reproduction is otherwise held constant. More formally, the manner in which selection modifies the schedule of reproduction will depend on the partial derivative of the rate of increase with respect to a given age-specific change. These partial derivatives for the rate of survival $p(x)$ and the rate of fecundity $m(x)$ at age $x$ are as follows:

$$\frac{\partial r}{\partial \ln p(x)} = \left(\frac{1}{T}\right) \sum_{x+1}^{\text{death}} e^{-ry} l(y) m(y)$$

$$\frac{\partial r}{\partial m(x)} = \left(\frac{1}{T}\right) e^{-rx} l(x)$$

where $l(x)$ is the probability of survival to age $x$, and $T$ is mean generation time. For any age within the reproductive span [any age $x$ for which $m(x) > 0$], the right-hand sides of these equations decrease as age $x$ increases. What these equations say, therefore, is that a given small change in survival or fecundity will have a greater effect on fitness and will, therefore, be selected more strongly if it is expressed earlier in life: that is, the rate of selection on any component of fitness decreases with age. The effect of a gene on survival or fecundity is discounted with increasing age by a factor $e^{-rx} l(x)$, or in other words by the rate of mortality [which decreases $l(x)$] and by the rate of population increase (which decreases $e^{-rx}$). This is the formal justification of the intuitive argument that I began with.

The implication of this conclusion is that selection will generally act so as to make the age at first reproduction as early as possible. This, however, is true only for microbes; in multicellular organisms, a period of growth occurs before reproduction. The weakening of the force of selection with age means that a special explanation is required for multicellularity and for somatic growth and delayed reproduction generally. A larger individual can produce more offspring, and if it grows fast enough, the increase in fecundity through growth may exceed the discounting factor $e^{-rx} l(x)$. Growth alone, however, cannot account for delayed reproduction because, on purely demographic grounds, there is no reason that an individual should not grow and reproduce at the same time. Reproduction should be delayed, therefore, only if future growth is retarded by reproduction more than future reproduction is discounted. This will only be the case if the virgins grow so fast that their discounted fecundity $e^{-rx} l(x) m(x)$ increases with age early in life. This seems likely to be true in many cases: in very small, young individuals, even a very few propagules will represent a large fraction of somatic mass, and reproduction must severely inhibit growth because it uses up most of the tissues on which future growth would have been based. It is thus the prospective cost of reproduction, in terms of a loss of potential fecundity later in life, that reduces the value of early reproduction and, in some organisms, may cause selection for a prolonged period of prereproductive growth.

## 99. *Senescence evolves because selection favors postponing deleterious gene expression.*

Mutations that increase fecundity early in reproductive life will spread more quickly than mutations with similar effects later in life; thus, selection will generally tend to increase early vigor. The converse is also true: selection will tend to reduce vigor later in life, causing a senescent decline in rates of survival and fecundity. Senescence is no doubt inevitable and would occur even if selection were suspended, because some of the injuries that the soma must inevitably sustain are irreversible and their effect cumulative. Nevertheless, the variation in rates of senescence among organisms implies variation in the effectiveness of repair mechanisms, and the weakness of selection against deleterious changes late in life explains why somatic maintenance is imperfect. Senescence may evolve as a byproduct of selection for early vigor in two ways, depending on whether or not prospective costs of reproduction are invoked.

**Mutation Accumulation.** The accumulation of deleterious mutations will be checked by selection at all ages, but countervailing selection is more effective in younger than in older individuals. The equilibrium level of mutational load will therefore increase with age: deleterious mutations will accumulate to a greater extent in later age classes. In the simplest case, different deleterious mutations will have different fixed ages of expression, and those that are expressed later in life will be maintained at higher frequencies in the population. It may also be that the age of expression of a deleterious mutation is itself a character that can be selected, in which case selection will favor shifting expression to later ages. This is the *mutation-accumulation* theory of aging, first suggested by Sir Peter Medawar.

**Pleiotropy.** Instead of having a fixed effect at a single age, mutations may be expressed differently in younger and older individuals. Mutations that are deleterious at all ages will be rapidly eliminated, of course, and those with beneficial effects at all ages will rapidly spread. The remainder will have opposed effects on fitness at different ages, increasing fitness at some ages but decreasing it at others. Those that increase survival or fecundity in younger individuals, at the expense of reducing survival or fecundity in older individuals, will be selected more strongly than those with the converse effect of benefitting older at the expense of younger individuals and will be more likely to spread or will spread more rapidly. Indeed, the greater effect of early vigor on overall fitness implies that a rather small increase in survival or fecundity early in life will be favorably selected, even if it has severely deleterious consequences later in life. Senescence then evolves because the expression of prospective costs of reproduction is biased by the general weakening of selection with age. This idea, originally due to George Williams of Stony Brook, is the *antagonistic pleiotropy* theory of aging.

**Soma and Germ.** Neither mutation accumulation nor antagonistic pleiotropy will cause senescence to evolve in unicells that reproduce by an equal fission. Reproduction in such organisms will be discounted by the rate of mortality and the rate of population increase, as in multicellular organisms, and they will generally evolve so as to reproduce as soon as possible. However, there can be no tendency for deleterious mutations, or the deleterious effects of mutations, to be expressed later in life, because there is no difference in age between the products of fission. If a lineage of unicells does senesce, it will be through some mechanism such as Muller's Ratchet (Sec. 52) and not because of the weakening of selection with individual age.

The conventional view of animal development, dating from August Weissman, is that hereditary transmission occurs solely through a special caste of germ cells, whereas the sterile vegetative tissues of the body, the soma, represent a temporary vessel for the germ cells that is rebuilt in every generation. The germ cells thus proliferate within bodies just as clones of unicellular algae or yeasts proliferate outside bodies; in the same way, they form lineages of dividing cells that do not senesce. The activity of the soma in supporting the proliferation of germ cells is discounted by mortality and population increase, but the persistence of the soma as a distinct physical entity through the lifetime of an individual implies that genes can be expressed at different ages during that lifetime and, therefore, that deleterious mutations, or the deleterious effects of mutations, can be expressed later in life. It is thus the soma that undergoes senescence. This is perfectly clear in the case of animals where the germ line is segregated from somatic cell lines very early in development, as in vertebrates, insects, and nematodes. It is much less clear in the case of organisms such as plants or cnidarians, where a population of stem cells continues to divide freely throughout life, giving rise both to germ cells and to somatic tissues. In the extreme case, a multicellular organism may reproduce vegetatively by dividing into two identical fragments, each of which regenerates the adult form. This would be equivalent to the binary fission of a unicell, and such organisms should not senesce. There is, indeed, some evidence that worms that reproduce by a nearly equal binary fission do not senesce, or senesce very slowly, and this supports the evolutionary view that senescence in organisms such as mice or flies is directed by the weakening of selection with age. In practice, fission is rarely perfectly equal. In vegetatively reproducing worms, for example, cells in the hind region of the body proliferate to form a new large offspring, whereas the tissues of the head region persist; successive head fragments might then senesce whereas tail fragments would not. In *Hydra*, a band of cells near the base of the trunk region divides continually to produce tissues that move upwards, differentiating as they do so, before eventually being shed from the body at the tips of the tentacles. It is not clear that a living fountain of proliferating stem cells should senesce at all. More generally, the evolution of senescence through mutation accumulation or antagonistic pleiotropy depends on the presence of a distinct and persistent soma that creates an unequivocal distinction between an old parent and a young offspring and, in life cycles based on different patterns of development, senescence may not occur, or may be greatly retarded, or if it does occur may require some different explanation.

**Selection for Delayed Senescence in *Drosophila*.** If senescence reflects the weakening of selection with age, then not only can it be accelerated by selecting for early reproduction, as the experiments described in Sec. 97 show, but it can also be delayed by selecting for reproduction late in life. A procedure that makes old individuals more likely to live is, of course, a more impressive testimonial to the theory than one that merely makes younger individuals more likely to die. The simplest procedure is to breed solely from older parents. This artificially opposes the discounting of reproduction with increasing age and should lead to an increase in fecundity late in life, with a correlated increase in longevity. (It necessarily creates direct selection for increased longevity, too, because only surviving individuals can breed.) Experiments of this sort have been conducted with *Drosophila* by Michael Rose of Dalhousie, Leo Luckinbill of Wayne State, and several other scientists. The design of such experiments is very straightforward: eggs are collected from vials containing females of known age. By setting up some lines recruited from females only a few days old and others from females 40 days or more in age, the effects of selection on early and late fecundity can be contrasted. Experiments that have been properly conducted, using outbred stocks maintained in the laboratory for several generations before the experiment, seem to give the expected result: lines maintained by reproduction late in life evolve greater longevity. Moreover, fecundity early in life often falls as a correlated response to increased reproduction later in life. The effect on longevity is heritable and seems to involve genes at several or many loci, acting more or less independently. These *Drosophila* selection experiments provide the strongest direct evidence for the evolutionary theory of senescence.

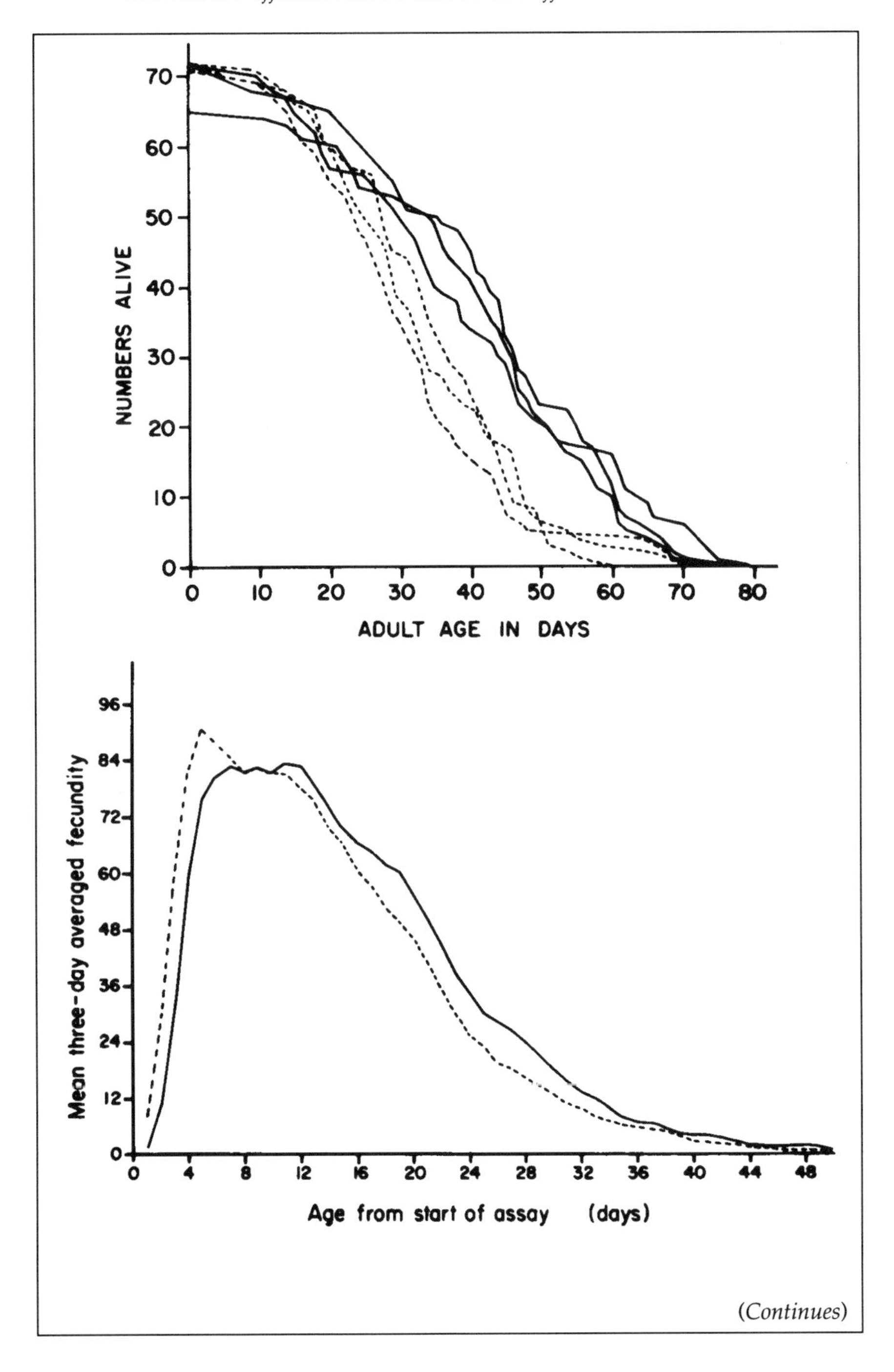

NUMBERS ALIVE
70
60
50
40
30
20
10
0
0
10
20
30
40
50
60
70
80
ADULT AGE IN DAYS
Mean three-day averaged fecundity
96
84
72
60
48
36
24
12
0
0
4
8
12
16
20
24
28
32
36
40
44
48
Age from start of assay (days)


(*Continues*)

(*Continued*)

Rose (1984; see also Rose & Charlesworth 1980, 1981) used a base population that had been maintained in the laboratory for over 100 generations on a 14-day cycle. This was selected in two ways: one set of lines was transferred on the same schedule, whereas another set was transferred at longer and longer intervals: at first 28 days, then 35, 42, 56, and finally 70 days. Eggs were collected only from the oldest females, so that only females that survive into old age are able to reproduce at all in the latter treatment. The upper figure shows the response of longevity: the populations transferred at longer intervals evolve longer lifespans. The lower figure shows the distribution of fecundity through life, averaged for all the selection lines. The effect of selecting for reproduction late in life is to increase late fecundity, at the expense of reducing fecundity early in life. Reverse selection increases early fecundity at the expense of accelerating senescence (Service et al. 1988). These results were broadly confirmed by an independent series of experiments run at about the same time by Luckinbill et al. (1984). Both research groups have continued to maintain and analyze these selection lines. They have reported morphological and physiological correlates of delayed senescence (Rose et al. 1984, Service et al. 1985, Service 1987, Graves et al. 1988, Service 1993) and investigated the genetic basis of the response to selection (Clare & Luckinbill 1985, Luckinbill et al. 1987, Luckinbill et al. 1988, Hutchinson & Rose 1991, Hutchinson et al. 1991). These and some earlier experiments are critically reviewed by Rose (1991). There are some discrepant results. Mueller (1987), Engstrom et al. (1992), and Roper et al. (1993) all obtained a direct response to selection for increased longevity, but failed to observe any substantial reduction in early fecundity. This may be attributable in part to differences in culture conditions, but may also reflect the different contributions of mutation accumulation and antagonistic pleiotropy in different experiments.

## 3.B. Selection in Several Environments

The world that I have thus far depicted is an extremely simple one, in which the environment is uniform and unchanging, unless it changes once to remain the same thereafter, and in which genes have straightforwardly beneficial or deleterious effects on characters that affect fitness. This is an essential pedagogical device; but, of course, the world in which organisms evolve is not really the calm immutable universe of the chemostat, but a more varied, turbulent and unbiddable place. Organisms must adapt, not merely to a given set of conditions, but to diverse and changeable conditions.

We can describe how organisms adapt to complex environments by an extension of the approach followed in previous sections of this part of the book. Characters such as (say) fecundity and wing length are clearly distinct, because they are measured in different ways, and these measurements are imperfectly correlated. The nature and magnitude of the correlation between them directs their joint evolution. Fecundity, however, need not be regarded as a single monolithic character; in the preceding sections, values of fecundity at different *ages* have been treated as different characters. The correlation between such components of fitness then directs the evolution of the life history. In a similar way, we can treat the values of fecundity (or of any other component of fitness) expressed in different *environments* as different characters. The response of a population to a complex environment can then be described in terms of the nature and magnitude of the correlation between the environment-specific values of components of fitness.

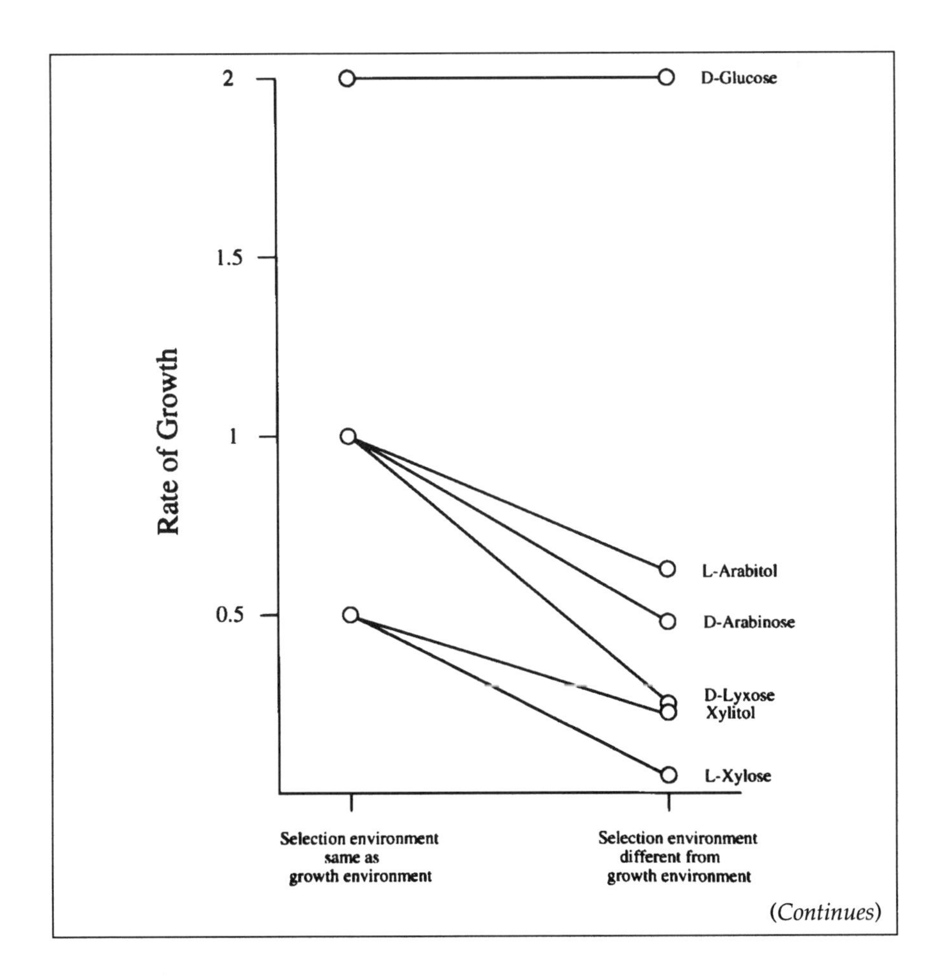


*(Continues)*

(*Continued*)

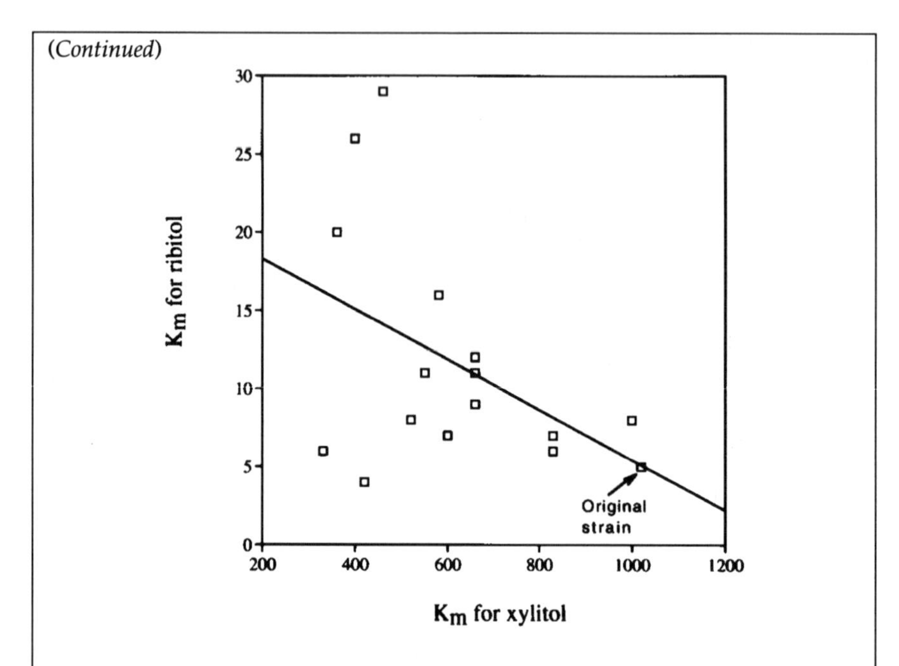


These two figures illustrate bacterial adaptation to different environments. The upper figure (drawn from data in Mortlock 1980, Table 1) shows the response of evolved and unevolved strains of *Klebsiella* to six different carbon sources. Values to the left are growth rates of populations that have previously grown for one cycle on a given sugar. Values to the right are the average growth rates of populations that have grown for one cycle on the five other sugars. For glucose, there is no effect of prior selection, showing that *Klebsiella* is so highly adapted to utilizing glucose that a single cycle of selection causes no appreciable further adaptation. For the other five compounds, growth rate is substantially increased by a single cycle of selection, demonstrating environment-specific adaptation. Growth rate is the reciprocal of the time in days needed to reach a standard turbidity.

The lower figure (drawn from data in Hartley 1984, Table III) shows the effect of continued selection of *Klebsiella* on xylitol. These experiments are described in Sec. 74. The plotted points are different strains isolated from xylitol chemostats. Selection causes a reduction in the $K_m$ for xylitol, but this is accompanied by an increase in the $K_m$ for ribitol. Thus, alterations in ribitol dehydrogenase are antagonistic, with an improved capacity to grow on xylitol causing a reduced capacity to grow on ribitol. The linear regression equation is $y = 21.5 - 0.016x$, with $r^2 = 0.21$.

The evolution of novel metabolic capacities in bacteria provides simple illustrations of principles that will be discussed at greater length subsequently. The crucial feature of laboratory environments is usually the limiting carbon source in the growth medium. We might grow bacteria on substrates ranging from glucose to more exotic sugars, such as xylitol or lyxose. Any culture will grow rapidly and immediately on glucose, but on other substrates cultures will grow slowly only after a lag of hours or days. In a glucose-limited chemostat, "growth rate" can be treated as a single character; in a more complex situation, it is easy to appreciate that "growth rate in a glucose environment" can be regarded as a *different character* from "growth rate in a xylitol environment". When we harvest these cultures after growth, each will be to some extent adapted to the substrate on which it has been growing, through selection during the lag period. If we were to grow these evolved strains on the whole range of substrates in the experiment, we would be likely to find that each grows better than the others on the substrate it has previously grown on. This shows that selection is capable of causing *specific adaptation* to a particular environment. Moreover, if we were to pursue the study further, we would probably discover that the evolution of a higher rate of growth on one substrate was accompanied by a lower rate of growth on others; in other words, that measures of fitness expressed in different environments tended to be *antagonistic,* in a fashion similar to measures of fitness expressed at different ages.

The main theme of the following sections is the extent to which specific adaptation is curtailed by the negative correlation, or antagonism, of performance in different environments. It is necessary, however, first to give a brief account of how environments vary in space and time and how organisms respond to this variation.

## 100. *The individual and the lineage have characteristic scales in space and time.*

**Individual Scale and Versatility.** A plant growing on the forest floor occupies a definite volume of space, in the soil and in the air above it. At any instant in time, conditions vary throughout this volume. The most obvious and important distinction is between soil and aerial conditions, which is reflected in the gross architectural design of the plant. But neither soil nor aerial conditions are uniform: some leaves will be receiving more light than others; some roots will be adequately supplied with water and some will not. Moreover, this pattern of variation changes through time. As sunflecks move across the plant, leaves will pass from light to dark and back again; roots will be wetted after rain and then dry out. An individual plant, thus, has a certain extension in space and time, and during its life it experiences a corresponding variation in conditions of growth. Its ability to cope with this variation may be called *versatility*. The more versatile the individual, the broader the range of conditions over which it can maintain growth. In order to do so, its physiological systems must be apt to accomodate themselves to changing circumstances, shifting in space and time as conditions dictate. The outcome of this underlying physiological variability, however, is to maintain components of fitness as nearly constant as possible: versatility implies the *stability* of survival and reproduction, despite environmental variation and because of physiological variability. All organisms must be versatile to some degree, although the amount of environmental variation that they experience, and the different contributions of spatial and temporal change, depend on their organization. A bacterial cell growing in favorable conditions has a spatial dimension of a few micrometres and a temporal dimension of a few hours; it is likely to be exposed to brief pulses of different nutrients and has inducible metabolic systems that switch the activity of the whole cell in order to exploit them. A tree has a spatial dimension of tens of metres and a temporal dimension of tens of years; although it must respond to temporal variation, it will also experience substantial spatial variation,and must be able to adjust light-harvesting or water-uptake systems to different levels in different parts of its body. Any particular organism, then, has a certain individual scale that represents its extension in space and time and must display the versatility appropriate to this scale. Large, motile, and long-lived organisms operate at larger scales and require greater versatility.

**Lineage Scale and Plasticity.** The seeds produced by an individual plant will be dispersed to a greater or lesser distance within the area occupied by the population and will germinate after a longer or shorter period of time. Each will grow in a different place from its parent and at a different time. Over and above the variation experienced by an individual during its lifetime, parents and offspring will experience different conditions of growth. Thus, the lineage has an extension in space, determined by the distance that propagules are dispersed, and an extension in time, determined by the time that elapses before the propagules germinate and grow. These constitute the characteristic scale of the lineage. In addition to their ability to accomodate changes in conditions during the lifetime of an individual, organisms must also adjust to the average change in conditions between one generation and the next. Insofar as this is possible, the same genotype will be able to express different phenotypes in different environments. Its ability to do so may be called *plasticity*. A more plastic genotype has a more variable developmental program that enables it to exhibit greater stability with respect to components of fitness over a broader range of environments.

I have defined versatility and plasticity in the context of plants that grow fixed in place at a given site. This makes it easy to visualize the environment as a patchwork of sites that differ in the physical conditions of growth and to set up trials in which environmental variation is represented by a series of treatments. Organisms that grow fixed in place and produce passively dispersed propagules may require some degree of versatility, but plasticity is likely to be more important when conditions vary more from site to site than they vary within a site over the lifetime of an individual. Motile organisms, on the other hand, may wander through many sites during their lives, and experience a broad range of physical conditions. To the extent that they can control their movements they may confine themselves to a single kind of site, or set of conditions, by eating only certain species of food plant, for example, or living always on the sea coast. By choosing their environment, they reduce the need for either versatility or plasticity. If they are merely swept passively along by wind or water, they must be able to deal with the full range of environmental variation within their lifetime and must be versatile rather than plastic. Much of what I have to say refers implicitly to plants, since they are most easily studied, and would have to be rephrased to apply to a planktonic diatom, for example, or a fish.

**Environmental Variance of Individuals and Lineages.** The distinction that I have made between versatility and plasticity is essentially the difference between reversible (versatile) and irreversible (plastic) developmental shifts. It is not a conventional usage: the ability of individuals to change their behavior from day to day, or even from minute to minute, is called plasticity by physiologists and ignored by geneticists, whereas the ability of the same genotype to develop into different kinds of individual is called plasticity by geneticists and ignored by physiologists. In this part of the book I am concerned with the evolutionary response of populations to selection in heterogeneous environments, and therefore primarily concerned with the average differences among individuals. I have therefore adopted a geneticist's definition of plasticity, while being careful to distinguish it from the equally important phenomenon studied by physiologists. Plasticity is thus the property of a genotype, not of an individual, and is measured by the environmental variance, that is, by the variance among individuals with the same genotype raised in different environments. It should be borne in mind that *more plastic genotypes are more stable*; that is, they display *less* variation for components of fitness as the result of displaying *more* variation for the physiological mechanisms that are responsible for maintaining survival and fecundity.

## 101. *The physical environment varies at all scales in space and time.*

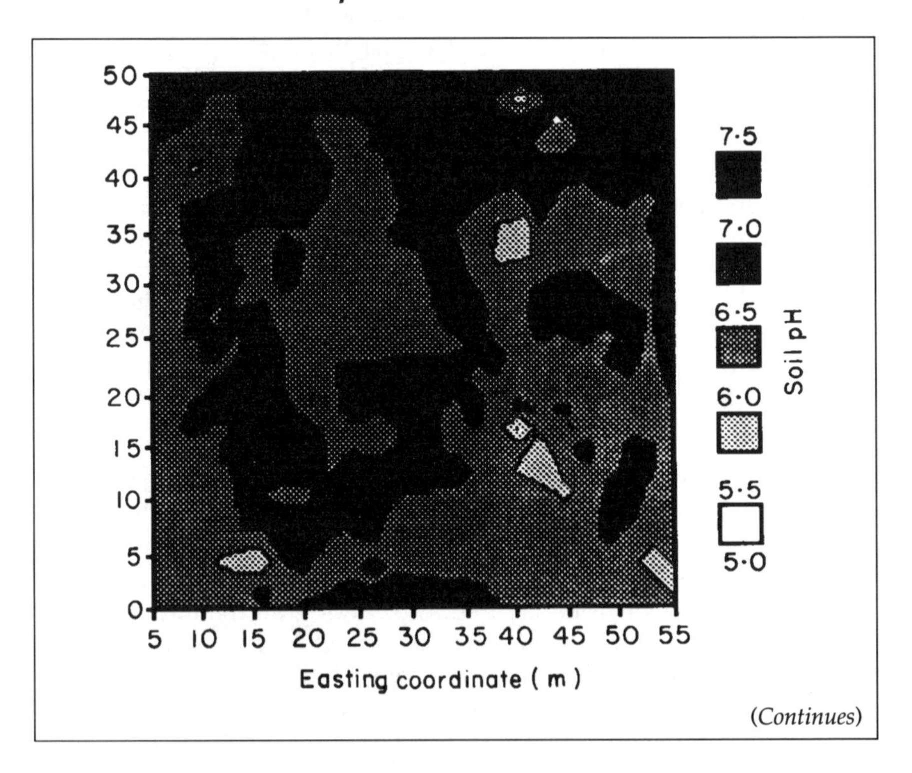


(*Continues*)

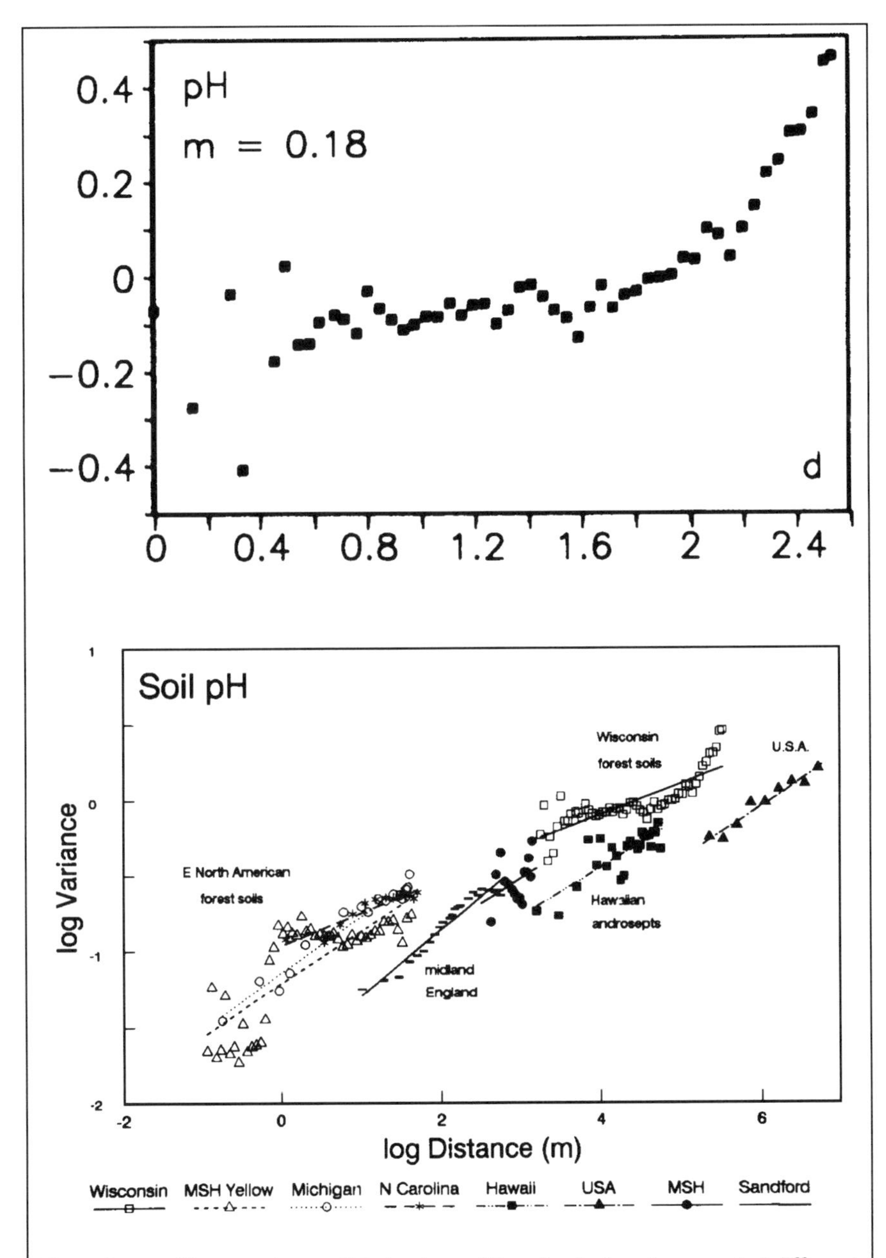


These figures illustrate the spatial structure of the physical environment at different scales, using data on soil pH. The upper figure is a contour diagram showing how

*(Continues)*

(*Continued*)

soil pH varies within a 50 m × 50 m quadrat in undisturbed beech-maple forest (Lechowicz & Bell 1991). It can be appreciated that the environment is not only variable, with soil pH ranging between about 5 and 7.5, but is also structured, with adjacent points in space tending to have similar pH. This is what gives the diagram its patchy appearance. Movement in any given direction is thus likely to take one from an area having some characteristic pH into another area with a somewhat different pH; over small distances (within the same patch, so to speak) this difference will be small, and over large distances (moving between patches) it may be more substantial. The difference in pH between two sites will therefore tend to increase with their distance apart. Another way of expressing this conclusion is to say that environmental variance increases with distance. The second figure shows how the variance of soil pH increases with distance at larger scales, in this case involving distances of up to $5 \times 10^5$ m in forests throughout the state of Wisconsin (Bell et al. 1993). It seems to be generally true that environmental variance increases indefinitely with distance, the plot of log variance on log distance being more or less linear. This relationship seems to hold at all scales; the bottom figure shows how soil pH varies over eight orders of magnitude of distance, from the few centimeters between survey points in the forest at our field station, to the thousands of kilometers separating localities in different parts of North America (Bell 1992c). The physical environment is therefore patchy at all scales: it would have the appearance of the uppermost diagram, whether viewed at a scale of centimeters or at a scale of kilometers. The literature of spatial heterogeneity is very extensive; my article (Bell et al. 1993) has a brief review, but the reader is warned that my views are unorthodox and have aroused considerable opposition.

**The Measurement of Environmental Heterogeneity.** The simplest way of measuring how the environment varies in space or time is to use instruments that report temperature, pH, reflectance, or some other physical variable. The advantage of this approach is that it provides standard and repeatable measures of single environmental factors. Its main disadvantage is that organisms do not necessarily respond to physical variation. It is reasonable to suppose, for example, that plant growth will respond to levels of nitrate and phosphate in the soil, so that if these vary the environment will be heterogeneous, in the sense that a single genotype will express different phenotypes in different places or at different times. Variation in physical factors, however, does not necessarily imply that organisms vary proportionately. Soil phosphate may vary widely from site to site, but if it is everywhere sufficient for growth, given the prevailing climate and the availability of water and other nutrients, then it will not cause any variation among the plants. Variation in (say) manganese levels would be even more difficult to interpret.

The best way out of this difficulty is to use plant response itself as the measure of environment. The variation among sites of the growth, or some other component of fitness, of a single genotype then provides a straightforward estimate of environmental variance, although, of course, it provides no information about the underlying physical factors responsible for this variation. There are two types of bioassay, one leaving the organisms undisturbed in situ and the other manipulating them in some way. Measurements in situ are easy to make, and it is possible to argue that they integrate all of the properties of natural populations. The cost of doing so is that they confound genetic with environmental variation. Within a population, this will lead to an overestimate of environmental variation. If different populations are included in the assay, the environmental variation is likely to be underestimated, because each population will tend to have become well adapted to its local environment. Studies should be carried out in situ only when they is good reason to suppose that these confounding sources of variation can be neglected.

More satisfactory estimates can be got by transplanting organisms to different sites. There are two basic transplant designs. In *explant* trials, samples of the environment, such as soil cores or vials of water, are brought into the laboratory and inoculated with a test organism. This is a very convenient procedure that is the basis of most assessments of environmental quality, but that necessarily underestimates the variance among the corresponding sampling sites because many of the features of the natural environment are not retained when small bits of it are extracted and brought into the laboratory. The alternative is the *implant* trial, in which organisms that have been bred in the laboratory are grown at different sites in the field. This is the phytometer technique, first used extensively by Frederic Clements of the Carnegie Institute and later revived by Janis Antonovics of Duke University. It is the most satisfactory method of estimating environmental heterogeneity, and if native species are used as test organisms in the environments in which they have evolved, it gives unbiased estimates of the variation of the environment as perceived by the organisms that live in it. Needless to say, it is also the most laborious technique, and extensive implant trials are very rare outside agriculture.

**Environments Are Not Uniform Even at Very Small Spatial Scales.** It is easy to appreciate that environments are often heterogeneous on the human scales of a meadow or a pond. We might expect, however, that on a smaller scale (different sites within a meadow or a pond, for example) the environment would be uniform. This is not the case. It is commonplace for agronomists to find that soil properties, such as water retention, particle size,or nutrient status vary on a scale of metres or less in cultivated fields, even though most agronomic practices tend to reduce environmental heterogeneity. These observations have been extended to natural environments by Antonovics and others at Duke, and by Martin Lechowicz and me at McGill. We found that physical factors, such as nitrate concentration and pH, varied substantially among soil samples within quadrats of undisturbed primary forest at all scales from 50 m down to about 0.1 m. These results were confirmed by explant trials, in which we grew genotypes of barley and *Arabidopsis* in soil cores taken from the forest and planted in the greenhouse. We also bred families of the annual herb *Impatiens*, collecting seedlings from the forest and planting their progeny back into the same area. *Impatiens* has both individual and lineage scales of about 1 m and 1 year, and our implant trial confirmed that there was substantial environmental variance at these scales within populations, showing that the environment varies on scales relevant to the growth of native organisms. Indeed, we have been unable to discover any scale, however small, at which forest soils are uniform. By collecting miniature soil cores and using them to prepare soil–water medium in which *Chlamydomonas* cultures can be grown, we have been able to detect heterogeneity at scales down to the limit of measurement at about 0.01 m.

**Environments Are Not Random Even at Very Large Spatial Scales.** As samples are taken from smaller and smaller areas, variance persists even at the smallest scales. Nevertheless, the quantity of variance decreases, as we would expect on the commonsense ground that larger areas are likely to contain a broader range of different conditions of growth. The increase in variance with distance reflects the fact that nearby sites are likely to be similar to one another and, thus, expresses the *pattern* of environmental heterogeneity. There appears to be no limit to this increase: as larger and larger areas are sampled, more and more environmental variance becomes apparent. As a rough general rule, the logarithm of environmental variance increases linearly with the logarithm of distance at all spatial scales. The slope of this graph is a measure of environmental pattern: a low slope indicates a fine-grained environment with little correlation between adjacent sites, whereas a steep slope indicates a coarse-grained environment in which adjacent sites are very similar. Pattern is the property of an environmental factor, not of an environment: different physical factors may show different patterns.

**Temporal Variation.** It is less convenient to study how environments vary through time, because investigations of year-to-year variation are necessarily lengthy, whereas patterns of change within years are dominated by diurnal and seasonal cycles. The only extensive data are supplied by agronomic surveys of yield in different years at different locations. These show that years and localities contribute about equally to environmental variance. They also show that the interaction between years and localities is an important source of variation, so that the effect of conditions in a given year is not the same in all localities, whereas the effect of conditions at a given locality differs from year to year. Temporal and spatial variation are thus not wholly independent.

Environmental heterogeneity seems to be the most important cause of phenotypic variation in quantitative surveys of plants. The environment in which they live is heterogeneous at all scales, and selection may therefore act differently from place to place, or from time to time, regardless of the size or longevity of the organism, and regardless of the distance that its progeny are dispersed in space and time.

## 102. *Genotypes vary in their response to the environment.*

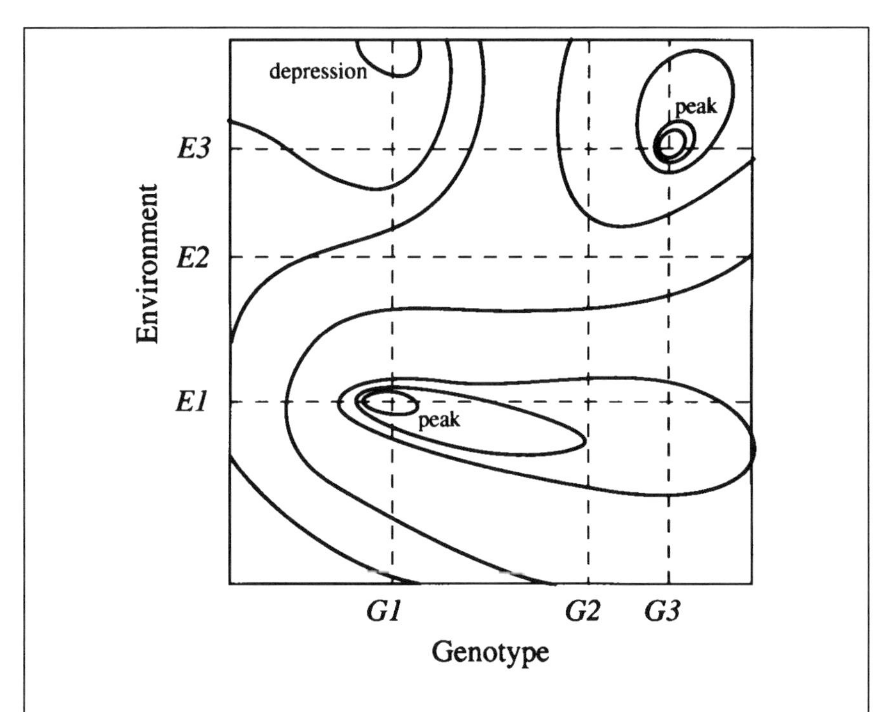


The interaction of genotype with environment may give rise to the complex sort of topology shown here. The relief (third dimension of the map) is fitness. A

(*Continues*)

vertical section represents environmental variance. Thus, G1 is a specialist type with very high fitness in some environments (near the peak at E1) but much lower fitness in others (near the depression at the top of the map). Conversely, G2 is a generalized or plastic type, with mediocre fitness everywhere. An horizontal section represents genetic variance, with some environments (such as E1) causing more genetic variance to be expressed than others (such as E2). Environmental change alters the relative fitness of genotypes and thus causes genetic change through selection. For example, a change from E1 to E3 will cause the replacement of G1 by G3.

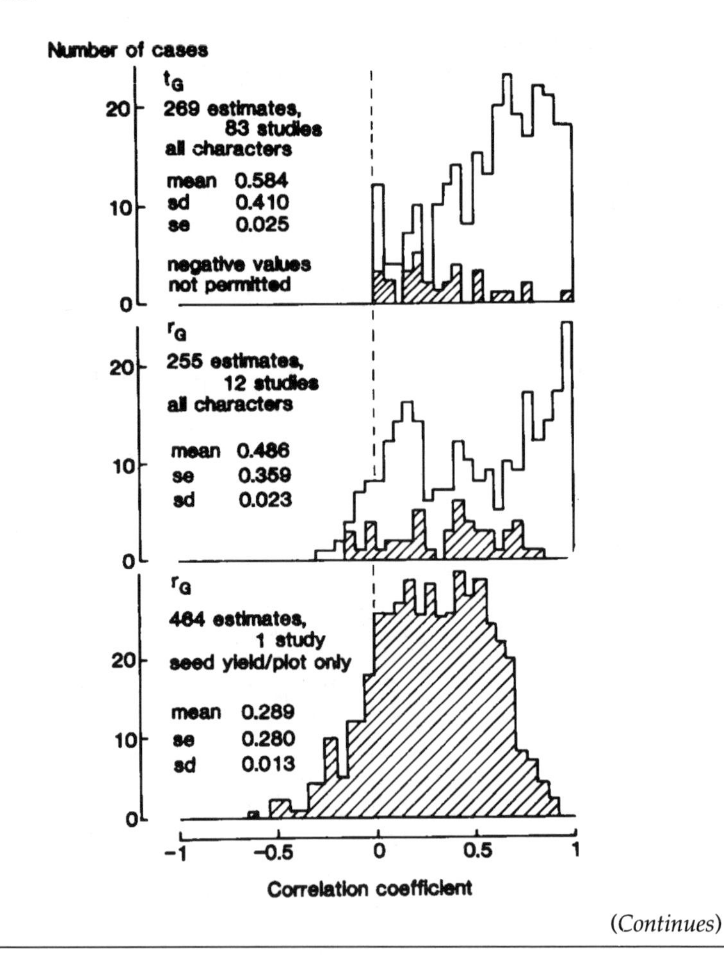


(Continues)

The magnitude of genotype-environment interaction, or in other words the precision of adaptation, is a central issue in evolutionary genetics. Our largest source of data is the results of crop trials carried out by agronomists. For a representative series of these trials, I have surveyed the journal *Crop Science* from 1978 to 1986, and abstracted all papers from which satisfactory estimates of sources of variation can be obtained. The data can be summarized in one of two ways. The first is to calculate the three components of phenotypic variation: the genetic variance $\sigma_G^2$, the environmental variance $\sigma_E^2$, and the genotype–environment variance $\sigma_{GE}^2$. Methods for doing this can be found in any elementary book on parametric statistics. The proportion of the genotypic variance contributed by straightforward genetic variance is then $t_G = \sigma_G^2/[\sigma_G^2 + \sigma_{GE}^2]$, an intraclass correlation coefficient. If this is close to unity, genotypes perform consistently over environments, the best or worst in one place being the best or worst everywhere else; that is, adaptation is very general. If $t_G$ has a value close to zero, then genotypes are very inconsistent, with fitness in one environment providing little if any information about its fitness in other circumstances; that is, adaptation is very specific. The survey yielded 269 estimates of $t_G$, from about 80 studies, whose frequency distribution is shown in the upper figure. The mean value is 0.602, with a standard error of 0.016. Further examination of the data tends to reduce this value. Thus, the 102 estimates from studies conducted over more than one year at more than one locality gave a mean value of 0.584. All these estimates refer to a very broad range of agronomic characters, from seed weight to milling quality; restricting the analysis to 30 or so studies dealing exclusively with seed yield (shaded), the character closest to Darwinian fitness gives a mean value of 0.35. Moreover, these estimates are not very sensitive to the nature of the source material: random selfed lines and outcrossed populations give similar results. In short, some 40% or more, say, a half, of genotypic variance, especially for characters close to Darwinian fitness, is attributable to the variation of relative performance among enviroments.

Another way of analyzing crop trials is to compute interclass genetic correlations, as described in the text. (In simple statistical models, the intraclass and interclass correlations have identical values, or some well-defined proportional relationship.) A second survey, from the same literature source, turned up 12 papers that supplied 255 estimates of genetic corelations between environments, either localities or years. These did not overlap with the first data set. The frequency distribution of genetic correlation coefficients is shown in the middle figure; its mean value is 0.486, with a standard error of 0.023.

The study of grain yield in winter wheat by Campbell and Lafever (1980), not included in either of the two previous data sets, is shown in the lower figure; this gave a mean genetic corelation coefficient of 0.289, with a standard error of 0.013.

(*Continues*)

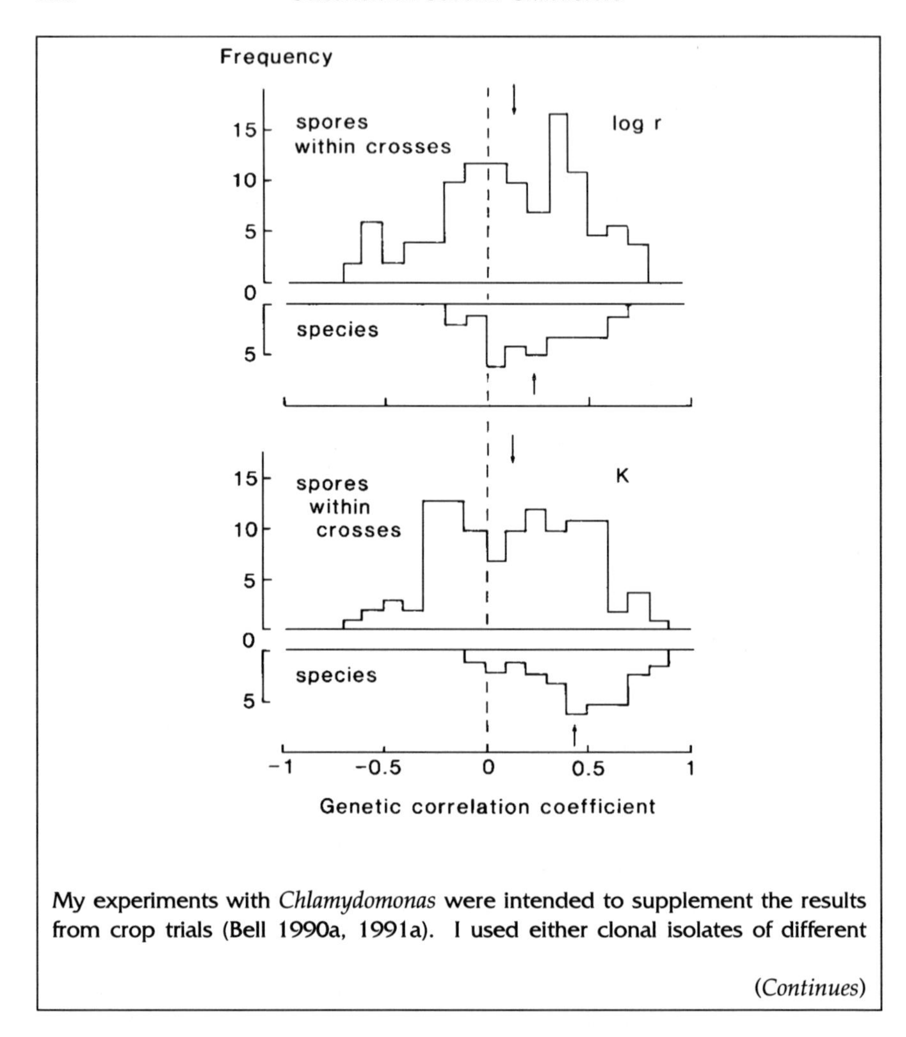


My experiments with *Chlamydomonas* were intended to supplement the results from crop trials (Bell 1990a, 1991a). I used either clonal isolates of different

(*Continues*)

The fact that fitness varies in time and space is an ecological principle that does not necessarily have any evolutionary consequences. If all genotypes vary in the same way and to the same extent over a given range of environments, then selection will not cause any change in their frequencies. It is only if *relative* fitness varies among environments that selection will act differently according to circumstances. This constitutes a third possible source of variation among individuals. If several genotypes are each tested in several environments, the variance of the average scores of genotypes represents genetic variance, and the variance of the average scores in species, or spores (clones) isolated from crosses within the single species *Chlamydomonas reinhardtii*. These were grown in a range of media with different concentrations of nitrate, phosphate, and bicarbonate. These histograms are the frequency distributions of the growth parameters $r_{max}$ and $K$ (see Part 5A) for different genotypes of *Chlamydomonas reinhardtii* (upper diagram of each pair) and for different species of *Chlamydomonas* (lower diagram). Arrows mark mean values.

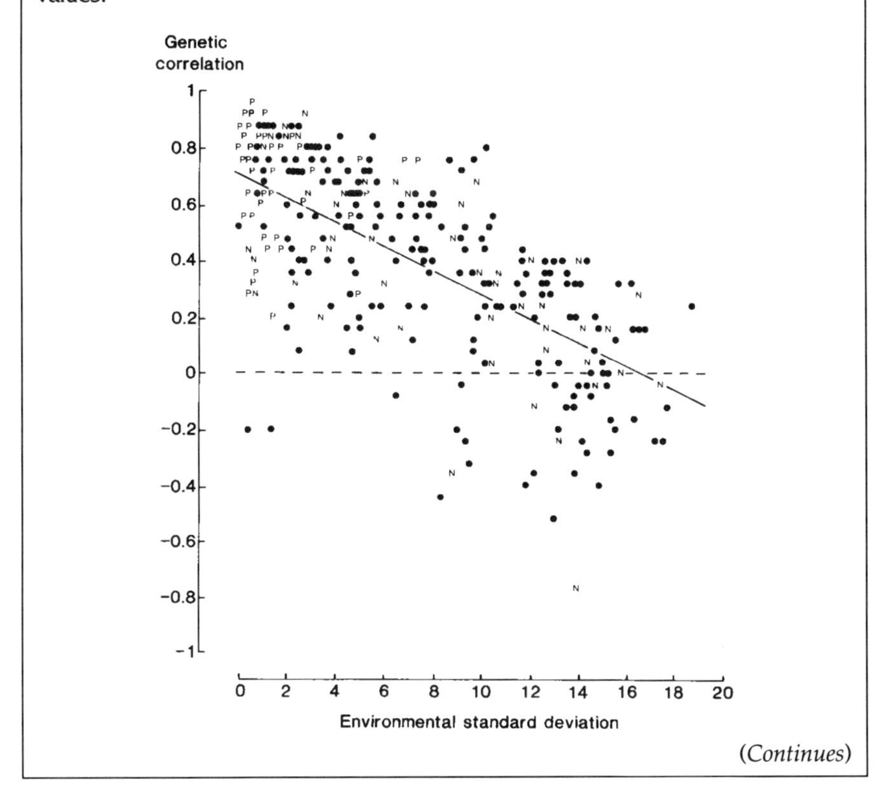


(*Continues*)

each environment represents environmental variance. If genotypes do not respond consistently to environmental variation, however, the score of a particular genotype in a particular environment may deviate from the score that would be expected on the basis of the average properties of that genotype and that environment. The variance represented by these deviations is attributable neither to purely genetic sources, nor to purely environmental sources, but rather to the interaction between genotype and environment G × E. This is the variance of relative fitness over environments.

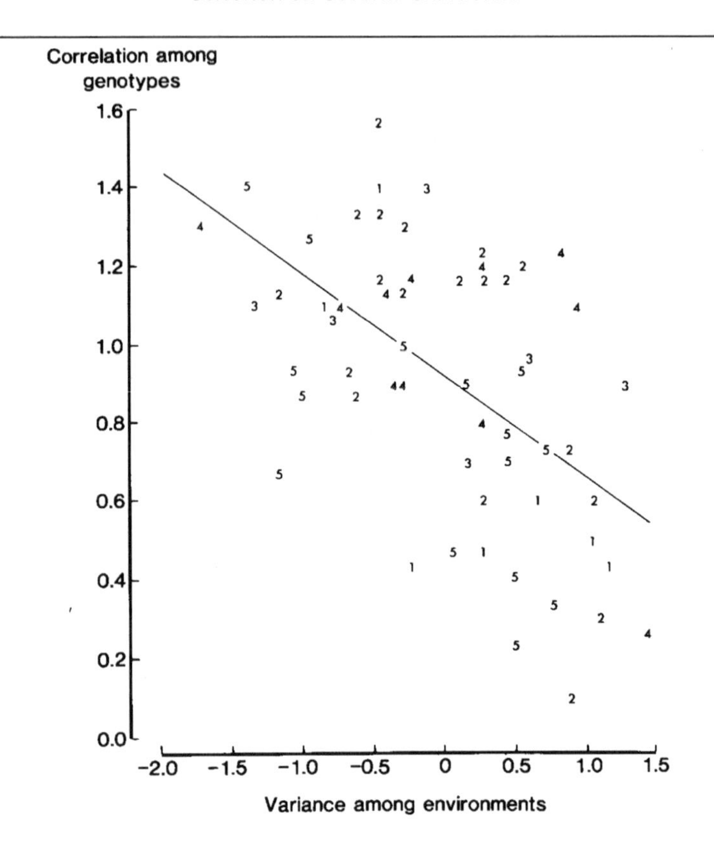


Both the *Chlamydomonas* system and crop trials can be used to define how relative fitnesses shift as environments become more different. This is done by calculating the genetic correlation and the environmental variance for each pair of environments. The environmental variance is simply the variance of mean fitness, although this is of course created by some underlying physical differences between the environments. In one *Chlamydomonas* experiment, shown in the upper diagram, 15 species were grown in media with five different levels of nitrate and of phosphate, allowing 300 pairwise comparisons between environments. As the environmental variance increased (primarily because of variation in overall concentration, especially of nitrogen), the genetic correlation decreased. For the most highly contrasted environments, the genetic correlation was on average about zero.

The lower figure shows a similar analysis of crop trials, although the diversity of the material made it necessary to transform the data in a rather complicated way. The correlation is the intraclass correlation $t_G$, transformed by arcsin square root (because it is a ratio). The variance is the residual value of environmental variance

(*Continues*)

(*Continued*)

regressed on genetic variance (to remove the autocorrelation of $\sigma_G^2$ with $t_G$), both variances being scaled on their means (because in different cases they are measured in different units). The plotted numbers are different kinds of character: 1, seed yield; 2, components of seed yield; 3, vegetative yield; 4, morphology; and 5, chemical composition. Again, genetic correlation declines as environmental variance increases.

The results from crop plants and from *Chlamydomonas* seem to be concordant. Relative fitnesses change when the environment changes, so that the range of environments to which any given genotype is well adapted is limited. Moreover, there seems to be a simple empirical rule describing these shifts in relative fitness, at least when environments differ only in the concentration of simple nutrients: that the genetic correlation declines from unity in identical environments towards zero in the most disparate environments.

**The Ecogenetic Landscape.** The interaction of genotype with environment might be visualized as a landscape, akin to the adaptive landscape used to represent interaction among genotypes (Sec. 54). I shall call it the ecogenetic landscape. Longitude represents a series of genotypes, as in the adaptive landscape; but latitude now represents a series of environments. The genotypes, which may be alleles at a single locus or clones that differ to any extent, are arranged from the least to the most fit; the longitudinal relief is genetic variation. Likewise, the environments are arranged from the least productive—that in which mean fitness is lowest—to the most productive, and the latitudinal relief is environmental variation. If we can arrange both genotypes and environments in this way, then the landscape will be a ramp, sloping smoothly upwards to the north-east. The same genotype will be selected, regardless of the state of the environment, because the same genotype is the most fit in all environments. However, it may not be possible to arrange either genotypes or environments in this way. If genotypes are ranked differently in different environments, then they cannot be unequivocally arranged from least to most fit. Consequently, rather than being a smooth ramp, the ecogenetic landscape will have a more or less complex topography of hills and valleys. This roughness of this topography is produced by the interaction between genotype and environment, with the height of the hills and the depth of the valleys representing the magnitude of G $\times$ E.

A landscape of this sort can be used to visualize how the population responds to environmental change in time or in space. Imagine that the environments are arranged in a temporal sequence, so that the landscape moves in a north-south plane past some line representing the present, like a conveyor belt. At any given time, the population occupies a small region of the map, represented by the prevailing genotype and its variants in a single environment. The fitness of the whole range of possible genotypes at this time is an east-west section through the landscape, with the hills and valleys seen in profile. If the environment has remained the same for long enough, the population will have climbed to the top of the hill or one of the hills. When the environment changes, the landscape moves north-south, and the profile of the hills and valleys alters. The population now moves in the east-west plane, as selection shifts the population uphill towards the newly optimal state. This represents directional selection, which always tends to increase the mean fitness of the population. When adaptedness has been restored, mean fitness is maximized, and the population stands at the top of the hill, with a few mutational variants straggling down the slopes. This is the familiar process by which the fit between population and environment continually deteriorates through external change and is continually restored through selection.

Now imagine that the latitudinal dimension is the range of different kinds of environment that are simultaneously available to the population, so that it represents spatial rather than temporal variation. The hilliness of the terrain now expresses the variation of relative fitness from site to site. At any given site, the population will move uphill to a local peak. The peak, however, represents different genotypes in different sites. Viewed from above, a population that is initially dispersed over the whole landscape would be seen as evacuating the valleys and moving in different directions up the slopes of different hills. Selection is directional at any given site, but because it acts in different directions at different sites it is *disruptive* at the level of the population as a whole. The disruptive selection engendered by G × E retards the fixation of any single genotype and thereby tends to conserve genetic diversity.

**The abcd Model.** The simplest version of the selective landscape involves two genotypes in two environments. I shall use this minimal fitness matrix in several different guises; in the present context it is a genotype–environment matrix that looks like this:

| | | Environment 1 | Environment 2 |
|---|---|---|---|
| Genotype | 1 | a | b |
| | 2 | c | d |

where abcd each represent the fitness of a genotype if it were growing in only one of the two environments; I shall usually call this the "performance" of a genotype in a given environment. The overall fitness of a genotype depends on its performance in both environments and on their frequency. Assuming the two environments to be equally frequent, there is genetic variance in fitness if the mean fitness of the genotypes differs when averaged over environments, that is, if $\frac{1}{2}(a+b)$ differs from $\frac{1}{2}(c+d)$. Likewise, there is environmental variance in fitness if average fitness varies among environments, that is, if $\frac{1}{2}(a+c)$ differs from $\frac{1}{2}(b+d)$, assuming the two genotypes to be equally frequent. It is, however, easy to construct an abcd matrix in which there is no consistent variation among genotypes or among environments despite any amount of variation in performance. Suppose that $a = d = 2$, for example, whereas $b = c = 1$. The two genotypes have the same mean fitness, and their average performance is the same in both environments. The overall variance in performance is wholly attributable to the difference in relative fitness between the environments: genotype 1 is twice as fit in environment 1, but genotype 2 is twice as fit in environment 2. Thus, there will be G × E whenever there is a difference between the diagonals of the genotype–environment matrix, that is, when $(a+d)$ differs from $(b+c)$. The quantity of G × E in this simple model is, in fact, $\frac{1}{4}[(a+d)-(b+c)]^2$.

**Generalists and Specialists.** There are two equally valid ways of interpreting G × E. On the one hand, it expresses the extent to which genetic variation is expressed differently in different environments. The quantity of genetic variation is not a fixed property of a population, but may vary according to the environment in which the population is living. There may, therefore, be greater opportunity for selection in some environments than in others. Moreover, part of this variation among environments in the quantity of genetic variation that is expressed in them arises because genotypes that have very high relative fitness in one environment are mediocre or even inferior in others. Thus, genotypes may be selected as *specialists* in particular environments. On the other hand, G × E expresses the extent to which environmental variation is expressed differently by different genotypes. Some genotypes may express very different phenotypes in different environments and, therefore, possess a large quantity of environmental variance; others will be less responsive and express more or less the same phenotype regardless of the environment in which they are raised. The quantity of environmental variance is itself a character and may vary among genotypes. Thus, genotypes may be selected as *generalists* over a range of environments. The main issue raised by G × E is the balance between generalization and specialization that should evolve in populations that live in a heterogeneous environment.

**The Magnitude of G X E.** Genetic variance is studied by geneticists; environmental variance is studied by ecologists; G × E has historically been neglected by both. There is thus no synthetic literature on the subject to which one can turn for an evaluation of the importance of G × E. However, G × E has long been of interest to agronomists, who usually wish to develop cultivars that will perform well over as broad a range of environments as possible, and the agronomic literature is replete with accounts of trials in which a series of cultivars are grown at several localities for two or three years. Such trials are by far the largest available source of information about how genotypes respond to environments, and they provide very extensive documentation of the magnitude and generality of G × E. As an example, the largest single survey that I have found was a yield trial conducted by Campbell and Lafever involving 19–30 cultivars of winter wheat scored at 12 localities over 7 years. How can we express the degree to which the relative fitness of these genotypes, measured as their seed yield, varied among the environmental conditions prevailing at different times and places? Perhaps the most straightforward way is to consider the variation of a series of genotypes in just two environments. The growth of a genotype in one environment could then be plotted against its growth in the other. This graph would then represent a genetic correlation; if the relative fitness of genotypes is similar in the two environments, the correlation is high, whereas if relative fitness in one environment is unrelated to relative fitness in the other, the correlation will be zero, and if high fitness in one environment is consistently associated with low fitness in the other, the correlation will be negative. The advantage of this approach is that the genetic correlation coefficient is related directly to the correlated response of one character to selection applied to another. It amounts to considering the different rates of growth expressed in the two environments as being two different characters; more generally, any attribute measured in several different environments can be treated as though it were as many different characters. The problem of how populations evolve in an heterogeneous environment can then be viewed in the wider context of how selection acts on several different characters simultaneously. In Campbell and Lafever's trial, the average of the 464 genetic correlations that can be calculated from their data was +0.29; half of the estimates fell between +0.09 and +0.51 and 15% of them were negative. The relative fitness of the genotypes they studied thus varied extensively from place to place, and from year to year, giving the impression of a selective landscape with a very rough and broken topography.

There are few comparable studies in non-agronomic situations. However, I have made extensive measurements of the growth of *Chlamydomonas* in the laboratory. These involve recording the density of clonal cultures throughout their growth in different media, providing an estimate of the rate of increase and final density of each genotype in each environment. In one experiment the clones were different named species; in another, different isolates of unknown taxa from a natural community; in a third, sibs from crosses within the same species. The results of all these trials were similar, yielding genetic correlations of 0.1–0.5 on average.

The large-scale *Impatiens* trials used to measure the heterogeneity of the forest environment also provide estimates of genotypic variances in a population growing in its native environment. The genotypes in this case are sibs; the environments are different sites in the forest, a few metres apart. What little genotypic variance of fitness and fitness components that was detectable in these circumstances was largely G × E.

**Relationship of G X E to Environmental Variance.** The main axis of the argument so far has been that environments are heterogeneous and that this heterogeneity causes variation in relative fitness. It will be very difficult to use these facts to interpret populations, however, unless there is a regular relationship between the extent of heterogeneity, measured as the variation in average fitness of a given set of genotypes, and the extent of variation in relative fitness. If the most trifling environmental shift may often cause a complete revolution in the ranking of genotypes, then it will be impossible either to predict or to explain evolutionary change in terms of environmental change. Rather, we might imagine that a very slight change in the environment will cause only very slight alterations in relative fitnesses and that greater environmental change will cause progressively more change in relative fitnesses. This amounts to saying that G × E should become larger, as a proportion of total genotypic variance, as environmental variance increases. In crop plants and in *Chlamydomonas*, I have found that the genetic correlation among environments decreases linearly as environmental variance increases. In very similar environments it is nearly unity; in the most dissimilar environments represented in the trials, it approaches zero. This supplies a simple general rule for relating ecological change to evolutionary change.

**The General Inconsistency of Genetic Response.** The crop trials are very extensive and prolonged; the *Chlamydomonas* trials allow precise measurements of growth in defined environments; the *Impatiens* trials were carried out on undisturbed natural populations. These very different systems all exhibit high levels of G × E for characters closely related to fitness, showing that, in general, genotypes respond rather inconsistently to environmental variation. This inconsistency is an essential part of the basic theory of selection: when the environment changes, the relative fitnesses of genotypes changes, causing some that were previously rare to increase in frequency through selection. Change in this model is rare, however, and at any particular time a genotype expresses a fixed fitness in a uniform environment.

This representation of the world, though convenient for many purposes, is evidently wrong. Instead of a world in which populations that have reached equilibrium under selection are occasionally disturbed by some abrupt perturbation, the surveys that I have described suggest that populations of all kinds of organisms, regardless of the scale on which they live, are kept in flux by continual change in the intensity and direction of selection. This is a very important change in perspective.

## 103. *The outcome of selection depends on the environment in which it is practised.*

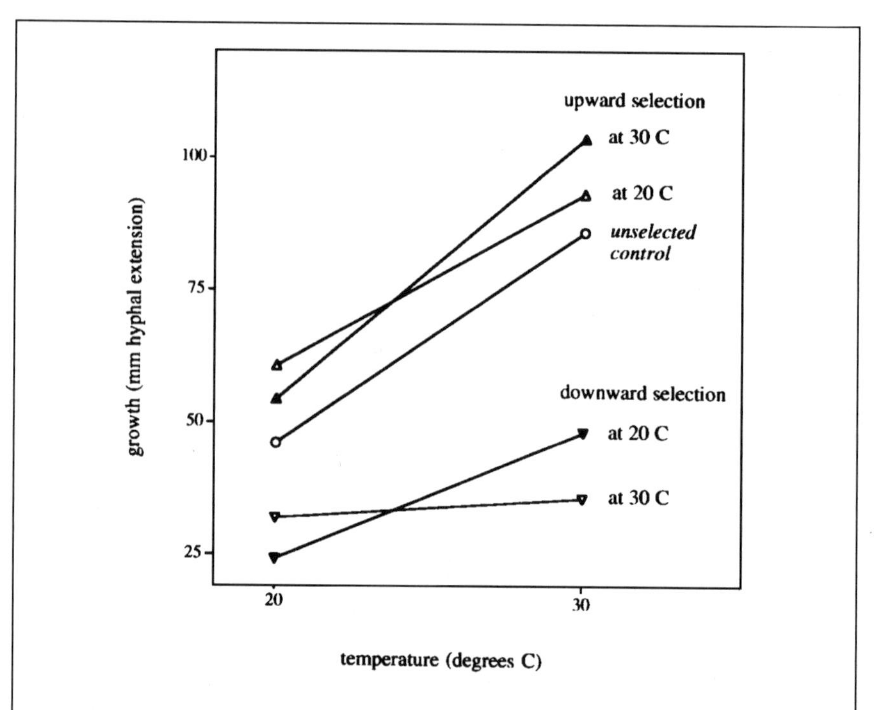


The direct and correlated response to selection in different environments can be seen in this diagram, drawn from data in Jinks and Connolly (1973; see also Jinks and Connolly 1975) on the growth of *Schizophyllum* selected upwards and downwards at two temperatures. The lines drawn here to represent the character

(*Continues*)

value of a genotype, or a selection line, over a range of environments is sometimes called a norm of reaction, or reaction norm. The direct response is the difference between a selection line, assayed in the environment of selection, and the unselected control. For example, the direct response to downward selection at 20 C is 24.25 − 46.25 = −22 mm. The correlated response is the difference in growth at a given temperature between the response of the line selected at the other temperature and the unselected control. Thus, the correlated response at 30 C to downward selection at 20 C is (32 − 46.5) = −14.25 mm. The full pattern of direct and correlated responses was as follows.

**Response to selection**

| | | *Direct* | *Correlated* |
|---|---|---|---|
| Upward selection | at 20 C | +14.5 | + 7.25 |
| | at 30 C | +17.75 | + 8.25 |
| Downward selection | at 20 C | −22 | −37.75 |
| | at 30 C | −50.25 | −14.25 |

Thus, the correlated response may exceed the direct response. This is because the reaction norm of the unselected control is not parallel to the reaction norms of the selection lines: selection is more effective in increasing growth at 20 C and more effective in reducing growth at 30 C. It is useful, however, to distinguish between the correlated response of lines (in a given test environment) and the correlated response in environments (of lines selected in another environment). In all cases, the line selected at a given temperature deviated more from the control than did the line selected at the other temperature. That is, in either environment the direct response (of the line selected in that environment) always exceeded the correlated response (of the line selected in the other environment). For example, at 30 C the direct response was −50.25 mm, and the correlated response −37.75 mm. This is why the reaction norms cross; this represents genotype–environment interaction caused by selection.

Selection for body size of mice in different nutritional environments was reported by Falconer and Latyszewski (1952) and Falconer (1960); see also Park et al. (1966) and the review by Hammond (1947). Frahm and Kojima (1966) and Cavicchi et al. (1989) selected for body size in *Drosophila* at different larval densities and temperatures. The work on barley Composite Cross I was reported by Suneson and Stevens (1953); for other examples of selection for agronomic characters in different environments, see, for example, Adair and Jones (1946) or Arboleda-Rivera and Compton (1974).

(*Continues*)

(*Continued*)

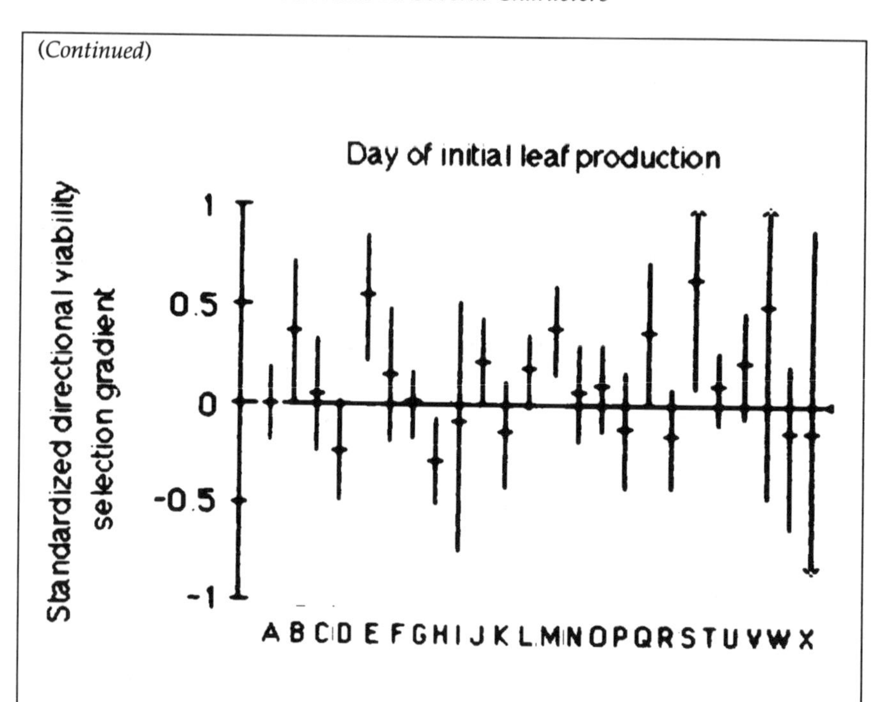


Such observations have rarely been extended to natural populations. This figure is an exception, showing the response of a phenological character, the day of initial leaf production, to very local differences in selection in a native population of *Impatiens*; Stewart and Schoen (1987) analyzed a range of other characters. The $x$-axis orders 24 sites within a 30 × 40 m area of forest floor. The $y$-axis is a standardized directional selection coefficient calculated from the partial regression of character values on viability. Note that the character is positively selected in some sites and negatively selected in others, despite the fact that the sites are only a few meters apart. This study is the best evidence currently available for variation in phenotypic selection at spatial scales comparable with the individual and lineage scales of the organism.

To the extent that the phenotype expressed by a given genotype varies with circumstances, selection for a given character in different environments will favor different genotypes. In the case of artificial selection, this implies that the response of a character to selection will depend on the conditions of culture; when the character is fitness, natural selection will produce different results in different environments.

**Artificial Selection for Body Size in Mice.** This argument leads to an important practical question: in selecting improved strains of crop plants or livestock, how should one choose the environment in which selection takes place? D.S. Falconer of Edinburgh investigated this problem by selecting for body size in mice, as a model or surrogate for attempts to increase carcase weight in larger mammals. The mice were reared either on the normal laboratory diet or on the same food diluted with a large quantity of indigestible fiber. There were thus two characters being selected in different lines: 'body size on a high plane of nutrition' and 'body size on a low plane of nutrition'. Both characters were selected upward and downward for about a dozen generations and, as expected, there was in every case a direct response to selection in the appropriate direction. It is the correlated responses to selection that were the main point of the experiment. We can ask three questions about these responses.

First is the question of whether the outcome of selection in either environment is completely reproducible in the other. If this is the case, the environments are not really different, with respect to evolutionary change, and it does not matter which of them we choose to select in. This is equivalent to saying that the genetic correlation between the two environments is unity. This was not invariably the case. The correlated responses to selection for increased growth on low plane (i.e., growth on high plane of the line selected for increased growth on low plane) and for reduced growth on high plane (i.e., growth on low plane of the line selected for reduced growth on high plane) were indeed comparable with the direct responses. However, although selection for reduced growth on the low plane caused a substantial direct response, there was no consistent correlated response (i.e., the line selected for reduced growth on a low plane showed no response when tested on the high plane). Moreover, there was a pronounced direct response to selection for increased growth on a high plane, but little if any correlated response (i.e., the line selected for increased growth on a high plane showed no response when tested on a low plane). The response to selection in one environment is thus not necessarily reproducible in another environment.

The second question is whether the correlated responses to selection can be predicted, by treating rates of growth in different environments as different characters and estimating their genetic variances and covariances. This is similar to asking whether the direct response can be predicted from estimates of genetic variances, and the answer is much the same. For the first few generations, the genetic correlation is a good predictor of the correlated response, but it is of little value in later generations, presumably because the genetic variances and covariances involved have themselves changed under selection.

Finally, we can ask which of the two planes of nutrition is preferable as an environment for selection. The answer depends on the direction of selection. For upward selection the low plane is better because the direct response achieved in this environment was reproducible in the other environment. Conversely, if the object is to reduce body size, it would be better to select on the high plane.

**Selection for Growth Rate at Different Temperatures in *Schizophyllum*.** John Jinks and his colleagues at Birmingham carried out several experiments to investigate the effects of selecting in different environments. *Schizophyllum* is a basidiomycete in which growth can be measured as the linear extension of hyphae in solid medium. High and low growth rate were selected by choosing the two most extreme of the 50 sexual progeny of the previous cycle. Both upward and downward lines were established at 20 C and 30 C; the growth of unselected lines is considerably greater at 30 C, which is thus the more favorable environment for growth, corresponding to the high plane of nutrition in Falconer's experiment. Despite the brevity of the experiment, the lines evolved specific adaptation to 20-C and 30-C conditions. Thus, at 20 C the lower of the two low lines and the higher of the two high lines were those selected at 20 C, whereas the lines selected at 30 C were the more extreme when tested at 30 C. The effect was nearly symmetrical, in that the difference between the direct and correlated responses was about the same at 20 C and at 30 C, in proportion to the growth of unselected lines.

**Natural Selection in Barley Composite Cross I at Different Localities.** H.V. Harlan of the U.S.D.A. mixed or crossed 11 barley cultivars to create diverse populations that were sown in several different widely-separated localities spanning the continental U.S.A. (see Sec. 42). For about a decade in the 1920s and 1930s each population was harvested without conscious selection to collect seed for resowing, while at the same time monitoring its genotypic composition. Some localities were used for only four or five years, but longer term records covering all 12 years,1925–1936,are available for stations in Montana, Idaho, and California. All the cultivars could grow reasonably well as pure stands at all the localities, but when they were sown together as a mixture, their frequencies changed rather rapidly through selection. Some cultivars seemed to be unequivocally inferior competitors and were quickly reduced to low frequencies at all localities. Others were more successful, but no cultivar increased in frequency at all localities; rather, the genotypic composition of the population changed in different directions in different populations. Composite Cross I was created by making 32 crosses among the 11 parental cultivars and was sown at the same localities as the mixture. Although its varietal composition can no longer be discerned directly, the frequency of distinct characters can be scored instead. Again, selection produced different results at different localities. *Black seed coat*, for example, was evidently nearly neutral at most localities, remaining in the population at about the same frequency for up to 12 years; at the California station, however, it declined from about 10% to about 1% within five years. *Two-rowed* plants increased in frequency somewhat at the Idaho and Montana stations, from about 20% to about 30%, but decreased to 5% in California; *smooth-awned* plants doubled in frequency in California and Montana, but showed no change in Idaho. All of these characters are attributable to single Mendelian genes, although, of course, selection acting at linked loci may have been responsible for some of the changes. These experimental populations of barley provide the best evidence that the G × E shown by cultivars when grown in isolation translates into differences in the direction of change in gene frequency within populations grown in different environments.

**Phenotypic Selection Within a Population of *Impatiens*.** The barley experiments show how selection will tend to produce local adaptation on a large spatial scale, using exotic plants. There have been very few attempts to work at smaller spatial scales or with native plants. Steve Stewart and Dan Schoen of McGill monitored groups of *Impatiens* from germination to death at 24 sites within a 1200-m$^2$ area of undisturbed forest, at the same locality where Martin Lechowicz and I had found substantial small-scale spatial heterogeneity for physical and biotic measures of the environment. They measured morphological and phenological characters, such as the rate and timing of leaf production, and showed that most of these characters affected the survival or fecundity of individuals at some of the sites. The effects were often quite strong, involving selection coefficients in excess of 0.2; moreover, the pattern of selection varied from site to site, sometimes favoring larger values of a character at some sites and smaller values at others. Spatial heterogeneity thus caused shifts in the direction of selection at scales of a few metres, corresponding to the lineage scale of these plants. It is doubtful whether natural environments are uniform with respect to selection at any scale.

## 104. *Specific adaptation causes a correlated increase of fitness in similar environments.*

**Selection for Growth Rate at Different Temperatures in *Schizophyllum*.** By testing the selection lines over a range of temperatures, the *Schizophyllum* experiment can be used to make a point that is often neglected, that is, whereas adaptation may be specific to a particular environment, it is not confined to that environment. Selection at 20 C specifically causes adaptation to 20 C; but this does not imply that at 20.1 C the line would be equivalent to an unselected line or to a line selected at 30 C. The effects of adaptation will be reproducible to a greater or lesser extent in similar environments. To investigate the breadth of adaptation associated with selection in a specific environment, the lines were tested at 2.5-C intervals between 15 C and 35 C after eight cycles of selection. Thus, the high line selected at 20 C had a higher growth rate than the unselected control at all except the highest temperature; it exceeded the 30-C line between 15 C and 25 C, but the 30-C line grew faster at temperatures of 27.5 C or more. If this were not the case, then the effects of selection could not cumulate because any trifling change in the environment would cause selection for a different set of genotypes.

**Selection for Growth Rate at Different Temperatures in *E. coli*.** The evolution of growth at different temperatures has been studied with larger populations over longer periods of time by using natural selection in

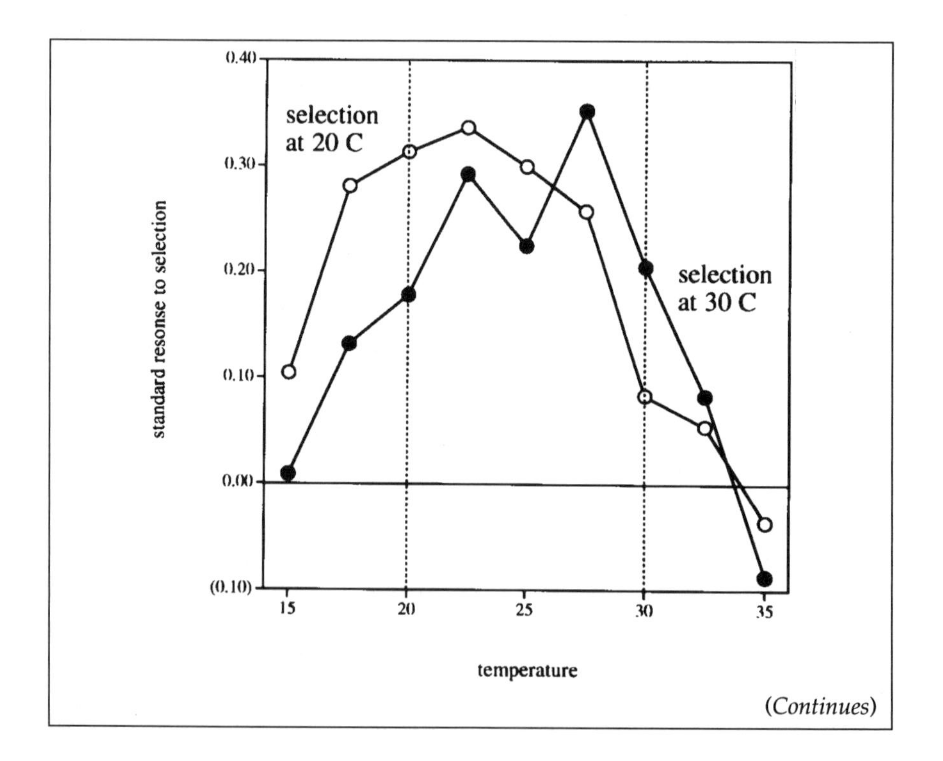


(*Continues*)

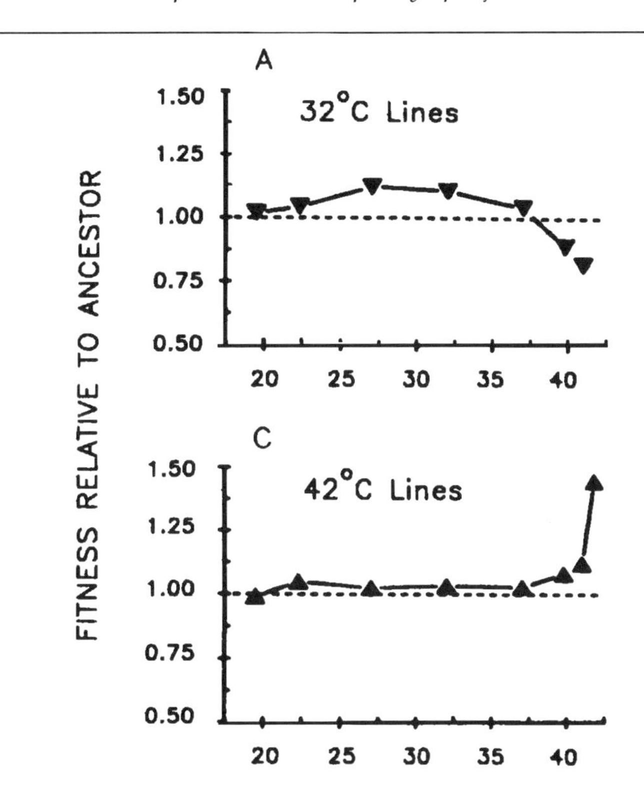


This upper diagram shows the results of testing the *Schizophyllum* upward selection lines over a wide range of temperatures. Because the unselected control line grew faster at higher temperatures, I have calculated a standardized response to selection as the difference between the selection line and the control, divided by the value of the control. There are three striking features of these results. Firstly, the response is positive in all cases, except at 35 C. There is therefore a very generalized correlated response to selection at either 20 C or 30 C, perhaps caused by adaptation to features of the laboratory environment common to both growth environments. Secondly, the response at either temperature seems to be a more or less smooth curve; the direct response is not qualitatively distinct from the correlated response in similar environments. Adaptation is not highly specific. Finally, the response curve for the 20-C line is shifted toward lower temperatures and that of the 30-C line towards higher temperatures. The specificity of adaptation is represented by the difference between the two curves.

The lower figure shows the fitness of *E. coli* lines selected at 32 C and 42 C over a range of temperatures, from Bennett and Lenski (1993). The bacteria are unable to persist in batch culture with a 100-fold daily dilution at temperatures below 19.5 C and die just above 42 C.

bacterial systems. The *E. coli* lines discussed in Sec. 64 had been maintained in the laboratory for 2000 generations, and had by this time become reasonably well adapted, both to general laboratory conditions and to a temperature of 37 C. Two sets of lines extracted from this population were then grown for a further 2000 generations, one at 32 C and one at 42 C. The effect of selection at a particular temperature on adaptation over a range of temperatures could then be assessed by competing the selection lines against their common ancestor. The results were similar to those got with *Schizophyllum*: the 32-C lines had greater fitness at temperatures around 30 C, and the 42-C lines had greater fitness at temperatures above 40 C.

A related issue addressed by this experiment is whether the quantitative increase of adaptedness in the environment of selection produces a qualitative shift in tolerance. It is possible to select strains of *E. coli* able to grow at temperatures in excess of the normal limit of 42 C, simply by isolating mutants at high temperatures, in the same way that one might select for resistance to an antibiotic. It is not clear, however, whether the usual limits to growth will be extended as a correlated response to selection within the normal range of tolerance. In these experiments, no such extension was observed. Although 42 C is very close to the upper limit to growth in the base population, continued selection at this temperature did not confer the ability to grow at 43 C. In a few cases, high-temperature lines grew at temperatures above 43 C, but only after an unusually long lag; this almost certainly involved the appearance and selection of novel highly tolerant strains not present at appreciable frequency in the 42-C lines.

## 105. *The evolution of specialization is obstructed by immigration.*

In an heterogeneous environment, the members of a given lineage will grow up in different conditions at different sites. An individual will survive and reproduce at rates characteristic of the site it lives in; but fitness is a property of lineages, not individuals, and the overall rate of increase of the lineage will be an aggregate property of the reproduction of its members. At any given site, we might measure the rate of increase of a certain clone, but although fitness has been defined as a rate of increase, it is a single rate of increase only in a uniform environment. The reproduction of a genotype at a particular site is not the fitness of that type, but rather a performance or *site-specific fitness* that represents a component of its fitness in the environment as a whole.

Site-specific selection will lead to site-specific adaptation; if organisms were completely sedentary, an heterogeneous environment would be populated by an intricate patchwork of locally adapted races at every scale down to that of the individual. However, all organisms must reproduce, and reproduction involves making propagules that disperse beyond the parent, into neighboring sites. Any site is therefore continually receiving immigrants whose parents have flourished elsewhere. To the extent that sites differ, and local adaptation tends to evolve, these immigrants will be well adapted to their natal site and, therefore, poorly adapted to the site that has received them, relative to the individuals resident at that site. The arrival of an immigrant in a locally adapted population, balanced by the death or emigration of a resident, is thus akin to the occurrence of a deleterious mutation. The immigration of poorly adapted individuals reduces the fit of population to environment, and local adaptation can be maintained only if selection is effective in restoring it. Site-specific selection is responsible for local adaptation; but at the same time, it threatens local adaptation when there is migration between sites. Indeed, immigration may often be a more potent force than mutation because immigrants are often much more frequent than mutants.

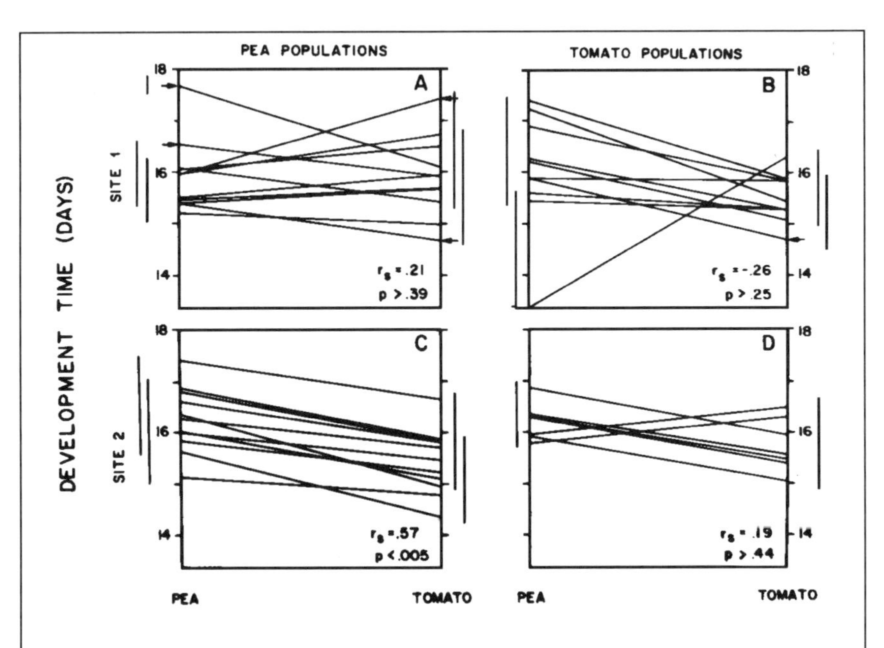


This figure shows the relationship between development time on cowpea and tomato food plants of full-sib families of the fly *Lyriomyza sativae*, from Via (1984).

(*Continues*)

(*Continued*)
The parental flies were collected from adjacent commercial plots of the two crops at two localities. The frequent crossing of the lines connecting performance on one host species with that on the other (norms of reaction) represents G × E caused by frequent changes in relative fitness between the two environments; note that in some cases a family that performs very poorly on one host performs very well on the other. Despite the opportunity for selection, however, no specialization has evolved: families collected from cowpea plots were not consistently superior on cowpea, nor were families collected on tomato consistently superior on tomato. Migration between adjacent plots of the two host species seems to have effectively obstructed selection for specialization.

The experimental work on *Drosophila* described in the text is from Wallace (1982); these experiments were continued in an attempt to detect sexual isolation between populations selected on the two salts, separately or together, and are discussed further in Sec. 171.

**Selection of Drosophila on Two Toxic Media.** Bruce Wallace of Cornell cultured *Drosophila* on medium containing either sodium chloride or copper sulphate. Populations cultured with either salt evolved some degree of resistance to it, as expected. These populations were transferred at intervals to cages in which both contaminants were present in separate vials. The flies in these two-salt cages were allowed to mate freely, and the larvae tested regularly to detect any tendency for those feeding on one salt to be more resistant to that salt than those feeding on the other. No such tendency was found, apart from the inevitable effects of the single episode of selection on either salt. The free migration and mating between the populations emerging from the two kinds of food vial was sufficient to prevent the evolution of specialized subpopulations.

## *106. Rare specialists will seldom evolve.*

The extent of specialization that can evolve in a particular kind of site is thus governed by the balance between site-specific selection and immigration. The rate of immigration is a function of the frequency of the site. An average kind of site that resembles most of the neighboring terrain and supports a large fraction of the population will in effect supply most of its own immigrants; the influx of poorly adapted genotypes from different kinds of site will be slight, and the maintenance of adaptation to ordinary or average conditions therefore requires only mild selection. An unusual kind of site is in the opposite position, because almost all the immigrants that it receives will come from different kinds of site. The maintenance of adaptation to unusual sites therefore requires intense selection in order to counteract this high rate of immigration. If the site is sufficiently rare, specialization is unlikely to evolve unless conditions there are lethal to immigrants.

The same argument also forbids the evolution of extremely precise adaptation to local conditions. If two sites differ only very slightly in their physical properties, it is nevertheless conceivable that one genotype may be best adapted to the first kind of site and a different genotype to the second. If the variation in relative fitness is correspondingly slight, however, it will be overwhelmed by even modest rates of immigration, and local adaptation will not be maintained. There is, in other words, a limit to the similarity of specialized types that can be maintained simultaneously in the same population.

Populations cannot, therefore, evolve an indefinite degree of specialization, and all individuals will be to some extent generalists, capable of growing successfully in a range of conditions. Indeed, the ideal type would be one that could grow successfully anywhere, combining in one genotype the attributes of every type of specialist. What is to prevent the evolution of a universal generalist?

## *107. The evolution of generalization is constrained by functional interference.*

The overall fitness of any type growing in a heterogeneous environment will be the average of rates of reproduction of members of that type at each site, weighted by the fraction of the population of the type that lives at each site. The concept that rates of reproduction in different conditions can be treated as different characters thus leads naturally to the concept that these rates are fitness components, just as survival and fecundity are fitness components in that both contribute to overall reproduction.

The evolution of fitness components, such as survival and fecundity, is constrained by a cost of reproduction. A similar principle of antagonism applies to the fitness components represented by the ability to reproduce successfully at different sites. It can be called the *cost of adaptation*: an increased ability to reproduce in any given conditions causes reduced performance in other conditions. It is almost self-evident that some such principle must apply. If it did not, the world would be occupied exclusively by a single type of organism, capable of expressing the optimal phenotype in any given circumstances. The diversity of all ecological communities shows that there are limits to the versatility and plasticity that organisms can display.

The primary constraint that leads to a cost of reproduction is the allocation of finite resources to alternative functions. This can scarcely be the source of the cost of adaptation: allocation is an individual concept, whereas the notion that a given type may grow poorly at one site as a consequence of being able to grow well at another site refers to a lineage. Antagonistic adaptation to different conditions requires, instead, a constraint imposed by *functional interference*. This is a very broadly applicable principle that is familiar in everyday life from the design of tools. A single tool may serve diverse purposes, as a combination pocket knife may include large and small blades, screwdriver, tweezers, and so forth. Nevertheless, surgeons or electricians do not usually rely on a Swiss Army knife; they have,instead,a battery of specialized implements, each of which is well suited to a particular task but of little use for others. A scalpel and a spoon are each apt for a particular task, but the features that fit them for one task disqualify them for the other, and an implement that combined these features would be effective for neither. Michael Ghiselin of Berkeley has emphasized the importance of this principle in the organization of the body, whose various functions are best served by specialized cells and tissues. It applies equally to different types of individual. The cryptic coloration of snail shells provides an example. Dark brown, unbanded shells are difficult to see against the sodden leaf litter of the woodland floor; yellow, banded shells are more likely to escape detection in rough grassland. What is cryptic at one site will be conspicuous at the other; better adaptation to one type of site can be achieved only at the expense of poorer adaptation to the other.

There is also an economic parallel to the corollary that populations can evolve specialization only for conditions that are sufficiently frequent or distinct. Whether or not it is economically worthwhile to manufacture a particular specialized implement depends on whether it serves an essential function for which no alternative exists. Soup cannot be eaten with a scalpel; if it were eaten only on one day in the year, it would still be profitable to make spoons, no matter how cheap scalpels were. The two implements are not *substitutable*. On the other hand, suppose that people rarely ate soft-boiled eggs. If teaspoons were cheap and plentiful, it would not be profitable to make egg spoons, because it would not be worthwhile to buy a special implement for eating eggs when a teaspoon is nearly as serviceable. Teaspoons and egg spoons are substitutable; their different characteristics, with respect to size, shape of bowl, and so forth, reflect functional interference, but it is in this case so slight that it is readily overcome by the much greater availability of one of the two implements. A sector that represents only a small fraction of the market cannot be exploited profitably if a substitutable product is widely available.

**Selection on Versatility.** The advantage of inducible enzyme expression is obvious: it reduces unnecessary metabolic activity. Nevertheless, it may also bear a cost. If the appropriate substrate (and inducer) is only intermittently present, then an inducible genotype will not respond immediately to the presence of the substrate, whereas a constitutive genotype will be able to take advantage of it immediately. The overall benefit, or cost, of inducibility will thus depend on the period of environmental fluctuations. Richard Lenski and his colleagues have studied this issue in the context of tetracycline resistance in *E. coli*. This system is encoded by an operon comprising a structural gene *tetA*, whose product flushes tetracycline from the cell, and a repressor *tetR*. The operon is derepressed by tetracycline, which binds to the repressor protein and thereby causes the structural gene to be transcribed. When tetracycline is absent, constitutive mutants with defects in *tetR* have appreciably lower fitness than inducible strains, confirming that the unnecessary metabolic activity represented by expression of *tetA* is indeed a handicap. In a medium containing tetracycline, however, the constitutive strains had the greater fitness, in part because they could begin growing sooner. Whether or not the versatility of the inducible operon is worthwhile will depend on the frequency, length, and severity of toxic episodes.

**Selection on Plasticity.** So far as I know, there are no directly comparable experiments on plasticity; no experiments, that is, that evaluate the fitness consequences of an irreversible response to the environment that is clearly appropriate, in the sense that the reversible response provided by an operon whose inducer is the substrate for its structural gene is appropriate. Instead, there have been several attempts to apply artificial selection to the environmental variance of a morphological character, such as bristle number, with the intention of evolving stocks in which the character is more or less stable among individuals raised in different conditions, whether or not they are better adapted to these conditions. The usual design involves some form of family selection, because an individual cannot be tested twice. Part of a family is raised in one environment and part in another, and the difference between the mean scores in the two environments is calculated; families with the least (or greatest) differences, that is, the least (or greatest) environmental variance, are selected. The difficulty with such experiments is that they may confound changes in the variance with changes in the mean—there are bigger differences between bigger things—and the main object of such experiments has often been to demonstrate that the environmental variance can be selected independently of the mean. For example, Marvin Druger of Syracuse selected for the variance of the number of scutellar bristles in *Drosophila*. There is a maximum of four normal bristles, with more being produced at 18 C (about 3) than at 28 C (about 1.5). Selection for families with a greater or lesser difference in bristle number at high and low temperatures was very effective, especially in the low-variance lines. Indeed, after about the 10th generation some families produced more bristles at 28 C than at 18 C, reversing the normal pattern of variation. At the same time, selection for low (or high) variance caused a decrease (or increase) in mean bristle number. The increased variance of the high lines was caused mainly by a tendency to produce more bristles at 18 C, without much change at 28 C. The two replicate low lines behaved differently. In one case, bristle number at 18 C decreased, whereas at 28 C it was unchanged; in the other line, there was both an increase at 28 C and a decrease at 18 C. Thus, the changes in mean bristle number do not by themselves explain the evolution of the environmental variance, which may have occurred in different ways in replicate lines. Whether or not such experiments imply that there is a distinct set of genes responsible for adjusting environmental variance, rather than simply modulating environment-specific performance, remains controversial.

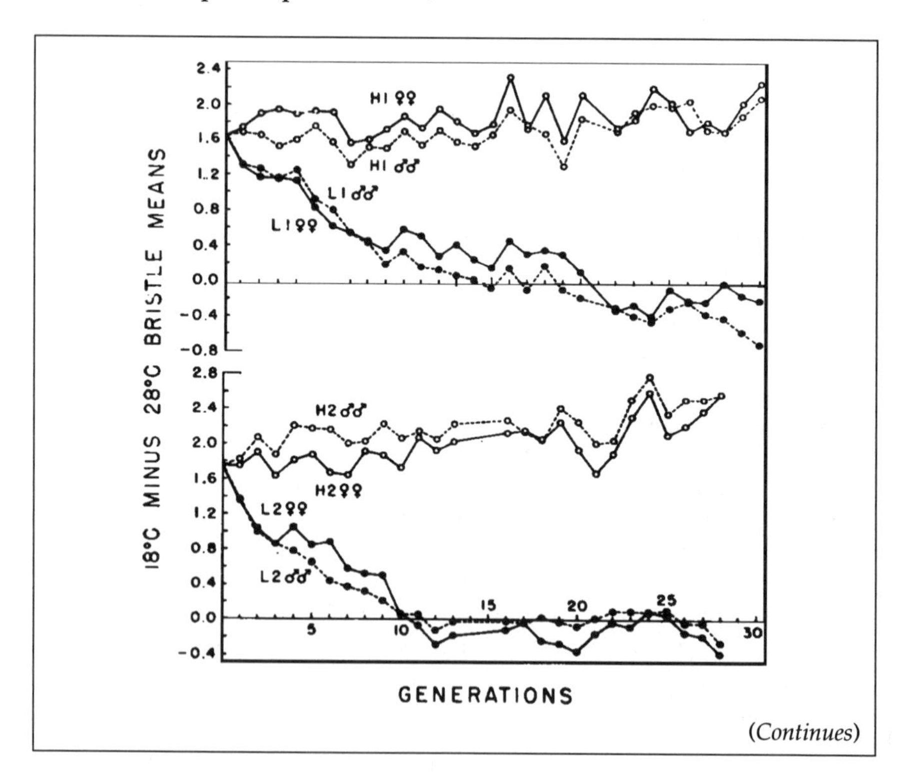


(*Continues*)

(*Continued*)
This diagram looks like the outcome of any other selection experiment, but rather than a mean value, it shows the difference in mean values in two environments. Lines selected upwards and downwards for the environmental variance of scutellar bristle number at high and low temperature diverge rapidly (from Druger 1967). There is a very similar experiment by Kindred (1965), and other work on *Drosophila* by Falconer (1957), Rendel and Sheldon (1960), Waddington (1960), Hillesheim and Stearns (1991), and Scheiner and Lyman (1991). Comparable experiments on plants were published by Brumpton et al. (1977) and Jinks et al. (1977). The topic was reviewed by Scheiner (1993).

The fitness consequences of inducibility of tetracycline resistance were investigated by Nguyen et al. (1989) and Lenski et al. (1994).

## 108. *Diversity is limited by the extent of the market.*

Functional interference explains why there should be a diversity of tissues within the body, or a diversity of types in the community. In either case, this diversity represents a *division of labor* between entities that are variously specialized for different tasks. The specialization of tissues within the body is a *cooperative* division of labor that acts to increase the performance of the body as a whole; it is akin to the division of labor among the employees of a single company. The specialization of different types in the population or community is a *competitive* division of labor that acts to increase the performance of each, not necessarily of the ensemble, by reducing the extent to which they compete with one another; it is akin to the division of labor among trades and professions in society. We are concerned here solely with the competitive division of labor.

The principle that regulates economic and ecological diversity was discovered before Darwin was born, by Adam Smith of Edinburgh: *the degree of division of labor is proportional to the extent of the market.* He gave the example of a nailsmith working in a small Scottish village, in the pre-railway society of the late eighteenth century. To make a nail requires a series of operations: drawing the wire, cutting the blanks, pointing and heading the nail. A single smith, working at nothing else, could make perhaps one-quarter of a million nails a year. But he could not possibly sell this quantity, because his market for nails would be limited to the houses of the village and a few outlying farms. In order to make a living, he must therefore be a general smith, for whom nail making is only a small part of his activities. A small market enforces generalization. In a large city, on the other hand, the market for nails is much greater, and the smith may well choose to make nails and nothing else. The advantage of doing so is that different activities often interfere with one another: turning from making nails to shoeing horses, for example, requires investment in a range of different tools and loses time in preparing the workplace for a different task. Indeed, if the market is sufficiently extensive it may greatly exceed the capacity of a single smith working with hand tools, who may then choose to specialize further by restricting his activities to a single operation, making blanks, for example, which are then sent to another workshop for finishing. A large market encourages specialization because if the market is large enough, even a small sector, comprising a very small fraction of total economic activity, can support specialized workers.

The simplest biological equivalent of this principle is that larger areas support more different types of organisms. Because environmental variance increases with distance, a larger area will display a larger overall quantity of environmental variance. This will be associated with a greater quantity of G × E, providing a greater opportunity for the maintenance of a diversity of specialized types. Thus, more species are found on larger islands. Within a given area, however, the extent of the market depends on productivity. In unproductive areas, generalists will be favored because a rare kind of habitat is incapable of supporting a substantial population. As productivity increases, the number of habitats that offer a viable way of life increases, and a more diverse community of specialists will evolve to exploit them. Thus, the species diversity of North American vertebrates is highly correlated with evapotranspiration, a measure of the energy absorbed by the land, and therefore of the quantity of work that can be done. Unfortunately, no experiments have been done to evaluate this principle.

## 109. *Specialized types may accumulate conditionally deleterious mutations.*

Zamenhof and Eichorn (1967) attributed the spread of auxotrophic mutants of *Bacillus* to energy conservation. Andrews and Hagerman (1976) found that *E. coli* strains constitutive for the *lac* operon were not consistently selected against when supplied with sugars other than lactose, even though the enzymes of the lactose pathway consitute 3–5% of the total cell protein. The elegant experiment summarized in the text is by Dykhuizen (1978).

Water loss from *Drosophila* is reduced by waxy epicuticular hydrocarbons. It increases in strains maintained for long periods of time in normal laboratory conditions, where humidity is high or where the food vials provide humid refugia. The total quantity of hydrocarbon is unchanged, but there is a shift from longer, branched molecules to shorter, simpler molecules (Toolson & Kuper-Simbrðn 1989). In consequence, the ability of the flies to endure changes in temperature

(*Continues*)

*(Continued)*
and humidity is reduced. It is not known whether this is attributable to pleiotropy (long branched hydrocarbons being more expensive to make) or to mutation accumulation.

A population cannot become perfectly adapted even to a uniform and unchanging environment, because deleterious mutations cannot be entirely eliminated, but are rather maintained in the population at a frequency that represents an equilibrium between the rate of mutation and the rate of selection (Sec. 25). Mutations that are nearly neutral may therefore be very frequent. The same principle applies to local adaptation. Mutations that reduce the fitness of a type in the site to which it is specifically adapted will be held at low frequency by selection; but mutations that reduce fitness in other sites will be nearly neutral in effect and will tend to accumulate. Long-continued selection in one site, in populations where migration is infrequent, for example, will for this reason lead to some loss of adaptedness to other kinds of sites. It is not necessary that such conditionally deleterious mutations be somehow entailed by increased adaptedness to local conditions. They will accumulate merely because adaptedness tends to break down unless it is actively and continually maintained by selection; in a heterogeneous environment with restricted migration, site-specific selection will lead to local adaptation, but cannot maintain adaptedness to the specific features of other sites.

This is the third context in which I have distinguished two types of process that contribute to adaptation. Selection may cause increased adaptedness to novel conditions through the spread of beneficial mutations, or it may maintain adaptedness through the elimination of deleterious mutations. The schedule of reproduction may respond to selection for early vigor, causing senescent decline in later life, or it may be molded by the greater frequency of deleterious mutations that act later in life. In the present context, the breadth of adaptation may be limited because specific adaptation to a given environment is accompanied by antagonistic effects in other environments, or because selection in one environment causes the accumulation of conditionally deleterious mutations. This dichotomy is a very old one, being stated explicitly by Darwin, who distinguished between the injurious effects of characters that were beneficial in other circumstances (to explain the reduction of wings in the insects of isolated islands) and the degeneration of organs that were not used (to explain the reduction of eyes in cave-dwelling animals).

In one sense, mutation accumulation it is not as severe a constraint as antagonistic pleiotropy because the selection of types that are well adapted to local conditions may proceed much more rapidly than the accumulation of nearly neutral mutations; some short time after the colonization of a site, therefore, a population may be reasonably well adapted but display little loss of fitness if it is then transferred to a new site. On the other hand, antagonistic pleiotropy requires some particular mechanism that will be different for different kinds of character, whereas the accumulation of conditionally deleterious mutations is a very general process that will occur whenever continued selection in one environment causes specific adaptation.

**Selection for Tryptophan Auxotrophs in *E. coli*.** Auxotrophic mutants often spread in chemostat populations of bacteria, when the substance they are unable to synthesize is supplied in the growth medium. This would be easy to understand if a metabolic pathway that is no longer necessary represents a metabolic burden, so that strains possessing the pathway reproduce less rapidly and are lost through selection. Such a theory of energy conservation would provide a general rationale for antagonistic pleiotropy and would explain why anabolic pathways are usually repressed by their end-products. For example, it is known that tryptophan auxotrophs often increase rapidly in frequency in glucose-limited chemostats, when tryptophan is supplied in excess. The *trp* operon of *E. coli* comprises five linked genes that encode the biosynthetic enzymes, controlled by an upstream promoter and an unlinked repressor. The pathway they form can be represented crudely as

$$\text{chorismate} \rightarrow \text{anthranilate} \rightarrow\rightarrow\rightarrow \text{indole} \rightarrow \text{tryptophan}$$

Daniel Dykhuizen, working at Chicago, used *trp* mutants to find out whether auxotrophs tend to spread in permissive conditions of growth because of the energy cost of maintaining redundant metabolic machinery. A mutation accumulation hypothesis was not even considered as an alternative. Neutral mutations certainly increase in frequency in chemostats—this provided much of the impetus for the early chemostat work—but the $trp^-$ mutants that Dykhuizen studied increased in frequency when competing against isogenic wild-type $trp^+$ strain thousands of times more rapidly than mutation could explain. However, an explanation in terms of energy conservation seemed almost as unappealing. Only about 1% of the total energy budget of the cell is used in making tryptophan; moreover, when inhibited by the presence of excess tryptophan in the medium, the level of tryptophan synthesis is only about 1% of normal, and so constitutes only about 0.01% of the total energy budget. The selection coefficients involved in the spread of $trp^-$ strains, on the other hand, are roughly 0.05–0.10. Thus, there is a substantial disproportion, amounting to about three orders of magnitude, between the amount of energy saved and the competitive advantage that this is supposed to create. It might be argued that very slight metabolic economies are somehow magnified into large differences in growth rate, but when auxotrophic and prototrophic strains are cultured separately, there is little if any difference in their rates of growth. It is also possible that the major expense of operating the pathway lies in making the enzymes involved, rather than tryptophan itself. Mutants in which these enzymes are not synthesized, however, have no greater advantage than otherwise isogenic mutants in which they are synthesized. Finally, selection should be weaker when using mutants in which only the first enzyme of the pathway is dysfunctional if indole is supplied in the place of tryptophan (because both mutant and wild-type strains would have to perform the final reaction in the pathway) and still weaker if anthranilate is supplied. However, this prediction could not be confirmed either. There is undoubtedly some pleiotropic effect in competitive conditions associated with the *trp*$^-$ genotype, but it is not merely an energy saving.

## 110. *Sorting or continued selection leads to negative genetic correlation for site-specific fitness.*

Functional interference leads either directly or indirectly to negative genetic correlation. If characters that influence site-specific fitnesses are encoded at different loci, it causes epistasis for site-specific fitness: color and banding are epistatic characters in snails, for example, because different combinations of color and banding genes are selected at different sites. If the characters are encoded at the same locus, interference causes antagonistic pleiotropy: a color allele is pleiotropic, for example, because its effects on reproduction vary among sites.

Furthermore, selection tends to create negative genetic correlations, as in the case of individual components of fitness. Variants that have high site-specific fitness at all sites will spread rapidly, whereas those that have low fitness everywhere will be lost. Selection will act to reduce genetic variance for overall fitness, while leaving the G $\times$ E variance intact. Thus, after a period of selection most genotypes will exhibit high fitness in some sites and low fitness in others.

Mutation accumulation will also tend to create negative genetic correlation. A strain will perform well in the conditions to which it has become adapted; in other conditions it may perform poorly because mutations that are nearly neutral in its native site are deleterious when expressed elsewhere. This tendency will strengthen as conditionally deleterious mutations become more frequent in the locally adapted strain, so that genetic correlation among sites will become increasingly negative with time.

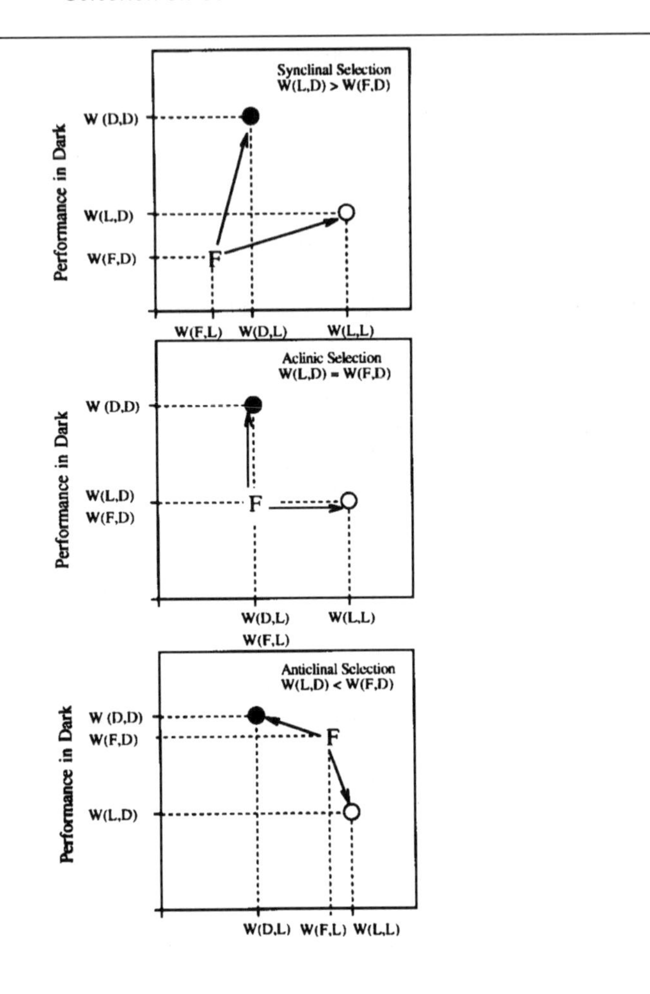


These diagrams illustrate the concepts of synclinal, aclinic, and anticlinal selection. The solid circle represents a line selected in the dark and the open circle a line selected in the light; both descend from the same ancestral genotype, the founder, F. The performance of a line selected in environment X and tested in environment Y is denoted W(X, Y): thus, W(D, L) is the performance in the light of a line selected in the dark. W(F, X) is the performance of the founder (stored or maintained as an unselected control) in environment X. Note that the degree of divergence of the selection lines, and the genetic correlation between them, is the same in every case, but is the outcome of functional interference only when selection is anticlinal.

(*Continues*)

*(Continued)*

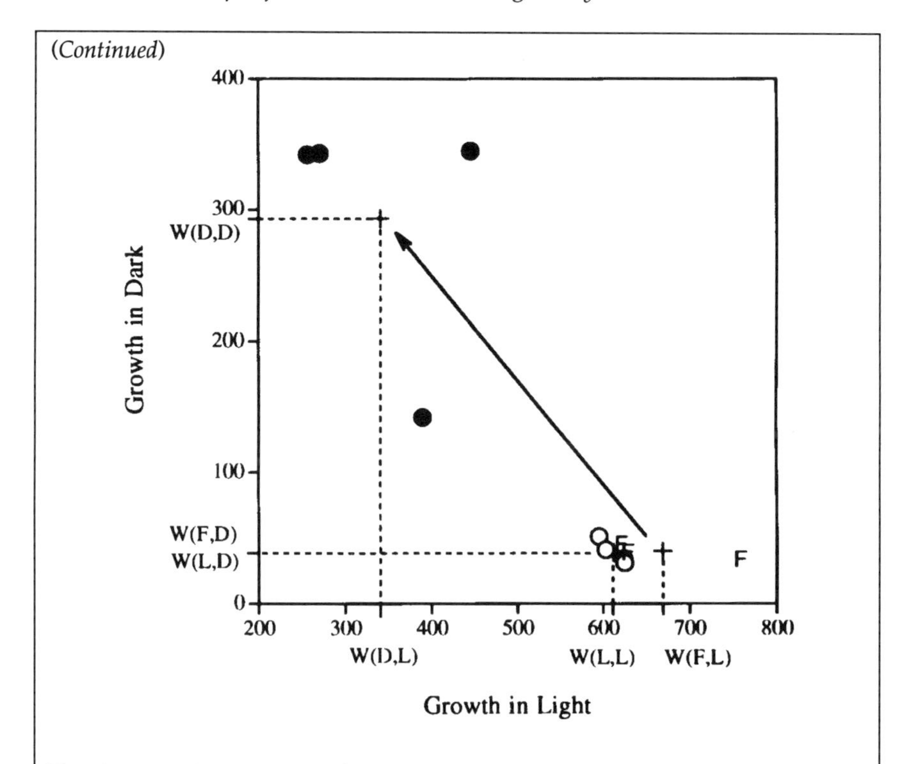


The diagram shows a case of asymmetrical anticlinal selection, from the experiment by Bell and Reboud (1996). Similar symbols are replicate spores from the same selection line. The founder grew well in the light but very poorly in the dark. Selection in the light was ineffective, W(L, L) $\sim$ W(F, L). Selection in the dark was highly effective, W(D, D) $\gg$ W(F, D) and anticlinal, W(D, L) $<$ W(F, L).

**The Farnham Sorting Experiment.** I have already described the sorting of variation among a large number of *Chlamydomonas*-like microbes isolated from soil samples taken along a single 25-m transect in a ploughed field by Hans Koelewijn, Patrick de Laguerie, and I (Sec. 42). When we measured their ability to grow in different types of environment in the laboratory (concentrated and dilute media, autotrophic and heterotrophic conditions, liquid and solid substrates) we found that there was a substantial amount of G $\times$ E, although no consistent tendency for strains that performed exceptionally well in one type of environment to perform poorly in others. This may well have been attributable, of course, to variation in the degree of exaptation to novel features of laboratory conditions of growth. The mixture of strains was then selected in each of the test environments for a few dozen generations. Clones isolated from the selected populations demonstrated the expected direct response to selection, growing better than the average of the base population in the environment in which they were selected; the correlated response to selection was generally antagonistic, causing reduced growth in the other environments.

**Continued Divergent Selection in *Chlamydomonas*.** If a single clone of *Chlamydomonas* is divided into two selection lines, one being grown autotrophically in the light in minimal medium and the other heterotrophically in the dark on supplemented medium, the two diverge. After a few hundred generations, the dark line shows increased growth in the dark, and the light line shows increased growth in the light, whereas each grows better than the other in the environment in which it has been selected. This cannot be attributed to the sorting of variation in site-specific fitness present in the base population because the base population was a clone; rather, negative genetic correlation between site-specific fitnesses has emerged in the course of selection.

However, the negative genetic correlation associated with adaptive divergence is not necessarily caused by functional interference. A line that is grown in the dark may evolve more rapid growth in the dark; but at the same time it may, to a lesser extent, evolve more rapid growth in the light, as the result of adaptation to common features of these two novel environments, such as suspension in liquid medium at rather high temperatures. It would then be superior to the base population in both environments. This may be called *synclinal* selection because the direction of the response to selection is the same in both environments. Provided that the response is greater in the environment of selection, this will nevertheless lead to adaptive divergence and the appearance of negative genetic correlation between the selection lines. On the other hand, the line grown in the dark may evolve more rapid growth in the dark, whereas the rate of growth in the light is reduced. It will then be superior to the base population in the environment of selection, but inferior in the other environment. This may be called *anticlinal* selection since the direct and correlated responses of site-specific fitnesses to selection are in different directions. This will equally lead to adaptive divergence and the appearance of negative genetic correlation. Indeed, from measurements of the selection lines alone it is impossible to distinguish synclinal from anticlinal responses. It is only anticlinal selection, however, that reflects functional interference between characters involved in site-specific adaptation.

The response to selection in the light was not very marked (green algae are apparently very good at photosynthesis after several billions of generations of selection) and involved little loss of ability to grow in the dark. The dark lines, on the other hand, responded markedly; two lines, in particular, were scarcely able to grow in the dark before selection, but evolved rapid dark growth after a year in dark culture. The response in the dark, especially in the lines that grew only very slowly before selection, was generally accompanied by reduced rates of growth in the light. Selection was thus sometimes markedly anticlinal, much more strongly so in the dark than in the light.

The source of the functional interference underlying the anticlinal response could not be identified in these experiments. If it gave rise to antagonistic pleiotropy, however, it could be detected because we would expect the enhancement of the ability to grow in one environment to be proportional to the reduction of growth in the other; the extent of regress should be correlated among lines with the extent of advance. Since this was found to be the case, some part of the cost of adaptation was probably attributable to antagonistic pleiotropy. This does not prove that mutation accumulation was not also in part responsible; the two processes are not mutually exclusive, and perhaps both often occur together. Indeed, in three lines we observed the spread of mutations. These were lesions of one of the two chlorophyll biosynthetic pathways and were easily detected because they caused cells to grow yellow in the dark. This is presumably nearly neutral when there is no light for photosynthesis; we do not know whether the mutations specifically reduced fitness in the light—there is a second chlorophyll biosynthetic pathway, active in the light, so the mutant cultures grew green when tested in the light—but they are certainly most unlikely to have increased it.

**Adaptation to Extreme Temperatures in *E. coli*.** The long-term *E. coli* lines selected at low and high temperature (section 103) showed clear anticlinal responses. A line selected at 32 C outcompeted its ancestor over a range of temperatures around 30 C, but was less fit than the base population at temperatures above about 35 C, and its thermal maximum contracted from 42 C to about 40 C. A line selected at 42 C showed increased fitness above 40 C, but reduced fitness below 32 C. However, these lines were exceptional; the other five replicate lines selected at either temperature all showed aclinic responses, becoming adapted to the environment in which they were selected without any evident loss of fitness at 37 C, or indeed at any other temperature, relative to the base population. Thus, both the *Chlamydomonas* and the *E. coli* experiments show that specific adaptation may cause antagonistic correlated responses in other environments, but they also show that this is by no means a universal rule, at least on timescales up to a thousand generations.

## 111. *Diversity can be maintained through selection in spatially heterogeneous environments.*

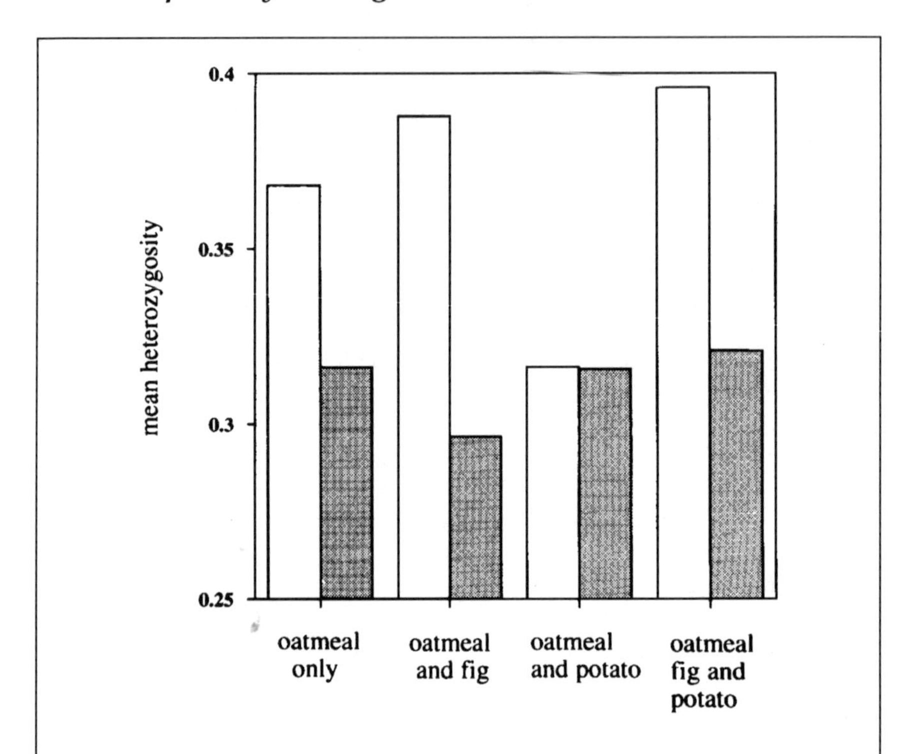


The experiment by Haley and Birley (1983), described in the text, seems to be the only one that tests the theory that genetic variance is supported through selection in spatially heterogeneous environments. The figure shows levels of heterozygosity in environments where one, two, or three types of food medium were available in separate vials. The two bars for each kind of environment represent the mean heterozygosity of replicate lines from two natural populations collected in South Australia (open bars) and the Netherlands (stipple bars). Allele frequencies were estimated for *Adh, Est-6, G-6pdh, a-Gpdh, Pgm, Lap-D*, and *Aph*, plus *6Pgdh* in the Australian lines; all the loci were essentially diallelic. Other experiments involve two different designs.

a) Fine-grained spatial heterogeneity with different food sources mixed together in the same medium and thus simultaneously available to all larvae, selecting for versatility; e.g., Minawa and Birley (1978).

(*Continues*)

*(Continued)*

b) Temporal heterogeneity, with different media available at different times, selecting for plasticity; e.g., Yamazaki et al. (1980).

Other studies include those by Powell (1971), McDonald and Ayala (1974), Powell and Wistrand (1978), and Oakeshott (1979). Birley and Haley (1987) measured linkage disequilibrium after selection in simple and heterogeneous environments. In no case was there a simple direct relationship between observed genetic diversity and environmental variability of any kind. The field has been reviewed by Hedrick (1986).

The negative correlations that may evolve between site-specific fitnesses when experimental populations are cultured under different conditions suggest that genetic diversity would be maintained if a single population were cultured in a heterogeneous environment. Curiously enough, there has been very little experimental investigation of this important point.

**The Maintenance of Enzyme Polymorphisms in *Drosophila*.** One approach to the maintenance of allelic diversity is to follow the frequency of specified alleles in different kinds of environment. Two Birmingham geneticists, C.S. Haley and A.J. Birley, set up *Drosophila* population cages to represent environments with different degrees of heterogeneity. The simplest environment included vials of a single standard food medium made up of oatmeal and molasses. In more complex environments, two types of medium were available, the standard recipe plus either fig or potato, in separate vials. The most complex environments contained all three media. Larvae would thus develop in a vial containing one of the three media, later emerging and mating with flies that in the more complex environments might have developed, and been selected, in a different food medium. They began their experiments with large populations that were segregating for alleles at several enzyme loci and monitored the frequency of these alleles at intervals for about 30 generations. Allele frequencies shifted substantially at most of these loci, almost certainly through selection, and mean heterozygosity also changed over time. There was no simple relationship, however, between genetic heterozygosity and ecological heterogeneity. The highest levels of heterozygosity were found in the most heterogeneous environments—those containing all three food media—but there was as much or more heterozygosity in the simplest environments as in those with two types of medium.

## 112. *Transplant experiments show that selection in spatially heterogeneous environments leads to local adaptation.*

Individuals belonging to the same species often have different forms in different localities, and it was long a matter of contention whether these were heritable or attributable to the direct action of the environment. This issue can best be resolved by the transplant experiment, introduced by Kerner in Austria in the 1870s, and later practised by Bonnier in France, Turesson in Sweden, Clausen in the United States, and many others. The simplest type of experiment is to bring plants from different localities into a *common garden*, where the differences they continue to express must be primarily heritable rather than environmental. Such experiments established that many of the distinctive forms of the same species found in different kinds of environment (ecotypes) are genetically different from one another. It was natural to speculate that each form has become adapted to the habitat in which it is found, through selection that varies from place to place. The possibility of local adaptation can be investigated by setting up several gardens, each representing a particular set of conditions;

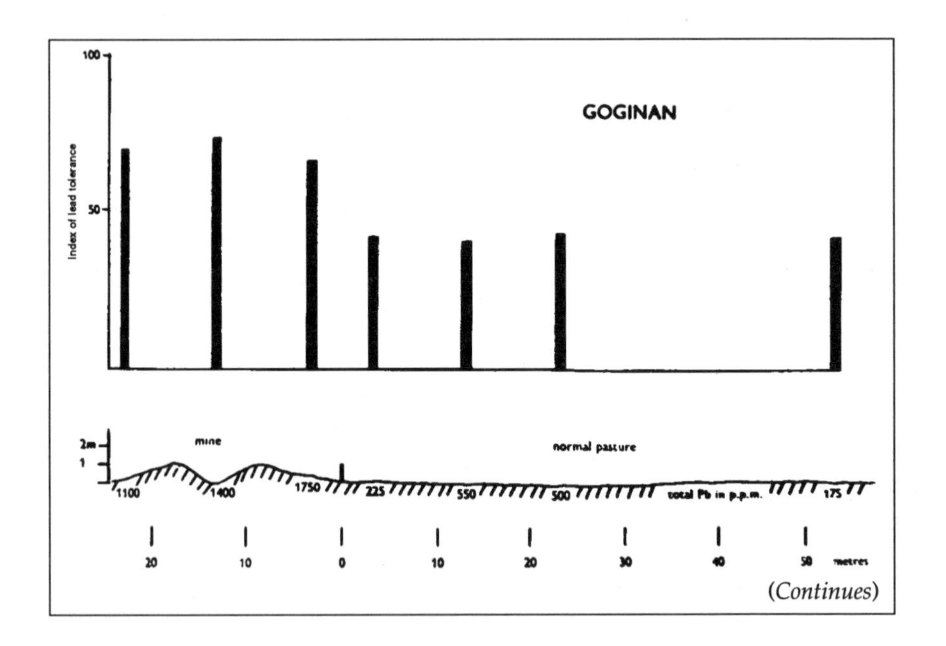


(*Continues*)

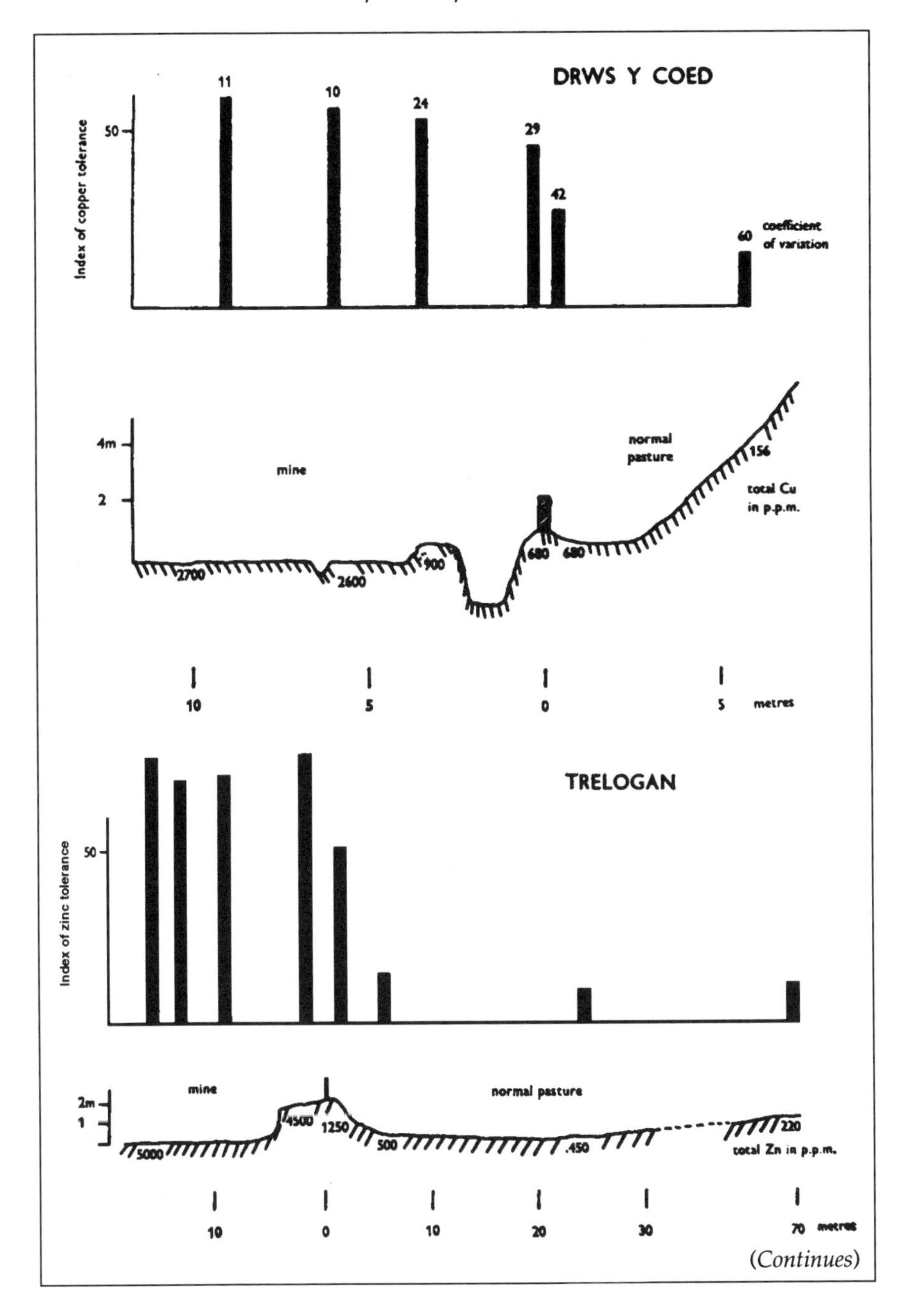
DRWS Y COED
Index of copper tolerance
50
11
10
24
29
42
60
coefficient of variation
4m
2
mine
normal pasture
156
total Cu in p.p.m.
2700
2600
900
680
680
10
5
0
5
metres
TRELOGAN
Index of zinc tolerance
50
2m
1
mine
normal pasture
5000
4500
1250
500
450
220
total Zn in p.p.m.
10
0
10
20
30
70
metres

(Continues)

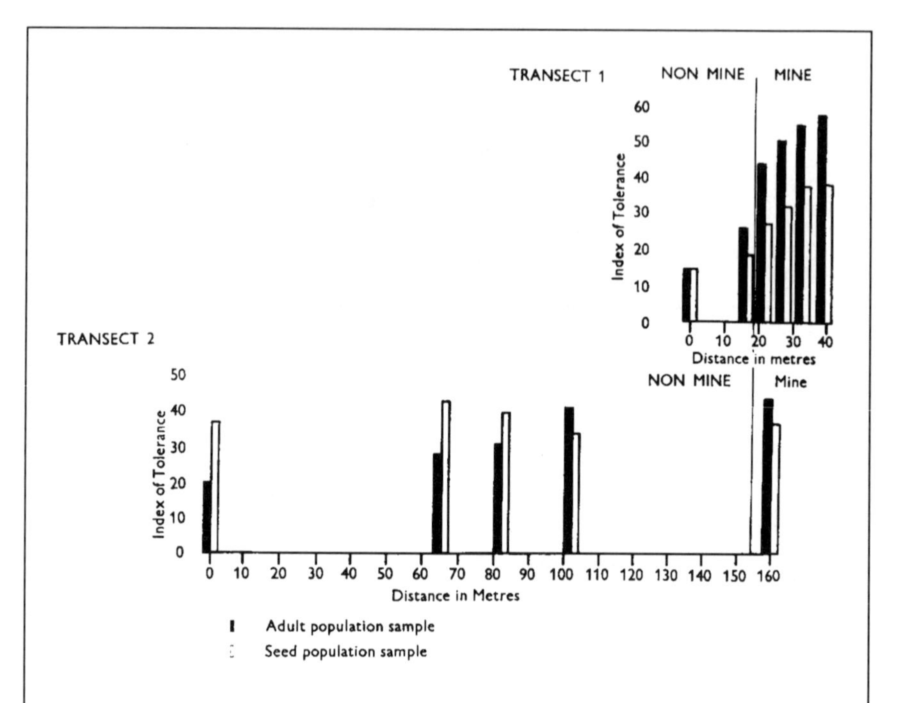


The upper figures show levels of tolerance to heavy metal contamination along transects from mine to pasture soils at three localities in North Wales, from Jain and Bradshaw (1966), who cite earlier references on the subject; see also McNeilly and Bradshaw (1968). Tillers were collected from sites along the transect and grown in permissive conditions in a common garden for two years. Their tolerance to the metal mined at the locality was then estimated from the growth of ramets grown in nutrient solutions containing standard concentrations of the same metal. Note the sharp increase in tolerance on a scale of 10 m or less at the mine boundary. The correlated changes in a variety of morphological and phenological characters are described by Antonovics and Bradshaw (1970).

The lower figure shows a detailed study by McNeilly (1968) of the upwind and downwind populations of *Agrostis* growing around a copper mine in a narrow valley. The solid bars represent samples of adult plants and the open bars samples of seed. At the upwind end of the mine (Transect 1) there is a sharp increase in tolerance at the mine boundary; at the downwind end (Transect 2) the gradient is much shallower, reflecting the asymmetry of intense selection against pasture genotypes on the mine, and the weaker selection against mine genotypes in the pasture. The difference between seed and adult samples reveals the direction and intensity of selection over a single generation.

(*Continues*)

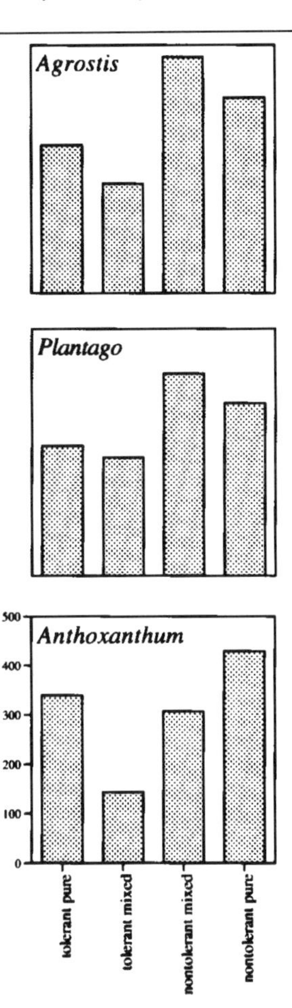


These diagrams illustrate the cost of adaptation to heavy metal pollution, from data in Cook et al. (1972). The plants were taken from mines in North Wales and tested in potting compost after an extended period of growth in a common garden. They were grown either as pure cultures or as mixtures. The non-tolerant plants are generally somewhat superior in pure culture, but this superiority is much more pronounced when tolerant and non-tolerant plants are grown in competition with one another. A similar diagram for growth on heavily contaminated soil would show the same effect in a more exaggerated form; the non-tolerant plants scarcely survive or grow in mine soil, although the additional effect of competition would be much less.

(*Continues*)

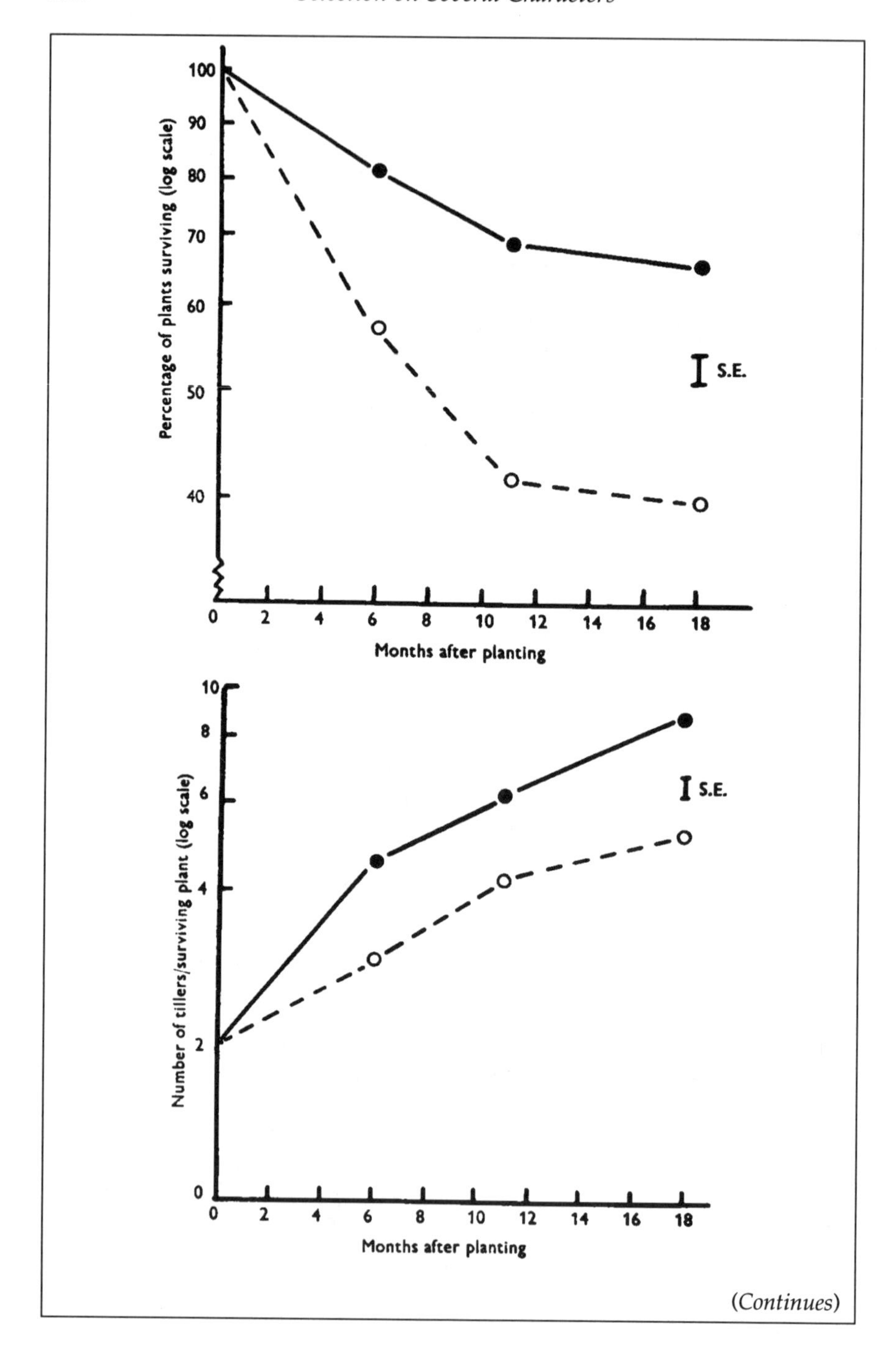

100
90
80
70
60
50
40
Percentage of plants surviving (log scale)
S.E.
0
2
4
6
8
10
12
14
16
18
Months after planting
10
8
6
4
2
0
Number of tillers/surviving plant (log scale)
S.E.
0
2
4
6
8
10
12
14
16
18
Months after planting


(Continues)

(*Continued*)

These diagrams show the survival (upper figure) and tiller number (lower figure) of transplants between home (solid line) and away (broken line) plots in limed and unlimed plots of the Rothamsted Park Grass Experiment, from Davies and Snaydon (1976); see also Snaydon and Davies (1972, 1982).

The pea aphid work is by Via (1984); recall that in another phytophygous insect the effect of selection was unable to overcome the effect of migration (Sec. 104).

The classic common garden transplant experiments include those by Bonnier (1895), Turesson (1922), Clausen et al. (1940); there is a good review by Hiesey (1940). Reciprocal transplant experiments in natural populations have been performed with *Ranunculus* (Lovett Doust 1981), *Plantago* (Antonovics and Primack 1982, van Tienderen and van der Toorn 1991, van Tienderen 1992), *Dryas* (McGraw and Antonovics 1983), *Phlox* (Schmidt and Levin 1985), *Amphicarpum* (Cheplick 1988) and *Diodia* (Jordan 1992). More general experiments involving unilateral transplanting into a range of sites have been reported by Cavers and Harper (1967). Schmitt and Gamble (1990) found that the fitness of inbred progeny declined away from the parental site in *Impatiens*. Outbreeding depression has been investigated by Price and Waser (1979), Bertin (1982), Sobrevila (1988), Newport (1989), Redmond et al. (1989), Waser and Price (1991) and McCall et al. (1991).

Kerner, for example, used alpine and lowland stations, and Clausen and his colleagues grew plants at a series of sites at different elevations. It is then possible to compare the growth of plants in their native locality with their growth elsewhere. In most cases, plants grow poorly in unfamiliar surroundings: lowland species rarely flower in the Alps, and lowland populations of widely distributed species of *Potentilla* or *Achillea* cannot survive at high altitudes, where there are flourishing resident populations of the same species. The most powerful version of such *transplant garden* experiments would involve following the changing frequencies of different types in populations established at different stations, but I have not found any studies of native vegetation comparable with the Composite Cross trials of barley. A nearly equivalent technique, however, is to move plants between localities, comparing the growth of resident plants, themselves dug up and replanted, with that of individuals from elsewhere. Such *reciprocal transplant* experiments are relatively easy to perform and allow local adaptation to be studied at any spatial scale.

**Mine and Pasture Populations of *Agrostis*.** I have already described the rapid evolution of tolerance to heavy metal contamination on old mine sites and spoil heaps in North Wales by *Agrostis*, *Anthoxanthum*, and other plants (Sec. 61). Even more remarkable than the rapid appearance of adaptation is its extreme specificity. At most sites, metal concentration drops off steeply at the edge of the old working, so that there is an abrupt change, often marked by a wall or a ditch, from the scanty vegetation of the mine to the much lusher pasture. Mine and pasture populations are genetically differentiated, with tolerance increasing steeply as one passes from pasture to mine, the frequency of tolerant individuals paralleling the concentration of metal in the soil. This differentiation is maintained despite the unrestricted movement of seed and pollen across the boundary. The mine site is often quite small and completely surrounded by pasture, so that a large proportion of new seedlings are incomers, bringing pasture genes onto the mine. The increase in frequency of non-tolerant genotypes caused by immigration must be opposed by intense selection against these genotypes, in order to account for the maintenance of tolerance on the mine. This is most clearly demonstrated by reciprocal transplants between mine and pasture soils which show that pasture plants are often almost incapable of growth on the mine, and seldom have fitnesses, relative to mine plants, of more than about 0.3.

The erosion of adaptation by immigration was very clearly demonstrated by Tom McNeilly, who studied a mine situated in a narrow glaciated valley down which the prevailing wind is funneled. At the upwind end of the mine, where pollen and seeds are blown from the pasture population on to the mine, an abrupt increase in tolerance occurred over less than 20 m at the boundary between pasture and mine. Adult plants from the mine (tested as tillers) were much more tolerant than seedlings, grown from the seed that the same adult plants produced in the field; the difference between tillers and seedlings represents the extent to which adaptation is broken down in every sexual generation by pollen from the pasture and is, therefore, a minimal estimate of the selection that must be acting in every sexual generation to restore adaptation. In this case, that selection is sufficient to keep mine and pasture populations distinct. At the downwind end of the site, the wind blows pollen and seeds from mine to pasture. The evolution of tolerance on the mine involves a cost of adaptation: the growth rate of tolerant plants in uncontaminated soil is less than that of non-tolerant plants, and reciprocal transfer experiments show that the fitness of mine plants on pasture, relative to the resident individuals, is usually about 0.7–0.9. Thus, there is rather strong selection against mine plants on the pasture (if this were not the case, of course, metal-tolerant plants would often be common even in uncontaminated pasture), but it is not nearly as strong as selection against pasture plants on the mine. Consequently, there is only a gradual decline in the frequency of tolerant adults, over a distance of about 200 m, as one passes from mine to pasture at the downwind end of the site.

As before, seedlings are less well adapted than adults, showing that local adaptation tends to be restored in every generation, but less effectively because selection is weaker and gene-flow stronger.

**The Rothamsted Park Grass Experiment.** The Park Grass experiment (Sec. 61) is another situation in which a high degree of local adaptation has evolved rather rapidly in adjacent populations. In this case, the differences between the physical environments in neighboring plots are caused by deliberately manipulating the nutrients supplied to them. The strength of selection involved in maintaining local adaptation was estimated by reciprocal transplants of *Anthoxanthum* tillers between comparable plots differing in macronutrient concentration or pH. For survival, the performance of incomers, relative to residents, was about 55% for transplants from unlimed plots (low pH) to limed plots (high pH), and about 85% for transplants in the reverse direction; for vegetative performance, the corresponding relative growth rates were about 65% and 75%. Overall, the relative fitness of incomers relative to residents is about 60–70%, and this rate of selection maintains the sharp distinctions in physiological characteristics between the populations on adjacent plots.

**Pea Aphids on Two Host Plants.** Pea aphids (*Acyrthosiphon*) live on legumes, reproducing parthenogenetically on the host during the growing season and producing a winged sexual phase in the fall. Sara Via of Cornell measured the fitness of local aphid clones on two host plants, alfalfa and red clover, by the extraordinarily thorough method of collecting all the progeny produced by isolated individuals confined to single host plants. By reciprocal transplants of clones from the host, on which they were found, to the alternative host, she showed that the relative fitness of incomers (clones collected from alfalfa and tested on clover, or vice versa) was about 0.25. An alternative way of expressing the result is as the genetic correlation of the fitness of clones between host plants, which was found to be $-0.7$. Divergent specialization to different species of host is thus maintained in the population by strong selection, despite annual episodes of dispersal.

**Plants in Natural Environments.** Reciprocal transplants between polluted and unpolluted sites, or between experimental plots, give a clear picture of local adaptation maintained by intense selection. A number of similar experiments in natural (or rather less drastically modified) environments have mostly supported the generalization that incomers are inferior to residents, but the results are weaker and more equivocal.

At large spatial scales, reciprocal transplants using seed or tillers as phytometers may provide good evidence for local adaptation. A survey of *Phlox* in Texas showed that incomers were less fit than residents at seven of eight localities, with an average relative fitness of about 0.55. A similar survey of *Plantago* in North Carolina included populations growing in pine stands on clay and sand, on shaded and unshaded lawns, amongst tall lush grass, and at an abandoned silver mine. Survival did not vary among populations—it seemed to depend on the good fortune of being sowed into a favorable microsite—and was no greater for residents than for incomers. Populations did differ in growth, however, and perhaps in fecundity; in general, residents grew more successfully than incomers, but there were many exceptions, including one population that was consistently inferior at its own site. The annual *Diodia* has morphologically different ecotypes in coastal and inland districts of eastern North America; at either kind of site, the resident type has the greater overall seed production, but is not necessarily superior in every component of reproduction.

At smaller spatial scales, the evidence is less extensive and less convincing. Transplanting *Plantago* between two nearby hayfield sites in Holland, one wet and the other dry, and a pasture site about 100 km distant gave little evidence for local adaptation. Total seed yield was negatively correlated between hayfields and pasture, but other characters, or comparisons of the two hayfields, gave low but positive genetic correlations. Reducing the scale still further, transplants were made among four sites about 50 m apart on a river bank, from a dry ridgetop to a wet hollow. There was a weak tendency for survival to be greater among residents than among incomers in two of the sites, but otherwise no consistent difference between residents and incomers for any component of fitness. A number of other studies have failed to find any evidence for local adaptation at small spatial scales.

Even at very small spatial scales, however, local adaptation may still be detectable, if the sites are sufficiently different. In one striking study in North Wales by Lesley Lovett Doust, ramets of *Ranunculus* were transplanted between adjacent grassland and woodland sites. The residents consistently produced more stolons and leaves. Interestingly, the effect was asymmetrical, woodland incomers to the grassland being much more severely handicapped than grassland incomers to the woodland: for total ramet production, for example, the relative performance of incomers was 0.22 in the grassland but 0.93 in the woodland. This may have been because the grassland population had evolved from the woodland population when the wood was cleared, hardly more than a decade previously. The experiment would then have censussed both current selection and the historical effects of past selection, the grassland population still being adapted in some degree to woodland conditions. Studying recently cleared woodland is perhaps uncomfortably close to studying abandoned mines; still, the intensity of selection against incomers is impressive.

**Outbreeding Depression.** In sedentary organisms, such as land plants, individuals that live close to one another are likely to be related. If they are diplonts, then when they mate their offspring are likely to be homozygous for deleterious recessive mutations and, thus, have low fitness because of inbreeding depression. On the other hand, individuals that live far apart are likely to be adapted to different conditions, and if this adaptedness depends on particular commbinations of genes, then their offspring will be unsuccessful because these locally adapted combinations have been broken up by recombination. The failure of wide crosses has been known for some time (in the *Drosophila* literature, it is attributed to "synthetic lethals"), and might generally be called "outbreeding depression". Granted that both crosses with close neighbors and crosses with very distant individuals are likely to be relatively unproductive, it follows that crosses will be most successful when they involve individuals at some intermediate distance. The existence of an optimal outcrossing distance is evidence for local adaptation based on epistasis, and the value of the distance itself is a measure of the spatial scale of local adaptation.

This quite different kind of experimental approach to local adaptation at very small spatial scales was pioneered by Mary Price and Nickolas Waser. They have used an outcrossed perennial *Delphinium* to show that crosses between individuals growing 10 m apart in the field produce more seed and more vigorous offspring than crosses between close neighbors 1 m apart and more than crosses between total strangers 100 m apart. It seems fair to note, however, that other biologists have found it difficult to confirm these results in other species, and whether an optimal outcrossing distance usually occurs within the area of distribution of an interbreeding population is not yet established.

**Specialization, Generalization, and the Pattern of the Environment.** Perhaps the best way of summarizing this pot-pourri of experimental results is to say that local adaptation in a heterogeneous environment seems to weaken with scale. Immigration to a given site by individuals from elsewhere is akin to mutation: it reduces the fit between the local population and the local environment, necessitating selection in every generation to restore local adaptation. At large spatial scales there is a great deal of environmental variance, supporting extensive G $\times$ E, whereas long-distance migration is likely to be a weak force. Divergent adaptation at these scales, perhaps involving distinctive ecotypes, should occur very frequently. At smaller and smaller scales the environmental variance, and thus G × E, decreases, whereas immigration increases; at some point, selection will be unable to maintain site-specific adaptation in the face of immigration. This interpretation of reciprocal transplant experiments in the field seems consistent with the *Chlamydomonas* experiments, where intense selection in very different conditions of growth is necessary to overcome high rates of immigration.

According to this view, the balance between specialization and generalization depends on the relationship between the lineage scale of an organism and the pattern of the environment. If environmental variance increases rapidly with distance, relative to the scale on which propagules are dispersed, progeny will tend to grow up in conditions similar to those experienced by their parents. Such environments are coarse-grained and can be thought of as being made up of large patches of similar types of habitat. Selection will be strong relative to immigration, and specialization is likely to evolve. If environmental variance increases only slowly with distance, offspring are likely to be dispersed to different kinds of site; such environments are fine-grained, an intricate patchwork of small areas representing different kinds of site. Site-specific selection is likely to be overwhelmed by immigration, so that local adaptation is obstructed, and the population is likely to be dominated by generalists. We can therefore make two predictions about the degree of specialization that is likely to evolve in spatially heterogeneous environments.

The first is that an increase in lineage scale, relative to environmental scale, will favor generalization. This might be caused by an increase in the production of dispersive propagules, or by an increase in their average distance of dispersal. Conversely, populations with little or limited dispersal are more likely to display specialization and, thus, to maintain large quantities of genetic variance.

Secondly, an increase in environmental scale, relative to lineage scale, will favor specialization. Within a fixed area, the diversity of specialized types that can be maintained will be proportional to the quantity of environmental variance. Moreover, specialization is favored when environmental variance increases steeply with distance; thus, populations are more likely to be specialized with respect to physical factors whose variance increases steeply with distance, and more likely to be specialized with respect to any given physical factor in environments where its variance increases with distance more steeply.

## 113. *Stable genotypes are favored when selection and environmental effects act in opposite directions.*

**The Regression Analysis of Stability.** A population may become locally adapted to any feature of the environment, even a feature as exotic as severe lead pollution. One universal source of environmental heterogeneity, however, is variation in productivity from site to site. I have already emphasized that the growth of a test organism is itself the most appropriate measure of the state of the environment. The average growth of a particular set of genotypes thereby defines the value of the environment at a given site, supplying a common scale that enables us to compare sites within any sort of environment. The behavior of a genotype can then be represented by plotting its growth at a series of sites as a function of their environmental value. This graph is usually linear and can be used to partition the environmental variance expressed by a genotype into two components. The first is the variance in growth of a particular genotype that is accounted for by variance in average growth or, in other words, that part of its environmental variance that is attributable to the regression of genotypic value on environmental value. The second part is the variance in growth expressed by a particular genotype at a fixed environmental value because of factors that affect the relative fitness of genotypes without affecting mean fitness; this part of its environmental variance is represented by deviations from the regression line. This regression analysis of G × E is useful because it isolates the response to one source of environmental variance—the availability of resources for growth—that can be studied in any organism in any kind of environment.

The slope of the regression represents the *responsiveness* of a genotype to improved conditions of growth. A shallow regression slope characterizes an unresponsive, or plastic, genotype, whose growth is not consistently affected by a general amelioration of the environment. This might be merely an unconditionally inferior genotype that is unable to grow well anywhere; or it might grow well in some conditions and poorly in others, but in a way that cannot be explained by variation in the general conditions of growth, in which case the regression analysis is uninformative. More interestingly, it might be a genotype that grows relatively well in unproductive conditions, but relatively poorly at more productive sites. Because its fitness does not vary much with general productivity, it might be regarded as a generalist that grows equally well anywhere. This is not necessarily the case, however, because its growth may vary widely with features of the

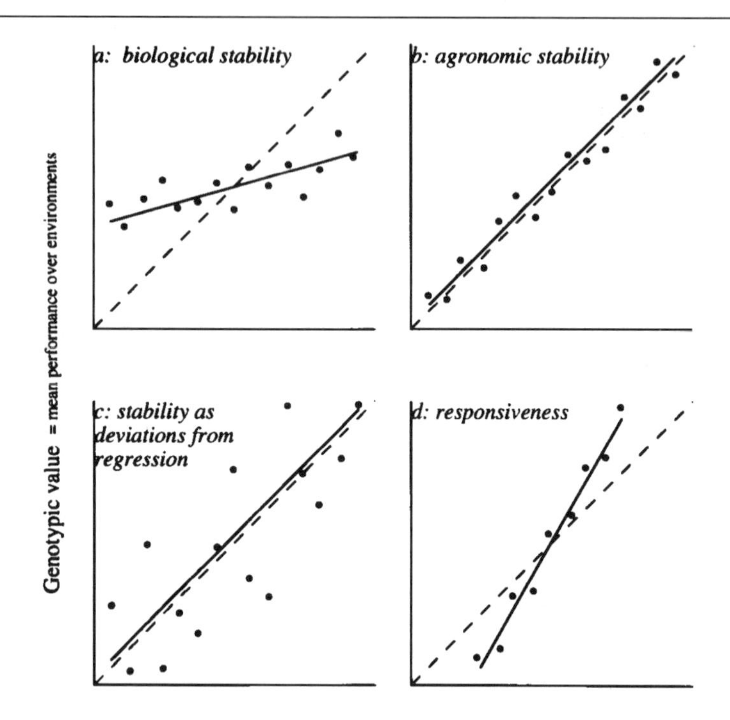


The regression analysis of G × E has been developed mainly by agronomists and has given rise to a large and rather confusing literature. Biologists are accustomed to defining a plastic type as one that displays little environmental variance for fitness, being everywhere intermediate between alternative specialists. Agronomists are also interested in plasticity, because their concern is to select cultivars that are broadly adapted and that can be grown in a wide range of localities, without being liable to suffer disastrous losses in poor years. The stable genotypes sought by agronomists, however, are not usually those with very low environmental variance because they would not be able to respond to increased nutrient input or improved management. Agronomic stability involves, not lack of response, but rather a predictable and regular response to environmental variation. A generalist in the agronomic sense is therefore a type whose response is as close as possible to the average. Regressing genotypic performance on mean

(*Continues*)

performance is a way of expressing stability as a single parameter, the slope of the regression. This can be used to identify genotypes that are generalists, in the sense of having a regression of close to unity. The object of selecting stability, in this sense, is therefore not to remove environmental variance, but rather to remove G × E. There are two potential sources of confusion in this procedure. The first is that the regression accounts for only the fraction of G × E that is consistently associated with environmental variance in average productivity. The remainder is associated with deviations from the regression; these deviations, indeed, have been proposed as an alternative or supplementary measure of stability that would be reduced to a minimum in desirable cultivars. The second, and more important, consideration is that agronomic stability is a highly relative concept. When plasticity is defined in terms of low environmental variance of fitness, it is necessarily defined with respect to a given range of conditions; a genotype that is plastic in one environment may not be in another. But when stability is defined in terms of low G × E, it is defined, not only for a given set of environments, but also for a given set of genotypes. A genotype that responds to variation in productivity close to the average of some given set of other genotypes is judged to be a generalist; but relative to some other set, it might be a specialist. Consequently, whereas the biological interpretation of plasticity can be analyzed rather straightforwardly, there are often difficulties in applying concepts such as heritability to the agronomically desirable trait of stability.

These figures illustrate the types of response that might be shown by different cultivars in a crop trial conducted at several localities or in several years. Environments may be evaluated by the mean performance of all genotypes, by the mean performance of all genotypes except the one whose response is being scored, or by the mean performance of a separate series of check cultivars. Four types of genotypic response may be found:

a) Low environmental variance, giving a shallow regression slope.

b) Low G × E variance, giving a regression slope close to unity.

c) A pronounced scatter of genotypic values around the regression line, showing that G × E is not consistently related to the mean environmental value.

d) High environmental variance, giving a steep regression slope.

Different concepts of stability are reviewed by Lin et al. (1986); see also Simmonds (1991).

(*Continues*)

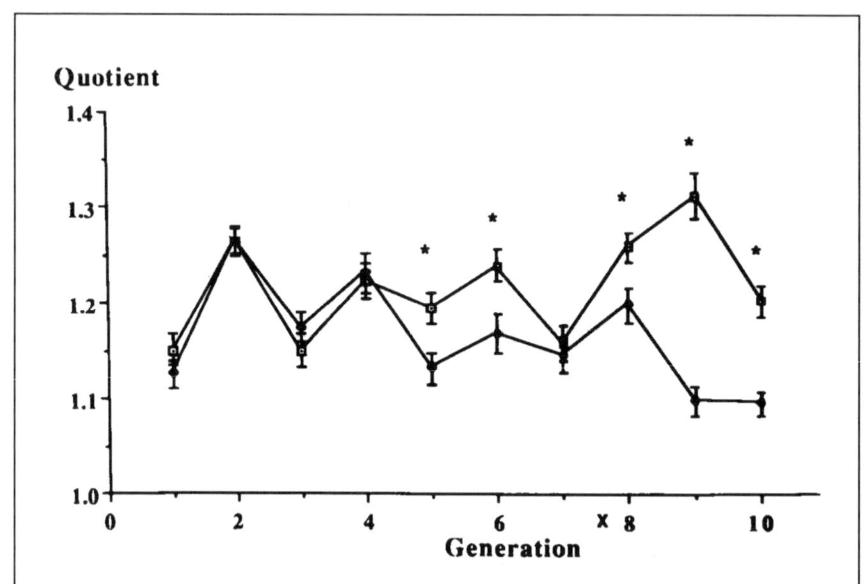


The measure of stability in this experiment (Hillesheim and Stearns 1991) is the ratio of mean body weight of lines selected in rich and poor conditions of growth (Quotient). This is equivalent to using linear regression, as earlier. Family selection for responsiveness (upper points) and biological stability (lower points) caused lines to diverge within 10 generations.

*(Continues)*

environment that are not consistently related to overall productivity. It is better interpreted as a type that is specialized for growth in unproductive sites. Conversely, a steep regression slope characterizes a highly responsive type that grows poorly in generally unproductive sites, but is able to exploit productive sites more effectively. Where we can recognize these two types of specialist, their regression lines will cross at some intermediate environmental value at which they have equal fitness. A generalist will exhibit nearly average growth at all sites, being inferior to one specialist and superior to the other. The regression of its growth on environmental value, which is defined as the mean growth of all genotypes, will therefore have a slope of unity. Consequently, provided that we exclude genotypes that are everywhere inferior, or everywhere superior, we can define a type specialized for growth in poor conditions as one whose regression on environmental value has a slope of less than unity, and a specialist for rich conditions as a type whose regression slope is greater than unity.

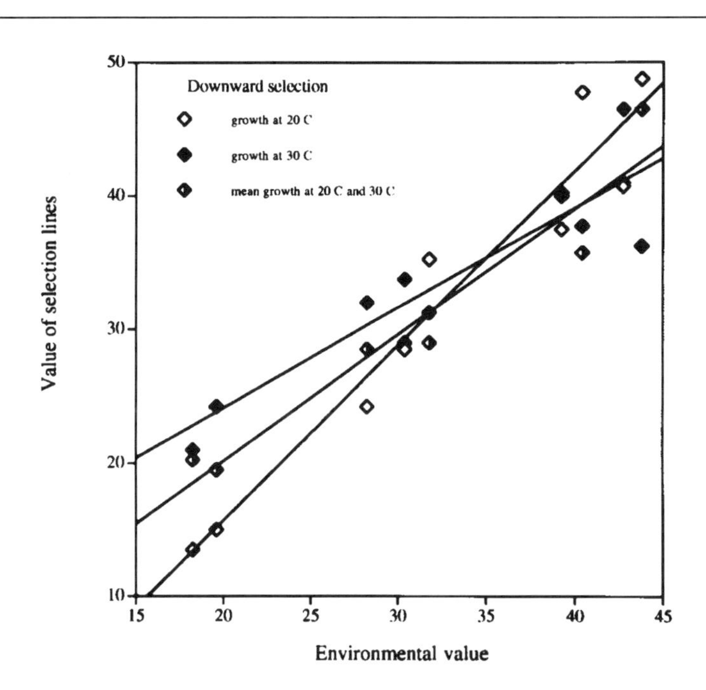


This figure shows the regression analysis of stability in the downward selection lines of the Jinks–Connolly experiment on *Schizophyllum* (see Sec. 103). Selection for low growth rate at 20 C increased responsiveness (regression slope = 1.31); selection at 30 C reduced environmental variation to produce lines with enhanced biological stability (slope = 0.75); selection for the mean value of growth at 20 C and 30 C reduced G × E variation to produce lines with enhanced agronomic stability (slope = 0.94). The effect of natural selection in fluctuating environments on stability is discussed in Sec. 113. For a general discussion of this effect, see Falconer (1990).

*(Continues)*

**The Direct Response to Selection.** Elke Hillesheim and Stephen Stearns of Basel selected for body size in *Drosophila* in rich and poor growth environments. Within a single family, the ratio of the body sizes of individuals tested in rich and poor environments is an estimate of plasticity; a high ratio is equivalent to a steep regression slope, indicating a responsive genotype, whereas a low ratio indicates stability. By selecting the families with the largest and the smallest ratios in each generation, they produced lines that were more or less responsive to environmental variation. Phenotypic plasticity is therefore a character that can evolve through selection in the short term.

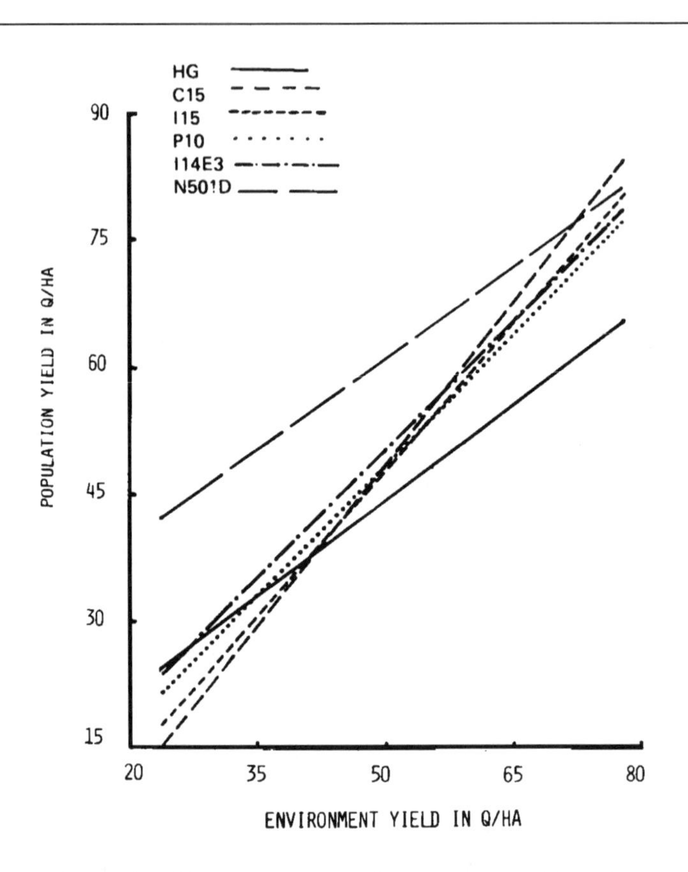


This figure shows the effect of mass selection for mean yield on the stability of yield in maize, from Mareck and Gardner (1979). The lower solid line is Hays Golden; the upper broken line is a check hybrid derived from Hays Golden. The intermediate lines are populations developed from Hays Golden by mass selection in various ways, tested at five localities in three years. Note that selection has greatly increased yield in good environments but somewhat decreased yield in poor environments, causing an increase in responsiveness. For a similar experiment, with an excellent discussion of the relationship between agronomic stability and G × E see Moll et al. (1978). Fatunla and Frey (1974) describe the effect of natural selection on stability in bulk-propagated populations of oats; regression stability was initially high then declined in normal populations, but began low then increased in irradiated lines. Langer et al. (1978) describe the stability of oat cultivars developed at different times in the past and grown in common gardens; they conclude that agronomic stability increased during the first decade of selection (to 1940) but changed little thereafter, perhaps because of genetic constraints

(*Continues*)

(*Continued*)
associated with the limited base of North American oat cultivars. See, also, Jinks and Pooni (1982), Powell and Phillips (1984), and Hoard and Crosby (1986).

**The Response of Plasticity to Selection in More and Less Stressful Environments.** Falconer's experiments with mice and Jinks' experiments with *Schizophyllum* had the same basic design, measuring the direct and correlated responses of growth rate in two environments, one of which was generally more favorable to growth than the other (Sec. 103). The Hillesheim–Stearns experiment, besides involving family selection on plasticity, also involved individual mass selection on body size in favorable and unfavorable environments. All three experiments yielded similar results: lines that were selected for increased growth in the more favorable environment grew faster in that environment than did lines selected in the less favorable environment, but they grew less well in the less favorable environment than did lines selected in that environment. Thus, the environmental variance of lines selected for increased growth in a favorable environment was greater than that of lines selected for increased growth in an unfavorable environment. In terms of the regression analysis of stability, the regression slope is steeper for the lines selected in the favorable environment, indicating that they were more responsive to environmental variation. The converse was true for lines selected for decreased growth: lines selected in the unfavorable environment had greater environmental variance or, in other words, were more responsive than lines selected in the favorable environment. We may summarize these results by saying that selection will favor individuals that are highly responsive to environmental variation when selection is applied in the same direction as the general effect of the environment.

This will be the outcome of selection whenever the correlated response is not as great as the direct response in either environment, as will usually be the case. Jinks interpreted this pattern in genetic terms (already alluded to in Sec. 103) as follows. Selection for high growth rate at 30 C will fix some alleles that increase growth regardless of temperature. It will, however, also favor alleles that have antagonistic effects on growth at high and low temperature, increasing growth at 30 C while decreasing growth at 20 C, and will thereby increase the responsiveness of growth rate to temperature. Selection for high growth rate at 20 C will likewise fix alleles that have the same effect on growth at all temperatures, and will also favor alleles that increase growth at 20 C while reducing it at 30 C, thereby reducing the responsiveness of growth to temperature. This interpretation posits two classes of genes. One includes genes that affect growth to the same extent in all environments, that will contribute equally to the direct and correlated responses to selection, and that do not affect plasticity. The other includes genes whose effect varies according to conditions, producing an antagonistic correlated response in conditions that are sufficiently different from the conditions in which selection is carried out because of some unidentified source of functional interference. The combined effect of these two kinds of genetic effects is that the correlated response will often be in the same direction as the direct response (because of genes in the first category), but will be less pronounced (because of genes in the second category). It follows that selection in more favorable environments will increase responsiveness and decrease plasticity, whereas selection in more stressful environments will decrease responsiveness and increase plasticity. Selection for reduced growth rate will produce the same results, bearing in mind that 30 C is the more stressful and 20 C the more favorable environment for the expression of low rates of growth.

Growth was measured in each generation of selection in the *Schizophyllum* experiment, so that the joint evolution of mean growth rate and responsiveness can be followed. Regardless of the environment in which selection was carried out, responsiveness declined in lines selected for low growth rate, and increased in lines selected for high growth rate. Responsiveness declined more steeply in lines selected for low growth at 30 C, however, and increased more steeply in lines selected for high growth at 30 C. This is the clearest experimental demonstration that the extent of plasticity or responsiveness depends on the environment in which selection is practised and that plasticity can be increased or reduced in a predictable manner as a correlated response to selection in different kinds of environment.

**The Response of Stability to Selection for Average Performance.** The *Schizophyllum* experiments also included lines that were selected on the basis of their average performance at 20 C and 30 C. Not surprisingly, these lines evolved intermediate growth rates when tested at either temperature. Their plasticity and responsiveness were therefore also intermediate: when selected upward, they were less responsive than the 30-C line but more responsive than the 20-C line; when selected downwards this relationship was reversed. These lines were thus more stable than either the 20-C or the 30-C lines, in the sense of showing less G $\times$ E, and therefore a regression slope of close to unity.

**The Response of Responsiveness to Selection for Increased Performance.** In view of the great interest in broad adaptability among agronomists, it is surprising that there does not appear to have been any attempt to select for stability in crop plants, beyond choosing the more stable varieties from a trial. There are, however, several descriptions of how stability changes during the course of a conventional program of selection for increased yield.

The selection of new crop varieties is generally carried out at an experimental station and the evolved lines then tested on farms under more or less normal agricultural conditions. Experimental lines are often grown as spaced plants, carefully nourished and protected, so that conditions of growth are usually more favorable at the experimental station than on the farm. Upward selection for yield is then expected to increase responsiveness. This has been demonstrated in several experiments with maize. For example, 15 cycles of mass selection at the Nebraska experimental station increased grain yield by about 14%. The base population was a cultivar called Hay's Golden, an open-pollinated variety developed during the drought years of the 1930s that is said to be resistant to high temperature and low water availability. When tested under modern conditions, it is low yielding and plastic, as expected: it yields reasonably well in the poorest environments, but shows little response to improved conditions. The selection lines are much more responsive: they are about equally productive, or perhaps a little less productive, in the poorest environments, but yield 20–25% more in the most favorable environments. Selection with the grain of the environment, so to speak, thus produces less plastic, more responsive genotypes, as we expect from elementary theory and the experiments with mice and *Schizophyllum*.

It follows from this argument that yield and responsiveness will usually be positively correlated, provided that selection for yield is practised in favorable environments. To put this in another way, plasticity will decrease as a correlated response to selection for increased yield. This is the general finding from agronomic trials and is attributable to the superior management usually practised at experimental stations. D.S. Falconer has pointed out that increased responsiveness will be a very general consequence of directional selection, even when lines are tested in the same environment that they were selected in. The conditions of growth experienced by individuals in a population will always vary somewhat, even when they are raised in the laboratory, or at a single location, and those that are selected will tend to be those that have experienced more favorable conditions (with respect to the expression of the character under selection) and are able to respond to them. Under stabilizing selection, on the other hand, individuals that express phenotypes close to the optimal value are likely to be those that are insensitive to environmental variation. Responsiveness should therefore increase under directional selection and decrease under stabilizing selection.

**There Is No Particular Environment That Is Best for Selection.** Agronomists have long argued whether selection should be practised under favorable or stressful conditions. On the one hand, favorable conditions might facilitate selection by allowing a greater range of heritable variation to be expressed; on the other hand, most farm environments are stressful in at least some respects. It will be evident from the preceding sections that no general answer can be given. Selection in favorable conditions will increase responsiveness, maximizing yield in good years, at good localities or on well-managed farms; selection in stressful conditions will increase plasticity, minimizing losses in poor years, and broadening the range of localities in which a cultivar can be profitably grown. Which procedure is preferable is determined by the subjective priorities of the breeding program. One might add that any attempt to select in stressful environments must take into account the possibility that plasticity with respect to one source of stress is not necessarily expressed when a different stress is experienced. The only sound general rule is that selection is best practised in the environment where the plants will eventually be cultivated because the direct response to selection will almost always exceed the correlated response. (This advice is not very practicable if it requires selecting independently at a great number of localities; and it cannot be applied to yearly variation at all.) Where the crop is grown in a range of environments, it is best to select for average performance over this range, if the object is to increase mean yield. The *Schizophyllum* experiments suggest that this will also enhance stability. If it is impracticable to select in several environments, the selection environment should approximate the average conditions that will be experienced by the commercial crop.

## 114. *Spatial variation supports specialization; temporal variation favors the evolution of generalization.*

**Geometric Mean Fitness in Variable Environments.** I have assumed so far that the members of a given lineage will broadcast their progeny haphazardly over the environment. A specialist type that thrives in a particular kind of site will distribute some of these progeny to suitable sites, where their success will compensate for the loss of their less fortunate sibs. The overall fitness of the lineage is the average of its fitness in different kinds of site, weighted by the frequency of those sites, and provided that

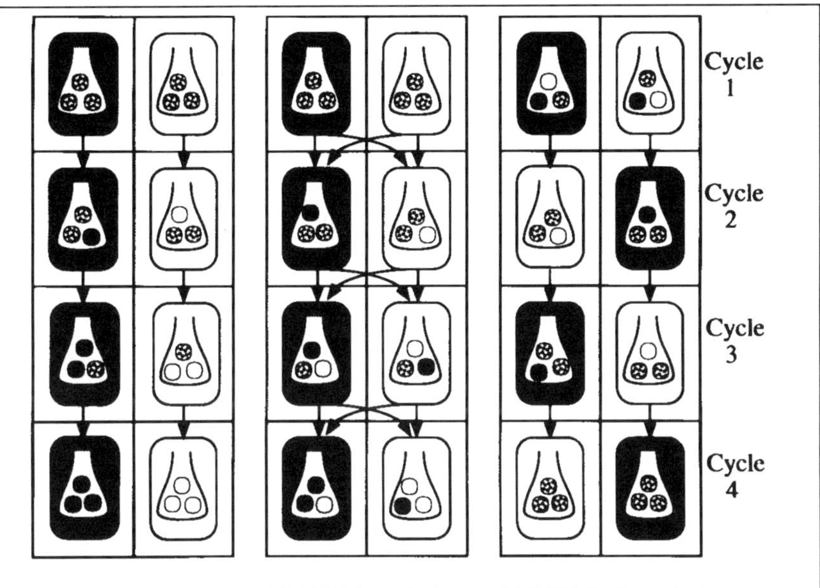


When cultures are maintained separately, the appropriate specialist will eventually become fixed in any given environment. Each culture will be uniform, but the population as a whole—the combined set of cultures—will be a diverse assemblage of different strains. If some degree of mixing occurs at each transfer, selection is less effective, and poorly-adapted genotypes—inappropriately-specialized strains and generalists—will be found in each culture. Thus, diversity will be maintained in each environment, as well as in the population as a whole. The amount of diversity within each culture will depend on the amount of mixing, relative to the intensity of selection. Finally, if the cultures are maintained separately, but each is subjected to a fluctuating environment, specialists will be lost, because each specialized strain is repeatedly exposed to conditions that it can neither flourish in nor escape from. Each culture will then come to comprise a uniform population of generalists. These concepts were tested experimentally by Reboud and Bell (1996).

suitable sites are not too rare, and that a fairly large number of progeny is produced, this average will remain almost constant from generation to generation. Imagine, however, that rather than being dispersed very broadly, the progeny instead all arrive in the same site, or the same kind of site. If parents have the ability to direct their progeny to a suitable site, for example, by laying eggs on the same food plant that they grew up on themselves, this will reinforce selection for specialization. Indeed, selection will favor habitat choice by specialists, and whenever it is effective the population is likely to comprise a diverse set of specialized types. On the other hand, suppose the progeny produced by a lineage are all dispersed to a single site, or a single kind of site, chosen at random. The expected fitness of the lineage in any given generation is the same as if its progeny were dispersed broadly, but this fitness will fluctuate from generation to generation, being high in generations when the lineage happens to occupy a suitable site and low when it does not. This will greatly impede the evolution of specialization. The fitness of a lineage depends on the factor by which its members increase in number; over several generations, this is equal to the product of the factors in each successive generation. This is no more than a restatement of the very general principle that a lineage will tend to increase exponentially, or geometrically, in numbers. The appropriate average fitness of a lineage in the long term is therefore the *geometric mean* of its fitness in each generation. (The geometric mean of a series of $n$ numbers is the $n$-th root of their product.) The more its fitness varies through time, the lower its geometric mean fitness will be. A small number does not greatly affect the total and, therefore, the arithmetic mean of a series; but it may greatly affect the product and, therefore, the geometric mean of the series. A specialized lineage may be maintained in a heterogeneous environment, even if it cannot grow at most sites; but if it is confined in each generation to a single kind of site, it will be eliminated once it encounters a site that it cannot tolerate, regardless of how successful it has been previously.

It is not very likely that any organism disperses by directing all the members of each lineage to a single random site. This, however, is essentially what happens as the environment changes from generation to generation. A given year (or some other appropriate period of time) is akin to a site that is colonized by all the members of a lineage, a site that none have encountered previously and to which none can be precisely adapted. A type that is highly specialized is likely to have very high fitness in some years at the cost of much lower fitness in others and will, therefore, have a high variance of fitness through time and a correspondingly low geometric mean fitness. A less highly specialized type will have the higher geometric mean fitness because of its lower variance over years. Functional interference thus constrains plasticity or, rather, makes it impossible for genotypes to have low environmental variance through being superior in all conditions.

In any given generation, a lineage will have some overall fitness $w$. Over a series of generations, the relationship between the geometric mean of $w$, $G_w$, and the arithmetic mean, $M_w$, in terms of the variance of $w$ through time is roughly $G_w = M_w - \frac{1}{2}\mathrm{Var}(w)/M_w$. John Gillespie was the first to point out clearly that because selection will tend to maximize geometric mean fitness, it will tend to minimize the environmental variance of fitness. The variance of fitness, like mean fitness, is a property of a genotype; G × E can be thought of as the genetic variance of environmental variance, genotypes with high variance being specialists and those with low variance being generalists. Selection can act on this variation, favoring specialists or generalists in different circumstances. In spatially heterogeneous environments whose composition does not vary much through time, the outcome of selection may be a more or less diverse assemblage of specialists adapted to local conditions, depending on the balance between functional interference and immigration. In uniform environments whose properties vary through time, a single plastic type with high geometric mean fitness is expected to prevail. The balance between specialists and generalists should reflect the relative magnitude of spatial and temporal variation in fitness.

**Selection of *Chlamydomonas* in Heterogeneous Environments.** The *Chlamydomonas* selection lines maintained for hundreds of generations in light or dark conditions can be used to investigate this generalization. Spatial heterogeneity can be simulated, as before, by mixing sister light and dark lines after each cycle of growth and redistributing the mixture to fresh light and dark flasks. Temporal heterogeneity is simulated by transferring the same line to light and to dark conditions in alternate growth cycles. The rationale behind the experiment was that specialized genotypes are guaranteed a refuge in the spatial treatment: a dark-adapted genotype, for example, might be rapidly eliminated in the light flask, but half the lineage would have been transferred to a dark flask, where they could as rapidly replace the light-adapted genotypes transferred with them. Under temporal variation, there is no refuge: a lineage must be able to reproduce with at least moderate success in both light and dark flasks, or it would quickly become extinct. We therefore expected lines exposed to spatial variation to remain specialized, whereas those exposed to temporal variation should evolve greater plasticity. We also expected that the generalist types evolving in the temporally varying selection lines would be inferior in either light or dark conditions to the specialized types in lines maintained permanently in light or dark flasks.

The outcome of the experiment, after about 100 generations of selection, was generally consistent with these predictions: spores isolated from the temporal lines expressed less genetic variation in either light or dark environments and expressed less G × E because their environmental variance was less than that of spores isolated from the spatial lines. The only surprising feature of the results was that the generalists from the temporal lines were nearly as fit in either environment as the specialists from lines maintained permanently in the light or in the dark, despite the cost of adaptation to these conditions that had been identified previously. Temporal variation seems to be an effective procedure for selecting superior generalists.

## 3.C. Selection Acting at Different Levels

### *115. Any entity with heritable properties can be selected.*

It is conventional to regard individuals as the objects whose variation causes selection and, in most cases, this usage is uncontroversial. The central problem of evolutionary biology is to explain the adaptedness of individuals and, in simple situations such as a culture of bacteria in a chemostat or a population of *Drosophila* in a cage, it is appropriate to treat adaptation as the result of selection among individuals. Nevertheless, this viewpoint has two limitations. The first, which I have already emphasized, is that genes, not individuals, are self-replicators and selection acts through the differential propagation of genetic lineages. Individuals reproduce, but do not replicate; each is a unique vessel constructed by genes and serving to transmit the information that they encode. The second is that many organisms do not possess distinct individuality, but nonetheless evolve. It is not possible to define individuals unequivocally in a slime mold, a rhizomatous plant or a filamentous alga; such creatures are not organized in the same way as fruit flies or people, and the definition that we choose depends on the use we intend to put it to.

More generally, therefore, any assemblage of genes can be selected on the basis of the heritable properties conferred on it by those genes and may as a result evolve distinctive adaptations. Such assemblages may represent components of individuals (such as cells), or they may themselves be assemblages of individuals (such as families).

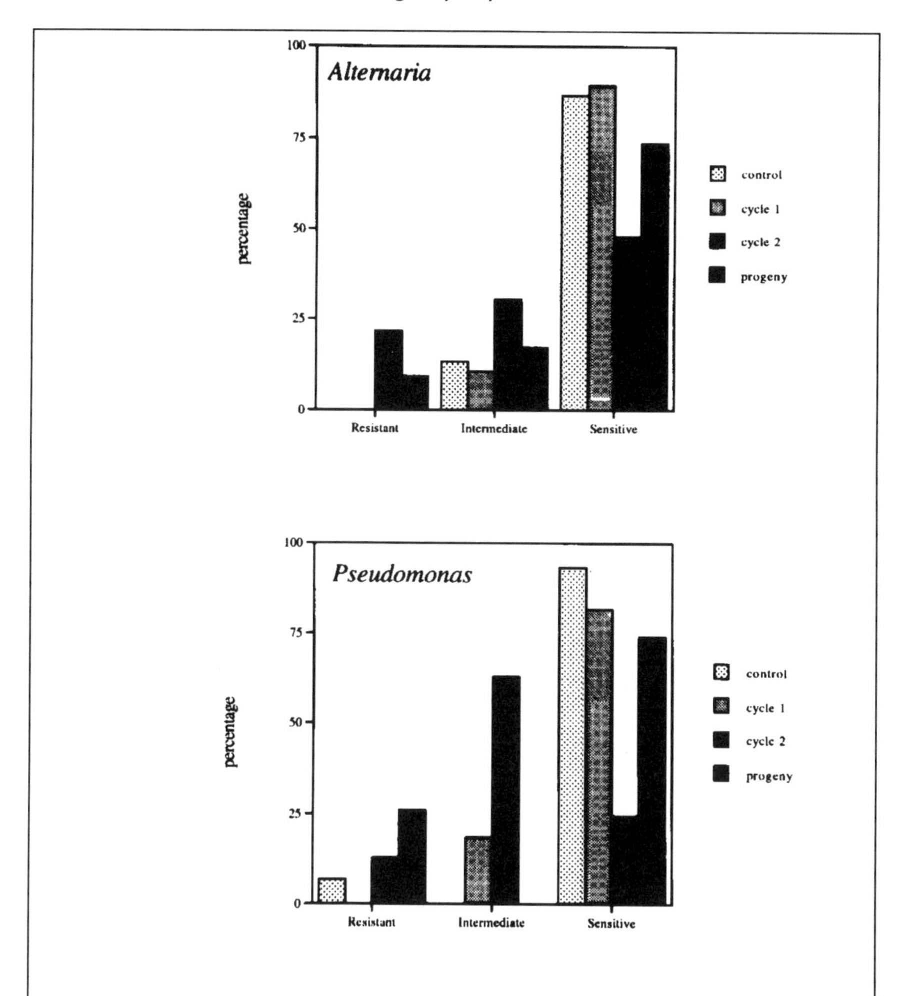


This figure illustrates the progress made by selection in cell suspensions of tobacco for resistance to the toxins produced by *Pseudomonas* and *Alternaria*, from data in Thanutong et al. (1983). Plants are scored as sensitive or resistant, with intermediate phenotypes distinguished in some cases. Control plants are an unselected line of a single cultivar, with a high level of sensitivity to both pathogens. Plants were selected by growing cells in suspension in permissive conditions, plating, transferring callus colonies to toxic medium, then transferring surviving colonies to fresh non-toxic plates for regeneration. After two cycles of selection, open-pollinated progeny were raised from seed produced by the regenerated plants. Note that selection is applied to calluses, not to cells. This is a common procedure, but greatly restricts the intensity of selection that can be applied; it

(*Continues*)

*(Continued)*
might be more effective to select first for resistance in cell suspensions and then for regeneration ability in callus cultures. A final feature of these experiments worth noting is that the calluses are often chimaeric, with different morphotypes expressed by replicate plants raised from the same callus. Similar experiments involving selection for resistance to various agents have been published by Carlson (1973), Maliga et al. (1975), Marton and Maliga (1975), Palacco and Palacco (1977), Gengenbach, Green and Donovan (1977), Widholm (1977), Chaleff and Parsons (1978), Mattern et al. (1978), and Sacristan (1982). There are reviews by Brettel and Ingram (1979) and Thomas et al. (1979). The work on phosphate starvation is by Goldstein (1991).

The importance of somatic selection in the evolution of developmental systems was emphasized by Buss (1987), but doubted by Slatkin (1985) and van Valen (1988). Michaelson (1987) has discussed cell selection as an integral part of processes such as limb development and pattern formation. The special case of tissue differentiation involving the death of some cells was reviewed by Ellis et al. (1991). The possibility of selection for rapidly proliferating variants in the germ line was raised by Hastings (1991). It has been shown that diploid cell lineages bearing deleterious mutations fall in frequency during the premeiotic development of the germ line in *Drosophila* (Abrahamson et al. 1966).

The concept of family selection was first extensively developed by Lush (1947a, 1947b). The efficacy of family selection relative to individual selection has been investigated experimentally by Kinney et al. (1970), Wilson (1974), Campo and Tagarro (1977), and Garwood et al. (1980).

In some cases, different assemblages replicate independently of one another, such as B-chromosomes and autosomes, but are likewise propagated through the reproduction of a common structure, such as the cell or individual. Such assemblages are composed of different genes that must to some extent compete with one another for the opportunity to propagate through reproduction. The evolution of competing assemblages of genes is discussed in Part 4.

In other cases, assemblages form nested hierarchies of reproducers, such as cells within meristems and meristems within plants. A given gene will be represented at all levels of such an hierarchy, but it is not necessarily expressed in the same way in each assemblage. Its effects at different levels can be viewed as different characters, so that selection at any given level is constrained by selection at other levels, in a way analogous to the way in which selection in one environment is constrained by selection in other environments. As in previous cases, we can anticipate a general tendency for characters expressed at different levels to have antagonistic effects on fitness. A gene that causes uncontrolled cell proliferation, for example, may be favored among cells within a meristem (because it will increase in frequency in the meristem), but not among meristems within a plant (because it will disrupt the normal function of the meristem and make it less able to form new shoots). The integrity of any hierarchy of replicators always tends to be compromised by the antagonism of effects at different levels.

## *Entities That Are Components of Individuals*

During the development of a multicellular organism, the tissues of the adult are formed by different cell lineages that in a remarkable display of phenotypic plasticity express widely varied morphologies from a common genotype. Every multicellular individual is a walking phylogeny, the end-product of a clonal tree of descent. The relationship among the component cell lineages of an individual is, of course, predominantly cooperative, in contrast to the predominantly antagonistic relationship among physiologically independent cells. Nevertheless, cell proliferation during development provides an opportunity for selection among cell lineages, independent of selection among individuals.

**Selection in Somatic Cultures.** A simple way of dramatizing the possibility of selection among cells within individuals is to grow cultures of somatic cells outside the body, either as colonies on plates or as cell suspensions in liquid medium. Cells from actively replicating lineages, such as fibroblasts or meristematic cells, can often be grown in this way and are treated essentially as if they were microbes. (The situation is often not quite so straightforward; cultured mammalian fibroblasts, for example, will cease dividing after a certain number of cell generations. But such complications do not affect the main point.) They can therefore be selected for particular properties, just as one can select bacteria for growth on a novel substrate. This approach has been particularly successful in selecting for resistance to toxins and pathogens. However, there are other possibilities: for example, A.H. Goldstein of California State University selected tomato cells for growth at very low levels of phosphate. He first spread cells on solid medium containing very little phosphate. Most cells starved to death, but one of the few colonies that was able to grow was transferred to liquid medium, at the same very low phosphate concentration, and maintained for several months. This selection line then grew faster than an unselected control in low-phosphate media. Selection had apparently enhanced the secretion of an extracellular acid phosphatase, and thereby greatly increased the rate of uptake of inorganic phosphorus. It seems likely that this response involves the derepression and constitutive expression of a normally inducible system, in a fashion reminiscent of many experiments in bacterial evolution.

The possibilities of selection in cell suspensions do not seem to have been very fully exploited yet. The technique has two chief limitations. One is, of course, that it can be applied only to organisms that are able to regenerate from single cells. The second is that selection among cells may be antagonistic to selection among individuals, so that the regenerated individuals may be abnormal, weak, or sterile. For example, P. Thanutong and his colleagues in Kyoto selected regenerated callus tissue of tobacco plants for resistance to *Pseudomonas* (which causes wildfire disease) and *Alternaria* (which causes brown spot). They were successful in obtaining resistant plants normal in appearance and karyotype by this technique, and the resistance was transmitted to sexual progeny. Many of the regenerated plants, however, were morphologically abnormal, or were partly or completely sterile. The proportion of sterile plants increased sharply in the second cycle of somatic selection, suggesting that selection among cells is strongly antagonistic to selection among individuals. It is my impression, although I may be mistaken, that general principles of selection have not yet been applied very rigorously to the production of variant plants from cell suspensions, so that the enormous possibilities of selecting in very large populations of cells have not yet been fully realized. Moreover, the breakdown of somatic integration in plants regenerated from cells selected for their ability to proliferate in suspension might provide a novel way of identifying genes concerned with developmental integration, although selection in cell culture does not seem to have been used for this purpose.

**Cancer.** The growth of cell lineages within the body is firmly repressed or controlled. These controls will grow weaker with age (Sec. 98), and any lineage that escapes them will of course proliferate at the expense of other tissue. A tumor thus represents a very simple case of selection, the spread of a novel variant able to replicate under restrictive conditions that suppress the replication of other lineages. Once a tumor has begun to form, it grows like a somatic cell culture and, of course, will undergo selection for variants with higher rates of growth; tumors will therefore tend to expand with increasing rapidity as they age. The interpretation of tumors as populations undergoing selection for increased rates of growth has only recently appeared in the medical literature.

**Somatic Selection.** Cancer is an extreme example of what should be a common pathology, the tendency of cell lineages to undergo selection for greater or more effective growth at the expense of the organism as a whole. Indeed, Leo Buss of Yale has founded a whole theory of development on the process of somatic selection. Somatic cells have renounced the possibility of reproducing indefinitely by committing themselves to a mortal individual soma, so any gene in a somatic cell lineage has only very limited opportunity to replicate. A variant gene that transgressed this limitation by re-entering the germ line would regain the possibility of indefinite replication, although it would to some extent thereby injure its own ability to replicate by disrupting development and reducing the rate of reproduction of the individual. It would at the same time injure the prospects of genes in other cell lineages that will be selected to inhibit individualistic variants of this sort. On the basis of this argument, Buss interprets phenomena such as the maternal control of early embryogenesis and the early segregation of a distinct lineage of germ line cells as evidence of the continued suppression of selfish somatic lineages. The role of somatic selection in the evolution of development has been widely questioned on the grounds that selection among individuals will effectively overpower any tendency for selfish lineages to spread, but there are some cases where it seems to be important. The cellular slime-mold *Dictyostelium* grows as a clone of independent amoeboid unicells feeding on bacteria. When local resources become depleted, the amoebae gather together to form a multicellular slug-like creature that differentiates into a stalk bearing a sporangium. The spores formed in the sporangium are dispersed, and each gives rise to a new clone of trophic amoebae, but the cells that constitute the stalk are reproductively sterile. When the organism is cultured on agar plates, the stalk is redundant, and variants that form sporangia with no stalk or with a much-reduced stalk commonly appear and spread. This appears to be an example of somatic selection, manifested when the normal countervailing force of selection among individuals is weakened.

**Germ-Line Selection.** Ian Hastings of Edinburgh has pointed out that a similar phenomenon may occur in the germ line. Cells capable of becoming spores or gametes form a proliferating clone within which selection will favor variants with high rates of growth. The frequency of such variants will increase in every generation, although if they are inferior as reproductive or sexual propagules, their spread will be opposed by selection among individuals. Germ-line selection has been widely discounted because in diplonts gametic types are normally produced in Mendelian proportions, suggesting that genotypes propagate at the same rate in the germ line. Modest differences in the rate of proliferation, however, would produce only slight departures from Mendelian proportions and would not be detected in most investigations.

**Selection in the Immune System.** The most remarkable example of selection among cell lineages is the way in which the vertebrate immune system works. The immune response involves a mechanism for generating extensive variation among lymphocytes through recombination and a mechanism for selecting lymphocyte clones that fulfil certain criteria, resulting in the production and spread of cell lineages able to attack a specific pathogen. The analogy with artificial selection is very striking.

## *Entities That Are Assemblages of Individuals*

Evolutionary theory often takes it for granted that organisms are sharply individualized, so that the outcome of selection can be evaluated by simply counting the numbers of different types of individual. Many organisms are constructed on a modular basis and comprise a more or less indefinite assemblage of repeated parts, such as ramets or zooids, that are in partial physiological contact with one another. These parts will no doubt vary to some extent through mutation, and different variants will compete, with the added complication that they will share resources. There must be a process akin to somatic selection, although in many cases each part possesses its own germ line and is capable of autonomous reproduction. Even highly individualized organisms, however, can form assemblages of related individuals, so that selection can act among lineages rather than among individuals within a lineage.

**Replica Plating.** A bacteriologist may want to select strains that are killed by a given treatment, strains that are sensitive to an antibiotic or a bacteriophage, for example, or that cannot grow when a particular nutrient is not supplied in the culture medium. A conventional screen is ineffective because it is only the undesired types that survive it. The usual way of finding the cryptic mutant clones hidden in the bacterial lawn is the technique of replica plating, developed by Joshua Lederberg in the 1950s. A velvet pad is pressed against the culture growing on the plate and then pressed, in turn, against two fresh plates. The pad acts as a multiple inoculator that transfers a sample of colonies from one plate to another, while retaining their spatial relationships. One of the two fresh plates is permissive, so that all colonies are able to grow; the other is restrictive, containing antibiotic or phage, or lacking a nutrient, so that sensitive or auxotrophic genotypes are unable to grow. The colonies that die on the restrictive plates can be detected but not selected; on the permissive plates they cannot be detected directly, but they will occupy the same position on the two plates, so that clonemates of the dead cells can be isolated from the corresponding position on the permissive plate.

**Sib Selection.** The corresponding technique for larger organisms is sib selection, the selection of sibs of individuals that display a particular property. This is useful when scoring a character kills many of the individuals, or makes it impracticable to breed from them, or when a character cannot be scored at all. The last possibility is the most common reason: the milk production of bulls cannot be measured directly, but bulls whose daughters are likely to have high milk yield can nevertheless be recognized by the superiority of their sisters.

**Family Selection.** The most familiar method of artificial selection is the mass selection of individuals, the breeder choosing the most extreme individuals in the population for propagating the line. Selection, however, may instead be applied to families, as I have already mentioned in the context of selection for recombination (Sec. 48). The reason for doing so is that the phenotype expressed by an individual depends on both genetic and environmental sources of variation, and if the most extreme individuals are predominantly those that happen to have experienced unusually favorable conditions of growth then mass selection will make little progress. It would, of course, be preferable to select solely on the basis of heritable differences, if these could be discerned. Roughly speaking, when offspring are reared separately, or when the offspring of all families are reared together as a group, the variance of family means represents genetic variance, and the variance of individuals around the mean of their family represents environmental variance. The breeder may choose, therefore, to select individuals from the most extreme families, rather than selecting the most extreme individuals in the population. This technique is particularly useful for characters whose variation is largely environmental, because the mean value of a family, being based on a number of similar individuals experiencing a range of different environments, is a more reliable measure of heritable superiority than the value of a single individual. Conversely, if all the members of a family are kept together, like the litter of a sow, the variance among families may be largely environmental; but the variance among individuals from the same family, growing in very similar conditions, will be largely genetic. In these circumstances, it may be most effective to select the best individuals in each family, disregarding the average value of the family as a whole. Mass selection can thus be thought of as having two components, one represented by selection among families, the other represented by selection among individuals within families. Each is of practical use in different circumstances.

Family selection works because the mean value of a character in a family is heritable: families, as well as individuals, can transmit qualities to their descendants. This is because sibs inherit copies of the same ancestral genes from their parents and therefore form part of the lineage of those genes. This is obvious enough in an asexual population, in which the families are clones, but it remains true in an outcrossed sexual population. Family selection is less effective in sexual organisms because a given gene is transmitted to only half the full-sib progeny: although the family as a whole is selected, only half are part of the lineage of the gene. The same principle applies, not only to sibs, but to any set of related individuals. Selecting sets of individuals that are related to one another in a given way is equivalent to selecting lineages of different degree, that is, lineages that extend backwards in time for the corresponding number of generations. In principle, selection can be applied to lineages of any degree; this idea is developed further in the next section. However, selection is less effective among lineages of higher degree. The response to selection among lineages, relative to the response to the mass selection of individuals, will be less when their members are less closely related; very roughly speaking, the response will decline with the square root of relatedness. In addition, as more and more generations are added to the lineage, selection can be applied only at longer and longer intervals. In practice, therefore, family selection is hardly ever used except for full-sib and half-sib families.

## 116. *Organisms may be selected to produce variation among their progeny on which selection can act.*

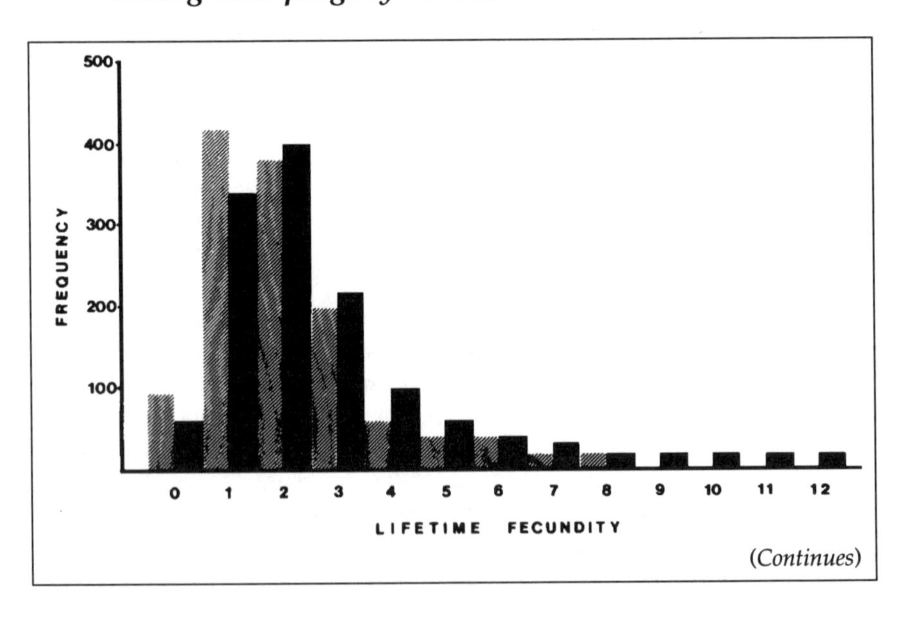


*(Continues)*

*(Continued)*
The lottery models originally developed by Williams (1975) were intended to explain the prevalence of sex: sexual families, being more diverse than clones, are more likely to include the very few highly superior genotypes able to flourish at any given site. Kelley (1989) collected individuals of *Anthoxanthum* from a semi-natural population and obtained clonal families by multiplying ramets and sexual half-sib families by germinating seeds (the plants are self-incompatible and wind-pollinated). He then planted mixtures of these two kinds of progeny, as vegetative ramets of comparable size, into mown grassland. Each clump of experimental plants thus included both uniform asexually derived ramets and diverse sexually derived ramets, descending from the same parental plant. The smallest clumps were single plants; the largest were groups of 16 plants, 8 of which were sexually derived and 8 asexually derived. The histogram is the frequency distribution of net reproductive rates for asexual plants (light bars) and for sexual plants (dark bars). The expected reproduction of plants descending from sexual seed is nearly 40% greater than that of plants descending from asexual tillers, and the design of the experiment suggests that their superiority is attributable to their greater diversity. The advantage of the sexual plants, however, was as marked among the isolated single plants as it was in the larger clumps. Moreover, there was little sign of the extreme truncation selection envisaged by lottery models. The experiment thus succeeded in establishing that selection will in some circumstances favor the diversification of progeny, but failed to discover why.

**Selection Arenas.** A breeder conducting a trial generally begins by assembling a highly variable base population that is likely to contain genotypes with the desired characteristics. Steve Stearns of Basel and Jan Koslowski of Cracow have suggested that a similar process may operate in nature. In a varying and uncertain environment, propagating a single specialized type would lead to low geometric mean fitness. It would then be preferable to produce phenotypically plastic offspring able to grow in a range of conditions. There are limits, however, to the evolution of plasticity (Sec. 107), and an alternative might be to transfer plasticity from the individual to the family as a whole. A diverse family of differently specialized offspring constitutes a *selection arena*, ensuring that in each generation at least some offspring will be produced that are well adapted to prevailing conditions. The success of the lineage is enhanced, not by consistently producing any given genotype, but rather by presenting the environment with a range of types, so that recurrent selection will maintain adaptedness. George Williams has developed a similar theory of *lottery selection*. Imagine that the environment is so variable that only a very few genotypes are capable of reproducing and that it cannot be predicted from a knowledge of previous generations which types will be successful in the coming generation. The situation is analogous to a lottery in which there are only a very few winning numbers. In such circumstances, it is futile to photocopy your lottery ticket; it is preferable to purchase tickets with different numbers, even if this means having fewer of them. These arguments have been used to explain the generation of genotypic diversity by recombination in sexual families.

**The Variability of Bacteria.** A single lineage of bacteria may produce a variety of distinct phenotypes when cultured for many generations. Some of these variants are classical mutations caused by some harmful external agency, such as irradiation or transposon insertion. Others, however, involve the rearrangement, duplication, insertion, or deletion of sections of the genome through recombinational processes of diverse kinds. In some cases, genetic rearrangements occur regularly in response to some particular signal and constitute systems for controlling gene expression. There remains a large category of rearrangements causing phenotypic changes, often with substantial effects on fitness, that seem to occur irregularly and irreproducibly. They may occur at very high frequency and, although inherited, may also revert at high frequency. Paul Rainey and his colleagues at Oxford have recently argued that the generation of high levels of phenotypic diversity within bacterial lineages by recombination may have evolved as a non-specific response to stress. When conditions of growth change or deteriorate, a lineage that is able to produce well adapted variants quickly would replace more conservative lineages. There is no suggestion that the variants are specifically adapted to the new conditions; only that a large quantity of random variation is likely to include a few types that will increase in frequency through selection. It is not yet known whether the rate of random genetic change generally increases in stressful conditions. If it does, our present concept of bacterial populations as being largely uniform and evolving through the stately substitution of a series of classical mutations will be supplemented, and to some extent supplanted, by a much more dynamic picture in which large amounts of variation may be generated in response to stress. Bacterial populations would then be much more variable, and would respond much more rapidly to selection, than has hitherto been thought likely.

## *117. Selection among lineages weakens with time scale.*

**Selection Among Lineages of Different Degree.** The ancestry of a group of organisms is familiarly represented as a bush or tree. Each line of the diagram indicates continuity of information and connects an entity with the descendent to which it gives rise by replication. The branching structure of the diagram traces genetic relationship extended in time, so that it can be used to show how entities that lived at different times in the past are related to one another. Strictly speaking, such a diagram can refer only to a gene, with branching caused by mutation; but with some loss of precision it can also refer to an asexual lineage (the lines connecting individuals) or even to a sexual or an asexual clade (the lines connecting taxa of equivalent rank, usually species). Imagine a tree so detailed that it represents the ancestry of every individual in a large clade of asexual organisms over millions of years. Now draw a horizontal line through the tree, representing an instant of time. This line will intersect every individual alive at that time. It will also intersect every group of related individuals each of which constitutes a bundle of adjacent lines. Looser and looser degrees of relationship are represented by larger and larger bundles (which could receive Linnean recognition as genus, family, order and so forth) until the most inclusive group is simply the entire clade. We could deliberately select individuals, or sibships, or groups of cousins, or any other sets of individuals that are related in a certain degree, however remotely. By doing so, we would be selecting among lineages that extend for different periods into the past. Individuals have no extension in time; sibships extend a single generation, and sets of cousins two generations; species or genera might extend for many thousands of generations into the past before the common ancestor of two competing taxa was reached. Thus, selection among groups of related individuals is equivalent to competition between lineages of corresponding extent. This raises no difficulty when the group is reduced to a single individual: selection among individuals leads to individual adaptedness, as the consequence of the differential proliferation of lineages whose individual members differ consistently in heritable properties. When the group is a set of related individuals, it raises the issue of whether, in nature, lineages rather than individuals are subject to selection.

A lineage of given degree is a component of lineages of higher degree, as sibships are components of sets of cousins, or species components of genera. Asking whether lineages are themselves subject to selection is equivalent to asking whether a lineage of given degree may have attributes that are selected through its proliferation within lineages of higher degree. It is the sets of individuals of given relatedness that are actually selected at any given time (the tips of the branches of lineages of given extent) and the differential proliferation of the lineages of higher degree founded by these sets that is the outcome of selection. Thus, we might ask whether a species may have attributes that cause its descendants, other species of which it is the ancestor, to become more numerous than the descendants of other species in the same clade. The answer depends on the sort of attribute that is being considered.

Any set of related individuals will display a certain mean value for any given character; this mean will vary among sets and will be inherited to some extent. Artificial selection can be applied to this mean, which will change through time as a result. Because artificial selection is evidently feasible (and routinely practised), natural selection is clearly conceivable. Whether or not this often happens in nature, and whether it produces results that are distinctively different from the outcome of mass selection on the same character, may be debatable; but there is no difficulty of principle involved.

The other kind of attribute is a character that is necessarily the attribute of a set of individuals and that cannot be reduced to the sum or average of individual character values. These are characters that can only be expressed by a set of individuals, living either contemporaneously or as a temporal sequence. The first category includes all characters that depend on interaction between individuals, such as the complementary use of resources in mixtures of genotypes, or the behavior of individuals towards one another in societies. These are discussed in Part 5. The second category includes characters that are responsible for long-term adaptations, that is, those that affect the ability of species (or other taxa) to survive and give rise to other species.

**Selection of Lineages for Specific Adaptation.** Any environment is likely to change abruptly at long and irregular intervals; this is part of the variation of the environment on all time scales. Organisms are thus liable to suffer infrequent catastrophes, as the result of the devastation caused by fire, flood, or some similar event. They might become adapted to resist this devastation in either of two ways. In the first place, they might become specifically adapted to a recurrent stress, for example, by producing seeds that are able to germinate after fire. If the return time of fires exceeds the lineage scale of the organism, then fire-resistant seeds will not, in most generations, increase the reproductive output of individuals. Such seeds will, rather, affect whether a lineage survives and proliferates. The degree of the lineage is the expected return time of fires, divided by the average generation period of the organism. The clade to which the organism belongs can be thought of as constituting a large number of lineages of this degree, in the same way that a lineage is made up of individuals. Fire-resistant seeds are a distinctive adaptation of lineages of a certain degree and evolve through the selection of such lineages. There is, in short, no difficulty in analyzing specific adaptation to rare events in terms of the selection of lineages of appropriate degree. Many people, to be sure, would regard this as unnecessary and would treat the evolution of fire-resistant seeds as an example of individual selection that occurs only occasionally. I think that this is less precise, but probably not seriously misleading.

Lineages of higher degree are less likely to evolve specific adaptations. This is because their direct response to selection is weaker, and the correlated response is more likely to be antagonistic. Events that happen less often, and therefore involve more extended lineages, will cause less intense selection within a given span of time. Any specific adaptation that is favored in this way will be opposed by shorter-term processes. In the first place, it will become degraded by mutation during the intervening period in which it is not being actively maintained by selection; the more extended the lineage, the longer the period involved and the greater the degree of degradation. Secondly, shorter-term selection acting through lineages of lower degree may often act antagonistically to longer-term selection. There may be some easily understandable reason for this: fire-resistant seeds, for example, may germinate less easily in normal years, so that selection among lineages, a few generations in extent, is opposed by selection among individuals within lineages. It will, however, by now be a familiar proposition that negative correlation between shorter-term and longer-term fitness will tend to evolve: genes that increase both will be fixed, and those that reduce both eliminated, leaving genes with antagonistic effects segregating in the population.

Longer-term adaptation is therefore less effectively wrought and more effectively opposed. There is no difficulty in explaining adaptation to circumstances that recur every few generations as being caused by selection among lineages of modest degree. It is much more difficult to identify specific adaptations of lineages thousands of generations deep; indeed, I do not know of any and doubt that selection at the level of species is ever effective in causing them to evolve. The field is cloudy, and there is no experimental work to guide us, but I am inclined to believe that the great bulk of adaptation is caused by the selection of individuals, or of lineages of low degree, and that it is a sound general rule to prefer interpretations of adaptation that invoke selection among lineages of lower degree.

It is not the point of this argument to deny that rare events may profoundly affect the history of lineages. It has been suggested that diatoms survived the end-Cretaceous mass extinction with relatively little loss because most species have dormant propagules that were able to survive the occlusion of sunlight caused by the dust and smoke following a major asteroid impact. This might be the case. It is, however, extremely unlikely that these resistant propagules represent a specific adaptation to an event that recurs at a time scale of tens of millions of years.

**Selection of Lineages for Adaptability.** A lineage that is able to adapt more readily to altered circumstances may be more likely to persist. This is the phylogenetic equivalent of plasticity; but what is meant is not the ability of a single genotype to express different phenotypes, but the ability of a population to express appropriately specialized genotypes. The response to selection will be proportional to the quantity of available genetic variation, and lineages that are able to store more variation, and release it on demand, so to speak, will be less likely to become extinct. This proposition was commonplace in evolutionary biology until quite recently and was the basis of the classical interpretation of genetic systems by the Oxford geneticist C.D. Darlington; it remains a staple of textbook accounts of evolution. There is nothing heretical about the idea: all things being equal, a lineage that is more apt to evolve is more likely to persist. There are, however, two reasons for doubting that adaptability will often evolve as an adaptation of lineages of high degree.

The first concerns the causes of extinction. A population can adapt to a changing environment only if the change occurs rather slowly (Sec. 56). The relevant physical features of the environment will change, at some long time scale, much faster than the population can respond through selection. When a pond dries up, the minnows die; they do not evolve into frogs. If the pond dries up every hundred years, then it must be resupplied with minnows by immigration. If all the ponds within the distribution of a species of minnow dry up in the same year, as might happen every million years, then the species becomes extinct. It is perfectly true to say that the minnows became extinct because they failed to adapt to a changing environment; but it might be more useful to say that they became extinct, not because the environment they lived in changed, but because the environment they could live in disappeared. It may be that extinction is usually caused by perturbations far beyond the range of adaptation available to the population through sorting. A population living in favorable conditions expresses a certain minimal range of genetic variation. If it is stressed, by high temperature or antibiotics or starvation, the genetic variance of fitness will increase, because some genotypes will be resistant to the stress although others will succumb. At some point, the stress will be so severe that no genotypes can survive, so that genetic variance collapses to zero. The expected persistence of a population depends on the relationship between this point and the time-scale of environmental perturbations of increasing magnitude. By storing more variation, a population extends the range of stresses that it can resist, but it is not clear that it can possibly store enough variation to prolong its existence substantially.

The second, and more important, ground for scepticism is that short-term processes will generally overpower longer-term processes. Variants that might survive some distant catastrophe are normally deleterious; if this were not the case, we should not have to explain their maintenance in the population in terms of their effect on the long-term fitness of the lineage. Their frequency will therefore be continually reduced by selection among individuals or lineages of low degree. In a trivial sense, extinction could always be avoided by more rapid adaptation; a population could survive any catastrophe by maintaining a few specialists for every imaginable contingency, like a vast tool kit. But if these specialists are less fit than average for most of the time, as they will be if there is a general cost of adaptation, then they will have been eliminated long before the rare catastrophe that they could resist occurs.

For these reasons, the view that characters such as the rate of mutation, or the balance between inbreeding and outbreeding, evolve so as to ensure the long-term evolutionary flexibility of the lineage have fallen out of favor among evolutionary geneticists. Nevertheless, there is one exception to this general scepticism: the prevalence of sex in most groups of eukaryotes is widely believed to be attributable to the greater adaptability of sexual lineages. This view dates back to the great German geneticist August Weismann, who, before the rediscovery of Mendelism, speculated that the sexual diversification of offspring provided the pool of variation that made possible a continuing response to natural selection. The simplest version of this argument is that the long-term geometric mean fitness of a clone is zero. No matter how well adapted, or how plastic, it will eventually be wiped out by a perturbation that exceeds its range of tolerance, whereas a sexual lineage will persist through the survival of a few rare recombinant offspring. Asexual taxa are thus evolutionary dead ends, incapable of further evolution, and doomed to early extinction. There is some experimental evidence for this view: asexual lineages of ciliates are demonstrably mortal, whereas sexual lineages are not (Sec. 52). This applies only in the special case of isolate culture, however, where irreversible mutation accumulation occurs fairly rapidly in the absence of recombination, and no comparable process has been reported from mass cultures. The strongest evidence is comparative: asexual taxa of metazoans such as arthropods or vertebrates are invariably derived from related sexual taxa, are never themselves ancestral to large clades, and are presumably quite young. It is not difficult to provide an explanation for this pattern: an asexual lineage derived from a sexual population may respond slowly to selection because the independent effects of genes are not effectively exposed to selection (Secs. 45–46), and being specialized in one way or another might be extinguished by a minor environmental change. This argument is much less convincing when population sizes are large enough for mutation to provide a substantial input of variation because selection is known to be effective in large clones (Sec. 59). In fact, bacteria and eukaryotic microbes, and even small metazoans such as rotifers, are very often asexual, or undergo incomplete or infrequent sexual processes.

**Assemblages Whose Properties Are Not Heritable Do Not Evolve.** Individuals are often associated together in groups that are not lineages. The members of a flock of birds, for example, may be unrelated to one another,

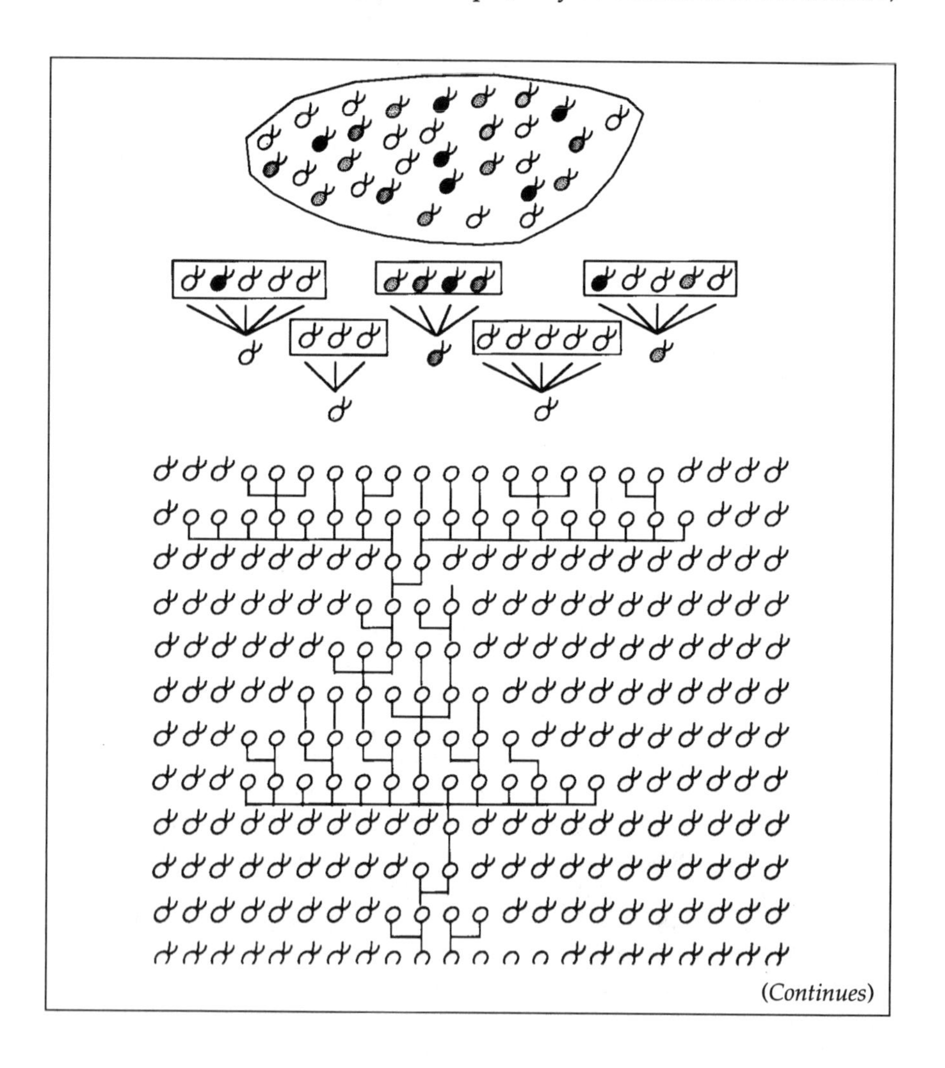

(*Continues*)

*(Continued)*

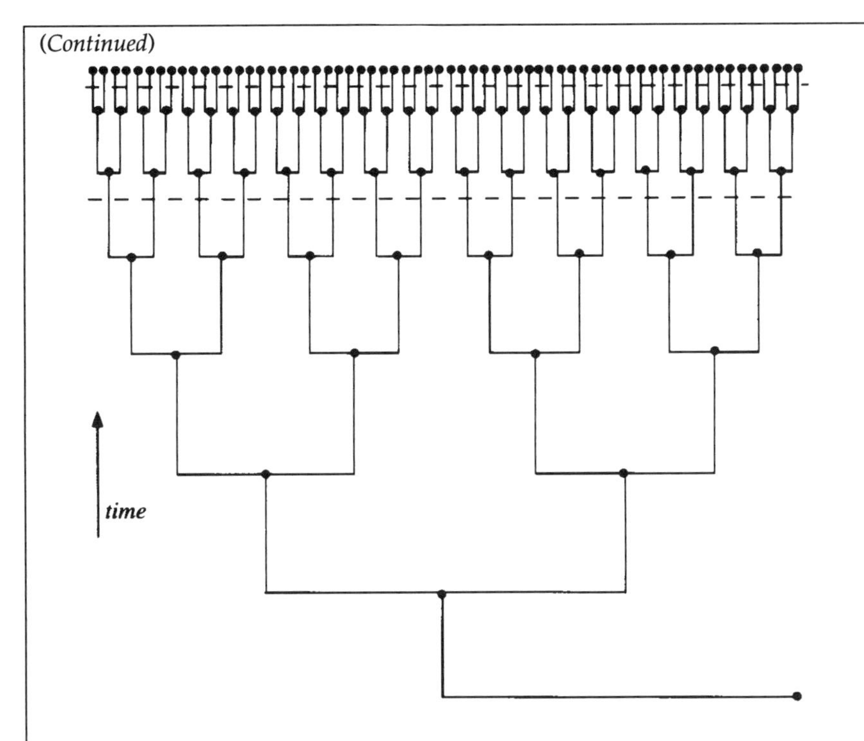


The first diagram illustrates three levels of selection. Individual mass selection for increased depth of pigmentation would favor the half-dozen or so darkest cells in the top picture. Family selection (middle) would favor the middle family with the highest mean score; individual selection would tend to favor some of the progeny of less heavily pigmented parents, which might not transmit the selected character value to their own progeny. The bottom picture shows a lineage of flagellaless mutants. Here, 12 generations are shown, with time increasing up the page. The character is usually disadvantageous, but occasional environmental changes give it a short-lived advantage over the flagellated type. It is thus maintained in the population by selection acting on lineages about 6 generations in extent. More generally, one can imagine selection acting at any level in a phylogeny. In the second diagram, the upper line intersects the phylogeny at the lowest level of classification, and represents selection among individuals. The lower line intersects the phylogeny at the level corresponding to monophyletic lineages two generations in extent; it represents selection among groups of individuals, such that the individuals in each group have their grandparent in common.

For contrasting views of the nature and occurrence of species selection, see Hoffman (1984) and Gilinsky (1986).

in the sense that two random members of the same flock are not more closely related than two birds from different flocks. Because flocks and similar groupings possess properties of their own, it has often been asked whether selection among such assemblages can cause specific group-level adaptations. The general answer is that this will not occur. Assemblages of unrelated individuals lack heritability and, therefore, changes in the frequency of assemblages with different properties cannot cumulate through time. If the fitness of an individual depends to some extent on interactions with other types of individuals, however, rather than on its own intrinsic properties, then it will vary according to the composition of the population. The outcome of selection when the environment includes other organisms is discussed in Part 5.

# 4

# *Autoselection*

## *118. The replication system itself is vulnerable to genetic parasites.*

Parasites such as helminths or rust fungi are organisms that make a living by exploiting the growth system of their host, diverting nutrients obtained or elaborated by the host to their own growth. All organisms have systems for reproduction that are based ultimately on the replication of nucleic acids; just as the growth system of an organism can be parasitized by other organisms, so the replication system of genes can be parasitized by other genes. We have already seen that a self-replicating system cannot be perfectly *precise*; nor can it be perfectly *specific*, replicating its own sequence and no other sequence whatsoever. This creates an opportunity for elements that are not autonomously self-replicating but that, rather, utilize the common replication system of the genome. I have already described a simple case of this sort in connection with Q$\beta$ (Sec. 6). The virus itself is a conventional parasite that uses the host cell as a source of raw materials. These can be assembled into new viral genomes only through the viral replicase, a diffusible protein encoded by the viral genome. The replicase, however, will also direct the replication of incomplete viral genomes that do not themselves encode the replicase. Such incomplete viruses (often called "defective interfering particles") are parasites of the viral replication system that often appear in the later stages of viral infection. The Q$\beta$ experiments, indeed, for the most part document the evolution and diversification of such defective interfering particles.

**Autoselection.** This principle applies with greater force to the genomes of cellular organisms that comprise a congeries of genes, most of which are incapable, separately or together, of self-replication but are, rather, replicated by a common genetic machinery. This machinery is not highly specific and will replicate a fairly broad range of genes; indeed, it will often replicate genes from very distantly-related organisms, which is why genetic engineering is feasible. Replicators that are able to utilize this machinery will therefore persist, even if they are not expressed in the phenotype so as to benefit the genome, or the organism, as a whole. Because they utilize the machinery of replication without contributing to its maintenance, they may be called selfish or parasitic elements. Their persistence, or spread, is not determined by the degree of adaptedness to external conditions that they confer, but solely by the extent of their ability to be replicated along with the rest of the genome. The process by which such elements evolve can thus be termed *autoselection*.

**The Selfish Behavior of Autoselected Elements.** A plasmid that is replicated along with host DNA runs one serious risk: if there is no means of ensuring regular segregation, some daughter cells will by chance receive no copies of the plasmid. These asexual lineages are permanently cured, because they are very unlikely to be re-infected. These cured lineages will continually increase, because there is a process that adds to their number, but none that necessarily reduces it, until the plasmid is extinct. The only plasmids we observe, therefore, are those that have evolved some means of preventing the accumulation of cured cells. One possibility is simply to maintain many copies, so that it is almost certain that every daughter cell will receive some (which can then supplement their numbers by additional replication); this is bad for the host to the extent that the maintenance and expression of plasmid genomes is metabolically expensive. A second possibility is to cause selection that favors cells that bear a plasmid, for example because the plasmid genome encodes an antibiotic resistance. Such plasmids are domesticated parasites that actually benefit their hosts. The third possibility is the least well known, but most clearly illustrates the selfish nature of many genetic elements and how they can be autoselected despite their antagonism to host function. Plasmid R1 is very stably maintained in *E. coli*. It includes a locus that encodes host killing (*hok*) and suppression of host killing (*sok*) mRNAs. The *hok* mRNA is translated into a small protein that collapses the proton gradient across the cell membrane, halting respiration and killing the cell. The *sok* mRNA is untranslated, but binds to *hok* mRNA and prevents it from being translated. The *sok* promoter is the more powerful, and so *hok* is effectively unexpressed and the cell is not harmed seriously by carrying the plasmid. Alas, if it should lose it. Although the daughter cell may not receive a copy of the plasmid, it will certainly receive some of the transcribed mRNAs in its cytoplasm. The *sok* mRNA is much less stable than *hok* mRNA, so that in a cell cured of the plasmid *hok* will be expressed, and the cell is killed by the *absence* of its parasite.

**Vestigial Genetic Elements.** In asexual organisms there will be little if any tendency for elements to be selected because they are able to replicate more rapidly than others in the same genome. If there is any cost of maintaining superfluous copies of an element in the genome, lineages that bear more over-replicating elements will proliferate less rapidly than lineages that bear fewer; thus, the frequency of selfish elements within lineages will be kept low through selection among lineages. Nevertheless, redundant elements that no longer contribute to the phenotype except as a burden will continue to be replicated by a machinery that is incapable of distinguishing between genes on the basis of their phenotypic effects. Genes that have long since lost their usefulness will persist for many generations, slowly degenerating as neutral mutations accumulate unchecked. The bits and pieces of genetic gibberish produced by recombinational excision and other processes will likewise continue to be replicated, until they are finally lost by the same sort of accident that created them. No doubt such elements are somewhat deleterious, insofar as energy and materials that could more profitably be deployed elsewhere are used to maintain them, but their effect on the cellular economy is probably very small. Incomplete and defective genes are quite abundant, especially in large genomes where their marginal effect will be less, and they constitute the largest category of selfish elements in asexual organisms. They are rudimentary or vestigial structures, akin to the pelvic girdle skeleton of whales and snakes, or the eyes of cave-dwelling shrimps and fish. Vestigial structures do not furnish very good evidence of selection, except that it is very weak in the later stages of the loss of a redundant feature, but they do give conclusive evidence of ancestry. The abundance of vestigial structures in the genome shows how mere passive replication can prevent for long periods the loss of elements no longer functional and, thus, emphatically suggests that the replication system might be actively exploited by elements that evolve for no other reason.

## 4.A. Elements That Utilize Existing Modes of Transmission

### *119. Mutator genes occasionally spread in asexual populations.*

High rates of mutation can be caused very simply by loss-of-function alleles of the loci encoding polymerases or other components of the replication

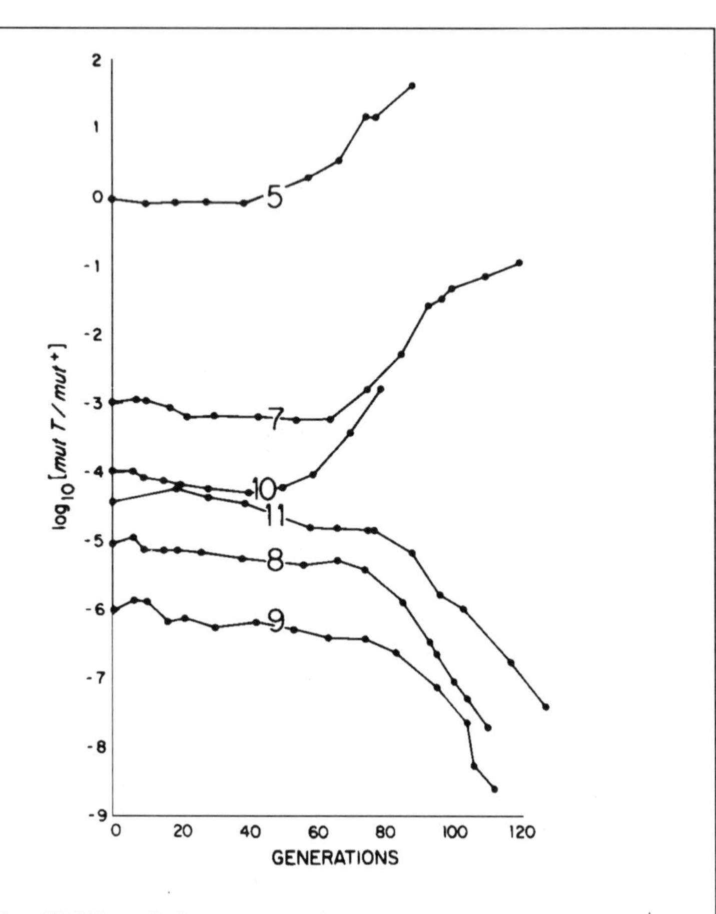


Chao and Cox (1983) studied competition between isogenic wild-type $mut^+$ and mutator *mutT* strains in glucose-limited chemostats. The *mutT* mutation increases the rate of AT → G C transversions by three or four orders of magnitude; for the genetics and mode of action of this and other mutator genes, see Cox (1976). Mixed populations in which mutator and wild-type strains were present in different proportions were made up from stationary batch cultures and used to inoculate the chemostat. There is an initial lag phase of about 60 generations, during which the frequency of *mutT* falls slightly but consistently. This represents the time taken for a beneficial mutation to arise and increase to appreciable frequency. The fall in the frequency of *mutT* during this period is caused by selection against the deleterious mutations that it induces and is necessarily linked to. If *mutT* is present at a sufficiently high frequency in the initial population, it is likely that the novel beneficial mutation will arise in a *mutT* cell, and that following the lag phase the *mutT* lineage will thereby spread. If *mutT* is very rare in the initial population,

(*Continues*)

(*Continued*)

the increase in mutation rate that it causes is insufficient to counter the greater abundance of the *mut*$^+$ strain, the first beneficial mutation is more likely to arise in a *mut*$^+$ background and, consequently, *mutT* is eliminated. See, also, Gibson et al. (1970), Nestmann and Hill (1973), and, especially, Cox and Gibson (1974).

system. To the extent that the rate of evolution is limited by the rate of mutation, some degree of infidelity of replication might be favored in the long term if it increases the rate of response to environmental change. Such selection is expected to be very weak (Sec. 117). Nevertheless, rather high rates of mutation may evolve in chemostat populations of bacteria. This is because mutator alleles, causing a high rate of errors during replication, can be autoselected in asexual populations through hitch-hiking (Sec. 45). Most of the mutations they induce will be deleterious; mutator frequency and mutation rate will therefore be low at equilibrium. In a novel environment, however, a mutator allele will occasionally cause a favorable mutation. As this spreads through the population, the mutator will spread with it because in the absence of recombination the two are transmitted together. Once the new mutation has been fixed, the mutator frequency will decline because more stable revertants that do not cause excess deleterious mutations will again be favored. Mutator genes will continue to wax and wane in this fashion, until the population has reached evolutionary equilibrium, new favorable mutations become very unlikely, and mutator genes are almost always very rare. Thus, periodic selection (Sec. 58) implies the periodic spread of mutator genes, despite the fact that they reduce fitness in nearly all the individuals that bear them.

The situation is quite different in sexual populations, where recombination separates mutator genes from the mutations they have caused. Selection is less effective in keeping mutation rates as low as possible because it can act only through a fraction of the progeny of a mutant individual; however, it will be completely ineffective in causing any substantial increase in mutator frequency because favorable mutations can be selected independently of the mutator genes that produced them. Mutation rates should therefore be somewhat higher on average in sexual populations, without displaying the extreme excursions that are liable to occur in asexual populations. I do not know whether this is, in fact, the case.

**Dynamics of Mutator Genes in Chemostats.** The best-known experiments on mutator genes were carried out by E.C. Cox and his colleagues at Princeton. They used a mutator allele of *E. coli* that increases the frequency of A : T to C : G transversions to about $u_m = 2 \times 10^{-6}$ per base pair per replication, causing about four additional base pair alterations per genome per generation. Most of these will be deleterious. If the initial number of mutator cells $N_m$ is rather small, so that $N_m u_m \ll 1$, only a small fraction of all possible mutations may occur in the population in every generation, and the mutator population is likely to dwindle to extinction before any beneficial mutation occurs. On the other hand, if $N_m u_m > 1$, almost all possible mutations will occur in every generation, including all possible beneficial mutations. Provided that the initial mutator inoculum is sufficiently large (more than about $5 \times 10^5$ cells in this case), it should tend to increase in frequency for as long as the population is not well adapted to the chemostat environment. This is what Cox observed: after a short lag of about 50 generations, a mutator strain tends to spread when inoculated at low frequency into a chemostat occupied by an isogenic non-mutator strain. The main reason for believing that this is attributable to a higher mutation rate, rather than to some directly stimulating effect of the mutator gene product, is that replicate cultures often behave in quite different ways. The mutator gene may spread slowly or quickly and sometimes fails to spread at all. The evolved populations may be changed in several ways: for example, they may adhere more readily to glass surfaces, reducing the rate at which they are washed out of the chemostat, or utilize novel carbon sources, or resist starvation more effectively; but different populations show different kinds of adaptation. Moreover, the superiority of the evolved population is usually retained even when the mutator gene is deleted. Thus, high mutation rates can evolve in chemostats because mutator alleles remain linked to the rare beneficial mutations that they cause.

## 120. *Autonomously replicating elements such as plasmids can spread among sexual lineages.*

The genome may replicate either completely or incompletely. Mitotic replication is complete, in the sense that any given gene in a parental genome will ordinarily be copied to every progeny genome. Meiotic replication is incomplete because any given gene is ordinarily copied to only half of all progeny genomes. Incomplete replication offers a much greater scope for autoselection because a gene that is transmitted to more than half the progeny will tend to spread among sexual lineages. Sexual organisms are thus expected to harbor a denser and more diverse fauna of selfish autoselected elements.

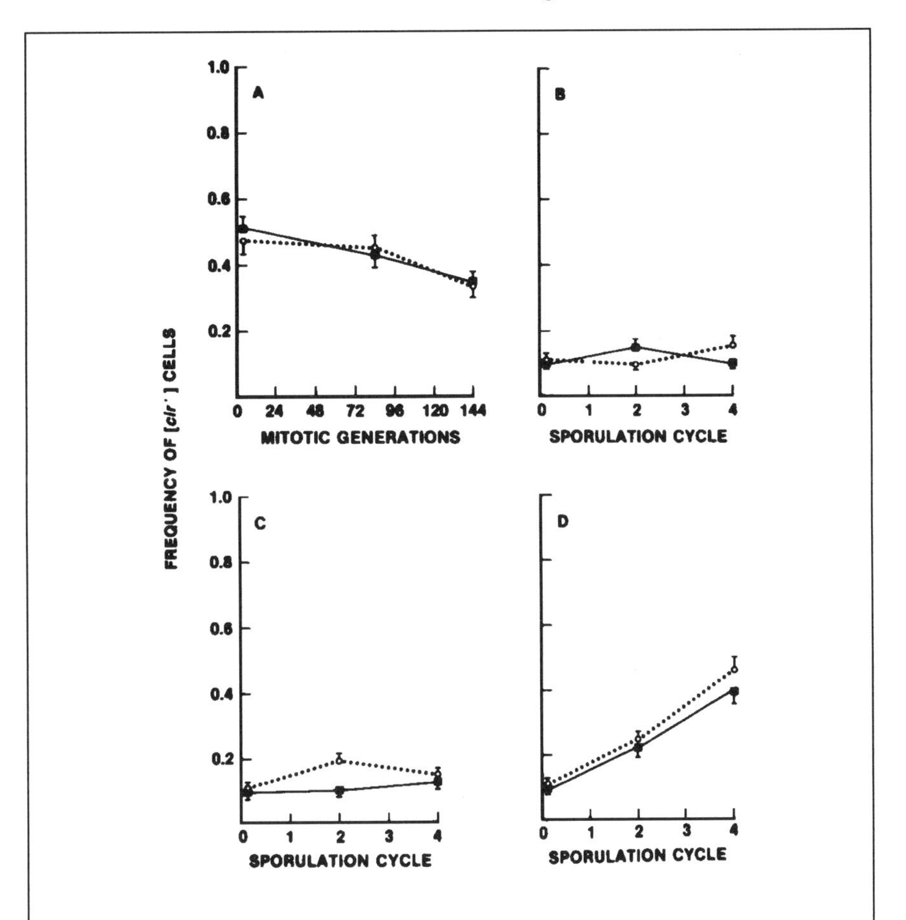


These figures show how the frequency of the 2-$\mu$m plasmid changes in experimental populations of yeast, from the paper by Futcher et al. (1988). The two lines in each figure are two different strains:

a) In asexual mitotic populations the frequency of the plasmid-bearing $cir^+$ strains declines slowly.

b) If cultures pass through successive sexual cycles, with gamete fusion occurring within the ascus, the plasmid does not increase appreciably in frequency.

c) The same result is obtained if unfused vegetative cells are killed with ether during the sexual cycle.

d) When the ascus is disrupted before gamete fusion, enforcing a high rate of outcrossing, the plasmid spreads rapidly through the population.

**Bacterial Plasmids.** There are usually several self-replicating structures in a bacterial cell: the large bacterial "chromosome" that encodes most of the vegetative activities of the cell and a number of much smaller plasmids that in most circumstances are not essential to the organism. Most bacterial lineages are genetically isolated almost all of the time, and cells maintain only one copy, or a very few copies, of any given kind of plasmid. Any plasmid that replicated more rapidly than the rest of the genome would be a burden on the cell and would tend to be eliminated from the population by selection among clones. Some plasmids, however, can be transmitted from cell to cell during conjugation. A plasmid that did not replicate during this process would merely exchange one genetic background for another. A transmissible, or mobilizable, plasmid, however, may replicate before transfer, so that one copy remains in the host cell, while the other is transmitted to its partner. Any plasmid that has this ability will tend to spread through the population by infecting all the lineages that are able to conjugate with one another, whereas a non-transmissible plasmid will always remain imprisoned with a single lineage. In the long term, any given clone is likely to become extinct, and a plasmid that is restricted to a single lineage will become extinct when its host does. Transmissible plasmids, as well as being more widely distributed at any given time, are therefore likely also to be longer-lived.

**The 2$\mu$ Plasmid of Yeast.** Most strains of yeast, *Saccharomyces*, contain a few dozen copies of a circular DNA molecule about 6 kb in length, the so-called 2$\mu$ circle. Although it is an unusual creature (free plasmids are rarely found in eukaryotic cells) we understand its maintenance and spread more clearly than that of conventional bacterial plasmids, thanks to the insights of Donal Hickey of Ottawa and his collaboration in experimental work with Bruce Futcher and his colleagues at McMaster University. In strictly asexual populations of yeast, the 2$\mu$ circle is slowly lost through two processes. The first is vegetative segregation. Because there is no mechanism that ensures an equal partition of copies between the mother cell and its budded daughter, some cells will receive more copies and others fewer. In each generation, a small proportion of cells, descending from parents who themselves possessed few copies, will receive no copies of the plasmid at all. The lineages they found are permanently cured of the plasmid, which cannot re-infect them as long as the population remains asexual. There is therefore a continual increase in the frequency of *cir*$^0$ cells, although in practice the average copy number is so high that this is only a weak force. Selection against *cir*$^+$ lineages is probably more important. The plasmid does not encode any vegetative function, and *cir*$^0$ strains are apparently normal in all respects. When isogenic *cir*$^+$ and *cir*$^0$ strains are mixed, the frequency of $cir^+$ falls by about 1% per generation, presumably because of the metabolic burden that 50 or 60 copies of the plasmid imposes on the host cell. Thus, the plasmid cannot spread in asexual populations, and even if it is initially present at high frequency it will be lost within a few hundred generations. It can be maintained permanently only in populations where cells go through a sexual cycle from time to time. When $cir^+$ and $cir^0$ cells mate, the plasmid is distributed to all four meiotic products, so that the sexual progeny of $cir^+$ and $cir^0$ parents are all $cir^+$. The lineages descending from the four meiospores all have a normal complement of plasmids, showing that the plasmid takes advantage of the opportunity for over-replication provided by the incomplete genome replication of the sexual cycle. If the plasmid is rare, $cir^+$ cells will double in frequency whenever the population as a whole goes through a sexual cycle. It must be noted, however, that this process demands, not merely sex, but outcrossing: mating between two $cir^+$ cells will not contribute to the spread of the plasmid. Yeast has two mating types, or genders, that define whether or not two haploid cells can mate. After meiosis in the diploid cell produced by sexual fusion, two spores are of one mating type and two are of the other; all four are either $cir^+$ (if at least one parent was $cir^+$) or $cir^0$ (if neither parent was $cir^+$). These cells are held within a thick-walled ascus, like seeds in a fruit, and if the ascus is not disrupted, sister cells will mate before being released. Experiments in which the ascus was allowed to remain intact after mating showed no increase in the frequency of $cir^+$ during six successive sexual cycles. It was only when the ascus was disrupted, and the spores from different asci allowed to mingle before the next round of mating, that the expected spread of the plasmid (roughly doubling in frequency after every sexual cycle) was observed. This is a clear example of a genetic element that spreads infectiously as a parasite of outcrossed sexual lineages.

**B-Chromosomes.** Eukaryotic cells contain a variety of parasitic and mutualistic replicators that, like plasmids, are physically separate from the host genome. A number of different kinds of micro-organisms and their domesticated descendants, the cell organelles, are found inside eukaryotic cells; they often interact strongly with the host genome, raising new issues that are discussed subsequently with respect to sex and gender and are treated more generally in Part 5.D. There is one type of genome parasite that deserves special mention, however, because it illustrates a different way in which sexual transmission can be biased. B-chromosomes, or accessory chromosomes, resemble normal autosomes, but do not encode any essential functions; they are present in some individuals but not in others, they may vary in number among individuals that bear them, and individuals that lack them develop normally. They are found in about 10% of all animals and plants and are in some respects analogous to plasmids. Like plasmids, they will tend to be eliminated from asexual or self-fertilizing populations, and consequently they are abundant only in outcrossed species. In multicellular organisms, however, they are not restricted to over-replicating at meiosis, but can also exploit the distinction between somatic and germ-line cells. They are transmitted through mitosis to all the tissues of the developing individual, but in many cases are known to become less frequent or numerous in differentiated somatic tissues, whereas they accumulate in meristematic or germ-line cells that alone will give rise to gametes. This makes good sense from the point of view of an autoselected genomic parasite: in somatic cells it would only be a hindrance, injuring the development and activity of its host, and thereby retarding the host reproduction on which its own transmission depends. A similar but more dramatic phenomenon occurs in ciliates, where parasitic genetic elements are excised from the sexual micronuclear genome before the vegetative macronuclear genome is formed.

## 121. *Transposable elements may be selected through the mutations they cause.*

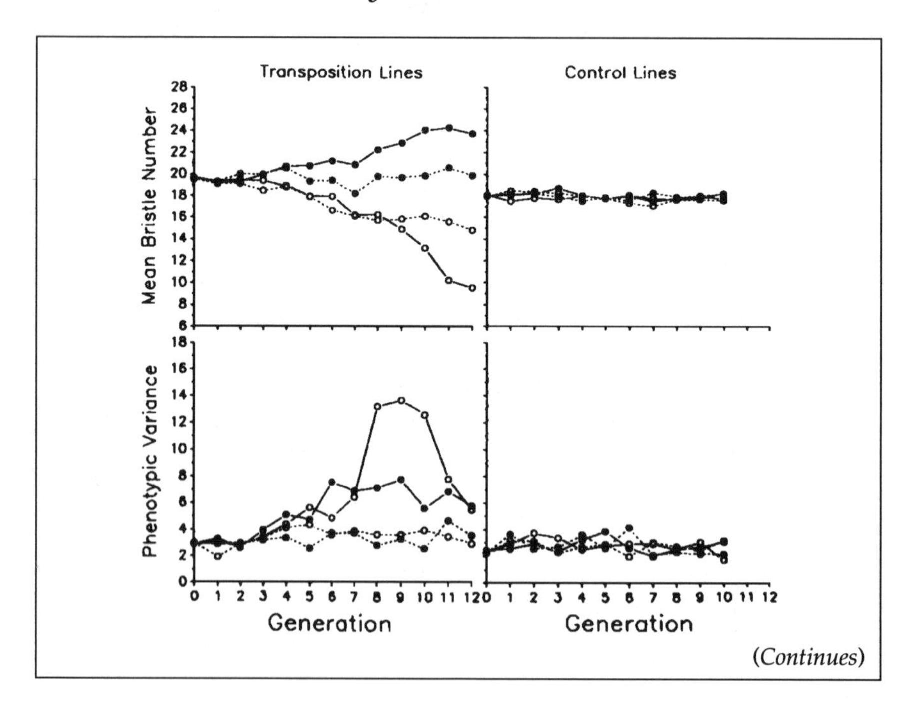


(*Continues*)

These graphs show how $P$-element transposition affects the response to artificial selection for bristle number in *Drosophila*, from Torkamenzahi et al. (1992). The two graphs on the right refer to inbred lines lacking $P$-elements, the two on the left to lines transformed by $P$ DNA. The lower pair of graphs give the phenotypic variance and the upper pair the mean of bristle number. The four plots on each graph are upward and downward selection lines founded from two base populations. The transposition lines are the more variable, and in three cases out of four respond to selection, whereas none of the control lines show any response. For related experiments, see Mackay (1985), Morton and Hall (1985), Torkamenzahi et al. (1988) and Pignatelli and Mackay (1989). The rescue of low-fitness lines of *Drosophila* through transposition was reported by Pasyukova et al. (1991).

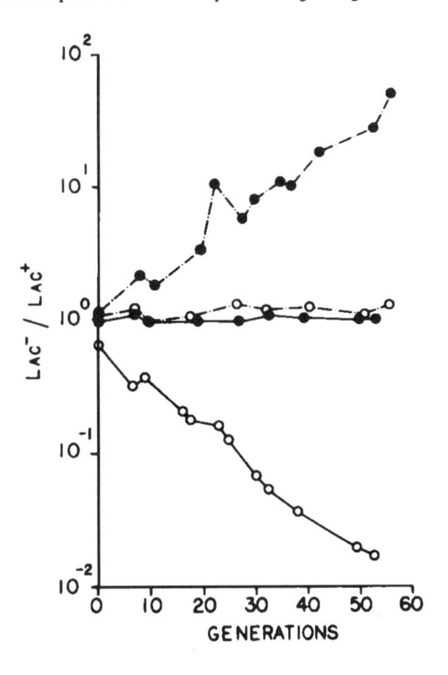


This figure demonstrates that new *IS*10 insertions increase fitness in *E. coli*, from Chao and McBroom (1985). The *IS*10 element is a component of the compound transposon *Tn*10, which consists of two *IS*10 elements flanking an internal sequence. Previous work by Chao et al. (1983) had shown that strains infected by *Tn*10 were competitively superior to uninfected strains if, and only if, they had experienced transposition by *IS*10 to a specific new site. Four strains were constructed, bearing all pairwise combinations of a selectable marker (*lac*$^-$ vs *lac*$^+$) and the presence or absence of the new *IS*10 insertion. The two middle plots show competition between two strains, both of which either do or do not bear

*(Continues)*

(*Continued*)

*IS*10; there is no appreciable change in frequency, demonstrating the neutality of the *lac* marker in glucose-limited conditions of growth. The strain bearing *IS*10 increases in frequency, relative to an isogenic strain, whether associated with $lac^-$ (upper plot) or with $lac^+$ (lower plot). The transposon *Tn*5 is likewise a compound structure, comprising two *IS*50 sequences and an internal region. The *IS*50 element increases host fitness even without transposition, possibly by increasing promoter activity (Biel and Hartl 1983, Hartl et al. 1983). Modi et al. (1992) set up chemostat populations of *E. coli* bearing a plasmid containing the transposon *Tn*3; after about 1000 generations, isolates from the evolved populations showed that the insertion of *Tn*3 into the bacterial genome was associated with substantial increases in fitness. Comparable results have been obtained with the *R*100 element (Condit 1990). The evolutionary genetics of the transposable elements of bacteria has been reviewed by Charleswoth and Charlesworth (1983), Charlesworth (1987), Charlesworth and Langley (1989) and Charlesworth et al. (1994). The *Ty* elements of yeast have been reviewed by Wilke et al. (1992); the experimental work summarized in the text is by Wilke and Adams (1983) and Adams and Oeller (1986).

Several kinds of genetic element are capable of moving from place to place in the genome, with or without replicating as they do so. Some are DNA creatures that move without replicating, but which may increase in numbers when they move from a region that has already been replicated to a region that has not yet been replicated; others produce RNA transcripts that can be used to make a new DNA copy that is inserted elsewhere in the genome. I shall call them all transposons, although different kinds are often distinguished by different names. There may be several or many copies of a given element in the genome. Copy number often varies among individuals and may increase within an asexual lineage through new transposition. It does not increase indefinitely, but is rather limited either by selection against clones with more transposon insertions than average, or by active regulation of copy number, either by the host genome or by the genome of the transposon itself. Transposons are akin to linear plasmids that have become incorporated with the host genome. Like plasmids, they are usually transmitted to all sexual progeny, through some kind of replicative transposition in the zygote. Bacterial transposons, like plasmids, often carry beneficial genes encoding functions useful to the organism, such as antibiotic resistance. Unlike plasmids, they often cause serious damage in eukaryotes, by inserting copies into functional genes, making it impossible to transcribe them accurately. Despite this, it is often argued that they are directly beneficial to the host individual or its lineage.

## *Specific Genetic Function*

Transposons might contribute to adaptedness in much the same way as ordinary genes, perhaps by acting as developmental regulators or switches. They would then be selected like any other gene. Despite the preference of some transposons for inserting into the upstream regulatory region of genes, the notion that they are part of a subtle system of gene control does not seem very compelling. Copy number and location vary among individuals but are substantially the same in different tissues of the same individual; if they behaved so as to regulate development, we would expect the opposite pattern. Transposons may be inactive in somatic cells, or even excluded from somatic cell lines, like B-chromosomes. Moreover, individuals that lack transposons seem perfectly normal. The only evidence in favor of this interpretation is that bursts of transposition in certain stocks of *Drosophila* are accompanied by a substantial increase in fitness. These stocks, however, have been deliberately selected for *low* fitness under natural selection; they are very sick flies, perhaps because of transposon insertions. It is scarcely surprising that the disease is alleviated when the pathogen releases its hold.

## *Generation of Variation*

A much more popular proposal has been that the mutations induced by transposons are beneficial in the long term, by permitting a more rapid response to selection. Long-term selection is likely to be weak and will rarely be effective when there is countervailing selection among individuals in the short term. Moreover, the mutations induced by transposition, involving massive disruption of gene sequences, seem unlikely to be beneficial in any circumstances (although if the transposon subsequently excises itself, the reconstituted gene may have changed only slightly). Nevertheless, there is some evidence that populations with active transposons evolve more rapidly.

**Effect of *P*-Elements on Response to Selection.** Most of the experiments of this kind have been performed by Trudy Mackay of Edinburgh, Adam Torkamenzehi of Sydney, and their colleagues, using artificial selection on bristle number in *Drosophila*. The transposons involved are the *P*-elements, which are known to be powerful mutagens. Crossing a *P*-bearing male with a female lacking *P*-elements causes a burst of transposition in the offspring and, consequently, a great deal of transposon-induced mutation. This dysgenic effect persists for several generations, until the number of copies of the element has approached its equilibrium value. The reciprocal cross (*P*-bearing female with non-*P*-bearing male) causes far less transposition, and the offspring are normal. If lines from dysgenic and non-dysgenic crosses are selected, response is much faster, either upwards or downwards, in the dysgenic lines. Moreover, in the dysgenic lines a very rapid initial response seems to approach a limit after a few generations of selection, in contrast to the more gradual and nearly linear response of the non-dysgenic lines. At the same time, phenotypic variance in the dysgenic lines remains much higher than in the non-dysgenic lines, even when the response to selection has become very slow. This suggests that the transposon-induced mutations that cause variation in bristle number have deleterious pleiotropic effects on viability or fertility, so that further advance under artificial selection is prevented by natural selection acting in the opposite direction (see Secs. 50 and 63). Indeed, one gene that was identified as having a large effect in reducing bristle number in one of the downward selection lines from the dysgenic cross causes sterility in females and cannot be maintained as a homozygote.

An alternative procedure is to compare inbred lines that lack *P*-elements with the lines developed by crossing them with nearly isogenic lines into which *P*-elements have been introduced. This has the advantage that the control lines (lacking *P*-elements) are expected to have very little genetic variance and, thus, little or no response to selection in the short term. An experiment of this sort has given a clear demonstration of the effect of transposition on the rate of evolution. Genetic variance for bristle number was generated by transposition about 30 times as rapidly as would ordinarily be expected from spontaneous mutation. These lines responded both to upward and to downward selection, whereas the control lines showed no reponse in either direction. The response in the transposition lines was greater for downward selection, where it was roughly equivalent to the response expected in a heterogeneous outbred population. Again, however, a large part of the response to downward selection was caused by a single mutation, causing the loss of bristles from many parts of the body, which had deleterious pleiotropic effects. Females carrying this mutation had very low fertility and were rapidly lost from the population when artificial selection was suspended.

These experiments show that transposon activity may increase the rate of response to selection by generating mutational variation in fitness. The severely deleterious side-effects often associated with transposition mutagenesis, however, seem inconsistent with its playing a major role in adaptive evolution.

## *Hitch-Hiking*

Studies of the population biology of transposons in cultures of asexual microbes have shown that they can persist or spread as a consequence of being linked to the mutations they cause. This is obviously similar to the idea that they serve to increase the rate of response to selection. Hitch-hiking, however, involves only selection among lineages within a population, a simpler and more plausible process than selection among populations, those that bear transposons replacing those that do not.

**The *Ty* 1 Element of Yeast.** The yeast genome usually contains 20 or 30 copies of the retrotransposon (RNA transposon) *Ty* 1, whose evolutionary biology has been investigated by Julian Adams and his colleagues at Ann Arbor. When a culture is founded by a strain bearing a single copy of *Ty* 1, copy number increases through time, to an average of about three copies after about 1000 generations. Sex is never induced in these cultures and probably never occurs. Moreover, adaptation to glucose-limited chemostats seemed to be associated with changes in copy number and location. It appears that *Ty* 1 transposition somehow benefits the cell. The most direct way of investigating this possibility is to allow strains that have undergone different numbers of transpositions to compete against one another; if transposition is sometimes beneficial to the cell as a whole, then lineages that have undergone appropriate transpositions will be selected, and the class with no transpositions will disappear from the population. The base population can be constructed by taking an exceptional strain that lacks *Ty* 1 and transforming it with a plasmid containing an active copy of the transposon under the control of a galactose promoter. In its new host, the transposon can be activated by growth on galactose, when it scatters between 1 and 20 copies of itself through the genome in an initial burst of transposition. If the cells are grown on glucose, the transposon is quiescent, remains on the plasmid, and few or no transpositions occur. In this way, strains that have experienced different numbers of transposition events can be made. In pure culture, transposition is on average deleterious: the galactose-grown strains in which about a dozen transpositions have occurred reach asymptotic densities about 8% lower than those of the isogenic glucose-grown strains. It seems reasonable to infer that transposition has reduced metabolic efficiency, presumably by disrupting active genes. Lethal disruptions would not be detected by this assay, of course, so transposition is probably more damaging than these results suggest. Strains in which transposition had occurred are more variable, as one would expect; more surprisingly, although the majority are inferior to those in which no transposition had occurred, a few actually reach a higher final population density, suggesting that the mutations caused by transposition can occasionally increase the fitness of the host cell. This was tested by allowing the strains to compete in mixed culture for about 100 generations. The diversity of these mixed populations decreased through time, and the average copy number was less after selection than before. Nevertheless, *Ty* 1 was not eliminated through selection; instead, strains that had not undergone transposition were rare or absent by the end of the experiment.

**The *IS*10 Element of *E. coli*.** Somewhat similar results have been reported by Lin Chao of Northwestern University for the bacterial transposon *IS*10. Strains that bear the transposon usually replace strains without the transposon in chemostat competition experiments. Moreover, if the *IS*10 strain loses, the element has not transposed; if it is excised from a successful strain, the strain loses its competitive superiority. Unlike *Ty* 1, however, *IS*10 inserts into a specific region of the genome in successful strains. This location is not far from a locus previously known to affect fitness in glucose-limited chemostats, and it is possible that *IS*10 disrupts a regulatory sequence and causes constitutive expression of some uncharacterized enzyme. A similar explanation may explain the dynamics of *Ty* 1, and would be consistent with what we know of the early stages of metabolic adaptation to novel environments (Sec. 70).

**Transposons as Mutators.** These experiments seem to establish that transposons can be maintained in populations through selection of the beneficial mutations that they sometimes cause. They seem entirely consistent with similar experiments on the mutator genes of bacteria. An element, whether a gene or a transposon, may cause a wide range of mutations, most of which are lethal or deleterious, although a few are beneficial in a novel environment. The superior variants produced in this way spread through the population under selection, and as they do so, both the mutant gene and the agent responsible for its mutation increase in frequency together. Mutator genes will behave like this only in asexual populations, where they remain linked to the mutations they cause; transposons normally cause mutations by being inserted into a gene and are, thus, tightly bound to the mutant gene whether the population is sexual or asexual. This is a process of hitch-hiking. Although it will cause the spread of mutation-inducing elements, in appropriate circumstances, however, it does not follow that the function of transposons is to induce mutations. After all, nobody would argue that the function of polymerase genes is to act as mutators by producing defective enzymes. Despite the clarity of the experimental work, it may be mistaken to interpret transposon structure as being specifically adapted to cause beneficial mutations in the host.

One weakness of the mutator theory of transposons is that it does not explain how a rare transposon can spread. If transposon-bearing cells are sufficiently abundant, a beneficial mutation induced by transposition is likely to appear sooner or later, and as it increases in frequency the transposon must rise too. However, if such cells are initially rare, however, they are likely to be eliminated, through selection against the deleterious mutations caused by transposition. An unstable equilibrium of this kind has, indeed, been demonstrated for the *IS*10 element of *E. coli*. Hitch-hiking may contribute to the dynamics of established transposons, but is unlikely to be responsible for their initial establishment in the population.

## 122. *Transposons spread infectiously in sexual populations.*

A quite different interpretation of transposons is that they are genomic parasites that are specialized to exploit the sexual cycle of the host by over-replication, transmitting copies to all four products of meiosis. They should, therefore, be most abundant in outcrossed sexual populations. In asexual populations they will be eliminated by selection among clones, because the clones that harbor the most transposons will suffer the highest rates of deleterious mutation.

Transposons are not rapidly eliminated from asexual populations, and mass cultures of microbes often seem to retain them with little change in copy number or location for long periods of time. Cliff Zeyl and I could find no appreciable decline in the abundance of *TOC*1 or *Gulliver* (transposons are often more imaginatively named than ordinary genes) in selection lines of *Chlamydomonas* that had been perpetuated vegetatively for about 1000 generations. However, this is not a very powerful test of the hypothesis. When transposons are rather quiescent, with transposition rates of the same order as mutation rates, selection against clones with above-average numbers of transposons will be correspondingly weak.

**Rapid Spread of Transposons in Sexual Populations.** The *P* element of *Drosophila* is a DNA transposon that causes hybrid dysgenesis, a condition characterized by sterility and chromosomal abnormalities, in the offspring of crosses between males that carry the element and females that do not.

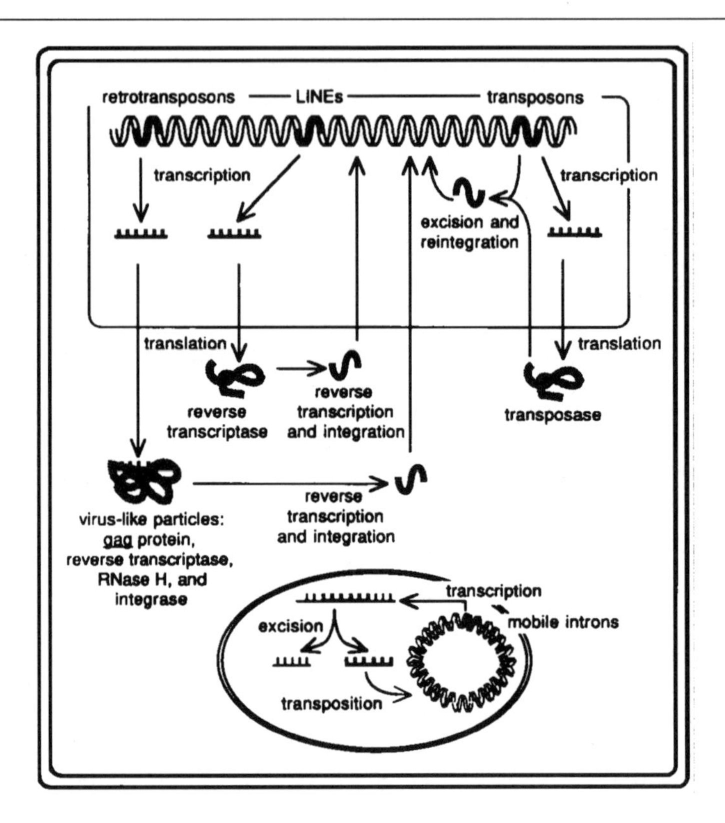


The nuclear and plastid genomes of eukaryotes harbour a host of mobile parasitic elements. Transposons (in the strict sense) are elements able to encode their own excision and re-insertion, so they move from site to site within the genome. If a transposon moves during DNA replication, from a site that has already been replicated to a site that has not yet been replicated, it creates two copies of the single initial element. Retrotransposons replicate more straightforwardly: the source copy is transcribed as RNA, and the transcript then inserted as DNA into another site. In structure and life cycle they are similar to retroviruses, except that they do not leave the cell. LINEs—Long Interspersed Elements—are broadly similar to retrotransposons, although they lack any detailed structural similarity to retroviruses. Mobile introns are found in the genomes of chloroplasts and mitochondria. A copy spliced from an RNA transcript can be inserted into the homologous site, when this is unoccupied. All of these kinds of element thus have the characteristic property of tending to increase in number within the genome.

(*Continues*)

(*Continued*)

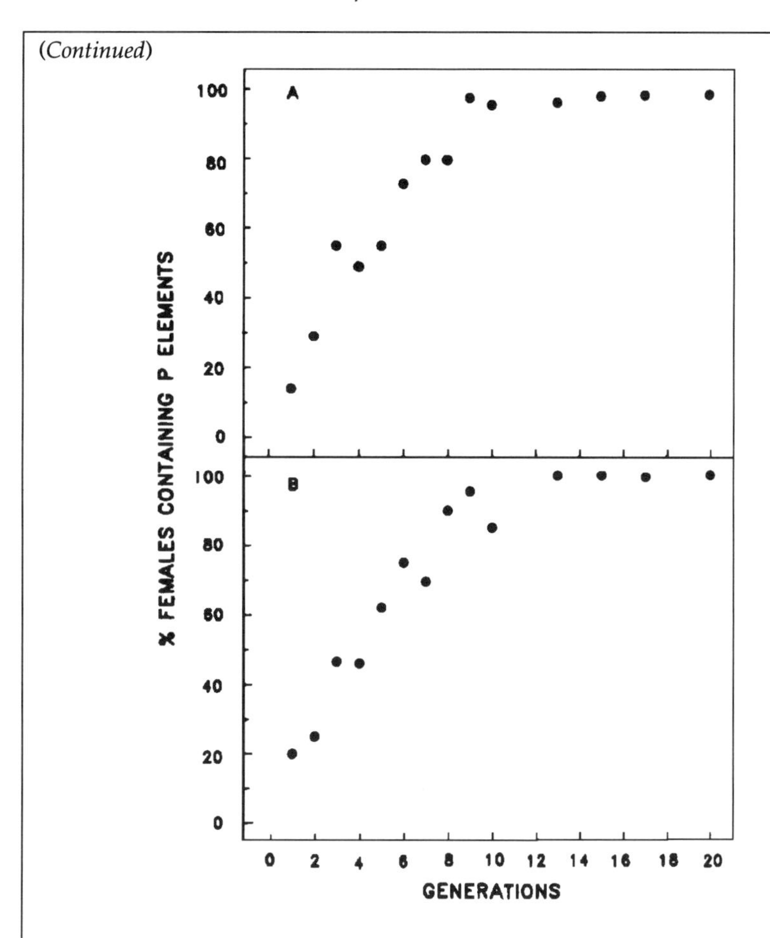


In asexual populations, transposons may be maintained stably, with little or no change in copy number or location, for hundreds or thousands of generations (Cameron et al. 1979, Zeyl et al. 1994). In sexual populations they can spread very rapidly. The figure shows the spread of *P* elements through experimental populations of *Drosophila* into which they had been introduced at a frequency of 0.05 (from Good et al. 1989). Note that the element nearly doubles in frequency each generation for the first few generations. Stevens and Wade (1990) have documented the rapid spread of a cytoplasmic element (probably a microbe) causing a dysgenic syndrome in laboratory populations of *Tribolium*.

It is present in all natural populations, so far as is known, but is absent from most laboratory lines that were founded and closed before about 1970. This curious fact is difficult to interpret except by a very recent, and very rapid, spread of *P* through the world population of *Drosophila melanogaster*. This can scarcely have taken place on such a scale, or at such a pace, if the *P* element merely increases the fitness of the flies that carry it. It must have been caused by replicative transmission, and the consequent infection of all sexual progeny of an infected fly and its uninfected partner. A group of Canadian scientists set out to imitate this process in the laboratory by establishing experimental populations with females from old, uninfected stocks, a few of which had been mated with *P*-bearing males. The result was unequivocal: the frequency of *P* elements among females increased from a few percent to 90% in 10 generations, and had reached fixation in 20. This can only have been an infectious process.

**Spread of Transposons in Sexual and Asexual Lines of Yeast.** The crucial issue is whether an active transposon will invade a sexual population, but not a comparable asexual population, when introduced at low frequency. The appropriate experiment is identical in concept with the study of the $2\mu$ circle of yeast, described earlier. Clifford Zeyl, working in my laboratory at McGill, took a transposon-free line of yeast and transformed it with a copy of *Ty*-3 linked to a $\beta$-galactosidase promotor. This created an infected line in which transposition could be induced (by growth on galactose) or repressed (by growth on glucose) at will. This line was then inoculated at low frequency into the isogenic uninfected line. The main result of the experiment, after eight sexual cycles, was quite clear: the transposon was lost from asexual lines, whereas it spread rapidly through sexual lines—but only if it was induced, and able to transpose. The only mechanism that could cause such rapid spread is the infectious transfer of the element to the genome of an uninfected sexual partner.

I have not attempted to conceal my preference for interpreting transposons, like transmissible plasmids and similar elements, as genome parasites that spread infectiously and are specialized for doing so. This is the natural interpretation of their structure and behavior, which is essentially that of an intracellular virus, and it explains how a new and rare transposon can spread through a population. Such elements may have other effects; in particular, if they are mutagenic then they may briefly increase the response to artificial selection, or spread through hitch-hiking. These, however, seem to be unselected consequences rather than adaptations. I expect this view to be strengthened as more experimental and comparative information becomes available in the next few years.

**Parasites of Transposons.** Defective copies of transposons are very common in eukaryotic genomes. In the limit, they may consist of little more than the repeated sequences that flank the central coding region of the active element. Although they are unable to produce a transposase, they can replicate and transpose, as long as they are accompanied by at least one active copy of the same element. They are therefore akin to defective interfering particles such as the evolved versions of Q$\beta$, living as parasites on the parasites.

## 4.B. Elements That Modify Existing Modes of Transmission

### 123. *The sexual cycle is often modified by elements able to elevate their own rate of transmission.*

Different features of the genetic system can be exploited by different kinds of parasitic elements. The genetic system, however, is not a fixed constraint; it may vary and evolve like any other character. The genes that encode the genetic system do not affect the phenotype directly but, rather, determine how it is inherited: with greater or less precision, by vegetative or sexual transmission, on nuclear or plastid genomes, and so forth. Selection may act on the genetic system itself in either of two ways. Firstly, any element that exploits some feature of the genetic system may be more effectively transmitted if it can modify or exaggerate that feature appropriately. Secondly, any alteration of the genetic system necessarily alters the nature or the rate of production of progeny genomes, and selection may act on individuals or lineages so as to favor one system over another.

**The Vulnerability of the Sexual Cycle.** The sexual cycle is much more vulnerable than the reproductive cycle to the first-order autoselection of genome parasites. There are two reasons for this. Firstly, the incomplete genome replication of the sexual cycle provides an opportunity for parasitic elements able to over-replicate; elements that over-replicate during mitosis will tend to be eliminated by selection among clones. Secondly, elements that modify the sexual cycle are unlinked to some extent from any deleterious effects that they have on the rest of the genome, whereas elements that modify clonal growth will be associated with the damage they cause. This is why most of the sections to follow are concerned with the induction, suppression, or manipulation of sexuality.

## 124. *Elements that encode sexual fusion will invade asexual populations.*

I have interpreted mobilizable plasmids as parasites of the sexual system of bacteria. Sexual fusion itself (bacterial conjugation) is induced by similar elements, the conjugative plasmids. The most familiar example is the $F$-plasmid of *E. coli*. In certain conditions (usually when the cells are starved) an $F^+$ cell (bearing an $F$-plasmid) will adhere to an $F^-$ cell (lacking an $F$-plasmid), forming a narrow cytoplasmic bridge between the two. A copy of the $F$-plasmid passes down this bridge, along with some part of the genome of the $F^+$ cell. When the two cells separate, the $F^-$ cell has become infected by the plasmid, and is now $F^+$, while part of its genome has been replaced by the homologous region from its partner. Conjugation thereby causes both infection and recombination. The production of recombinant genomes in this way may have profound evolutionary consequences, and conjugative plasmids may be selected in part through the properties of the recombinant genomes with which they are associated. The best evidence for this is that the plasmid is sometimes integrated into the host genome, where it will continue to elicit fusion, but is very rarely itself transmitted. Nevertheless, the most straightforward interpretation of conjugative plasmids is that like more conventional parasites they have evolved to encode the behavior that promotes their own transmission. However, I do not know of any attempt to demonstrate experimentally that $F$-plasmids will invade an asexual population of bacteria. Natural populations are predominantly $F^-$, suggesting that the plasmid severely reduces the fitness of the host genome, but the constraints on the spread of conjugative plasmids have yet to be investigated.

Nothing strictly comparable to conjugative plasmids occurs in eukaryotes, although an element in the mitochondrial genome of slime molds has been identified that behaves in a similar way, by causing mitochondrial fusion within the cell. I have suggested that the genes that determine mating type in eukaryotic microbes are the descendants of parasitic elements, encoding cell fusion, that have become integrated into the genome; but this is not yet widely accepted.

## 125. *In many sexual populations, autoselection favors genes that suppress sex.*

Sex usually reduces the short-term rate of replication to some extent, because the complicated processes of cell fusion, chromosome rearrangement, and meiosis consume time and material that could otherwise be used for growth. In multicellular organisms that have highly differentiated male and female gametes, sex greatly retards vegetative proliferation. Imagine a population that consists of (say) 50 males and 50 females; the argument works just as well for outcrossed hermaphrodites, but is slightly more complicated. If each inseminated female produces 10 offspring, then the population as a whole produces a total of 500 new individuals. But a comparable population that consisted of 100 parthenogenetic females, whose eggs did not need to be fertilized, would produce 1000 new individuals. An asexual lineage that produces unreduced spores by mitosis will therefore proliferate twice as fast as a sexual lineage that produces reduced gametes by meiosis. Consequently, any element that suppresses meiosis, and allows offspring to develop without fertilization, will tend to spread through a sexual population; indeed, as long as it is rare it will double in frequency in every generation. This simple conclusion will hold whenever the male contribution to the offspring is physiologically negligible.

It is disturbing that so simple an argument leads to the conclusion that outcrossed sexual systems with differentiated male and female gametes should not exist. It is always possible to argue that sex is so deeply implicated with reproduction in organisms, such as mammals and birds, that parthenogenetic variants never arise. This is not a very convincing assertion; there are parthenogenetic beetles, fish, and lilies, so it does not seem likely that any general law forbids the development of complex multicellular creatures without sex. In any event, sex is only an occasional interruption for the large number of organisms whose life cycle consists for the most part of repeated episodes of asexual proliferation, and they could certainly abandon sex entirely.

The sexual cycle thus presents us with two puzzles about how selection acts. The first is the rarity of sex in microbes that lack differentiated gametes or fusion partners. Autoselection will favor the infectious spread of elements able to encode fusion; it is presumably opposed by natural selection among clones, but we do not know how. The second, which has received much more attention, is the prevalence of sex among organisms with differentiated gametes. Autoselection will favor the suppression of sex by genes that direct the production of unreduced eggs by mitosis. It is presumably opposed by natural selection that favors sexual lineages, but it has been difficult to say why sexual lineages should be superior. It is possible that they possess a long-term advantage through greater evolutionary flexibility (Sec. 117), but the two-fold advantage of asexual lineages in the short term strongly suggests that sex must be associated with some equally compelling short-term superiority.

The efforts to identify this short-term effect are discussed in Secs. 152 and 160.

## 126. *Some genes distort meiotic segregation so as to favor their own transmission.*

The segregation of chromosomes during meiosis is generally "fair", that is to say, Mendelian, in that the two alleles present at a heterozygous locus in the diploid genome are transmitted with equal frequency to the haploid products of meiosis. Meiosis, however, offers an opportunity for autoselection because any gene that is transmitted to a majority of haploids will necessarily tend to increase in frequency. Plasmids and transposons achieve this by over-replication, but genes that are replicated in concert with the rest of the genome can also enhance their own transmission. In most cases, they succeed in skewing transmission by destroying haploids that bear their allelic partner. This phenomenon is called *meiotic drive.*

### *Meiotic Drive of Nuclear Genes*

**The Segregation-Distorter of *Drosophila*.** The best-known example is the segregation distorter of *Drosophila*. It consists of a cluster of tightly linked genes, the most important of which are *SD*, the gene responsible for biased transmission, and *Rsp* (responder) that modulates and can oppose the effect of *SD*. In $SD/SD^+$ heterozygotes, sperm bearing the susceptible allele of the responder locus, $Rsp^s$, develop abnormally, with a syndrome of inadequate chromatin condensation, the failure of spermatids to become individual cells, and defective sperm maturation. Sperm bearing the insensitive allele $Rsp^i$ develop normally. If there is no recombination between the two loci in $SD\ Rsp^i/SD^+\ Rsp^s$ heterozygotes, then the $SD\ Rsp^i$ chromosome will spread because by destroying $SD^+\ Rsp^s$ sperm it is transmitted to 95% or more of the progeny. Thus, if *SD* and $Rsp^i$ occur in circumstances that prevent recombination between them (usually by being linked in the same inversion), autoselection will cause an increase in segregation distortion. *SD* chromosomes occur in most natural populations of *Drosophila*, but at frequencies of only a few percent. Their spread is checked by countervailing natural selection. The suppression of recombination near the *SD* locus causes the accumulation of deleterious recessive mutations under Muller's Ratchet (Sec. 52). When two flies bearing *SD* mate, the *SD/SD* homozygotes they produce are therefore likely to be inviable or sterile. Furthermore, males in which half the sperm are killed may simply have reduced virility; this is a more serious possibility in *Drosophila*, where the sperm are extremely large and few in number, than it would be in most other animals.

**The *t*-Locus of Mice.** A very similar system is found in mice, where a series of distorter genes acting on a responder occur in a region where inversions suppress crossing-over. About 25% of wild mice are heterozygous for a *t* haplotype. Males that are heterozygous for the *t* complex transmit it to 90% or more of their offspring; as with *SD*, however, the homozygotes are severely handicapped.

**The Spore-Killer of Neurospora.** Meiosis in the filamentous fungus *Neurospora* leads to the production of an ascus bearing eight haploid spores; these germinate to form a haploid but multinucleate mycelium. Spore-killer *Sk* genes kill the four spores from the non-*Sk* parent. *Sk* genes do not kill themselves: a cross between allelic killers yields eight viable spores. This self-protection seems to be attributable to some diffusible substance, since sensitive nuclei are rescued in giant multinucleate ascospores that also contain *Sk* nuclei. It is not known what limits the spread of *Sk*, although it is possible that *Sk* mycelia are vegetatively inferior.

## *The Darwinian Stability of Mendelian Transmission*

The list of elements that subvert the Mendelian rules could be lengthened; but it could not be lengthened indefinitely. The rules remain rules; despite the opportunities for disobedience, there are few transgressors. The stability of Mendelian inheritance, so often taken for granted, may be viewed instead as one of the least understood features of biology. One possibility is as follows. Meiotic segregation is often subverted by elements able to bias their own transmission. If successful, such elements become fixed, despite any opposing force of natural selection; but having been fixed, they then exert no net effect on segregation, and the Mendelian rules are restored. Only in those cases where there is some strongly antagonistic pleiotropic effects of transmission bias, as there is in the case of *Drosophila SD* or mouse *t*-alleles, do we observe any continuing departure from the Mendelian norm. From this perspective, segregation distortion is not exceptional, but commonplace, and the genome is choked with elements that were formerly selected through their ability to skew transmission in their own favor, but that now consent to Mendelian rules because their more pliable competitors have been eliminated. This is entirely speculative. We still lack any experimental demonstration of how selection operates to perpetuate the normal pattern of Mendelian inheritance.

### *The Uniparental Transmission of Plastids*

Chloroplasts and mitochondria are very often (but not always) transmitted to sexual progeny by only one parent; this parent is usually (but not always) the mother. This may cause no surprise, in view of the fact that the mother normally contributes an enormously larger quantity of material to the egg and might simply swamp any male contribution. However, there are too many exceptions to the rule for this lazy explanation to be accepted; in conifers, for example, it is the male that transmits the chloroplast. Moreover, the exclusion of plastids from sperm and pollen seems to be an active and presumably adapted process. The issue is most acute in microbes such as *Chlamydomonas*, where the plastids are usually (but not always) transmitted uniparentally, even though the fusing gametes are equal in size. In *Chlamydomonas reinhardtii*, the single chloroplast is transmitted to sexual progeny through the *mt+* parent, because the genome of the chloroplast from the *mt-* parent is destroyed by exonucleases shortly after gamete fusion. The mitochondria, on the other hand, are transmitted through the *mt-* parent. It is not known how this system is maintained under selection. One guess is that it avoids costly internecine strife between rival plastids battling for transmission to the haploid spores. Another is that there is a beneficial interaction between chloroplast genes and nuclear genes located near the mating-type region. Even the most elementary experiments, however, such as selecting for reversed transmission, have yet to be attempted.

## 127. *Genes that are transmitted by only one gamete gender may be selected to bias sexual development.*

The equal numbers of male and female offspring produced by organisms with chromosomal sex determination is a classical instance of Mendelian segregation. The response of the sex ratio to natural selection depends on how males and females interact so as to determine the fitness of the lineage as a whole; this is discussed in Sec. 167. The sex ratio will also be affected by the autoselection of segregation distorters. Grossly unbalanced sex ratios may be merely the consequence of segregation distortion, but they may be caused by the distorting element so as to enhance its transmission.

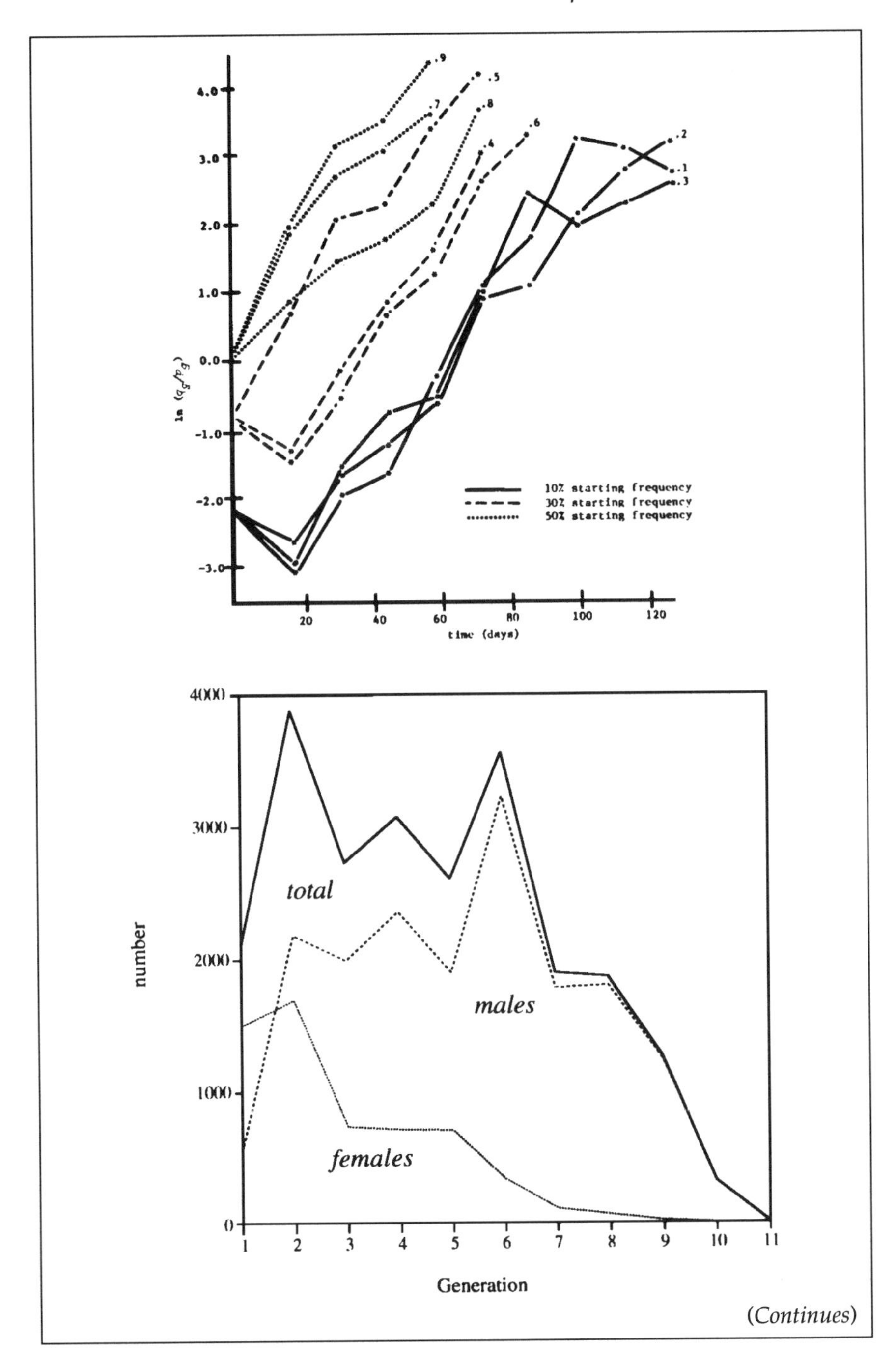
4.0
3.0
2.0
1.0
0.0
-1.0
-2.0
-3.0
ln $(q_g/p_g)$
20
40
60
80
100
120
time (days)
10% starting frequency
30% starting frequency
50% starting frequency
4000
3000
2000
1000
0
number
total
males
females
1
2
3
4
5
6
7
8
9
10
11
Generation

(Continues)

The upper graph shows the spread of Y chromosomes to which the *segregation-distorter* gene had been translocated through large cage populations of *Drosophila*, from Lyttle (1977). The lines represent three replicate populations at three initial frequencies of SD; note that the log frequency ratio increases more or less linearly over time until the SD; Y chromosome is effectively fixed. The generation time is about 12 days. The lower graph shows the concomitant changes in population size. Total population number is roughly conserved for the first six generations, but the number of females continually decreases until none remained in the 10th generation, and the population declines to extinction.

(*Continues*)

## *Meiotic Drive on Sex Chromosomes*

When genes that bias transmission are located on a sex chromosome, they will cause dramatic changes in the sex ratio of progeny. In many cases, broods may consist entirely of daughters or entirely of sons.

**Sex Ratio of *Drosophila*.** The *sex-ratio* gene *SR* of *Drosophila* is an X-linked segregation distorter. It acts by destroying Y-bearing spermatids during spermatogenesis; consequently, the offspring of *SR* males are all daughters. Its spread is opposed by natural selection, primarily because females homozygous for *SR* have only about half the viability of wild-type flies. There is also a long-term effect of *SR*, however, since if it were ever fixed there would be no males in the next generation, and the population would die out. It is possible that this plays a part in curtailing the spread of *SR* chromosomes.

**Sex Ratio Distortion in Aedes.** A similar distorter *D* is known from the mosquito *Aedes*, but is Y-linked: the offspring of *D* males are all sons. The *D* gene causes breaks in X chromatids. As *D* spreads through a population, the proportion of males increases, and thus the rate of reproduction falls, leading to the decline and eventually the extinction of the population. This has led to the use of *D* to control (that is, to eradicate) mosquito populations. The simplest scheme is to release very large numbers of *D*-bearing males, so that in the following year females (which alone take a blood meal) are scarce. More ingeniously still, it is possible to construct chromosomes that bear both *D* and a conditional lethal, for example, a mutation that is lethal only when it is very cold or when low concentrations of insecticide are present. This chromosome increases to high frequency in the population, which is then highly vulnerable to a cold snap or chemical treatment.

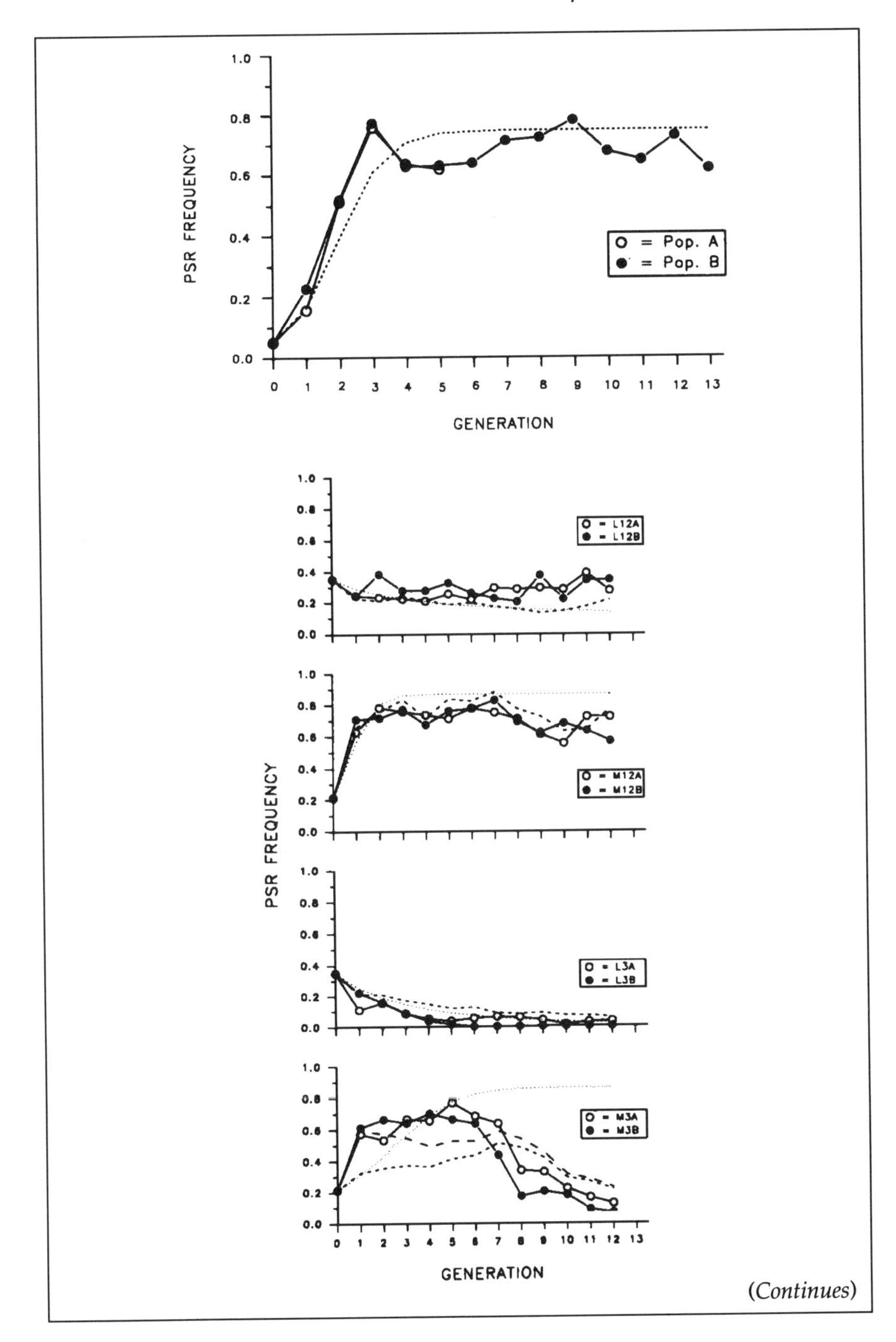
PSR FREQUENCY
GENERATION
O = Pop. A
● = Pop. B
O = L12A
● = L12B
O = M12A
● = M12B
O = L3A
● = L3B
O = M3A
● = M3B

(Continues)

Autoselection of the Paternal Sex Ratio (PSR) element of *Nasonia* has been studied experimentally by Beukeboom and Werren (1992); for a description of the element, see Werren et al. (1987). It is transmitted only through fertilized eggs that would normally develop into female offspring. Its dynamics therefore depend on the fraction of eggs that is fertilized; in fact, if that fraction is $f$, then PSR should move asymptotically toward a frequency of $(2 - f)/f$. Thus, if all eggs are fertilized, it will become fixed; if fewer than half are fertilized, it will be lost. The upper diagram shows the spread of PSR when it is introduced at low frequency into a large cage population of *Nasonia*. The dotted line is the predicted response to selection, given the rate of fertilization observed in the population; with a fertilization rate of about 0.8, the predicted frequency of PSR at equilibrium is 0.75, reasonably close to the observed value of 0.68. The frequency of the element in natural populations will depend in part on this powerful autoselection, and in part on two other processes.

a) The rate of fertilization will itself depend on the frequency of PSR. Moreover, it will vary with the presence of other genetic elements, such as Maternal Sex Ratio (MSR): its presence will catalyze the spread of PSR because it increases the supply of fertilizable females. These considerations anticipate the discussion of frequency-dependent selection in Part 5.B.

b) As PSR spreads, it has the same effect as a Y-acting segregation distorter: the frequency of males increases and the productivity of the population is likely to fall. Thus, natural selection among groups is antagonistic to autoselection within groups. The spread of PSR may be constrained by this antagonism when the population as a whole is broken up into a large number of breeding groups (as seems to be the case for *Nasonia* in nature) because groups with a high frequency of PSR will contribute fewer progeny to the population as a whole. This anticipates the discussion of group selection in Part 5.C.

These predictions were tested experimentally by setting up four types of population, founded from stocks infected (M) or uninfected (L) by MSR, and broken up into a large number of small groups (3, three foundress wasps for each group) or a smaller number of large groups (12, twelve foundresses per group). The lower graphs show the observed changes in two replicate populations of each type, compared with the predicted response, represented as before by a dotted line. The prediction is reasonably good for the L12 and L3 treatments: the element is maintained in the population with larger and fewer groups, although as expected its frequency is much lower than in the large undivided cage population and is strongly reduced or eliminated when there are more, smaller groups. The difference between L and M populations is also correctly predicted: PSR frequency is higher in populations infected with MSR. Although the observed response is reasonably close to prediction in M12, however, it is clearly far astray in M3. This was almost certainly because fertilization rates in the experimental populations diverged from those in control populations, on which the predictions were based.

(*Continues*)

(*Continued*)

The broken lines show attempts to update the predicted response using the actual fertilization rates in the experimental populations.

There is an extensive review of sex-ratio distorters in animals by Hurst (1993). The *Trichogramma* system is descibed by Stouthamer et al. (1990). There is a large literature on *Armadillidium*; for a recent discussion see Juchault and Legrand (1989). The dynamics of the X-linked *sex-ratio* element of *Drosophila* have been studied experimentally by Moriwaki and Kitagawa (1957), Ikeda (1970) and Fitz-Earle and Sakaguchi (1986).

**Dynamics of Pseudo-Y Distortion in *Drosophila*.** There is no natural Y-linked distorter in *Drosophila*, but Terrence Lyttle, working in Madison, manufactured one by translocating a part of the second chromosome containing the *segregation-distorter* locus to the Y-chromosome. In its new location, it is selected more effectively than when borne on an autosome, mainly because it is permanently heterozygous. In experimental populations, it behaves in much the same way as an allele with a two-fold to four-fold advantage in a haploid population, spreading rapidly when introduced at moderate frequency and becoming nearly fixed within a dozen or so generations. Because males bearing the translocation produce only sons, the frequency of females drops as the *SD* males become more abundant. This eventually brings about first the decline and then the extinction of the population. The system cannot be used straightforwardly to control natural populations, but it is an impressive laboratory demonstration of the effects of sex chromosome drive.

## *Infectious Agents That Control Sexual Development*

*SR* and *D* are nuclear gene whose effect on progeny sex ratio is incidental and, in fact, tends to cause countervailing natural selection. More interesting are elements outside the nucleus that not only cause changes in the gender or sexuality of offspring but that are autoselected because of these changes. Most of them are bacteria or, less often, eukaryotic microbes that are vertically transmitted in egg cytoplasm and are, therefore, selected to induce their host to develop as a female.

**Induction of Parthenogenesis in *Trichogramma*.** Several species of the small wasp *Trichogramma* are wholly asexual, producing unreduced eggs by mitosis. R. Stouthamer has made the remarkable discovery that these species will produce sexual males and females if treated with antibiotics. If the treatment is continued over several generations, the lineage becomes obligately sexual, like most hymenopterans. It seems almost certain that parthenogenetic development is induced by a bacterium transmitted in the egg. In a sexual species its host might be a male, and it can therefore enhance its own transmission by suppressing its host's sexuality and forcing it to develop as an asexual egg-producing individual. Several cases are now known in which sexuality can be induced in parthenogenetic organisms by treatment with antibiotics or prolonged exposure to high temperature, and the suppression of sex by intracellular microbes may be quite widespread.

**Feminizing Agents.** Other agents cause the individuals that bear them to develop as sexual females. These have been described from a wide range of organisms. One classic case, described early in this century by Alfred Vandel and recently studied in depth by P. Juchault, J.J. Legrand, and their colleagues at Nice, concerns the isopod *Armadillidium vulgare*. The nuclear sex determination system is female heterogamety: uninfected XY individuals develop as females. However, XX individuals that are genetically male develop as females if they are infected with a rickettsia-like bacterium. This is transmitted through egg cytoplasm, so that the broods of infected XX neofemales consist largely or entirely of daughters. The condition can be induced by tissue transplantation and cured by high temperature or antibiotics. Some cases of feminization, however, cannot be cured. They are thought to be caused by a transposable element transferred from the bacterial genome to a host X-chromosome that is thus converted into a functional Y. It is subsequently inherited in a more or less Mendelian fashion. These baroque genetics are typical of populations where sex determination is influenced by infectious elements and, indeed, the details of the system, involving not only the behavior of the microbe but also the response of its host, are much more complicated than this brief account suggests.

**Male Killers.** Infectious agents transmitted through the female line can be autoselected if instead of feminizing their host they kill its male offspring. This will increase their rate of transmission if maternal resources that would otherwise support the growth of male offspring are redirected to increase the number or quality of daughters. Such agents may act in female offspring to kill their male sibs. Alternatively, they may act in the males, committing suicide by killing their hosts, but in doing so benefitting copies of the same agent in the sisters of their hosts; this would be an example of kin selection (Sec. 147). The best-known case is the *sex ratio* condition of *Drosophila* that is caused by a spiroplasma that kills male offspring very early in embryogenesis. (This is not related to the *SR* system described previously.) The organism can be cultured outside the host and used to infect new hosts. When infected flies are introduced into laboratory populations, the agent may increase in frequency, but its spread is usually checked, and at equilibrium a nearly equal sex ratio is restored. The nature of the selection opposing the spread of the spiroplasma has not been identified, although at high population density infected females may be less fecund.

**Paternal Sex Ratio in *Nasonia*.** John Werren of Utah has described an unusual sex ratio distorter in *Nasonia*. This small wasp has the usual hymenopteran genetic system, in which fertilized diploid eggs develop as females, whereas unfertilized haploid eggs develop as males. It harbors several distorters, including *son killer* , a maternally inherited bacterium that is lethal to the male eggs of infected females, and *maternal sex ratio* , a feminizing cytoplasmic agent that also causes the production of all-female broods. *Paternal sex ratio* causes the production of all-male broods. It is not a microbe, but a B-chromosome that is transmitted solely through sperm. It acts by destroying all paternal chromosomes, except itself, in the zygote. This causes fertilized eggs to become haploid and, thus, to develop into males. These males then transmit the B-chromosome to all their progeny, whereas a diploid female would transmit it to only half her progeny. The total destruction of the genome with which an element is transmitted is a forceful example of the Darwinian logic of autoselection.

**Male-Sterile Elements in Plants.** Most angiosperms have hermaphroditic flowers that bear both male and female sexual structures. In male-sterile individuals the male structures are aborted and the flowers are female; the population then consists of a mixture of females and hermaphrodites. The genes responsible for male sterility are often (although not always, or exclusively) borne on the chloroplast genome. Chloroplasts are usually transmitted only through the female line: a plastid gene that suppresses male function will therefore be autoselected, provided that some of the resources otherwise committed to male function can be diverted to seed production.

# 5

# *Social Selection*

The processes of selection that I have described so far are atomistic. The characteristics that are selected are assumed to be expressed by isolated individuals, whose reproduction varies according to the relationship between their phenotype and the state of the external environment. The virtue of this approach is that it makes a clear distinction between the population that is undergoing selection and the environment in which it is selected. It is then possible to investigate selection and predict the course of evolution by studying isolated individuals, or by treating individuals as though they were isolated, and inferring the dynamics of populations from the aggregate properties of their members. This simple distinction between organism and environment, however, although clear, may be misleading. Individuals seldom, if ever, live wholly apart from all other organisms, passively transforming physical resources into offspring; rather, they live embedded in a community of other organisms, whose activities will affect their own. The *physical* environment, indeed, provides little more than the context for a much richer *biotic* environment, in which other individuals may be rivals, partners, dangers, or opportunities. These various interactions among individuals give rise to the *social selection* of attributes whose effect on fitness varies according to the composition of the community. The effect of social selection depends on the type of social interactions that are involved.

In the first place, reproduction will generally be affected by the *density* of the population. Types that have high fitness as solitary individuals will not necessarily retain their superiority when many individuals are growing together. Even when these neighbors are similar, in the sense of having similar effects on one another's growth, the relative fitness of different types may vary with population density. This gives rise to *density-dependent selection*.

Individuals of different kinds do not necessarily have similar effects on one another's growth; they may be either more or less severely affected by the presence of their own kind than by that of others. Reproduction will then depend on the composition of the population, giving rise to *frequency-dependent selection.*

When reproduction depends on the density or composition of the population, groups of individuals will have properties that are not displayed by isolated individuals of the same kinds. These social properties will be selected, regardless of whether the individuals concerned belong to the same lineage. Social interactions among individuals may thereby lead to competition between social groups, a process of *group selection.*

Finally, social interactions are not by any means restricted to individuals of the same species, or even to individuals following the same general way of life. On the contrary, quite unrelated organisms that are parasites, predators, or partners are a major component of the biotic environment. The selection mutually imposed by dissimilar species leads to a process of *coevolution.*

## 5.A. Selection Within a Single Uniform Population: Density-Dependent Selection

The biotic environment is not merely an additional source of selection. The social selection that it creates is different in kind to selection that is mediated by purely physical factors. The most important difference between biotic and physical factors is not the difference between living and inorganic agents but, rather, the difference between *depletable and non-depletable resources.*

The simplest examples of selection involve non-depletable resources, using the word in a very broad sense: the evolution of tolerance to heavy metals, adaptation to high temperature, the response to selection for high bristle number. In all of these cases, the modification of the population through selection has no effect on the environmental agent responsible for selection. When *Agrostis* evolves metal tolerance, the metal content of the soil does not change; *E. coli* is powerless to alter the temperature of the chemostat; the experimenter is unmoved by the hairiness of the flies. In such cases, adaptation can be understood by imagining each individual to be separately exposed to selection in its own small vial, being evaluated against some fixed standard and reproducing more or less successfully

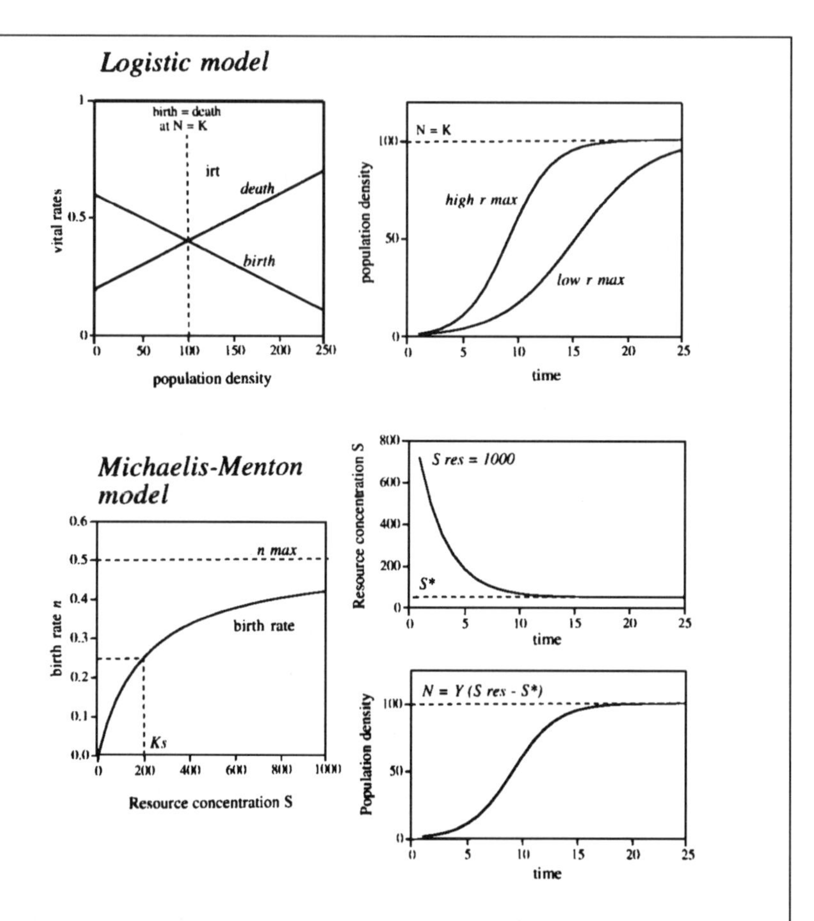


**Logistic Growth.** In exponential populations, such as batch cultures, population number $N$ grows as $dN/dt = rN$, so that $N_t = e^{rt} N_0$. The logarithmic or "instantaneous" rate of increase $r$ is the difference between the (logarithmic) birth-rate $n$ and the (logarithmic) death-rate $m$. The population will approach a steady state only if $r < 0$, implying $n < m$, when $N$ exceeds some critical value. The simplest assumption is that birth-rates and death-rates are linear functions of population density, $n = n_0 - \nu N$ and $m = m_0 + \mu N$. Substituting these forms into the exponential population equation, $dN/dt = [(n_0 - \nu N) - (m_0 + \mu N)]N$. By defining $r_{\max} = n_0 - m_0$ and $K = r/(\nu + \mu)$, this can be rewritten more concisely as $dN/dt = r_{\max} N(1 - N/K)$.

**Michaelis–Menton Growth.** In continuous cultures, the death-rate $m$ is the fraction of the population removed per unit time by outflow; a population that did not reproduce would have $dN/dt = -mN$. The birth-rate $n$ is a rate of division that falls as $N$ increases and the concentration of the limiting resource in the

(*Continues*)

(*Continued*)

nutrient medium $S$ falls. By analogy with enzyme kinetics, $n = n_{max}[S/(S + K_s)]$, where the half-saturation constant $K_s$ is the resource concentration at which $n = n_{max}/2$. A population that did not die would have $dN/dt = nN$. Hence, the overall rate of change in population number caused by birth and death is $dN/dt = (n - m)N = n_{max}N[S/(S + K_s)] - mN$. The quantity of resource in the vessel will change through time because it is augmented by input from the reservoir, but depleted by outflow and consumption: $dS/dt = mS_{res} - mS - rN/Y$, where $Y$ is a yield constant expressing the efficiency with which the resource is used by the organisms. When the culture is in steady-state there is no further change in resource concentration, and $S = S^* = mK_s/(r_{max} - m)$, where $r_{max} = n_{max} - m$; at this point, population density is $N = N^* = Y(S_{res} - S^*)$.

as a result. Many environmental factors are non-depletable in this sense. The temperature of a hot spring, the partial pressure of oxygen in the modern atmosphere, the pH of calcareous soil, the hydrostatic pressure in the deep ocean: all of these will be unaffected by whether or not a population succeeds in adapting to them.

Other kinds of resources are different, because they can potentially be used up faster than they can be supplied. When a few bacteria are inoculated into a vessel containing lactose as a carbon source, they will at first find themselves in a world of plenty and will proceed to increase rapidly in numbers. The type with the highest rate of increase will, of course, increase in frequency as well as in abundance. A larger population, however, has a larger appetite for lactose, and as the population increases the availability of lactose will fall. The same principle of selection will continue to hold: the type with the greatest rate of increase will spread. The type that proliferates the most rapidly when lactose is present in excess of the capacity of the population to metabolize it, however, is not necessarily the type that will be the most successful when the population has grown and lactose is scarce. When a resource is depletable, the environment will change through time, as a consequence of the adaptedness of the population, and the direction of selection may change as a result.

## Density Regulation

As long as the supply of resources is sufficient to support growth, the population will tend to increase exponentially, according to the simple law relating the future population $N_{t+1}$ to the present population $N_t$ as $N_{t+1} = e^r N_t$, where $r$ is the exponential rate of increase. If individuals are rare and widely scattered, then the rate of increase $r$ is as large as it can be, given the rate of supply of resources; call this value $r_{max}$.

As the population grows, the environment will deteriorate. It will become increasingly impoverished because the ration of depletable resources will fall, and it will become increasingly polluted because the concentration of toxic metabolites will increase. For either of these reasons, or for both, the rate of increase will decline, until at some point $r = 0$. The population is at equilibrium at this point, because any further addition of individuals will cause $r$ to become negative, so that the population declines, whereas any decrease in numbers will restore a positive value of $r$, so that the population increases. This process of density-regulation can be represented mathematically in any number of ways, but the two simplest depend on whether one is thinking in terms of serial transfer or continuous culture.

**Logistic Growth.** In the serial transfer of batch cultures, a small inoculum is allowed to grow for a period before the transfer of a sample to fresh medium. From any number of possible assumptions about how the increasing population density affects the growth rate, the simplest is that birth rates fall and death rates rise linearly with the logarithm of population density. The per-capita growth rate at any population $N$ is then $r = r_{\max}(1 - N/K)$, where $K$ is the maximum number of individuals that can be supported indefinitely in a given environment. This formulation is the so-called logistic equation that has been used extensively by ecologists.

**Michaelis–Menton Kinetics.** In continuous culture, a fresh medium is supplied continuously at a given rate. The rate of increase of the population will depend on the concentration of the limiting medium in the culture vessel $S$ that will be determined in part by the rate of inflow and in part by the population itself. By analogy with the Michaelis–Menton equation describing the kinetics of enzyme action, the per-capita rate of increase can be represented as $r = [n_{\max}S/(S + K_s)] - M$. In this more complicated formulation, $n_{\max}$ is the maximal rate of division at very low population density; $K_s$ is the so-called half-saturation constant, the concentration of substrate at which the rate of division is equal to half its maximal value $n_{\max}$; and $M$ is the dilution rate of the chemostat. The maximal rate of increase is $r_{\max} = n_{\max} - M$. This formulation tends to be used by experimentalists who work with microbes in chemostats. The parameter $r_{\max}$ has the same meaning as in the logistic formulation. It is unfortunate that $K_s$ and the $K$ of the logistic equation should look so similar; they stand for completely different quantities, but their usage is too deeply entrenched for the symbols to be changed. It should also be borne in mind that whereas the two parameters $r_{\max}$ and $K$ of the logistic are independent, the two parameters $r_{\max}$ and $K_s$ of the chemostat equation are not.

Either of these formulations leads to a partition of the rate of increase $r$ at a given population density $N$ into two components. The first expresses the maximal rate of increase, attained when the population is so sparse that it is not regulated by its own density. This introduces no new principle: if density can be neglected, types with greater values of $r_{max}$ will spread. The second expresses the way in which population growth rate is regulated through population density or through the decrease in the availability of resources caused by increasing population density. This introduces the new problem of how selection will act on $K$ or $K_s$ so as to maximize the realized rate of increase $r$.

## 128. *Relative fitness may change with population density.*

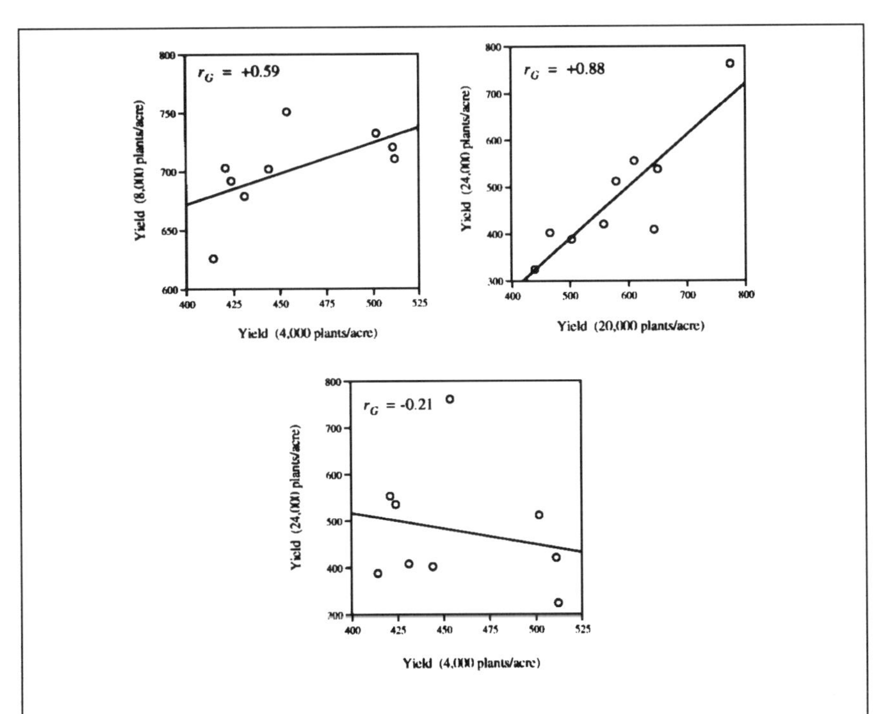


These figures show the yield of a series of single-cross corn hybrids varies with planting density, from data published by Lang et al. (1956; Table 1, low nitrogen treatment). Units of yield are bushels/acre. The upper two graphs show that

(*Continues*)

(*Continued*)
genotypes behave consistently at similar planting densities, whether low (4,000 plants/acre) or high (20,000 plants/acre); in either case, the genetic correlation is positive and moderately high. When very dissimilar densities (in this case, 4,000 vs 24,000 plants/acre) are compared, however, there is little if any consistent relationship between the yield of a genotype at one density and its yield at another. There are many comparable studies in the agronomic literature; for example, El-Lakany and Russell (1971, maize), Baker and Briggs (1982, barley) and Westermann and Crothers (1977, beans). The *Drosophila* experiment is by Lewontin (1955); other laboratory studies include Sokal and Sullivan (1963, house-fly) and Sokal and Huber (1963, *Tribolium*). Muellar and Ayala (1981, Fig. 1) show explicitly how the genetic correlation decays as the difference in density increases in *Drosophila*; c.f. Sec. 102.

If all genotypes respond in the same way to population density, then one will have the greatest rate of increase at all densities and will become fixed, no matter what the density or how density changes through time. Variation in density affects the outcome of selection only if it affects relative fitness, with some genotypes being superior at low density and others at high density. We can, then, treat density as if it were a physical factor and investigate the magnitude of the genotype-density interaction in experiments where a series of genotypes is scored at a number of different densities (see Sec. 102).

**Larval Viability in Drosophila.** Richard Lewontin reared from 1 to 40 larvae of 19 different inbred lines of *Drosophila* in standard culture vials. Survival was generally rather low at the highest densities, either because the larvae were starving or because they were being poisoned by their own waste products. More surprisingly, solitary larvae also fared poorly in many cases, their survival being lower than that of larvae growing up with one or a few companions: social interactions may be beneficial. The pattern of response to density, however, varied among the lines; some showed poor survival at low density, surviving best with 4 or 8 larvae per vial, whereas in other lines the solitary larvae survived well, and survival decreased regularly as density increased. This variation can be summarized as a genetic correlation of survival at different densities, in the same way that such genetic correlations can be calculated for other kinds of environmental variable. Lewontin presented his data as a table, each row being an inbred line and each column a density, so that the entry in each cell is the fraction of the larvae of a given strain surviving at a given density. The intraclass correlation coefficient is quite low, at about $t_G = 0.30$, but because the observations are unreplicated, this reflects stochastic variation from vial to vial, as well as genotype–environment interaction. The interclass genetic correlation is probably more informative; if we take all pairwise combinations of densities, the average genetic correlation of larval survival is $r_G = +0.39$, with a standard error of 0.06. The most interesting comparison is that between the density that was, on average, optimal (4 larvae per vial) and that which produced the lowest survival (40 larvae per vial): this correlation is close to zero. In short, genotypes seem to respond to larval density in this experiment much as they generally do to other features of the environment: there is substantial genotype–environment interaction, reflecting variation in relative fitness over environments, and if the environments are sufficiently different, the genotypes are uncorrelated.

**Crop Yield.** Agronomists have two reasons to be concerned about density regulation and, in particular, about how cultivars differ in their response to density. Firstly, new cultivars are often selected and tested as spaced plants, or in short rows, where their promise may not reflect their performance in commercial conditions, at higher densities in much larger plots. Secondly, the farmer wishes to plant as economically as possible. Increasing the planting rate will increase yield only up to a certain point; beyond that point, yield remains nearly constant, because more means worse: the mature plants are more numerous but individually less fruitful. Different cultivars, with different responses to density, may have different optimal planting rates. An experiment by a group of agronomists from the Urbana experimental station in Illinois will serve as an example. They planted nine corn hybrids at densities from 4,000 to 24,000 plants per acre, and then used a low, moderate, or high level of nitrate fertilizer. As the density of plants increased, the average weight of the cobs fell, and an increasing fraction of stems were barren. Total yield thus increased with planting rate to a certain point, but then fell. The density at which increased planting no longer causes increased yield depends on the amount of nitrate applied (corn has a very high demand for nitrogen): when little nitrate is applied, yield is maximized by planting about 12,000 plants per acre, but with heavy nitrate application the maximum yield is got from planting 20,000 plants per acre. This response, however, varies among varieties. The hybrid that yieldest most at 24,000 plants per acre was only a mediocre performer at low density, whereas the hybrid that was the poorest at high density was the second best in the group at the lowest density of 4,000 plants per acre. There is again an indication that the change in relative fitness with density is such that, if performance is compared at two extreme densities, genotypes are virtually uncorrelated.

## 129. *Characteristic genotypes evolve in starving populations.*

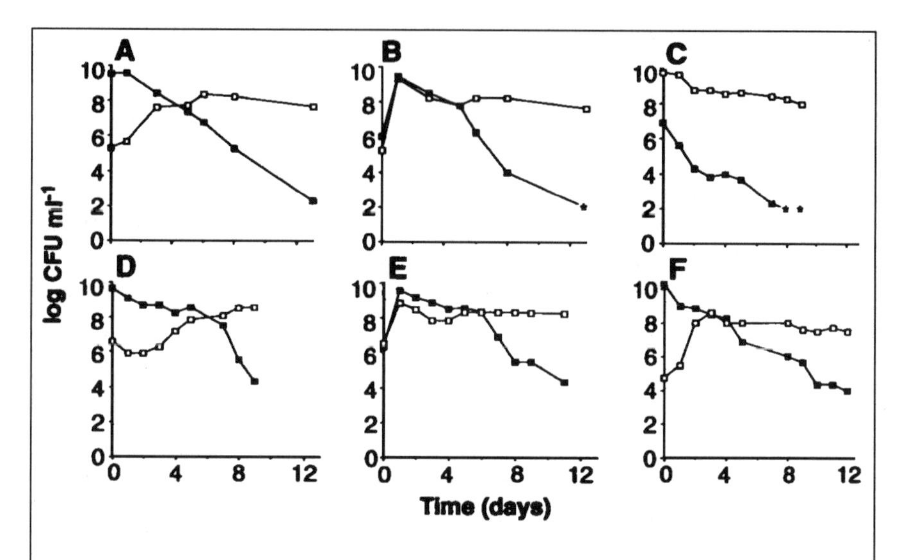


Selection in aged cultures of *E. coli*, from Zambrano et al. (1993):

a) Cells from an aged culture (open symbols) spread in young populations (solid symbols) when inoculated at low frequency.

b) When young and old populations are mixed in equal proportions, the young population is eliminated.

c) Young cultures (solid and open symbols represent two replicates) decline through time when cultured on their own.

d) The mutation *rpoS819* (open symbols) spreads in an isogenic *rpoS* (solid symbols) population when inoculated at low frequency. This is the mutation isolated from aged cultures and then inserted into a wild-type background.

e) The *rpoS819* mutation eliminates *rpoS* when both are initially equally frequent.

f) Cells from an aged culture of an *rpoS819* strain (open circles) spread in a population of the same strain from a young culture.

The analysis of selection in batch culture refers to Vasi et al. (1994).

If a culture of microbes is allowed to grow in a fixed volume of medium, it will at first increase rapidly, then grow more slowly, and finally cease growth altogether at some characteristic population density. The limiting nutrients are now almost completely exhausted, although there may be some recycling from dead cells. Nevertheless, cells may remain viable for long periods of time, often becoming characteristically specialized for a dormant or semi-dormant lifestyle. Cultures of *E. coli* can remain viable for several days or even weeks; I have got viable *Chlamydomonas* from liquid cultures that have sat forgotten in the laboratory for nearly two years. These stationary or declining populations live in highly stressful conditions of extreme famine and are usually regarded as being quite inert. Nevertheless, there are some signs of life. Long-neglected plates of *Chlamydomonas*, covered with a yellow or white field of dead or moribund cells occasionally show one or two green colonies that have regained some vitality. Recent work with bacteria shows that stationary populations are more dynamic than has previously been thought, and they evolve genotypes specifically adapted to a starvation regime.

**Stationary Cultures of *E. coli*.** A group of geneticists at Harvard Medical School led by Maria Mercedes Zambrano kept *E. coli* cultures for up to two weeks, long after they would normally have been discarded. These cultures enter stationary phase after about a day, when they are still viable but have stopped dividing. After three days, most of the cells are dead. A week later, however, the fraction of viable cells increases. Moreover, these cells are dividing; this can be proven by treating the cultures with an antibiotic that inhibits cell division but not cell growth, so that long filaments form in growing cultures. If a small quantity of these aged cultures is inoculated into a one-day-old stationary culture, the cells from the aged culture rapidly spread, eliminating the younger cells. This is not caused by any inhibitory effect of the aged culture medium itself, because young cells resuspended in a cell-free filtrate of an aged culture remain viable. Moreover, the superiority of the aged cells is maintained through several successive cycles of growth. Mutation in the stationary phase has led the selection of genotypes able to proliferate after the growth of the original strain has long since completely ceased.

A mutation responsible for this behavior has been identified. One of the proteins induced in the stationary phase, and responsible for continued viability, is a product of the *rpoS* gene. Viable cells from aged cultures bear a frameshift mutation in *rpoS* that causes the final four amino-acid residues at the 3′ end to be replaced by 39 new residues. When this mutant gene is transduced into young cells, it confers the ability to spread in stationary phase. Moreover, if the mutant cultures are themselves aged, further, unlinked mutations occur that enhance the ability to grow in stationary phase and that replace the original mutant strain in mixed cultures. Selection is not, then, confined to young, rapidly growing populations; the characteristic process of sequential substitution also occurs in old and numerically static cultures, causing the evolution of strains adapted to crowded conditions.

**Selection in Batch Culture.** In serial transfer experiments, the population repeatedly experiences an alternation between rich, uncrowded conditions and a depleted environment with a high population density. The overall increase in fitness that is routinely observed could be caused by adaptation to any of the four main density regimes of batch culture. The first is the lag that occurs when cells from an aging culture are transferred to fresh growth medium that they are temporarily unable to exploit. The second is the maximal rate of growth of induced cells at very low density. The third is the affinity of the cells for the substrate that will be important once the rise in population density has made resource acquisition competitive. The fourth is the death rate of starving cells in the exhausted medium, before transfer. The relative importance of each of these phases will depend on how the culture is cycled; however, it is easy to see that under normal operating conditions there will be much more opportunity for selection to act early in the growth cycle, by reducing the lag or increasing maximal growth rate. This turns out to be the case: the lag and maximal growth rate are substantially altered by selection, with little if any change in substrate affinity or death rate. Selection should modify the later stages of growth only if the cultures are maintained for long periods of time in stationary phase, or if very large inocula are used to transfer the cultures from cycle to cycle.

## 130. *The genotype able to subsist on the lowest ration prevails in density-regulated populations.*

It is evident that the maximal rate of increase at very low population density will not necessarily predict the outcome of selection when populations continue to grow to the point where they begin to become severely inhibited by their own density. It should again be emphasized that the types with the greatest rates of increase will spread through selection at any density; but the realized rate of increase is specific to a given density, and the outcome of selection will depend on $K$ or $K_s$, as well as on $r_{max}$.

### *Batch Culture*

A clone inoculated into a fixed volume of medium grows rapidly at first, with $r = r_{max}$, and then more slowly, until finally $r = 0$ and growth ceases.

At this point the population has reached its carrying capacity, with $N = K$. The individual ration of any resource is thus proportional to $1/K$. If two clones are inoculated simultaneously, they will begin growing at different rates, the type with the greater value of $r_{\max}$ increasing in frequency. After some time, the combined population of the two clones will attain the carrying capacity of the clone with the smaller value of $K$. This clone will therefore cease growing. The other clone continues to grow, further reducing the ration for the clone with lower $K$ and thus forcing it to decline in frequency. Eventually only the clone with the greater value of $K$ will remain. This simple argument is perfectly general and applies to any number of competing clones or genotypes: in a density-regulated population, the type with the greatest value of $K$ will become fixed. Another way of

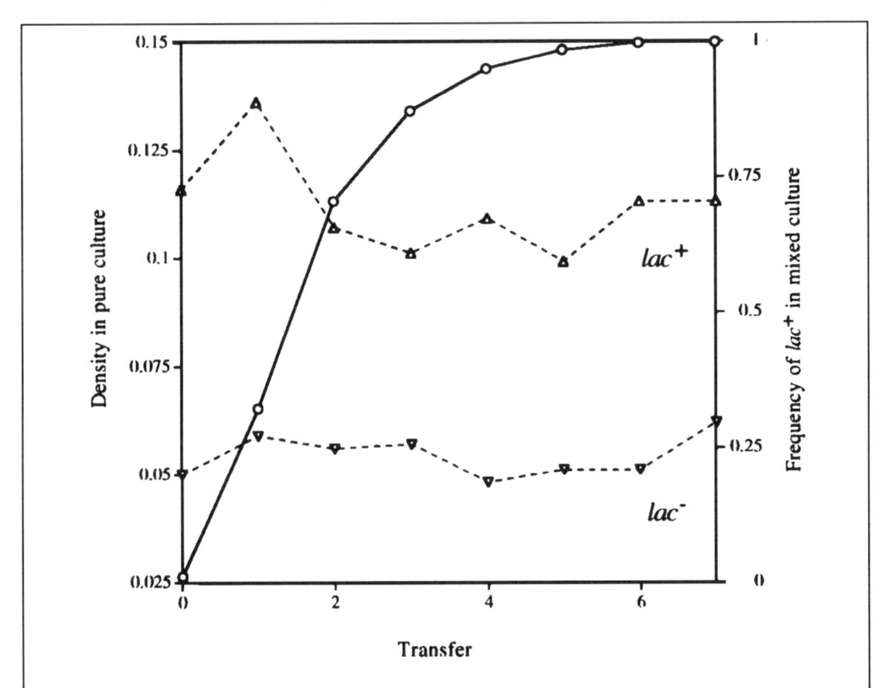


In the experiment by Smouse and Kosuda (1977), the carrying capacity of *lac*$^+$ strains of *E. coli* consistently exceeded that of *lac*$^-$ strains in pure culture (triangles and broken lines) in media containing lactose and other sugars. When the strains are mixed (circles and solid line), *lac*$^+$ eliminates *lac*$^-$ after about half a dozen transfers. Gause's work on *Paramecium* and other systems is summarized in his classic book (Gause 1934).

(*Continues*)

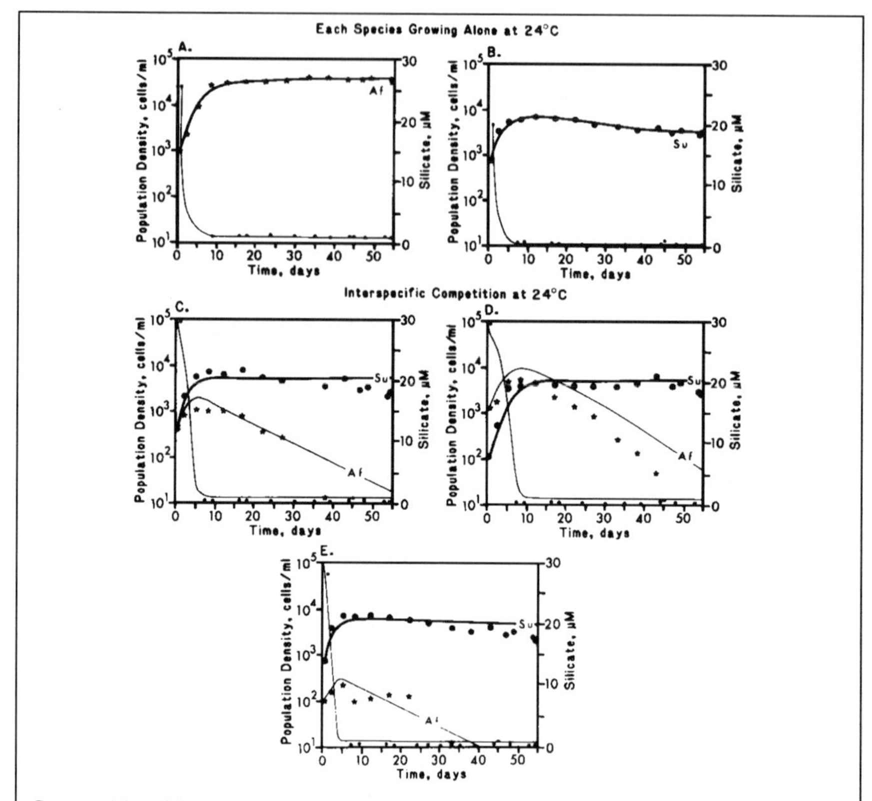


Competition between diatoms for silicate has been studied experimentally by Tilman (1977, 1981; Tilman et al. 1981; reviewed by Tilman 1982). The upper two graphs show the dynamics of population density and resource abundance in pure cultures of a) *Asterionella* (Af) and b) *Synedra* (Su). The lower three graphs show that *Asterionella* is eliminated through competition with *Synedra* in mixed cultures, regardless of initial frequency, as predicted from the resource dynamics of the pure cultures. The work on *Aphanizomenon* is by Zevenboom et al. (1981).

(*Continues*)

expressing this is that the successful clone will be that which can live on the lowest ration. Genetic variance in $r_{max}$ does not affect the outcome of selection, provided that the population is allowed to remain for long periods close to carrying capacity. It is required, however, that all types have the same effect on one another; that is, that any single individual depresses the per capita rate of growth of all types, including its own type, to the same degree. If this is not the case, selection will depend on frequency rather than (or as well as) on density.

(*Continued*)

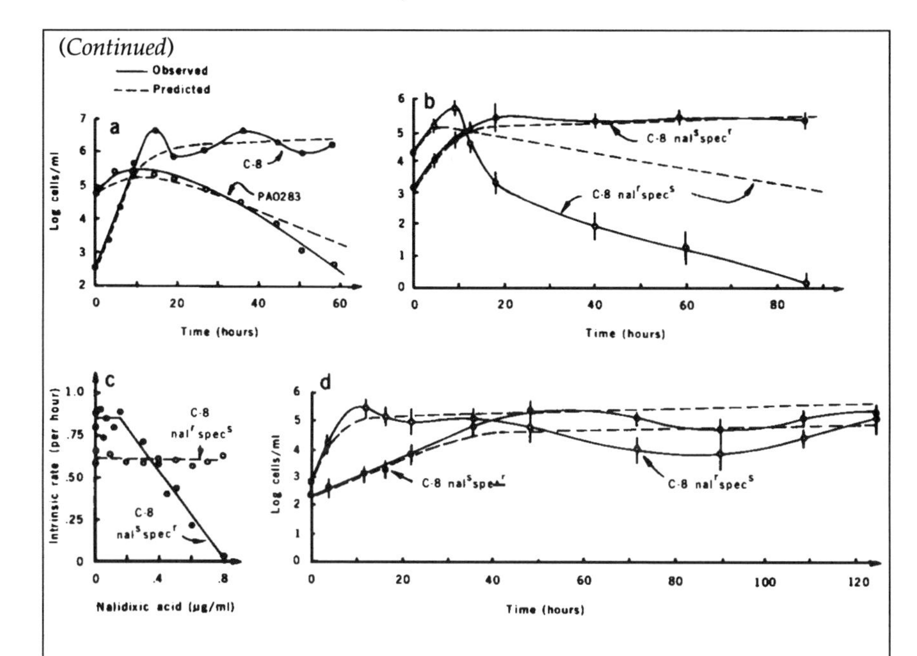


These figures show the outcome of the chemostat experiments by Hansen and Hubbell (1980).

a) *E. coli* (C8) replaces *Pseudomonas* (PAO283) in tryptophan-limited chemostats.

b) The $nal^S spec^R$ strain of *E. coli* has the greater $r_{\max}$, and eliminates the $nal^R$ $spec^S$ strain from mixed cultures.

c) The $r_{\max}$ of the $nal^S$ strain is reduced by nalidixic acid, such that at some concentation the $nal^S$ and $nal^R$ lines have the same $r_{\max}$.

d) At this critical concentration of nalidixic acid, the $nal^S$ and $nal^R$ lines coexist indefinitely.

**Classical Mixed-Species Competition Experiments.** The point of this simple theory is that it should be possible to predict the outcome of selection from the behavior of pure cultures: a genotype that grows to high density as a pure culture will supplant in mixed culture any genotype whose carrying capacity is lower, regardless of their maximal rates of increase and regardless of their initial frequencies. A great many experiments were conducted by ecologists to test this principle between the 1930s and the 1960s. These experiments are basically simple sorting procedures, in which the frequency of two species is followed through time. In most cases, however, two species are so different that the requirement that they have equal effects on one another is unlikely to be true. Perhaps the closest approach to a purely density-dependent process was G.F. Gause's famous *Paramecium* experiment. He cultured *Paramecium aurelia* and *Paramecium caudatum* separately in batch, replacing 10% of the medium daily. *P. aurelia* is the smaller animal (about 40% as large as *P. caudatum*) but is much more numerous at equilibrium. Consequently, the total volume of *P. aurelia* consistently exceeded that of *P. caudatum*. In mixed cultures, only *P. aurelia* persisted, with *P. caudatum* being driven down to very low frequencies after two or three weeks of competition. This is consistent with simple theory, although it is not a very critical test because *P. aurelia* also had the greater maximal rate of increase. The reason for the superiority of *P. aurelia* is not known; Gause himself suggested that it was more resistant to a toxin produced by the bacteria that were supplied as food.

**Lactose Variants in Lactose-Supplemented Chemostats.** Similar experiments involving different variants of the same species are not easy to find. Perhaps the point is too elementary. Peter Smouse of Michigan and Kazuhiko Kosuda of Josai University grew a $lac^-$ mutant of the K12 strain of *E. coli* in competition with the normal $lac^+$ type, in medium containing lactose, arabinose, and glucose. Not surprisingly, the $lac^+$ clone had the higher carrying capacity in pure culture, being able to utilize part of the medium that was not available to the $lac^-$ clone. In mixed culture, the $lac^+$ clone quickly became fixed, regardless of the concentration of sugars in the medium or the initial frequencies of the two types.

## Continuous Culture

The analysis of continuous cultures is slightly more complex. It has often been suggested that the type with the lower $K_s$, and therefore the greater affinity for the substrate, should prevail. There is no reason, however, to suppose that the population corresponding to the substrate concentration $K_s$ has any special significance. The correct procedure is to calculate the substrate concentration representing an equilibrium at which the population no longer increases and the substrate itself is no longer depleted. This critical concentration is $S^* = K_s(M/r_{\max})$. The dilution rate $M$ is set by the experimenter, but types may vary in $K_s$ or $r_{\max}$, or both, and thereby in the critical concentration $S^*$ at which they can no longer increase. The rule is similar to that for logistic growth: the type with the lowest value of $S^*$ can subsist on the smallest ration and will, therefore, exclude all other types. It should therefore be possible to predict the dynamics of mixed cultures from a knowledge of the kinetic properties of the component types grown in isolation.

**Competition Experiments with Diatoms.** Diatoms are unicellular photosynthetic protists that are very abundant in the open water of lakes. They construct a box-like test of silica and are often limited by the supply of silicate in nature. David Tilman of Minnesota grew *Synedra* and *Asterionella* in silicate-limited chemostats at a dilution rate of $M = 0.11$ per day. By estimating the $r_{max}$ and $K_s$ of pure cultures, the critical concentrations of silica were calculated to be $S^* = 1.0\ \mu M$ for *Synedra* and 2.8 $\mu$M for *Asterionella*. The actual concentrations of silica in the chemostats when the cultures had reached stationary phase were, in fact, lower than this (0.4 $\mu$M for *Synedra* and 1.0 $\mu$M for *Asterionella*), but in the same proportion. It can, therefore, be predicted straightforwardly that *Synedra* will displace *Asterionella* from mixed cultures, regardless of their initial frequencies, which is exactly what happened: mixtures of the two types reduced the concentration of silica in the medium to about 0.4 $\mu$M, at which *Synedra* can maintain itself but *Asterionella* cannot. The criterion that is used to predict this outcome is quite general; the particular outcome that is predicted, however, is specific to the conditions in which the experiment was run because when these are altered $r_{max}$ and $K_s$ are likely to change. In fact, the superiority of *Synedra* holds only above 20 C for the growth medium, light level, and dilution rate used in these experiments; below 20 C, *Asterionella* has the lower value of $S^*$, and, as expected, excludes *Synedra* from mixed cultures.

**A Spontaneous Mutant of a Cyanobacterium.** Cyanobacteria are photosynthetic prokaryotes. Some species that grow as filaments possess specialized cells, the heterocysts, that are able to fix nitrogen. W. Zevenboom and his colleagues were growing clonal cultures of *Aphanizomenon* in light-limited chemostats when they noticed that one culture had suddenly begun to increase in biomass, while at the same time changing morphologically. This turned out to be a spontaneous mutant that had lost the ability to develop heterocysts and therefore could not fix nitrogen, but that was able to grow at lower light levels in the chemostat. Therefore, it excluded the normal heterocyst-bearing type from chemostats where nitrate was supplied in excess but light was limiting, by virtue of its lower $S^*$ for light.

**Bacterial Competition Experiments.** Stephen Hansen and Stephen Hubbell of Iowa used bacterial systems to investigate competitive exclusion in chemostats. Strains of *E. coli* and *Pseudomonas* that are unable to synthesize tryptophan have similar $r_{max}$ but very different $K_s$ in tryptophan-limited chemostats. *E. coli* has the smaller $K_s$ and therefore the smaller $S^*$ and rapidly excludes *Pseudomonas* from mixed cultures. Selection in this strain of *E. coli* for resistance to nalidixic acid and streptomycin gave rise to two lines with distinctive resistance phenotypes that had very similar $K_s$ in tryptophan-limited chemostats, but quite different $r_{max}$. The selection line that had the greater $r_{max}$ necessarily had the lower $S^*$ and eliminated the other line from mixed cultures. The successful line happened to be the one that was sensitive to nalidixic acid. When nalidixic acid was added to the growth medium, both $r_{max}$ and $K_s$ of this strain were reduced, whereas the resistant strain was not appreciably affected. Nalidixic acid could thus be used to titrate the growth medium so that the two lines had very nearly the same $S^*$, although their $K_s$ and $r_{max}$ were widely different. When mixtures were cultured in these conditions, both types persisted in the chemostat, at roughly equal frequencies, for at least 100 hours.

The process of sorting in density-regulated populations is thus quite straightforward, as long as the competing strains have similar effects on one another. Density-dependent selection will favor the type with the smallest ration because it will reduce the concentration of the limiting nutrient below the point at which other types are able to grow. This is a familiar economic principle: in the same way, unrestricted competition among workers at a particular trade in a saturated economy will drive down the rate of wages to the lowest level at which subsistence is possible.

## 131. *Efficient and profligate resource use are antagonistic adaptations.*

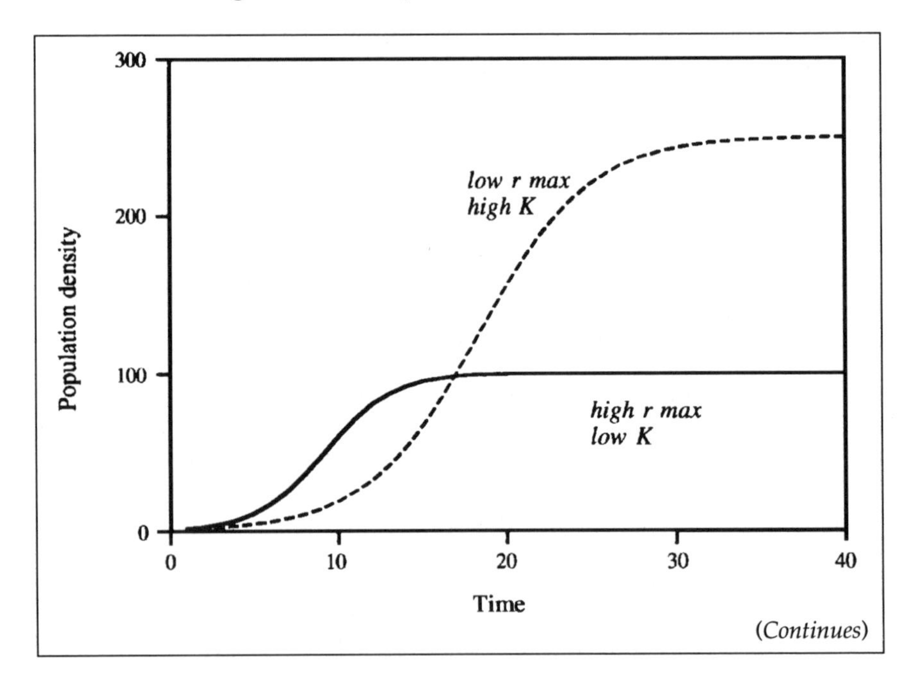


(*Continues*)

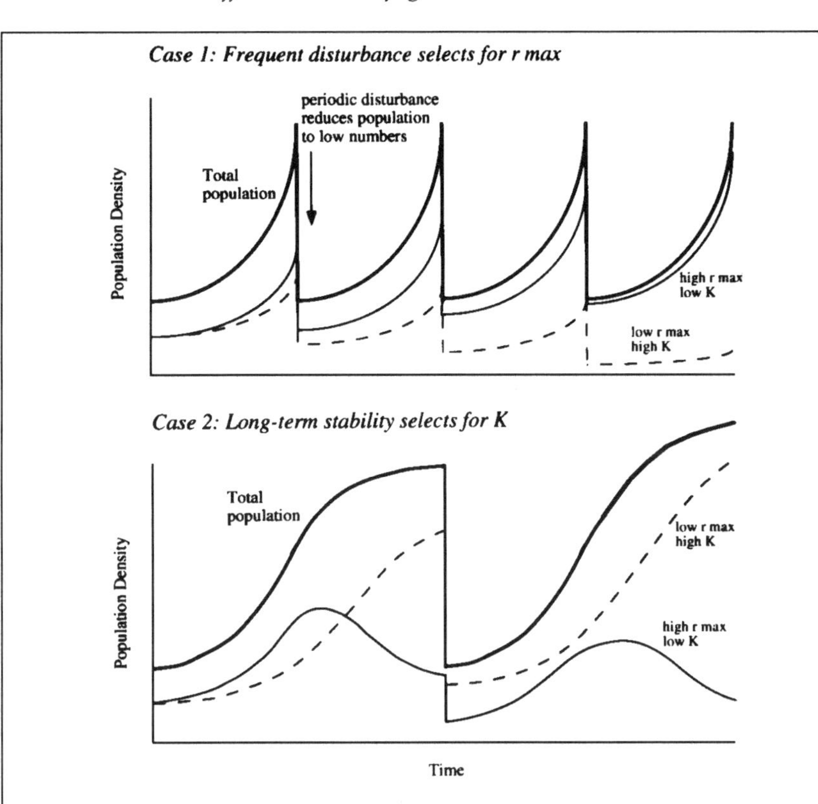


The outcome of competition between strains with different demographic characteristics depends on the frequency of disturbance.

a) Two strains when grown in isolation may exhibit contrasting demographies, one having a high maximal rate of increase $r_{max}$ but a low carrying capacity $K$, whereas the other increases slowly at low density but attains a greater density at equilibrium.

b) If the environment is disturbed frequently—the population as a whole being abruptly reduced in numbers—then the population is kept in a state of nearly exponential growth. The strain with the greater $r_{max}$ will then increase in frequency (refer back to Sec. 4).

c) If the environment is seldom disturbed, the population is density-regulated for most of the time. Total population density may then exceed the carrying capacity of the strain with lower $K$. In these circumstances, the strain with the higher $K$ will increase in frequency, despite its lower value of $r_{max}$.

*(Continues)*

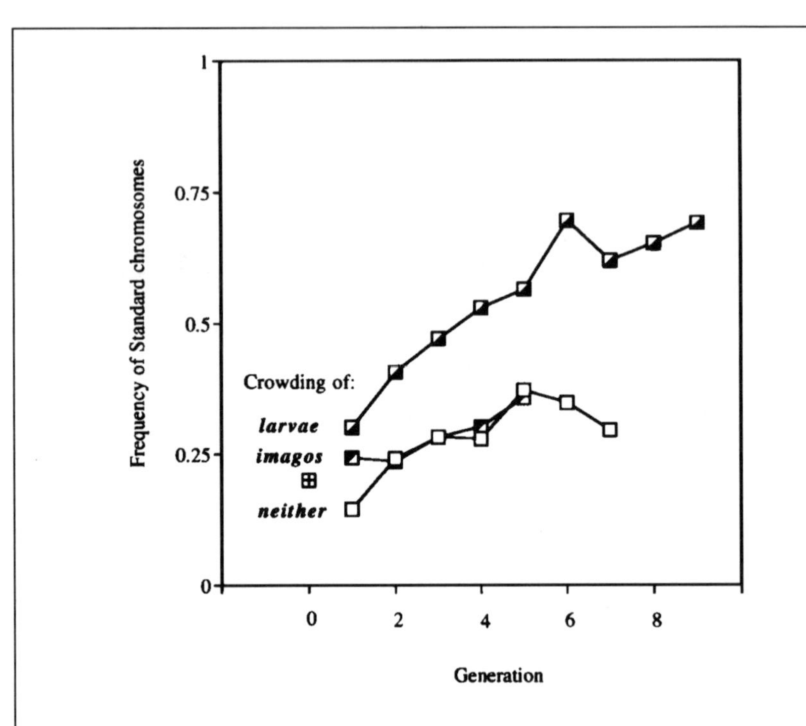


Birch (1955) found that the outcome of selection in mixtures of Standard and Chiricahua karyotypes depended on density; if the larvae were crowded, the population became about 70% Standard, and if uncrowded about 30% Standard. The base population (square with cross) was 20% Standard. For other experimental treatments of the effect of crowding on the outcome of selection in *Drosophila*, see Taylor and Condra (1980) and Barclay and Gregory (1981). The ciliate experiments are by Luckinbill (1979). For a background to $r$–$K$ selection theory, see Boyce (1984).

(*Continues*)

If a population is continually rarefied, so that reproduction is essentially unchecked by density, selection will favor the types that have the highest maximal rates of increase. On the other hand, in populations that are perennially close to carrying capacity, the types that can subsist on the smallest ration will be the most successful. The reproductive characteristics that evolve will thus depend on the demographic history of a population, which will in turn depend on the pattern of ecological change that it experiences. If the rate of supply of resources is nearly constant through time, as it is in a chemostat, the ability to reproduce at the lowest possible

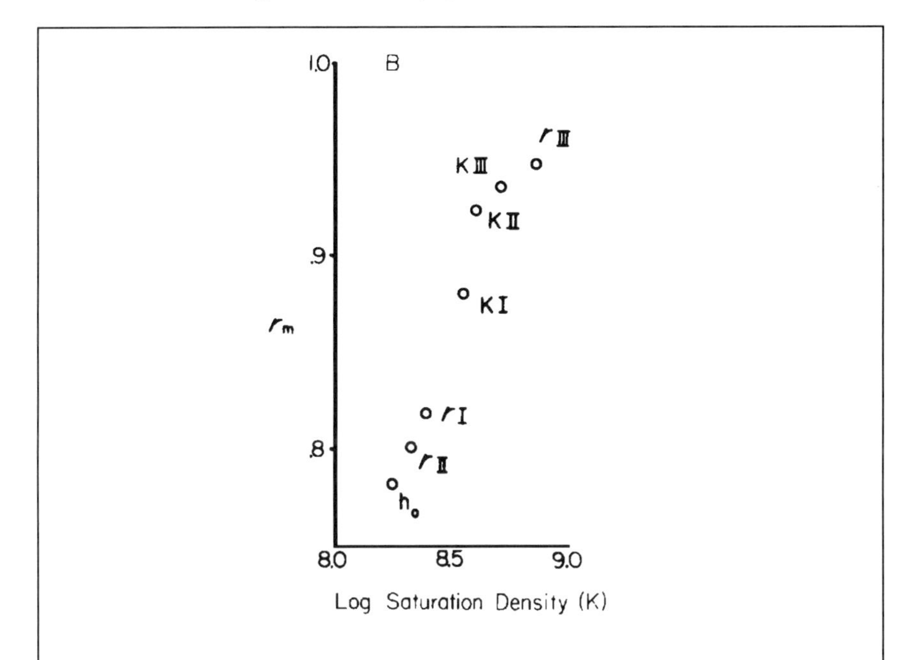


This figure, from Luckinbill (1984; see also Luckinbill 1978) shows the characteristics of novel mutants from $r$-selected (transferred at short intervals in log phase) and $K$-selected (transferred at longer intervals in stationary phase) lines of *E. coli*, when tested in pure culture. The base population is $h_0$; the $r$-selected lines are $r$I, $r$II, and $r$III; the $K$-selected lines are $K$I, $K$II, and $K$III. Note the strongly positive genetic correlation of $r$ and $K$. In mixed cultures of $r$-selected and $K$-selected lines, pure-culture performance reliably predicts the outcome of selection: any of the $K$ lines eliminates $r$I or $r$II, but $r$III eliminates any of the $K$ lines.

(*Continues*)

ration will be the crucial adaptation. If the rate of supply of resources fluctuates widely in time, as it does when batch cultures are transferred frequently, population density will often be small relative to resource concentration, and the rate of unrestricted growth will alone determine fitness. We therefore expect different kinds of adaptation in stable and in disturbed environments.

## r–K Selection

A theory of this kind was first advanced by Theodosius Dobzhansky, to contrast the expected outcome of selection in the relatively stable

(*Continued*)

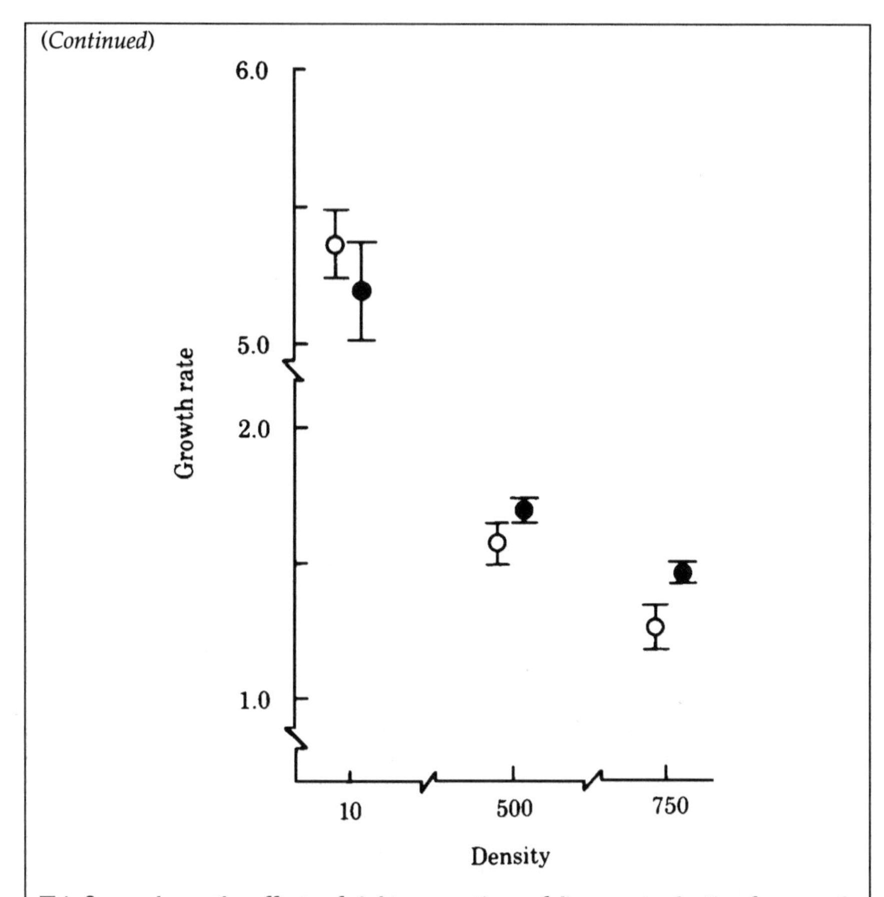


This figure shows the effects of eight generations of divergent selection for growth at high and low density in *Drosophila*, from Mueller and Ayala (1981): a modest degree of antagonism, in the expected direction, between maximal growth rate and population density in $r$-selected (open circles) and $K$-selected (solid circles) lines. For subsequent results in this system, see Bierbaum et al. (1989), Mueller (1991), and especially Mueller et al. (1991); also Joshi and Mueller (1988) on larval feeding rate, and Mueller and Sweet (1986) on pupation height. There is another large-scale experiment in the same general area by Bakker (1969).

economies of tropical regions with that in the more changeable conditions of temperate latitudes. It was later elaborated by Robert MacArthur, who, being an ecologist rather than a microbiologist, thought in terms of logistic population growth and distinguished selection acting through $r_{max}$ in disturbed environments from selection acting through $K$ in stable environments. Types that have high values of $r_{max}$ will make profligate use of abundant resources, being unchecked by any severe constraint on the rate at which resources can be acquired. Types that have high values of $K$, on the other hand, will use scarce resources as efficiently as possible, being able to subsist and reproduce on the lowest possible ration. This is an intuitively attractive dichotomy: when fuel is cheap, vehicles tend to be large, over-powered, and inefficient, but if it becomes expensive, these types are driven out of the market by smaller cars that get better mileage. It seems unlikely that adaptation will be indifferent to whether or not rates of uptake are a severe constraint on rates of reproduction, and so adaptations that increase $r_{max}$ will generally be antagonistic to those that increase $K$. This is the basic concept underlying the theory of *r–K selection*, leading to the conclusion that selection in density-independent and density-regulated populations will lead to characteristically different kinds of adaptations.

The term "$r$–$K$ selection" is, of course, a misnomer. Selection will always favor types with greater realized rates of increase $r$, and the contrast is properly between $r_{max}$ and $K$. Moreover, the dichotomy between populations that are unconstrained by resource supply and those that are perpetually starving, although useful to organize ideas, is too sharply drawn: most populations will spend most of the time at densities intermediate between zero and carrying capacity. Nevertheless, the realized rate of increase at densities that are kept low relative to resource supply may be more sensitive to changes in $r_{max}$ than to changes in $K$, whereas in populations that reproduce for much of the time at densities close to carrying capacity the reverse is likely to be true.

As an account of how selection will act differently on components of the realized rate of increase in different circumstances, the theory, as I have so far summarized it, is plausible and straightforward. It is, however, by no means straightforward to estimate the parameters of logistic growth in density-regulated populations in the field. An alternative approach is to argue that selection on $r_{max}$ will lead to a correlated response of other components of fitness that are easier to measure. In particular, density-independent rates of increase may be elevated by rapid development and the production early in life of a large number of small offspring. In density-regulated populations, on the other hand, selection for high values of $K$ is likely to favor long-lived, slowly developing types that produce a few large offspring. The extensive development of $r$–$K$ selection theory during the 1960s and 1970s thus became increasingly concerned with the interpretation of life histories and, in particular, with the interpretation of age-specific schedules of reproduction in different ecological circumstances. It therefore came into conflict with the parallel development of theories based on costs of reproduction that I have described in

Part 3.A. These cost-based theories could be formulated more rigorously and tested more easily, and as they became widely understood during the 1980s, the theory of $r$–$K$ selection, as an interpretation of life histories, was largely abandoned. Indeed, the two most authoritative recent accounts of the evolution of life-histories, by Derek Roff of McGill and Steve Stearns of Basel, relegate $r$–$K$ selection to a few paragraphs largely concerned with exposing its deficiencies.

In my view, this rejection is premature. The central idea that the profligate and the efficient use of resources are antagonistic is at least as clear and compelling as the notion that present and future reproduction are antagonistic. The two approaches are different: $r$–$K$ selection generally neglects age structure, just as cost-based theories neglect resource supply. However, they are not mutually exclusive. It is true that selection acting on schedules of reproduction may be independent of density; for example, earlier reproduction will be favored (other things being equal) whether or not the population is density regulated. This does not mean, however, that the partial effect on fitness of a change in the age at first reproduction, relative to the partial effect of a comparable change in some other feature of the life history, will be insensitive to resource supply or population density. The costs of reproduction on which the balance of selective forces acting on the life history depends derive eventually from a shortage of resources and will be modulated by population density. The converse is also true: the effects of resource supply on life-history characters will depend on the relationships between them. Moreover, these relationships may themselves change with population density. It is has often been found in crop trials that the correlation structure of components of yield changes with planting density: total yield may be almost independent of a particular component of yield at low density, yet highly correlated with it when the plants are more crowded. Density-dependent selection and age-specific selection should be regarded as complementary rather than competitive interpretations of life history.

## *r and K in Unselected Populations*

The estimation of $r_{max}$ and $K$ requires measuring populations rather than individuals, and has rarely been attempted on a large scale. So far as I know, the most extensive series of observations is supplied by the thousands of cultures of *Chlamydomonas* whose growth has been measured in my laboratory over the last few years. These include different species, different isolates of the same species, and the progeny of sexual crosses among unrelated strains, all cultured in liquid minimal medium without previous selection. There appears to be no strong or consistent correlation between $r_{max}$ and $K$ in these arbitrary populations.

## *Sorting of r and K in Random Populations*

**Artificial Selection for $r_{max}$ in Ciliates.** Leo Luckinbill of Wayne State grew replicated mixtures of four strains of *Paramecium*. After each growth cycle, the $r_{max}$ of each replicate was estimated, and those replicates with the highest values were subdivided and transferred to fresh medium. This procedure was successful in increasing $r_{max}$; however, it also increased $K$. When the strains were cultured separately, it was found that there was a positive correlation between $r_{max}$ and $K$ in the base population, and the effect of short-term selection had therefore been merely to fix the strain which had both the greatest $r_{max}$ and the greatest $K$. With so few strains being employed, this experiment is relatively uninformative.

**Natural Selection in *Drosophila*.** An experiment by Charles Taylor and Cindra Condra of Riverside illustrates the difficulty of testing $r$–$K$ selection theory independently of age-specific effects. They set up $K$-selection lines of freshly collected *Drosophila* stocks in population cages where food vials were renewed on a four-week cycle, causing obvious and severe crowding of the larvae. They felt it to be impracticable to set up comparable $r_{max}$-selection lines by rarefaction, and instead allowed only the first 100 flies emerging to lay eggs for two days on a relatively large amount of medium, so that the larvae were uncrowded. After about 10 months of selection, the life histories of the selection lines had diverged, when tested in a common environment. The $r_{max}$ lines developed more rapidly, and began oviposition about one day younger than females from the $K$ lines. This seemed consistent with theory, if early reproduction is crucial to increasing $r_{max}$. There was, however, little difference in the number of eggs produced at this time, and the $K$-line females actually produced more eggs later in life and lived longer. The earlier reproduction of the $r_{max}$ lines and the greater fecundity of the $K$ lines roughly balanced, so that both had about the same rate of increase overall. These results are difficult to understand in terms of density-dependent selection, but the design of the experiment introduced age-specific as well as density-dependent effects. By taking the first females to emerge and allowing them to lay eggs for only two days, the experimenters selected strongly for increased reproduction early in life. We know from other experiments that this is likely to have the pleiotropic effect of reducing expected fecundity in later life. Thus, any density-dependent effects that occurred were overshadowed by the response to age-specific selection.

**Age-Specificity of Density-Dependent Selection.** Some experiments have deliberately incorporated age-specific density effects. L.C. Birch, the well-known Sydney ecologist, set up an experiment while visiting Dobzhansky's laboratory to investigate the reasons for the seasonal fluctuation in the frequency of inversion types in some *Drosophila* populations. In population cages under normal conditions of culture, a mixture of Standard and Chiricahua karyotypes reaches an equilibrium of about 70% Standard, because heterozygotes are the fittest genotype, and Standard homozygotes are superior to Chiricahua homozygotes. If both adults and larvae are thinned, the relative fitness of the homozygotes reverses, and at equilibrium about 70% of the chromosomes are Chiricahua; thus, selection on the two types is density dependent. It probably acted through differences in mating or fecundity, rather than mortality. Thinning the larvae, but crowding the adults, gave the same result; but if the larvae were crowded and the adults thinned, the Standard type was again in a majority at equilibrium. Thus, Chiricahua is selected, not merely at low population density, but specifically at a low density of larvae.

Hugh Barclay and Patrick Gregory of Victoria measured the effects of larval and adult crowding on the life history. Their very complicated experiments, which I shall summarize very briefly, involved removing adults, larvae, or both, from the cultures at frequent intervals for the six months of selection, before assaying the selection lines and the controls (in which adults and larvae were both crowded) in a common environment. When both adults and larvae were removed, maintaining low densities of both, the lines evolved a shorter lifespan and greater early fecundity, which is consistent with density dependent selection. When only one of the two stages was removed, the experiments gave erratic results that are difficult to interpret in terms of either density-dependent or age-specific selection.

## *Continued Density-Dependent Selection*

The experiments that I have just described seem inconclusive; they seem to involve density-dependent effects, but because of limits of material or time, or because age-specific effects were predominant, they do not clearly demonstrate an antagonism between the components of realized rates of increase in density-regulated populations. Very few well-designed long-term experiments have been attempted.

**Spontaneous Mutants in Density-Regulated Bacterial Populations.** Perhaps the most elegant experiments are those reported by Leo Luckinbill, who selected *E. coli* populations in batch culture with glucose as a carbon source. $K$-selected lines were allowed to grow into stationary phase before being transferred; lines selected for $r_{max}$ were transferred more frequently, before the supply of glucose was exhausted. The base populations were mixtures of histidine auxotrophs and prototrophs; when the frequency of the auxotrophs changed abruptly, it signaled the passage of an advantageous mutation in the process of spreading through the population. This generally occurred after a few hundred generations of selection and was used as a signal that exactly one mutant adapted to frequent or infrequent transfer had become fixed. This mutant strain could then be isolated and tested against other strains, under both frequent and infrequent transfer.

The outcome of selection was unambiguous. The single-mutant selection lines invariably showed the appropriate direct response, the $r_{max}$ lines having a higher $r_{max}$ than the parental strain, and the $K$ lines a greater $K$. When tested in the environment of selection, the selection lines displaced the parental strain. There was, however, no evidence of any antagonism between $r_{max}$ and $K$; in every case, the selection lines were superior to the parental strain in both $r_{max}$ and $K$. Moreover, when the lines were mixed and allowed to compete for about 30 generations, the outcome of selection was independent of environment: when an $r_{max}$ strain and a $K$ strain competed, the same strain always won, regardless of whether transfers were frequent or infrequent. This was not an artefact of the particular system of batch culture used in the experiment. A second experiment that used semicontinuous culture in an anaerobic fermenting vessel gave the same result. Luckinbill concluded that his experiments gave no support to the theory that selection for $r_{max}$ and selection for $K$ gave rise to antagonistic adaptations.

The only flaw in this experiment is the deliberate attempt to select the first adaptive mutations that spread successfully. Glucose-limited liquid culture may be a novel environment for the strain used as a base population. The results can then be explained by supposing that the first advantageous mutations to appear are simply those that are unequivocally superior in this environment, with greater $r_{max}$ and greater $K$ than the unselected initial population. For this reason, the experiments do not seem to me to be fatal to the basic concept of $r$–$K$ selection.

**Long-Term Selection Lines of *Drosophila*.** The most extensive experimental study of density-dependent selection is that carried out over the last decade by Laurence Mueller of Irvine, with his colleagues and students.

His experimental material was derived from a set of strains, homozygous for the second chromosome, derived from recently captured wild flies. Growth at different densities was positively correlated among these strains: those that grew well at one density grew well at other densities, although this correlation was small when the strains were cultured at very different densities (Sec. 128). These sets were crossed, and the progeny selected divergently for $r_{max}$ and $K$. Lines selected for $r_{max}$ were rarefied by transferring a predetermined number of adults to fresh vials every week, thus keeping larval density quite low; in the $K$-selected lines, all surviving adults were transferred every four weeks, so that densities were much higher.

After eight generations, the lines were assayed after having been cultured in a common environment for two generations. At low density the $r_{max}$ lines had the greater rates of growth, although only by a small margin; at high density, the $K$ lines had appreciably faster growth. This is the expected result of sorting a diverse population in which there is initially little correlation between $r_{max}$ and $K$ at very low and very high densities.

The life history of the flies was studied later in the experiment, after about 20–40 generations of selection. There seemed to be no difference between the lines in adult survival or fecundity at any density, but larval characteristics had been modified: at high larval densities, the $K$-selected larvae survived better, although at low densities there seemed to be little difference between the lines. The $K$-selected larvae also grew into somewhat larger adults. The differences in population growth rate were thus generated largely through larval adaptations.

The nature of these adaptations was investigated subsequently, after more than 100 generations of selection. Resource competition among *Drosophila* larvae under laboratory conditions seems to be a fairly simple process: they pump the medium through themselves, and those that pump faster will grow faster and be more likely to survive. Larvae from the $K$ lines pump faster: they move their mouthparts in and out about 15% more quickly than larvae from the $r_{max}$ lines. They also behave differently at the end of larval life. *Drosophila* larvae may pupate directly on the surface of the food medium, or may first crawl some way up the wall of the vial. In uncrowded cultures, it is safe to pupate on the medium, and unnecessary to spend energy crawling up the vial. In crowded cultures, however, the medium becomes as wet and churned as a farmyard in November, and pupae on the surface of the medium are likely to drown. Even those that crawl a little way above the surface are likely to be dislodged by more active larvae crawling past them. In the $r_{max}$ lines, the average pupation height was about 5 mm, and 40% of the larvae pupated directly on the surface of the medium. In the $K$ lines, the larvae crawled nearly twice as high to pupate, and only 15% remained on the surface. Density-dependent selection seems to have modified larval feeding rates and pupation behavior to increase larval survival and adult size at high larval density.

These experiments demonstrate that characteristic adaptations evolve at high density. They do not directly demonstrate any antagonism between low-density and high-density adaptation, because there seemed to be little response to selection at low density, and in any case, the selection lines were not compared with the base population. A form of back selection was used to resolve this point. After about 200 generations of selection, the $r_{max}$ lines were divided, some remaining at low density, whereas others were transferred to the high-density conditions of the $K$ lines. After 25 generations, the growth-rate and adult productivity of the new $K$ lines had dropped, relative to the lines retained at low density. The low-density conditions therefore seem to require adaptations that are lost when selection is relaxed or changed, although it is not known what these are.

These experiments have succeeded in documenting density-dependent selection, by showing that lines selected at high and low density evolve divergent and antagonistic adaptations. However, they do not seem to be the adaptations anticipated by $r$–$K$ selection theory. The $K$-selected larvae feed faster, but they do not seem to feed more efficiently: larvae selected at low and at high density grow at about the same rate at any given food level. There is as yet little experimental support for any general principle of density-dependent selection.

## 5.B. Selection Within a Single Diverse Population: Frequency-Dependent Selection

### 132. *Genotypes themselves constitute environmental factors.*

Genotypes may sometimes have the same effect on one another, as I have so far assumed; but this must be a special case. It is more likely that some genotypes will have stronger and others weaker effects on their neighbors or on the individuals with whom they are temporarily interacting. Growth and reproduction will then depend on what kind of neighbors an individual lives alongside; neighbors are one of the features of a particular site, or way of life, and can be treated as an environmental factor like nutrient supply or temperature. The basic **abcd** model that was used to represent the response of genotypes to physical environments can be pressed into

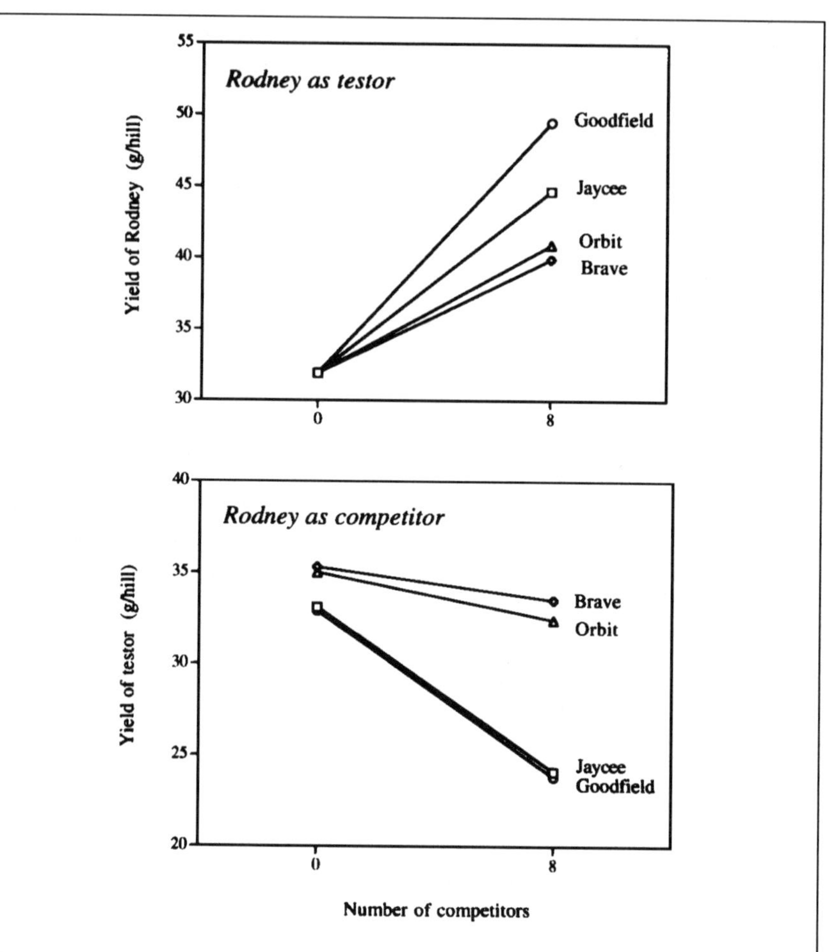


These two graphs illustrate some of the interactions among oat varieties described by Smith et al. (1970). The upper graph shows the yield of Rodney when grown with different kinds of neighbor on the same hill plot. There were four treatments, involving the growth of the central plant in a $3 \times 3$ hill with zero, two, four or eight neighbors of a different variety (thus, the zero level is eight neighbors of the same variety); the lines on the graph are linear regressions. In all four cases, a Rodney plant yields more when presented with different neighbors than when grown with other Rodney individuals. The lower graph shows the effect of Rodney on the yield of other varieties. There are many similar experiments with a variety of crops, for example Hinson and Hanson (1962, soybean), Doney et al. (1965, potatoes), Kannenberg and Hunter (1972, maize) and Beg et al. (1975, peanuts). The classic experiments with rice mentioned in Sec. 132 are by Roy (1961). Mather and

(*Continues*)

> *(Continued)*
> Caligari (1983) have developed a useful method for expressing the results of such experiments in terms of how individuals exert competitive effects on others, and in turn themselves respond to competition, and use it to interpret experiments with *Lolium* and *Drosophila*.

service to show how genotypes respond to one another.

| | | Neighbor | |
|---|---|---|---|
| | | Type 1 | Type 2 |
| Self | Type 1 | a | b |
| | Type 2 | c | d |

Thus, b is the performance of a Type 1 individual (as Self), when growing together with a Type 2 individual (as Neighbor). The important difference between this situation and the response to physical environment is that the relationship between Self and Neighbor is mutual—genotypes act as neighbors to one another. The clear distinction between genotype and environment that is the basis of classical genetic analysis begins to break down at this point; as it does so, the analysis itself becomes less apt to deal with the dynamics of the system. The rest of this chapter will deal with the new concepts that are necessary to understand the process of selection in self-referential systems where genotypes mutually influence one another.

**Stronger and Weaker Competitors.** A classical design in agronomy for testing the performance of cultivars is the bordered hill plot. A single self plant is grown in the middle of a small clump of neighbors, to simulate the commercial situation, in which it would be growing within a field of other plants. For example, a $3 \times 3$ arrangement provides a single central test plant with a border of eight others. This can be used to compare the effect of a border of similar plants with that of a border of different plants. For example, the arrangements

```
S S S      N N N
S S S      N S N
S S S      N N N
```

can be used to identify the environmental effect of similar and dissimilar neighbors on the performance of the central test plant S. Olin Smith and his colleagues at St Paul, Minnesota, used this design to investigate the effect of competition in mixed culture on the seed yield of five oat cultivars. They found that the yield of the test plant was substantially affected by the identity of its neighbors; some neighbors enhanced yield, relative to growth in pure culture, whereas others depressed it. One cultivar in particular (Rodney) grew much better in the company of other cultivars than it did when its neighbors were other Rodney plants. At the same time, other cultivars were suppressed by Rodney neighbors. In this sense, Rodney is a strong competitor, perhaps because it was the tallest of the cultivars tested and may have been able to shade its neighbors. Taking Rodney as type 1, we can express this in terms of the abcd model by saying that $a < b$ and $c < d$. This constitutes environmental variance, supplied by the ecological effect of different neighboring genotypes.

**Larval Viability in *Drosophila*.** The experiment in which Lewontin measured larval viability as a function of the density of pure cultures (Sec. 128) was supplemented by a second experiment in which the same inbred lines were reared as mixtures. The mixtures comprised equal numbers of the test strain and of a strain carrying the sex-linked mutation *white*. If genotypes have the same effect on one another, then the viability of the *white* larvae will be the same at a given total density, regardless of the genotype with which it is paired. This, however, was not the case: at any density, different strains had different effects on the viability of the *white* larvae. From the point of view of *white*, therefore, different neighbors represent consistently different environments.

## 133. *Neighbors affect relative fitness.*

Lewontin's experiment can be viewed from another perspective by comparing the viability of strains in pure culture with their viability when cultured with *white*. As an average over all strains, viability was about the same at a given total density, whether or not *white* were present. The strains, however, responded in different ways: in some, viability was enhanced by the presence of *white* (rather than the same number of flies of the same strain), whereas in others it was reduced. This is a special case of genotype–environment interaction. The two environments are the presence of other larvae of the same strain, and the presence of *white* larvae. Genotypes respond differently to these environments, so that performance in one is poorly correlated with performance in the other. Because the environments concerned are genotypes, we can symbolize the interaction, not as G × E, but rather as G × **G**, the **G** being the genotype of the neighbor acting as the environment.

**Interaction Among Rice Cultivars.** In Lewontin's experiment, the *white* strain was used as a neighbor for all the test strains. A still more interesting design is to use every strain both as self and as neighbor, so that all the mutual relationships among the strains can be described. Subodh Roy, of the Indian Statistical Institute at Calcutta, grew three rice cultivars in pairwise combinations in small hill plots. The cultivars do not have euphonious names, and can be referred to simply as 2A, BK, and RP. There were four plants on each plot, two of one sort and two of another, and Roy measured the seed yield of each of them. His results were thus the yield of a given type, when grown in combination with a given neighbor.

| | | Neighbor | | |
|---|---|---|---|---|
| | | 2A | BK | RP |
| | 2A | 570 | 574 | 743 |
| Self | BK | 717 | 553 | 822 |
| | RP | 76 | 148 | 321 |

There are two sorts of effect here. In the first place, there are stronger and weaker competitors. RP is clearly a weak competitor: the other cultivars perform well when they have RP as a neighbor, and RP itself performs poorly when its neighbors are 2A or BK. This is a straightforward environmental effect. It is impossible, however, to characterize 2A and BK as being either strong or weak competitors. BK yields better when its neighbors are 2A than it does when grown in pure culture; but 2A, in turn, yields better when its neighbors are BK than it does when grown as a pure stand. This constitutes G × **G**: it is impossible to define the relative fitnesses of 2A and BK, unless the neighbors with which they are growing are specified.

## 134. *Social interactions lead to frequency-dependent selection.*

Whenever there are unequal competitive interactions among genotypes, the fitness of a genotype must depend on the frequency of different kinds of neighbor. In the simplest case, neighbors have a straightforward environmental effect, as strong or weak competitors. The fitness of a genotype is then reduced when the frequency of stronger competitors is increased. If there are only two types in the population, this implies that the fitness of either type is directly proportional to its frequency. This effect occurs in Roy's hill-plots of rice. Because RP is a weak competitor, its fitness will

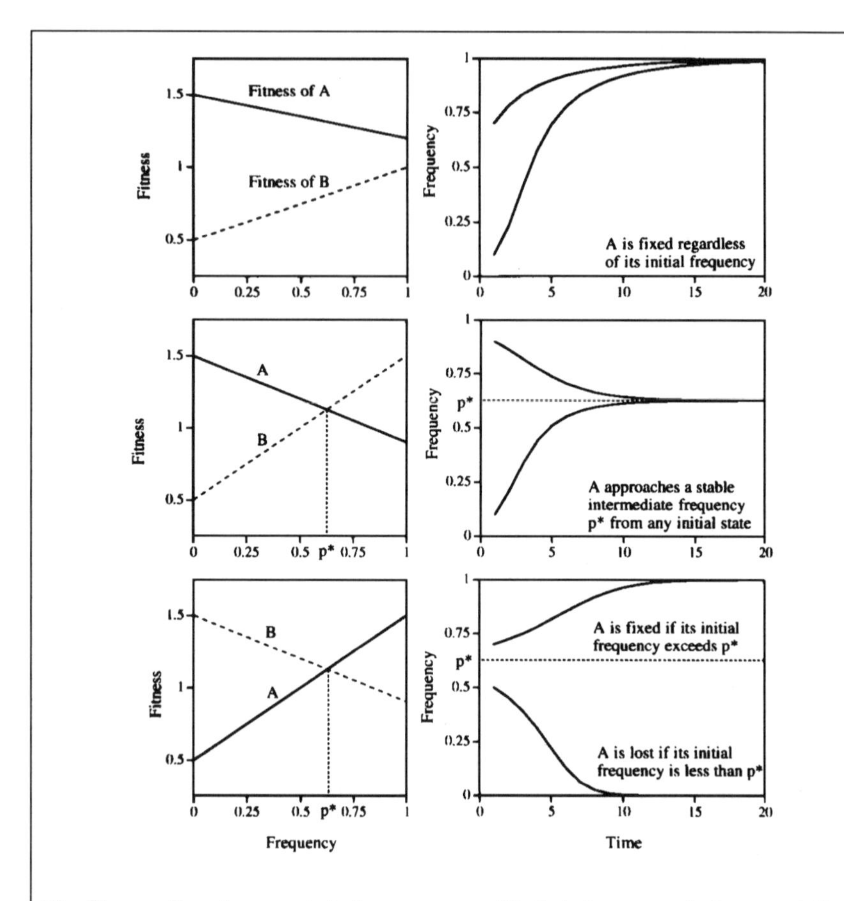


The fitness of two (or more) strains may vary with their frequency in the population.

- If one strain nevertheless has the greater fitness at any frequency, it will become fixed, regardless of its initial frequency.
- However, if the relative fitness of either strain declines as its frequency increases, there may be an intermediate frequency $p^*$ at which the two have the same fitness. This frequency is a stable equilibrium towards which the population will evolve from any initial state; note that at equilibrium the fitnesses of the strains are equal, but their frequencies may be dissimilar.
- If the fitness of either strain increases with its frequency, the outcome of selection depends on its initial frequency. In this case, $p^*$ is an unstable equilibrium.

(*Continues*)

*(Continued)*

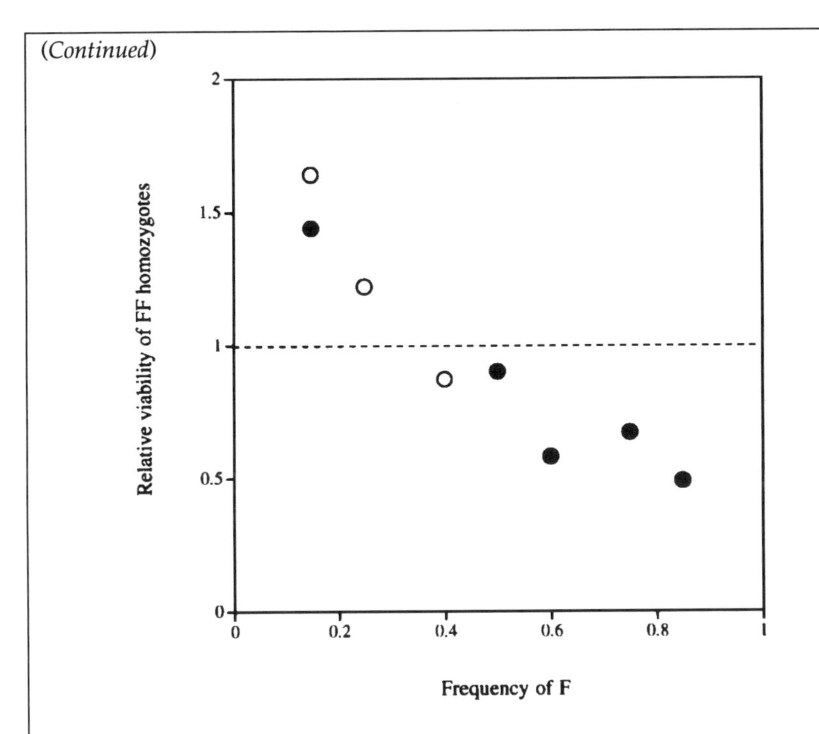


Morgan (1976) set up lines homozygous for the fast F and slow S alleles at the *Est-6* locus from large cage populations of *Drosophila melanogaster*. The populations had been maintained in the laboratory for about 18 months before the experiments were conducted. Four replicate cultures of 200 larvae each were established at frequencies of the F allele ranging from 0.15 to 0.85. The FF and SS flies emerging from these cultures were counted, and I have calculated the relative fitness of FF as the ratio of the fraction of FF to the fraction of SS surviving. The relative viability of FF declines with the frequency of FF in the initial population of larvae, such that FF larvae are superior to SS at frequencies less than about 0.4, but inferior at higher frequencies. Other reports of frequency-dependent selection at enzyme loci include Kojima and Yarborough (1967), Yarborough and Kojima (1967), Kojima and Tobari (1969), Huang et al. (1971), and Yamazaki (1971). Comparable experiments with karyotypic variation in *Drosophila* have been conducted by Levene et al. (1954) and Nassar et al. (1973); there are also similar experiments by Anxolabehere (1971) on a variety of loci, and by Bundgaard and Christiansen (1972) on two marked fourth-chromosome stocks. Harding et al. (1966) described the frequency-dependent advantage of heterozygotes at a seed-color locus in lima beans. The field has been reviewed by Kojima (1971) and by Ayala and Campbell (1974).

depend on the frequency of stronger competitors such as 2A and BK: the more frequent they are, the lower will be the relative fitness of RP. RP, however, has a lower fitness than the other two at any frequency, and its fitness simply decreases as the other two types become more abundant.

The relationship between 2A and BK is different: either may have the greater fitness, depending on the frequency of the other. I have discussed these results in terms of the yield obtained by a given type; however, they can be expressed in a different way, in terms of the effect of a given type on its neighbors' yield. In a 2A environment—that is, at a site where most neighbors are 2A plants—BK produces more seed than 2A does itself. However, the converse is also true: in a BK environment, 2A has the greater fitness. The relative fitness of either type will depend on its frequency in the population. The occurrence of G × **G** thus leads to selection that changes in direction, and not merely in magnitude, according to the frequency of competing types.

**Frequency-Dependent Selection at Enzyme Loci in *Drosophila*.** Some enzyme loci in *Drosophila* have two (or more) alleles that reach characteristic frequencies in cage populations that are not deliberately selected; that is, replicate populations with different initial gene frequencies will usually evolve towards the same frequencies. In the interesting cases, these frequencies are much too large or much too rapidly attained for the maintenance of diversity to be explicable in terms of random processes acting on functionally equivalent genes. These have attracted some attention, because they might reveal a mechanism of selection capable of explaining the allelic diversity typical of many loci in natural populations. The most extensive work was done by Ken-Ichi Kojima of Austin and his collaborators, shortly after the introduction of protein gel electrophoresis as a routine procedure into population genetics. It had until then been widely accepted that a substantial part of genetic variation was maintained through heterozygote advantage, or heterosis. Once the extent of allelic diversity in natural populations was uncovered, it was immediately realized that it was unlikely to be maintained by heterosis, since no population could sustain the cost of selection incurred by the elimination of so many homozygotes. The focus of Kojima's work was thus to show that frequency-dependent selection was an acceptable alternative to heterosis as a Darwinian explanation of allelic diversity. He worked principally on an esterase locus, *Est-6*, and the alcohol dehydrogenase locus *Adh*, both of which have two alleles that coexist more or less indefinitely in cage populations, with the rarer allele at a frequency of about 0.25.

There are two ways of finding out how such alleles are maintained. The first is to use the results of the selection experiment itself, by using a particular hypothesis of selection to predict the dynamics of the approach to equilibrium. The main difficulty with this approach is that when genotype frequencies are the only information available, fitness will always appear to be frequency-dependent. Even if the locus is heterotic, the allele or the homozygote that is present in excess will fall in frequency, so that its fitness, as estimated by the ratio of frequencies in successive generations, will be low so long as its frequency remains high. It is possible to correct for this effect, but the differences between populations evolving under heterosis and those evolving under frequency-dependent selection are then rather small, unless gene frequencies are extreme. It is also regrettably true that most of the simplifying assumptions that one must make in order to calculate the expected generational changes produce a spurious impression of frequency-dependent selection if the population does not, in fact, conform to them. The initial claims for frequency-dependent selection at *Est-6* were for these reasons not completely convincing. The second approach is to measure fitness components directly in separate tests where the genotypes are deliberately reared together at different frequencies. Although this seems straightforward, it can also be problematic. For example, the overall viability can be estimated by using homozygous lines and their hybrids to set up a population of premated females that will give rise to progeny genotypes in known proportions and then score the genotypes of adults in the next generation. Experiments of this sort showed that the viability of either homozygote appeared to be greater when it was less abundant. The procedure works, however, only if the genotypes are equally fecund; if the heterozygote is the most fecund, it will produce additional homozygous progeny of each kind in equal numbers, but the effect of this will be to increase the frequency of the rarer homozygote by a larger proportion, giving a misleading impression that viability is frequency-dependent. In fact, viability, but not fecundity, was found to be rather strongly frequency-dependent. In addition, Phillip Morgan of Monash University subsequently measured viability directly in larval populations made up of the two homozygotes in different proportions and found that the viability of either type declined linearly with its frequency, at least for *Est-6*. Although Kojima's experiments have been criticized, they seem to me to show fairly satisfactorily that the evolution of a characteristic allele frequency at these two loci is driven primarily by frequency-dependent selection.

## 135. *The outcome of selection in mixed cultures may not be predictable from the behavior of pure cultures.*

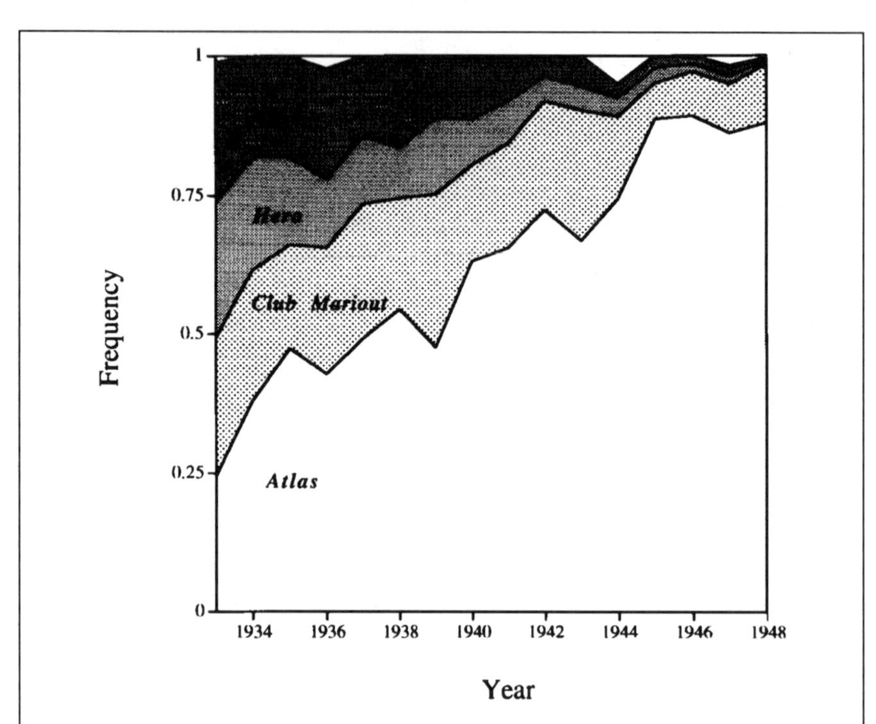


The diagram shows the outcome of selection in a mixture of four barley cultivars, from Suneson (1949): Vaughn is the most productive in pure stand, but Atlas spreads in the mixed population. The small proportion of unassigned plants in some years is made up of hybrids or contaminants. Blijenburg and Sneep (1975) have conducted a comparable experiment with barley. Similar but briefer experiments were reported by soybeans by Mumaw and Weber (1957), with the grass *Dactylis* by Eagles (1983), and for species of duckweed, *Lemna*, by Clatworthy and Harper (1962). The general outcome of such experiments is that there is not necessarily a large positive correlation between yield in pure culture and success in mixtures. I have found that in mixtures of two closely related genotypes of *Chlamydomonas*, however, the genotype with the more rapid growth in pure culture rapidly excludes the other (Bell 1992). Moreover, Austin (1982) has published a careful quantitative account of the relationship between yielding ability and competitive ability in five species of pasture grasses, concluding that more than half the variance in competitive ability was associated with variation in yield in pure culture.

In a physically heterogeneous environment, the fitness of a genotype will depend on the frequency of the kind of site to which it is well-adapted. Likewise, in a diverse population, the fitness of a genotype will depend on the frequency of different kinds of neighbor. The dynamics of the two situations, however, are quite different. The frequency of a particular kind of site is regarded as a fixed feature of the environment. The frequency of different kinds of neighbor, on the other hand, will change through selection. In a 2A environment, BK has the greater fitness. Because of this, it will increase in frequency. But as it does so, it transforms the environment by increasing the probability that a given plant will have BK neighbors. If BK becomes very abundant, most plants will have BK neighbors; 2A now has the greater fitness and will in turn tend to increase in frequency. The characteristic property of frequency-dependent selection is that the outcome of selection itself transforms the environmental factors responsible for differential fitness. It is a familiar proposition that selection changes the genetic composition of populations; the new element is that a change in the composition of the population may in turn change the direction of selection.

If relative fitness varies with frequency, then the performance of individuals in pure cultures of their own genotype may not reflect their behavior in mixed cultures. This introduces a distinction between the concepts of adaptedness and fitness that until now I have used more or less interchangeably . Adaptedness means the ability of a genotype to proliferate as a pure culture in a given environment. The environment may or may not vary in space or time, and its important features may be physical or biotic, but it is static in the sense that it does not change in response to the degree of adaptation that the population evolves. Fitness means the ability of a genotype to proliferate in competition with others in a diverse population. The competitive environment may well be dynamic, in the sense that it is changed as the result of selection. In many cases, the two concepts really are interchangeable. Clones of bacteria could be ranked in the order of their rates of growth in the presence of high concentrations of an antibiotic. This ranking represents their degree of adaptedness to an environment in which the antibiotic is thought to be the main agent of selection. If we now mix the clones and grow the mixture in the presence of the antibiotic, the more resistant clones will increase in frequency, and the clone that becomes fixed will probably be the clone that has the highest rate of growth in pure culture. It is usually assumed by theoreticians that adaptedness and fitness are equivalent in this way. However, this is not necessarily the case: if there are genetic interactions that create frequency-dependent selection, growth in pure culture may be an unreliable guide to growth in mixtures. There are thus two different approaches to the study of selection: the comparison of kinetic parameters in pure culture and the measurement of changes of frequency in mixtures. Which of these approaches is preferred usually depends on whether the investigation is seen as studying adaptation to a fixed, externalized physical environment or to a dynamic, responsive biotic environment.

**Yield and Persistence of Barley Cultivars.** These two approaches may give conflicting results. Coit Suneson has described the behavior of four varieties of barley—Vaughn, Atlas, Hero, and Club Mariout—in California. When they are grown in the usual way, as pure cultures several acres in extent, their yields per unit area are ranked as: Vaughn > Hero > Atlas > Club Mariout. These yields vary from year to year, and from place to place; but certainly Vaughn was superior, exceeding Atlas and Club Mariout for yield in 12 out of 15 years. Suneson grew the four as an annually resown mixture between 1933 and 1948, without conscious selection. Barley is almost entirely self-fertilized, so that the varieties remained distinct, and the experiment involves only simple sorting. The experiment was begun with equal frequencies of all four varieties. After 16 years, Atlas had risen to a frequency of nearly 90%; Club Mariout at first declined, but appeared to stabilize at about 10% of the population; Hero and Vaughn had almost entirely disappeared, dropping to frequencies of less than 1%. Clearly, pure-culture yield would have been a very poor guide to the evolution of the mixed population; in this case, indeed, yield in pure culture and success in mixture were negatively correlated.

**Yield and Persistence of Soybean Cultivars.** A similar but briefer experiments was reorted by C.R. Mumaw and C.R. Weber from the experimental station at Ames, Iowa. They prepared three different mixtures, each of three strongly contrasted soybean varieties, and propagated them for five years. In all three cases, the frequencies of the component varieties shifted quite rapidly, in one case from the initial value of one-third up to about 80% for the most successful type and down to less than 5% for the least. In two of the three experiments, the most frequent variety after five years was the variety most productive in pure stand. In the third case, however, the second most productive variety dominated the mixture, and the most productive had fallen to a frequency of less than 10%. In these experiments, then, there was no consistent relationship between yield in pure stand and success in mixture. All of the genotypes that increased in frequency had a branched habit and might have spread because they were able to shade out their neighbors.

## *136. Selection for yield in pure culture may cause enhanced self-facilitation.*

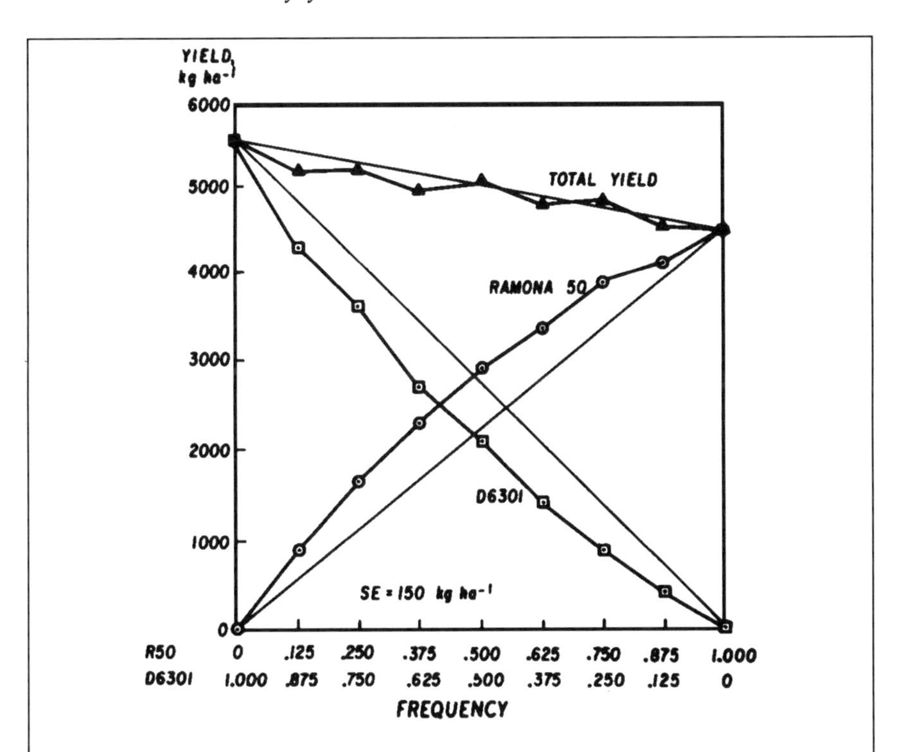


Although there is no necessary relationship between yield in pure culture and competitive ability in mixtures among arbitrary genotypes, it is often possible to identify antagonistic relationships among particular combinations of genotypes. This is illustrated by the replacement-series experiments involving tall and dwarf wheats described by Khalifa and Qualset (1974). This design is equivalent to single-generation selection experiments in which the output of seed from a given genotype is compared with its input frequency in mixtures of varied composition. In this case, a dwarf cultivar D6301 has the greater yield in pure culture, but always decreases in frequency in mixture with the tall cultivar Ramona 50. Similar results have been reported from barley by Sakai and Gotoh (1955), from rice by Jennings and de Jesus (1968), Jennings and Herrera (1968) and Kawano et al. (1974), and from cassava by Kawano and Thung (1982). The evolution of antagonism between pure-culture yield and competitive ability is reviewed for annual seed crops by Donald and Hamblin (1976), who cite a large number of experiments on cereals and legumes.

Some crop plants are selected as individuals, others as communities. Crops such as fruit trees and cabbages are normally grown as rather widely spaced individuals that are harvested separately. They are therefore selected as individuals, typically for large size; the modern plants (or at least the harvested parts) are much larger than their wild progenitors. Annual seed crops such as wheat, oats, and barley are harvested as communities, the object being to obtain as large as possible a yield from a given area of land, without regard to the production of individuals. Nevertheless, although they are harvested as communities they have often been selected as individuals. Particularly tall and luxuriant plants with large and abundant seed tend to be preferred as stocks for breeding. Two Australian agronomists, C.M. Donald and J. Hamblin, have argued that selecting seed crops on the basis of individual performance is inconsistent. Tall plants with broad horizontal leaves that tiller or branch profusely are superior competitors that will yield well because they are unlikely to be repressed by less vigorous neighbors. Being superior competitors, however, they are themselves likely to repress the growth of their neighbors. When grown as pure stands they may not be especially productive in terms of yield per unit area, precisely because they compete intensely among themselves. Selection for total yield per unit area is more likely to favor weak competitors, whose modest yield as individuals is accompanied by less adverse effects on their neighbors. I have already pointed out that yield in pure culture may be unrelated, or even negatively related, to success in resown mixtures. If there is an antagonism between total yield and individual yield, then the modern practice of selecting cultivars for their yield as pure stands will cause a correlated decline in their competitive ability, as measured by their loss from deliberately contrived mixtures.

**Competition Between Tall and Dwarf Wheat.** M.A. Khalifa and C.O. Qualset of Davis made a series of mechanical mixtures in which a tall cultivar (about 120 cm in height at maturity) and a dwarf cultivar (about 80 cm in height) were present in different proportions. The dwarf type yields about 20% more per unit area, when grown in pure stand. In mixtures, however, the tall cultivar is always the stronger competitor, presumably because it shades the shorter type. In mixtures where the tall type was abundant, the dwarf plants had fewer kernels per spike; when the dwarf variety was abundant, the tall plants had heavier kernels. Thus, dwarf neighbors benefited the tall plants, whereas tall neighbors injured the dwarf plants. Consequently, the yield of the dwarf plants, relative to their yield in pure culture, fell as their frequency decreased.

**Semidwarf Varieties of Rice as Weak Competitors.** Traditional rice cultivars are rather tall, leafy plants with high seed yield as individuals. They have now been largely replaced by semi-dwarf varieties that produce a heavier crop on a given area of land. P.R. Jennings and J. de Jesus planted two tall and three semi-dwarf varieties in equal proportions on experimental plots in the Philippines. Within four generations, the high-yielding semi-dwarf varieties had been almost eliminated from resown populations that became dominated by the lower yielding of the two tall varieties. In subsequent experiments, heterogeneous mixtures of plants from crosses between tall and semi-dwarf varieties became dominated by taller plants. Selection for yield in in pure culture thus tends to reduce fitness in mixed populations.

## 137. *The social relations between strains can be altered through selection.*

The distinction between adaptedness and fitness implies that neighbors mutually modify one another's environment. This will lead directly to selection that causes changes in the frequency of lineages in the population. However, it may lead indirectly to selection for altered or improved competitive abilities. That is to say, a lineage might evolve so as to become adapted to neighboring or interacting individuals of other lineages, just as it might become adapted to any other sort of environmental modification.

There are three sorts of experiment that might address this issue. In the first kind, a population is cultured repeatedly in the presence of some stock genotype. In this case, the neighbor to whose presence the population is adapting is always the same, being drawn in every generation from some uniform line, maintained separately. This is essentially the same as adapting to some feature of the physical environment, the only novelty being to discover whether different genotypes do, in fact, modify the environment in substantially different ways.

The second kind of experiment is to perpetuate a mixed population for several generations, without conscious selection, and then to compare the competitive relationships among its component lines with those among the equivalent lines in the base population, or in concurrent pure cultures. This design allows lineages to become mutually adjusted to one another, provided that the population remains diverse for long enough for selection to be effective. This can readily be enforced, by transferring fixed numbers of types in each generation, regardless of population frequencies. The experimental population thus consists of a definite number of types (in practice, two) that are deliberately maintained in the same proportions.

Finally, the population can simply be propagated by bulk transfer, if it is expected that its diversity will be maintained through natural selection.

The experimental population is in this case an indefinite mixture, whose composition is not directly controlled. Selection can act in two ways in such populations. It may, as before, modify the properties of the various types within the population. It may also, however, modify their frequencies, thereby causing a change in the average behavior of the population, regardless of whether any types have been individually modified.

## *138. A population may become specifically adapted to the presence of another species.*

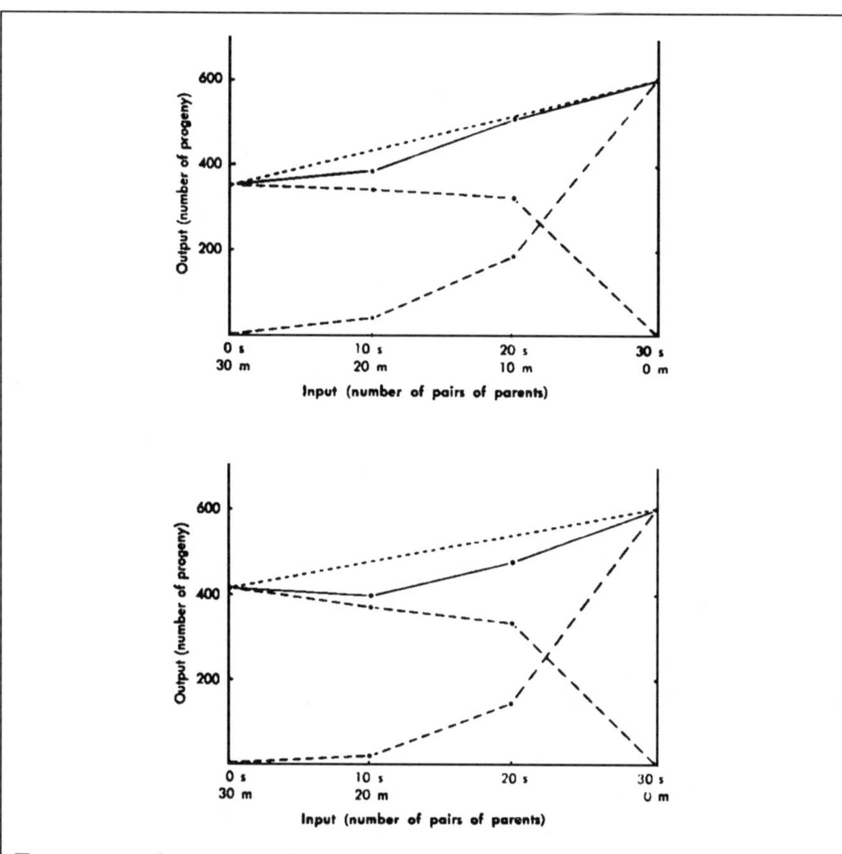


These are replacement-series diagrams for one replicate in Futuyma's experiment (Futuyma 1970), involving a selection line (upper graph) and its control (lower graph), after five generations of selection. There is no increase in output

(*Continues*)

(*Continued*)
of *melanogaster* (m) attributable to co-culture with unselected *simulans* (s), although the initial competitive superiority of *melanogaster* means that the scope for improvement is rather limited. There is some suggestion that total output is depressed less severely by competition in the selection line. For a similar experiment with *Tribolium*, see Park and Lloyd (1955).

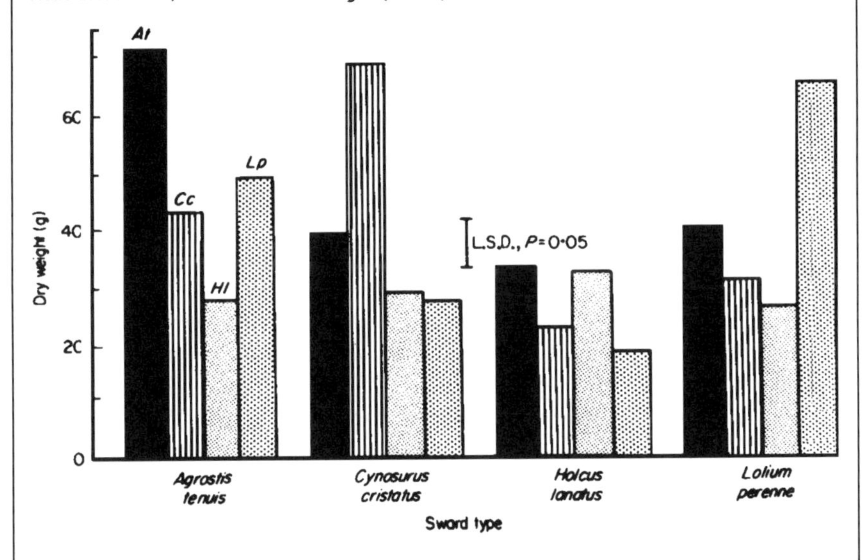


This diagram from Turkington and Harper (1979) shows the yield of clover ramets sown into artificial pure stands of the four grass species that occurred in the same meadow: At, *Agrostis tenuis*, Cc *Cynosurus cristatus*, Hl *Holcus lanatus*, and Lp *Lolium perrenne*. Note the tendency for ramets to grow better in a sward of the grass species with which it was associated in the field.

The stickleback experiment described in the text was published by Schluter (1994).

Social relations may be very one-sided . A rare grass in a meadow will be grazed along with the rest of the herbage by a herd of cattle. The relative fitness of different types of the grass may vary according to the intensity of grazing, whereas the cattle are indifferent to whether such a minor item in their diet occurs or not. From the point of view of a species that is perennially rare, generalized grazers are a fixed component of the environment, akin to frost or drought. Their effects may vary through time, but they vary independently of any evolutionary modification of the rare prey. In the same way, many species of minnow may feed on an abundant species of cladoceran in a lake. A rare species of minnow may become better adapted for capturing cladocerans, but so long as it remains rare it will cause no appreciable selective mortality in its prey, which will retain its former characteristics. This situation can be represented in the laboratory by culturing a population generation after generation with neighbors that are freshly recruited in every generation from a stock that is itself maintained as a pure culture.

The experimental population consists of a number of lines that can readily be distinguished by some marker and that do not interbreed, or that are kept from interbreeding. In an outcrossing population such lines would not remain distinct and could not evolve as units. Competitive ability can be modified only among different species, or among different lineages of asexual or self-fertilizing organisms. Although the lines must be stable, they must nevertheless also be sufficiently variable to respond to the selection caused by their neighbors. They must therefore be populations and not merely different mutants. The interesting question is one of scale: how different must these interacting populations be in order to elicit specific adaptation? It would not be surprising if experimental populations of flies evolved in response to geckos on the walls of the cage or nematodes in the food medium (although I do not know of any such experiments). It would be more surprising if similar neighbors of the same species generally modified the environment so distinctively that specific adaptations evolved as a response.

**Competition Between Similar Species of *Drosophila*.** A favorite system for studying the evolution of social interactions between related species is the sibling pair *Drosophila melanogaster* and *Drosophila simulans*, that are morphologically almost indistinguishable but do not interbreed. Doug Futuyma of Stony Brook reared 10 generations of a highly heterogeneous *melanogaster* population in the company of an equal number of *simulans* recruited in every generation from a stock that was marked, for ease of identification, with a mutation causing dark body color. Control *melanogaster* populations were maintained at the same time as pure cultures. After 10 generations, flies from the control or the selected *melanogaster* lines were mixed with equal numbers of stock *simulans*. The offspring of these flies grew up together and eventually emerged as adults. If social selection had been effective, the proportion of *melanogaster* emerging from the mixed cultures should be greater for the selection lines than for the controls. In fact, there appeared to be little if any difference. In one of the replicate selection lines, an excess of *melanogaster* emerged; but in most lines there was no appreciable difference, and in two cases the control lines actually yielded more *melanogaster* than the corresponding selection lines. When Futuyma tested the individual strains (kept in mass culture during the selection experiments) from which his base population had been constructed, he could find no evidence that they differed in fitness when cultured with equal numbers of *simulans*. Selection for social adaptation to a closely related species was therefore ineffective: the base population may have contained no genetic variance for specific social behavior, and none emerged through sorting.

**Local Adaptation to Neighbors.** A lineage of plants in a natural population may remain associated with the same set of neighbors for many generations. It may then become adapted to a particular species, just as it may become adapted to a particular soil type. Roy Turkington and John Harper of Bangor sampled clover from a field in which it was growing among a patchwork of four species of grass. They multiplied each ramet in the greenhouse, so that they were able to plant out replicates of a given genotype back into its home site or into any other kind of site. There was a striking tendency for ramets planted back into their home site to yield more and survive better than those planted into other sites. Moreover, this local adaptation continued to be expressed when the clover was planted into pure stands of one of the four species of grass grown in the greenhouse: genotypes performed better in swards of the grass species with which they had been associated in nature. This provides very strong evidence that a population can become specifically adapted to the presence of another species, although the ecological differences among the grasses must be much greater than the difference between two closely related species of *Drosophila*.

**Character Displacement.** It has long been anticipated that two similar species, or strains of the same species, will evolve so as to become more different because types that are most different from the bulk of the combined population will experience less competition. For example, where two species of the stickleback *Gasterosteus aculeatus* complex occur in the same lake in western Canada, one is usually specialized for feeding on benthic invertebrates, having few gill-rakers and a wide gape, whereas the other is specialized for limnetic life, having a narrow gape and many more gill-rakers, and feeding on planktonic crustaceans. If either species is present alone, it will usually have an intermediate morphology and feed on both benthic and planktonic prey. Dolph Schluter of the University of British Columbia ran a single-generation selection experiment by stocking artificial ponds with fish from a lake where only a single species occurred, and which thus had an intermediate phenotype, and then introducing limnetic individuals to some of the ponds. He found that individuals most dissimilar from the limnetic form, with more benthic appearance and habits, had the higher growth rates when the limnetic form was present, although the effect was small.

## 139. *Two interacting species may become mutually modified.*

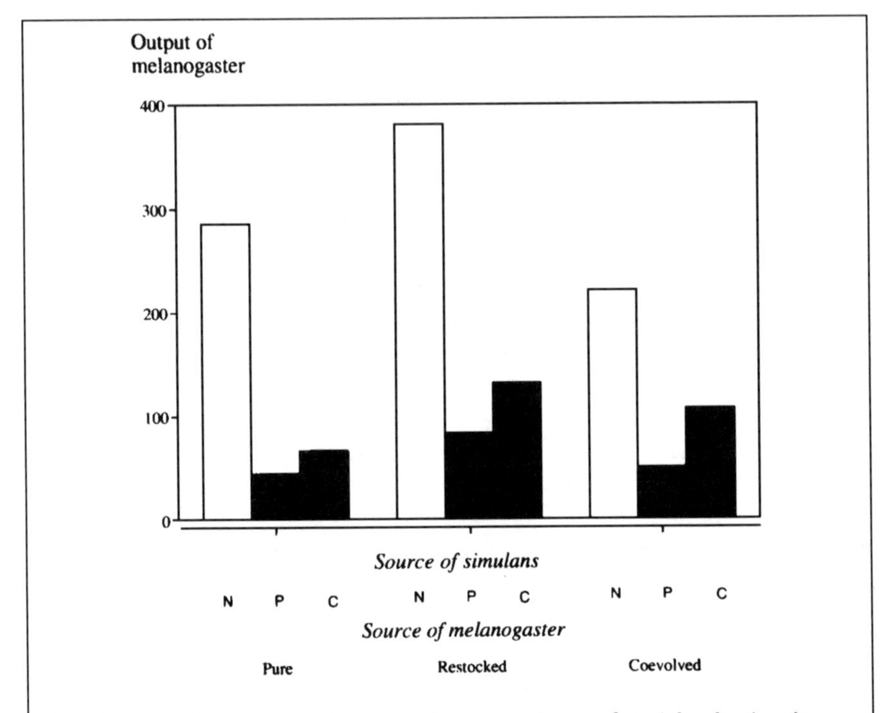


This diagram shows the outcome of 77 generations of social selection in van Delden's experiment (van Delden 1970). The data are the output of *melanogaster* from competition vials. The parental flies were evolved in pure cultures of *melanogaster*; restocked cultures, supplied with *simulans* from a control pure culture of simulans in each generation; and coevolved cultures, both the *melanogaster* and *simulans* being propagated, at fixed proportions, from generation to generation.

The competing *simulans* flies came from: None, N, no simulans present, i.e., the output of pure cultures of *melanogaster*; Pure, P, the pure culture control line; and Coevolved, C, the same treatment as that for *melanogaster*.

The main result is the increased productivity of *melanogaster* from the two competition lines, especially the Restocked line. The productivity of *simulans* varied in the same direction: parents from the Coevolved line produced more offspring (242 per vial) than parents from the Pure culture (201 per vial) when competing with *melanogaster* (presumably, though the text is not clear, from the Coevolved line). No standard errors are available for these estimates. Similar experiments have been reported by Moore (1952) and Barker (1973). There is a general review of selection in mixed-species culture by Arthur (1982).

When two species are both moderately abundant, each will form part of the other's environment. They will then behave quite differently from physical features of the environment, because each will repond to the other's presence and, in responding, will change the social environment that the other experiences. This may result in specific mutual modification through a dynamic process of social selection.

**Mutual Modification of Similar Species of *Drosophila*.** The same species pair of *Drosophila melanogaster* and *Drosophila simulans* was used in this kind of experiment by J.S.F. Barker of Sydney and W. van Delden of Groningen. The basic procedure is to set up pure and mixed cultures from laboratory strains of the two species that can be distinguished because they carry genes conferring different eye color.

Barker transferred 10 pairs of *melanogaster* and 10 of *simulans* in every generation in his mixed cultures, and 20 pairs in each of the control pure cultures. The flies could be extracted from the lines at intervals and tested to see whether the competitive ability of the selection lines had changed relative to that of the controls, as judged by the proportions of the two species emerging from mixed cultures. The experiment was continued for about 60 generations, but was bedevilled by the usual difficulties of long-term studies: the author moved his laboratory from Chicago to Sydney during the experiment, the culture medium was changed twice, some of the mixed cultures had to be discontinued because there were not enough flies from one of the two species, and at least one *melanogaster* line was contaminated by immigrants. Perhaps as a result of these misadventures, the behavior of the pure-culture stocks changed through time, and the behavior of the selection lines is difficult to interpret. It does seem that there was little if any change in social behavior for the first 30 generations or so. After 40 generations, there was some indication that one of the *simulans* lines had evolved somewhat different social properties. These seemed to involve the production of greater numbers of *simulans*, without a corresponding reduction in the production of *melanogaster*. It is possible that in this one line the two species had diverged somewhat so as to compete less intensely than they did at first. Whatever changes took place, however, were certainly slight and difficult to identify with certainty.

Van Delden's experiment was even longer, being carried out to about 80 generations. It combines the two designs I have discussed so far. In a restocked selection line, the *simulans* were supplied in every generation from the control pure culture, whereas in a coevolved line they were transferred from the previous generation of the mixed culture. There appeared to be no response to selection after 10 generations, but quite pronounced effects began to emerge after about 60 generations. In either kind of selection line, the production of both *melanogaster* and *simulans* increased. The effect in *melanogaster* was more marked for the line in which it coevolved with *simulans*. However, the production of pure cultures set up from the selection lines also increased, relative to that of the controls. Van Delden's results are unfortunately described too tersely for one to be certain of what happened, but it seems as though there was a change in general social behavior (as shown by the increased productivity of the coevolved selection lines in pure culture) and a more specific response (as shown by the improved performance of *melanogaster* from the restocked line when tested against stock *simulans*). In both cases, the response seemed to involve some sort of mutual facilitation, similar to that which may have occurred in Barker's experiment, with flies of one species increasing in abundance without causing a corresponding loss to the other species.

## *140. It is doubtful whether arbitrary strains of the same species often become mutually modified.*

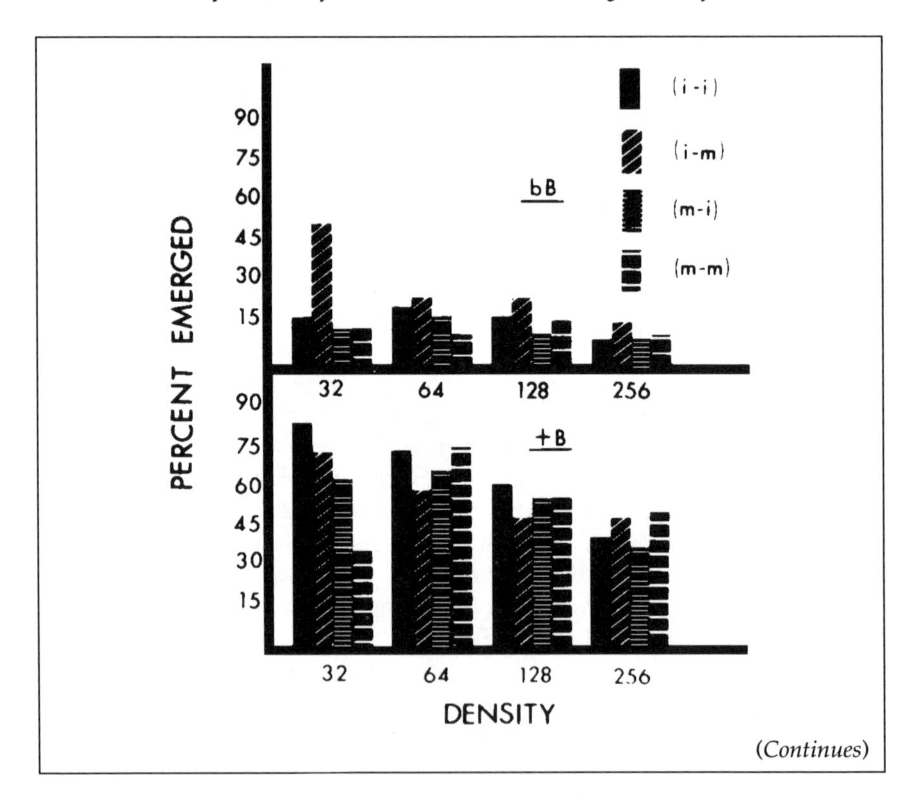


(*Continues*)

*(Continued)*
Pruzan-Hotchkiss et al. (1980) propagated pure and mixed cultures of the mutant bB and wild-type +B strains of *Drosophila melanogaster*, which can be distinguished by body color. In each generation the mixed lines were reconstituted with 7 pairs of +B and 10 pairs of bB. After eight years, flies were tested in the following conditions:

*i–i* flies from pure (isolated, i) cultures of one stain tested in the presence of flies from the pure culture of the other strain;

*i–m* flies from pure cultures of one strain tested in the presence of flies from mixed (m) cultures of the other strain;

*m–i* flies of a given strain from mixed cultures tested in the presence of flies from pure cultures of the other strain;

*m–m* flies of a given strain from mixed cultures tested in the presence of flies of the other strain from mixed cultures.

The testing procedure involved inoculating medium with equal numbers of eggs from each strain, at four densities, and counting the flies of each strain emerging from the vials. If selection in mixture improves the ability of flies to survive in the presence of the competing strain, we expect that $m–m > m–i > i–i > i–m$ for both strains. This outcome was not consistently observed at any density. For other experiments with similar designs, see Seaton and Antonovics (1967, *Drosophila*), Sokal et al. (1970, *Musca* and *Tribolium*), Bryant and Turner (1972, *Musca*), Dawson (1972, *Tribolium*), Sulzbach and Emen (1979, *Drosophila*) and Sulzbach (1980, *Drosophila*). Hartl and Jungen (1979) reported several instances of reversal of competitive dominance (see next section) between strains of *Drosophila melanogaster* cultured together.

**Competitive Ability in Different Strains of *Drosophila*.** The differences between strains of the same species may be relatively slight, and selection for altered competitive abilities correspondingly weak and non-specific. Nevertheless, a well-known experiment by A.P.C. Seaton and Janis Antonovics seemed to demonstrate rather strong and specific selection of this kind. They studied competition between wild-type flies and a stock carrying the wing mutation *dumpy*. The experiment was founded with progeny from crosses within the wild-type and *dumpy* stocks. These were cultured together until they eclosed as adult flies. While still virgin, these flies were removed and the two stocks, distinguishable from the *dumpy* marker, separated. In this way, two sexually isolated populations could be maintained in competition with one another. In every generation, the selected flies from one of the two stocks were tested against unselected flies of the other stocks, reared as pure cultures. After only three or four generations of selection, there was a remarkable increase in the specific competitive ability of the selected flies, demonstrated by an increase in the proportion of flies from the selected strain emerging from mixtures with the unselected stock of the other strain. This was caused by an increase in the number of the selected flies, rather than by a decrease in the number coming from the unselected stock. Hence, it seemed that the main effect of selection in mixture was a reduction in the intensity of competition between the strains.

**Competitive Ability in *Musca*.** Robert Sokal and his colleagues at Stony Brook set out to repeat Seaton and Antonovics' experiment, using strains of the housefly, *Musca*. After about ten generations of selection, however, neither of the two strains they used seemed to have responded to social selection in mixed culture. Edwin Bryant, one of Sokal's collaborators, argued that one reason for this discrepancy might be that Seaton and Antonovics had given *dumpy* a head start in mixed cultures by putting its eggs into the vials a couple of days before adding wild-type eggs; *dumpy* is so sickly that unless it is given this head start it is quickly eliminated from the mixture. He therefore set up another experiment, with Carl Turner at Houston, giving a similar head start to a rather weak green-eyed strain of *Musca*. This time, the survival of the mutant strain increased markedly in mixed culture within four generations, whereas it remained about the same in pure culture. The wild-type strain that it was competing with did not change. The improvement in the mutant strain was traced to an acceleration of hatching caused by selection against late-hatching larvae in the mixtures, although it was not clear why the same improvement should not also occur in the pure cultures. At all events, this second experiment seemed to confirm Seaton and Antonovics' original result.

**Negative Results from *Drosophila*.** This result has not been quantitatively confirmed in subsequent experiments. For example, David Sulzbach of Wesleyan University created two highly heterogeneous strains by matings among different sets of unrelated isolates from widely separated localities, one set being marked by the eye-color mutants *vermilion* and *brown*. After about 20 generations of selection in mixture, there were no signs of any substantial modification of competitive ability, although some replicate lines may have evolved weak and rather erratic tendencies to perform better in the presence of their neighbor. Like Bryant, Sulzbach reasoned that the head start given to a weak competitor might be the critical factor in getting a rapid improvement in competitive ability, so he tried the effect of giving a two-day start to the brown-eyed strain for the first eight generations of selection. In one replicate (but not in other lines) the survival of the brown-eyed flies increased sharply during the early generations of selection. It seems doubtful whether this improvement was caused by social selection; the line was originally weaker than the control line, for unknown reasons, and merely regained the same level of performance as the control.

The longest social selection experiment was reported by a group of biologists from small New England colleges led by Anita Pruzan-Hotchkiss who maintained a wild type and a mutant in pure culture and in restocked mixture for eight years. Both strains were rather unusual: the wild-type line was founded by a single fertilized female, and the mutant was a multiply-marked compound autosome construct. They were used in the experiment because the hybrids are inviable, so that no special procedures are necessary to keep the two strains isolated. They seemed to show no specific adaptation to each other's company. It seems that 200 generations was not sufficient for these two strains to evolve any specific response to one another.

**Competition Between Species of *Tribolium*.** Flour beetles in the genus *Tribolium* were used in several classic ecological experiments on interspecific competition by Thomas Park at Chicago in the 1930s and 1940s. Peter Dawson of Oregon State subsequently used *Tribolium castaneum* and *Tribolium confusum* to investigate the evolution of competitive ability. This was a conscious attempt to repeat the *Drosophila* experiments, using the same design to culture coevolving mixtures over 5 or 10 generations. The beetles showed the same range of social behavior as the flies—an unequivocally stronger competitor in one case, frequency-dependent fitnesses in another—but there was no tendency at all for this behavior to change through time.

**The Interpretation of Negative Results.** There is a very strong tendency to publish experiments that work. The phrase itself is revealing: an experiment that works provides striking support for an hypothesis. Everyone is aware that experiments should be attempts to falsify hypotheses, but everyone nevertheless organizes their experimental work in the hope that it will support some hypothesis they favor. After all, it is easy to produce negative results through laziness or lack of skill, whereas a clear demonstration of some new or hitherto obscure phenomenon is a great deal more difficult to accomplish. The proportion of negative results in experiments designed to detect the evolution of specific social interactions between different strains therefore carries some weight. It is most unusual for selection experiments to fail repeatedly to procure a response, and the most likely explanation for this plethora of negative or marginal results is that social selection of the kind envisaged in the experiments simply does not occur very often. Even the positive results are not very convincing—ironically, because they are too strikingly positive. They all seem to involve mutant strains that are feeble in pure culture, and it seems likely these instances of rapid improvement following selection in mixture merely represent the immediate effects of selection on populations whose mean fitness is low, either because the base population carries unconditionally deleterious mutations or because it is at first poorly adapted to the environment in which the experiment is conducted.

The rationale for these experiments is that strains of the same species are likely to have very similar ecological properties, and will therefore compete intensely. This may be so; but when the base populations are very similar, or highly heterogeneous, or both, the difference between self and neighbor is slight or variable, and there will rarely be strong or consistent selection for a *specific* response to the competing strain. In the limit, the two strains are identical, and only a general density-dependent response is possible. The lack of a specific response has been attributed to epistasis, the social interaction between related strains involving combinations of genes that cannot be selected effectively in sexual populations. This may be true, but it seems equally likely that genes causing a specific response towards similar neighbors are just very rare. Moreover, the prevalence of one species or another in mixed-species cultures is usually strongly affected by the physical environment, and very slight changes in culture conditions may cause the frequency of related strains of the same species to change markedly from one trial to the next. The character that is measured in these experiments—percentage emergence of one type—is thus likely to show a high degree of genotype–environment interaction, to have a correspondingly low heritability, and therefore to respond only very slowly to selection.

However, the generally negative outcome of the experiments is not uninformative. Arbitrary pairs of closely related species, or strains of the same species, seems to represent a scale at which social adjustments evolve only very feebly or slowly. This is perfectly consistent with the widely held view that in sexual organisms the species is a natural ecological unit constituting the minimal level of distinctiveness that competition can effectively enforce.

## 141. *Social relations may evolve in self-perpetuating mixtures.*

Experiments using enforced mixtures test whether arbitrary strains can adapt to one another's presence. Arbitrary strains, however, do not necessarily form stable mixtures. Whether or not two strains will continue to interact as neighbors depends on the nature of the social relationship between them, and the relationships that evolve in a population where diversity is maintained through selection may be quite different from the behavior of the sole remaining genotype in a uniform population, to whom an exotic neighbor is suddenly introduced. In self-perpetuating mixtures, therefore, the average behavior of genotypes isolated from the population may change through time because social selection increases the frequency of types that interact in a certain way with their neighbors.

**Reversal of Social Dominance.** Pairs of related species can sometimes be maintained as self-perpetuating mixtures in which the two species coexist in roughly constant proportions for long periods of time. In some cases, these proportions have been seen to shift abruptly, so that one species,

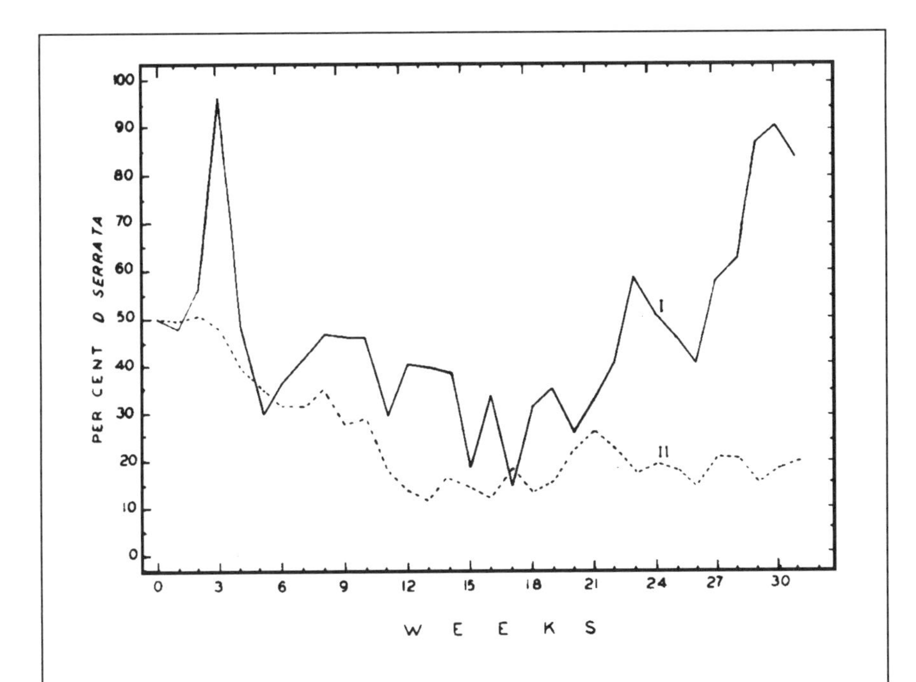


This figure illustrates a reversal of dominance in mixed cultures of *Drosophila nebulosa* and *Drosophila serrata*, from Ayala (1969). *Drosophila serrata* normally declines to a frequency of about 20%, as in line II; in line I, however, it begins to increase after about 20 generations of coculture, eventually attaining a frequency of about 90%. The superiority of the evolved line I flies was confirmed by extracting genotypes and retesting in competitive cultures. For comparable observations on *Musca*, see Pimentel et al. (1975).

(*Continues*)

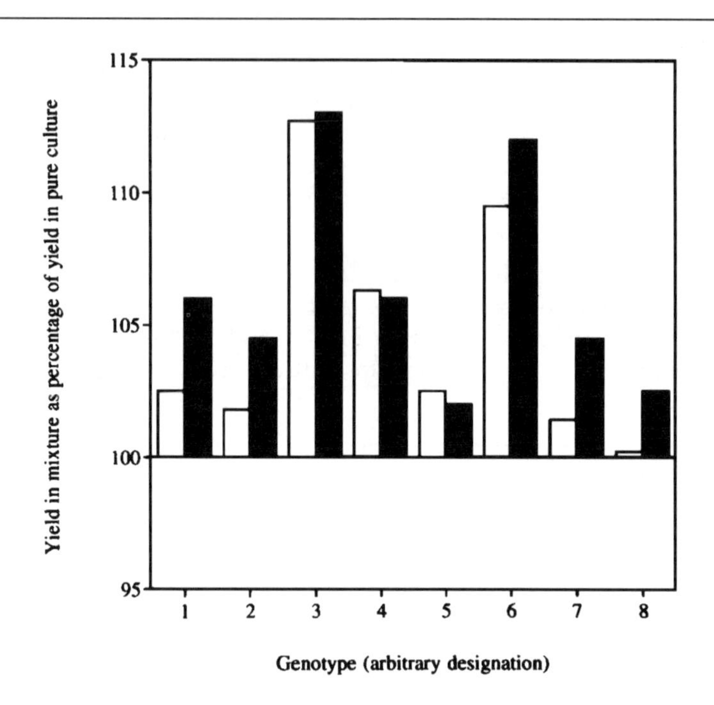


Allard and Adams (1969) extracted eight genotypes from barley Composite Cross V after 18 generations of selection and measured their yield in pure and mixed stands. This diagram shows the yield of each genotype, as a percentage of its yield in pure stand, when surrounded by neighbors of a single different genotype (open bars, mean of all seven mixtures with all seven neighbors) and when surrounded by all eight genotypes (solid bars). For a comparable but briefer study in soybeans, see Boerma and Cooper (1975).

These experiments are relevant to the practical issue of whether it is more efficient to use a process analogous to natural selection (bulk or mass selection) or to artificial selection (pedigree selection) to improve crop plants. Broadly speaking, pedigree selection is usually more effective, but bulk selection is much cheaper. Experiments bearing on this issue have been published by Taylor and Atkins (1954), Suneson (1956), Jain (1961), Hallauer and Sears (1969), Compton et al. (1979) and Feaster et al. (1980), among others.

(*Continues*)

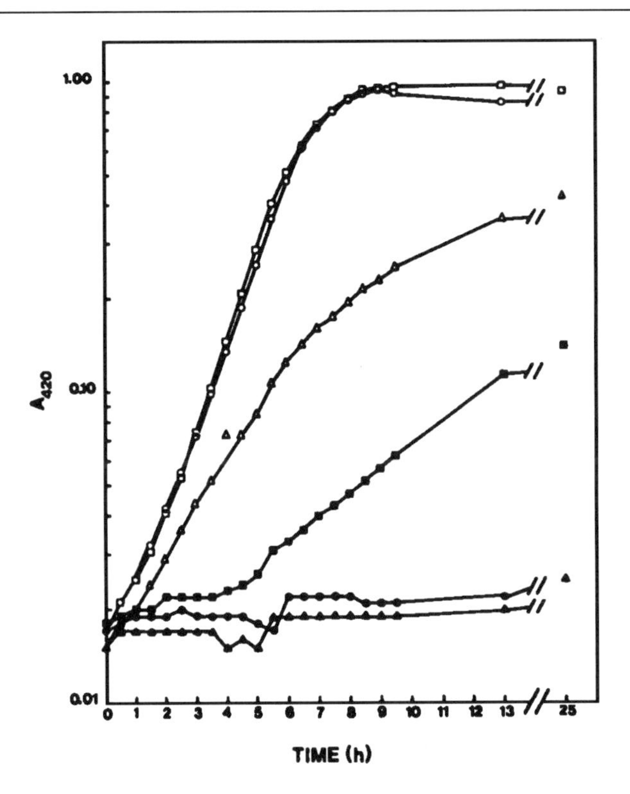


This diagram shows the growth of large-colony (CV101) and small-colony (CV103) variants isolated from long-term *E. coli* chemostats, in medium previously exhausted by the growth of one or the other type; from Helling et al. (1987).

a) Open circles and squares: growth of CV101 in medium previously exhausted by CV101 (circles) or CV103 (squares), but supplemented with additional glucose. The growth curves are normal, so there is no substantial effect of growth-inhibiting substances.

b) Open triangles: growth of CV103 in medium exhausted by CV103 but supplemented by glucose. Growth is normal for this strain.

c) Solid squares: growth of CV101 in medium exhausted by CV103, without glucose supplementation. Because growth is possible, CV101 must be capable of growing on substances excreted by CV101, in the absence of glucose.

d) Solid circles and triangles: growth of CV101 in medium exhausted by CV101 (circles) and of CV103 in medium exhausted by CV103 (triangles). Neither strain can grow in medium that it has itself exhausted.

*(Continues)*

(*Continued*)

For a complicated experiment measuring the growth of species of *Chlamydomonas* in medium exhausted by the same or different species, see Bell (1990). Levin (1972) also reports the coexistence of two strains of *E. coli* in a simple chemostat system, although the mechanism involved is unknown.

formerly rare, becomes prevalent. For example, David Pimentel and his colleagues cultured the housefly *Musca* and the blowfly *Phaenicia* together. One always eventually eliminated the other, but the species that seemed to be prevailing was sometimes rapidly replaced by its competitor. In one experiment, for example, *Phaenicia* struggled along for about 40 weeks at low frequency and eventually became very rare. At this point, however, it began to increase in numbers, and 20 weeks later had completely eliminated *Musca* from the community. This may have been caused by social selection within the *Phaenicia* population. While it was rare, social selection would have favored individuals that succeeded in competition with *Musca*, rather than those successful in intraspecific competition, and a variant appearing in about the 40th generation was so well-adapted to competing against *Musca* that it drove it to extinction. Other explanations are possible. A subtle shift in laboratory environment or handling procedures might have tipped the balance decisively against the type that was formerly predominant. This possibility can be investigated by comparing the supposedly evolved population with the base population or with a population maintained in pure culture in mixtures with the other species. This is convincing only if the environment in which the lines are tested is the same as that before the shift took place, and this may be difficult to establish. In Pimentel's experiment, the selection lines were deliberately supplemented from time to time by unselected flies taken from wild population, with the intention of reducing the level of inbreeding, and the successful strain of *Phaenicia* may simply have been introduced in this way. Reversals in social dominance are thus often difficult to interpret as unequivocal evidence for social evolution in one of the competing species, although the circumstantial evidence is sometimes quite suggestive.

**Evolution of G x *G* in Defined Mixtures.** In Roy's rice plots, social relations were dominated by the environmental effect of genotypes as neighbors and, in particular, by the weakness of competition from RP individuals. If these plots were self-perpetuating populations, it is easy to appreciate that RP would quickly be eliminated: it performed less well than either of the other two genotypes, no matter who its neighbors were. This would leave only 2A and BK. The social matrix would now be as follows.

| | | Neighbor | |
|---|---|---|---|
| | | 2A | BK |
| Self | 2A | 570 | 574 |
| | BK | 717 | 553 |

The purely environmental effect of neighbors is now much reduced. The 2A strain does not represent a consistently rich or poor environment for growth; instead, it represses itself more than it represses BK. Likewise, BK represses itself more than it does 2A. Thus, the elimination of an unequivocally weak competitor has selectively retained self-inhibiting types. These are likely to be maintained indefinitely through negative frequency-dependent selection. One effect of selection on an arbitrary population of interacting genotypes is thus to transform their environmental effects into G × **G** by retaining self-inhibiting types that form a stable mixture. There might instead be a single unequivocally superior competitor, or an unstable mixture of self-facilitating types; but in that case there would be no social interactions to observe.

**Mutual Facilitation in a Population of Barley.** Barley Composite Cross V was created by intercrossing 31 disparate varieties and afterward propagated in bulk without conscious selection. The population was thus a highly heterogeneous mixture of many different self-fertilizing lineages. R.W. Allard and Julian Adams extracted genotypes from this population at Davis 18 generations after its foundation, and measured their response to one another as neighbors in 3 × 3 hill plots. In most cases, the seed yield of the central plant was greater when its neighbors were dissimilar than when they were members of the same selfed line. In terms of the abcd model, this implies that most pairs of genotypes expressed $a < b$ and $d < c$. Moreover, neighbors of a given genotype tended to repress the growth of the same genotype more strongly than they repressed other genotypes, so that $a < c$ and $d < b$. This was in contrast to experiments with arbitrary pairs of cultivars, where they repeated Suneson's work with more or less the same result: some cultivars, such as Vaughn, were unequivocally inferior competitors that usually suffer in mixture with other varieties. Natural selection in Composite Cross V seems to have caused the evolution of mutual facilitation of growth among genotypes.

**Evolution of Novel Social Relations in Bacterial Populations.** The classical view that populations are effectively uniform for most of the time, with continued selection causing the occasional spread of a beneficial mutation, underlies the conventional interpretation of periodic selection in chemostat populations. It is challenged by the unexpectedly high flux of mutants in some experiments. It is also challenged by the unexpected diversity of at least some chemostat populations. Robert Helling and his colleagues at Ann Arbor found that nearly neutral genotypes of *E. coli* conferring resistance to phage T5 fluctuated in frequency in long-term glucose-limited chemostats, flagging the spread of novel favorable mutations in the usual way. Less expectedly, they also found that variants giving small colonies on agar also appeared in all cultures that were maintained for more than about 100 generations. These variants, which bore all the markers of the initial inoculum, and therefore could not be dismissed as recurrent contaminants, never took over the population but, instead, persisted indefinitely at intermediate frequencies. Further experiments showed that the small-colony types processed glucose much more rapidly than the large-colony types. One might expect that they would quickly eliminate their competitors. The large-colony types, however, can grow in medium that has been exhausted by the small-colony types. It seems that the large-colony types evolve as scavengers that are able to grow on substances excreted by the rapidly growing but inefficient small-colony types. Quite novel kinds of social relationship may evolve after several hundred generations of culture, even in physically constant and uniform conditions, and in doing so support diversity in a system where only uniformity might have been anticipated.

## 142. *Whether mixed cultures become uniform or diverse depends on the relative strength of self-inhibition.*

In natural populations, the social relationships between sexually isolated lineages can be modified only if they continue to interact with one another over long periods of time, which requires that the diversity of the population be maintained. However, whether or not selection itself tends to maintain diversity, in turn, depends on the nature of competitive relationships within the population.

I have so far used terms like "competition" and "competitive ability" in a purely phenomenological way: competition is the way in which the reproduction of one type is altered by the presence of another. The ecological and physiological mechanisms that underlie this effect are important, however, because they can determine how the direction of selection is related to frequency.

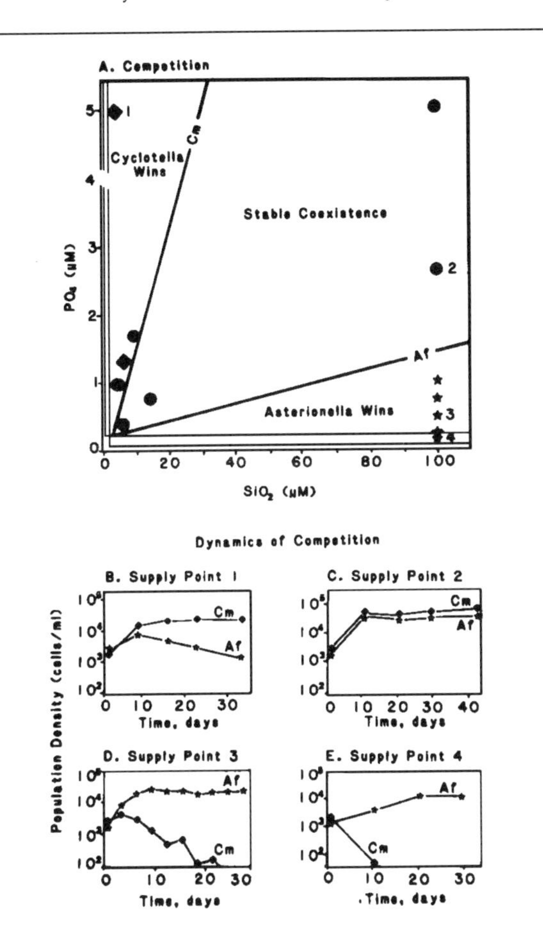


These diagrams illustrate the experiment by Tilman (1977, Tilman and Kilham 1976) described in the text. From growth in pure culture, whether one species eliminates the other, or both coexist, can be predicted for given concentations of phosphate and silicate. The combinations tried experimentally (upper diagram) were:

a) diamonds, *Cyclotella* wins.

b) stars, *Asterionella* wins.

c) circles, coexistence.

The lower four diagrams show the population dynamics for the four numbered points.

*(Continues)*

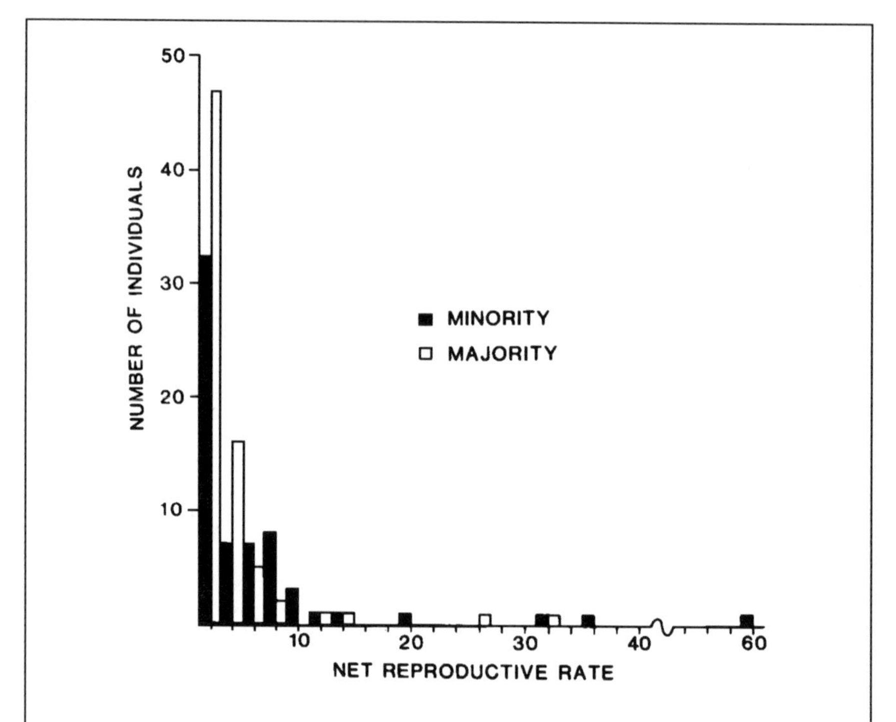


This figure shows the reproduction of minority plants (with dissimilar neighbors) and majority plants (with similar neighbors) in the experiment on *Anthoxanthum* by Antonovics and Ellstrand (1984). The net reproductive rate is the product of survival and fecundity, measured after three years' growth. Plants growing with dissimilar neighbors tend to have higher rates of reproduction, suggesting a general advantage for rare types. Similar experiments have been reported by Schmitt and Ehrhardt (1987, *Impatiens*), Willson et al. (1987, four herbs), Kelley (1989, *Anthoxanthum*), Tonsor (1989, *Plantago*), McCall et al. (1989, *Impatiens*) and Burt and Bell (1992, *Impatiens*). In general, these later studies do not support a large and consistent advantage for rare types when outcrossed families are compared with inbred families or clones.

For experiments on apostatic selection caused by predators, see Allen and Clarke (1968), Allen (1972, 1974, 1975), Manly et al. (1972), Soane and Clarke (1973) and Cook and Miller (1977). The topic is reviewed by Clarke (1979).

(*Continues*)

(*Continued*)

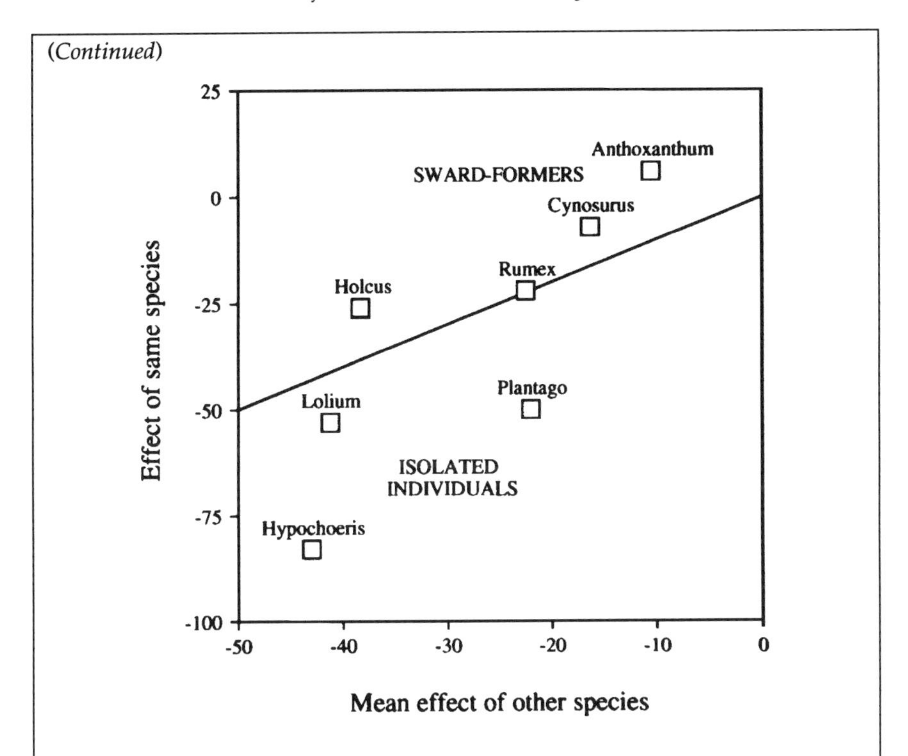


This diagram illustrates the outcome of one experiment from the large and often confusing literature on allelopathy, from data in Newton and Rovira (1975), as described in the text. Units are g/pot as an excess over pots receiving no leachate. The experiment compares the effect of leachate from pots holding the same species with that of leachate from pots holding seven other species of pasture herbs. These effects are highly correlated ($r = 0.80$). However, sward-producers tend to be repressed more by other species by themselves, whereas plants that usually grow as isolated individuals are repressed to a greater extent by leachate from the same species. Line represents equal effect.

## *Negative Frequency Dependence Through Resource Competition*

If different types have the same resource requirements, they will compete intensely, because what is gained by one is necessarily lost by the others. The outcome of such density-dependent selection is the fixation of the type that can subsist on the smallest ration (Sec. 130). It is unlikely, however, that several types will each require the same resources in the same proportions, unless they differ from one another only by a few mutations affecting the efficiency of resource utilization. More generally, different types deplete different resources, or they deplete the same resources at different rates. The more different they are in size, shape, physiology, ecology, or behavior, the less their resource requirements are likely to overlap. To put this in another way, types that are more similar are likely to compete more intensely.

To express this in terms of the **abcd** model, grant that in the absence of competition the two types have equal fitness. The **abcd** then represent the effect of one type on the other or on itself. If more similar types compete more intensely, neighbors of given type will repress the growth of other individuals of the same type more severely and the growth of different individuals less severely. Thus with resource competition we might expect $a < c$ and $d < b$. When either type is rare, almost all its neighbors will be of the other type; its relative fitness will be the greater, and it will increase in frequency. As it becomes more abundant, an increasing fraction of its neighbors will be of the same type, its growth will be more and more severely repressed, and its fitness will fall. Resource competition is thus likely to generate *negative frequency-dependent selection*, the fitness of either type falling as its frequency increases.

Because we have assumed that the two types are equivalent in all other respects, each is superior to the other, when it is itself rare. Either type, if introduced at low frequency into a population dominated by the other type, would tend to spread under selection. Clearly, neither type can be eliminated from the population. The population must therefore remain mixed at equilibrium. This conclusion does not necessarily follow if the two types are in other ways unequal. One type might be much better adapted, in the absence of competition, so that it always retains the greater fitness in competitive situations, even though its fitness declines with its frequency. Nevertheless, negative frequency-dependent selection always tends to retard the loss of diversity.

This kind of selection may contribute to the great diversity of many populations and communities. I have emphasized that the usual effect of selection is to reduce diversity, because mutants and migrants are less likely to reproduce successfully. In special cases, such as the greater fitness of heterozygotes, the outcome of selection is a mixed population; but in such cases the mixture is maintained through the elimination of large numbers of homozygotes in each generation, and it may be doubted that the population could sustain such heavy losses at very many loci. Negative frequency-dependent selection avoids this difficulty, because it will maintain diversity with little or no cost. Either type will tend to decrease in frequency when it is abundant, until its fitness is equal to that of the other. The two types thus have equal fitness at equilibrium, and there is no cost of selection.

**Frequency-Dependence and Environmental Heterogeneity.** In a homogeneous environment defined by two limiting resources, at most two types can coexist permanently through frequency-dependent selection. More generally, the number of types that can be maintained in this way cannot exceed the number of limiting resources. This may appear to set a very low ceiling on the amount of diversity that can be sustained through selection. This is not necessarily true. Different types may be limited, not by different resources, considered as chemical elements or compounds, but by different forms and ratios of these resources. Resources take different forms when their availability is modified by physical or biotic factors. All plants require phosphate, but some may be more efficient in taking it up from solution at high soil pH and others at low soil pH. Nematodes and owls also require phosphate, but the one is adapted to extracting it from plant roots and the other from mice. Even if resources are supplied in solution in an environment that is otherwise completely specified, the relative fitness of different types may vary with resource supply ratios, as Tilman's diatom cultures show. Thus, heterogeneous environments where the form or ratio of resources varies from place to place may provide a substrate for the maintenance of highly diverse communities through selection.

I have already introduced the idea that selection in heterogeneous environments may sustain diversity (Sec. 111). However, it does not necessarily do so. Imagine an environment whose heterogeneity is reflected in a large quantity of G $\times$ E for site-specific reproduction, so that the relative fitness of types varies to any extent from place to place. At the beginning of each cycle, each site is seeded with a fixed number of individuals, representing a random sample from the population as a whole. The descendants of these individuals may occupy the site for any number of generations before eventually dispersing, but throughout this time they continue to increase exponentially in numbers; the population is regulated at a certain density only after the dispersing individuals leave their sites and before they are resettled into new sites at the beginning of the next cycle. The total productivity of a site then depends on the mean rate of increase of its inhabitants; few types will be well-adapted to rare or extreme kind of site, which will thus contribute few individuals to the next generation. In these conditions, the abundance of a given type at the end of a cycle of growth within sites will be determined solely by its rate of growth in each kind of site, weighted by the frequency of the different kinds of site. The type with the greatest weighted average rate of increase will spread to fixation. Diversity is lost during this process, regardless of how heterogeneous the environment may be.

It is probably more realistic to suppose that each site provides a finite supply of resources, so that density will be regulated at the level of the site, rather than the population as a whole. If growth continues within the site for long enough, the sites will reach their carrying capacity, contributing this fixed number of emigrants to the population at the end of the cycle. In the simplest case, all sites support an equal number of individuals, and in each kind of site the best-adapted type becomes fixed before the end of the cycle. It is easy to appreciate that each kind of site now provides a refuge for the type best-adapted to it. The overall fitness of any given type depends not so much on its mean performance over sites—it is lethal in most of them—but on the frequency of the site to which it is the best-adapted type in the population. Even types that are specialized for very rare combinations of conditions cannot be eliminated, because these sites represent a refuge from which they cannot be dislodged. More generally, local density regulation will create frequency-dependent selection. When a type is rare, relative to the availability of the sites to which it is the best adapted, its fitness will be high. This is because nearly all its neighbors in these sites will be inferior, and each individual that migrates to such a site at the beginning of the cycle will have produced a large number of descendents by the end. If the type becomes very abundant, relative to the availability of its preferred sites, then several or many individuals will colonize each of these sites. Because the total productivity of the sites is fixed, each of these founders will leave relatively few descendents; thus, as the frequency of the type rises, its fitness falls, as members of the type increasingly compete among themselves. In this way, local density regulation in heterogeneous environments causes negative frequency-dependent selection. This provides the main reason for supposing frequency-dependence to be so ubiquitous that it sustains a large proportion of the diversity of natural communities.

**Depletable Resources Provide a General Advantage for Rare Types.** The spatial heterogeneity that is envisaged in arguments of this kind is a fixed property of the environment, unaffected by the activities of the organisms that inhabit it. A given type will have high fitness as long as it remains rare, relative to the availability of sites to which it is well-adapted . It will, however, often be the case that resources are depleted by being consumed; a type that is well-adapted to consuming a particular kind of resource will then become less successful as it becomes more abundant because in becoming more abundant it necessarily despoils the sites where it has been successful. Variation in how depletable resources are utilized would thus provide a very general advantage for rarity. Furthermore, it might explain why offspring are produced sexually, because every sexual offspring has a rare genotype and might thus compete less intensely with its sibs than would a comparable individual in a clonal brood.

Janis Antonovics and Norman Ellstrand used the famous *Anthoxanthum* sward at Duke to conduct a test of this idea. They collected tillers from various parts of the field, multiplied them up clonally to obtain large numbers of identical tillers, and then planted them out in a hexagonal design, as follows:

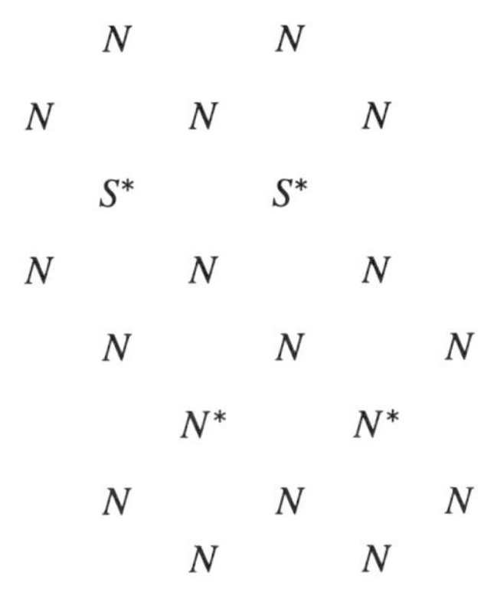


The test plants are marked with an asterisk; their growth was monitored for up to three years. The $N$ testers are surrounded with neighbors from the same clone, and represent a locally abundant type; the $S$ testers are from a second clone, have dissimilar $N$ neighbors, and thus represent a locally rare type. Rare genotypes tended to survive better and to be more fecund than common types; indeed, the relative fitness of the common type was on average only about 0.66. This experiment aroused a great deal of interest, because it demonstrated a substantial advantage for rarity in quasi-natural conditions. Subsequent attempts to repeat it have not yielded such striking results, although there has been some indication that rare types are often fitter when the sward is attacked by herbivores or pathogens.

**Competition for Two Resources in Diatoms.** When two types are limited by a common depletable resource, density-dependent selection will cause the single type that can subsist on the lowest ration to become fixed. When there are two depletable resources, frequency-dependent selection may sustain both types. The simplest situation involves resources that are essential and non-substitutable; that is, a certain supply of both is essential for growth, growth is not increased by supply in excess of this minimum, and no supply of one, however great, can compensate for a shortfall in the other. If one type can subsist on a lower ration of both resources, then it will become fixed. However, the requirements of the two types may differ. One type will then be able to subsist on a lower ration of one resource, whereas the second type requires less of the other resource. This implies that the two types will consume the two resources in different proportions. The growth of both types will drive down resource supply to the point where either type is barely able to subsist on the resource for which the other type has the lower requirement. At this point, the two types are limited by different resources: each will be limited by the resource of which it requires the greater ration. This is an equilibrium, because neither type is able to increase. This mixture will be stable, preserving diversity, provided that either type represses its own growth more severely than it represses the growth of the other type. This is equivalent to saying that both types must consume proportionately more of the resource that is limiting their own growth. The condition for the indefinite coexistence of two types in a mixture is thus that both should consume proportionately more of the resource for which they require the greater ration.

This argument was developed by David Tilman, who tested it with mixtures of different species of diatoms in semi-continuous cultures maintained for about 50 generations. At a certain death rate (i.e., outflow rate), in pure culture the minimal subsistence levels of silicate and phosphate for *Asterionella* were 1.9 $\mu$M silicate and 0.01 $\mu$M phosphate. The corresponding levels for *Cyclotella* were 0.6 $\mu$M silicate and 0.2 $\mu$M phosphate. The growth of a mixed population will drive silicate down to 1.9 $\mu$M (at which point *Asterionella* is limited by silicate, but not by phosphate), and phosphate down to 0.2 $\mu$M (at which point *Cyclotella* is limited by phosphate, but not by silicate). At this point, *Asterionella* consumes proportionately more silicate ($1.9/0.01 = 190$ for *Asterionella*, whereas $0.6/0.2 = 3$ for *Cyclotella*). *Asterionella* therefore represses its own growth, through depletion of silicate, more severely than it represses the growth of *Cyclotella*, through depletion of phosphate. The converse applies to *Cyclotella*: both types are inhibiting their own growth more effectively than they are inhibiting the growth of their competitor. These predictions can be tested by manipulating rates of nutrient supply. If the rate of supply of phosphate relative to that of silicate exceeds the ratio in which the two are consumed by *Cyclotella*, the species limited by phosphate at equilibrium, then *Cyclotella* is released from its limitation by phosphate, whereas *Asterionella* remains limited by silicate. Consequently, *Cyclotella* will spread to fixation. Conversely, *Asterionella* will become fixed if the supply ratio of silicate exceeds its consumption ratio by *Asterionella*. The results of competition experiments with different supply rates of the two resources agreed well with the predictions from pure-culture dynamics.

**Apostatic Selection.** Predators and other antagonists are, in a sense, the negative image of growth sites: a genotype may prosper through avoiding predation, just as it may prosper by finding suitable conditions for growth. Bryan Clarke of Nottingham has repeatedly urged that predation may cause negative freqency-dependent selection, through the preference of predators for more abundant types of prey. This has been dubbed "apostatic selection", the rare types being the apostates. One simple (and rather enjoyable) experiment is to lay out pastry baits in different colors on university lawns. These are eaten by university birds, who display a marked preference for the more common color. Extrapolated to less contrived conditions, for example, snails in hedgerows, this supplies a simple mechanism for the maintenance of diversity through selection favoring rare types.

## *Positive Frequency-Dependence Through Pollution*

Individuals may compete indirectly through a scramble for resources, those who are more able impoverishing the less able; they may also compete more directly, by blocking access to resources that would otherwise be available. The main cause of interference is pollution: the suppression of growth by metabolites or other substances released into the medium or administered to competitors.

**Allelopathy.** Inhibitors that are produced during the course of growth can be thought of as negative resources. If a single inhibitor is present, the type that can subsist at the highest dose will be fixed. If two types produce different inhibitors, then the one that can subsist at the higher dose of both will be fixed. Otherwise, there is an equilibrium that is stable if either type is more strongly repressed by the inhibitor it produces itself. On the other hand, if either type is relatively immune to its own product, the equilibrium is unstable, and the type that is initially the less strongly repressed (because it is the more abundant, for example) will become fixed. I do not know whether there is any physiological reason that individuals should be either more or less strongly affected by their own waste products than by those of others. There is, however, a compelling evolutionary reason: types that are specifically poisoned by their own waste products cannot increase in frequency, and if they exist at all, they must remain rare. Most of the cases we observe, therefore, should involve specific inhibitors that repress the growth of unlike types more effectively than they repress the growth of their own type. This will cause positive frequency-dependent selection, with fitness rising as frequency increases. The outcome will be the fixation of a single type and the loss of diversity.

Some plants produce substances that inhibit the growth of neighbors different from themselves; this phenomenon is known as allelopathy. Whether or not specific growth inhibition often occurs has been controversial, primarily because of the difficulty of interpreting experiments that use macerated or dying plant tissues. A careful experiment by E.I. Newman and

A.D. Rovira of Bristol, however, uncovered allelopathic interactions between most of the pasture herbs that they investigated. They grew eight common species in separate pots, collecting the liquid that percolated through the growth medium and applying this leachate to other individuals of the same eight species. By measuring the growth of the recipient plants, they could construct the 8 × 8 matrix defining the effect of the leachate of each of the donors on each of the recipients. On average, this leachate repressed growth; the substances excreted by plants therefore have a general inhibitory effect on growth. There was no overall tendency for plants to be less strongly repressed by leachate from their own species. However, some species were less strongly and others more strongly repressed by self-leachate. Species such as *Anthoxanthum* and *Cynosurus* were less strongly repressed by leachate from their own species than by leachate from other species; on the other hand, species such as *Plantago* and *Hypochoeris* were actually more strongly repressed by their own species. This seems to be related to their growth habit in natural populations. *Anthoxanthum* and *Cynosurus* form more or less continuous swards that can dominate permanent grassland, whereas *Plantago* and *Hypochoeris* are found as isolated individuals or small groups, never as pure stands. Thus, although there is no general tendency for self-inhibition to be weaker than inhibition by different types, there does seem to be a tendency for self-inhibition to be weaker in species that usually grow with neighbors of the same species. Whether a similar pattern is displayed by different genotypes of the same species is not known. Nor is it possible to infer cause and effect from these data: plants with relatively weak self-inhibition may for this reason grow as pure stands, or plants that grow as pure stands may evolve weak self-inhibition. There is a fertile and so far neglected field for evolutionary experiments here.

**Simultaneous Resource Competition and Inhibition.** Two types may coexist on a single resource if at the same time they produce a single inhibitor. If one type can subsist on a lower ration of the resource and a higher dose of the inhibitor, then it will become fixed. One type, however, may be able to subsist on a lower ration of the resource while being repressed by a lower dose of the inhibitor. The two will then coexist if the type that is resource limited at equilibrium produces proportionately more inhibitor.

The classic experiment involving a situation like this is Gause's study of yeast mixtures. He used the common budding yeast *Saccharomyces* and a smaller yeast that he refers to (incorrectly) as "*Schizosaccharomyces*." The limiting resource was sugar; the inhibitor was the ethanol produced by fermentation. In pure culture, *Saccharomyces* has the higher $r_{max}$ and the higher $K$. Nevertheless, both species persist in mixed cultures for at least 20 or 30 generations, though *Schizosaccharomyces* is the less abundant. Gause explained the maintenance of diversity in mixed cultures by the greater inhibitory effect of *Schizosaccharomyces*, which produces about twice as much ethanol per unit volume as does *Saccharomyces*.

## 143. *The selection of social behavior leads to an Evolutionarily Stable State.*

The best introduction to evolutionary game theory is the book by Maynard Smith (1982), who analyzes more complex versions of the Hawk-Dove game as well as a range of other games. The game discussed in the following section is usually known as the Prisoner's Dilemma and has a large literature; see Axelrod and Hamilton (1981) and Axelrod (1984).

The experimental evidence suggests that selection is unlikely to modify social relations between arbitrary strains of the same species. When strains exhibit strongly marked differences in behavior, however, selection will readily alter their frequencies in the population. How can we predict the outcome of selection from a knowledge of social behavior?

I have given two different and apparently incompatible accounts of evolution in a mixture of socially interacting individuals. On the one hand, tall, aggressive individuals exclude shorter plants from mixtures. On the other hand, natural selection in mixtures favors an harmonious accomodation between types. Is there any general rule that will predict whether certain social characteristics will be favored, and whether at equilibrium the population will be uniform or diverse? The problem is quite subtle. At first site, it might seem obvious that aggressive plants able to repress their neighbors will spread through a population. On reflection, however, when such individuals become abundant, they will for the most part be repressing themselves, whereas less aggressive types might share resources to their mutual profit.

**Warfare in Hydrozoans.** Leo Buss of Yale and Rick Grosberg of Davis discovered a beautiful example of dichotomous social behavior in *Hydractinia*, a marine hydrozoan that forms colonies of zooids on gastropod shells. When two founding zooids settle on the same shell, they will proliferate until the edges of the two colonies come into contact. Colonies that have the same genotype at a locus that determines somatic compatibility will then fuse. If they are dissimilar, they will respond in one of two ways. The first is to lay down a fibrous matrix that forms a stable boundary at which growth by both colonies ceases. The two colonies then share the resource provided by the limited area of the shell. The second is to proliferate a stolon, into which nematocysts from nearby zooids migrate. These nematocysts are discharged into the foreign tissue of the neighbor, forming a necrotic zone of dying zooids. This continues until the neighboring colony is destroyed. Some genotypes (which can be replicated by dividing the growing colony into portions) are always stoloniferous; some are never stoloniferous; others are able to switch from one behavior to the other. When a stoloniferous colony meets a matrix-forming colony, it invariably destroys it after a more or less prolonged struggle. When two stoloniferous colonies meet, one (the faster-growing colony) destroys the other. The stoloniferous colonies are certainly the stronger competitors, in the sense of repressing the growth of their neighbor more effectively. Yet both types coexist in *Hydractinia* populations, so aggression cannot always pay. How are the pacific matrix-forming colonies maintained in the population?

**The Hawk-Dove Game.** The situation in *Hydractinia* is a particularly dramatic example of the general tendency for social behavior to be less aggressive in many cases than a naive interpretation of Darwinism would lead one to expect. This was a fertile source of misunderstanding in studies of animal behavior for many years before the development of evolutionary game theory by John Maynard Smith in the 1970s.

Game theory was originally developed by John von Neumann to explain how human conflicts are resolved. The central concept of game analyses is that individuals will come to behave in such a fashion that no alternative behavior will profit them more, given that their opponents are following the same rule. It has been only partially successful as a theory of human social behavior, largely because it assumes that the outcome of a conflict can be expressed on a linear scale of utilities, and that players always behave rationally in their own interests so as to obtain the greatest possible utility. It scarcely needs pointing out that people often behave irrationally and that the different utilities at stake in a conflict may be incommensurable. It is much more plausible, however, to assert that in a population whose members have long interacted with one another in similar circumstances, Darwinian fitness provides a single linear scale of utility and that the population will have become modified through selection acting strictly on the differences in fitness caused by different behaviors. We can then substitute an evolutionary prediction for the outcome of social conflict: social behavior in a population will evolve towards a state such that no type expressing a behavior different from that prevailing in the population can spread. The population must then be at an equilibrium that can be called the Evolutionarily Stable State, or, to introduce one of the few useful acronyms in evolutionary biology, the ESS.

The contrast between aggressive and pacific individuals was the subject of the original application of evolutionary game theory to social behavior. In a very simplified form, the argument is as follows. Imagine that individuals compete in pairs for some resource whose value can be represented as $V$: this value is fundamentally an increment in fitness, although of course it will be physically and immediately apparent as calories, space, access to water, or some such benefit. When two Doves meet, they behave in such a way that the resource is shared between them, like the matrix-forming types of *Hydractinia*. Hawks are aggressive, like the stoloniferous types of *Hydractinia*. When a Dove meets a Hawk it immediately retreats, so that the Hawk gets all of the resource and the Dove gets none. When two Hawks meet, they fight for possession of the resource. The victor gets all of the resource and the loser none; moreover, the loser suffers damage (such as destruction by the nematocysts of the winning colony) that is equivalent to losing $C$ units of resource. We can then represent these interactions in terms of the abcd model, with one modification: the abcd in this case do not represent fitness, but instead represent the effect on fitness of an interaction with a given type of neighbor. The force of this assumption is that Hawks and Doves are equally successful in acquiring uncontested resources and thus have some equivalent baseline fitness; the difference in fitness between the two types arises solely through social encounters in which they must compete for additional shares of resource. Type 1 is Dove, Type 2 is Hawk. When two Doves meet, they share the resource, each obtaining $\mathsf{a} = V/2$. When two Hawks meet, each has an equal chance of obtaining the resource of value $V$, or sustaining an injury equivalent to a loss in value $C$; the expected value of the contest to either individual is thus $\mathsf{d} = (V - C)/2$. A Hawk that meets a Dove is always successful, so that $\mathsf{c} = V$ and, conversely, $\mathsf{b} = 0$.

A behavior is an ESS if, in a population in which almost all individuals display that behavior, no other behavior causes a greater increment in fitness to the individual displaying it. Another way of expressing the same principle is to say that a behavior represents an ESS if, when competing with the type displaying that behavior, individuals with the same behavior gain more than individuals with any different behavior. In terms of the abcd model, this requires that $\mathsf{d} > \mathsf{b}$ (Hawk is an ESS) or $\mathsf{a} > \mathsf{c}$ (Dove is an ESS). As the game has been defined, Dove cannot be an ESS, because if nearly all of the population were Doves, a Dove would obtain $V/2$ from each contest, whereas a rare Hawk (all of whose contests would be with Doves) would obtain $V$. Thus, $\mathsf{c} > \mathsf{a}$ for the game, and hawks would therefore tend to spread in a population of Doves. Doves, however, will not spread in a population of Hawks if they gain less from an encounter with a Hawk than a Hawk itself gains from an encounter with another

Hawk. Hawk is thus an ESS, provided that $d > b$, that is, $(V - C)/2 > 0$. This will be true if $V > C$, that is, if the chance of gaining the resource by fighting for it is worth the risk of being injured.

This, however, may not be true; in *Hydractinia*, for example, the cost of losing a conflict with a stoloniferous colony is complete destruction. If $C > V$, then neither Hawk nor Dove is an ESS. The ESS is, instead, a mixture of the two behaviors, in proportions such that either behavior would be more profitable if it were expressed less frequently. The frequency of Hawk in this stable mixture is

$$f_{ESS} = (c - a)/[(c - a) + (b - d)]$$

The ESS is thus a mixture if $c > a$ and $b > d$, that is, if either behavior is more profitable when rare, because neighbors suppress similar individuals more severely.

Whether or not aggressive behavior prevails in a population will, then, depend on how severely aggressive individuals injure one another when they become abundant. The argument is entirely phenotypic and is rather insensitive to assumptions about genetics. Aggression may be a fixed and heritable behavior of a certain group of individuals, in which case the ESS is the frequency of individuals of this type in the population at equilibrium. Alternatively, every individual may display aggressive or pacific behavior at random in any given encounter: in this case, the ESS is the frequency with which any given individual is aggressive when the population is at equilibrium. In part because of its freedom from any particular genetic model, evolutionary game theory can be applied very broadly to any situation involving social interactions, and its initial introduction led to an efflorescence of theoretical and comparative studies. Experimental work, however, has concentrated on demonstrating adaptedness, usually by showing how the behavior of individuals changes appropriately when their environment is manipulated, and there seem as yet to be no selection experiments that trace the evolution of the ESS.

## 144. *Cooperative behavior may be selected during repeated contests.*

This very simple idea explains why populations are not always dominated by the most aggressive kinds of individual. It may be as much as we need to understand the rudimentary social relations between barley plants or fly larvae, but it scarcely provides a clear explanation for societies in which individuals routinely assist others, let alone for communities organized for mutual benefit, such as lichens or plant-mycorrhizal associations, where individuals depend absolutely on the continued goodwill of their partners.

Hawk, after all, may not always be an ESS, but Dove never is. It has been difficult to understand how cooperative social behavior can evolve in a Darwinian world that seems to reward only selfishness.

To show how the problem has been solved, it helps to define a game in which selfish and cooperative individuals compete. It is essentially a trading game. Two individuals each possess a good that the other requires; say, one grows corn and the other makes plows. They arrange to meet at some rendezvous, each bearing a large box, and then exchange boxes and leave. If both are honest, the agriculturalist will receive a box containing a plow, and the industrialist will receive a box containing corn; both will benefit. But what prevents dishonesty? After all, either may be tempted to bring an empty box, obtaining what they require (they hope) without supplying anything in return. If both succumb to temptation, then neither benefits, and the social arrangement breaks down. The equivalent of Dove, then, is an honest trader who always cooperates with his neighbor; the equivalent of Hawk is the dishonest trader who defects by providing an empty box. The social matrix is as follows.

| | | Neighbor: Cooperate | Neighbor: Defect |
|---|---|---|---|
| Self | Cooperate | Award for honesty | Booby prize for cooperating with a dishonest neighbor |
| | Defect | Cheating gets something for nothing | Deadlock of two dishonest traders |

It will be cheaper to bring an empty box, so we can assume that $c > a$ and $d > b$. On the other hand, if both traders are honest, each gains more than if both are dishonest: $a > d$. The game is therefore defined by $c > a > d > b$. The solution is obvious from the preceding section: the only ESS is the Hawk-like behavior of always defecting. That is the dilemma.

Either trader would benefit if they behaved honestly, provided that the other reciprocated. If they meet only once, no such arrangement will be honored, because neither can know whether or not their neighbor will defect. If they were to meet repeatedly, however, their past behavior may indicate what they are likely to do in the future. The expected future behavior of one's neighbor could then serve to guide one's own. To be sure, the calculation is a delicate one. If one's neighbor has a fixed behavior (always cooperate, or always defect, or any alternation of cooperating and defecting), then it is obviously profitable to defect in every encounter. It is only if one's neighbor's behavior depends in some appropriate way on one's own response that any sustained cooperation will be possible. The appropriate way, however, is not easy to define. Complete dishonesty is clearly an ESS: if all one's neighbors always defect, it is folly to cooperate. This does not rule out the possibility that some less adamantly uncooperative behavior might also be an ESS. To identify this behavior, however, required the unlikely collaboration between a political economist, Robert Axelrod, and an evolutionary biologist, W.D. Hamilton, and involved one of the most extraordinary selection experiments ever devised.

Axelrod approached the problem empirically by inviting people who might reasonably be supposed to be experts, for example, game theoreticians in economics and sociology, to specify the best possible social behavior, by defining the response to one's neighbor in each of a long series of encounters. There was no limit on the complexity of such behaviors, which were expressed as computer programs of any length. Whether one cooperated or defected in a given encounter could involve remembering the outcome of every previous encounter with one's neighbor, weighting more and less recent outcomes appropriately, inferring their likely future behavior, projecting its consequences for one's own behavior, expressing the result as a probability distribution, and then drawing a random number to decide how to act. Or, of course, you could just always defect. The failure of previous attempts to solve the problem was confirmed by the fact that all 14 programs submitted were different. They were entered into a tournament, along the lines of a hill-plot crop trial. Each program was matched with a partner for a long series of encounters, accumulating a score from their outcome. After all the pairwise combinations had been tried, the program with the best average score was declared the winner. Rather surprisingly, this was the shortest program in the tournament, devised by Anatol Rapoport, a philosopher at the University of Toronto. Its recipe for social behavior was very simple, involving none of the labored ingenuity of other entries: always cooperate at the first meeting and thereafter do what your neighbor did at the previous meeting. Rapoport called this pattern of behavior "Tit-for-tat." It is friendly on first encounter, promptly and invariably punishes any dishonesty, and when honest trading is resumed immediately forgives—and forgets. Although these straightforward virtues might seem attractive, however, it was not easy to explain why they should be so successful. Axelrod therefore organized a second tournament, in which more than 60 contestants, profiting from the outcome of the first, invented yet more ingenious methods of exploiting one's neighbor. Rapoport, with really superb aplomb, simply submitted Tit-for-tat again, unrevised. It won.

For an evolutionary biologist, this is not quite conclusive: it is perfectly conceivable that a behavior might have a high average score over pairwise contests, and yet fail to spread in a heterogeneous population. (I shall discuss whether this is likely to be true in the next section.) The programs were next entered in a selection experiment. The programs constituted a population in which each was at first equally frequent, there being equal numbers of copies of all of them. Each copy was paired with a random partner, and its score over a long series of encounters used as a measure of fitness to adjust the frequency of copies of that program in the population entering the next cycle. As the experiment progressed, different programs waxed and waned. Aggressive programs able to exploit foolishly cooperative neighbors at first began to spread rapidly; but as they became more abundant, they became more likely to meet one another and fail together in deadlock. After 1000 generations, Tit-for-tat was the clear winner, not only more abundant but still spreading more rapidly than any of its competitors.

With the benefit of hindsight, it was subsequently possible to prove that Tit-for-tat is an ESS, provided that neighbors interact for long enough. It nonetheless remains as surprising as it is instructive that such an agreeably social pattern of behavior should be so successful. After all, Tit-for-tat never wins. At best, it will draw (if its neighbor always cooperates); more often (if its neighbor ever defects, even once) it will lose, though not by much. It is successful because more aggressive behaviors end up by punishing themselves. By rewarding similar patterns of behavior, Tit-for-tat gradually creates a social environment in which it prospers. Moreover, it does not require a capacious memory or advanced mental powers, but merely the ability to follow a very simple rule that can evolve as readily in bacteria as in apes. The mutual profitability of reciprocating past favors will lead to the evolution of cooperative social behavior between partners who interact repeatedly, without involving any process more onerous than straightforward Darwinian selection.

## 145. *Fitness in mixed cultures is not necessarily transitive.*

Most familiar games are *transitive* in the sense that players can be ranked unequivocally in terms of their ability: if $A$ beats $B$, and $B$ beats $C$, then $A$ will beat $C$. It is generally assumed that populations behave in the same

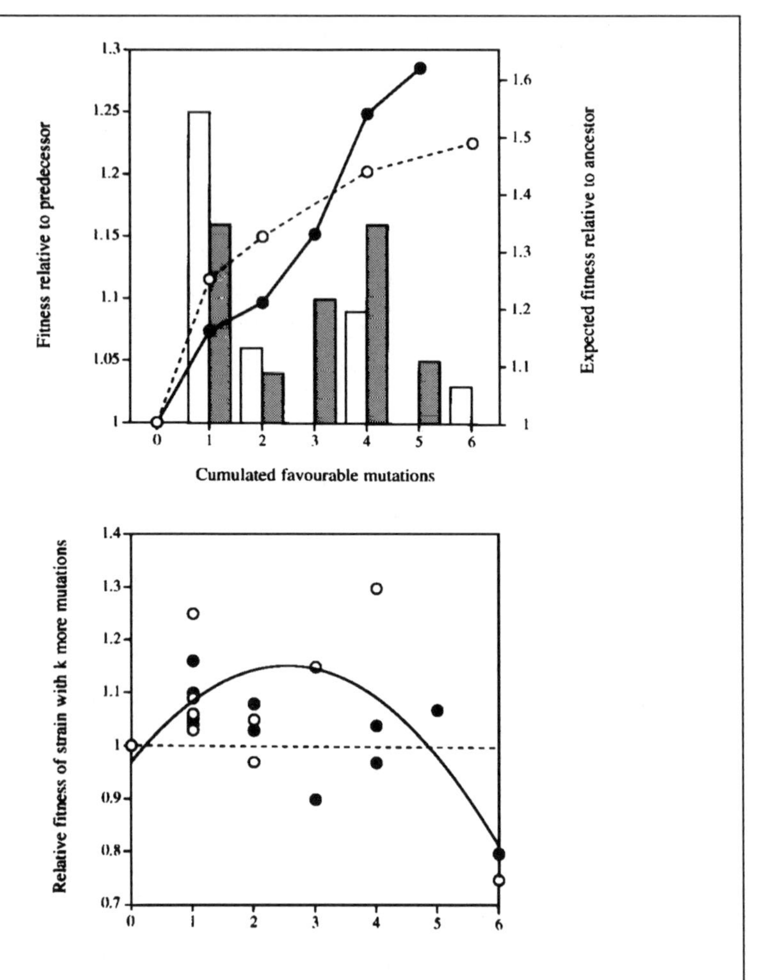


These diagrams show the outcome of competing strains evolved in the chemostat against their more or less remote ancestors, from the diploid series in the experiments on yeast by Paquin and Adams (1983). There are two sets of trials, differing only in the state of a 5-fluouracil resistance marker, distinguished by open bars and broken lines, or stippled bars and solid lines, in the upper diagram, and by open or solid symbols in the lower diagram. The passage of successive favorable mutations was signaled by fluctuations in the frequencies of markers, in the usual way, enabling strains that had accumulated different numbers of favorable mutations to be isolated and stored. The histogram bars in the upper diagram show the fitness of successive mutant strains relative to their immediate ancestor; in all cases, a single additional mutation increases fitness, as expected. The lines show

(*Continues*)

(*Continued*)

the expected fitness of strains with successively more mutations relative to the original strain, assuming that each mutation contributes independently to fitness; because each mutation causes an increase in fitness, relative to its immediate ancestor, expected fitness relative to the initial strain increases monotonically. The lower diagram shows the actual relative fitness of the evolved strains, relative to ancestors with up to six fewer mutations. Strains with a single extra mutation always have greater fitness; this is the result already shown in the upper diagram. However, there is no consistent tendency for strains with two or more additional mutations, for example, one isolated after the passage of five successive mutations, compared with one isolated earlier after the passage of three successive mutations, to display greater fitness. Indeed, there is some indication that after about 300 generations and the passage of six mutations, each superior to its predecessor, the evolved strains are actually inferior to the original inoculum. The solid line is a fitted quadratic equation ($y = -0.028x^2 + 0.144x + 0.97, r^2 = 0.44$), showing that relative fitness does not increase monotonically relative to distant ancestors. Thus, fitnesses in this experiment appear to be intransitive.

The general transitivity of fitness among different species of *Drosophila* in competition experiments has been documented by Richmond et al. (1975) and Goodman (1979).

way, with the type that is on average most successful in pairwise encounters spreading through a mixed population under selection. Furthermore, the same rule can be applied to successive gene substitutions. A population placed in a novel environment will adapt through a succession of substitutions, and if genotypes are extracted from the population at intervals and afterward made to compete in mixed cultures, those that evolved later will eliminate those that evolved earlier. However, games, and perhaps evolutionary processes, are not necessarily transitive. The standard example is the Rock–Scissors–Paper game played by children. Rock (a clenched fist) blunts Scissors (two spread fingers), and Scissors cut Paper (the open hand); however, Paper wraps Rock, so the profitability of the three behaviors cannot be expressed on a single scale. If evolutionary processes are likewise intransitive, their analysis will be greatly complicated. In socialized environments, the concept of adaptedness must be replaced by a stricter concept of fitness; but if social relations are intransitive then there is no fixed scale on which fitness can be expressed.

Intransitivity represents a higher order interaction among genotypes, which might be symbolized as G × **G** × G: the outcome of competition between two types depends on whether a third is present. Because neighbors may modify their environment, while the outcome of competition between two types is often sensitive to the state of the environment, it is quite possible that intransitivity is widespread. The empirical evidence is not yet sufficient to know whether this is really the case.

**Social Hierarchies in Mixtures of *Drosophila* Species.** The simplest approach is to measure the outcome of selection in pairwise mixtures of a large number of types. The degree of intransitivity can be expressed by considering all the possible combinations of three types. The pairwise interactions of these three types are either transitive or intransitive; the frequency of transitive triplets is then a measure of the overall transitivity of the social matrix. A large experiment of this kind was reported by D. Goodman of the Scripps Institute, who set up all the pairwise combinations of 19 strains of *Drosophila*, representing 15 different species. The pairs were set up with five females of each kind, the proliferating population being transferred from vial to vial for nine weeks, when it was counted. The type that was then in a majority can be regarded as the winner; in many cases, one type was almost or completely eliminated. Assessed in this way, the social matrix was strikingly transitive. The strains could be arranged in an almost perfect linear sequence, with the common laboratory species *melanogaster, simulans*, and *pseudoobscura* at the head, and the less familiar *paulistorum, pallidipennis, funebris*, and others at the foot of the ladder. About 95% of triplets were transitive, and the few exceptions almost all concern two species of similar competitive ability, both of which were still present in substantial numbers at the end of the experiment.

**The Outcome of Selection in Complex Mixtures.** The outcome of selection in pairwise mixtures leads naturally to selection experiments involving three or more strains: if social relations are transitive, then the outcome of selection in mixtures of three or more types can be predicted from the outcome of pairwise competition. Rollin Richmond of Indiana and a group of collaborators investigated this prediction in the triplet of *Drosophila pseudoobscura, D. willistoni*, and *D. nebulosa*. In three of four cases, *pseudoobscura* eliminated *nebulosa* from mixtures, the exception being highly unstable, so *pseudoobscura* > *nebulosa*. *Pseudoobscura* and *willistoni* usually coexisted for at least 200 or 300 days, although in one case *willistoni* was eliminated; thus, *pseudoobscura* >= *willistoni*. *Willistoni* and *nebulosa* also usually continued to coexist, but in one case *nebulosa* was eliminated, so *willistoni* >= *nebulosa*. These pairwise experiments thus suggest that *pseudoobscura* is the strongest competitor and *nebulosa* the weakest. When the mixture of all three species was tested, *nebulosa* was eliminated in all six replicates, and *willistoni* from three; *pseudoobscura* persisted in all. The actual social hierarchy was thus *pseudoobscura* > *willistoni* > *nebulosa*, which is consistent with the results of the pairwise trials.

**Transitivity in Evolved Mixtures.** Experiments such as these seem to show that as a general rule the social hierarchy of arbitrary mixtures of related species is transitive, or nearly so. Whether this applies to strains that have evolved together in asexual or self-fertilized organisms is not known. The reason for thinking that it might not apply is that any transitive hierarchy will be reduced by selection, leaving only the strongest competitor. The elements of the hierarchy that will be the most difficult for selection to digest will be those involving intransitive relationships, leaving these as a more prominent feature of evolved populations than experiments with arbitrary strains would suggest. However, I do not know of any demonstration that transitivity decreases through sorting in heterogeneous mixtures.

**Intransitive Fitnesses of Successive Substitutions in Chemostat Populations of Yeast.** The successive substitution of adaptive mutations in asexual populations can be followed by the fluctuations in frequency of neutral markers (Sec. 58), allowing strains that have accumulated a known number of adaptive mutations to be isolated from the chemostat. Charlotte Paquin and Julian Adams of Ann Arbor were apparently the first to think of testing whether the sequential substitution of these mutants represented a hierarchy of types of increasing fitness that would behave transitively in pairwise competition with one another. They used haploid and diploid lines maintained in glucose-limited chemostats for about 250 generations (the same experiments are described from a different point of view in Sec. 163). Isolates that differed by a single mutation always behaved consistently: the strain with one more mutation replaced the other. This was expected, since the strain with one more mutation had, in fact, replaced the other during the evolution of the population in the chemostat. Isolates that differed by several mutations behaved less consistently. They were always less fit, relative to the original strain, than would be expected from the separate increments in fitness caused by each successive new mutation. Fitness did not, then, simply cumulate through time. Although the difference in fitness between a given strain and the strain that it had just replaced remained about the same throughout the experiment, this difference represented a smaller and smaller advantage over earlier strains. Indeed, in some cases, strains that had accumulated five or six adaptive mutations were *less* fit than the original type; in consequence, fitness in the diploid population at least actually *decreased* during the experiment. Clearly, this yeast population was not merely adapting to a novel physical environment; it was also responding to the social environment created by the currently prevalent strain, in such a way as to create an intransitive succession of new types.

# 146. *Intransitive social relations destabilize genotype frequencies through time-lagged frequency-dependent selection.*

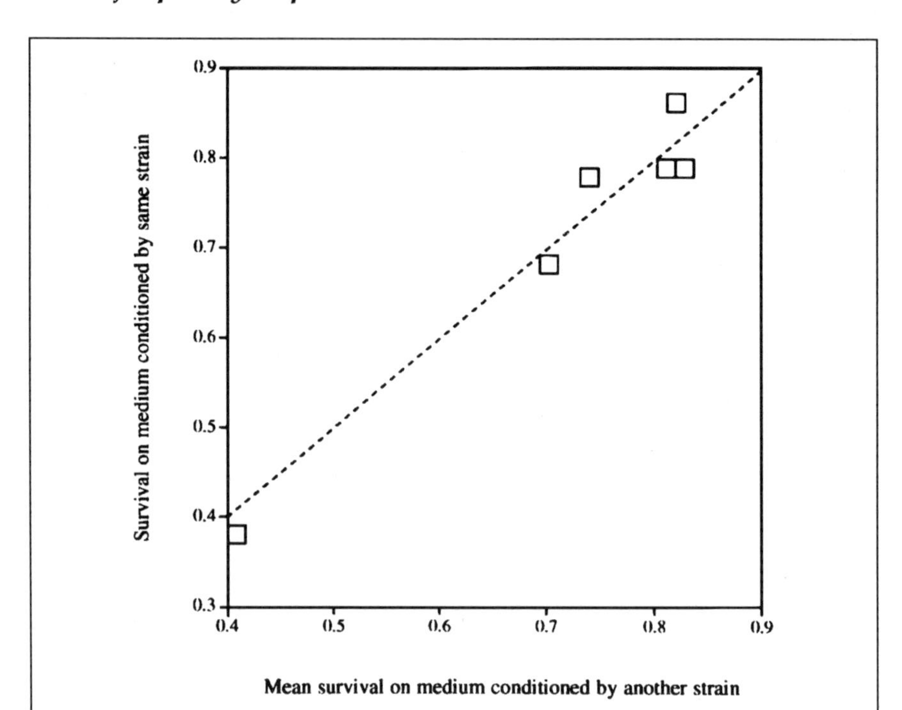


This diagram shows the effect of previous conditioning of the food medium by strains of *Drosophila melanogaster*, from an experiment by Dolan and Robertson (1975). They used five strains derived from the Kaduna isolate and a sixth line from the Oregon R stock. Each point is based on the number of adults emerging from a total of 1500 larvae. The broken line represents equal survival whether the medium has been previously conditioned by the same or by another strain. The outlier, with low survival, was from a line selected for low bristle number. There was no consistent tendency for conditioning by the same strain to depress survival, although there is a weak suggestion that self-conditioning may be beneficial in some cases and deleterious in others. Similar ambiguous results were obtained by Weisbrot (1966) and Dawood and Strickberger (1969); a strong negative effect of self-conditioning was reported by Huang et al. (1971). This result can be compared with Newton and Rovira's experiment in Sec. 141.

(*Continues*)

*(Continued)*

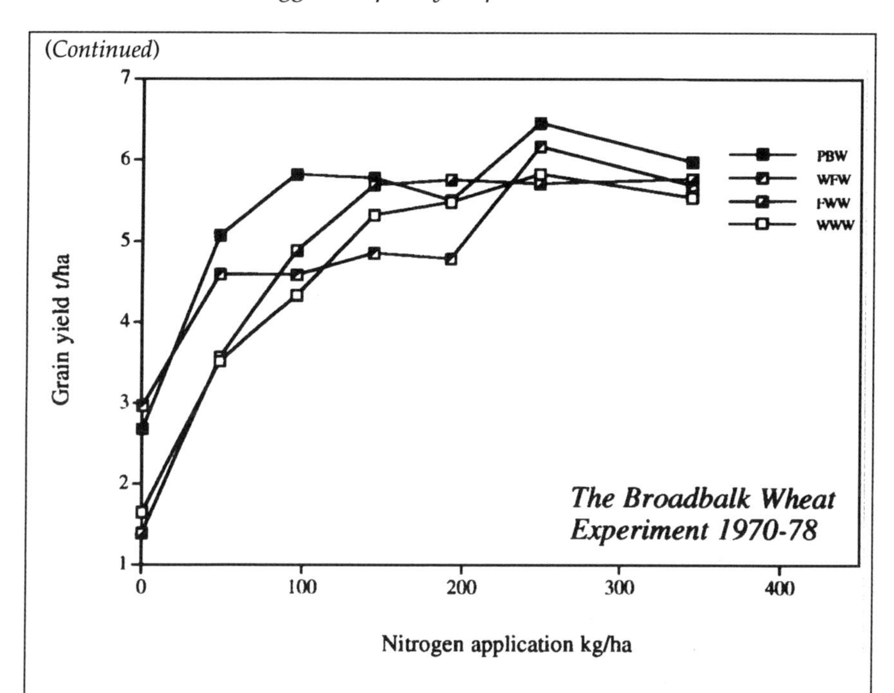


This diagram shows recent results from the Broadbalk Wheat Experiment, from Dyke et al. (1983). The four rotation treatments are: WWW, continuous wheat; FWW, second wheat crop after fallow; WFW, first wheat crop after fallow; and PBW, wheat crop following potatoes and beans. Continuous wheat yield is calculated as the average of the two sections on which this sequence was followed. Yields at a given level of nitrogen application are unweighted means over all treatments (involving other elements) representing this level of nitrogen. The results are for three rotations in 1970–1978. Ripley (1941) describes a variety of experiments and gives a thorough historical review.

The basic dynamic consequence of intransitive interactions can be appreciated by imagining the outcome of the Rock–Scissors–Paper game. It is probably easy to understand that when individuals are free to change their behavior, the best that can be done in any given encounter is to choose one of the three possibilities at random, each with one-third probability. Children soon learn this. But what if Rock, Scissors and Paper are three different genotypes, with fixed behavior resulting in the intransitive social relations that their names suggest? To settle draws, I shall assume that if two similar individuals meet they behave in a Dove-like manner and share whatever resource is being contested. Suppose that one type, say, Paper, is initially somewhat more abundant than the others, which are about equally frequent. The premium given to draws means that Paper has a slightly greater fitness, since more of its encounters will be with Paper neighbors and it will increase in frequency. As it does so, its fitness will increase—because there are more draws—and it will eventually become very abundant, displacing the other two types. But at some point, a largely Paper population will be invaded by Scissors, since they will win almost all their encounters. A population consisting largely of Scissors is no more stable, however, because it will be invaded by Rocks. Not only is there no equilibrium, there is no end-point: the population will cycle endlessly between its three possible states.

A population under selection in a novel physical environment will increase asymptotically in adaptedness, with the interval between successive beneficial mutations lengthening through time. On the other hand, if selection is caused largely by intransitive social interactions among genotypes, the population may never reach any definite end point, fitness will not tend to increase, and genotypes will replace one another in an endless sequence, with a more or less constant interval between the spread of successive clones. There is a fundamental point at issue here. The conventional view is that evolution through selection is elicited by exogenous changes in the physical environment, and if these were to cease any directional evolution would soon run down, with the fixation of the best-adapted type. The alternative view is that social selection based on largely intransitive interactions will drive an endless process of change, even if the physical environment remains precisely the same. Whether this alternative view has much substance is not yet known: there have simply not been enough careful long-term experiments. Lenski and Bennett's bacterial populations (see Sec. 64) seem to evolve more and more slowly through time, but in Paquin and Adams' yeast populations, the interval between successive substitutions did not lengthen appreciably during their experiment.

**Time Lags.** When social relations are intransitive, any genotype will, by increasing in frequency, depreciate its own social environment in the future. The spread of Scissors through a Paper population is checked, not by self-inhibition (draws are profitable), nor by any increase in the ability of its opponent (Scissors always wins against Paper), but because its very abundance creates a social environment in which Rock is a superior competitor: its present success ensures its long-term ruin. The three types all experience negative frequency-dependent selection, but the negative effect of increasing frequency on fitness is postponed. The effect of this time lag is to destabilize the genetic equilibrium that is normally associated with negative frequency-dependent selection. When a given increase in frequency will only cause a reduction of fitness several generations hence, the frequency of a type will overshoot its equilibrium value before being restrained by selection. Instead of settling down to a stable mixture, therefore, the different types endlessly cycle or fluctuate in frequency.

**The Conditioning of Food Medium by Larvae.** A population of *Drosophila* growing in a vial will soon exhaust the food available for the larvae and must be transferred to a fresh vial. If the larvae were killed or removed before the food was used up, however, the vial could be restocked with new larvae. It would not be surprising to find that these new larvae did not grow as well on medium that had been conditioned by the growth of a previous cohort of larvae, if only because the food supply will have been depleted to some extent; but it would be interesting to know whether the effect of conditioning were in any way specific, with different types having more or less severe effects on succeeding generations. D.S. Weisbrot of Berkeley grew various strains of *Drosophila* as larvae for three days before killing them and restocking the vials with the same or with different strains. The survival of a Californian strain of *melanogaster* was actually enhanced in medium conditioned by *pseudoobscura*, whereas *pseudoobscura* was strongly repressed by the previous activity of *melanogaster*. This Californian wild-type also repressed other strains of *melanogaster*. The social effects of species or strains may thus be propagated through time, with one generation of larvae affecting the next, in a way that varies among genotypes. The effect did not seem to be caused by the depletion of yeast or by the presence of dead larvae in the medium, so the accumulation of metabolites seems the most likely explanation. In Weisbrot's experiment, the four *melanogaster* strains tested were generally repressed less by the previous activity of their own strain than they were by that of other strains. This would lead to a positive frequency-dependent effect on fitness and the rapid fixation of one type. This, however, is probably not a general phenomenon: the main lesson of this and similar experiments is that they show how specific social effects can persist through time.

**Crop Rotation.** The cause of time-lagged negative frequency-dependent selection is the projection of self-inhibition through time. The belief that any single type, continually resown, will sour its own environment is the basis of crop rotation, a very ancient practice (Virgil mentions it) whose modern literature is rather scanty. P.O. Ripley, working at the Ottawa experimental station, published a series of experiments in the 1940s to show that cereal-cereal rotations generally increased seed yield. Most of these were done in agronomic conditions, but one particularly simple experiment involved growing corn, rye, and oats in 2-gal earthenware jars. After the crop had been harvested, the jar was resown, either with the same cereal or with a different one. It is interesting to compare his results with Roy's hill-plots of rice cultivars. When the yield of a particular temporal sequence is expressed in terms of the yield of a sequential monoculture (the same cereal sown in successive seasons), the social matrix was as follows.

| | | Neighbor | | |
|---|---|---|---|---|
| | | Corn | Rye | Oats |
| | Corn | 1 | 1.27 | 1.21 |
| Self | Rye | 1.29 | 1 | 1.59 |
| | Oats | 2.11 | 1.98 | 1 |

In Roy's experiment, neighbors were grown on the same plot at the same time; in Ripley's experiment, neighbors were grown on the same plot but in the preceding season. There is clearly a tendency for self-inhibition to extend through time: each cereal yields more when its plot was previously occupied by a different type.

**The Broadbalk Wheat Experiment.** The classic crop-rotation experiment is the Broadbalk Wheat Experiment, originally designed to investigate the performance of continuous wheat monocultures under different soil treatments, and pursued, though with several changes in design, at the Rothamsted station since 1843. The design followed in 1968–1978 included a third-year fallow and a sequence of potatoes–beans–wheat, and allows continuous wheat to be compared with two sorts of rotation. There were also different levels of nitrogen application. The mean yield of wheat (over all levels of nitrogen) were 4.53 tons/ha for continuous wheat, 4.69 tons/ha for the second wheat crop after fallow, 4.81 tons/ha for the first wheat crop after fallow, and 5.34 tons/ha for wheat after potatoes and beans. Continuous culture of the same crop thus depresses seed yield, in the sense that yield is about 20% greater when the two preceding crops were different species of plant. The difference is most pronounced, amounting to almost a doubling of yield, at the lowest level of nitrogen application. One possible explanation for this is that the different crops have complementary resource requirements and that their effects on relative levels of soil nutrients persist from one growing season to the next, being reduced or removed by large inputs of extraneous nutrients.

The Rock–Scissors–Paper game is admittedly fanciful, and the rotation of different species of crops is rather remote from conventional issues in evolutionary biology. Nevertheless, the occurrence of intransitive social relations and the propagation of social effects through time raise the possibility of a continuously dynamic process of selection. This becomes more substantial when we consider how selection within one population will affect the evolution of completely different kinds of organisms in the surrounding community, which is the subject of Part 5.D.

## 5.C. Several Populations: Kin Selection and Group Selection

There is a very large and diverse literature, much of it polemical in tone, on the topics of kin selection and group selection. I can do no more than refer to some of the basic documents and reviews. The modern discussion of group selection has two roots. One is a review of Simpson's book on long-term evolution (Simpson 1945) by Sewall Wright (1945): Simpson argued that selection among groups could not cause the evolution of social characters, in reply to which Sewall Wright sketched a theory of group selection resembling his "shifting-balance" models of individual selection (see Sec. 54). The topic lay dormant until the critiques by Maynard Smith (1964) and Williams (1966) of the treatise by Wynne-Edwards (1962), which elaborated an already familiar line of reasoning: Dunbar (1960), for example, had earlier argued explicitly that whole communities or ecosystems could be effectively selected. Differing points of view about the subsequent course of the controversy are given by Maynard Smith (1976) and Wilson (1983). The modern discussion of kin selection began with two articles by Hamilton (1964a, 1964b). The book by Wilson (1980) is the most extensive development of a modern theory of group selection, comparing it with kin selection and showing that it gives essentially isomorphic results. The field has recently been revisited by Williams (1992).

The social behavior of individuals defines the society that they constitute. This society has itself some of the attributes of an individual: it may grow, divide, and die. These attributes may vary with the behavior of its members: cooperative individuals are more likely to form a stable society than those who generally attempt to exploit or suppress their neighbors. This will lead to a political process of selection, in which more vigorous, stable, and rapidly proliferating societies will tend to supplant neighboring societies in which social behavior is less effective in enhancing the fitness of the group as a whole. Indeed, this political selection, through which one kind of society is replaced by another, has been very widely regarded as being the chief process by which social behavior evolves.

The notion that characters evolve through the selection of more or less extensive groups of individuals, because such characters affect the fitness of the population or the species, or even the local community or the ecosystem, has a long history in evolutionary biology and in related disciplines, such as ecology and animal behavior. It was chrystallized in a famous treatise on social behavior published in 1964 by V.C. Wynne-Edwards of Aberdeen. To put it very briefly, Wynne-Edwards argued that a society in which every individual was adapted to harvesting resources as rapidly as possible would soon exhaust the local resource supply, and all would then starve. The efficacy of selection acting at the level of individuals would thus encompass the ruin of the population as a whole. On the other hand, if individuals were to regulate their behavior, refraining from growth and reproduction when resources were scarce, the local resource supply would be self-sustaining, and the population could survive indefinitely. This self-denial would evolve because prudent and frugal societies would eventually replace their spendthrift neighbors. Arguments of this kind, less explicitly and comprehensively developed, were commonplace at the time, and can still be found in elementary textbooks of biology. They have a very serious weakness. Granted that societies where individuals conserve resources by refraining from reproduction will arise, they will continually be invaded by types that reproduce at the greatest possible rate, even when resources are scarce. Selection among groups is therefore opposed by the selection of individuals within groups. The latter is expected to be the more effective, because individuals are more numerous and reproduce more rapidly than societies and can therefore be selected more intensely. These and other objections were marshalled by George Williams, in a counterblast to Wynne-Edwards' thesis so convincing that the interpretation of social behavior as evolving through the selection of groups was utterly discredited. It was this revulsion of feeling that led to a search for interpretations based on the selection of individuals, and I have already described how an individualistic theory of social cooperation was eventually developed. Nevertheless, it is undeniable that societies have distinctive properties that cause some to flourish while others fail, and in the sections to come I shall consider how a political process of selection may contribute to the evolution of social attributes.

These sections have been foreshadowed in part by the discussion of selection at different levels in Part 3.C. There are, however, two fundamentally different points at issue. The selection of lineages of different extent involves the differential proliferation of groups of individuals, but the characters involved may be expressed solely by individuals and have no effect on neighbors. The rate of proliferation of a lineage, in such cases, is merely the mechanical consequence of the adaptedness of its members. The distinctive feature of social behavior is that it will affect the proliferation of a group through its effect on neighbors, apart from its effect on the reproduction of the individual that expresses it. It is only in this case that the selection of groups of individuals, through their distinctive characteristics as groups, constitutes a new kind of process, whose outcome may differ from that of the selection of individuals considered in isolation.

## 147. *Selection among clones favors nepotism.*

The simplest society is that formed by the clonal proliferation of individuals from a single founder. Having similar phenotypes, they are likely to compete intensely amongst themselves, so that the more abundant they become the less successful each individual is likely to be. The most obvious response to social selection is to avoid one another. There is, however, a second possibility. If members of the same clone are often associated with one another, they will be selected to reduce the degree to which they repress the growth of their neighbors. The characteristics that evolve in this way may go beyond simple cooperation. A type that reproduces very slowly or that even does not reproduce at all may be selected, provided that it enhances the growth of its neighbors. The reason is almost self-evident. Imagine a genotype that represses the growth of the individual in which it is expressed, but enhances the growth of its neighbors. If the positive effect on its neighbors exceeds in sum its negative effect on the individual concerned, the reproduction of the local group as a whole will be increased. If, in addition, its neighbors are members of the same clone, they will bear the same genotype, and the number of copies of this genotype will increase, relative to the number of copies of an alternative genotype that neither repressed the growth of the individual nor enhanced the growth of its neighbors. A genotype may increase in frequency even if it is lethal, provided that it has the pleiotropic effect of adding more than a single individual to the combined reproduction of its clone mates.

I have previously explained how the concept of fitness refers to a lineage and expresses its capacity to proliferate. This capacity was regarded as being completely reducible to the reproduction of independent individuals, so that one could speak loosely of the fitness of an individual. When we drop the assumption that individuals reproduce independently of one another, and recognize instead that social relations may exist among members of a clone, we do not dilute in any way the basic principle that the future representation of that clonal lineage in a wider population of such lineages will depend solely on its rate of proliferation. We must, however, retreat from the view that the rate of proliferation of a clone is reducible to individual reproduction and must, therefore, modify the concept of individual fitness. The equivalent concept in the situation where social relations exist between members of the same clone extends beyond the individual expressing a particular behavior to include the effects of its behavior on its clonemates. The individual has a certain fitness by virtue of the reproduction that it is able to accomplish alone, corresponding to the concept of individual fitness in an asocial context. But we must also take into account the effect of its behavior on the combined reproduction of other members of the same clone. It is this *inclusive fitness* that determines the overall rate of proliferation of the clone and, thus, its future representation in the population.

Inclusive fitness = Reproduction achieved by individual, independently of any aid given by relatives
+
Reproduction achieved by clonemates through assistance given by individual.

A trait may be selected, even though it reduces the reproductive capacity of the individuals that express it, provided that it increases inclusive fitness through enhancing the reproduction of other individuals of the same type. Selection that favors nepotic behavior that aids relatives is called *kin selection*.

**Artificial Selection for Clonal Productivity.** The object of selection in cereal crops is to increase seed yield per unit land area. Genotypes that are highly productive as individuals and that tend to spread to dominate mixed populations do not necessarily give high agronomic yield, because when grown as clones each plant may repress the growth of its neighbors (Sec. 136). It is therefore preferable, when it is economically feasible, to select among clones or selfed lines grown as extensive pure cultures, rather than selecting among individuals. The outcome of such selection (such as the modern semidwarf rice cultivars) is often a weak competitor, rapidly eliminated from mixed cultures, but highly productive as a pure sward. This is an example of artificial kin selection.

**The Development of Multicellular Individuals.** The most familiar and highly organized clonal aggregations are the bodies of multicellular organisms. The development of any but the simplest multicellular creature involves the death or sterilization of many of the cells involved. The germ line may be segregated early in development (as in *Volvox* or vertebrates), or it may be continually recruited from a population of stem cells (as in polyps or plants), but in either case, a cell that has become differentiated for a particular somatic function very rarely gives rise to germ cells. Some cells, like Claudius, even aid further development by dying. At a coarser scale, the zooids of some colonial hydrozoans and ectoprocts develop as purely somatic creatures, serving the reproductive zooids of the colony; in highly differentiated colonies, such as siphonophores, a zooid may become little more than a bract or scale that protects its neighbors. The genotype that directs the terminal somatic differentiation of cells or zooids is favored through kin selection, the fitness of the clone as a whole being enhanced by the nepotic behavior of these non-reproductive components.

Vassiliki Koufopanou and I obtained some experimental evidence for the utility of somatic tissue in *Volvox*. This creature is a hollow sphere of several hundred cells, the great majority of which are small somatic cells resembling *Chlamydomonas*; a dozen or so much larger germ cells are set aside early in development by an unequal division. The germ cells develop into new colonies that are eventually released from the parental colony, leaving it as a hulk of somatic cells that soon die. We macerated colonies to isolate germ cells, which are capable of growing as unicells in nutrient medium, in the absence of a soma. The specific growth rate of these isolated germ cells, however, was less than that of the intact colonies. The rate of proliferation of the clone as a whole is thus increased when most of its members are unable to reproduce, but serve instead to accelerate the reproduction of a small set of clone mates.

**Kin Selection in Dispersed Clones.** Kin selection usually involves social interactions among neighbors, who can be assumed to be members of the same clone because of their proximity. It is much less likely to happen when the members of a clone are widely dispersed among other clones, unless there is some special mechanism by which clone mates can identify one another. A possible instance of kin selection in dispersed clones is the production of colicin by certain bacteria. Colicins are highly toxic antibiotics encoded by a plasmid gene. This gene is normally switched off, while another gene, whose product protects the cell against colicins, is switched on. Cells that bear the plasmid are thus immune to colicins. When the colicin gene is expressed, it kills the cell and all its neighbors, except those that bear a copy of the colicin plasmid. This suicidal behavior may evolve through kin selection. If nutrients are so scarce that the host cell begins to starve, a plasmid can make a fresh supply of nutrients available to members of the same clone (and thereby used to replicate copies of the plasmid) by killing neighboring competitors that lack the plasmid.

## *148. Selection among sexual kin is depreciated by their partial relatedness.*

The same principle applies to sexual relatives. A gene that directs co-operative behavior will spread even if it harms its bearer, by benefitting related individuals that bear a copy of the same gene. In outcrossed sexual lineages, however, there is the complication that only the gene itself is transmitted clonally, the genome being diluted and recombined in every generation with genes from other lines of descent. Consequently, the probability that two individuals in a sexual lineage both possess copies of the same gene varies with their coancestry. For example, the probability that two full sibs have both inherited a copy of a nuclear gene for nepotic behavior present in one of their parents is one-half. In more remote relatives, this probability falls off geometrically: for an aunt and a nephew it will be one-quarter, for two first cousins one-eighth, and so forth. This probability is the coefficient of relatedness $r$. Nepotic behavior will be favored by kin selection only if it is directed towards individuals who bear a copy of the gene that directs it, so among sexual kin there is a substantial risk that it will be misdirected and benefit, instead, individuals bearing a different allele. In sexual populations, therefore, the benefit $b$ of a nepotic act, in terms of its effect on the rate of increase of the genes that direct the act, must be discounted by the relatedness $r$ of the recipient. Kin selection will favor such acts only if the cost $c$ to the individual expressing them is sufficiently small, such that $c < rb$. This is the basis of J.B.S. Haldane's famous remark, to the effect that he could reasonably be expected to lay down his life to save eight cousins. The principle was codified by W.D. Hamilton in the 1960s, and forms the basis of an extensive theory of the evolution of social behavior through kin selection. There appears, however, to be little if any experimental work.

**Social Organization in Hymenopterans.** On the other hand, there have been many comparative studies, of which perhaps the best-known concern the remarkable social organization of certain ants, bees, and wasps. Although the details of hymenopteran social organization are very complicated, the basic problem is very simple: the daughters of the queen help in various ways to raise the queen's offspring, while largely refraining from reproduction themselves. The phenomenon is similar to the polymorphism of zooids in colonial invertebrates, but is the more striking in a mobile, individualized, sexual organism. Darwin himself realized the difficulty that non-reproductive social castes posed for the theory of natural selection and suggested the correct answer—the selection of family groups. The peculiar genetics of hymenopterans suggests a more precise solution. The fertilized eggs of hymenopterans develop into diploid females, the unfertilized eggs into haploid males. The diploid females make haploid sexual eggs by meiosis; the haploid males make haploid sperm by mitosis. Consequently, a fertilized egg receives half of its mother's chromosomes, but all its father's chromosomes. Mother and daughter share half their genes, so that $r = \frac{1}{2}$, as in more conventional animals. Two full sisters, however, share half their mother's genes, but all their father's genes; their relatedness is therefore $r = \frac{3}{4}$. In other words, a female ant is more closely related to her sisters than she is to her daughters. Kin selection will therefore favor aiding sisters rather than daughters, for acts involving a comparable sacrifice of personal fitness; and aiding sisters rather than daughters is the distinctive feature of hymenopteran sociality.

## 149. *Selection among groups of unrelated individuals may favor cooperation.*

It is easy to picture kin selection working through an extended self-interest in which a gene expresses a phenotype that aids other copies of the same gene in the bodies of relatives. It is more difficult to explain acts aiding unrelated individuals that hinder the reproduction of the individual expressing them. Indeed, it is logically impossible to explain such acts as evolving through selection, if they are directed toward a random sample of the population. There is only one loophole. Selection will favor genes that direct cooperative or altruistic behavior, provided that this behavior aids individuals that bear a copy of the same gene; the degree of relatedness of these individuals may be unspecified, or very distant, or even non-existent, in the (improbable) event that two individuals bear genes with the same sequence that have arisen independently by mutation. This might happen in a spatially structured population, where social interactions that affect reproduction occur within restricted groups or clusters of individuals. In the simplest case, these groups are simply constituted as random samples from the general population. They will nevertheless differ in composition to some extent, because more individuals of a given type will be allocated to some groups, and fewer to others, merely by chance. There will therefore be some degree of variation in social behavior among groups, corresponding to the binomial variance of random sampling. This variance among groups can be selected, because groups that flourish because of the social tendencies of some of their members will expand at the expense of competing groups. The regeneration of the binomial variance among groups in every cycle will continually renew the process of selection, with some degree of social cooperation arising as a consequence.

Recall the trading game that I described in Sec. 144. Although the cooperative behavior Tit-for-tat is an ESS, a single Tit-for-tat mutant would not spread in a population where every other individual always defected. Tit-for-tat could spread in the selection experiment only because it was originally present at an appreciable frequency and, therefore, occasionally encountered itself, or a program with similar tendencies. The situation simulated by the experiment, indeed, was a population divided into random groups of two individuals. In any given group, Tit-for-tat was nearly neutral: it did no better than its partner, and if paired by an uncooperative program, would do a little worse. It spread because of the greater productivity of cooperative pairs. The same principle applies more broadly. Imagine any behavior that is cooperative, in the sense that it benefits others (by trading honestly, for example), without harming appeciably the individual expressing it. Within a group of any size, the trait is neutral, because the benefit is conferred on cooperative and uncooperative individuals alike, and will not tend to spread. Within the population at large, however, the trait will be selected, because those groups that have a higher frequency of cooprative individuals reproduce more successfully. At the level of the population, of course, cooperative behavior spreads because cooperative individuals have the greater relative fitness. Nevertheless, this would not be apparent from a study of the consequences of cooperation at the level of the local group in which the behavior is actually expressed. A character may be selected in a structured population because it increases absolute fitness, although it has no effect on relative fitness within the local group.

## 150. *Altruism evolves through group selection only if alternative social types are overdispersed among groups.*

### *Selection Among Temporary and Persistent Local Groups*

If social interactions are unique, with individuals parting after a single encounter, then cooperative behavior cannot evolve. In these circumstances, Tit-for-tat amounts to a single cooperative act extended to a partner who invariably profits by defecting. Altruistic behavior that benefits others while actually harming the individual that expresses it is not, in general, favored by selection among groups in a structured population. There is only one circumstance in which it can evolve. Imagine that there are two types in the population, one of which behaves altruistically, whereas the other behaves selfishly. If the population is divided at random into groups of any size, genotypes that express selfish behavior will spread. We can imagine, however, that altruistic individuals can by some means seek one another out, so that in the limit the population will consist of two sorts of group, one comprising altruistic individuals and the other only selfish individuals.

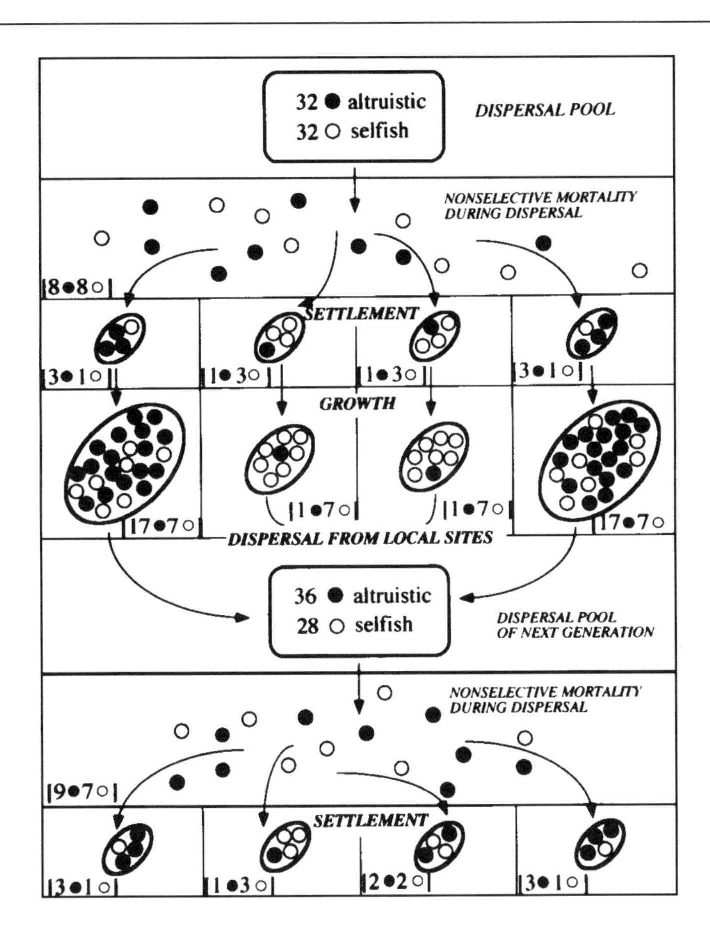


"Altruistic" or cooperative individuals behave in some way so as to promote the growth of the local group, enhancing the reproductive success of all members of the group equally. "Selfish" individuals have no such effect on the group as a whole, and always reproduce more than altruists. Consequently, the frequency of altruism always falls within any local group, regardless of its composition. However, the greater growth of groups with a high frequency of altruists may cause altruism to increase in frequency in the population as a whole.

(*Continues*)

(*Continued*)

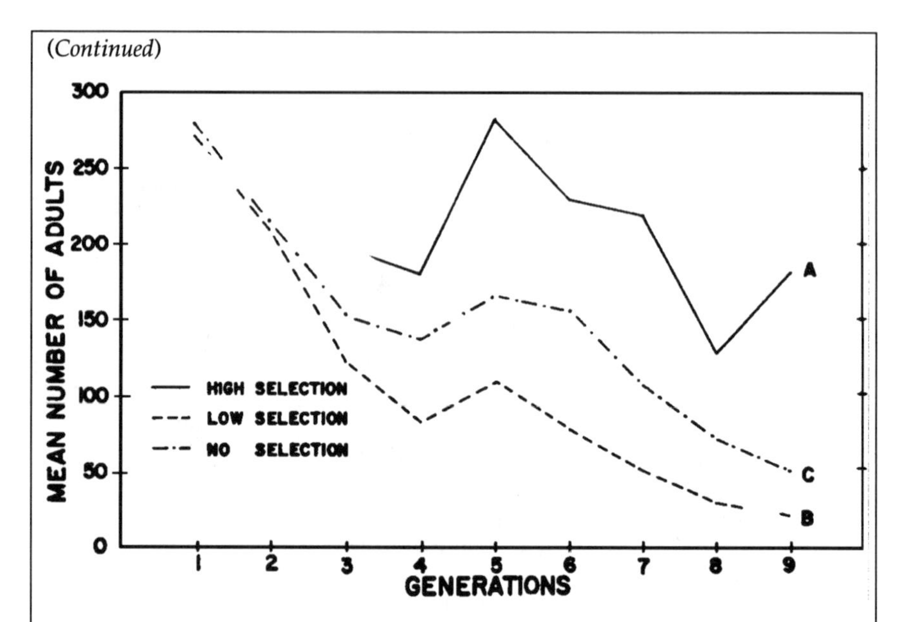


This figure shows the oucome of upward and downward artificial group selection for population number, in an early *Tribolium* experiment by Wade (1977). Other papers in this series concern the correlated response of competitive ability (Wade 1980), the genetic and demographic basis of the response (McCauley & Wade 1980), the effect of different rates of migration (Wade 1982), the response to relaxation of group selection (Wade 1984) and the interaction of group size and migration (Wade & McCauley 1984). A parallel experiment was reported by Craig (1982). Group selection in two-species communities of *Tribolium* has been studied on similar lines by Goodnight (1990a, 1990b).

Altruistic acts will then benefit other altruistic individuals. Because the altruistic groups will expand at the expense of the selfish groups, the frequency of altruism in the population must increase, and the genotype responsible will eventually become fixed. More generally, altruism can be selected, provided that the variance among groups is greater than the binomial variance supplied by random sampling. The greater the variance, the more extreme the degree of altruism that can evolve. There have been two attempts to show how such variance can arise: one, developed by David Sloan Wilson, refers to temporary aggregations of individuals, whereas the other, originally suggested by Sewall Wright, involves more or less permanent local populations.

**Temporary Groups.** The condition that altruistic genotypes should be overdispersed is quite onerous, especially if social groups are temporary aggregations that re-form in every generation. Altruistic individuals might be able to recognize altruistic tendencies in others, and choose to associate with altruistic partners; but this requires considerable powers of discrimination and is vulnerable to a dishonest pretence of altruism, when this cannot be tested by repeated encounters. It is also conceivable that altruism might be associated with other behaviors that cause altruistic individuals to live in similar places; but this can scarcely be the general case and, again, seems likely to be circumvented by types that express the habitat preference but not the altruism. The most likely reason for altruists to be consistently associated in the same social groups is that they are relatives who grow up close to one another. This amounts to nepotism evolving through kin selection, however, rather than to altruism evolving through group selection.

**Persistent Groups.** There are other possibilities if the social groups are separate populations that persist for many generations. Small, isolated populations will diverge through genetic drift (Sec. 37): roughly speaking, if population size $N$ and the rate of migration among populations $m$ are sufficiently low, such that $Nm < 1$, populations will tend to become fixed for one or another type. The variance that is generated in this way is available for selection among populations. Those with a high proportion of cooperative or altruistic individuals may well flourish, sending out migrants that can occupy vacant sites or providing a continual stream of altruistic immigrants for predominantly selfish populations. Altruistic behavior may thus spread infectiously through the whole assemblage of populations. This process of selection among populations, however, will be opposed by selection for selfish behavior within populations, which will be most effective when the populations are large (so that the outcome of selection within each is not jeopardized by drift) and migration is frequent (so that any altruistic population receives a continuous influx of selfish individuals). The population does not have to be very large, nor migration very frequent, for selection of individuals within populations to be much the more effective process: this is a good reason to doubt that characters often evolve through benefiting the group rather than the individual.

**Group Selection and Kin Selection.** The distinction between group selection and kin selection, though conventional, seems somewhat artificial. Small persistent populations are likely to be highly inbred, of course, and altruistic acts will then be directed largely towards kin. The subdivision of the general population into small, nearly isolated groups serves to create a situation where altruistic acts are likely to be directed towards kin (or where the altruistic behavior associated with a given genotype is likely to be directed towards individuals with the same genotype) and are likely to be selected for that reason. Selection will favor lineages in which social acts tend to benefit members of the same lineage; it cannot favor lineages that promote the reproduction of other lineages. Group selection is a fallacious concept when it is held to cause the evolution of characteristics that benefit populations, or species, or communities, as a whole, without any distinction of ancestry.

## *Group Selection in* Tribolium

The exception to the general paucity of experimental work on kin selection and group selection is the remarkable series of selection experiments reported during the last 20 years by Michael Wade and his collaborators at Chicago. They are not concerned directly with social behavior, but rather with the "population phenotype" resulting from the aggregate of individual interactions. In most cases, this phenotype is population number, reflecting the traditional concern of group-selection theorists with the evolution of density regulation. A population founded by a small number of adult flour beetles, *Tribolium*, is put into a glass vial containing the nutrient medium, a mixture of flour and yeast. The characteristics of the population that develops in this vial can then be selected: that is, the most extreme vials can be chosen to propagate the line, just as in most experiments the most extreme individuals are chosen. This process of selecting vials rather than individuals amounts to artificial group selection, and the main purpose of the experiments is to find out whether it is effective in modifying the population phenotype.

**Divergence Through Genetic Drift.** The experiments typically involve a large number (of the order of 100) small populations. The populations really are small, usually being founded by only 16 adult beetles, which give rise after a few weeks to 100 or 200 descendents. When a small proportion of the populations is chosen at random to reconstitute the population array in each cycle, there is no consistent selection for any particular phenotype, and so the mean productivity, the number of beetles alive in each vial at the end of the cycle, is not expected to change. Each population, however, will drift independently, when each is a separate unit that propagates itself from cycle to cycle, and the variance among populations will increase through time. By choosing a few populations at random to reconstitute the whole array, a process of populational drift is superimposed on the drift within each population. Just as the effect of drift within a population is to reduce the genetic variance among individuals by fixing one or another lineage or allele, the effect of populational drift will be to reduce the genetic variance among populations within the array. This will not necessarily happen if each population is undergoing a similar process of individual selection: this will tend to make populations more similar, and the random selection of populations will retard their convergence. If we neglect individual selection, the divergence of populations will be promoted by the random sampling of individuals within each population and will be restrained by the random sampling of populations within the array. The rate of divergence will be increased by low population number and a low rate of population extinction. Broadly speaking, these expectations are borne out by the experiments, although small populations founded by 6 or 12 adults, after diverging rapidly during the first few cycles, were no more variable after 10 or 11 cycles than larger populations founded by 24 or 48 adults.

After about 10 cycles of random selection, there was about a fourfold variation in productivity among populations. These differences were heritable, in the sense that daughter populations tended to resemble one another, and resembled the parental population from which their founders were drawn. The genetic variance of population productivity arising in this way is then the basis for selection among populations.

**Response to Group Selection.** The random selection of populations can be contrasted with two other modes of selection. The first is no selection at the group level, with all populations being propagated in the same way. There will be selection among individuals within each population, of course, but this need not necessarily lead to any change in mean productivity. The second is directional group selection, for increased or decreased productivity, by propagating the whole array from the most extreme vials. Randomly selected and unselected populations diverged through time, as expected; indeed, the randomly selected populations diverged more, suggesting that unselected populations were experiencing strong individual selection of some sort. This variance was harvested by selection among vials: after nine cycles, the upward selection line was producing on average nearly 200 adults per vial, and the downward line only 20. The cause of the divergence was that the upward line had retained more or less the same productivity as the base population, whereas the productivity of the downward line had declined steeply. There was also, however, a substantial decline in the productivity of the randomly selected and unselected lines. The most straightforward explanation of these results is that productivity was generally reduced in these small populations by inbreeding depression, which could be countered by group selection for high productivity or exacerbated by group selection for low productivity.

Both the divergence of randomly selected populations and the response to selection are reduced if the selected populations are supplemented by a few migrants from the discarded populations. Nevertheless, even when

25% of the individuals transferred in each cycle were migrants, there was still a detectable response to directional group selection.

**The Contingent Response to Group Selection.** Populations were selected for productivity without regard to the particular individual and social behaviors underlying the population phenotype. When these behaviors were investigated, it was found that they had evolved differently in populations selected in the same direction. Low productivity, for example, can evolve through group selection by reduced fecundity early in life or by increased rates of cannibalism. The outcome of group selection seems to be as contingent as the outcome of individual selection.

**Antagonism of Group Selection and Individual Selection.** Group selection leads to characteristic adaptations only if it is opposed to individual selection and thus causes a different outcome. When the upward and downward selection lines were reared as mixtures with another species of *Tribolium*, the downward lines were considerably more successful. Group selection for low productivity had thus enhanced interspecific competitive ability, selecting for aggressive types that performed well in mixtures. Selection for high productivity had the converse effect of enhancing social behavior that could be exploited by another species. The beetles thus respond in the same way as crop plants selected for high yield per unit area as pure stands.

Wade's experiments do not directly address the evolution of nepotic or altruistic behavior. Nevertheless, they have the usual force of experiments in artificial selection. If conditions are such that small populations with particular characteristics are favored, these characteristics can be modified by selection among groups, and the social behavior of individuals may evolve as a correlated response.

## 151. *The most productive mixture is not necessarily the ESS.*

Because the type that spreads in mixed cultures is not necessarily the type that is the most productive in pure stand, the outcome of selection among individuals may differ from the outcome of selection among groups. I have assumed so far that selection favors a single type with a particular social behavior. However, a similar principle applies to mixtures of types. This is an important point, because the composition of a mixture is necessarily the property of a population, and selection for mixtures is necessarily selection among populations or groups of some sort.

**The Most Productive Population.** A population is mixed at equilibrium if either type is more fit than the other or others, when rare, so that it tends to spread when its neighbors are predominantly of the other type; in terms of the abcd model, the ESS is mixed if $c > a$ and $b > d$ (Sec. 143). The overall productivity of a mixture, however, is not necessarily maximized at the ESS. A mixture is more productive than either pure stand if either type is on average more productive when its neighbors are different than it is as a pure stand. The conditions for this to be the case are thus $(b+c) > 2a$ and $(b + c) > 2d$. When both conditions are satisfied, the Most Productive Population (MPP) is a mixture in which the frequency of type 1 is

$$f_{MPP} = \tfrac{1}{2}[(b - d) + (c - d)]/[(c - a) + (b - d)]$$

The ESS and the MPP are obviously closely related, inasmuch as both involve mixtures if there is less competition (or more cooperation) between unlike than between like types. These mixtures, however, do not have the same composition; they are related as

$$f_{MPP} = \tfrac{1}{2} f_{ESS}[1 + (b - d)/(c - d)]$$

The ESS is the outcome of pure individual selection in a single homogeneous population; the MPP can be thought of as the outcome of pure group selection in a spatially structured population. One important practical consequence of this distinction is that agronomists cannot rely on individual mass selection (harvesting and resowing a population without conscious selection) to produce a mixture with maximal yield.

## 152. *Arbitrary mixtures tend to be more productive than the mean of their components but not as productive as the best component.*

These arguments are of substantial practical importance, because if the MPP is often a mixture they suggest that agronomic trials might be directed toward selecting the best mixture rather than the best single variety.

**The Productivity of Equal Pairwise Mixtures.** The simplest experiment is to measure the productivity of equal mixtures of several arbitrary types in all pairwise combinations. This procedure is equivalent to the diallel cross of conventional genetics, but generates combinations of individuals rather than combinations of genes. Such mixtures may be superior in one

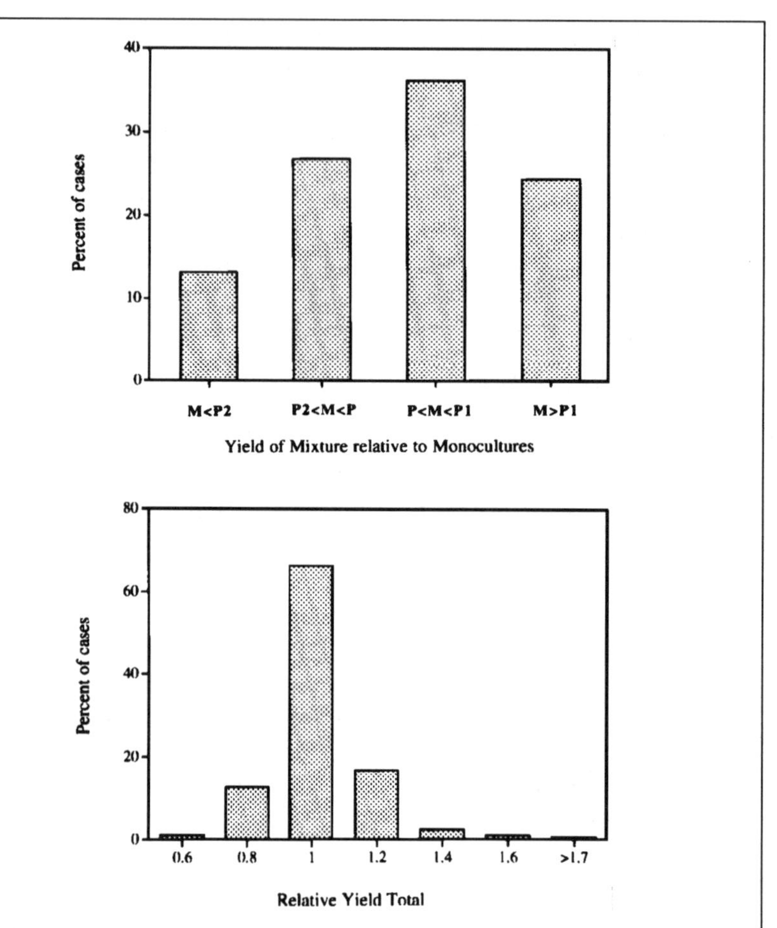


These diagrams summarize Trenbath's review of vegetative production (dry weight of stems) in equal binary mixtures of cultivars or species, mostly of grasses (Trenbath 1974). Combinations of legumes and non-legumes are excluded. The upper diagram compares the yield of the mixture **M** with the yield of the lower-yielding **P2** and the higher-yielding **P1** components grown in pure culture at the same total density, and with the mean of the two pure cultures **P**. There were 344 cases, from 16 studies. The lower diagram is an alternative representation from de Wit replacement-series designs. If the mean yield per plant of type $i$ in mixture with type $j$ is $Y_{ij}$, then the Relative Yield Total is defined as $\frac{1}{2}(Y_{ij}/Y_{ii}+Y_{ji}/Y_{jj})$. A value of unity indicates that the yield of the mixture will be intermediate between that of the two monocultures, i.e., $\mathbf{P2} < \mathbf{M} < \mathbf{P1}$ in terms of the upper diagram. In fact, two-thirds of all cases had a Relative Yield Total of between 0.9 and 1.1,

(*Continues*)

emphasizing that transgressive mixtures are rare. There were 572 cases from 16 studies, overlapping broadly with those included in the upper diagram.

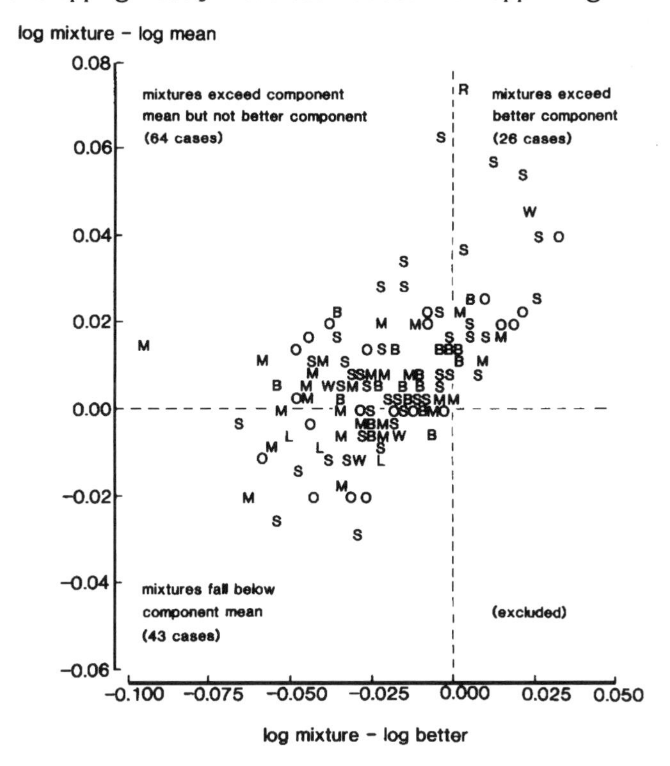


This diagram summarizes my own review of seed yield in equal binary mixtures of cereal cultivars. I collated 169 experiments from 21 studies published in *Crop Science*, involving seven species. Yields vary greatly with the crop, of course; however, the log ratio of mixture and monoculture yield is Normally distributed and does not vary among species of crop. The $y$-axis if this graph is (log mixture yield − log mean yield of components in pure culture), whereas the $x$-axis is (log mixture yield − log yield of better component). The symbols identify different crops: B barley, L lima bean, M maize, O oats, R rye, S soybean, W wheat. Points falling in the upper right quadrant are 27 cases in which the mixture outyields the better component; the 64 cases in the upper left quadrant are those in which the mixture exceeds the mean of the components, but not the better component; in 43 cases, in the lower left quadrant, the mixture is inferior to the component mean. Overall, mixtures yielded more than the mean of their components, but the difference was, on average, only 1.8%. Moreover, they were generally not as productive as the better monoculture, the mixture yield being, on average, only 95.6% as great as that of the better component.

(*Continues*)

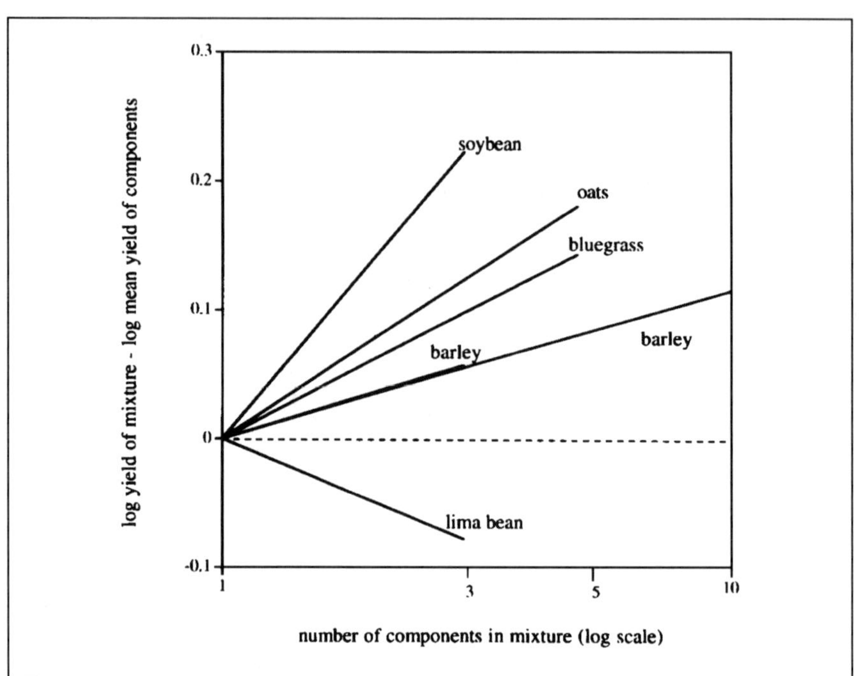


This diagram summarizes the results of six studies in which mixtures of 3–10 cultivars were constructed. The lines are linear regressions of the log ratio of mixture yield to mean monoculture yield, on log number of cultivars in mixture. All but one (bluegrass) are seed crops. There is a tendency for the superiority of mixtures to become more pronounced as they become diverse, although the effect is generally weak, with many exceptions to the rule. The most extensive experiment, with 10 cultivars of barley, is by Clay and Allard (1969). The careful experiments on mixtures of two and three cultivars of barley by Qualset (1981) are also worth consulting. There is also an experiment involving 14 soybean cultivars by Walker and Fehr (1978), not included here because the results were not expressed in comparable form; it showed a very weak tendency for mixture superiority to increase with diversity.

(*Continues*)

of two respects. Firstly, they may be more productive than the average of their components in pure culture. This demonstrates some degree of social interaction between the components of the mixture, but the MPP is nonetheless a pure stand of the more productive variety. In extensive trials, however, the more productive variety is not known in advance, and sowing a mixture may be less risky. Secondly, the mixture may be more productive than the better component. Mixtures that constitute an MPP

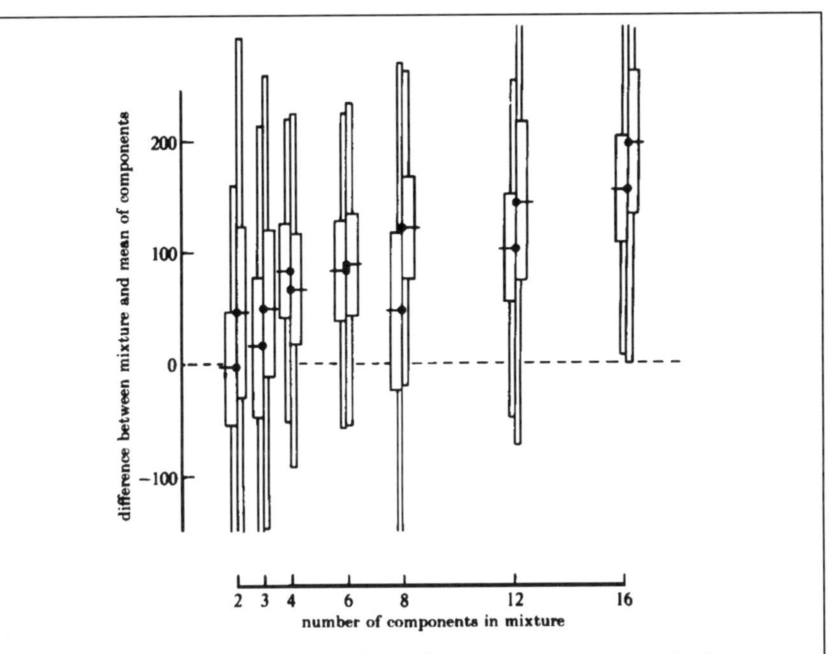


This diagram shows the results of a *Chlamydomonas* experiment in which mixtures were constructed by choosing 2, 3, 4, 6, 8, 12 or 16 strains at random from a collection of 68 different strains. The right- and left-hand bars show the 10 random mixtures grown in one of two media, and the graph shows the mean (dot and horizontal line), the 95% limits of the mean (broad box) and the 95% limits of the mixtures (narrow box). The productivity of each mixture could then be compared with the mean productivity of its components. This difference increases with the diversity of the mixture. Although this experiment shows very clearly that mixtures are generally superior to the mean of their components, it does not exclude the possibility that this is caused by an increase in the frequency of more productive strains during a cycle of growth. Few if any of these mixtures, even at the highest level of complexity, were transgressive.

(*Continues*)

are said to be *transgressive* and, if they could be identified and selected, they would be unequivocally superior in agronomic practice.

**Vegetative Yield of Crops.** There is an extensive literature on the yield of mixtures of lines or cultivars in crop plants, and rather than cite individual cases I shall refer to two extensive surveys. B.R. Trenbath of Adelaide reviewed crops that are harvested for their vegetative parts, covering over 300 trials from 15 studies. Mixtures outyielded the average of their

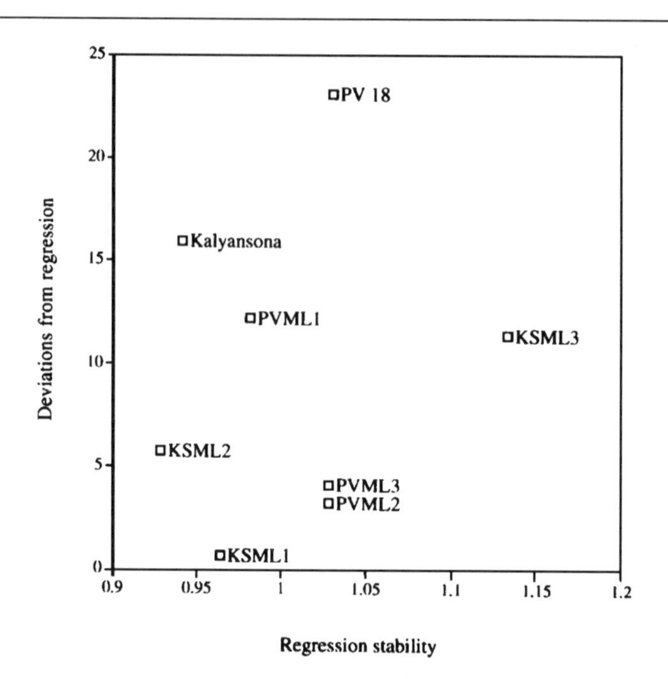


This diagram compares the stability of mixed populations (multiline cultivars) with that of pure cultures (the recurrent parents), from trials by Gill et al. (1984). The parental genotypes were Kalyansona and PV 18, two Indian cultivars of wheat, from which the two multiline series KSML and PVML were developed by repeated backcrossing and selection. Three multilines from each series, together with their recurrent parents, were tested at nine localities in the Punjab over three years. There were very substantial G × E interactions involving years, localities and year–locality combinations. The stability of the eight lines over all environments was assessed by regression analysis, as described in Sec. 112, so as to separate the systematic effect of variation in the mean productivity of environments from other sources of G × E. In this case, the regression stability of the multilines is not markedly different from that of their parents: the multilines show neither a consistently lower regression slope (biological stability through lower environmental variance) nor a slope consistently closer to unity (agronomic stability through less G × E). Indeed, one of the multilines, KSML3, appears to be the most responsive line. There is a marked difference between the multilines and their parents, however, in the magnitude of deviations from regression. These multilines were thus more stable than their parents because they were not as severely affected by particular sources of local environmental variation. Since the development of the multilines involved selection for disease resistance, it is likely that they were more stable in part because they were less vulnerable to short-lived local epidemics.

(*Continues*)

(*Continued*)
For large-scale comparisons of the stability of arbitrary mixtures with that of their components, see Shorter and Frey (1979) and Frey and Maldonado (1967).

components in about 60% of these trials. About a quarter of all mixtures were transgressive, yielding more than the better component, although about half as many were transgressive in the opposite direction, being inferior to the poorer component. There is thus a rather pronounced tendency for mixtures to be more productive than expected. However, the results are erratic and highly variable; there are few, if any, cases in which a specified mixture has been shown by repeated trials to be consistently overyielding.

**Seed Yield of Crops.** I have carried out a similar survey of seed yield, collating nearly 200 trials from 20 studies. It leads to similar conclusions. Mixtures were more productive than the average of their components in nearly 70% of all cases, but outyielded the better component in only about 20% of cases. The superiority of mixtures seems to be very general (it was reported in 19 of the 20 studies) but it is also very slight: on average, mixtures yielded only about 2% more than the mean of their components and yielded nearly 5% less than the better component.

**The Effect of Increasing Diversity.** The natural extension of work on equal binary mixtures has been to ask whether a more pronounced increase of yield can be obtained from mixtures with several or many components. Experiments of this sort have given vague and conflicting results that are difficult to summarize briefly; certainly there is no regular and predictable increase in yield with diversity. It seems that different species or cultivars of crop plants may interact quite strongly with one another, but these interactions are as likely to be antagonistic as to be cooperative. When the plants are grown close together, different mixtures vary widely, some being overyielding and others underyielding; as spacing increases, both antagonistic and cooperative tendencies weaken, the mixtures become more alike, and eventually any mixture effect disappears.

**Mixed Cultures of *Chlamydomonas*.** The results that I have obtained with mixed cultures of *Chlamydomonas* are consistent with general agronomic experience. If different types are grown together over one growth cycle of about 10 generations, the density of the culture may differ from the mean density of its components when grown separately for either of two reasons. The types may interact antagonistically or cooperatively; however, over 10 generations the composition of the population may also change through selection. When the types are different species, pairwise mixtures reliably exceed the average of their components, and this excess increases steadily with diversity. There is no evidence, however, that these mixtures, however diverse, regularly exceed the productivity of their best component. There are quite strong interactions among different species, because cell-free filtrates of single-species cultures affect the growth of other species. When the same experiment is tried with sibs from crosses within a species, pairwise mixtures are roughly as productive as the better of the two components. In this case, however, there is no doubt that the effect of mixture is caused simply by the displacement of the less productive genotype because the genetic variance of growth rate is greatly attenuated by a single cycle of growth.

The *Chlamydomonas* results suggest that the properties of mixtures are dependent on their genetic scale. Different species may interact strongly, sometimes constituting consistently overyielding mixtures. Different varieties of the same species may show the same effect, although on average it is very weak; in mixtures of random genotypes from the same family, any more subtle social effects are erased as one genotype comes to dominate the mixture. If this is the case, it will be possible to select overyielding mixtures only when their components have sufficiently distinct ecological properties.

**The Tangled Bank.** The offspring of a sexual cross constitutes an arbitrary mixture of recombinant genotypes. I have suggested that sexual families might be more successful than clones because of the general tendency of mixtures to outyield the average of the components. The argument can be understood through a simple economic analogy. In a small, closed community, some professions will be more profitable than others; the dentist (say) is likely to earn more than the shop assistant. Nevertheless, parents raising a family in the community would be ill advised to have all their children trained as dentists, because they would then compete among themselves, and their average income would fall. It would clearly be preferable to enter them into a variety of different professions. The diversification of offspring genotypes through outcrossing might serve the same function of reducing competition among sibs and thus increasing the success of the brood as a whole. I called this theory the Tangled Bank, after an expression in the final paragraph of the Origin of Species, referring to the diversity of natural communities.

It may be helpful at this point to distinguish among three ways in which the diversification of sibs might be favored by selection. The first is the lottery principle (Sec. 116): a diverse sexual brood is more likely to include one of the few genotypes that happen to be well adapted to current conditions in a continually changing environment. The second is a general advantage of rarity (Sec. 141): any given type is more likely to have higher fitness while it remains rare, so selection favors the breaking up of genotypes that are currently abundant. The third is the Tangled Bank that asserts that a *collection* of rare types will have greater overall productivity than any uniform group. Because all three involve some form of competition among sibs, they are often confused.

The Tangled Bank is an attractive explanation for the prevalence of sex in relatively stable and saturated communities in tropical and marine environments where competition for resources is likely to be continual and intense. Nevertheless, the evidence from crops and *Chlamydomonas* has persuaded me that it is unlikely to be correct. Arbitrary mixtures do indeed seem to be more productive than the average of their components, but the effect is too slight at the scale of offspring within a single sexual brood for it to overcome the manifest inefficiency of sex (Sec. 125).

**The Stability of Mixtures.** It has been repeatedly suggested that apart from any effect on average yield mixtures are likely to vary less among environments than are pure cultures of a single genotype. This might be of some practical importance because it would suggest that farmers might be less likely to suffer severe losses by using mixtures, even if their yield often fell short of what could be obtained in a particular season from the best available variety in pure stand. From a more academic point of view, it would suggest that arbitrary mixtures, such as sexual sibships, might have greater geometric mean fitness than clonal lineages. The most extensive experiment that bears on this issue was a trial involving 28 oat cultivars and their 378 equal binary mixtures, grown at two locations in two successive years. The genetic variance of seed yield among mixtures is expected to be half that among pure lines, but was instead only about one-fifth as great. The mixtures therefore show less environmental variance, presumably because lines that were low yielding in pure culture had enhanced yields in mixture, and vice-versa. Moreover, G $\times$ E was much less for mixtures than for pure lines, presumably because when the yield of one component of a mixture is reduced, that of the other component was quite likely to be increased. Both the absolute yield and the relative yield of mixtures were much more stable for mixtures than for pure lines. This result seems to be quite general: the effect of mixture on stability seems to be much more consistent than its effect on arithmetic mean yield. This effect will be most pronounced when genotypes differ markedly in performance from one environment to another, that is, when G $\times$ E is a major source of variation.

**Evolving Mixtures.** I have emphasized that the mixture that evolves through frequency-dependent selection does not generally have the most productive composition. Nevertheless, the conditions that the ESS and the

MPP be mixtures rather than pure stands are similar, in that both require that b and c should be large relative to a and d. It is reasonable to expect that the components of a mixture whose composition has evolved through individual selection should be more productive in combination than sets of genotypes chosen at random. There are some indications that this may be the case: Allard and Adams found that lines from Composite Cross V (described in Sec. 141) showed a much stronger tendency for elevated yield in mixture after 18 generations of mass selection. Allard has reported a somewhat similar experiment with lima beans propagated for up to seven generations. He compared seed yield in five different kinds of population: the two parental lines $\mathbf{P_1}$ and $\mathbf{P_2}$ (both, of course, previously selected independently for high yield), with mean $\mathbf{P}$; the equal mechanical mixture of these lines, $\mathbf{M_{12}}$; a mechanical mixture of equal amounts of seed from 30 $F_4$ families, $\mathbf{M_{F4}}$; the mean of the same $F_4$ families grown in pure culture, $\mathbf{F_4}$; and an unselected $\mathbf{F_7}$ bulk, $\mathbf{B_{F7}}$. The general result was that yields were ranked in this way:

$$\mathbf{B_{F7}} = \mathbf{P_1} > \mathbf{M_{F4}} > \mathbf{P} > \mathbf{M_{12}} > \mathbf{F_4} > \mathbf{P_2}$$

Because $\mathbf{M_{F4}} > \mathbf{F_4}$ even though $\mathbf{M_{12}} < \mathbf{P}$, the greater diversity of the $F_4$ mixture may be responsible for its superiority. However, $\mathbf{B_{F7}} > \mathbf{M_{F4}}$ suggests a further increase in yield attributable to individual selection acting during bulk propagation, although it is possible that the type resembling the higher yielding parent had increased in frequency through selection in the bulk.

**Selection of Mixtures.** A more effective way of producing high-yielding mixtures is to select directly on mixture composition. Despite a considerable amount of interest in social evolution and in the agronomic properties of mixtures, however, there seems to have been no attempt to prolong the artificial group selection of mixture composition beyond the single generation of a crop trial. This leaves a wide opportunity for interesting experiments.

## 5.D. Coevolution

### 153. *There are powerful and highly specific interactions between organisms that are not ecologically equivalent.*

The social relations I have described in previous sections have concerned similar kinds of organisms: closely related species, or genotypes within the

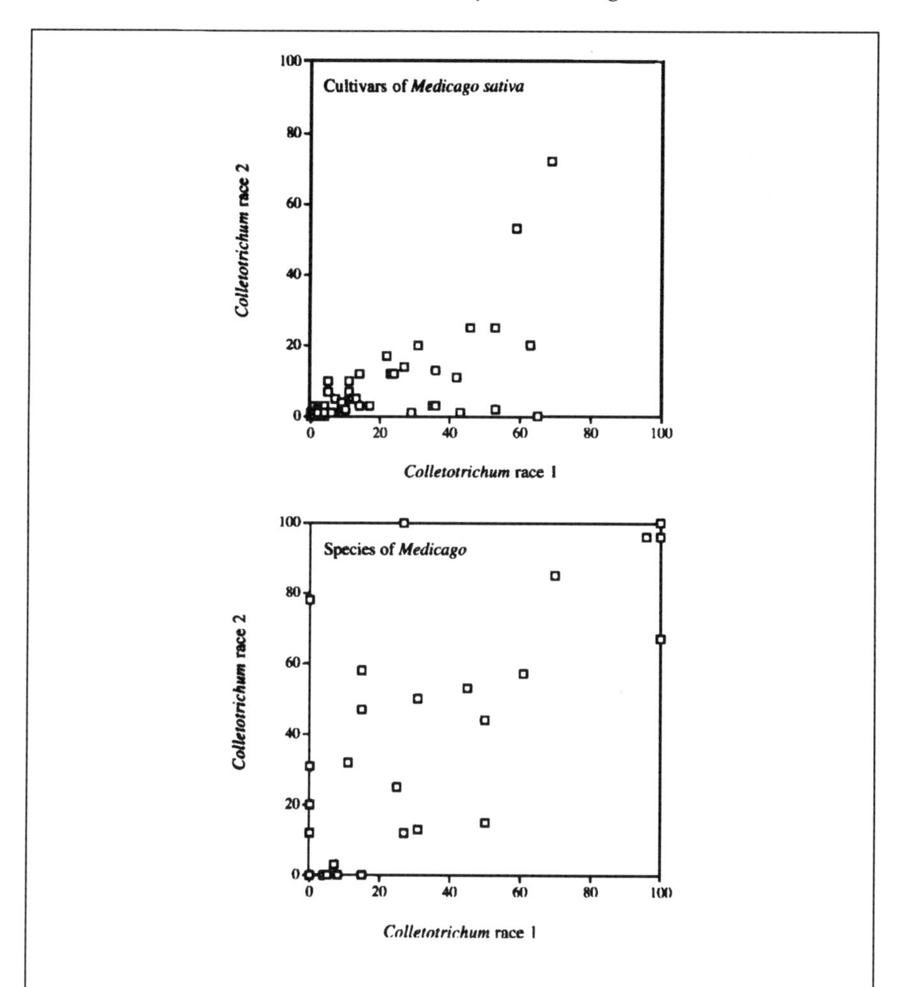


These two diagrams show the resistance (as a percentage) of plant genotypes to two races of the fungus *Colletotrichum trifolii*, which causes anthracnose in alfalfa and other crops. The genotypes in the upper graph are cultivars of alfalfa; those in the lower graph are different species in the same genus. Susceptibility to the two fungal genotypes is positively correlated among plant genotypes: for cultivars, the genetic correlation coeffificient is $r_G = +0.63$, and for species $r_G = 0.78$. These data are from Elgin and Ostazewski (1982).

These results are quite normal. I collated the disease response of crop genotypes to pathogen genotypes from a dozen studies, including the two illustrated, and found the overall mean genetic correlation was $r_G = +0.46$. Nevertheless, a glance at the figures will show that interactions are often very specific: some

(*Continues*)

*(Continued)*

plant genotypes are highly susceptible to one fungal genotype, while being highly resistant to the other (or, from the other point of view, a fungal genotype may be virulent on one plant genotype and benign on another). Other experiments measure the yield response of the crop and appear to give a similar picture. The literature on the ecology, dynamics, and genetics of plant–pathogen interactions is far too large to summarize here, however briefly, but there is an excellent account from a population biology perspective by Burdon (1987). The study of wheat soilborne mosaic virus discussed in the text is by Kayasthar & Heyne (1978).

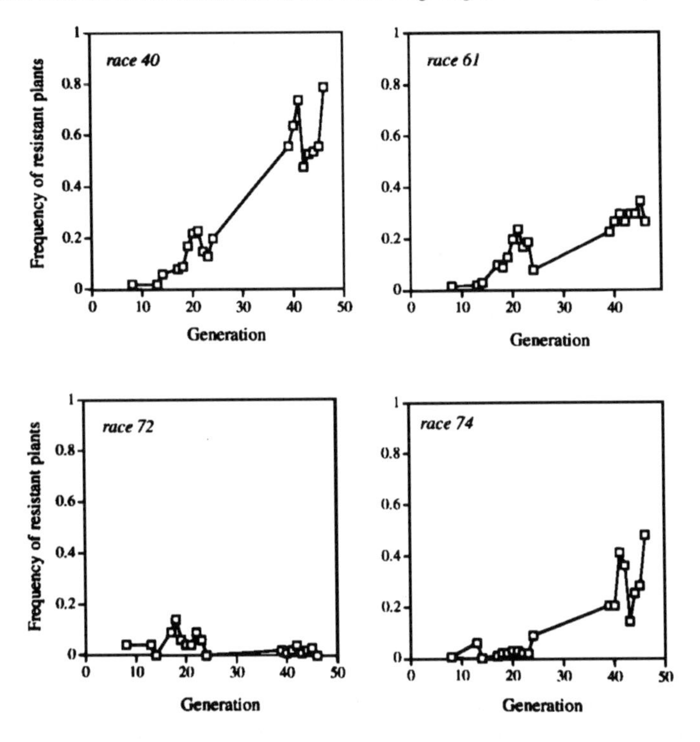


This diagram shows the frequency of genotypes resistant to four races of the scald fungus, *Rhynchosporium secalis*, in barley Composite Cross II during 45 generations of propagation in seminatural conditions (data from Webster et al. 1986). Note the pronounced increase in resistance to three of the four races during the later generations. The virulence of the pathogen population is described by McDonald et al. (1989). For other studies of the evolution of pathogen resistance in this population, see Jackson et al (1978), Muona et al. (1982) and Sagai-Maroof et al. (1983).

same species. Entirely different kinds of organisms, however, often form very close associations, either as partners or as antagonists. The selection that is generated by such associations has been studied most intensively in agriculture, where the fungal and bacterial communities that live on crop plants are of enormous economic importance.

**The General Effect of Other Organisms on Relative Fitness.** Crop trials that investigate how relative fitness varies in different physical environments are usually conducted so that other organisms, especially competitors, herbivores, and pathogens, have a minimal effect on the results. If genotypes vary in their ability to withstand competition or attack, the routine application of herbicides and pesticides to the experimental plots will cause the amount of environmental and G × E variance to be underestimated. There is no doubt that this is the case. Briefly, two main kinds of experiment have been attempted. The first involves measuring the impact of disease on cultivars grown in a range of more or less infested environments. The literature shows that disease scores vary in much the same way as other agronomic characters, the relative health of genotypes varying in different localities and years. The second kind of experiment is to measure the yield of cultivars that are either exposed to or protected from pathogens. This can be done by manipulating the environment, so that the plants are grown either in infested or in uninfested conditions; alternatively, it can be done by manipulating the genotype, so that lines known to be resistant or susceptible to a particular pathogen are grown over a series of more or less infested environments. An example of an experiment that combines the two approaches was reported by B.N. Kayasthar and E.G. Heyne. They constructed five pairs of nearly isogenic lines of wheat, one member of each pair being resistant and the other susceptible to the soil-borne mosaic virus of wheat. Some of the plots in which the wheat was sown were known to be clear of the virus, whereas others were infested. The genotype–environment abcd matrix for mean yield was as follows.

| | | Environment | |
|---|---|---|---|
| | | Uninfested | Infested |
| Genotype | Resistant | 2672 | 2222 |
| | Susceptible | 2705 | 1737 |

This neatly illustrates the usual results of such experiments. There is a pronounced environmental effect, yield being reduced by disease in the infested plots. There is also a distinctive pattern of G × E: the resistant plants produce more seed on the infested plots, as expected, but the susceptible plants have the higher yield in the absence of disease.

**Specific Genetic Interactions.** The interaction between the wheat and the virus is not only substantial, it is also highly specific, and is caused largely by a single gene for resistance to this disease in the wheat. Such genes are not uncommon; moreover, they are often matched by specific genes in the pathogen that determine whether or not it is virulent on a given strain of host. In contrast to the rather vague genetic basis for social relations among strains of the same species there is often a simple Mendelian basis for the relationships between resistant or susceptible host plants and the virulent or avirulent fungi and bacteria that live on them. These relationships can be explored in trials where lines or cultivars of the crop are exposed to defined clonal isolates of the pathogen. The analysis of this genotype–genotype matrix introduces a new and important complication.

We might measure the seed yield of each plant strain when exposed to each pathogen strain. This would enable us to express how the relative fitness of hosts varied according to the pathogen isolate they were exposed to. It is, however, only for economic reasons that we are primarily interested in the host; from a biological standpoint, we will be just as interested in how the relative fitness of pathogen strains varies according to the strain of host they grow on. We would then measure the productivity of the pathogen, although this is never done in agronomic trials, except as the intensity of disease symptoms expressed by the host. There are thus two social matrices shown as follows.

| | | Pathogen | |
|---|---|---|---|
| | | Type 1 | Type 2 |
| Host | Type 1 | a | b |
| | Type 2 | c | d |

| | | Host | |
|---|---|---|---|
| | | Type 1 | Type 2 |
| Pathogen | Type 1 | *a* | *b* |
| | Type 2 | *c* | *d* |

Here, the abcd represent the fitness of a host type when exposed to a particular pathogen type (or the effect on host fitness of that type), whereas the *abcd* represent the fitness of a pathogen type when exposed to a particular host type. These matrices are irreducible: they cannot be expressed, as all previous cases have been, as a single matrix. The reason is that the abcd cannot be interpreted as fitnesses (or effects on fitness) relative to the *abcd*. Thus, if a is large and *c* small, we cannot infer anything about the outcome of selection in either population. It is in this sense that the two components of the system, such as a plant host and a fungal pathogen, are not ecologically equivalent. When we are dealing with ecologically non-equivalent organisms, we can no longer describe the social evolution of a single population or set of populations, but must instead think in terms of the *coevolution* of two distinct entities. This consideration extends, of course, to any number of host and pathogen genotypes, and, what is more important, to any number of interacting but ecologically non-equivalent organisms.

I have expressed this argument in terms of the interaction between two antagonistic organisms because, for excellent reasons, a concern with pathogens engrosses the agronomic literature. Nevertheless, it applies with equal force to organisms that interact cooperatively and can be called mutualists or partners. There are, for example, several studies of the response of crop plants to different strains of the nitrogen-fixing root-nodule bacterium *Rhizobium*, showing that partners also have specific mutual responses.

**Coevolution of Barley and Scald.** The Composite Cross barley populations set up 40 or 50 years ago in California (see Sec. 103) have been monitored at intervals for various agronomic characters, including disease resistance. They are attacked by a variety of pathogenic fungi, including *Rhynchosporium*, which causes scald. Both host and pathogen are variable: different barley genotypes are resistant to different races of the fungus. Most of the varieties used to set up the initial hybrid populations were highly susceptible to most or all races of scald. After 45 generations of propagation, however, the frequency of resistance to most races had increased substantially. Moreover, many plants bore genes at different loci conferring resistance to several races of scald. A originally rather susceptible population, then, that is exposed to pathogens under more or less natural conditions evolves higher levels of disease resistance, through selection sorting both the initial variation and the variation arising subsequently from recombination. The plants in turn drive evolution in the pathogen populations. Composite Crosses II and V are closed populations that have been grown on isolated plots for many years, and each has evolved a characteristic community of pathogens. The *Rhynchosporium* populations are genetically quite diverse, as assessed by enzyme electrophoresis, with three or four common genotypes making up the bulk of the population and a great many less common types. The same two genotypes together made up about 60% of the fungal population on both barley populations. There was, however, no concordance among the remaining genotypes, some of which occurred at frequencies of 10–20% on one population but were rare or absent on the other. Selection on host populations that differ in genetic composition thus gives rise to pathogen populations that differ in composition. The virulence of the *Rhynchosporium* genotypes was assessed by inoculating a standard series of barley cultivars. The most virulent genotypes, attacking the widest range of test cultivars, were more frequent in Composite Cross II, which is the more resistant of the two barley populations. Thus, the interaction between pathogen and host causes the coupled evolution of resistance and virulence.

### 154. *Interactions between ecologically non-equivalent organisms are the main source of time-lagged frequency-dependent selection.*

The social matrices of non-equivalent organisms are coupled, so that selection in one will drive evolution in the other. The dynamics of these coupled systems are quite complex, but for asexual organisms, or alleles of a single gene, we can make a basic distinction between two types of systems.

**Isomorphic Social Matrices.** First, the two matrices may be similar in form, in the sense that they can be arranged so that the greatest fitness in either is found in the same cell, or along the same diagonal. For example, suppose that the two matrices can be arranged so that $\mathsf{a} > \mathsf{d} > \mathsf{b} > \mathsf{c}$ and $a > d > b > c$. We can symbolize the two types of one species as H1 and H2, and those of the other species as P1 and P2. The joint population may have any composition to begin with; say that it is dominated by H1 and P2 and so can be represented as H1P2. Because one population is dominated by H1, the frequency of P1 will rise in the other, since $a > c$, leading to a H1P1 population. This is stable against the spread of either H2 (because $\mathsf{a} > \mathsf{c}$) or P2 (because $a > c$). A similar argument shows that if the population were initially H2P1, it would evolve to a different stable state, H2P2. The greater is the frequency of H1, the greater the fitness of P1, and vice versa; the greater is the frequency of H2, the greater the fitness of P2, and vice versa. This type of system thus generates positive frequency-dependent selection, leading to the fixation of a single type in both populations. Because the two matrices have the same form, they can be said to be isomorphic. The condition for isomorphism is that $(\mathsf{a} - \mathsf{c})(a - c) > 0$, or $(\mathsf{b} - \mathsf{d})(b - d) > 0$, or both.

**Anisomorphic Social Matrices.** If the two matrices are different in form, the dynamics of the coupled system are very different. Anisomorphic matrices have both $(\mathsf{a} - \mathsf{c})(a - c) < 0$ and $(\mathsf{b} - \mathsf{d})(b - d) < 0$. For example, $\mathsf{d} > \mathsf{a} > \mathsf{b} > \mathsf{c}$ and $b > c > a > d$ are anisomorphic. If the joint population is initially H1P2, as before, then it will evolve toward H2P2 (because $\mathsf{d} > \mathsf{b}$), which will in turn evolve toward H2P1 (because $b > d$), subsequently to H1P1 (because $\mathsf{a} > \mathsf{c}$), and finally to H1P2 again (because $c > a$). The system

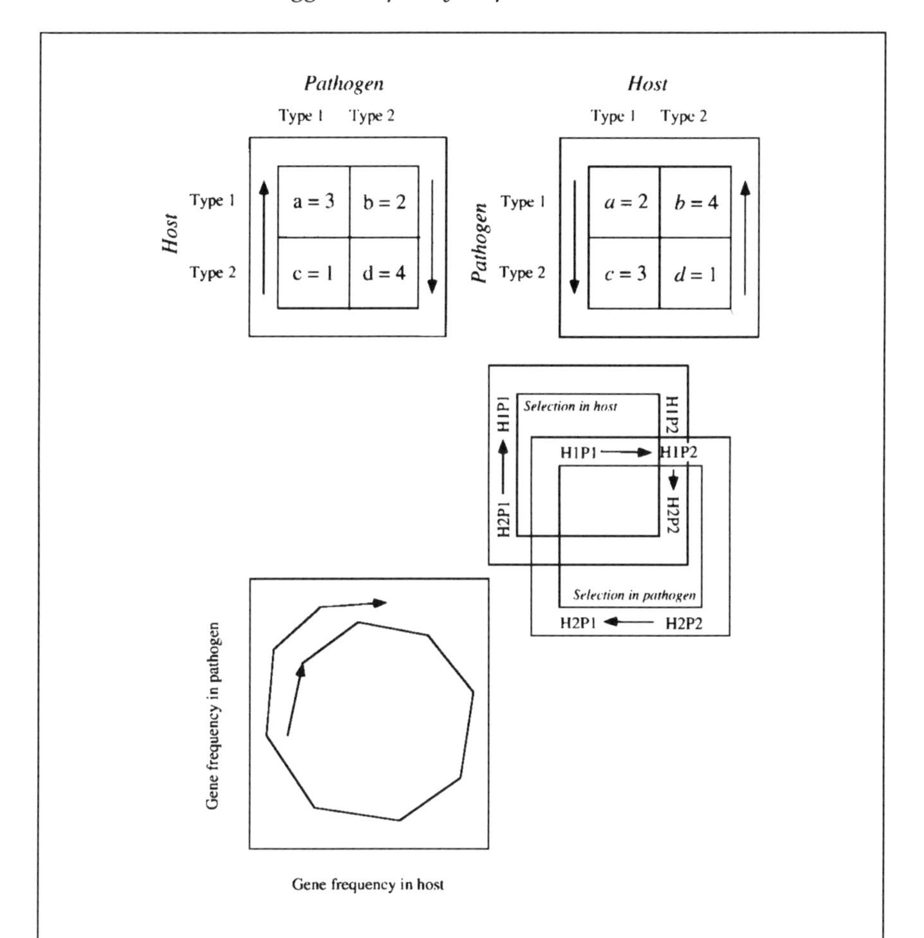


The two figures at the top are the anisomorphic social matrices discussed in the text, showing the evolutionary transitions in host and pathogen separately. Superimposing these matrices, so that given host–pathogen combinations occupy the same positions, makes the cyclical nature of the interaction clear. This can be expressed in the form of a phase diagram, a plot of gene frequency in the pathogen on gene frequency in the host, which will cycle indefinitely. (In practice, large fitness differences tend to force the cycle out to extreme frequencies that would be likely to cause the fixation of one type or the other in either population.)

*(Continues)*

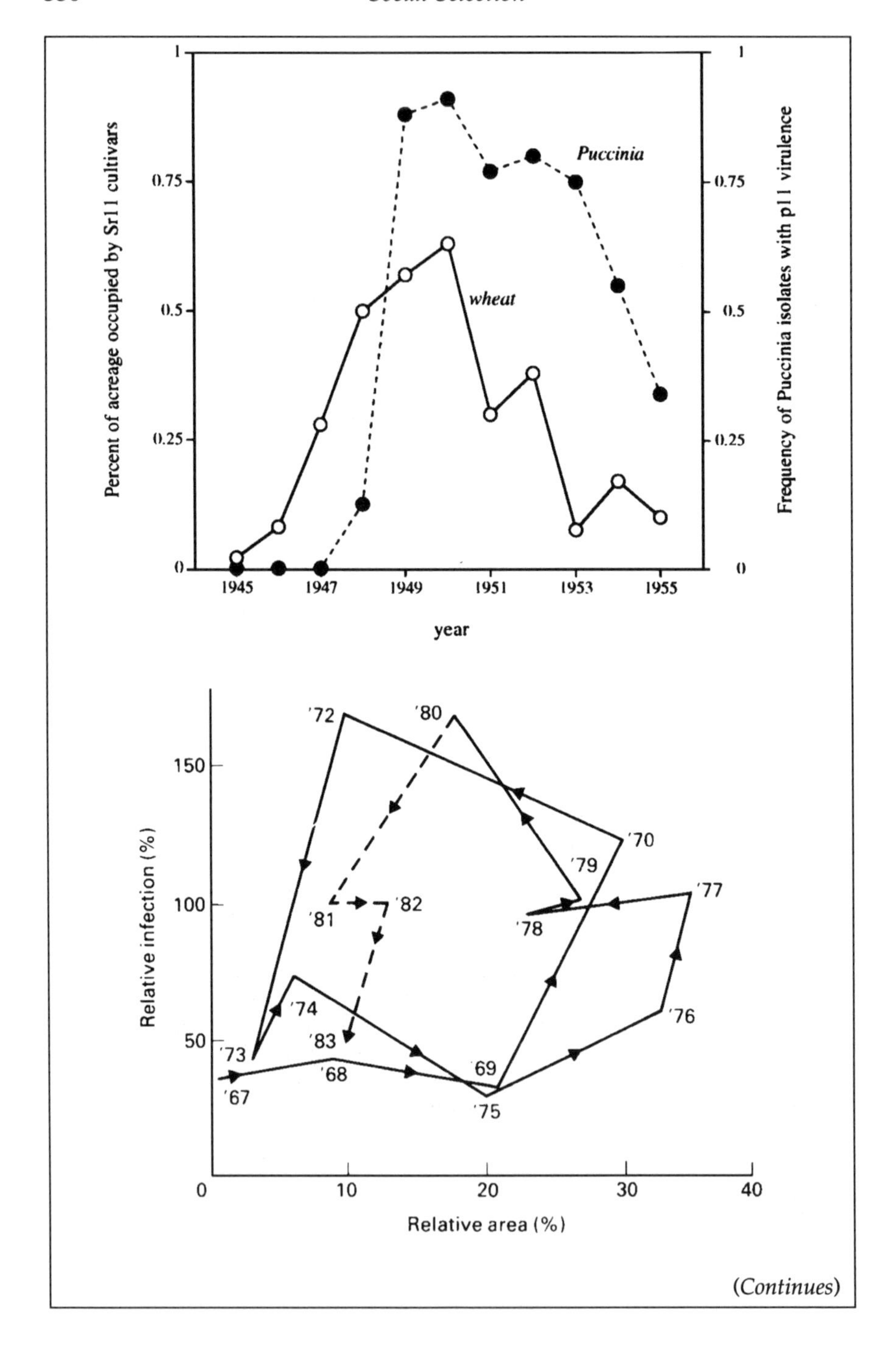

Percent of acreage occupied by Sr11 cultivars
Frequency of Puccinia isolates with p11 virulence
Puccinia
wheat
1
0.75
0.5
0.25
0
1945
1947
1949
1951
1953
1955
year
Relative infection (%)
Relative area (%)
150
100
50
0
10
20
30
40
'67
'68
'69
'70
'72
'73
'74
'75
'76
'77
'78
'79
'80
'81
'82
'83


(Continues)

*(Continued)*

The top diagram shows coupled changes in Australian populations of wheat carrying the resistance allele *Sr*11 and populations of the rust fungus *Puccinia graminis* carrying the corresponding virulence factor *p*11, from data in Watson and Luig (1968). This gives a clear picture of the spread of a pathogen genotype following the rise of a host genotype, followed by the decline of the host genotype and then the decline of the pathogen genotype. In this and similar cases, of course, it is the response of the farmers, rather than the direct response of the crop plant, that determines host dynamics. I have chosen this data so as to give a clear illustration of the process. In many cases the dynamics are more complex; for example, virulent races of the pathogen often persist at high frequency even when susceptible hosts have been driven out of cultivation (see Johnson 1987).

The lower diagram shows the coupled oscillations of barley cultivars carrying *Mla*12 resistance and mildew (*Erisyphe graminis*) races carrying *Va*12 virulence, as described in the text (from Wolfe 1987). For a general discussion of barley–*Erisyphe* genetics, see Wolfe and Barrett (1977).

Finckh and Mundt (1993) followed the frequency of four wheat cultivars in mixtures that were either inoculated with stripe rust or protected by a fungicide. There were substantial shifts in frequency over a period of three years, apparently unrelated to seed yield in pure culture, but similar shifts occurred in the infested and uninfested populations. In all cases, the rate of change in frequency of the cultivars in the evolving mixtures fell as their frequency increased; however, it is not clear whether fitnesses were frequency dependent, and the experiment was too brief to detect any time-lagged frequency dependence. Murphy et al. (1982) described changes in the composition of oat mixtures in plots infested or uninfested by crown rust. Kilan and Keeling (1990) described changes in the frequency of resistance genes in soybean populations on plots infested with the rot fungus *Phytophthora*.

is therefore never at rest, but continually cycles H1P1 → H1P2 → H2P2 → H2P1 → H1P1 → ⋯.

**Evolutionary Time Lags.** At some point in this process, the joint population is predominantly H1P1. Because P1 is common, the fitness of H1 is high, and if anything it will tend to increase in frequency at the expense of H2. The abundance of H1, however, selects for the spread of P2 that as it occurs increases the fitness of H2, until it eventually exceeds that of H1. The same argument applies to both genotypes in either population: the fitness of an abundant genotype will fall, because its abundance selects a well-adapted antagonist that some alternative genotype is better adapted to resist. Thus, when a genotype increases in frequency, its fitness falls, after a lag required for its specific antagonist to spread in the other population. This time-lag destabilizes the composition of either population, causing a coupled oscillation in the frequency of genotypes in both.

A similar process occurs when the matrices are isomorphic, except that in this case the fitness of a genotype increases with its frequency, so that when it overshoots its equilibrium point it continues to spread with increasing rapidity, until it has become fixed.

The generation of time-lagged frequency-dependent selection through the prior conditioning of growth medium may have seemed a rather marginal or artificial concept when it was introduced in Sec. 146. However, the evolutionary lag that characterizes the coupled selection of ecologically non-equivalent organisms implies that such processes occur very generally.

**Coupled Evolution of Barley and Powdery Mildew.** In the late 1960s barley cultivars carrying a new gene *Mla*12 for resistance to powdery mildew began to be used on a large scale, occupying about a quarter of the acreage by the end of the decade. At this point they began to lose their effectiveness as the corresponding virulence genes *Va*12 spread in the fungal population, and were largely withdrawn from cultivation. Nevertheless, varieties bearing *Mla*12 were still being used in breeding programs, where they were necessarily selected in the presence of virulent populations of mildew. The outcome was the re-emergence of *Mla*12, combined in new ways with other sources of resistance. This new range of cultivars were planted extensively during the 1970s, occupying about a third of the acreage by 1977. Their resistance to mildew, however, was eroded year by year, until by the end of the decade it had been overtaken by the evolution of virulent strains of mildew. Naturally, these cultivars were in turn largely abandoned, and a third range developed. The coupled evolution of virulence in the pathogen and resistance in the host thus drives a cyclical process in which the response of fungal strains carrying *Va*12 virulence first checks and then reverses the spread of barley varieties carrying *Mla*12 resistance.

In a simple system of this sort, history continually repeats itself. There is, however, no need to think that history ever *precisely* repeats itself, except in models. The expression of virulence or resistance is often epistatic, depending not only on the state of other loci with similar effects, but also on genes that normally play no part in host–pathogen interactions. At every turn of the cycle, continued directional selection on virulence in the pathogen (or resistance in the host) is likely to be a contingent process whose outcome depends in part on interacting genes that appear during the course of selection. Moreover, to the extent that pathogen virulence is a specific response to the particular mechanism of host resistance, and vice versa, the contingent outcome of selection in the pathogen will drive a contingent response by the host. At the level of overall virulence or resistance, host–pathogen systems may converge to regular oscillations, but the underlying genetic and physiological mechanisms of resistance and virulence may nonetheless diverge rapidly. Indeed, this is what happens in the barley–mildew system, where *Mla*12 is deliberately combined with other genes in order to overcome the existing sources of fungal virulence.

## 155. *Continued selection may lead to an arms race.*

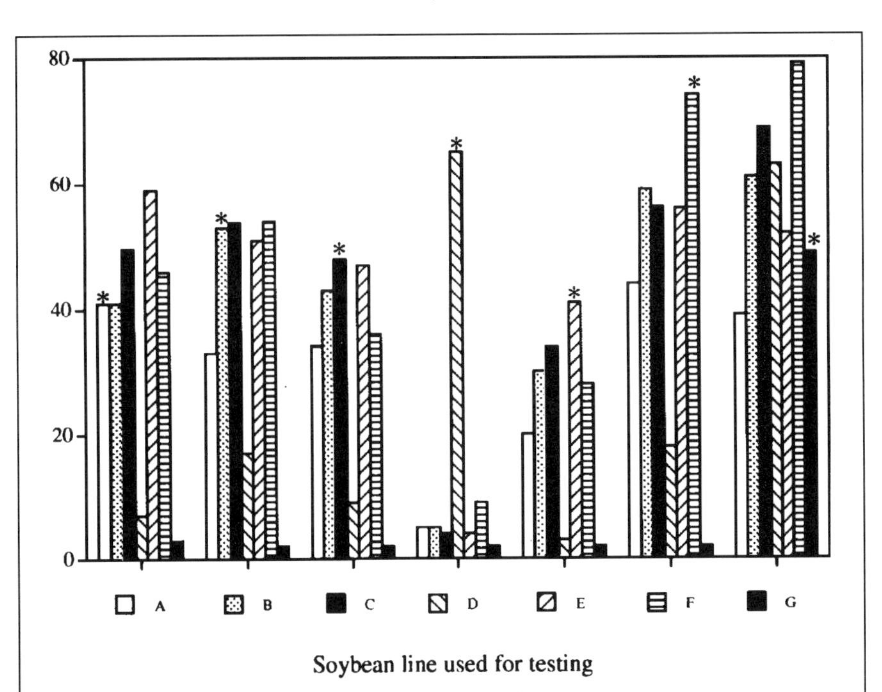


This diagram shows how the soybean cyst nematode, *Heterodera glycines*, evolves specific virulence toward soybean cultivars, using data from McCann et al. (1982). Resistant varieties of soybeans were grown in an agricultural field to collect nematodes capable of growing on them. The nematodes isolated from a particular variety were then grown with that variety in the greenhouse for two years. Each nematode population was then tested on each soybean variety, by counting the number of female nematodes produced when a given soybean variety is inoculated with a given nematode population. The diagram is grouped by the soybean variety used for testing (for simplicity, I have relabeled these varieties A–G, retaining the order in the original paper). For each of the seven test varieties, there were seven nematode populations, each selected on one of the soybean varieties. I have marked the column representing the performance of a nematode population with the soybean variety on which it was selected with an asterisk. The trial shows two patterns.

(*Continues*)

a) A given nematode population tends to perform better with the host on which it was selected. This is especially evident for populations that perform poorly on most hosts, such as the population selected on G.

b) On a given soybean variety, the population selected on it performs better than any other population. This is especially evident for resistant varieties, such as D.

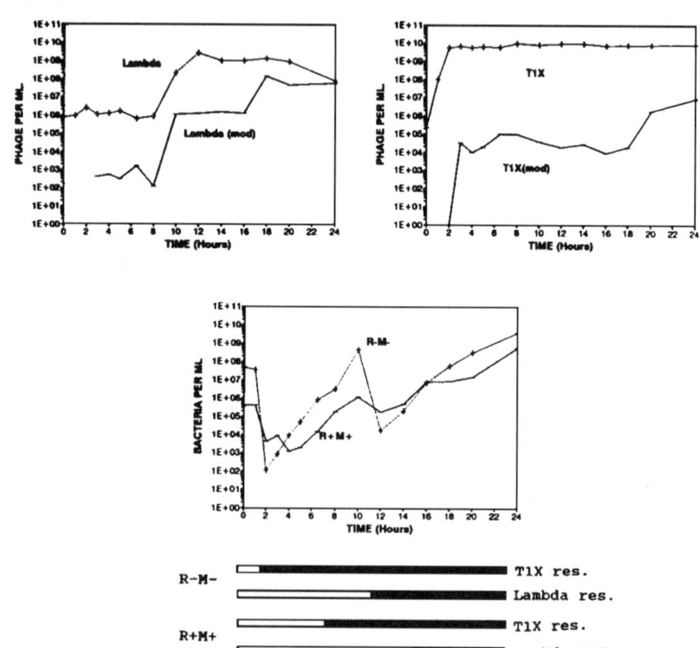


This figure shows details of the initial dynamics of a bacteria–phage system, from Korona and Levin (1993; see also Levin 1988). An *R+M+* strain was inoculated at low frequency into a liquid culture of *R-M- E. coli* in the presence of two types of phage, T1X and Lambda. T1X is the more virulent and rapidly increases in abundance, driving down the bacterial population. During the first two hours of the infection, *R+M+* cells decline in abundance less rapidly and thus increase in frequency. Types resistant to T1X then spread through the *R-M-* population (as shown by the filled section of the bar at the bottom of the figure), as modified types spread in the phage population. *R-M-* then increases, whereas *R+M+* continues to decrease until resistant *R+M+* types appear. After about 12 hours, the same succession of events occurs with Lambda: a general decline in abundance, with *R+M+* increasing in frequency, the subsequent spread of resistant types in *R-M-* and of modified types in Lambda, and the rebound of *R-M-*. In this experiment, *R+M+* increased in frequency by the end of the first cycle of growth; this was not

(*Continues*)

(*Continued*)
generally the case in experiments where only a single type of phage was present, and the increase was not sustained in subsequent growth cycles even when two or more phage types were present.

Adaptation to a new host by *Tetranychus* was described by Gould (1979). Host-range extension is of course often observed when crops are introduced into a new area; see Sec. 173.

The evolution of *SD*-infected stocks of *Drosophila* is described by Lyttle (1979). He also describes a line in which suppression was achieved through the selection of XXY and XYY aneuploids (Lyttle 1981), emphasizing the contingent outcome of coevolutionary struggles. There are comparable accounts of the coevolution of the *Drosophila* genome and the transposable *P* element in experimental populations by Daniels et al. (1987) and Kidwell et al. (1988).

The barley–mildew system is an isolated case. I do not know of any other clear example of coupled genetic oscillations. They will be detected only in long-term studies of systems whose genetics are accessible, and this may in large part explain their rarity in the literature. It is also possible, however, that the abcd model cannot be applied straightforwardly to most natural populations. There are two major complications. The first is that any given population may interact with many partners and antagonists, so that its genetic dynamics may be much more complicated than a simple cycle, giving instead the impression of aimless fluctuations of rather small magnitude; it is only in isolated couples of strongly and specifically interacting organisms that the underlying phenomenon will be easily detected. Second, there is built into the abcd model an assumption that the component populations each comprise a small set of defined genotypes, whose evolution occurs through a process of sorting. It is entirely possible that there is more often a continued selection of novel genotypes arising through mutation. The evolution of the system will still be driven by time-lagged frequency-dependent selection, but instead of the endless alternation of the same range of types, there will instead be an endless succession of different types, the outcome of an arms race between a population and its antagonists.

**Experimental Extension of Host Range in a Mite.** Fred Gould of Stony Brook collected *Tetranychus*, a polyphagous spider mite, from a natural population in a pear orchard, and reared them on lima beans in the laboratory for eight months. The mites grow well on lima beans, with high fecundity and low mortality. A second line was then established in a cabinet containing both lima beans and cucumber plants. The mites will feed on both the beans and the cucumber, but juvenile mites usually die on cucumber, so the bean plants are destroyed more rapidly. The remaining mites must then feed on cucumber alone until transferred to a new cabinet. After nearly two years of selection, the mites had evolved much higher survival rates on cucumber. The evolution of new sources of virulence can therefore occur quite rapidly, within less than 20 generations. The evolved populations were also tested on exotic hosts that are normally resistant to the mites. The correlated response to selection on cucumber included an increased survival rate on potato and tobacco, although not on plantain. Adaptation to one antagonist is thus not completely specific and may have manifold consequences for interaction with others.

**Evolution of Restriction-Modification Systems.** Bacteria produce restriction enzymes that cut DNA molecules wherever a particular sequence of a half-dozen or so bases occurs; different enzymes recognize different sequences. They also, necessarily, produce modification enzymes that, in most circumstances, prevent their own DNA from being cut, usually by methylating a particular sequence of bases; different enzymes modify different sequences. The combination of a restriction enzyme and an appropriate modification enzyme constitutes a restriction-modification system that will destroy foreign DNA, such as invading viral genomes, without endangering the bacterium itself. It is not quite a foolproof system; occasionally, the viral DNA will be modified by the host enzyme, and the virus will then be able to reproduce. Its progeny inherit the modified DNA, and this virulent genotype can then spread through, and destroy, the bacterial population. Consequently, there should always be selection for new restriction-modification types that will provide immunity against bacteria as long as they remain rare. This would account for the diversity of bacterial restriction-modification systems.

Some experiments by Ryszard Korona and Bruce Levin, working at Amherst, cast doubt on this simple interpretation of restriction-modification as a generally effective source of resistance to pathogens. They introduced an *R+M+* strain bearing a functional system into liquid cultures of an isogenic *R-M- E. coli* population lacking restriction-modification function. In the absence of phage, *R+M+* is nearly neutral, and continues to persist at low frequency. When phage is present, the *R+M+* strain increases in frequency, especially if there are two or more types of phage in the culture. However, this advantage is transient; within a few generations, resistant genotypes with altered cell membranes that prevent phage adsorption appear and subsequently spread through the bacterial population, so that after one or two transfers the population is almost completely resistant. Because these mutations are likely to occur in the more abundant *R-M-* genotype, *R+M+* cannot spread when introduced at low frequency. Experiments on solid medium, however, where each viable cell gives rise to a discrete colony, had a different outcome. The growing colony will eventually make contact with a phage particle that initially infects cells at the margin and then proliferates inward to destroy the whole colony. In *R-M-* colonies, there are too few cells for resistant mutants to arise before the colony is destroyed. *R+M+* colonies, on the other hand, are usually able to digest the invading phage before they acquire the appropriate modification, preventing the infection from taking hold, provided that phage density is not too high. Even at high density, restriction-modification slows down the pace of the initial attack, giving time for envelope-based resistance to evolve. The course of the arms race in bacteria-phage systems is thus unexpectedly sensitive to bacterial ecology.

**Modifiers of Meiotic Drive.** The Segregation-distorter element *Sd* of *Drosophila* (Sec. 125) is harmful to most of the genome because of the sterility or inviability of *SD/SD* homozygotes. Consequently, genes unlinked to *Sd* will tend to evolve so as to suppress its effects. Conversely, genes linked to *SD* would be selected to overcome suppression, because they benefit by being transmitted along with *SD*. There is therefore an antagonism between different kinds of element within the same genome; a similar antagonism will arise in the case of other autoselected elements, such as B-chromosomes or transposons. In many cases, suppression is caused either by alleles at the *Responder* locus or by dominant genes of large effect on other chromosomes, whereas enhancement of *SD* is caused by genes on the same chromosome arm coupled to *Sd* by inversions. The evolution of suppression has been studied experimentally by Terrence Lyttle of Honolulu, using stocks in which the second chromosome, bearing *SD*, is linked to the Y chromosome by a translocation. This causes extreme sex-ratio distortion, because males transmit only the driven Y, so that all their progeny are male. The usual sources of suppression do not evolve in such populations, because major genes of large effect are absent from the base population, and insensitive *Responder* alleles cannot be transferred to the *SD* chromosome, because it is linked to Y by translocation, and crossing-over does not occur in male *Drosophila*. Instead, the experimental populations evolved over about 90 generations to suppress distortion through the selection of recessive genes of small effect on the third and fourth chromosomes. It seems likely that many of these genes arose by mutation during the course of the experiment, and therefore that different lines tended to accumulate different sources of suppression. At the same time, genetic enhancers of drive were detected on the compound Y and second chromosome. Autoselected elements may thus inaugurate a complex coevolutionary succession of antagonistic elements within the same genome.

## 156. *Neighbors that constitute mutually renewable resources evolve to become partners.*

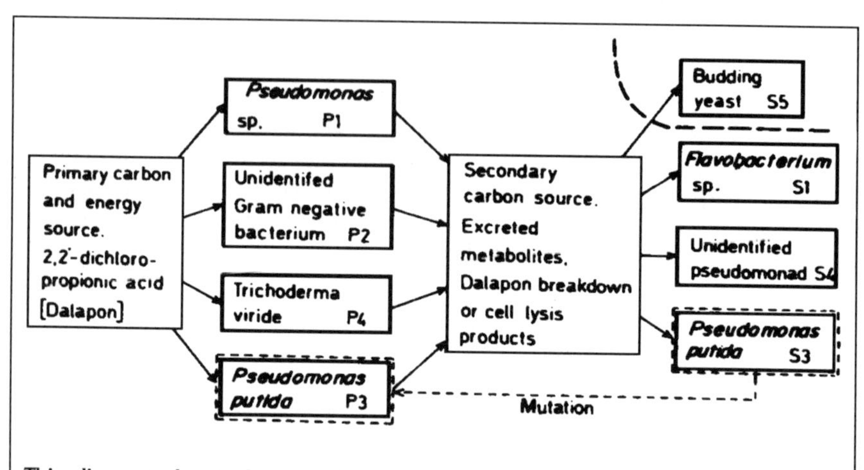


This diagram shows the microbial community consistently emerging from soil samples cultured in chemostats with the herbicide Dalapon as sole carbon and energy source, from Senior et al. (1976).

There are several well-known examples of partnerships between phototrophic and heterotrophic organisms: lichens, or angiosperms and mycorrhizal fungi, or *Hydra* and intracellular *Chlorella*, for example. The basis for their cooperation is (roughly speaking) the exchange of fixed carbon for nitrogen or other nutrients. These partnerships can be dissolved, and sometimes are, yet in most circumstances remain remarkably stable. Their stability can be understood through an extension of the trading game used to illustrate the evolution of cooperation between members of the same population (Sec. 144). A pine tree and its mycorrhizal fungi are associated physiologically, although not genetically or reproductively; each produces independent propagules, and any two young individuals must establish a new partnership. This partnership lasts for life. The two partners normally grow continuously, but we can imagine that their growth occurs as a large number of short episodes. In each episode, the tree offers organic carbon compounds to the fungus, and the fungus offers an increased flow of mineral nutrients to the tree.

**Partners Have Isomorphic Social Matrices.** If the two were ecologically equivalent, this long series of encounters between the same two individuals would tend to select cooperative partners, as I have already explained. Plants and fungi are not ecologically equivalent, however: the cooperative behavior of a mycorrhizal symbiont cannot be explained in terms of the interactions between different fungal genotypes. Nevertheless, the logic of Tit-for-tat behavior continues to hold. Plant and fungus are permanently associated and are unlikely to change partners. Each has something valuable to trade and would suffer if its partner defected; each, so to speak, constitutes a renewable resource for the other. The symmetry of their interests implies that their social matrices are isomorphic, so that the outcome of selection will be the fixation of a single type in both populations. Both plant and fungus, however, will prosper if they cooperate, and neither will prosper if both defect. Cooperative types will therefore spread in both populations. Indeed, cooperation is more likely to evolve than if the two organisms were in direct reproductive competition, because quantitative symmetries are irrelevant: a plant will benefit from cooperating with its mycorrhizal associate if it thereby increases its fitness, if only very slightly, even if as a consequence it provides an enormous benefit to the fungus.

These arguments remain speculative; there has been little or no experimental work on partners who reproduce separately, and the evidence for the interpretation that I have just described remains circumstantial. It might pay either partner to defect if the terms of trade change, so that a partner can more profitably be treated as a non-renewable resource: a green hydra growing in the dark, for example, draws no benefit from its photosynthetic symbionts and is likely to digest them. It will also pay to defect if the association is coming to an end; if your partner is sick and likely to die, it is better to realize the capital, rather than have the investment fail. There is clearly an opportunity for selection experiments on highly integrated systems, such as lichens, to test ideas such as these.

**The Dalapon Community.** The extent of evolved partnerships among different kinds of organisms and the way in which they influence the structure of communities remain poorly understood. An interesting line of approach is suggested by the experiments of E. Senior and his colleagues at Canterbury and Warwick on the community of microbes that develops in soils treated with the herbicide Dalapon, a chlorinated propionic acid. When soil samples are cultured in chemostats with Dalapon as the only source of carbon and energy, they quickly and consistently give rise to a characteristic community of seven different microbes. Three of these are able to grow as pure cultures in Dalapon medium: a species of *Pseudomonas*, an unidentified Gram-negative bacterium, and the filamentous fungus *Trichoderma*. The other four members of the community were *Pseudomonas putida*, an unidentified pseudomonad, a *Flavobacterium*, and a budding yeast. They were unable to grow on Dalapon as pure cultures, but formed a conspicuous part of the Dalapon community: they presumably fed on metabolites excreted by the primary Dalapon users. The composition of the community changes somewhat with conditions of culture: when the dilution rate rises above 0.2 per hour the yeast drops out, and if it exceeds 0.45 per hour *Pseudomonas* displaces the Gram-negative bacterium as the most abundant primary user. Nevertheless, the community is remarkably stable, showing no qualitative change in composition over several thousand generations of continuous culture. This represents a sorting experiment that selects a group of six or seven interacting species from the teeming microbial community of the soil. Unfortunately, the nature of their interactions has not yet been ascertained. The primary users are clearly supporting the secondary users, but whether the secondary community enhances the growth of the primary users by removing metabolic wastes, and what the relationships among the species in either community are, remains unknown. Interestingly, the experiments also showed that the community can continue to evolve: a strain of *Pseudomonas putida* with a modified dehalogenase able to hydrolyze Dalapon to pyruvic acid appeared and spread after a few hundred generations, adding a new component to the community of primary users.

## 157. *Vertically transmitted symbionts evolve to become closely integrated partners.*

Cooperative partnerships are likely to evolve when the partners are necessarily associated for extended periods of time, whether through lack of mobility, long-continued culture in the same chemostat, or for any other reason. The most durable associations are formed between organisms that are not only physiologically but also reproductively dependent. When the offspring of the partners themselves remain associated, it is not only individuals but lineages that continue to interact, and selection will favor combinations of lineages that enhance one another's productivity.

**A Novel Protist-Bacterium Partnership.** K.W. Jeon and his colleagues in Paris observed a remarkable instance of the evolution of a novel partnership in the laboratory. Their cultures of *Amoeba* became accidentally infested by an unknown bacterium able to multiply inside the host cells. The bacterium was initially pathogenic, killing most of the amoebas it infected. The bacteria released by the lysis of the dead host would then infect other amoebas. A few hosts survived this infection, though retaining a residual population of bacteria in their cytoplasm. When these amoebas divided, both daughters inherited an intracellular population of bacteria, so the bacteria were necessarily vertically transmitted with their host. The evolution of a specific relationship between the amoebas and the bacteria could be demonstrated by reciprocal transplant experiments: bacteria extracted from the perennially infected host lines were pathogenic in naive hosts, and unevolved bacteria were pathogenic to evolved hosts whose resident bacteria had been cleared out by antibiotics. Moreover, in some lines this relationship became obligate after about 100 host generations:

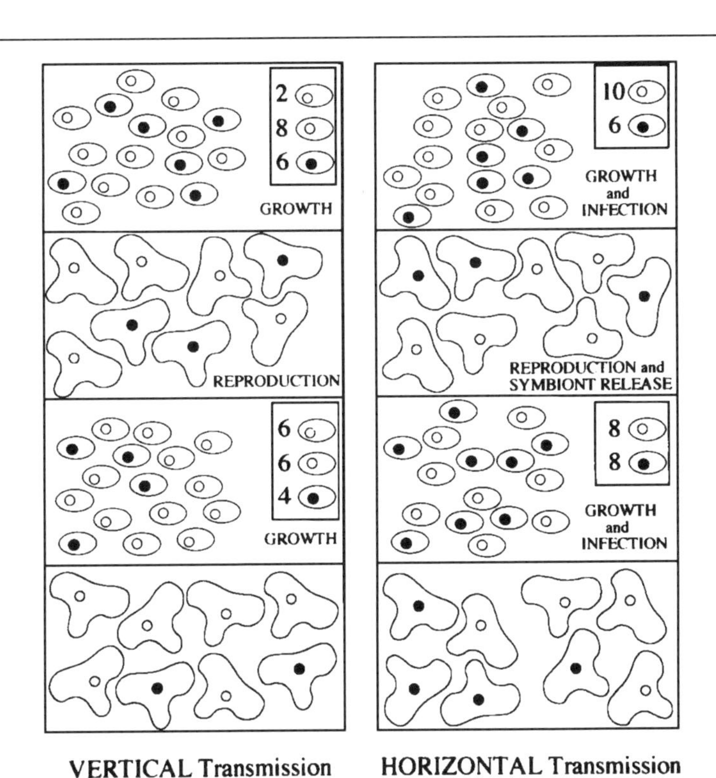


When symbionts are transmitted vertically, so that offspring inherit their parents' symbionts, strains that promote the reproduction of their host are favoured, because infected lineages of susceptible hosts (marked by an open circle) will increase in frequency at the expense of both uninfected lineages of susceptible hosts, and resistant lineages (solid circle). Horizontally-transmitted symbionts are released and infect new, unrelated hosts in every generation. They will be selected for their ability to infect, even if they damage the host individual with which they are associated. Susceptible host lineages will decrease in frequency, and resistance spreads in the population.

*(Continues)*

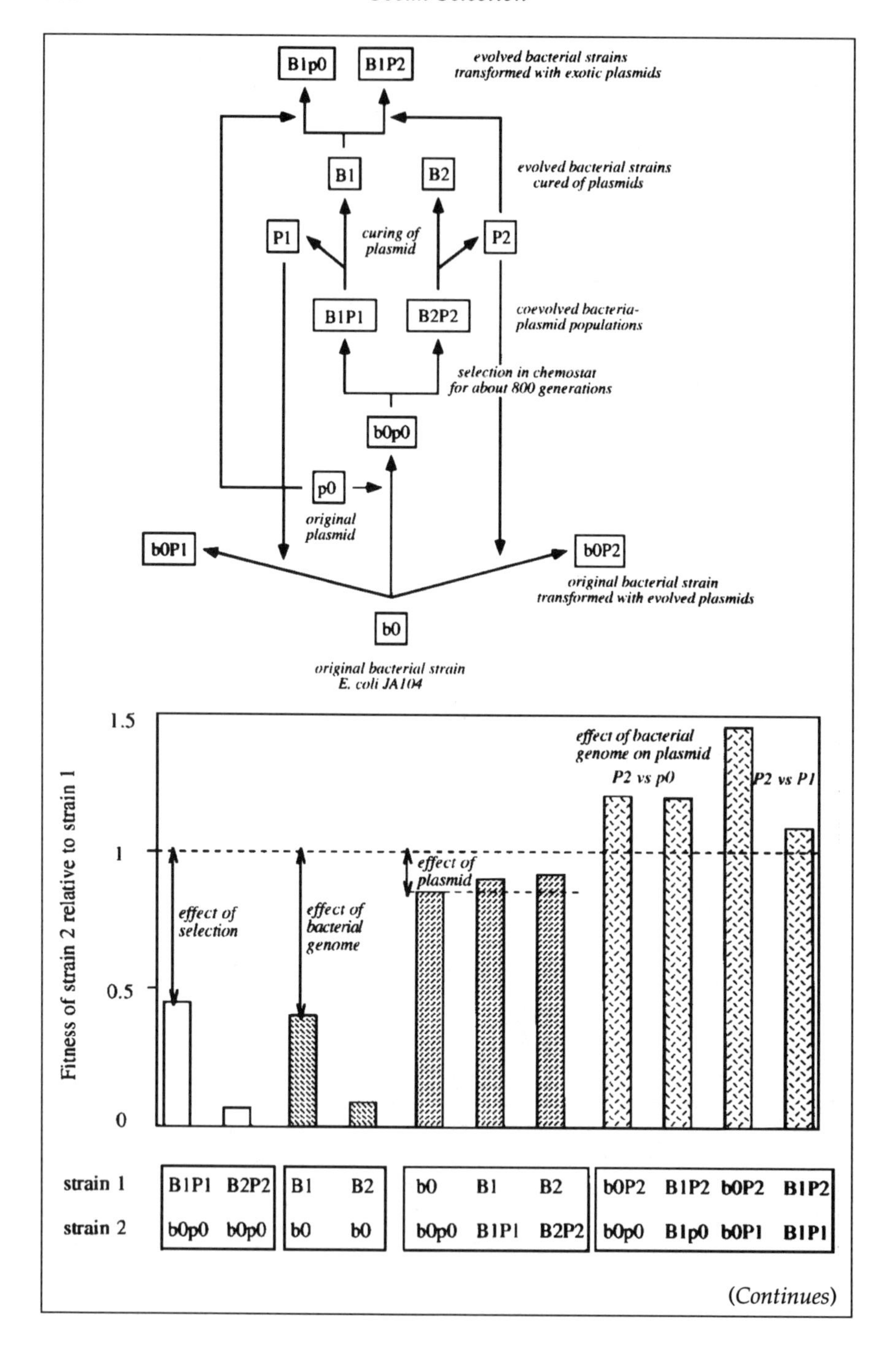
B1p0
B1P2
evolved bacterial strains transformed with exotic plasmids
B1
B2
evolved bacterial strains cured of plasmids
P1
curing of plasmid
P2
B1P1
B2P2
coevolved bacteria-plasmid populations
selection in chemostat for about 800 generations
b0p0
p0
original plasmid
b0P1
b0P2
original bacterial strain transformed with evolved plasmids
b0
original bacterial strain E. coli JA104
Fitness of strain 2 relative to strain 1
1.5
1
0.5
0
effect of selection
effect of bacterial genome
effect of plasmid
effect of bacterial genome on plasmid
P2 vs p0
P2 vs P1
strain 1
strain 2
B1P1 B2P2
b0p0 b0p0
B1 B2
b0 b0
b0 B1 B2
b0p0 B1P1 B2P2
b0P2 B1P2 b0P2 B1P2
b0p0 B1p0 b0P1 B1P1

(*Continues*)

(*Continued*)

These diagrams illustrate the coevolution of bacteria and plasmids, from the study by Modi and Adams (1991). The original bacterial and plasmid strains are designated b0 and p0, respectively, and the evolved strains B1, B2 and P1, P2. The flow chart shows how the various bacterial and plasmid lines were selected and tested. When different combinations of bacteria and plasmids were constructed, the main results shown in the bar diagram were as follows:

a) Selection causes adaptation to chemostat conditions (e.g., B1P1 vs b0p0).

b) This adaptation is largely attributable to changes in the bacterial genome (e.g., B1 vs b0).

c) The plasmid is deleterious, but its effect on the bacterium is less in the evolved lines (e.g., B1P1 vs B1, compared with b0p0 vs b0).

d) The effect of the plasmid depends on the bacterial genome with which it is associated (the effect of p0 vs P2 is the same in b0 and B1, but P2 is more deleterious in b0 than in B1).

For similar studies, see Bouma and Lenski (1988) and Lenski et al. (1994). Comparable experiments on the coevolution of bacteria and phage were reported by Bull and Molineux (1992) and Bull et al. (1991). The work on the *Amoeba*–bacterium system is by Jeon and Lorch (1967) and Jeon and Jeon (1976).

the amoebas were no longer able to grow successfully without the bacteria. This shows how it is possible to reconstruct the initial stages of the evolution of highly modified bacterial endosymbionts, such as chloroplasts and mitochondria, in the laboratory. Such burgeoning relationships can be found in a range of extant protists, such as the bacterial community of *Pelomyxa* or the photosynthetic entities of *Glaucocystophora*.

**Experimental Evolution of Cooperation in a Bacterium-Phage System.** The best evidence for the evolution of cooperation through the selection of vertically transmitted combinations of lineages comes from an elegant experiment by Jim Bull and Ian Molineux of Austin, in collaboration with Bill Rice at Santa Cruz. They used a filamentous phage that infects *E. coli*. Filamentous phages are relatively benign parasites whose progeny can pass out through the host cell envelope without disrupting it. The bacterium thus continues to divide, distributing phage to both daughters, although at a diminished rate, because of the metabolic burden of maintaining 100 or so replicating copies of the phage. Infected cells are immune to further infection. This is important, because if the phage has no monopoly on a host, it will have less interest in maintaining host viability. The phage used by Bull and his colleagues had additional DNA inserted into the region between its two origins of replication; this DNA encoded antibiotic resistance and was used as a selectable marker to ensure that all bacteria carried copies of the original phage or one of its evolved derivatives. Selection for cooperative phage (phage genotypes that interfered less with the reproduction of their hosts) was imposed simply by serial transfer in medium containing antibiotics. This ensured a high degree of fidelity between partners (as Bull expresses it), because all surviving cells carried the phage and were thus immune from further infection; thus, genotypes able to infect new hosts would gain no advantage, whereas types that enhanced the reproduction of their hosts (relative to the usual deleterious effect of infection) would increase in frequency through the increased frequency of the host lineages that bore them. Selection was applied in the opposite direction by collecting free phage from the culture and inoculating them into a fresh culture of uninfected bacteria. This enforced low partner fidelity by separating phage and host lineages, selecting for infectious phage that should evolve so as to maximize the production of infectious progeny, regardless of their deleterious effect on host reproduction. The high-fidelity and low-fidelity lines were propagated for 15 growth cycles, each comprising about 10 bacterial generations.

The hypothesis that cooperation is favored by the permanent association of lineages makes several predictions about the consequences of mixing cells (and the phage they contain) from the high-fidelity and low-fidelity lines. If the mixture is simply inoculated into fresh medium, the high-fidelity phage should increase in frequency, by virtue of the higher rate of growth of bacteria whose phage are cooperative. The same thing should happen if the culture also includes a large number of other bacteria that are resistant to the phage; but if there is a large proportion of uninfected susceptible bacteria in the culture, then the low-fidelity phage, with their greater infective ability, will be the more successful and might even spread. All of these predictions were confirmed. Either the bacteria or the phage, or both, had evolved so that bacterial lines maintained in close association with their resident phage were no longer harmed as severely by infection.

The simplicity of the phage genome—the creature is only about 7000 nucleotides in length, comprising 10 genes—made it feasible to investigate in detail the genetic changes associated with the evolution of cooperative behavior. These turned out to be of great interest because different changes occurred in different selection lines. In one case, there were loss-of-function mutations that created non-infectious phage; these were infrequent because most mutations that knock out phage genes are lethal to the host cell. In other cases, the phage genome becomes integrated with the host genome. This presumably enhanced host reproduction by reducing the number of phage copies from 100 or so down to 2 or 3. Finally, several lines evolved highly defective phage-like plasmids from which almost all of the phage genome had been eliminated: these consisted of the two phage origins of replication, the antibiotic resistance factor and a short length of DNA presumably derived from the original phage genome. These were incapable of autonomous self-replication, depending on a functional integrated copy of the phage. Almost as notable as the divers adaptations that did evolve were those that did not. For example, replication-deficient phage did not appear, although they might have been expected, perhaps because there is no mechanism to regulate the distribution of phage between daughter cells, so that lines with few phage copies would continually segregate lineages that lacked phage altogether. Social evolution in this system is thus highly contingent, leading to similar phenotypes by quite different genetic routes.

**Coevolution of Bacteria and Plasmids.** Plasmids are more highly integrated than phage with their bacterial hosts. Conjugative and mobilizable plasmids can spread through autoselection despite their deleterious effect on the growth of their hosts (Sec. 120). Some plasmids encode functions beneficial (in certain environments) to their host, such as antibiotic resistance, and will be selected for this reason. Others, however, confer no distinctive phenotype, cannot be transferred from cell to cell, and are mildly deleterious to their host. In such cases, a more cooperative relationship should evolve.

Judith Bouma and Richard Lenski studied a 4-kb plasmid of *E. coli* that encodes multiple antibiotic resistance but reduces bacterial growth rate in the absence of antibiotics. A cell line into which the plasmid had been inserted was cultured for about 500 generations in the presence of antibiotics; this ensured the continued association of bacterial and plasmid lineages because the plasmid cannot be transmitted from cell to cell, and any cells that had lost the plasmid would be killed by the antibiotic. At the end of the experiment, the mutual adaptation of bacterial and plasmid genomes was measured by constructing all combinations of evolved and unevolved bacteria and plasmids. This showed that the bacterial genome had been modified through selection for types in which the deleterious effect of the plasmid was reduced; no modication of the plasmid could be detected. More surprisingly, the evolved bacterial strain outcompeted the original plasmid-free strain, even in growth medium containing no antibiotics. This new cooperative relationship had been established through a unilateral modification of the host genome because the same advantage was realized by the evolved bacterial strain when transformed with the original, unmodified plasmid.

A similar experiment was conducted by Rajiv Modi and Julian Adams at Ann Arbor, who cultured a population founded by a single plasmid-infected clone for about 800 generations in a chemostat. They did not select against cured cells which were generated by unequal segregation and increased in frequency throughout the experiment because of their higher rate of growth. The decline in the frequency of infected cells was very erratic: mutations occurring either in the bacterial or in the plasmid genome that enhance bacterial growth rate occasionally spread through the infected sector of the population, causing it to increase briefly in frequency. The fitness of the evolved population (under the conditions of glucose-limited continuous culture) was greatly increased through the periodic selection of the beneficial mutations whose passage was marked in this way; the fitness of the original strain, with its original plasmid, was only $\frac{1}{10}$–$\frac{1}{2}$ that of the evolved strain, with its evolved plasmid, when the two are put into competition with one another. By comparing cured lines from the original and the evolved strains, it could be shown that much of this advance was caused by the modification of the bacterial genome, as would be expected. Exchanging plasmids between lines, however, showed that it was also attributable in part to changes in the plasmid and to altered interactions between bacterial and plasmid genomes. The adaptation of plasmid-infected lineages to a novel environment thus involves some degree of coevolution toward a more cooperative relationship. The changes in the plasmid genome that contributed to this are not known precisely, but resulted in fewer copies per cell, presumably through reduced rates of replication.

## 158. *Resistance and virulence are costly.*

Whereas vertically transmitted symbionts evolve as partners, the converse is also true: whenever symbionts have the opportunity to change partners, they are less likely to evolve cooperative relationships, and horizontally transmitted symbionts tend to evolve as antagonists. Ecologically equivalent species become competitors; non-equivalent species evolve asymmetrical relationships as pathogen and host or predator and prey. The virulence of a pathogen indicates its ability to make use of a particular environmental opportunity provided by the host; the resistance of the host indicates its ability to grow in the hostile environment created by the pathogen. As with any other environmental factor, the precision of adaptation is constrained by its cost: the cost of resistance in hosts and the cost of virulence in pathogens.

**The Cost of Resistance.** The enfeeblement of resistant genotypes in uninfested environments has been studied carefully in two contexts, the development of crops resistant to disease and the relationship between bacteria

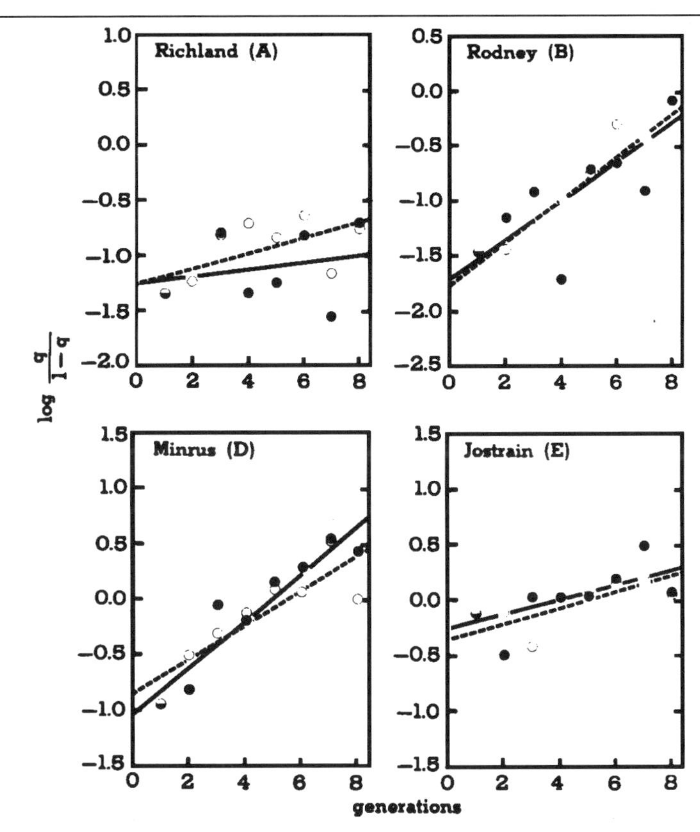


This figure shows the correlated response to selection of oat stem rust *Puccinia graminis* on two oat cultivars for eight sexual generations, from Leonard (1969). The $y$-axis is $\log [q/(1-q)]$, where $q$ is the frequency of avirulent spores in the rust population. This is the standard procedure for estimating fitness in experiments with microbes, the slope of the regression being a relative fitness (Secs. 40–41). The frequency of avirulent spores, presumably a very heterogeneous collection of genotypes, was estimated as the frequency of resistant-type lesions on the infected host plants. The rust population was selected on either Craig (broken line and filled circles) or Clintland A (solid line and dots) before being tested on eight other cultivars. This figure shows the results for four of the test cultivars. A positive regression indicates that selection on Craig or Clintland had favored genes that are avirulent on the test cultivar. Since all the regressions—including those not shown here—were positive, there is a general tendency for genes causing virulence toward host genotypes that have not recently been encountered to be lost from the population. For a similar experiment, see Alexander et al. (1985). For a general perspective, see Leonard (1987).

(*Continues*)

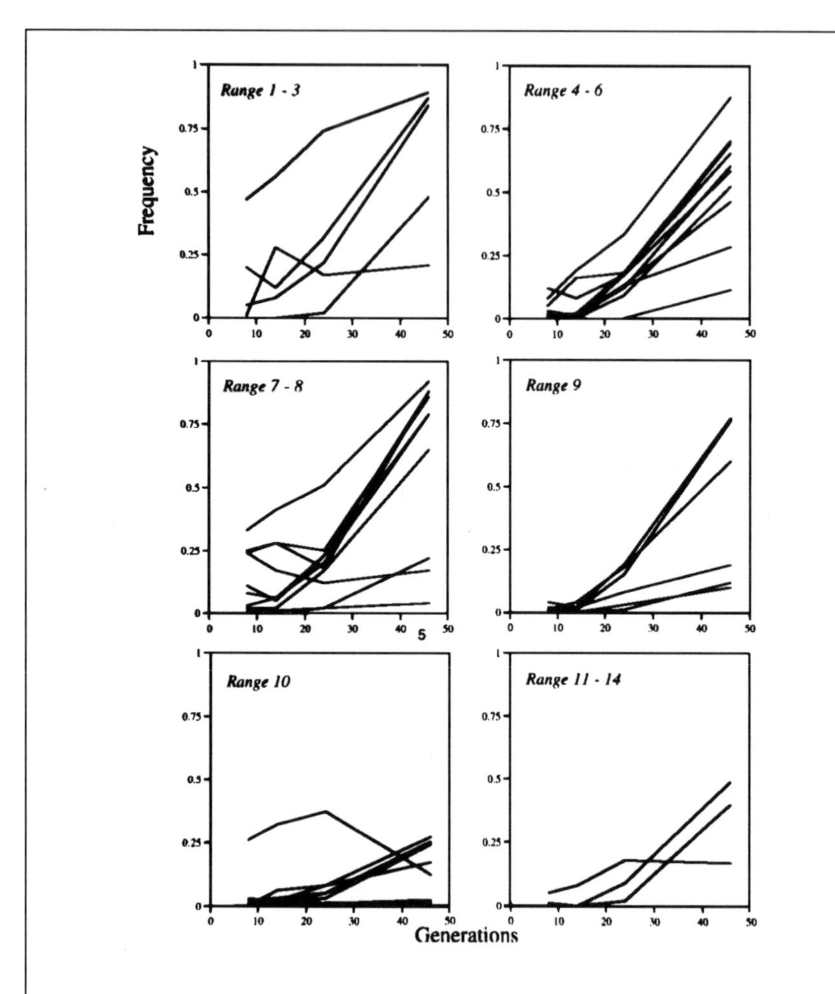


This diagram illustrates the evolution of resistance to pathogen strains with different host ranges in barley Composite Cross II, from data in Webster et al. (1986). The $y$-axis of each graph is the frequency of individuals in the barley population that are resistant to strains of the scald fungus, *Rhynchosporium secalis*, whose breadth of adaptation is assessed as the number of a series of 14 test cultivars they are able to exploit. The $x$-axis is the number of generations of propagation, the experiment being scored in generations 8, 14, 24, and 46. The simplest strains can attack only a few cultivars (range 1–3 or 4–6 meaning virulent on 1–3 or 4–6 of the test cultivars). A higher level of resistance to almost all these strains evolves in the barley population, with resistance in many cases increasing from very low frequencies to dominate the population. The more complex strains can

(*Continues*)

(*Continued*)

attack many of the test cultivars. Resistance to these strains evolves much more slowly, and in many cases does not evolve at all; in the highest category (range 11–14) resistance was demonstrated for only three of the nine strains.

The investigations of the cost of resistance to phage T4 in *E. coli* are by Lenski (1988a, 1988b).

and phage. Disease resistance in crops is discussed subsequently. To illustrate the evolution of phage resistance in bacteria, I shall use some recent experiments by Richard Lenski on *E. coli* and phage T4. Unlike the filamentous phages, T4 is lytic: rather than farming its host, it consumes all available cellular resources and then kills the cell when the progeny phage are released.

*E. coli* has two means of preventing phage from entering the cell. The first is to develop a mucoid capsule that prevents phage from gaining access to the cell envelope. Lenski studied a second mechanism of resistance, involving mutations that alter the lipopolysaccharide core of the cell envelope. There are three loci, or closely linked sets of loci, that encode the structure of the core. A series of independent isolates resistant to T4 have mutations at one or another of these loci: there are thus several different routes to resistance. All were associated with reduced fitness when grown in competition with susceptible phage in media where no phage were present, demonstrating a primary cost of resistance. The main point of Lenski's experiment was to investigate how this cost might be modified through selection. There are two possibilities. In the first place, the different isolates did not all express the same cost of resistance. There were two groups, probably representing mutations at different loci. One was greatly enfeebled in the absence of phage, with a fitness, relative to susceptible strains, of only about 0.55; the other was less severely handicapped, having a relative fitness of about 0.85. Provided that these fitnesses are transitive, selection will favor the isolates whose resistance was less costly, thereby reducing the average cost of resistance. The second possibility is that the cost might be further reduced by mutations at other loci that compensated for the physiological defects associated with resistance. This possibility was investigated by growing pure cultures of susceptible and resistant strains by serial transfer in batch culture for about 400 generations. Selection during this period increased the fitness of the susceptible strains by about 10%, the result of adaptation to the novel conditions of culture; the fitness of the evolved susceptible strains relative to that of the initial clones was thus about 1.10. The relative fitness of the resistant strains, however, increased during the same period from 0.66 to 1.03. These evolved resistant strains had thus restored fitness to a level comparable with that of the base population and were little inferior to the evolved susceptible strains. This advance was not caused or accompanied by a reversion to sensitivity; the evolved lines were all still fully resistant to T4. It must instead have been caused by selection for genes that ameliorated the initially deleterious side effects of the resistant phenotype.

Resistance is always to some extent specific: a host genotype is resistant to a specific category of pathogens, not to all. Most of the strains resistant to T4 were also resistant to T7, and it might be expected that the increased fitness of T4-resistant strains in permissive conditions would be accompanied by a loss of resistance to T7. This was not the case: resistance to T7 was not appreciably lower in the evolved populations. It remains possible, however, that selection in the presence of T4 would eventually lead to highly specific modifications causing increased susceptibility to T7.

If this experiment were tried with types that secrete a mucoid capsule to hinder infection by the virus, it would be anticipated that selection in a permissive environment, with no virus present, would cause an increase in fitness through selection for the loss of capsule synthesis. The surprising feature of Lenski's experiment is that relative fitness increased without any concomitant loss of resistance. Mutations that modify the cell envelope, however, have no substantial economic effects on the cell; it is probably no more expensive to make a defective envelope than it is to make a normal one. Selection can therefore favor mutations that compensate in some way for this defect, without affecting the resistance that it confers. Moreover, because resistance is in this case associated with the partial loss of normal function, it will not be degraded through time in permissive environments by mutation accumulation in the way that a gain-of-function mutation would be. Thus, selection may exacerbate or ameliorate the side effects of disease resistance, depending on the genetic and economic consequences of a particular resistance mechanism.

**The Cost of Virulence.** Crops that are grown in novel environments where an otherwise endemic pathogen does not occur often lose their resistance, although curiously enough there seem to be few careful quantitative studies of this phenomenon and no deliberate selection experiments, despite its obvious importance. There is a larger literature on the cost of virulence in pathogens. K.J. Leonard of North Carolina State collected sexual spores of oat stem rust from barberry (the alternate host) and then recycled asexual generations on single oat cultivars for eight generations, scoring pathogenicity to a set of test cultivars in each generation by inoculating them with spores from the experimental plants. He found that virulence against the test cultivars decreased through time, presumably through selection against unnecessary virulence in populations evolving on a single host cultvar. The selection coefficients involved were quite large (of the order of 10%), so that the evolution of virulence directed specifically against one cultivar may result in the rapid loss of virulence toward others.

**Resistance and Virulence in Barley Composite Cross II.** The cost of resistance limits the degree to which hosts can become adapted to the local community of pathogens; likewise, the cost of virulence limits the degree of adaptedness of the pathogen. This is why most host individuals are susceptible to most pathogen genotypes, and most pathogen individuals avirulent on most host genotypes. The range of adaptation of the scald fungus *Rhynchosporium* was estimated by inoculating a test series of 14 barley cultivars. This defined 75 races by the different combinations of cultivars they could attack; only one of the 75 was virulent on all 14 hosts. The cultivars, in turn, varied in the extent of their resistance; the most susceptible resisted 21, the most resistant 56 of the fungal genotypes. Now, suppose that a heterogeneous barley population such as Composite Cross II is propagated in the presence of the native population of *Rhynchosporium*; what breadth of adaptation will evolve? Fungal strains that have a narrow host range are rather abundant and will select for resistance among the genotypes susceptible to them. Strains with a very broad host range are rare (because multiple virulence is costly to maintain), and the weak selection they impose is unlikely to cause hosts with a broad range of pathogen resistance to evolve (because multiple resistance is costly). Broadly speaking, the barley population behaved as expected: resistance to simple races of fungi, with narrow host ranges (as gauged by the 14 test cultivars), evolved rapidly, whereas plants remained susceptible to fungal strains with broad host ranges. There were many exceptions to these generalizations, however, and the argument I have given is no more than a crude approximation.

**The Canonical Form of the Social Matrices of Antagonists.** On balance, it seems likely that the correlated response to selection imposed by other organisms is a special case of the cost of adaptation to any kind of novel environment. In most cases, the evolution of specific resistance by hosts and virulence by parasites will be associated with lower relative fitness in environments where the particular pathogen or host involved does not occur. The social matrices defining the relationship between host and pathogen genotypes can then be represented as follows.

| | | Pathogen | |
|---|---|---|---|
| | | Virulent | Avirulent |
| Host | Resistant | a | b |
| | Susceptible | c | d |

| | | Host | |
|---|---|---|---|
| | | Resistant | Susceptible |
| Pathogen | Virulent | $a$ | $b$ |
| | Avirulent | $c$ | $d$ |

These matrices represent the interaction of a particular pair of host and pathogen genotypes, the virulent pathogen type and the resistant host type being defined relative to one another. The avirulent and susceptible categories, on the other hand, may be very heterogeneous mixtures of genotypes displaying varying degrees of resistance and virulence toward one another. A cost of resistance implies that c < a and b < d; a cost of virulence implies that $a < c$ and $d < b$. This might be called the canonical form of the anisomorphic matrices that govern the coevolution of antagonistic organisms. It will tend to cause sustained oscillations of genotype frequency in both populations.

## 159. *Genetically uniform populations of hosts elicit epidemics of short-lived pathogens.*

These fluctuations can be abolished by deliberately enforcing uniformity in the host population. We might, for example, choose to culture a single resistant genotype, perhaps in the hope of avoiding disease. Whether or not this hope is realized depends on the size and the rate of turnover of the host population and its pathogens. Any uniform population of hosts will cause the selection of the pathogen genotypes best able to exploit the particular host type. If the host population is very small, this selection will be inappreciable; any tendency for a specific virulence to be selected will be opposed by its cost. But if the host population represents a large resource, relative to other available hosts, then selection among the pathogens for

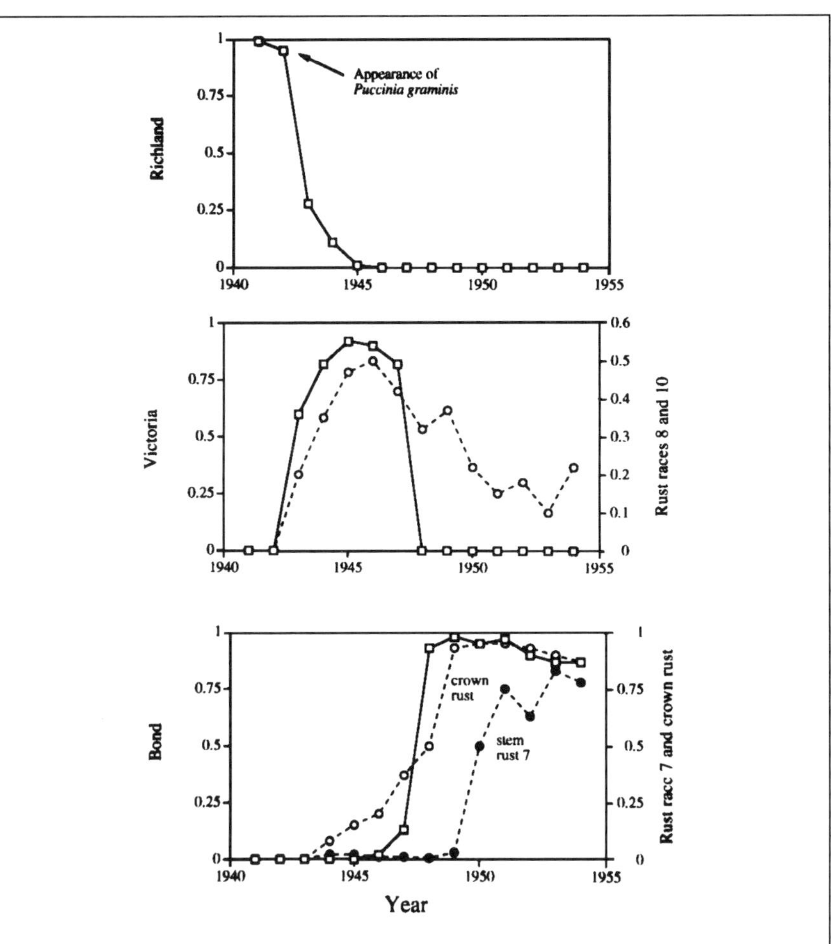


These diagrams shows the coupled changes in oat cultivars and oat rust genotypes in Iowa during the 1940s and 1950s (from Browning & Frey 1969; the figure is redrawn from their Fig. 1). The $y$ axis (left) is the fraction of the total acreage occupied by different groups of cultivars; the $yy$ axis (right) is the frequency of different races of rust fungus is isolates from the crop. In 1941, almost the whole acreage of oats was occupied by Richland and related cultivars. These were highly susceptible to oat stem rust (*Puccinia graminis*) and rapidly disappeared from cultivation after its appearance in 1942. They were replaced by the hybrid Victoria cultivars that were resistant to stem rust races 1, 2, and 5, but susceptible to races 8 and 10. These latter races, rare in 1942, rapidly increased in frequency. At the same time, an epidemic of crown rust (*Helminthosporium victoriae*) began to develop. With the rise of these new pathogens, Victoria was

(*Continues*)

(*Continued*)

forced out of cultivation, being replaced by Bond derivatives. These were resistant to races 8 and 10, but susceptible to race 7, which thereupon began to spread. At the same time, races of crown rust virulent on Bond also appeared. This continual coevolutionary arms race is a very common, perhaps an invariable, outcome of planting very large acreages to uniform, or very similar, crops: at the time when Browning and Frey were writing, the lifespan of a new oat cultivar was only about 5 years; wheat cultivars lasted 15 years in Canada, but only 5 in Mexico; barley varieties are often overwhelmed in 3 or 4 years. The literature of disease epidemics in uniform stands, and on the possibility of using mixtures and multilines to control disease, is very large, but there are excellent reviews by Browning and Frey (1969), Marshall (1977), Barrett (1981), Wolfe and Barrett (1981) and Wolfe (1985). Selection on powdery mildew *Erisyphe graminis* in pure and mixed stands of barley was studied by Chin and Wolfe (1984). The productivity of barley mixtures at infested and unifested sites was described by Wolfe and Barrett (1980). The aphid–*Anthoxanthum* experiment was by Schmitt and Antonovics (1986).

specific virulence will be very strong. If the pathogen population is in turn very small, or reproduces very slowly, this selection will again be ineffective; if specific virulence were to evolve in the pathogen, specific resistance would arise more rapidly and spread more quickly in the host. However, if the pathogen population is very large and reproduces very rapidly, then virulent types will arise quickly and spread through the pathogen population before the host population is able to respond to selection for the new source of resistance now required. The result is a devastating epidemic that may destroy the host population completely.

The scenario of a very large uniform host population and an even larger and more rapidly cycling pathogen population is supplied by many crop plants and some breeds of livestock. Cereal crops, in particular, are often grown as populations of hundreds of millions of individuals of the same cultivar. One of the main preoccupations of agronomists is the essentially evolutionary problem of how to prevent epidemics in such extensive pure stands; I shall illustrate some of the issues involved with barley and its fungal pathogen *Erisyphe* (powdery mildew) drawing mostly on the work of Martin Wolfe and John Barrett at Cambridge.

**Variability and Disease.** The ancestral populations of *Hordeum spontaneum* from which cultivated barley was selected still grow wild in the Middle East, in the company of other grasses and their herbivores and pathogens, including *Erisyphe*. A large proportion of plants in these populations is resistant to heterogeneous bulk inoculates of *Erisyphe* and must bear many different sources of resistance, encoded by different genes. On the other hand, an equally large proportion is susceptible to at least some of the *Erisyphe* strains in the inoculum. The host population is thus a mixture of resistant and susceptible plants, although whether this mixture is stable or fluctuates in composition over time is not known. Likewise, the populations of *Erisyphe* infesting these wild grasses have a wide host range that matches the broad resistance shown by their hosts. In these conditions, disease is endemic, and most plants (perhaps nearly all) show some symptoms of disease, but epidemics do not seem to occur.

The interdependence of host and pathogen variation can be studied in cultivated populations. The cultivars Hassan, Midas, and Wing incorporate three different sources of resistance to powdery mildew, each of which opposes a specific source of virulence in the fungus. If the cultivars are grown as pure stands, the pathogen populations that infest them are dominated by the corresponding simple races, each with a single source of virulence. In binary mixtures of Hassan and Wing, mildew strains able to grow on both components are abundant early in the season, although later they tend to be replaced by the two simple races, each able to grow on only one component of the mixture. In the three-way mixture, strains of *Erisyphe* with all three sources of virulence become more abundant. The host range of the mildew thus increases with the diversity of the barley. The effect is rather slight, however: strains with all three sources of virulence, for example, increase in frequency from about 5% in pure stands to about 12% in the three-way mixture.

**Monoculture and the Agronomic Arms Race.** Powdery mildew was reported as a minor pest in the 19th century, but the first epidemics occurred in the early years of the 20th century. Its rise coincides with the changeover from genetically heterogeneous landraces of barley to the extensive monocultures of single varieties that characterize modern agriculture. By the 1930s, many different sources of virulence had been identified in the fungus, and almost all the commercial varieties then available were susceptible to one or another of them. One variety, however, possessed a source of resistance to all known strains of *Erisyphe* that turned out to be attributable to a single major gene and was, therefore, readily transferred to other high-yielding cultivars. These first began to be planted extensively in the late 1940s; 10 years later, they occupied hundreds of thousands of hectares, including 70% of the barley area in Germany. By this time, however, all were susceptible to some strains of mildew, and shortly afterwards they were withdrawn from commercial use. This pattern of a highly resistant variety selecting virulent strains of pathogen when planted over very large areas has become familiar in barley and other grain crops as the single most damaging side effect of monoculture farming. The conventional response has been to engage in an arms race with the pathogen by selecting for resistance to current pathogen strains. The resources of the pathogen are so great, however—10 generations for every barley generation, and a million spores for every barley plant—that sequential selection for resistance on the experimental farm may be too slow to counter selection for virulence in the commercial fields. An alternative approach is to check the evolution of virulence by incorporating several different sources of resistance into the barley population.

**Multilines and Mixtures in Disease Control.** There are three ways in which this could be done. The most obvious is to incorporate several different sources of resistance into a single line. The main problem with this approach is the cost of resistance: multiply-resistant varieties are likely to have low yields in most years, which is why wild populations contain a substantial proportion of susceptible plants. A more sophisticated alternative, to which there is no close analog in natural populations, is to introduce different sources of resistance into nearly isogenic lines of a high-yielding strain. The result is a multiline, in which individuals vary in specific resistance, but are nearly identical in yield when virulent races of the pathogen are absent. The main drawback of developing multilines is that by the time the back-crossing program has been completed, the high-yielding recurrent parent has been overtaken by even higher-yielding varieties. The third strategy overcomes this problem by simply using a mechanical mixture of current high-yielding selections into which divers single sources of resistance have been introduced by a single generation of crossing.

Mixtures reduce the rate of spread of disease chiefly through the lower density of plants susceptible to particular strains of pathogen, so that fewer pathogen spores from any given lesion are likely to reach a susceptible host. Plants that are resistant to a particular strain act like flypaper, trapping fungal spores and thereby reducing the effective rate of spore dispersal. For example, in mixed stands of barley and wheat, the wheat is of course completely resistant to powdery mildew; the rate at which disease spreads in the barley is directly proportional to the percentage of barley in the mixture. Wolfe and Barrett conducted a large-scale trial of 47 mixtures of 25 barley cultivars representing 12 resistance phenotypes. They were unusually productive: 39 exceeded their component means and 26 exceeded the highest yielding component. The effect of mixture varied with the intensity of mildew infestation. In the least infested sites, mixtures yielded about 3% above their component means, which is consistent with general experience (Sec. 152). In heavily infested sites, however, the mixtures exceeded their component means by nearly 10%. Mixtures can thus be effective in suppressing disease and preventing epidemics. Moreover, the identification of specific sources of resistance provides a rational basis for the artificial selection of productive mixtures, although this has not yet been attempted. However, there have been two objections to the widespread utilization of varietal mixtures. The first is merely the legal difficulty of registering mixtures for commercial use. The second is the argument that presenting the pathogen population with several different sources of resistance simultaneously might result in the evolution of a super-race capable of overcoming them all. Whether or not this is a very likely outcome of selection has been controversial. From an evolutionary standpoint, however, it is clearly preferable to challenge the pathogen population with several sources of resistance simultaneously than to deploy them sequentially.

**Randomly Reconstituted Mixtures.** An evolutionary solution to the problem of disease epidemics in monocultures might be to sow mixtures that are newly constituted in every year as random samples from a large pool of high-yielding cultivars each bearing a single source of resistance. So far as I know, this has been neither advocated nor tried.

**Sexual Sibships as Disease-Resistant Mixtures.** The natural analog of a cereal monoculture is a clone or a selfed line. Outcrossed sexual sibships in which diverse sources of resistance are randomly combined may be less vulnerable to disease epidemics. Whether or not this would represent a substantial advantage for outcrossing seems again to be a question of scale, depending on whether or not sexual sibships present a sufficiently broad resistance spectrum to inhibit the spread of disease when growing as neighbors. There is little experimental work that directly addresses the issue. Joanna Schmitt and Janis Antonovics, working on the intensively studied *Anthoxanthum* population at Duke University, set up a hill-plot design in which the central plant was surrounded by four neighbors that were either full sibs of the central plant or unrelated plants from random seed; other plants grew singly. The experiment was fortuitously attacked by aphids, with infestations approaching 50%. The single plants suffered the most damage, showing that a fringe of neighbors may dilute an infestation. Plants surrounded by sibs or by unrelated plants survived equally well in groups that were not attacked by aphids, but when aphids were present the plants with unrelated neighbors survived better. This suggests that disease is less debilitating in mixtures of dissimilar individuals. Unfortunately, clonal groups were not used in the experiment, and so it is not known whether the more limited diversity of the sib groups might have had any effect on the severity of damage by aphids.

## 160. *The environment always tends to deteriorate.*

Parasites and pathogens will evolve so as to be best able to exploit the most abundant types of host, whose fitness will decline as a result. A similar generalization holds for the pathogens themselves: hosts will be selected for resistance to the most abundant types of pathogen, whose fitness will likewise decline. In either case, the average individual has relatively low fitness. From the point of view of the most abundant lineages in the population, the environment continually tends to get worse.

It might be thought that this is unjustifiably pessimistic; after all, when partners evolve so as to cooperate more closely, the environment will appear to both of them to be improving. Once a partnership has evolved, however, the genotypes responsible will be fixed in the population, with perhaps some slight further accomodation being made from time to time. Mutualistic relationships, especially highly integrated ones, do not change much over time. Most of the change that organisms experience is therefore contributed by antagonistic relationships, which are continually changing for the worse.

**The Darwinian Theory of the Environment.** In one of the most famous metaphors of ecology, G.E. Hutchinson spoke of the ecological theatre and the evolutionary play. The environment is thought of as a fixed frame of reference, within which the evolutionary action is contained. This may be the case for geological or meteorological processes that may be adapted to but cannot be deflected. However, it does not apply to neighbors: to the host of partners, competitors, and antagonists that are the most important features of the environment. Social selection will readily cause a population to become adapted to its neighbors; but it will with equal facility enable its neighbors to respond, leading to closer cooperation or to a ceaseless struggle. Once the importance of biotic features of the environment is recognized, the distinction between the theatre and the play is blurred; once the coevolutionary response of antagonistic neighbors is seen as the most important source of continuing selection, the distinction all but disappears. There is no fixed frame of reference, but only local and temporary structures that are continually recast by the action of the play itself.

Indeed, the imagery of a play enacted on an unresponsive and unaltering stage is one that might appeal more to an ecologist that to an evolutionist. A better analogy might be a street market, crowded, noisy, and intensely competitive. The merchants, in family groups that might extend over four generations, use any device to sell at the highest price their wares will command. They may even conspire with one another to maintain the price—but only if they know their neighbors well. Their customers attempt to buy at the lowest price they can persuade the merchants to accept. A merchant will occasionally discover some new trick of selling, and grow fat for a while, but when it is seized on by others it will eventually become known to the customers and lose its effectiveness. There is ceaseless maneuvring, constant bargaining, an endless succession of local triumphs and failures, all organized by the common theme of buying cheap and selling dear. But although there is a common theme, there is no plot. The action does not move in a predestined way to a foreseeable end, but instead unfolds as an historical sequence of events, shaped by simple forces, but unpredictable in detail. The market, of course, is affected by physical events: it may slacken when it rains, will be disrupted by an earthquake, may be abandoned altogether if the sea reclaims the sandspit it is built on. But the evolution of the market, the rise and fall of lineages of merchants and customers, the development of techniques for buying and selling, cannot be understood in terms of geology or meteorology; the springs of action are provided by the actors themselves.

The shift from the analogy of a play, scripted and confined, to the analogy of a market, represents the shift from viewing the environment as an unresponsive external constraint, causing the selection of certain phenotypes, to appreciating that adaptation is seldom final or conclusive, but instead tends to procure its own overthrow by selecting compensatory adaptations among neighbors. General Darwinism is not only a theory of how populations respond to the environment in which they live; it is also a theory of the environment itself. The lack of fit between the population and its social environment is not only restored by selection, but is also caused by selection.

**The Red Queen.** The Red Queen famously remarked to Alice that in her country it was necessary to run very fast just to stay in the same place. Mutually antagonistic organisms live in the same country. It has a Darwinian environment in which adaptation is merely provisional, and evolution a continual process of adaptation and counter-adaptation that would continue even if the physical environment were no more changeable than a chemostat. We do not know whether the world is really like this or not: the theory has yet to be seriously investigated. It can be tested quite easily, however, because it puts an arrow on ecological time. For example, suppose that we were to maintain large clonal populations, perhaps of sedges or duckweed, both in the field and in the greenhouse. The field populations are continually replenished from the surplus reproduction of the greenhouse stocks. In the field, the clones will offer new resources to the local community of herbivores and pathogens, who will evolve strains capable of exploiting them. The host population cannot evolve resistance, because the plants are continually being replaced from greenhouse stocks; the rate of reproduction in these populations will then simply decline through time as individuals become increasingly vulnerable to the increasingly sophisticated adaptation of local antagonists. No such process will occur in the greenhouse, where the plants can be protected from pests; thus, the fitness of a given clone growing in the field will continually decline relative to the fitness of the same clone maintained at the same time in the greenhouse. A program of experiments such as this could determine whether phenomena such as the breakdown of resistance in widely planted cultivars also apply generally in natural environments.

Until such a program is mounted, the most persuasive argument for a Red Queen interpretation of the world is the prevalence of sex. There are several processes that may give sexual organisms an advantage of some sort over asexual rivals, which I have described piecemeal throughout the text: mutation clearance (Sec. 87), the prevention of mutation accumulation (Sec. 52), selection that often changes direction (Sec. 87), extreme truncation selection (Sec. 116), and the greater productivity of mixtures (Sec. 152). All of these theories identify certain consequences of sex, but none have been very convincing as explanations for the routine maintenance of so elaborate and expensive a process in most eukaryote lineages. The Red Queen, however, makes sense of the riddle. The most puzzling feature of sex is that it breaks up combinations of genes that work and replaces them with new and untried combinations. This makes no sense if the world is more or less constant; it still makes no sense if environments change randomly from generation to generation. But it will make sense if the present success of a genotype brings about its future failure; only then will there be a systematic advantage for producing progeny different from their parents.

The way in which sex is selected through the coevolution of antagonists can be represented by one final elaboration of the abcd model. Imagine that genes for virulence are segregating at two loci in the pathogen and are specifically opposed by genes for resistance at two loci in the host. Both social matrices thus involve four genotypes—I shall not attempt to draw them. The gene frequency at both loci in both populations will, of course, tend to oscillate through time. The correlation between the loci, however, will also oscillate through time, with the sign of linkage disequilibrium changing from positive to negative and back again, as first one and then another combination of genes is favored by selection. Sex always tends to reduce the magnitude of linkage disequilibrium, and a sexual lineage will pass more easily and rapidly from positive to negative disequilibrium, or the reverse. The coevolution of antagonists is the only process that will naturally and necessarily create the only situation in which sex will increase the rate of evolution.

# 6

# *Sexual Selection*

Natural selection is caused by differential reproduction; in an asexual world, this concept would be sufficient to account for adaptedness. However, in many organisms—perhaps, fundamentally, in all eukaryotes—the vegetative processes of growth and reproduction are occasionally interrupted by sexual episodes.

Although the nature of sexuality was worked out before the end of the 19th century, misconceptions about the process continue to cloud thinking about evolution. This is largely attributable to the fact that sex and reproduction are intimately connected in large and familiar animals, especially in vertebrates, and to a lesser extent in flowering plants. Respectable textbooks of biology still refer to sexual reproduction (I probably have somewhere), as though, in Michael Ghiselin's phrase, sex were a kind of reproduction. To understand what sex is, and why it introduces new kinds of evolutionary principle, it is helpful to put dogs and daisies on one side, and think instead of ciliates, seaweeds, foraminiferans, or yeasts; the bulk of living diversity, in fact.

Ciliates may as well serve as an example, although I shall suppress the details of their rather imaginative genetic and developmental systems. They are large predatory unicells that are abundant in freshwater and marine environments. The vegetative cells are diploid. After growing to a sufficient size, they reproduce exclusively by binary fission, dividing by mitosis to form two nearly equal daughters. In certain conditions (usually when they are crowded or starving) a pair of cells will fuse; the nucleus of each cell goes through meiosis to form two surviving haploid nuclei, one of which remains in the parental cell, while the other migrates into the its partner, there to fuse with the corresponding product of meiosis. Diploidy having been restored, the two cells separate and swim away. The complementary processes of reduction and fusion together constitute the sexual cycle. The point is that *sex and reproduction are completely distinct*. They are,

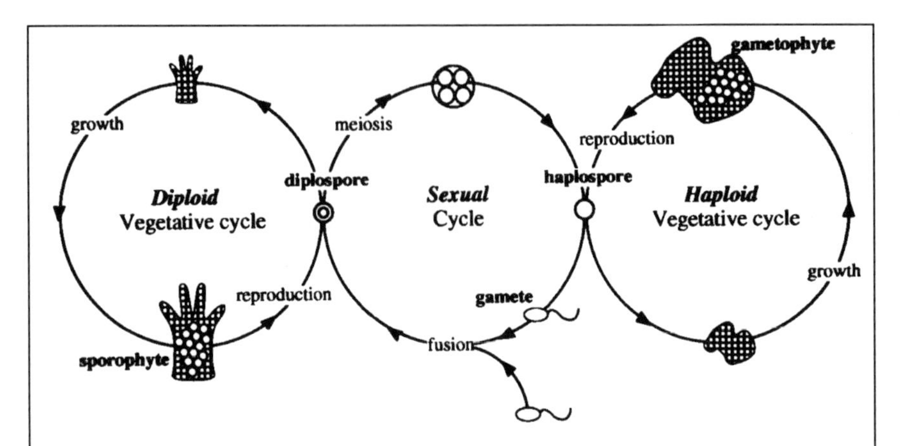


This picture gives the basic elements of the eukaryotic life cycle. There are two cycles of vegetative growth and reproduction, in haploid and in diploid individuals, linked by the sexual cycle. The three cycles pivot on two types of spore, haplospore and diplospore. The cycle as drawn here differs from the conventional representation (as a single circle) but makes clear the relationship between sexual and vegetative processes (see Bell 1994).

The sexual cycle is much more difficult to understand than the vegetative cycles, and the experimental basis of knowledge is narrow. Consequently, many of the sections in this part of the book are skeletal accounts of general principles, skimming the surface of a very large theoretical, speculative, and comparative literature. The best way of gaining an introduction to this literature is from edited volumes of papers by the leading researchers; the most recent are Campbell (1972), Halvorson and Monroy (1985), Stearns (1987), Michod and Levin (1988), Otto and Endler (1989), Harvey et al. (1991), Kirkpatrick (1994) and Thornhill (1993).

indeed, opposed: growth and reproduction generate two similar cells from one parent, whereas sex changes the nature but not the number of individuals.

All unicellular and many multicellular eukaryotes follow the same pattern that is obscured only when new individuals develop from fertilized eggs exclusively. The eukaryotic life cycle thus comprehends two quite different processes. The spore—any uncommitted cell—may develop either vegetatively or sexually. Vegetative development involves growth, with or without development as a multicellular individual, followed by reproduction that regenerates similar spores. Sexual development involves differentiation as a gamete, followed by fusion with another gamete, after which the spore is regenerated by meiosis. The complete life cycle of sexual organisms involves both vegetative and sexual processes, and natural selection, acting through vegetative development, is thus incomplete as a description of how they evolve. Selection acting on the sexual cycle differs from natural selection in that it acts through sexual fusion, rather than through vegetative fission.

## 161. *Sexual selection is competition among gametes for fusion.*

Sexual selection is ineradicably associated with peacocks. In order to understand why it is different from natural selection, it is necessary to return to microbes. During the vegetative cycle, individuals developing from spores compete to acquire resources. The surviving individuals each comprehend a greater quantity of matter and are enabled to reproduce by transforming captured resources into spores. Types that can do this more rapidly increase in frequency through natural selection. The organisms involved may be obligately outcrossing, so that progeny can be produced only by a combination of two parents, but the theory would require only the most trivial amendment to accomodate this curious restriction on reproduction. It remains a vegetative theory, grounded on the concept of differential growth.

Suppose, however, that the spore differentiates into a gamete, switching development from the vegetative cycle to the sexual cycle. It will not then compete for resources with other gametes in order to grow and reproduce more rapidly. Gametes do not grow or reproduce at all. They do nevertheless give rise to other spores. When two gametes fuse, the fusion product is a diploid spore, the zygote. This may develop directly, if vegetative growth occurs in the diploid phase, or it may first go through a meiosis to yield haploid spores, if growth occurs in the haploid phase. In either case, gametes that succeed in fusing give rise to a new generation of spores that are capable of vegetative growth. Gametes that do not fuse have one of two fates: they may redifferentiate into spores and re-enter the vegetative cycle, or they may die. In either case, they are unable to proceed further in the sexual cycle. Just as the vegetative cycle is completed by the fission of a single individual, so the sexual cycle is completed by the fusion of two gametes.

The subject of competition in the sexual cycle is, then, not growth, but fusion. Gametes compete for fusion partners. The availability of fusion partners is often limited, although it need not necessarily be, and it is more severely limited in some circumstances than in others. Competition among gametes for fusion partners gives rise to sexual selection. Types that are more apt to fuse increase in frequency because they are incorporated disproportionately into the spores that will enter the succeeding vegetative cycle of growth and reproduction.

In many microbes, the distinction between sexual selection and natural selection is perfectly clear, because cells that are morphologically alike have contrasting fates, either fusing in pairs to form a single spore or growing to divide into two equal spores. In other forms, the distinction is blurred because gamete fusion is mediated by the fusion or copulation of individuals that bear gametes. These gamete-bearing individuals are called gamonts. In most fungi and ciliates, for example, it is gamonts that fuse, prior to the fusion of the gametes they bear. In multicellular animals and plants, gametes may fuse inside or outside the bodies of gamonts, and it is very often the behavior of the gamonts (such as peacocks) that determines the success of gametes, rather than the behavior of the gametes themselves. In these cases, sexual selection modifies the properties of gamonts. Nevertheless, it does so by causing changes in the frequencies of different types in the population of competing gametes, through their success in completing the sexual cycle. Natural selection in these cases acts through the relative productivity of gametes, sexual selection through the relative success of the gametes that are produced. The life cycles of animals and plants are often so highly modified from the simpler cycles of microbes that the distinction between sex and reproduction, and thus between sexual selection and natural selection, is easily misapprehended. The underlying principle, however, remains the same.

## 162. *Sexual selection and natural selection are antagonistic.*

The overall fitness of any type with a complete life cycle has thus two components: its vegetative success through growth and its sexual success through fusion. If the two were always proportionate, there would be no special problem of sexual selection: more vigorous types would both reproduce more profusely and fuse more readily. A special theory of sexual selection is required only if vegetative and sexual success are antagonistic. It has long been felt that this is the case; indeed, the initial impulse for the development of a theory of sexual selection by Darwin was the perceived inutility—from a vegetative point of view—of peacocks' tails and similar extravagances.

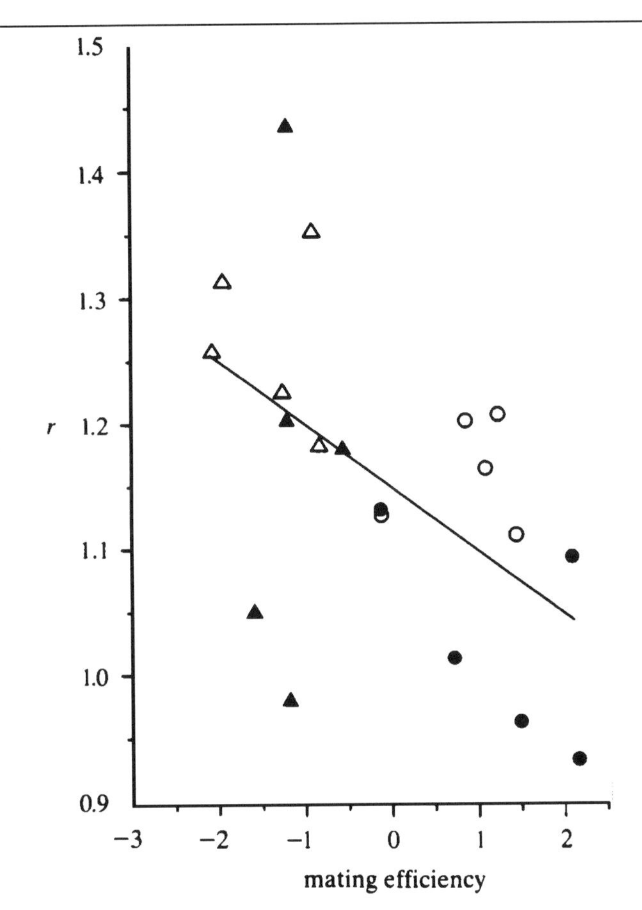


This figure shows the antagonism between natural selection and sexual selection evolving in experimental populations of *Chlamydomonas*, from da Silva and Bell (1992). The $y$-axis is the vegetative rate of increase, $r_{max}$; the $x$-axis is a measure of mating efficiency. The selection lines were as follows:

a) Filled circles: fast mating, zygotes selected after culture allowed to mate for a few hours only.

b) Open circles: slow mating, zygotes selected after culture allowed to mate for 24 hours.

c) Triangles: no mating, vegetative cultures of mt+ (solid triangles) or mt- (open triangles) cells only.

Note that the vegetative cultures have greater $r_{max}$ but lower mating efficiency, whereas selection for mating increases mating efficiency but is associated with lower $r_{max}$. For examples of the loss of sex in long-continued vegetative cultures, see Bell (1992).

**Sexual Enfeeblement of Vegetative Lineages.** The antagonism of sexual selection and natural selection will cause sexual function to deteriorate as a correlated response to natural selection and vegetative function to deteriorate as a correlated response to sexual selection. The most extensive, if unconscious, selection experiment has been the long-continued propagation of certain crop plants without sex. Root crops such as potatoes, sweet potatoes, and yams, for example, are propagated as tubers, or tuberous roots; citrus and other fruit trees are grown as buds grafted onto rootstocks; many berries are propagated as suckers, runners, or cuttings. In most cases, such crops are difficult to induce to flower, or the flowers are defective, and if seeds are produced their viability is often low. It seems that selection for vegetative characters has caused the sexual system to decay, through antagonistic pleiotropy or mutation accumulation. In parthenogenetic animals that have evolved recently from sexual stocks, secondary sexual structures, such as seminal receptacles, are often variable, defective, or absent.

A similar phenomenon has been noticed by microbiologists. Isolates of algae or fungi are often identified from their sexual behavior, mating with some known species in the laboratory, and are afterward propagated asexually for long periods of time. It is often then discovered that the strain mates reluctantly, if at all. Sexual function can often be fully restored, provided that a few zygotes can be obtained from the old asexual culture, within three or four sexual generations of intense selection for zygotes.

**Sexual and Natural Selection in *Chlamydomonas*.** *Chlamydomonas* is a haplont that reproduces indefinitely by mitosis, unless a sexual cycle is induced by starving the cells for nitrogen, when they transform into gametes. Jack da Silva and I set up selection lines, some of which were induced periodically to go through a sexual cycle, whereas others were kept perennially asexual. In the sexual lines, we exposed the mated cultures to chloroform vapor, which kills any unmated gametes but allows the thick-walled zygotes to survive. This enabled us to select for the readiness of gametes to fuse, by allowing the cultures to mate for only a certain time before killing all those that had not yet fused. This time was steadily reduced during the experiment, from six hours down to an hour and a half. The asexual lines evolved the higher rates of increase, but their mating ability declined; after about 100 asexual generations, indeed, their zygote production fell to only a few percent of its original value. This confirmed the casual observation that sexual ability declines in asexual cultures. The sexual lines mated rapidly and vigorously, as one would expect in a situation where failure to do so was lethal, but grew less rapidly as vegetative cultures. This seems to be the clearest experimental evidence that sexual and natural selection are fundamentally antagonistic.

## 163. *The life cycle is balanced between sexual selection and natural selection.*

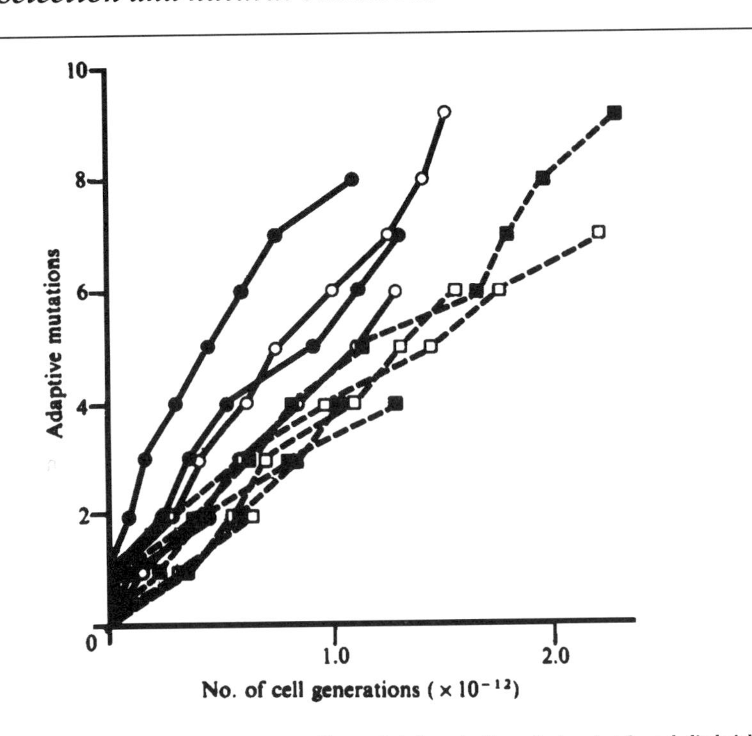


This figure shows the rate of fixation of beneficial mutations in haploid and diploid chemostat populations of yeast, from Paquin and Adams (1983). Square symbols represent haploid populations, circles diploid populations; filled and open symbols are two strains. The growth of haploid and diploid cultures in different environments was described by Adams and Hansche (1974).

Selection has been applied to populations of pollen grains in two ways. The first is to increase competition among pollen grains by providing larger or more heterogeneous pollen loads. Several authors have reported that heavier pollen loads lead to more vigorous progeny (e.g., Ter-Avanesian 1978, Mulcahy et al. 1978, Lee and Hartgerink 1986, Stephenson et al. 1986, Schlichting et al. 1987). Others have varied the distance that the pollen tube must grow, sometimes finding that longer styles produce more vigorous progeny (e.g., Mulcahy and Mulcahy 1975, Ottaviano et al. 1982, 1988). The second technique is to select pollen grains in stressful environments. Most studies apply the stress to the sporophyte and score resistance in the gametophyte. Hodgkin (1990), however, has reported that when pollen is selected in the presence of *Alternaria* toxin, resistance to the toxin is increased in the pollen of the subsequent generation. The topic, which

(*Continues*)

(*Continued*)

remains somewhat contentious, has been reviewed by Ottaviano and Mulcahy (1989) and Hormaza and Herrero (1992).

The life cycle of sexual eukaryotes comprises two growth cycles, representing the proliferation of haploid and diploid individuals, linked by the sexual cycle. Haploid individuals (gametophytes) produce haploid gametes. These fuse to form a zygote that develops as a diploid individual (sporophyte). The sexual germ cells undergo meiosis to form haploid spores that close the cycle by developing as haploid individuals. Because sexual fusion must be balanced by a reduction division, this alternation of haploid and diploid generations is a necessary and fundamental feature of sexual lineages. However, the relative importance of haploid and diploid growth varies enormously. In some organisms, such as charophytes and many unicellular and filamentous algae and fungi, the zygote is the only diploid structure in the life cycle, and only haploid individuals grow and reproduce. Conversely, in groups such as animals and flowering plants, it is the diploid generation that is responsible for growth; indeed, in animals, the gametes are the only haploid structures in the life cycle. Other organisms, including many seaweeds, grow both as haploid and as diploid individuals. The evolutionary consequences of haploid and diploid growth, and the way in which the balance between haploid and diploid growth is maintained by selection, is still unclear. This is mainly because we lack a convenient experimental system in which this balance can be manipulated. Consequently, this section is highly speculative.

## *Vegetative Theories of the Life Cycle*

Most attempts to explain the evolution of the life cycle refer to the different effects of natural selection in haploid and diploid populations growing vegetatively.

**The Ecophysiology of the Genome.** The simplest difference between haploids and diploids is gene dosage. The consequences of this difference are well known in yeast: diploid cells are bigger; contain twice as much DNA, RNA, and protein; and express higher specific activities for a range of enzymes. This need not lead to any difference in fitness: if the rate of work is twice as high in diploid cells, the amount of machinery needed to sustain it is twice as great. Elementary geometry dictates, however, that when the volume of a cell is doubled, without change in shape, its surface area does not increase proportionately, so that the ratio of surface area to volume decreases. If growth is limited by some externally supplied nutrient, diploid cells may have a lower rate of growth because the rate of flux through the cell surface is less per unit volume. Julian Adams and Paul Hansche investigated this possibility by growing isogenic haploid and diploid strains of yeast in different media. When all nutrients were supplied in excess, there appeared to be little if any difference in growth rate related to ploidy. (This was surprising, because one of the commonplaces of yeast genetics is that diploids almost always outgrow haploids in normal laboratory conditions.) When organic phosphate was limiting to growth, the haploid strain had a selective advantage of about 7% and displaced the diploid in competition trials. Yeast cells cannot take up organic phosphate directly, but first hydrolyze it with an extracellular acid phosphatase. It seems likely, then, that selection for haploids in this experiment is attributable to their greater relative uptake rate.

One weakness of this interpretation is that it defines an advantage for haploidy, but not for diploidy. It might be that the greater rate of transcription in diploid cells is advantageous when nutrients are present in excess, by permitting the evolution of rapid but wasteful resource utilization. In dilute media, haploids would grow faster because of their greater relative surface area and also because the genome itself would not represent as great a drain on limited resources. Thus, the interaction of ploidy with enviroment would favor diploid growth when resources are plentiful, but haploid growth when they are scarce.

**Mutational Load.** The most popular theory of ploidy is genetic redundancy: diploidy provides a protection against deleterious mutations by supplying a spare copy of every gene. (It is said that H.J. Muller, dusting his food liberally with pepper, was warned that it was mutagenic: "That's why we're diploid," he responded.) If a population with haploid growth were to switch abruptly to diploid growth, its mean fitness would increase because many of the deleterious mutations maintained in the population would be shielded by normal alleles. Diploid growth is then unconditionally advantageous; the theory does not explain why so many creatures are haplontic.

This advantage is transient, because new deleterious mutations would then proceed to accumulate in the diploid population. Indeed, the mutational load at equilibrium will be greater in the diploid population because the selection acting against deleterious mutations is weakened when they are complemented by dominant alleles. If the population were now to revert abruptly to haploidy, however, all the deleterious recessive mutations that had accumulated would now be expressed, causing a substantial reduction in mean fitness. The evolution of diploidy is then virtually irreversible.

This process is quite different from historical contingency, where the course of evolution is essentially irreversible because it is highly improbable that selection will cause a population to retrace precisely the same pathway of genetic change. There is, nevertheless, no objection in principle to it doing so: contingency operates precisely because the many different pathways to the same outcome are all more or less equally navigable. Diploidy may be irreversible because it blocks the path behind it.

**Variation and the Response to Selection.** Vegetative ploidy will affect the rate at which populations respond to natural selection. If a diverse population is sorted by selection, a haploid population will respond more quickly than a comparable diploid population because every gene is expressed in haploids, whereas the heritability of fitness in diploids is reduced by dominance. If the population is so large, or the time period so long, that novel mutations are the primary source of adaptation to a novel environment, however, then a diploid population will evolve more rapidly. This is not only because it contains twice as many targets for mutation, but also because a mutation that represents the acquisition of a new activity, at the expense of losing its original activity, can be selected because the original activity is still expressed by the unmutated allele. Diploidy thereby facilitates adaptation to novel circumstances through two aspects of redundancy: there are more opportunities for mutation to occur and less chance that a mutation will cause a fatal loss of function.

Charlotte Paquin and Julian Adams cultured haploid and diploid populations of yeast in chemostats. These populations undergo periodic selection, as successive mutations that are beneficial in the novel chemostat environment sweep through the population (Secs. 58 and 145). After the population is first set up, neutral mutations that are originally very rare will increase in frequency at a rate equal to the rate of mutation. A beneficial mutation that occurs within a few dozen generations will almost certainly occur in a genome that does not bear a neutral mutation at some convenient marker locus; therefore, the frequency of neutral mutations at the marker locus will fall as the beneficial mutation spreads. When it has become fixed in the population, neutral mutations at the marker locus begin to accumulate again; thus the successive substitution of uncharacterized adaptive mutations can be followed by the sawtooth fluctuation of neutral markers. The rate of fixation of beneficial mutations in these experiments was $3.6 \times 10^{-12}$ per cell per generation for the haploid cultures, but $5.7 \times 10^{-12}$ for the diploids. The diploid populations were thus evolving about 60% faster than the haploids, at least on a per-cell basis. This is consistent with the occurrence of partly dominant mutations at about twice the rate in diploids. It is offset, however, by the greater abundance of the haploids: because they are smaller, they were about 40% more dense than the diploids. The overall rate of fixation per population was thus only slightly greater for the diploids.

**Pollen Selection.** In flowering plants haploid growth is very restricted, and the male gametophyte comprises only three haploid nuclei. Dennis Mulcahy has suggested that large populations of pollen grains could supply a very effective selective screen for characters that are subsequently expressed in the sporophyte. This is an attractive idea, because a substantial fraction of the genome is expressed in the gametophyte, including many loci that are also expressed in the sporophyte. (It would not work in animals, where the properties of sperm seem to be controlled by the diploid genome of the gamont, rather than by the haploid genome of the gamete.) Moreover, it is sometimes possible to grow haploid plants from germinated pollen grains. The technique has not been used extensively so far, largely because of technical difficulties in recovering the selected pollen and using it to fertilize female plants. It will be effective only if there is a positive genetic correlation between gametophytic and sporophytic expression, so that sporophytic performance advances as a correlated response to gametophytic selection. It may be doubted that this will be generally true. There is some evidence for a positive correlation, especially with regard to tolerance of stresses, such as high temperature, salinity, or toxins. A positive correlation, however, is to be expected when plants are exposed to novel conditions of growth and may be reversed under continued selection (Sec. 92). Moreover, one would also expect that selection of gametophytes for vegetative performance would induce a deterioration in their sexual performance. Whether or not pollen selection will turn out to be useful in agronomy may be doubted, but it will certainly be interesting in terms of how selection acts on the components of the life cycle.

## *The Sexual Theory of the Life Cycle*

Theories that attempt to explain sexual life cycles solely in terms of natural selection acting on vegetative characteristics seem to me unsatisfactory. An alternative point of view is that the life cycle comprises two cycles of vegetative growth linked by the sexual cycle and interprets the relative extent of haploid and diploid growth as representing a balance between sexual selection and natural selection.

The only clear-cut genetic consequence of growing as a haploid or a diploid concerns the nature of the gametes. When growth is haploid, identical gametes are produced by mitosis; when growth is exclusively diploid, diverse gametes are produced by meiosis. In organisms with highly differentiated male and female gametes, sexual competition among male gametes will be very severe because so many more are produced than can possibly succeed in fusing. Any variant that has a slightly greater chance of fusing will be strongly selected, and so the meiotic diversification of gametes will be worthwhile if it creates a few highly superior gametes, even at the expense of creating a large proportion of inferior types that have little chance of success. Moreover, the characteristics that contribute to the sexual success of a male gamete are unlikely to be the same as those that enhance vegetative performance. The opposed effects of sexual selection and natural selection will exacerbate this antagonistic pleiotropy over the life cycle as a whole because alleles that enhance sexual ability, even at the expense of vegetative ability, will be favored during the sexual cycle; whereas those that enhance vegetative performance, even at the expense of sexual ability, will be favored during the vegetative cycle. Sexual selection will therefore continue to act on variation in gametic performance, and this will indirectly select for diploid growth.

The situation is quite different if gametes of different gender are nearly the same size because most will succeed in fusing, and sexual selection will be weak. Moreover, because the gametes are not strongly differentiated, they may be selected for nearly the same qualities as vegetative cells; in many microbes, especially, gametes and vegetative spores are very similar in structure and behavior. Sexual selection and natural selection will then act predominantly in the same direction and will jointly favor a single type that is successful both as a gamete and as a spore. This will indirectly select for a haploid growth cycle that proliferates uniform gametes mitotically.

The strength of sexual selection and its opposition to natural selection in different circumstances explains the general concordance of the structure of the life cycle with vegetative size and complexity and with the extent of gamete dimorphism: diploid growth tends to characterize large, complex organisms with highly differentiated male and female gametes. However, there has been no experimental work in this area.

## *164. Gender and species are the contexts for sexual selection and natural selection.*

Selection of any kind acts through the differential proliferation of lineages; how it acts depends on the structure of these lineages, that is, on the way in which alternative genes are associated during development and transmission. This intertwining of genealogies is expressed through the concepts of gender and species. The two concepts are complementary and together define the limits to sexual fusion. Gametes are unable to fuse if they belong to the same gender or to different species; they may fuse only if they have different gender and belong to the same species. Genes, or other entities, may compete only if they belong to a common category, within which one may displace another. Natural selection is the consequence of competition within the species; sexual selection is the consequence of competition within a gender.

This is a clear and useful point of view for thinking about most familiar animals and plants. For other organisms the situation is less straightforward, as the concept of gender becomes extended, and the concept of species breaks down.

## *The Nature of Gender*

**Asexual Systems.** A sexual system must possess at least two genders; if there were only one, mating could not occur and the species would be regarded as being asexual. All microbes are capable of more or less indefinite vegetative reproduction. When a diverse population is introduced into a new environment, and proceeds to reproduce, one genotype will be the fittest and will increase in frequency towards fixation. By chance, it will belong to one mating type (the equivalent of gender, when gametes of different gender are not readily distinguishable) or another. This mating type will thus come to dominate the population. Equal proportions are restored after meiosis in sexual progeny, but if there is long-continued vegetative growth, the slower-growing mating type may be eliminated from the population through natural selection. The population would then comprise a single mating type; it could not mate, and perhaps in time the genes directing fusion would deteriorate, with the accumulation of mutations that are neutral so long as mating does not occur, to the point where mating would not be possible even if the other mating type should re-enter the population. It is not yet known whether asexual microbes can be interpreted as having a single gender, rather than none.

**Bipolar Heterothallic Systems.** Sexual populations of many algae and fungi are bipolar, with two mating types. In the simplest cases, such as *Chlamydomonas* and *Neurospora*, each mating type is controlled by a single gene: each individual bears a single copy of the gene, which is necessary and sufficient for the expression of mating type, and individuals do not bear copies of both mating-type genes. They are thus heterothallic: all the members of a given clone express the same mating type. One of the most intriguing recent discoveries of molecular genetics is the extraordinary nature of the mating-type genes themselves. They occupy homologous regions of the genome (and therefore segregate in Mendelian ratios at meiosis), but they are not allelic. In all the bipolar fungi that have been investigated so far, they are of different size, incorporate coding regions that are in different positions, may be transcribed in different directions, and encode substantially different products. They are so different that it is difficult to believe that they have diverged from a single common ancestor. Such genes are said to be *idiomorphic*. It is my view that idiomorphic mating-type genes arose independently as transposable elements able to encode fusion, behaving in a manner similar to the conjugative plasmids of bacteria. Gender would then initially evolve through the autoselection of parasitic genetic elements. This view is not widely accepted.

**Multipolar Systems.** Other microbes, including many fungi and ciliates, have multipolar heterothallic systems. The population includes several or many genders—hundreds in some fungi. A common arrangement is for two loci to control sexual compatibility, with each locus having a series of alleles. One locus regulates whether or not fusion with a given partner is permissible; the other regulates whether meiosis will be successful if fusion occurs. Each locus may itself be compound, a region comprising two or more closely linked loci. In the simplest case, both loci are bipolar. There are then four genders: the rule is that cells will mate, producing viable sexual progeny, if they have different genes at the locus controlling fusion compatibility and the same gene at the locus controlling meiotic compatibility. The fusion genes are idiomorphs; the meiotic genes are allelic. With more loci, and more alternative genes at each locus, gender becomes riotous. In these highly multipolar systems, the fusion genes are generally allelic, but perhaps represent variants of two or more idiomorphs.

**Homothallism.** Finally, many microbes are homothallic: mating is permitted between any two cells, even if they are members of the same clone. Homothallism is quite common among *Chlamydomonas*-like organisms isolated from soil samples. This could be interpreted as an indefinite number of genders, each individual being sexually distinct. The genetic basis of homothallism is not known, however, and it may often represent a fundamentally bipolar system in which all individuals have both mating-type genes. Other microbes, such as yeast, are definitely homothallic but bipolar. Each cell bears a copy of each mating-type gene, but only one is located at the locus responsible for mating-type expression. When a daughter cell is formed by budding, the resident copy moves out of the expression locus and the alternative gene is inserted. The cell therefore has a single sexual identity, but the clone expresses both. Heterothallic clones are easily produced by jamming the resident copy in the expression locus. It is remarkable that yeast mating-type genes are idiomorphic and transposable.

**Other Restrictions on Fusion.** In some microbes, and most multicellular organisms, gametes with different gender differ in morphology, physiology, and behavior. Such differentiated systems are invariably bipolar, at least on a morphological level. However, further restrictions on gamete fusion may evolve. Many hermaphroditic organisms are incapable of self-fertilization. The incompatibility systems of flowering plants, such as the style polymorphism studied by Darwin in *Primula* and other plants, may prevent fertile matings, even between morphologically dissimilar gametes from unrelated parents. All such restrictions on gamete fusion are equivalent to gender.

**Individual Gender of Multicellular Organisms.** Multicellular individuals produce many gametes, and can themselves be assigned a gender, on the basis of the gametes they bear. In dioecious populations, each individual bears gametes of a single gender, male or female; hermaphrodites bear gametes of both genders. The successful transmission of male and female gametes often requires male and female individuals to behave differently, so that sexual selection may cause the secondary sexual structures of male and female individuals to diverge markedly. Gender is not always a permanent attribute of individuals, as it is of gametes; an individual may change from male to female or vice versa, according to circumstances.

## *The Nature of Species*

The term "species" has always haunted evolutionary biology. Unlike all other Linnean categories, which are clearly artificial and can be rearranged at will, the species has been widely regarded as a natural category that exists independently of any particular scheme of classification. It is for this reason that the process of *speciation* has been regarded as an important component—sometimes as the central component—of evolution. In practice, however, it has been surprisingly difficult to agree how the species should be defined.

There have been three major attempts to identify a property of the species that makes it more than a formal morphological category. The *ecological* species concept is economic: a species is a set of individuals that compete together, such that what is gained by one is lost to others. The *phylogenetic* species concept is historical: a species comprises the set of populations between two successive speciation events—species are the lines on phylogenetic trees whose terminal taxa are species. The *sexual* species concept recognizes mating as the criterion of species membership: a species is a set of interbreeding sexual individuals. (This is often called the *biological* species concept, but this seems absurdly pretentious.) The sexual view of species has dominated the field for the last 40 or 50 years, but the issue remains contentious. This is in large part because the controversialists have usually worked on strongly individualized, obligately sexual animals such as vertebrates and insects which are in practice almost invariably classified according to morphological similarity and would be classified in the same way no matter what species concept were employed. The striking coagulation of lineages among such organisms into bundles called *species* is caused by the intimate association of sex with reproduction. When the two basic processes of the life cycle are separated, the bundles loosen. Plants are less strongly individualized, may reproduce vegetatively, and often hybridize. Among protists, more complex systems of gender create incomplete or asymmetrical sexual relationships between lineages, leading to the indirect exchange of genes. Thus, if A mates with B, B with C, and C with D, then in organisms with differentiated bipolar gametes A would be the same gender as C, and A could itself mate with D. This is not necessarily the case among protists, where sexual compatibility may be asymmetrical. A and D may be unable to mate, but they are nevertheless able to exchange genes through a series of intermediaries. As sex becomes less frequent these links become more tenuous, until in asexual microbes they disappear almost entirely. There is thus a continuum from asexual organisms whose lineages ramify independently, through those in which sex and reproduction are distinct and whose lineages are connected as a loose network of sexual compatibility, to those in which sex and reproduction are confounded, where certain lineages are completely intermingled within the bounds set by gender and by the same token completely separated from other lineages. No single species concept can be applied universally, and I suggest that the attempt to do so should be abandoned. There is then no single or central problem of speciation. There are, instead, two less sonorous but more readily soluble problems: how sexual compatibility and incompatibility evolve and how this affects the adaptive divergence of lineages.

## 165. *Fixed permanent gender prevents self-fertilization.*

**Infectious Spread of Gender in Microbes.** The dynamics of gender in natural populations of eukaryotic microbes is not well understood. Populations with strongly polar gender may occasionally become asexual by the loss of mating types through drift or natural selection, and the sexual properties of the remaining type may change or deteriorate through mutation accumulation (Sec. 110). A sexually compatible type that enters the population from outside, or that arises within the population by mutation, will then spread infectiously through autoselection, in a manner analogous to the spread of rare conjugative plasmids in a bacterial population. Whether this simple process is responsible for the maintenance or diversification of gender in microbial populations is not known. Mutation does not create different genders in bipolar microbes with idiomorphic mating-type genes, but merely reduces sexual competence. It is not implausible, however, that new mating types might arise by mutation in multipolar systems where many mating-type genes are allelic.

**Gender and Outcrossing.** Gender promotes outcrossing. Heterothallism, dioecy, and self-incompatibility will at least serve to prevent fusion between gametes from the same clone or the same individual. The importance of this ban depends on whether there is haploid growth. When this is the case, identical gametes are produced mitotically by a large gametophyte. If they fuse, the diploid genome will be a doubled haploid, homozygous (barring recent mutations) at all loci, and meiosis will return the original haploid genome. This seems a waste of time, but is otherwise harmless. With diploid growth and no freeliving gametophyte, meiotic segregation diversifies the gametes. A deleterious recessive mutation will be partitioned into many gametes, and fusion among them will create homozygous genomes in which such deleterious mutations are expressed. Inbreeding depression is a very common phenomenon in outcrossed organisms and will effectively prevent them from evolving self-fertilization. A new mutation that caused high levels of selfing would be selected against because of the lower fitness of the homozygous genomes with which it is associated. However, this is a transient state of affairs: within a very few generations, most deleterious recessive mutations will have been cleared from the population, and the mutant will no longer be consistently selected against.

## 166. *The opposition of sexual and natural selection causes the evolution of the male–female distinction.*

If gametes encounter one another largely by chance, the number of fusions involving a particular type of gamete will be proportional to its abundance. Producing more gametes from a more or less fixed mass of material implies that the gametes will be smaller (see Sec. 92). Sexual selection thus favors genes that direct the production of numerous small gametes. Natural selection pulls in the opposite direction. The competitive ability of vegetative organisms is generally enhanced by large size. A large zygote is more likely to give rise to a successful vegetative lineage; this is especially true for multicellular organisms whose early embryonic development is fueled by zygotic reserves. Natural selection thus favors large gametes that form large zygotes when they fuse.

The opposed tendencies of sexual and natural selection may cause gametes to evolve toward an optimal intermediate size. On the other hand, it may lead to disruptive selection if medium-sized gametes are mediocre both in sexual and in vegetative abilities. For this to be the case, increasing gamete size must have a disproportionate effect on zygote survival, as it will do (over some range of sizes) if the zygote must use stored resources for development before it can feed. The most successful types of gamete will then be very large or very small cells.

In a population containing gametes of different sizes, natural selection will act against the zygotes produced by fusion between two small gametes. Small gametes will therefore be selected to fuse preferentially with larger gametes. This will lead to a correlation between gamete size and gamete gender, small gametes being of one gender and large gametes another.

Small gametes being the more numerous, they will compete for fusion with the larger gametes. Sexual selection among the smaller gametes will thus favor yet smaller and more numerous gametes. This process, however, will itself intensify competition for fusion; as the small gametes become increasingly numerous, fusion partners become increasingly scarce, and sexual selection acts ever more powerfully to favor smaller size. Very small gametes will, however, give rise to small zygotes; natural selection will then favor an increase in the size of the larger gametes so as to maintain the viability of the individuals developing from the zygotes. The opposed tendencies of sexual and natural selection will in this way lead to an increasingly exaggerated distinction between large and small gametes of different gender.

The same process of specialization will extend to other properties of gametes, especially their motility. Small gametes competing for fusion partners will often evolve or maintain a high degree of motility because the rate of encounter between gametes will be proportional to their relative velocities. Large gametes whose fitness is determined largely by the cytoplasmic resources they donate to the zygote will be selected in the opposite direction, because being more or less assured of fusion by the superabundance of small gametes of different gender they can conserve resources by evolving or maintaining immotility.

The evolution of small motile gametes and large immotile gametes of different gender establishes the basic male–female distinction that characterizes most large multicellular organisms and some protists. The interpretation of male and female that I have given here, in terms of the opposed tendencies of sexual and natural selection, is supported fairly convincingly by comparative evidence, but remains, unfortunately, untested by experimentation.

## 167. *Sexual selection favors the minority gender.*

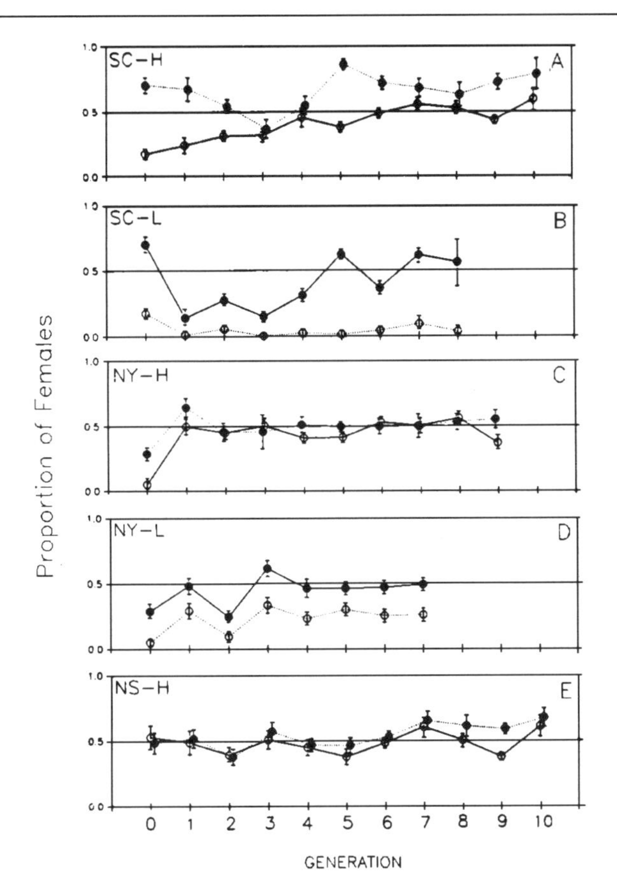


This diagram shows the outcome of propagating experimental populations of *Menidia* at different temperatures from Conover et al. (1992; see also Conover & Van Voorhees 1990). Fish were collected from localities in South Carolina (SC), New York (NY) and Nova Scotia (NS), and subsequently raised at a constant temperature of either 28 C (high temperature, H) or 17 C (low temperature, L) during the sensitive period of larval development during which gender is determined. The solid lines show sex ratios during 10 generations of propagation, tested at the same temperature as the selection environment. The broken lines show the sex ratio when tested at the other temperature. Note that the difference between the solid and broken lines is a measure of the extent to which sex ratio is determined by environment rather than genotype. At high rearing temperaturees the level of environmental determination seems to decrease, suggesting that

(*Continues*)

(*Continued*)
sex-determining genes insensitive to temperature are being selected; however, this was not generally the case in lines reared at low temperature.

**Mating Types in Microbes.** In a population of bipolar sexual microbes producing gametes of a single fixed size, the minority mating type will be sexually the more successful. Because fusion involves gametes of different gender, the proportion of the minority mating type that achieves fusion will necessarily exceed the proportion of the majority type that does so. Gametes that fail to fuse have little future if they are unable to develop vegetatively or if the zygospore is a resistant stage able to survive conditions that kill vegetative cells. The frequency of the minority mating type will then tend to increase. If as a result it becomes eventually the majority type, then it will suffer a corresponding disadvantage and will tend to decrease in frequency. Clearly, the two mating types will be equally frequent at equilibrium. In the simplest case, when mating type is Mendelian and only zygospores can survive the winter, equality is restored in a single sexual generation, regardless of any disparity in frequency caused through natural selection by the differential vegetative proliferation of the mating types. In multipolar systems, this is not necessarily the case because the more numerous mating types will participate in more fusions. A very abundant mating type will be relatively unsuccessful because a large fraction of its gametes will fail to find a compatible partner; nevertheless, the frequencies of mating types at equilibrium will be determined in part by their sexual and vegetative abilities and will not necessarily be equal.

**The Population Sex Ratio.** In organisms with strongly differentiated gametes, individuals possess gender by virtue of bearing only one type of gamete. The proportion of male and female individuals in the population (the "sex ratio") is again governed by the sexual advantage of the minority gender. If females are more abundant than males, a male will on average mate with more than one female and will therefore produce more offspring than the average female. A gene that causes more sons than daughters to be produced will be selected; the reason is not that individuals bearing the gene will produce more children—the total number of sons and daughters may be fixed—but rather that they will produce more grandchildren, through their sons. The spread of a son-producing gene will increase the proportion of males in the population. If they become more abundant than females, the converse argument applies, and genes that cause the overproduction of daughters will be favored. The survival rates of males and females may differ, just as the mating types of bipolar protists may have different rates of vegetative propagation, but selection will nevertheless lead to an equal sex ratio at birth. This important argument was advanced by Sir Ronald Fisher.

**The Population Sex Ratio and the Progeny Sex Ratio.** If the population sex ratio is not equal, selection will favor individuals that produce a greater number of offspring of the gender that is in a minority. Once it has reached equality, however, selection ceases to act. A population in which males and females are equally frequent may be composed of individuals all of which produce equal numbers of sons and daughters, or it may consist of equal numbers of individuals that produce sons or daughters exclusively.

**Sex Allocation in Hermaphrodites.** Although not as obvious, a parallel argument is valid for outcrossed hermaphrodites: at equilibrium, the expenditure of resources on male and female gametes will be equal. The principle can indeed be extended very broadly to cover the proportion of hermaphroditic and unisexual individuals in mixed populations, for example, or the age at which sequential hermaphrodites change sex.

**Genetic Constraints on Sex Ratio Evolution.** The evolution of sex ratios and sex allocation is a classic case of frequency-dependent selection, generated by competition for sexual resources. In organisms with chromosomal sex determination, however, it can be argued that an evolutionary principle is not required: Mendelian segregation will ensure that the sexes are equally numerous at birth. This does not explain why gender should be inherited in so equitable a fashion and does not apply to non-chromosomal mechanisms of sex determination. Nevertheless, it is certainly genetics rather than selection that is the proximate cause of equal sex ratios. Moreover, there are powerful constraints that prevent selection from modifying the outcome of chromosomal sex determination. The sex ratio is one of the few characters that does not respond to artificial selection in *Drosophila*. The best evidence for the general principle that equal sex ratios are established through the advantage of the minority gender comes, ironically, from situations in which the balance of the genders is disturbed: in particular, from biased sex allocation in self-fertilizing organisms, and from biased sex ratios in structured populations.

**Experimental Evolution of Balanced Sex Ratios in a Fish.** Gender in the Atlantic silverside *Menidia* is influenced by temperature; broods raised at low temperatures usually develop as females, whereas those raised at high temperatures develop as males. However, there also appear to be some genetic effects that modify the effect of temperature. David Conover and his colleagues at the State University of New York established laboratory populations at different temperatures, so that their sex ratios were initially highly skewed in one direction or the other. In almost all cases the minority sex increased in frequency in the following generation. After a few generations of selection, the sex ratio in all populations had approached equality, where it remained for the duration of the experiment. Moreover, when selection lines were tested at a different temperature, sex ratios often became skewed again; selection for equal sex ratios was thus specific to the environment in which the selection was practiced. So far as I know, this is the only experimental demonstration of the general principle that frequency-dependent selection leads to equal investment in male and female offspring.

## 168. Competition among sons or daughters yields diminishing returns for allocation to that gender.

### *Skewed Sex Ratios in Structured Populations*

**Sib Competition and the Sex Ratio of Progeny.** The general Fisherian proposition that sex ratios will be equal at equilibrium holds within a single interbreeding population. Suppose, however, that all social interactions occur between neighbors in rather small subpopulations, or districts, with progeny dispersing at longer or shorter intervals throughout the area occupied by the population as a whole. The progeny of any given individual make up a substantial proportion of their group, and they will compete largely among themselves, rather than competing equally with all other individuals, as they would in a single well-mixed population. This must often be the case: sibs will tend to live nearby unless they are promptly and widely dispersed by wind or ocean currents. Sib competition inevitably leads to diminishing returns because the more offspring are produced, the more intensely they will compete, and the more effectively each will suppress the other. The reproductive success of an individual will increase with the number of progeny it produces, of course, but it will not increase proportionately when sibs compete among themselves rather than competing in the population at large. As the number of progeny produced increases, then, the quality of individual progeny declines: more means worse. Sib competition affects how selection acts on the sex ratio if it differs according to gender. If sons compete (whereas daughters do not), then producing sons will lead to diminishing returns, such that the more sons are produced, the less valuable each becomes. This must alter the value of a son, relative to the value of a daughter. The overall value of sons, or daughters, will thus depend on their representation in the brood, as well as on their frequency in the population.

**Local Mate Competition.** If your neighbors are few, it is wasteful to produce many sons, who will compete among themselves to fertilize your neighbors' daughters. A very few males, on the other hand, are often able to fertilize many females; there is less likely to be sexual competition among daughters. When there is local competition for mates, therefore, selection will favor individuals that produce an excess of daughters.

**Local Resource Competition.** Conversely, an excess of males will evolve if group selection, rather than rewarding profligacy, instead favors reproductive restraint. When there is severe local competition for resources among growing progeny, for example, seedlings growing up thickly around their mother, producing too many may mean that all will starve. Local resource competition thereby favors male-biased sex ratios.

**Sexual Selection and Group Selection.** Another way of expressing this process is that in structured populations sexual selection and group selection will favor different sex ratios, and the sex ratio that evolves represents a compromise between these two opposed tendencies. The Fisherian principle legislates an equal sex ratio in interbreeding groups of any size; thus, if the subpopulations were completely isolated, an equal sex ratio would evolve in each. When each subpopulation contributes offspring at intervals to a common pool, however, those with more females will usually contribute more offspring: subpopulations consisting entirely of females, except for a single male to fertilize them, would contribute the most. The sex ratio is thus driven towards equality by sexual selection and towards a large preponderance of females by group selection. The outcome is a compromise between these two, the value of which depends on how many sexual generations pass within each subpopulation before the offspring are dispersed.

**Sex Ratios in Hymenopterans.** In hymenopterans, a female is able to manipulate the sex ratio of her offspring by choosing to fertilize eggs or not. There is much more opportunity to adjust the sex ratio in such a system, and sex ratios are often far from equal. This is sometimes attributable to autoselected genetic parasites that bias sexual development (Sec. 127), but may also represent an adaptive response based on nuclear genes that influence female behavior.

## *Sex Allocation in Hermaphrodites*

The proportion of sons and daughters produced by an individual affects its fitness through the number of grandchildren that descend from it. The brood can be seen as a compound individual that may have a single gender (all sons or all daughters) or a mixed gender; the gender of this compound individual affects the number of offspring it produces. A single individual may also have compound gender, producing both male and female gametes. Selection acts on the mixed gender of hermaphroditic individuals in the same way that it acts on the mixed gender of broods.

**Self-Fertilization.** The most obvious consequence of hermaphroditism is the possibility of self-fertilization. This may be a very useful ability in itself if sexual partners are rarely available; most annual plants that frequently found new populations in disturbed sites are self-fertilizing hermaphrodites. There is also a more subtle advantage of self-fertilization. An outcrossed hermaphrodite is likely (all things being equal) to allocate its resources equally to male and female gametes. If it then turns to selfing, a very small proportion of the sperm it produces will suffice to fertilize all its eggs, and the remainder can still be used for outcrossing. A mutation that causes this switch to self-fertilization will be transmitted to three-quarters of the self-fertilized eggs (as a homozygote in one-quarter of them) and to half the cross-fertilized eggs. The number of copies of the mutation transmitted to progeny is thus 50% greater than the number of copies of the normal gene transmitted by an outcrossed individual. The mutation will therefore be autoselected and will increase to fixation unless prevented by some other agent of selection, such as natural selection through inbreeding depression in diplonts.

In a population of selfed hermaphrodites, selection will usually drive down male allocation. Outcrossing is a chancy business when pollen or sperm are cast to wind or water, whereas an extra egg is nearly certain to be fertilized. The production of male gametes will be reduced to the point where they are just sufficient to ensure the fertilization of all the female gametes, outcrossing being abandoned. If giving up the production of outcrossing sperm makes it possible to produce an equivalent mass of eggs, then a selfed hermaphrodite with minimal male function will transmit genes, including those encoding its sexual behavior, nearly twice as fast as an outcrossed hermaphrodite with equal allocation to male and female gametes. Obligate selfing is therefore associated with the same automatic advantage as asexuality.

**Functional Interference in Sexual Structures.** In outcrossed organisms, there is no automatic advantage in being an hermaphrodite. Selection favors a division of sexual labor within an individual in the same circumstances that it favors a similar division between individuals within a brood, or a division of vegetative labor between different kinds of cell: labor should be divided when similar structures interfere with one another, causing diminishing returns on further investment.

Functional interference is easy to appreciate in selfed hermaphrodites: the pollen grains produced by an individual are competing with one another for the chance to fertilize. In structured populations, selection will favor female-biased sex allocation (through local mate competition) or male-biased allocation (through local resource competition) in hermaphrodites, just as it will favor biased sex ratios. Moreover, whenever the male gametes, or the female gametes, produced by the same individual compete with one another, creating diminishing returns in terms of sexual or vegetative success, selection will favor hermaphrodites over single-sexed individuals, even in outcrossed species. Secondary sexual structures, such as the anthers and styles crowded into the middle of a flower, may also show this kind of functional interference; if a given mass of tissue promotes sexual performance more effectively when it is divided between anthers and styles, the flower itself, not only the individual, should be hermaphroditic. In organisms that have a unitary rather than a disseminated system of sexual structures, the reverse is more likely to be the case. Animals must make a substantial minimum investment in oviducts, shell glands, uterus, and the like before they can function properly as females; males require a similar list of equipment; an hermaphrodite in which neither system was adequately constructed would be unsuccessful either as a male or as a female. These large fixed costs thus favor the separation of the sexes.

## *169. Sexual selection modifies the type whose gametes are present in excess.*

The form of secondary sexual structures can also be modified through sexual selection, either in organisms with separate sexes or in hermaphrodites. There are many cases, especially among organisms with separate sexes, where these modifications reach bizarre proportions. It was, indeed, the difficulty of accounting for characters, such as the plumage of male peacocks, in terms of natural selection that led Darwin to develop a special theory of sexual selection.

The extreme modification of structures and behaviors associated with mating is often confined to one gender; this is, indeed, the main circumstantial evidence that they evolve through sexual selection. This gender is usually, but not always, the male. The principle that governs this process is that sexual selection modifies the type whose gametes are present in excess.

**Gender and Sexual Competition.** Sexual selection is caused by sexual competition, in which each gender is a resource for the other. The limiting resource is the gender that is the less abundant. Gametes of the gender that is present in excess must compete for access to the limiting gender and will thereby be modified through sexual selection. In multicellular organisms, gamete success is often greatly enhanced because the gametes are brought into contact by specialized structures or behaviors displayed by vegetative individuals. These vegetative individuals will then compete and are liable to be secondarily modified through sexual selection. The same principle applies: individuals bearing gametes of the gender present in excess will compete for access to individuals bearing gametes of the minority gender. Male gametes are usually much more numerous than female gametes, and therefore males will usually compete for access to females, and are more likely to respond to sexual selection. The principle extends to hermaphrodites: hermaphroditic individuals acting as males will compete for access to those acting as females, even though the same individual may act as male and female at different times, or even, when copulation is reciprocal, at the same time.

**Bateman's Principle.** This principle has been expressed in different ways by a variety of biologists, reflecting the context of the problem they set out to investigate and the organism they were studying. The classical restatement was made by A.J. Bateman of Cambridge, as the result of experiments in which he used genetic markers to estimate the number of progeny produced by individual male and female *Drosophila* when several flies of each gender were cultured together and allowed to mate freely. Both male and female flies may copulate several times, but the number of copulations has a different effect on the number of offspring they produce. Females require one copulation in order to produce a batch of fertilized eggs, but their output of eggs was not increased by additional copulations. The number of offspring fathered by a male, on the other hand, increased in proportion to the number of females he mated with. Almost all females produced offspring; the number each produced would depend primarily, not on sexual success, because they are almost certain to be courted and inseminated, but on the quantity of resources they have been able to garner and turn into eggs. Variation among females in resource-gathering ability will evolve through natural selection. A substantial proportion (nearly a quarter) of males, on the other hand, failed to copulate successfully and thus fathered no offspring at all. The remainder copulated more or less frequently and gave rise to varying numbers of descendents as a result. Reproductive success thus varied more among males than among females. The greater variance of male reproductive success (strictly speaking, sexual success) is often called *Bateman's Principle*. It is the outcome of the more intense sexual competition among males, arising from the greater abundance of male gametes.

This need not necessarily have any evolutionary consequences. If we were to choose pairs of flies at random, allowing them to mate before returning them to the population and repeating the procedure, the variance of the number of copulations would be the same for males and females, but the variance of reproductive success would nevertheless be greater for the males than for the females. Because the flies were chosen at random, the fact that some male flies were much more successful than others would cause no consistent change in male attributes. It is only if the greater variance among males is caused in part by heritable variation in sexual ability that the response to sexual selection will be greater among males than among females.

**Sexual Competition and Parental Investment.** Male gametes being more numerous, they are also smaller. Although their genetic contribution to the zygote is the same as that of the female gamete, their cytoplasmic contribution is much less. Another way of expressing the general principle of sexual competition, suggested by Robert Trivers, is that the response to sexual selection will be greater in the sex that invests less in the progeny. When the male contributes no more than a gamete to each of his offspring, this point of view involves no new insights. In some cases, however, the male may provide resources that can be used by the developing embryo: a nuptial meal attached to the spermatophore in some insects, brooding of the eggs by seahorses, or feeding the growing nestlings in many birds. When the parental investment of the male approaches or exceeds that of the female, male gametes, however abundant, may be no more readily available than female gametes. Females must then compete among themselves for access to males and, as a result, may be strongly modified by sexual selection.

## *170. Mating success can be increased or decreased through artificial sexual selection.*

There have been many experimental demonstrations that artificial selection can alter the vegetative characteristics of populations. I presume that it would be just as easy to modify sexual characteristics, but there have been very few attempts to study sexual selection in the same way.

**Mating Speed in *Drosophila*.** Inseminated female flies usually discourage courting males, and other things being equal males that succeed in mating quickly will have a considerable sexual advantage. Aubrey Manning of Edinburgh selected for mating speed by choosing the ten fastest and the ten slowest pairs from samples of fifty, establishing two replicate

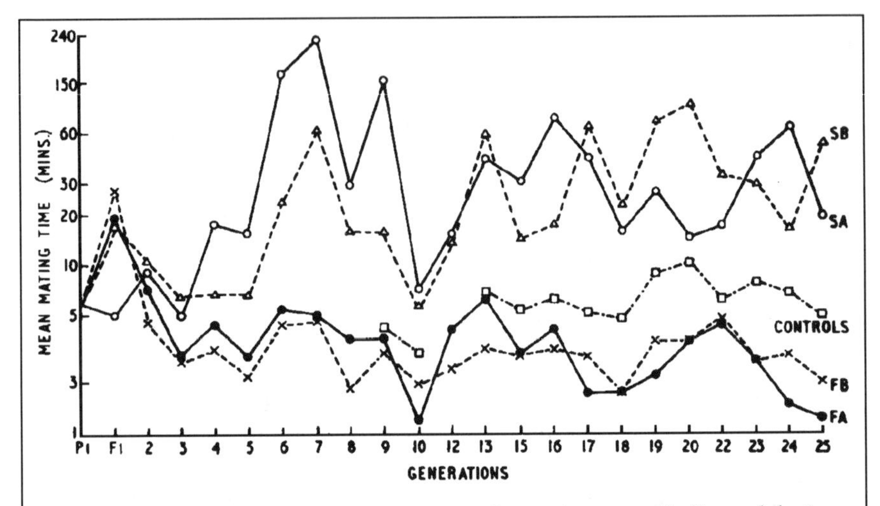


This diagram shows the response to selection for mating speed in *Drosophila*, from Manning (1961). There were two fast lines (FA, FB), two slow lines (SA, SB) and an unselected control line. The *y*-axis is a probit transform of mean mating time. For comparable experiments, see Manning (1963, 1968), Manning and Hirsch (1971), and Spiess and Stankevych (1973). Gromko et al. (1991) reported a response to divergent selection for copulation duration in *Drosophila*.

For a positive response to sexual selection among male gametophytes, see Ottaviano et al. (1983); conflicting results in radish *Raphanus* were reported by Marshall and Ellstrand (1986) and by Snow and Mazer (1988). Selection on flower size and pollen production was described by Stanton and Young (1994).

fast lines and two slow. After about twenty generations, the fast and slow lines had diverged substantially from the unselected controls. Half the unselected flies had mated after about 5 minutes; in the fast lines this time had been reduced to less than 2 minutes, whereas in the slow lines it had increased to more than 15 minutes. Most of the fast flies, indeed, had mated before the first of the slow flies had done so. The base population, a large cage population that had been maintained in the laboratory for several years, must have contained a large amount of genetic variance for mating behavior that was readily harvested by selection. The realized heritability in the first few generations of selection was about 30%, after which the response was much slower.

Crossing the lines showed that the differences in mating speed were heritable. They were transmitted both by males and by females, although crosses between the lines suggested that the males had been more affected by selection. A subsequent experiment in which only one sex was selected showed that selection for slow mating was much more effective in males than in females, although selection among males only was not nearly as effective as the selection of pairs practiced in the first experiment. The behavioral changes that caused the response of mating speed to selection were most interesting. When pairs were selected, there was a change in general levels of activity, but in an unexpected direction: the fast maters were relatively inactive flies. The reason is that rapidly moving females cannot be successfully courted. In the slow-mating lines the flies would move around for several minutes when introduced into the mating chamber and would rarely court or mate during the period. The flies from the fast-mating lines, on the other hand, crawled only a short distance before beginning courtship. However, this outcome depends on an interaction between the sexes. When flies are selected through the speed with which they mate with an unselected control stock, as in the second experiment, the situation is different. Generally speaking, the first females to mate are inactive, slowly moving individuals, but the first males to mate are active flies that are the first to discover the females. Selection for slow mating applied to males alone therefore led to a decrease in male activity: many of the slow males remained almost motionless when introduced into the mating chamber, and could not mate until a female happened to walk by within range. Interestingly, the same phenotype was produced by back selecting the fast lines in the pair-selection experiment, extremely inactive flies that mated very slowly because they were unable to find partners. The behavioral causes of mating success thus differed from experiment to experiment. In particular, slow mating can evolve through either very high or very low levels of general activity, depending primarily on whether selection is applied to one sex or to both.

**Gametophytic Selection in Flowering Plants.** I have already referred to suggestions that selection on gametophytes may produce a correlated response in sporophytes (Sec. 163). More straightforwardly, sexual selection of gametophytes should enhance their own sexual abilities. Thus, pollen grains can be selected for their ability to fertilize by placing many pollen grains on the stigma, creating intense competition among them, and harvesting the seeds produced. The results of such experiments have so far been equivocal. Several authors have reported success, but a careful study of pollination in the wild radish, by Allison Snow and Susan Mazer of Davis, yielded no reponse to selection. One explanation of the variable experimental results is that fertilization success depends primarily not on the male gametophytic genome but rather on the interaction between the male gametophyte and female sporophytic tissue.

**Flowers.** The form, coloration, and scent of flowers evolve through sexual selection for an adequate level of pollen import, to ensure the fertilization of the plant's own ovules and, more importantly, for a maximal rate of pollen export, to fertilize the ovules of other plants. These characters have also been deliberately selected, yielding cultivars to decorate the garden or the table. In most cases rather little has been accomplished by selection, beyond a general increase in the size of the bloom and sometimes brighter or different coloration. I have the impression that the very large blooms developed by horticulturalists lack scent, suggesting some antagonistic pleiotropy among sexually selected floral characters; but this is merely an impression.

One of the few experimental attempts to manipulate the form of flowers by selection was made by Maureen Stanton and Helen Young, again working on wild radish at Davis. In material from natural populations, petal size increases with pollen and nectar production, but not with ovule or seed number, as one would expect from the predominantly male role of flowers in serving pollen export rather than pollen import. Artificial selection for the greatest and smallest ratios of petal size to pollen production for only two generations removed the correlation between secondary sex allocation to petals and the primary allocation to pollen, suggesting that in natural populations this correlation is maintained by selection.

## 171. *Sexual competition among members of the same gender leads to the exaggeration of secondary sexual structures.*

Broadly speaking, there are two schools of thought regarding the evolution of secondary sexual characteristics. The first is that sexual and vegetative performance are positively correlated, so that sexually successful individuals give rise to vegetatively superior progeny. Natural selection and sexual selection then act in the same direction. In most cases, this implies that females prefer to mate with the more vigorous males or at least with males that will yield more vigorous progeny. The second point of view is that sexual and vegetative performance are negatively correlated, so that sexually successful individuals have relatively low viability or fecundity. Natural selection and sexual selection then act in different directions. In most cases, this implies that males compete directly among themselves for access to females.

### *Concordant Natural and Sexual Selection*

Females will be selected to mate preferentially with males whose progeny are likely to be vegetatively superior. Sexual selection will thus enhance the discriminatory abilities of females. Linda Partridge of Edinburgh compared the offspring of females that were allowed to mate freely with those of females that were constrained to mate with a single partner chosen at random by the experimenter. The offspring of the females that were given the opportunity to choose their partners were on average about 4% more viable than those from the arranged marriages.

There are two principal difficulties in accepting this process as a general basis for sexual modification. First, males may falsely advertise their genetic worth. Females should therefore discriminate among possible mates only on the basis of characteristics that cannot readily be misrepresented, such as body size. It is not clear that many interesting sexual modifications, such as bright patches of color or conspicuous crests, sacs, or wattles, are unequivocal signals of heritable vegetative superiority. Second, genes that confer a vegetative advantage and that are thus appropriate criteria for the choice of mates will be strongly selected and will pass quickly to fixation; there is then no variation left on which a choice can be based. The various theories of sexual selection through female choice have been developed largely in response to this second objection. They fall into two categories, depending on whether the lack of fit between population and environment is caused by genetic deterioration or by environmental change.

**Mutation and Inbreeding.** Recurrent deleterious mutation will maintain genetic variance for vegetative fitness, and females may avoid males whose appearance or behavior signals a high mutational load. Some major mutations in *Drosophila* are indeed associated with poor male mating success. Inbred flies from an outbred population suffer inbreeding depression and may be avoided by females. Courtship in *Drosophila* involves a dance in which the partners, face to face, perform a series of rapid shuffles from side to side. John Maynard Smith found that inbred males were unable to keep up the rapid tempo of the dance and suggested that females rejected dancing partners whose inferior athletic abilities might indicate a heritable defect.

**Environmental Change.** Females might discriminate against poorly adapted males in a novel environment. These might be immigrants in a patchy environment, or males adapted to the previous state of a changing environment. If the environment is changing through time, female discrimination will continue to be selected only if genetic variation continues to be maintained among the males. This has led W.D. Hamilton to argue that disease resistance is an appropriate criterion for female choice, because alternative resistance genes will cycle in frequency through time in response to the delayed frequency-dependent selection induced by pathogens and parasites (Sec. 154). Male displays are then viewed as medical examinations that allow females to scrutinize potential sexual partners for infection. This attractive idea continues to be controversial.

**Nonheritable Male Handicaps.** Amotz Zahavi of Tel Aviv has put forward the ingenious notion that females should choose to mate with males whose phenotype is clearly maladapted, on the grounds that having survived despite their handicap they must be exceptionally vigorous in other respects. If the handicap is not heritable, for example, a leg broken by accident and subsequently healed, then it is possible that handicapped males might sire unusually well-endowed offspring, although it is not clear that females use accidental injuries of this sort as criteria in choosing mates. If the handicap is heritable, such as long tail plumes that impede flight, the superior genes that allow males to survive will be transmitted to daughters, where they can be expressed without the handicap. Whether this compensates for the production of handicapped sons, however, is highly debatable.

## *Antagonistic Natural and Sexual Selection*

In some cases, it seems clear that secondary sexual characters have been exaggerated to the point where they must impair vegetative fitness. It is equally clear that such characters can evolve despite natural selection if females prefer to mate with bizarre males. A series of experiments over the last decade or so have established that in strongly dimorphic species females do indeed prefer the extremes of male showiness. Malte Andersson of Uppsala demonstrated this kind of preference in a wild population of African widow-birds. Male widow-birds have enormously long tail feathers that appear to be used in courtship displays but are aerodynamically unsound; the females have more conventional tails. By cutting the tail feathers, it is possible to create males with very short tails, with normal long tails (by reattaching the cut section with glue) or with extremely long tails (by gluing a long piece onto a tail that had not been shortened much). The males with very short tails attracted fewer females than the normal birds, and those with extremely long tails attracted more females than did males of normal appearance. Highly exaggerated structures may, then, evolve because they are favored by sexual selection, despite their disadvantage under natural selection, as Darwin realized long ago. It is the aesthetics that is more puzzling: why should females possess such apparently inappropriate tastes?

**Unselected Female Choice.** The simplest answer is that they do because they do. Some quite unrelated area of activity, such as gathering food or building nests, happens to make a certain kind of structure or a certain sound or movement more apparent or more attractive to females. A variant male that develops a long sword-like tail or croaks at a certain pitch is favored as a mate because of the pre-existing aesthetic or sensory bias of the female, and such variants may spread because of their sexual advantage, even if they reduce male viability. This was Darwin's original notion, and it has recently received some experimental support.

**Selected Female Choice.** Sir Ronald Fisher proposed a more general explanation of bizarre sexual modification. To begin with, female preference is based on some trait that is favored through natural selection, for any of the reasons already mentioned. Females that are better able to choose the more fit males will produce more fit offspring, and selection will thereby enhance the discrimination that females exercise among males. Males in which the character is more clearly and strongly developed will be more likely to be chosen as mates and will thereby receive a double advantage, having greater viability and greater sexual success. As the character becomes more exaggerated, its sexual effect will increase, but it will become less well adapted for its vegetative function. Sexual selection and natural selection now pull in opposite directions. By this time, however, most females have inherited a tendency to use this character as a criterion in choosing mates. Those who choose males in which it is exaggerated beyond usefulness will continue to be be favored, because their sons will be more likely to obtain mates, and their sexual prowess will more than outweigh their loss of viability. The success of the offspring of such matings will create a correlation between the tendency, expressed in males, to develop an exaggerated version of the character, and the tendency, expressed in females, to prefer to mate with such males. The process now feeds upon itself: a more pronounced female preference favors a greater exaggeration of the male character, which in turn favors the most discriminating females. It will be brought to a halt only when natural selection against the modified males becomes overwhelmingly strong.

This truly beautiful theory explains the extreme contingency of bizarre sexual modification. Characters that evolve through natural selection are often highly convergent. The mottled plumage of ground-nesting birds, the countershading of pelagic fish, the stripe passing through the eye in frogs; such characters evolve because they render animals less conspicuous, and they evolve, in more or less the same form, in many independent lines of descent. Conspicuous warning or frightening patterns, such as fake eyespots or bands of highly contrasted colors have likewise evolved independently on many occasions. Sexual modifications, on the other hand, are usually peculiar to a small clade and involve different structures, or different kinds of modification, in nearly every case. This is to be expected if the ultimate exaggeration of a male character depends on the prior establishment of one particular female preference from the very large number that might have arisen.

**Direct Sexual Competition Among Males.** Female preferences create indirect sexual competition among males. This competition is sometimes more overt, with males struggling among themselves for the physical control of access to females. Male red deer are much larger than the females, bear antlers that grow only during the rut, and engage one another in grim combat where the prize is an harem of hinds and the penalty may be a lingering death. Such contests among males provide a vivid enactment of sexual competition, and even in less theatrical cases, the evolution of male structures and behaviors through sexual selection is universally accepted. Moreover, combat ability has on occasion been deliberately selected, especially in fighting cocks and Siamese fighting fish, although I have not, unfortunately, come across any careful analysis of these unintentional selection experiments.

Sexual modifications that are useful in combat, such as large size and specialized weapons, are quite widespread, at least among animals, and have received a great deal of attention from naturalists and theoreticians. This has tended to obscure the fact that the usual outcome of sexual competition among males is in the reverse direction. Male gametes evolve to become small and numerous through sexual selection, and male individuals are usually modified in the same way and for the same reasons. When females mature a bulky crop of large gametes, they are likely to become sedentary and specialized through natural selection for gathering resources. A large male with an equally bulky testis would be carrying around far more male gametes than could be used in a single mating and would be likely to find and inseminate fewer females than would a large number of small males amounting to the same total mass. Producing numerous small sons will be favored by sexual selection in most circumstances. In most invertebrates, not to mention a host of other organisms, from filamentous algae to ferns, males are smaller than females. In some cases, this reduction in size is carried to extremes that are more bizarre than antlers or wattles. A male rotifer, for example, is little more than a testis equipped with a locomotor organ. The dwarf males of some barnacles and angler-fishes settle and live parasitically on, or in, the females they discover and are, in some cases, so highly reduced as adults that the females in which they embedded themselves were originally described as hermaphrodites.

## 172. *Adaptive divergence is hindered by outcrossing.*

Adaptedness to a particular site or way of life is seldom based on a single gene, but rather depends on a constellation of genes at many loci. It is therefore hindered by mating with differently adapted individuals because recombination will break up a successful combination of genes to produce inferior progeny. This is simply the converse of the view that recombination enables sexual lineages to adapt quickly to continually changing circumstances. Sex allows mutations that have arisen independently in different lines of descent to be combined into the same line of descent. But the other side of the coin is that sex causes genes that are combined in the same line of descent to be distributed among different lines of descent.

Adaptive divergence would be facilitated if individuals that are similarly adapted mate with one another. If there are genes that can specify appropriate mating preferences, they will tend to increase in frequency. Sexual isolation will evolve through a coupled process of natural selection and sexual selection: sexual selection will favor individuals that mate appropriately, because natural selection favors their progeny, in whom parental adaptations are conserved. Genes that confer local adaptedness will thereby become correlated with genes for sexual preference, and when this correlation is complete the outcome is an ecologically distinctive, sexually isolated new species.

Unfortunately, the same argument that applies to combinations of genes that confer local adaptedness also applies to combinations of genes, some of which affect vegetative performance and others sexual preference. The correlation between the two will be broken up by recombination. Selection against hybrids may favor genes that restrict mating, but this selection must be very intense if it is to overcome recombination. Indeed, it has been very difficult to see how this obstacle to speciation can be surmounted and even more difficult to demonstrate experimentally the processes involved in the differentiation of sexually isolated lineages from an outcrossing population.

The strength of selection needed to overcome the dissipative tendency of recombination explains the lumpiness of obligately sexual clades. A minor specialization that confers some slight advantage will scarcely ever become established as a distinct lineage because the advantage of mating assortatively will be too slight to counteract the effect of recombination in an outcrossing population. It is only major modifications that entail large adaptive differences that can engender selection powerful enough to drive the evolution of sexual preference, and outcrossing species therefore tend to be widely separated and highly distinctive. In asexual or self-fertilizing organisms this constraint does not apply, and instead of a limited number of clearly distinct types there is often instead an indefinite assemblage of "microspecies" that can be only vaguely demarcated. There is a curious echo here of the old argument of Fleeming Jenkin that troubled Darwin so much in his revision of the *Origin*. Darwin's theory requires the accumulation of slight variations over long periods of time, and yet (Fleeming Jenkin argued) these slight variations will quickly be lost through crossing with the parental stock. This argument was discredited with the development of particulate genetics: genes are not blended out of existence, and genetic variance is conserved indefinitely in outcrossing populations in the absence of selection and sampling error. Nevertheless, its ghost is still with us, whispering that it is very difficult to explain how slight modifications of sexual preference can accumulate when they are continually unlinked from the characters on which the preference is based.

## 173. *Sexual isolation may evolve directly through selection for specialization or habitat choice.*

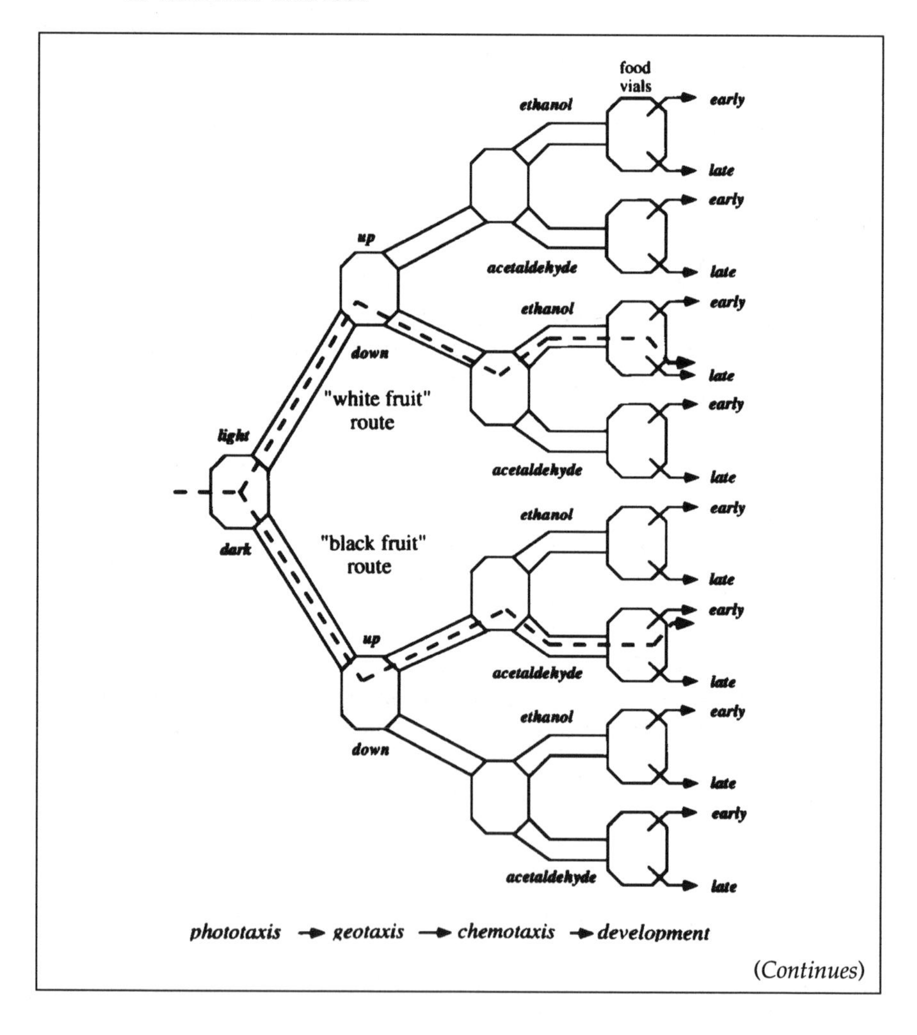


(*Continues*)

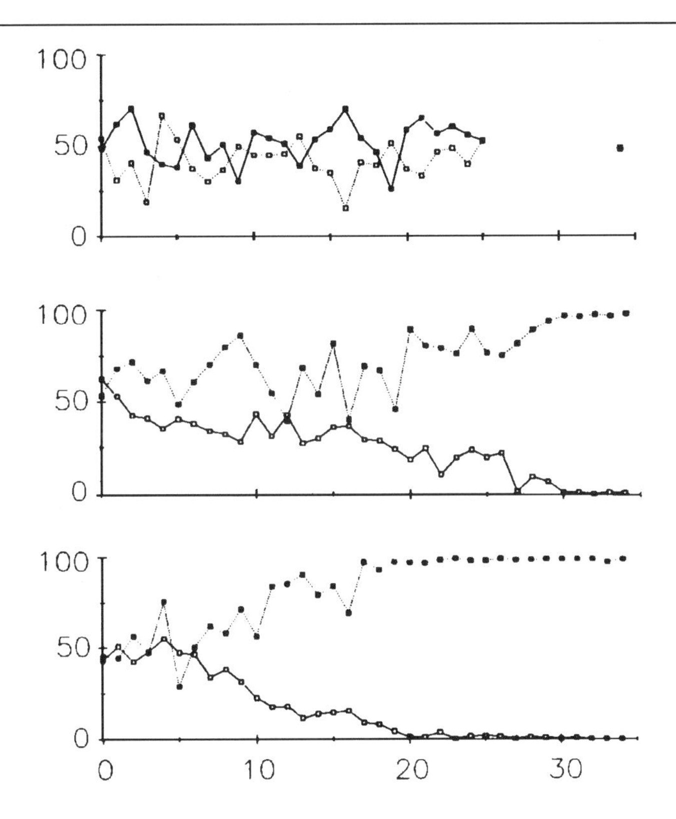


These diagrams show the evolution of sexual isolation through selection for habitat choice in the experiment by Rice and Salt (1990). The top figure shows the experimental design: only flies choosing the "white fruit" (light, downwards, ethanol, late) or "black fruit" (dark, upwards, acetaldehyde, early) sites survived. Flies making different decisions in the maze were identified by an ingenious phenocopy technique involving adding a supplement to the food medium in the vials representing one habitat that suppressed the expression of an eye-color mutation, providing a permanent but non-heritable marker. The diagrams here show the frequency of this marker in each of the two preferred habitats; an absolute preference for one of the two habitats would be indicated by a frequency of unity in one habitat and zero in the other. There are three types of line. The upper panel shows unselected control lines that showed no consistent preference. The middle panel shows selection lines in which only flies moving to one of the two preferred types of habitat were allowed to breed. The response to selection is rather slow and erratic, but ends in an almost complete sorting by habitat preference. The lower panel shows selection lines in which flies were retained not only if they moved

(*Continues*)

to one of the preferred habitats, but also if their parents had displayed the same preference. The response to selection is here much smoother and more rapid.

For the evolution of sexual isolation in *Rhagoletis*, see Bush (1969).

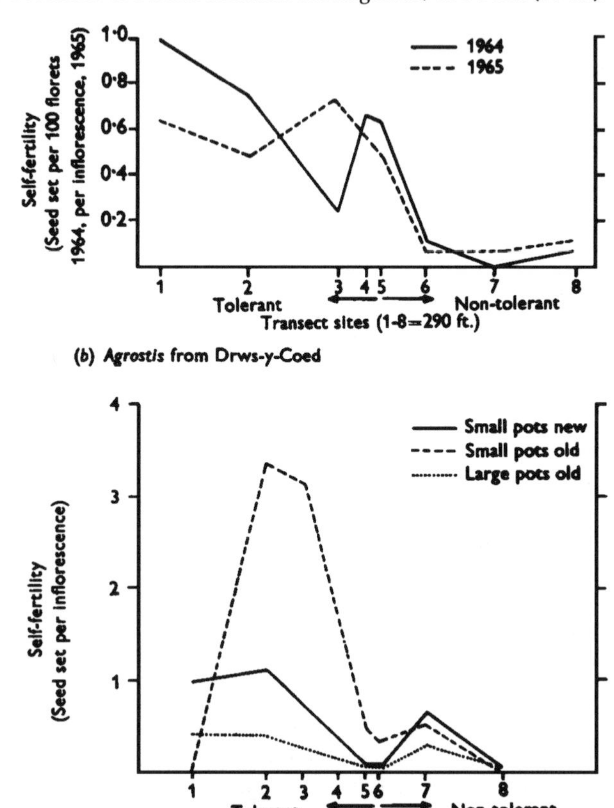


These diagrams show the self-fertility of plants sampled across the boundary of a lead mine at Trelogan and a copper mine at Drws-y-Coed, North Wales, from Antonovics (1968). They should be compared with the pattern of change in metal tolerance along similar transects (Sec. 111). McNeilly and Bradshaw (1968) found that mine populations also flower about a week earlier than pasture populations; this difference was retained in common-garden experiments and may have arisen through selection for sexual isolation under outcrossing. The difference was especially marked at the mine boundary or on small mines. Similarly, Snaydon and Davies (1976) found that plants collected near the boundary of plots in the Rothamsted Park Grass Experiment (Secs. 61 and 112) flowered several days earlier than their neighbors on either side.

(*Continues*)

*(Continued)*

Ehrman et al. (1991) selected *Drosophila* in heterogeneous environments to investigate the evolution of sexual isolation as a correlated response to divergent adaptation. They adulterated the food medium with sodium chloride or copper sulphate. Where only one of the two salts was present, the flies evolved some degree of resistance to it. When both were present, it was hoped that specialized resistance would evolve, with hybrids at a disadvantage and consequently selection for assortative mating. This did not occur. When the flies were tested after 11 years, or more than 250 generations, of selection, there was no tendency for flies developing from eggs laid on either salt to mate preferentially with other such flies.

In populations that remain outcrossing, the breakdown of the correlation between adaptedness and mating preference is caused by the two basic processes of the sexual cycle, fusion and meiosis. It can be prevented if certain types of gamete do not meet; or if, having met, they do not recombine. Genes that hinder migration or recombination will be favored by natural selection because they will tend to remain associated with the specialized genotypes they conserve. They will incidentally facilitate the development of associations between genes responsible for specialized adaptation and genes regulating mating preferences. The evolution of ecological or genetic constraints on mating is thus a first step towards the unfettered mating preferences that characterize sexual species.

## *Restricted Migration*

Different ways of life may prevent contact between strains almost entirely. The strains may live in different places; but they may equally breed at different times, feed on different fruit, or parasitize different species of fish. Disruptive natural selection for activity schedules or diet or searching behavior will then lead to a certain degree of sexual isolation. The evolution of characters that cause sexual isolation as a byproduct is likely to be self-reinforcing. Divergent specialization is opposed by recombination between differently specialized genotypes. If some slight degree of specialization is nevertheless acquired, it will incidentally hinder mating between the specialized types. Because the rate of recombination is less, the rate of selection needed to cause further divergence will be proportionately less. Further divergence will further isolate the strains; thus divergence will progressively lower the barrier to further divergence. Characters that affect the rate of sexual contact between individuals are thus most likely to show divergent adaptation; in doing so they will facilitate the subsequent divergence of other characters.

**Selection for Habitat Choice in *Drosophila*.** Bill Rice and George Salt ran a particularly colorful experiment to find out how fast habitat choice could evolve and how complete would be the sexual isolation it caused. Young flies entered a maze that presented them with a sequence of different kinds of choice. The first chamber had a light and a dark exit; passing through either, they could then move upwards or downwards; finally, they could enter chambers scented with either acetaldehyde or ethanol. They thus sorted themselves among eight vials, depending on their phototactic, geotactic, and chemotactic behavior. They mated and laid eggs in these vials, and the eggs were collected either at the beginning, in the middle, or at the end of the oviposition period. The combination of spatial and temporal criteria thus provided a total of 24 different habitats, or ways of life. All but two were lethal. Flies were reared only from eggs laid late in the light, downward, ethanol habitat, or laid early in the dark, upward, acetaldehyde habitat. These were deemed to represent two types of fruit that would support larval growth; other choices led to inappropriate oviposition sites, where the larvae could not survive. The larvae from the two habitable vials were reared separately, and after eclosion the young flies were reintroduced to the maze for the next generation of selection. After about 30 generations of selection, the separation between flies utilizing the two habitable vials was almost complete: those born in one of these vials almost invariably returned to the same vial as adults. Because mating occurred almost exclusively within the vials, rather than at intermediate points in the maze, habitat choice enforced sexual isolation, so that by the end of the experiment the population comprised two sexually independent strains.

**Habitat Expansion in *Rhagoletis*.** Similar experiments have been done unintentionally whenever crop plants have been introduced into a new region. The clearest example of isolation caused by the evolution of habitat choice is the tephritid fly *Rhagoletis*, studied by Guy Bush. Its native host plant in North America is hawthorn. When orchards of exotic fruits were planted in its range during the nineteenth century, some strains were able to exploit them. It was reported from apple in 1864 and has since established itself on pear and cherry. The adults have a strong tendency to mate on the host they fed on as larvae. Moreover, the rate of larval development depends on the host species—cultivated fruits provide a richer pasture—so that the flies emerge from different hosts at different times of the year. About 100 generations of selection has produced distinct strains specializing on different host species that are sexually isolated by virtue of their evolved habitat preferences.

**Habitat Choice and Mating Preference.** Strains that mate in different places or at different times are necessarily sexually isolated because their members rarely meet. It seems reasonable to suppose that they will proceed to evolve distinctive mating structures or behaviors, so that members of different strains will not mate with one another, should they ever come into contact. However, this has not yet been observed. The *Drosophila* selection lines from Rice and Salt's experiment mate randomly when mingled in the same vial; and different host-specific strains of *Rhagoletis* readily mate among one another in the laboratory. There is as yet no experimental evidence that mating preferences evolve as a correlated response to selection for habitat preferences.

## *Restricted Recombination*

**Loss of Outcrossed Sex.** Local adaptation that is broken down by immigration and outcrossing can be conserved by asexuality or self-fertilization. Moreover, a gene that directs parthenogenesis will tend to spread because by suppressing recombination it causes its own transmission to be associated with that of locally adapted genotypes. A shift from a sexual to an asexual life cycle may evolve only rarely in some groups, from the paucity of genetic variation for so radical a change. Self-fertilization, on the other hand, evolves very readily in hermaphroditic organisms, such as flowering plants. It will be selected only if inbreeding depression is moderate and local adaptation does not depend on heterosis, but it is so easily accessible that it should be the most frequent response of the genetic system to divergent selection.

**Self-Fertility in Metal-Tolerant Plants.** The local adaptation of grasses such as *Agrostis* and *Anthoxanthum* to the polluted soil of former mines is continually diluted by pollen from non-tolerant plants on the adjacent pastures (Secs. 61 and 112). Janis Antonovics found that metal tolerance was associated with self-fertility: all non-tolerant plants produced very few or no seeds in isolation, whereas a minority of the tolerant plants were highly self-fertile. Moreover, average self-fertility fell abruptly at the mine boundary, in parallel with the decline in metal tolerance. Selection against the fertilization of pasture plants by pollen from the mine is much weaker, and there was no indication that selfing was selected on the pasture.

**Linkage and Pleiotropy in Outcrossing Populations.** Self-fertilization is an extreme case in which one's preferred partner is oneself. This effectively retains epistatic combinations of genes. Even in outcrossing populations, however, such combinations can be retained if the genes are fairly tightly linked, because the intensity of selection required to maintain adaptedness will be less when the rate of recombination is less. Genes affecting mating preferences will thus be more effectively selected if they are brought into linkage with the genes responsible for adaptedness. Closer linkage will thus itself be favored by selection, but the second-order selection involved is likely to be very weak. Bill Rice has emphasized that the limit of this process is when genes have pleiotropic effects both on vegetative fitness and on sexual preference. The correlation between adaptedness and preference is then no longer dispersed by recombination, and sexual selection and natural selection can operate in parallel. This kind of mutualistic pleiotropy seems quite likely to evolve. Genes at different loci that cause mating preferences will be selected only weakly in the presence of recombination and are unlikely to increase in frequency. Pleiotropic genes that cause inappropriate matings, resulting in the production of inferior hybrid offspring, will be selected against. The few genes that direct both appropriate specialization and appropriate mating are thus likely to be chiefly responsible for effective sexual isolation.

**Abrupt Origin of an Incompatible Strain.** The difficulty of evolving sexual isolation between outcrossing lines in a heterogeneous environment can be evaded if it is supposed that isolation can arise abruptly and is then followed by divergence. There is no doubt that this can sometimes happen. Polyploid lineages are established in a single generation, either by the failure of meiosis after fusion or as the result of a vegetative doubling. The diploid gametes produced by a tetraploid parent will produce triploid plants on crossing with the original strain, and these are likely to be inviable or sexually sterile. The tetraploid will therefore not persist unless it is asexual, or can self, or forms a separate interbreeding population.

## 174. *Mating preferences may evolve as a correlated response to powerful divergent selection.*

When migration or recombination are constrained, a certain degree of isolation has already arisen. It is characteristic of sexual species, however, that this isolation is maintained, even when individuals from different lineages are intermingled at all stages during their lives. Mating preferences may evolve through directional sexual selection, if females choose to mate with superior males. Mating preferences will lead to complete sexual isolation only if they evolve through disruptive sexual selection, with differently adapted individuals choosing to mate with appropriate partners. The evolution of sexual isolation therefore hinges on disruptive sexual selection being elicited as a response to disruptive natural selection.

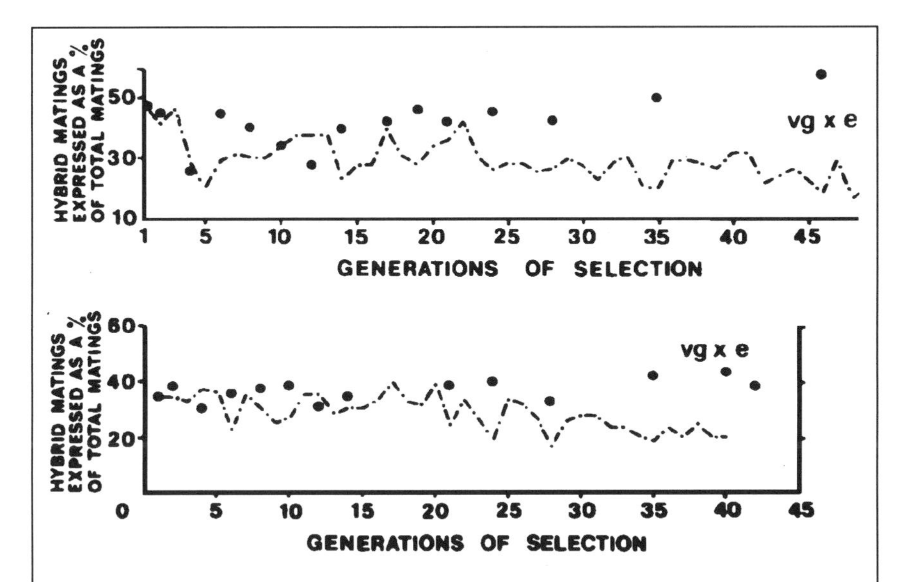


This diagram shows how the frequency of matings between two strains of *Drosophila* evolves when hybrids are eliminated, from Crossley (1974). The line is the frequency of matings between Ebony *e* and Vestigial *vg* individuals in the light and dark selection lines; the plotted points are the corresponding estimates for unselected control lines. The decline in heterogamous matings, relative to the control, was caused by an increase in both $e \times e$ and $vg \times vg$ matings in both light and dark conditions.

The experimental modification of sexual isolation between *Drosophila pseudoobscura* and *Drosophila persimilis* was reported by Kessler (1966); see also Koopman (1950). Dobzhansky and Pavlovsky (1971) conducted similar experiments with races of *Drosophila paulistorum*. These races are to a large extent sexually isolated, both because they are reluctant to mate with one another and because male hybrids are sterile. One isolate, originally classified as belonging to the Orinocan race, was found to produce sterile male progeny with Orinocan mates when retested after four or five years in culture, although the flies mated freely. The cause of the change is unknown, although infectious agents of some kind might have been responsible. This line was used to investigate whether behavioral isolation could be superimposed on the newly arisen sexual incompatibility through artificial selection. By eliminating hybrid progeny, some degree of behavioral isolation evolved after 70 generations of selection.

The original study describing sexual isolation evolving as a consequence of disruptive selection for bristle number was reported by Thoday and Gibson (1962). Subsequent experiments by Scharloo, den Boer and Hoogmoed (1967), Chabora (1968), Barker and Cummins (1969a, 1969b) and Spiess and Wilke (1984) gave negative results, although some degree of assortative mating was noted by Barker

(*Continues*)

(*Continued*)

and Karlsson (1974), in the experiment described in the text. A number of other experiments are reviewed by Thoday and Gibson (1970). Positive results, however, were obtained by Soans et al. (1974) and Hurd and Eisenberg (1975), who selected for positive and negative geotaxis in houseflies. In the latter experiment, three lines were established: selection for positive geotaxis in isolation; selection for negative geotaxis in isolation; and selection for positive and negative geotaxis, the selected individuals being mingled in equal numbers before mating. Mating trials after 16 generations of selection suggested that a rather strong tendency for individuals to mate with partners that had been selected in the same direction had evolved in all three lines.

If large populations often contain genetic variance for female mating preferences, it should be possible to modify these preferences through selection. There have been few experiments in which mating preferences are selected directly, by identifying individuals that prefer to mate with some specified kind of partner. There are, however, many experiments in which mating preferences are selected indirectly, as a correlated response to disruptive selection. The basic concept behind these experiments is that adaptive divergence through natural selection creates sexual selection for preferential mating as a consequence of the superiority of the offspring from matings betwen similar parents. There are two basic designs, differing in the reason that hybrid or intermediate types should be inferior. Firstly, progeny from wide crosses, between lines that have long been apart, might be completely inviable or sterile, even in conditions where either parent could flourish. Secondly, the progeny might be inferior to either parent because a particular scheme of artificial selection penalizes intermediate phenotypes, and yet be completely viable and fertile in permissive conditions.

## *Disruptive Selection with Inviable Hybrids*

Experiments of the first sort generally involve discrete strains, marked by major genes; their object is to reduce the frequency of mating between individuals bearing different markers. The standard design is to permit males and females of two different strains to mingle freely, subsequently destroying the hybrid progeny (recognized as recombinants for the markers), so that the lines are perpetuated only through matings between parents of the same strain. Artificial selection is thus applied directly to the markers and acts indirectly through sexual selection on the mating preferences. Such experiments (almost all with *Drosophila*) usually succeed in modifying mating preferences, creating or enhancing some degree of sexual isolation between the strains.

**Sexual Isolation Between Strains of the Same Species.** Stella Crossley of Oxford used *Ebony* and *Vestigial* mutants extracted from the same laboratory population. The mutants are interesting because their phenotypes are likely to affect mating behavior. In both cases, males are less successful in mating than wild-type flies. *Ebony* males have poor vision and are likely to lose sight of their partner; *Vestigial* males have highly reduced, non-functional wings and are unable to produce the buzz that is an important part of normal courtship behavior. In the light, *Vestigial* females mate randomly, whereas *Ebony* females, for unknown reasons, prefer *Ebony* males. In the dark, the *Vestigial* males cannot buzz and cannot see their partners either: *Ebony* males are thus preferred by both types of female. These initial preferences can be modified through selection, by discarding recombinant wild-type progeny. About 40 generations of selection caused a pronounced reduction in the frequency of mating between *Ebony* and *Vestigial*, both in the light and in the dark. Sexual selection increased the mating abilities of *Vestigial* males in the dark and of *Ebony* males in the light. This, however, did not enhance sexual isolation and may have reduced it. It was instead female behavior that was responsible for the decrease in mating between the two strains. In the dark, *Ebony* females repelled *Vestigial* males, either by wing-flicking (which halts male courtship) or simply by jumping away, so that most of their matings were with *Ebony* males. *Vestigial* females did not modify their behavior in the dark, but in the light both *Vestigial* and *Ebony* females actively repelled males of the other strain. The complex courtship behavior of female *Drosophila* can thus be tuned by sexual selection when powerful artificial selection is enforced against hybrids.

**Sexual Isolation Between Closely Related Species.** *Drosophila pseudoobscura* and *D. persimilis* are very similar species that can hybridize in nature. Seymour Kessler of Columbia selected for both an increased and a decreased degree of isolation between them. To select for low isolation, females from the low line of one species were mingled with unselected males of the other species: the first females to mate were chosen to perpetuate the line by removing them and allowing them to remate with males that had been selected in a similar manner. To select for high isolation, all the flies from the selection lines that mated with the other species were discarded, the surviving conspecific males and females were reintroduced to one another, and the first pairs to mate were selected. The response to selection was erratic, but after 18 generations the high and low lines had diverged in both male and female descent. The effect was again largely attributable to the behavior of the females. *D. persimilis* females are reluctant to mate with *D. pseudoobscura* males, even in the unselected lines, and selection had little effect on this behavior; if their partners are *D. pseudoobscura* males, there is vitually no contact between them. The behavior of *D. pseudoobscura* females was more malleable: in particular, the low females became much more receptive to *D. persimilis* males. Thus, it is possible either to enhance or to diminish the degree of isolation between two species that have only recently become distinct, through selection in the short term.

## *Disruptive Selection with Viable Hybrids*

In the second kind of experiment, no particular notice is taken of the hybrids. Individuals with extreme phenotypes are chosen as parents to generate disruptive selection, but the selected adults are mated as a group, with no control being exercised over their behavior. Hybrid offspring are thus penalized only because they are likely to have intermediate phenotypes that will be discarded by the experimenter.

**Disruptive Selection on Bristle Number in *Drosophila*.** Over 30 years ago, a remarkable experiment was published by J.M. Thoday and J.B. Gibson of Cambridge. In each generation, they extracted 80 females and 80 males from the experimental population, and selected the 8 females and 8 males with the most sternopleural bristles and an equal number with the fewest. The 32 selected flies were then mingled together in the same vial and allowed to mate. The males were then discarded, and the females moved to rearing vials to lay their eggs, keeping the high and low females separate. The variance of bristle number increased through time, so that by the 12th generation of selection the frequency distribution was bimodal. One mode consisted entirely of the offspring of the low-selected females, the other of the offspring of the high-selected females. When high and low individuals were deliberately crossed, however, their offspring were intermediate. Since few or no intermediate individuals were at this point present in the experimental population, it seems likely that flies with many bristles and those with few had become separated into two sexually isolated groups. This process occurs, as the authors remarked, "with astonishing rapidity."

Alas, it has not recurred. The dramatic result of the experiment spurred another dozen or more laboratories to repeat it, but they were uniformly unsuccessful. Indeed, the one generalization that may safely be made from these experiments is precisely the opposite of the original conclusion: sexual isolation does not generally evolve as a correlated response to disruptive selection of moderate intensity. This is not in itself surprising, in view of the effect that recombination is likely to have in uncoupling mating preference from bristle number; what remains surprising is the original result. J.S.F. Barker of Sydney ran a more elaborate version of the experiment, using the same original base population. He obtained a comparable direct response to selection for bristle number only when using much more intense selection on a larger number of flies in each generation. In these circumstances (but not with weaker selection, or smaller samples) some degree of sexual isolation did evolve, at least in the earliest generations. In this case, however, selection was so intense that after the third generation all matings were between high flies or between low flies, so that the procedure was equivalent to discarding all the hybrid offspring. This will not explain Thoday and Gibson's result because in their experiment selection was relatively weak. It seems to me more likely that the very small number of individuals selected created by chance a correlation between bristle number and mating preference. This does not appear to have been present in the base population (the evidence is equivocal) but could have arisen in subsequent generations. This would account for the rapid divergence of the high and low lines and would be consistent with Powell's boom-and-bust experiments, described in the next section. Whatever the truth of this may be, the important point is that the many attempts to repeat Thoday and Gibson's experiment have demonstrated that mating preferences do not usually evolve as a correlated response to disruptive selection of moderate intensity on an arbitrary character.

## *175. Sexual isolation may evolve as a correlated response to divergent selection in separate lines.*

Evidently, sexual isolation is likely to evolve only if disruptive natural selection is extremely strong—so strong, indeed, that the hybrids are unable to survive at all. Hybrid lethality might be the consequence of long isolation, but does not seem very promising as a first step in the evolution of isolation. In order to start up a coupled process of natural and sexual selection, it seems necessary, early in the history of diverging populations, to protect them from recombination. The most obvious way in which this might happen in nature is if they are in different places, with little exchange of migrants between them. Each can then evolve local adaptation through natural selection. As they do so, their sexual apparatus or behavior may become modified in different ways, so that the contingent outcome of sexual selection will prevent their interbreeding, should they subsequently come into contact again. It is not necessary that this separation should have evolved through selection for habitat preference; it is usually supposed to have arisen quite accidentally, with populations being spatially

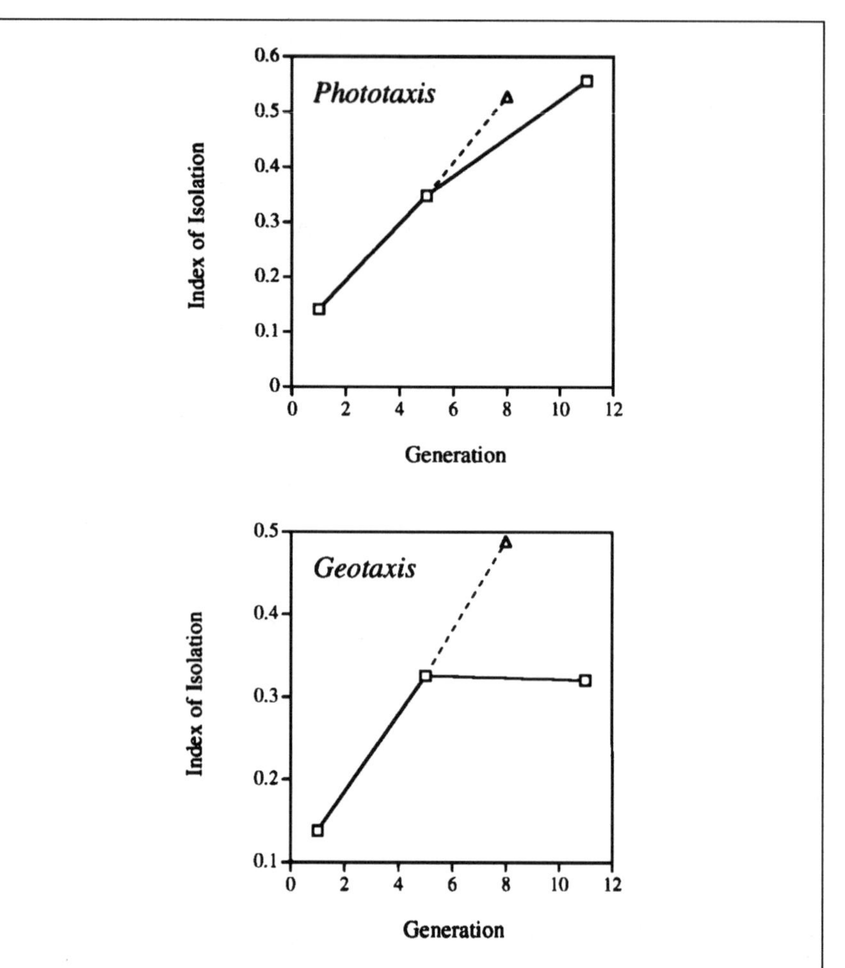


These diagrams show the outcome of mating trials between lines of *Drosophila* selected for positive and negative phototaxis (upper diagram) or for positive and negative geotaxis (lower diagram) from del Solar (1966). The index of isolation is an increasing measure of the frequency of matings between individuals from lineages selected in the same direction. The broken line and triangle represent lines in which selection was relaxed for three generations; it is somewhat surprising that isolation seems to increase at the same rate, or an even greater rate, as in the selection lines. Recall the experiments by Soans et al. (1974) and Hurd and Eisenberg (1975), demonstrating the evolution of partial sexual isolation between isolated lines selected in opposite directions for geotaxis (Sec. 172). Burnet and Connolly (1974) found that lines selected for increased and decreased locomotor activity had evolved mating preferences after about 100 generations. MacArthur

(*Continues*)

*(Continued)*
(1949) found no sexual isolation between mice selected upward and downward for body size. Similar experiments summarized in the text were published by Robertson (1966), van Dijken and Scharloo (1979a, 1979b) and de Oliveira and Cordeiro (1980). Replicated natural selection in different environments was described by Kilias et al. (1980); there is a comparable experiment by Dodd (1989), who obtained similar results. The experiments involving lines passed repeatedly through bottlenecks of very low population number are reported by Powell (1978; see also Dodd & Powell 1985) and Galiana et al. (1993); there are comparable studies by Ringo et al. (1985) and Meffert and Bryant (1991). An excellent review of these and related experiments has been written by Rice and Hostert (1993).

separated because the territory between them has become unsuitable for growth and impassable to migrants. The role of geographical isolation in facilitating the evolution of sexual isolation has been championed by Ernst Mayr of Harvard. The experimental issue is whether separate divergent selection lines eventually exhibit some degree of sexual isolation. There are about a dozen experiments of this sort; none have involved the evolution of complete sexual isolation, but several have reported some degree of assortative mating.

## *Divergent Selection in Complete Isolation*

The simplest experiments are those in which there is no contact between selection lines for some considerable period of time, after which their sexual compatibility is assessed. This procedure is very similar to destroying all the hybrids between the lines, and it gives similar results.

**Morphology.** As well as the disruptive selection experiment just described, J.S.F. Barker also ran two directional selection lines, for increased and decreased bristle number. They were tested after about 30 generations by mingling flies from the two lines and scoring their mating behavior. The morphological selection seemed to have caused changes in mating behavior; in particular, flies from the high line mated much more quickly than flies from the low line. This meant that most pairs that formed within a few minutes of the flies being introduced to one another comprised a high male and a high female, the low pairs being rare. This was a consequence of mating speed, however, not of any preference of one type over the other. After a few hours, all types of pairs were equally frequent. Divergent selection had caused no tendency for flies selected in the same direction to prefer to mate with one another.

**Physiology.** Two Brazilian biologists, Alice Kalisz de Oliveira and Antonio Rodriguez Cordeiro, selected *Drosophila* on excessively acid and basic food media, as well as on medium of normal composition. After about 100 generations of selection the lines had diverged considerably, with the number of offspring surviving to adulthood being much greater on the medium on which their parents had been selected. This adaptation was most pronounced for the lines on acid medium, which produced about five times as many offspring as the basic lines when tested on acid media; the basic lines produced about twice as many offspring as the acid lines when tested on basic media. When the lines were crossed, the hybrids were notably inferior to their adapted parent on the acid medium, but there was little difference on the basic medium. Some degree of hybrid inferiority had thus arisen, although this cannot have affected the mating behavior of the flies, the selection lines being maintained quite separately. Mating behavior was assessed by putting males from one line into a vial containing females from the same line and from one of the other two lines. There was an overall tendency for matings between flies from the same selection treatment to occur in excess. The design of the assay seems to imply that this was caused by male preferences; however, there was a pronounced bias only when females from the acid lines were available as mates, so female choice may have contributed to the expression of the male preference. The biases that were displayed were rather weak and erratic and did not seem to increase during the course of the experiment.

Forbes Robertson of Edinburgh added EDTA to the growth medium to select for resistant lines. Crosses between the selection lines and the unselected controls showed varying degrees of hybrid inferiority. This was largely attributable to genes on the third chromosome; when third chromosomes from the control population were introduced into the selection lines they caused sterility or death. The lines mated randomly, however, and attempts to select for assortative mating were unsuccessful.

**Behavior.** One advantage of using physiological rather than morphological characters is that natural selection can be used in place of artificial selection, greatly increasing the size of the selected sample. Behavior is even more awkward to study than morphology, because selection may require a lengthy scrutiny of each individual. This difficulty has been circumvented for geotaxis and phototaxis by the construction of ingenious mazes, designed so that the flies sort themselves out into different compartments according to their behavior, making large-scale selection experiments feasible. Eduardo del Solar of Santiago scored mating behavior in lines that had been selected for phototaxis and geotaxis, positive and negative, by Boris Spassky at the Rockefeller University. One generation of selection produced a very weak tendency for flies from the same line to mate with one another; this tendency was much stronger after 5 and 11 generations of selection and was maintained after 3 generations of reduced selection.

Several experimenters have found that the modification of behavior through selection is soon accompanied by the appearance of preferential mating. However, this is not always the case. F.R. van Dijken selected for high and low general activity in Wim Scharloo's laboratory at Utrecht. Flies were selected on the locometer, a series of conical compartments that allow passage in only one direction, so that the most active flies accumulate in the furthest compartments. When the mating behavior of the flies was examined after 30 or 40 generations of selection, it was found to be highly non- random; but this was almost entirely attributable to the fact that the males selected for high activity were quicker off the mark; in trials where one kind of female was mingled with both kinds of male, the high-activity males mated much more quickly, regardless of which kind of female was offered. Like selection for phototaxis or geotaxis, selection for general activity led indirectly to the modification of sexual behavior, but in this case one selection line acquired an unconditional sexual advantage.

## *Sexual Divergence of Small Populations*

Hampton Carson has suggested that very small populations may evolve distinctively different sexual preferences, even if they inhabit similar environments. He envisaged a series of populations colonizing disturbed sites, in which they repeatedly increased in numbers because of adaptation or lack of competition, before abruptly declining to very low numbers because the environment changes or because they have exhausted local resources. The population is thus repeatedly sampled from a very few individuals, who might by chance be sexually idiosyncrastic. After a number of cycles of increase and decrease, some populations, while still outcrossing, might have lost the ability to cross with members of other populations.

A number of selection experiments have attempted to identify a process of this sort. Jeffrey Powell of Yale assembled a highly diverse population by crossing isofemale lines of *Drosophila* collected from several localities. This was allowed to increase for two or three generations to about 10,000 individuals. Eight pairs of flies were then chosen to found new populations, each of which again increased in numbers before being reduced to a single mated pair. After four such population cycles, the mating preferences of flies from different lines were compared with those of flies from control lines that had been maintained more or less constant in size. Flies from these control lines showed no particular preference for mates from the same line. The experimental lines, on the other hand, often showed departures from random mating when flies from different lines were mixed together. When a substantial preference was observed, it always involved flies from the same line preferring to mate with one another. About a third of all the lines behaved like this. The strongest preferences persisted for some time after the end of the experiment, when the flies had been cultured at more or less constant density for about 10 generations.

At least three other experiments, with similar protocols, have found a similar but less pronounced tendency for mating preferences to evolve. The most extensive was run by a team led by Agusti Galiana of Valencia, who used large numbers of replicate lines, set up controls for inbreeding and other complications, and used between one and nine pairs to found new populations after each crash. After four or more population cycles, about 10% of crosses between lines showed some evidence of assortative mating. This seemed more likely to happen when the number of founders was very small. The mating preferences shown by these lines, however, were rather weak. In short, severe fluctuations in population number do lead to the appearance of mating preferences, provided that the residual populations are extremely small. It is not important that the effect occurs in only a minority, perhaps a small minority, of experimental lines: nobody expects speciation to occur in every isolated group. It is, however, cause for concern that the mating preferences that have been observed are, for the most part, rather weak. They certainly do not even approach sexual isolation in any case.

The severe bottlenecks in population number that the experimental populations passed through would have caused high levels of inbreeding, and the chance fixation of alleles at many loci. It is unlikely, however, that inbreeding itself caused the mating preferences that were observed. Powell also tested random inbred lines, maintained without a periodic reduction in numbers, and failed to find any consistent deviation from random mating. Indeed, crosses between inbred lines of *Drosophila* usually produce superior F1 progeny, and selection would rapidly disperse any preference that had arisen. It is rather the combination of inbreeding with rapid population expansion that seems to be responsible for the divergence of mating preferences. The argument usually made is that the increase in variation during population expansion coupled with the opportunity for intense selection during the subsequent crash can lead to a "genetic revolution", and the establishment of novel genotypes unique to each line. I have not been able to understand clearly what constitutes a genetic revolution, but if one were to occur I would expect it to lead to hybrid breakdown, rather than to mating preferences. Instead, crosses among the experimental lines give normal offspring in both the experiments just described. The experiments might instead be interpreted in terms of sexual selection. If the original population were segregating for genes that affected the choice of males by females, the single pair chosen to perpetuate the line might be highly unrepresentative. The particular female preference established by chance would then induce selection, during the subsequent expansion, for the preferred male type. Fisherian sexual selection on female preference would then become more intense, and lines that had by chance acquired somewhat different preferences would begin to diverge rapidly. The preferences exhibited during these experiments were relatively modest and might soon be dissipated if the lines were allowed to interbreed freely; their relevance to general theory is that they suggest how Fisherian sexual selection may participate in the rapid evolution of sexual isolation.

## *Secondary Contact between Divergent Lines*

These experiments demonstrate that some degree of isolation can evolve between lines that are kept completely separate. They suggest that hybrid inferiority is likely to arise through selection on physiological characters, and, rather more strongly, that mating preferences may evolve as a correlated response to selection on behavioral characters. Neither generalization is based on a very extensive body of experiments, but both are what common sense might lead us to expect. Theodosius Dobzhansky has suggested that the hybrid inferiority arising while the lines are separate might cause the evolution of mating preferences when they are mingled. Any tendency toward mating within lines will be reinforced by selection against the inferior offspring produced by mating between lines. The difficulty with this attractive idea is, of course, that the association between genes for specific adaptedness and genes for preferential mating will be disrupted by recombination. The experimental evidence for reinforcement remains sketchy. The experiments in artificial disruptive selection that I have described seem to show that preferences rarely evolve in freely mating populations that are initially undifferentiated. They do evolve when hybrids are deliberately eliminated, but there is then no mutual reinforcement of adaptedness and sexual isolation, because the isolation is already complete. There are very few experiments that investigate the reinforcement of partial isolation. One attempt was made by Forbes Robertson, who re-established contact between the lines that he had selected with and without EDTA added to their food, by the simple device of conecting the population cages with glass tubes. There was no sign that any substantial mating preferences had evolved 20 generations later.

## *Contingency and Pleiotropy*

There are two leading explanations for the mass of experimental work that I have summarized. One is that sexual selection leads to a highly contingent process of evolution, exaggerating whatever character first pokes its head above the parapet, so to speak. The other is that sexual isolation will evolve only when the characters involved in divergent adaptation have the pleiotropic effect of hindering some kinds of matings while facilitating others. The two are not mutually exclusive.

**Replicate Selection Lines.** If contingency is mainly responsible for sexual divergence, then it should not require divergent natural selection; independent lines selected in similar environments should show the same sort of behavior. The plainest evidence for contingency is thus the modest degrees of isolation that develop between small populations. The plainest evidence against it is that (on the scanty evidence available) replicate selection lines do not become isolated. For example, G. Kilias and his colleagues collected *Drosophila* from two localities in Greece, and then cultured their decscendants either in cold, dry conditions in the dark or in warm, humid conditions in the light. After five years in the laboratory, flies from either locality showed a rather strong preference for partners from the same line, so that, as in many other experiments, divergent selection led to some degree of sexual isolation. Flies from different localities selected in the same conditions mated freely with one another, however, showing no preference for partners from their own line. The sexual preferences displayed by flies cultured in different environments therefore represent a correlated response to selection, rather than purely historical divergence.

**Pleiotropy.** Sexual isolation is certainly a pleiotropic outcome of various behaviors that might be exposed to natural selection, and Rice and Salt's experiment provides a simple and convincing realization of this process. Whether or not these behaviors will in themselves lead to mating preferences when partners from different divergent lines are made freely available to one another is less clear. No such preferences evolved in Rice and Salt's experiment; on the other hand, a few generations of selection for phototaxis created strong mating preferences in del Solar's material. In most cases where noticeable preferences have developed, it has been through selection for behavioral characters. Divergent selection for physiological characters, on the other hand, sometimes leads to hybrid inferiority. The partial isolation caused pleiotropically through selection on behavior might reduce recombination to levels where it would permit reinforcement based on physiological characters to occur. This is perhaps too complex a theory to carry much conviction.

Experimental studies have so far failed to yield a convincing account of speciation. Although some degree of sexual isolation has been observed in several experiments, the characteristic boundaries of sexual species have not yet been created. It is my belief that we have been looking in the wrong place: the sexual habits of large and complex creatures such as *Drosophila* are simply too fixed to be altered radically in the short term. I expect that more rapid progress will be made only when we turn to organisms whose sex lives are more malleable and try to create new species of sexual microbes.

# *General Bibliography*

This is a list of some general works that are useful introductions to the theoretical literature largely unacknowledged in the text. It is not intended to be complete. It is restricted to single-author monographs, and I have not included anything earlier than Fisher and Haldane.


## General concepts

Dawkins, R. (1976). *The Selfish Gene*. Oxford University Press, Oxford, England.

Dawkins, R. (1986). *The Blind Watchmaker*. Norton, New York; Longman, London.

Gould, S.J. (1980). *The Panda's Thumb*. Norton, New York.

Gould, S.J. (1983). *Hen's Teeth and Horses' Toes*. Norton, New York.

## Selection

Endler, J.A. (1986). *Natural Selection in the Wild*. Princeton University Press, Princeton, NJ.

Sober, E. (1984). *The Nature of Selection*. MIT Press, Cambridge, MA.

Williams, G.C. (1966). *Adaptation and Natural Selection*. Princeton University Press, Princeton, NJ.

Williams, G.C. (1992). *Natural Selection. Domains, Levels, and Challenges*. Oxford University Press, Oxford, England.

## General Evolutionary Biology

Futuyma, D.J. (1986). *Evolutionary Biology*. 2nd Ed., Sinauer, Sunderland, MA.

Haldane, J.B.S. (1932). *The Causes of Evolution*. Longman, London.

Huxley, J.S. (1942). *Evolution: the Modern Synthesis*. Allen & Unwin, London.

Ridley, M. (1993). *Evolution*. Blackwell Scientific, Oxford, England.

Simpson, G.G. (1944). *Tempo and Mode in Evolution*. Columbia University Press, New York.

Simpson, G.G. (1953). *The Major Features of Evolution*. Columbia University Press, New York.

## Evolutionary Genetics

Dobzhansky, T. (1970). *Genetics of the Evolutionary Process*. Columbia University Press, New York.

Fisher, R.A., (1930). *The Genetical Theory of Natural Selection*. Oxford University Press, Oxford, England.

Ford, E.B. (1965). *Ecological Genetics*. 2nd Ed., Methuen, London; Wiley, New York.

Lewontin, R.C. (1974). *The Genetic Basis of Evolutionary Change*. Columbia University Press, New York.

Maynard Smith, J. (1989). *Evolutionary Genetics*. Oxford University Press, Oxford, England.

Nei, M. (1987). *Molecular Evolutionary Genetics*. Columbia University Press, New York.

## Theoretical Population Genetics

Bulmer, M.G. (1985). *The Mathematical Theory of Quantitative Genetics*. Clarendon Press, Oxford, England.

Crow, J.F. (1986). *Basic Concepts in Population, Quantitative and Evolutionary Genetics*. Freeman, New York.

Crow, J.F., & Kimura, M. (1970). *An Introduction to Population Genetics Theory*. Harper & Row, New York.

Falconer, D.S. (1981). *Introduction to Quantitative Genetics*. 2nd Ed., Longman, London.

Hartl, D.L. (1980). *Principles of Population Genetics*. Sinauer, Sunderland, MA.

Kimura, M. (1983). *The Neutral Theory of Molecular Evolution*. Cambridge University Press, Cambridge, England.

Lush, J.L. (1945). *Animal Breeding Plans*. The Collegiate Press, Ames, IA.

Roughgarden, J. (1979). *Theory of Population Genetics and Evolutionary Ecology*. Macmillan, New York.

Sewall Wright (1968–1977). *Evolution and the Genetics of Populations* (A Treatise in Three Volumes). The University of Chicago Press, Chicago, IL.

## Special Topics

Charlesworth, B. (1980). *Evolution in Age-structured Populations*. Cambridge University Press, Cambridge, England.

Charnov, E.L. (1982). *The Theory of Sex Allocation*. Princeton University Press, Princeton, NJ.

Maynard Smith, J. (1982). *Evolution and the Theory of Games*. Cambridge University Press, Cambridge, England.

Mayr, E. (1966). *Animal Species and Evolution*. Harvard University Press, Cambridge, MA.

Roff, D.A. (1992). *The Evolution of Life Histories. Theory and Analysis*. Chapman & Hall, London.

Endler, J.A. (1977). *Geographical Vatriation, Speciation and Clines*. Princeton University Press, Princeton, NJ.

Rose, M.R. (1991). *The Evolutionary Biology of Aging*. Oxford University Press, Oxford, England.

Stearns, S.C. (1992). *The Evolution of Life Histories*. Oxford University Press, Oxford, England.

Trivers, R. (1985). *Social Evolution*. Benjamin/Cummings, Menlo Park, CA.

Wilson, D.S. (1980). *The Natural Selection of Populations and Communities*. Benjamin/Cummings, Menlo Park, CA.

# *References*

Note: Entries are arranged in alphabetical order ignoring lower-case prefixes such as de (as in de Vilmorin), del, di or van; upper-case prefixes El-, Mac, Mc and O' are retained, Mac and Mc being separated.


Abdullah, N.F., & Charlesworth, B. (1974). Selection for reduced crossing-over in *Drosophila melanogaster*. *Genetics*, 76: 447–451.

Abrahamson, S., Meyer, H.U., Himoe, E., & Daniel, G. (1966). Further evidence demonstrating germinal selection in early premeiotic germ cells of *Drosophila* males. *Genetics*, 54: 687–696.

Acton, A.B. (1961). An unsuccessful attempt to reduce recombination by selection. *The American Naturalist*, 95: 119–120.

Adair, C.R., & Jones, J.W. (1946). Effect of environment on the characteristics of plants surviving in bulk hybrid populations of rice. *Journal of the American Society of Agronomy*, 38: 708–716.

Adams, J., & Hansche, P.E. (1974). Population studies in microorganisms. I. Evolution of diploidy in *Saccharomyces cerevisiae*. *Genetics*, 76: 327–338.

Adams, J., & Oeller, P.W. (1986). Structure of evolving populations of *Saccharomyces cerevisiae*: Adaptive changes are frequently associated with sequence alterations involving mobile elements belonging to the Ty family. *Proceedings of the National Academy of Sciences, USA*, 83: 7124–7127.

Adams, J., Paquin, C., Oeller, P.W., & Lee, L.W. (1985). Physiological characterization of adaptive clones in evolving populations of the yeast, *Saccharomyces cerevisiae*. *Genetics*, 110: 175–185.

Agar, W.E. (1913). Transmission of environmental effects from parent to offspring in *Simocephalus vetulus*. *Philosophical Transactions of the Royal Society of London Series B*, 203: 319–351.

Agar, W.E. (1914). Experiments on inheritance in parthenogenesis. *Philosophical Transactions of the Royal Society of London Series B*, 205: 421–489.

Alexander, H.M., Groth, J.V., & Roelfs, A.P. (1985). Virulence changes in *Uromyces appendiculatus* after five asexual generations on a partially resistant cultivar of *Phaseolus vulgaris*. *Phytopathology*, 75: 449–453.

Allard, R.W. (1963). Evidence for genetic restriction of recombination in the lima bean. *Genetics*, 48: 1389–1395.

Allard, R.W., & Adams, J. (1969). Population studies in predominantly self-pollinating species. XIII. Intergenotypic competition and population structure in barley and wheat. *The American Naturalist*, 103: 621–645.

Allard, R.W., Zhang, Q., Saghai Maroof, M.A., & Muona, O.M. (1992). Evolution of multilocus genetic structure in an experimental barley population. *Genetics*, 131: 957–969.

Allen, J.A. (1972). Evidence for stabilizing and apostatic selection by wild blackbirds. *Nature*, 237: 348–349.

Allen, J.A. (1974). Further evidence for apostatic selection by wild passerine birds: Training experiments. *Heredity*, 33: 361–372.

Allen, J.A. (1975). Further evidence for apostatic selection by wild passerine birds: 9:1 experiments. *Heredity*, 36: 173–180.

Allen, J.A., & Clarke, B. (1968). Evidence for apostatic selection by wild passerines. *Nature* 220: 501–502.

Andrews, K.J., & Hageman, D.G. (1976). Selective disadvantage of non-functional protein synthesis in *Escherichia coli*. *Journal of Molecular Evolution*, 8: 317–328.

Antonovics, J. (1968). Evolution in closely adjacent plant populations. V. Evolution of self-fertility. *Heredity*, 23: 219–238.

Antonovics, J., & Bradshaw, A.D. (1970). Evolution in closely adjacent plant populations. VIII. Clinal patterns at a mine boundary. *Heredity*, 23: 349–362.

Antonovics, J., & Ellstrand, N.C. (1984). Experimental sudies of the evolutionary significance of sexual reproduction. I. A test of the frequency-dependent selection hypothesis. *Evolution*, 38: 103–115.

Antonovics, J., & Primack, R.B. (1982). Experimental ecological genetics in *Plantago*. VI. The demography of seedling transplants of *P. lanceolata*. *Journal of Ecology*, 70: 55–75.

Anxolabehère, D. (1971). Sélection larvaire et fréquence génique chez *Drosophila melanogaster*. *Heredity*, 26: 9–18.

Arboleda-Rivera, F., & Compton, W.A., (1974). Differential response of maize (*Zea mays* L.) to mass selection in diverse selection environments. *Theoretical and Applied Genetics*, 44: 77–81.

Árnason, E. (1991). Perturbation-reperturbation test of selection vs. hitch-hiking of the two major alleles of *Esterase-5* in *Drosophila pseudoobscura*. *Genetics*, 129: 145–168.

Arthur, W. (1982). The evolutionary consequences of interspecific competition. *Advances in Ecological Research*, 12: 161–187.

Atchley, W.R., Rutledge, J.J., & Cowley, D.E. (1982). A multivariate statistical account of direct and correlated response to selection in the rat. *Evolution*, 36: 677–698.

Atwood, K.C., Schneider, L.K., & Ryan, F.J. (1951a). Periodic selection in *Escherichia coli*. *Proceedings of the National Academy of Sciences, USA*, 37: 146–155.

Atwood, K.C., Schneider, L.K., & Ryan, F.J. (1951b). Selective mechanisms in bacteria. *Cold Spring Harbor Symposia on Quantitative Biology*, 16: 345–355.

Austin, M.P. (1982). Use of a relative physiological performance value in the prediction of performance in multispecies mixtures from monoculture performance. *Journal of Ecology*, 70: 559–570.

Axelrod, R. (1984). *The Evolution of Cooperation*. Basic Books, New York.

Axelrod, R., & Hamilton, W.D. (1981). The evolution of cooperation. *Science*, 211: 1390–1396.

Ayala, F.J. (1966). Evolution of fitness. I. Improvement of the productivity and size of irradiated populations of *Drosophila serrata* and *Drosophila birchii*. *Genetics*, 53: 883–895.

Ayala, F.J. (1968). Genotype, environment and population numbers. *Science*, 162: 1453–1459.

Ayala, F.J. (1969). Evolution of fitness. IV. Genetic evolution of interspecific competitive ability in *Drosophila*. *Genetics*, 61: 737–747.

Ayala, F.J., & Campbell, C.A. (1976). Frequency-dependent selection. *Annual Review of Ecology and Systematics*, 5: 115–138.

Baker, G.A., Huberty, M.R., & Veihmeyer, F.J. (1950). A uniformity trial on unirrigated barley of ten years' duration. *Agronomy Journal*, 42: 267–270.

Baker, R.J., & Briggs, K.G. (1982). Effects of plant density on the performance of 10 barley cultivars. *Crop Science*, 22: 1164–1167.

Bakker, K. (1969). Selection for rate of growth and its influence on competitive ability of larvae of *Drosophila melanogaster*. *Netherlands Journal of Zoology*, 19: 541–595.

Baptist, R., & Robertson, A. (1976). Asymmetrical responses to automatic selection for body size in *Drosophila melanogaster*. *Theoretical and Applied Genetics*, 47: 209–213.

Barclay, H.J., & Gregory, P.T. (1981). An experimental test of models predicting life-history characteristics. *The American Naturalist*, 117: 944–961.

Barker, J.S.F. (1973). Natural selection for coexistence or competitive ability in laboratory populations of *Drosophila*. *Egyptian Journal of Genetics and Cytology*, 2: 288–315.

Barker, J.S.F. (1988). Population structure. In W.G. Hill & T.F.C. Mackay (eds.), *Evolution and Animal Breeding* (pp. 75–80). Commonwealth Agricultural Bureaux International.

Barker, J.S.F., & Cummins, L.J. (1969a). Disruptive selection for sternopleural bristle number in *Drosophila melanogaster*. *Genetics*, 61: 697–712.

Barker, J.S.F., & Cummins, L.J. (1969b). The effect of selection for sternopleural bristle number on mating behaviour in *Drosophila melanogaster*. *Genetics*, 61: 713–719.

Barker, J.S.F., & East, P.D. (1980). Evidence for selection following perturbation of allozyme frequencies in a natural population of *Drosophila*. *Nature*, 284: 166–168.

Barker, J.S.F., & Karlsson, L.J.E. (1974). Effects of population size and selection intensity on responses to disruptive selection in *Drosophila melanogaster*. *Genetics*, 78: 715–735.

Barnes, B.W. (1968). Stabilizing selection in *Drosophila melanogaster*. *Heredity*, 23: 433–442.

Barrett, J.A. (1981). The evolutionary consequences of monoculture. In J.A. Bishop & L.M. Cook (eds.), *Genetic Consequences of Man-made Change* (pp. 209–248). Academic Press, London.

Bateman, A.J. (1948). Intra-sexual selection in *Drosophila*. *Heredity*, 2: 349–368.

Bauer, G.J., McCaskill, J.S., & Otten, H. (1989). Travelling waves of *in vitro* evolving RNA. *Proceedings of the National Academy of Sciences, USA*, 86: 7937–7941.

Beaudry, A.A., & Joyce, G.F. (1992). Directed evolution of an RNA enzyme. *Science*, 257: 635-641.

Beg, A., Emery, D.A., & Wynne, J.C. (1975). Estimation and utilization of inter-cultivar competition in peanuts. *Crop Science*, 15: 633–637.

Bell, A.E., & Burris, M.J. (1973). Simultaneous selection for two correlated traits in *Tribolium*. *Genetical Research*, 21: 29–46.

Bell, A.E., & McNary, H.W. (1963). Genetic correlation and asymmetry of the correlated response from selection for increased body weight of *Tribolium* in two environments. *Proceedings of the XI International Congress of Genetics* (p. 256).

Bell, A.E., & Moore, C.H. (1972). Reciprocal recurrent selection for pupal weight in *Tribolium* in comparison with conventional methods. *Egyptian Journal of Genetics and Cytology,* 1: 92–119.

Bell, A.E., Moore, C.H., Bohren, B.B., & Warren, D.C. (1952). Systems of breeding designed to utilize heterosis in the domestic fowl. *Poultry Science,* 31: 11–22.

Bell, G. (1978). Further observations on the fate of morphological variation in a population of smooth newt larvae (*Triturus vulgaris*). *Journal of Zoology,* 185: 511–518.

Bell, G. (1984). Measuring the cost of reproduction. II. The correlation structure of the life tables of five freshwater invertebrates. *Evolution,* 38: 314–326.

Bell, G. (1988). *Sex and Death in Protozoa: The History of an Obsession.* Cambridge University Press, Cambridge, England.

Bell, G. (1990). The ecology and genetics of fitness in *Chlamydomonas.* I. Genotype-by-environment interaction among pure strains. *Proceedings of the Royal Society of London Series B,* 240: 295–321.

Bell, G. (1990b). The ecology and genetics of fitness in *Chlamydomonas.* II. The properties of mixtures of strains. *Proceedings of the Royal Society of London Series B,* 240: 323–350.

Bell, G. (1991a). The ecology and genetics of fitness in *Chlamydomonas.* III. Genotype-by environment interaction within strains. *Evolution,* 45: 668–679.

Bell, G. (1991b). The ecology and genetics of fitness in *Chlamydomonas.* IV. The properties of mixtures of genotypes of the same species. *Evolution,* 45: 1036–1046.

Bell, G. (1992a). The ecology and genetics of fitness in *Chlamydomonas.* V. The relationship between genetic correlation and environmental variance. *Evolution,* 46: 561–566.

Bell, G. (1992b). The emergence of gender and the nature of species in eukaryotic microbes. *Verhandlung der Deutschen Zoologischen Gesellschaft,* 85: 161–175.

Bell, G. (1992c). Five properties of environments. In P.R. Grant & H.S. Horn (eds.), *Molds, Molecules and Metazoa* (pp. 33–56). Princeton University Press, Princeton, NJ.

Bell, G. (1996a). Experimental Evolution in *Chlamydomonas.* I. Short-term selection in uniform and diverse environments. Unpublished.

Bell, G. (1996b). Experimental Evolution in *Chlamydomonas.* IV. Artificial selection for cell size in large clonal populations. Unpublished.

Bell, G., Lechowicz, M.J., Appenzeller, A., Chandler, M., Deblois, E., Jackson, L., Mackenzie, B., Preziosi, R., Schallenberg, M., & Tinker, N. (1993). The spatial structure of the physical environment. *Oecologia,* 96: 114–121.

Bell, G. and Rebound, X. (1996). Experimental Evolution in *Chlamydomonas.* II. Adaptive divergence in strongly contrasted environments. Unpublished.

Bennett, A.F., Dao, K.M., & Lenski, R.E. (1990). Rapid evolution in response to high-temperature selection. *Nature,* 346: 79–81.

Bennett, A.F., & Lenski, R.E. (1993). Evolutionary adaptation to temperature. II. Thermal niches of experimental lines of *Escherichia coli. Evolution,* 47: 1–12.

Bennett, A.F., Lenski, R.E., & Mittler, J.E. (1992). Evolutionary adaptation to temperature. I. Fitness responses of *Escherichia coli* to changes in its thermal environment. *Evolution,* 46: 16–30.

Berger, P.J. (1977). Multiple-trait selection experiments: current status, problem areas and experimental approaches. In E. Pollak, O. Kempthorne, & T.B. Bailey (eds.), *Proceedings of the International Conference on Quantitative Genetics* (pp. 191–204). Iowa State University Press, Ames, IA.

Bernard, C., & Fahmy, M.H. (1970). Effect of selection on feed utilization and carcass score in swine. *Canadian Journal of Animal Science*, 50: 575–584.

Berry, R.J., & Crothers, J.H. (1970). Stabilizing selection in the dog-whelk (*Nucella lapilus*). *Journal of Zoology*, 155: 5–17.

Bertin, R.I. (1982). Paternity and fruit production in trumpet creeper (*Campsis radicans*). *The American Naturalist*, 119: 694–709.

Beukeboom, L.W., & Werren, J.H. (1992). Population genetics of a parasitic chromosome: experimental analysis of PSR in subdivided populations. *Evolution*, 46: 1257–1268.

Biebricher, C.K. (1983). Darwinian selection of self-replicating RNA molecules. *Evolutionary Biology*, 16: 1–51.

Biebricher, C.K., Eigen, M., & Luce, R. (1981a). Kinetic analysis of template-instructed and *de novo* RNA synthesis by Q$\beta$ replicase. *Journal of Molecular Biology*, 148: 391–410.

Biebricher, C.K., Eigen, M., & Luce, R. (1981b). Product analysis of RNA generated *de novo* by Q$\beta$ replicase. *Journal of Molecular Biology*, 148: 369–380.

Biebricher, C.K., Eigen, M., & McCaskill, J.S. (1993). Template-directed and template-free RNA synthesis by Q$\beta$ replicase. *Journal of Molecular Biology*, 160: 175–179.

Biebricher, C.K., & Luce, R. (1993). Sequence analysis of RNA species synthesized by Q$\beta$ replicase without template. *Biochemistry*, 32: 321–327.

Biebricher, C.K., & Orgel, L.E. (1973). An RNA that multiplies indefinitely with DNA-dependent RNA polymerase: Selection from a random copolymer. *Proceedings of the National Academy of Sciences, USA*, 70: 934–938.

Biel, S.W., & Hartl, D.L. (1983). Evolution of transposons: natural selection for *Tn5* in *Escherichia coli* K12. *Genetics*, 103: 581–592.

Bierbaum, T.J., Mueller, L.D., & Ayala, F.J. (1989). Density-dependent evolution of life-history traits in *Drosophila melanogaster*. *Evolution*, 43: 382–392.

Birch, L.C. (1955). Selection in *Drosophila pseudoobscura* in relation to crowding. *Evolution*, 9: 389–399.

Birley, A.J., & Haley, C.S. (1987). The genetical response to natural selection by varied environments. IV. Gametic disequilibrium in spatially varied environments. *Genetics*, 115: 295–303.

Blijenburg, J.G., & Sneep, J. (1975). Natural selection in a mixture of eight barley varieties grown in six successive years. I. Competition between the varieties. *Euphytica*, 24: 305–315.

Boerma, H.R., & Cooper, R.L. (1975). Performance of pure lines obtained from superior-yielding heterogeneous lines in soybeans. *Crop Science*, 15: 300–302.

Bonnier, G. (1895). Recherches expérimentales sur l'adaptation des plantes au climat alpin. *Revue Générale du Botanie*, 20: 217–358.

Borisov, V.M. (1978). The selective effect of fishing on the population structure of species with a long life cycle. *Journal of Ichthyology*, 18: 896–904.

Bos, M., & Scharloo, W. (1973a). The effects of disruptive and stabilizing selection on body size in *Drosophila melanogaster*. I. Mean values and variances. *Genetics*, 75: 679–693.

Bos, M., & Scharloo, W. (1973b). The effects of disruptive and stabilizing selection on body size in *Drosophila melanogaster*. II. Analysis of responses in the thorax selection lines. *Genetics*, 75: 695–708.

Bos, M., & Scharloo, W. (1974). The effects of disruptive and stabilizing selection on body size in *Drosophila melanogaster*. III. Genetic analysis of two lines with different reactions to disruptive selection with mating of opposite extremes. *Genetica*, 45: 71–90.

Bouma, J.E., & Lenski, R.E. (1988). Evolution of a bacteria/plasmid association. *Nature*, 335: 351–352.

Boyce, M.S. (1984). Restitution of r- and K-selection as a model of density-dependent natural selection. *Annual Reviews of Ecology and Systematics*, 15: 427–447.

Boyce, M.S., & Perrins, C.M. (1987). Optimizing Great Tit clutch size in a fluctuating environment. *Ecology*, 68: 142–153.

Brettel, R.I.S., & Ingram, D.S. (1979). Tissue culture in the production of novel disease resistant crop plants. *Biological Reviews*, 54: 329–345.

Bridge, R.R., & Meredith, W.R. (1983). Comparative performance of obsolete and current cotton cultivars. *Crop Science*, 23: 231–237.

Bridge, R.R., Meredith, W.R., & Chism, J.F. (1971). Comparative performance of obsolete varieties and current varieties of upland cotton. *Crop Science*, 11: 29–32.

Brown, W.P., & Bell, A.E. (1961). Genetic analysis of a "plateaued" population of *Drosophila melanogaster*. *Genetics*, 46: 407–425.

Browning, J.A., & Frey, K.J. (1969). Multiline cultivars as a means of disease control. *Annual Reviews of Phytopathology*, 7: 355–382.

Brumpton, R.J., Boughey, H., & Jinks, J.L. (1977). Joint selection for both extremes of mean performance and of sensitivity to a macroenvironmental variable. I. Family selection. *Heredity*, 38: 219–226.

Bryant, E.H., & Turner, C.R. (1972). Rapid evolution of competitive ability in larval mixtures of the housefly. *Evolution*, 26: 161–170.

Bull, J.J., & Molineux, I.J. (1992). Molecular genetics of adaptation in an experimental model of cooperation. *Evolution*, 46: 882–895.

Bull, J.J., Molineux, I.J., & Rice, W.R. (1991). Selection of benevolence in a host-parasite system. *Evolution*, 45: 875–882.

Bumpus, H. (1899). The elimination of the unfit as illustrated by the introduced sparrow, *Passer domesticus*. *Marine Biological Laboratory (Wood's Hole) Biology Lectures* 1898: 209–228.

Bundegaard, J., & Christiansen, F.B. (1972). Dynamics of polymorphisms. I. Selection components in an experimental population of *Drosophila melanogaster*. *Genetics*, 71: 439–460.

Burdon, J.J. (1987). *Diseases and Plant Population Biology*. Cambridge University Press, Cambridge, England.

Buri, P. (1956). Gene frequency in small populations of mutant *Drosophila*. *Evolution*, 10: 367–402.

Burleigh, B.D., Rigby, P.W.J., & Hartley, B.S. (1974). A comparison of wild-type and mutant ribitol dehydrogenases from *Klebsiella aerogenes*. *Biochemical Journal*, 143: 341–352.

Burnet, B., & Connolly, K. (1974). Activity and sexual behaviour in *Drosophila melanogaster*. In J.H.F. van Abeelen (ed.), *The Genetics of Behaviour* (pp. 201–258). North Holland, Amsterdam.

Burt, A., & Bell, G. (1992). Tests of sib diversification theories of outcrossing in *Impatiens capensis*: effects of inbreeding and neighbour relatedness on production and infestation. *Journal of Evolutionary Biology*, 5: 575–588.

Bush, G.L. (1969). Sympatric host race formation and speciation in frugivorous flies of the genus *Rhagoletis*. *Evolution*, 23: 237–251.

Buss, L. (1987). *The Evolution of Individuality*. Princeton University Press, Princeton, NJ.

Buss, L., & Grosberg R.K. (1990). Morphogenetic basis for phenotypic differences in hydroid competitive behavior. *Nature*, 343: 63–66.

Buzzati-Traverso, A.A. (1955). Evolutionary changes in components of fitness and other polygenic traits in *Drosophila melanogaster* populations. *Heredity*, 9: 153–186.

Cain, A.J. & Sheppard, P.M. (1954). Natural selection in *Cepaea*. *Genetics*, 39: 89–116.

Cairns, J., Overbaugh, J., & Miller, S. (1988). The origin of mutants. *Nature*, 335: 142–145.

Calef, E. (1957). Effects on linkage maps of selection of crossovers between closely linked markers. *Heredity*, 11: 265–279.

Calhoon, R.E., & Bohren, B.B. (1974). Genetic gains from reciprocal recurrent and within-line selection for egg production in the fowl. *Theoretical and Applied Genetics*, 44: 364–372.

Calkins, G.F. (1919). *Uroleptus mobilis* Engelm. II. Renewal of vitality through conjugation. *Journal of Experimental Zoology*, 29: 121–156.

Calkins, G.F. (1920). *Uroleptus mobilis* Engelm. III. A study in vitality. *Journal of Experimental Zoology*, 31: 287–305.

Campbell, B. (ed.). (1972). *Sexual Selection and the Descent of Man*. Aldine, Chicago, IL.

Campo, J.L., & Tagarro, P. (1977). Comparison of three selection methods for pupal weight of *Tribolium castaneum*. *Annals of Genetics and Selection in Animals*, 9: 259–268.

Campo, J.L., & Velasco, T. (1989). An experimental test of optimum desired-gains indexes in *Tribolium*. *Journal of Heredity*, 80: 48–52.

Carlson, P.S. (1973). Methionine-sulfoximine-resistant mutants of tobacco. *Science*, 180: 1366–1368.

Carson, H.L. (1964). Population size and genetic load in irradiated populations of *Drosophila melanogaster*. *Genetics*, 49: 521–528.

Castle, W.E., & Phillips, J.C. (1914). Piebald rats and selection. An experimental test of selection and of the theory of gametic purity in Mendelian crosses. *Carnegie Institute of Washington Publication*, 195: 1–31.

Castleberry, R.M., Crum, C.W., & Krull, C.F. (1984). Genetic yield improvement of U.S. maize cultivars under varying fertility and climatic environments. *Crop Science*, 24: 33–36.

Cavers, P.B., & Harper, J.L. (1967). Studies in the dynamics of plant populations. I. The fate of seed and transplants introduced into various habitats. *Journal of Ecology*, 55: 59–71.

Cavicchi, S., Guerra, D., Natali, V., Pezzoli, C., & Giorgi, G. (1989). Temperature-related divergence in experimental populations of *Drosophila melanogaster*. II. Correlation between fitness and body dimensions. *Journal of Evolutionary Biology*, 2: 235–251.

di Cesnola, A.P. (1907). A first study of natural selection in *Helix arbustorum* (Helicogena). *Biometrika*, 5: 387–399.

Chabora, A.J. (1968). Disruptive selection for sternopleural chaeta number in various strains of *Drosophila melanogaster*. *The American Naturalist*, 102: 525–532.

Chaleff, R.S., & Parsons, M.F. (1978). Direct selection *in vitro* for herbicide resistant mutants of *Nicotiana tabacum*. *Proceedings of the National Academy of Sciences, USA*, 75: 5104–5107.

Chambers, G.K. (1988). The *Drosophila* alcohol dehydrogenase gene-enzyme system. *Advances in Genetics*, 25: 39–107.

Chao, L., & Cox, E.C. (1983). Competition between high and low mutating strains of *Escherichia coli*. *Evolution*, 37: 125–134.

Chao, L., & McBroom, S.M. (1985). Evolution of transposable elements: an *IS*10 insertion increases fitness in *E. coli*. *Molecular Biology and Evolution*, 2: 359–369.

Chao, L., Vargas, C., Spear, B.B., & Cox, E.C. (1983). Transposable elements as mutator genes in evolution. *Nature*, 303: 633–635.

Charlesworth, B. (1987). The population biology of transposable elements. *Trends in Ecology and Evolution*, 2: 21–23.

Charlesworth, B., & Charlesworth, D. (1983). The population dynamics of transposable elements. *Genetical Research*, 42: 1–27.

Charlesworth, B., & Charlesworth, D. (1985). Genetic variation in recombination in *Drosophila*. I. Responses to selection and preliminary genetic analysis. *Heredity*, 54: 71–83.

Charlesworth, B., & Langley, C. (1989). The population genetics of *Drosophila* transposable elements. *Annual Reviews of Genetics*, 23: 251–287.

Charlesworth, B., Sniegowski, P., & Stephan, W. (1994). The evolutionary dynamics of repetitive DNA in eukaryotes. *Nature*, 371: 215–220.

Cheplick, G.P. (1988). Influence of environment and population origin on survivorship and reproduction in reciprocal transplants of amphicarpic peanutgrass (*Amphicarpum purshii*). *American Journal of Botany*, 75: 1048–1056.

Chetverin, A.B., Chetverina, H.V., & Munishkin, A.V. (1991). On the nature of spontaneous RNA synthesis by Q$\beta$ replicase. *Journal of Molecular Biology*, 222: 3–9.

Cheung, T.K., & Parker, R.J. (1974). Effect of selection on heritability and genetic correlation of two quantitative traits in mice. *Canadian Journal of Genetics and Cytology*, 16: 599–609.

Chin, K.M., & Wolfe, M.S. (1984). Selection on *Erisyphe graminis* in pure and mixed stands of barley. *Plant Pathology*, 33: 535–546.

Chinnici, J.P. (1971). Modification of recombination frequency in *Drosophila*. I. Selection for increased and decreased crossing-over. *Genetics*, 69: 71–83.

Chippindale, A.K., Leroi, A.M., Kim, S.B., & Rose, M.R. (1993). Phenotypic plasticity and selection in *Drosophila* life-history evolution. I. Nutrition and the cost of reproduction. *Journal of Evolutionary Biology*, 6: 171–193.

Chung, C.S., & Chapman, A.B. (1958). Comparisons of the predicted with actual gains in selection of parents of inbred progeny of rats. *Genetics*, 43: 594–600.

Clare, M.J., & Luckinbill, L.S. (1985). The effect of gene-environment interaction on the expression of longevity. *Heredity*, 55: 19–29.

Clarke, B. (1979). The evolution of genetic diversity. *Proceedings of the Royal Society of London Series B*, 205: 453–474.

Clarke, J.M., Maynard Smith, J., & Sondhi, K.C. (1961). Asymmetrical response to selection for rate of development in *Drosophila subobscura*. *Genetical Research*, 2: 70–81.

Clarke, P.H. (1978). Experiments in microbial evolution. In T.C. Gunsalos, L.N. Ornston, & J.R. Sokatch (eds.), *The Bacteria* (Vol. VI, pp. 137–218). Academic Press, New York.

Clarke, P.H. (1983). Experimental evolution. In D.S. Bendall (ed.), *Evolution from Molecules to Men* (pp. 235–252). Cambridge University Press, Cambridge, England.

Clarke, P.H. (1984). Amidases of *Pseudomonas aeruginosa*. In R.P. Mortlock (ed.), *Microorganisms as Model Systems for Studying Evolution* (pp. 187–232). Plenum Press, New York.

Clatworthy, J.N., & Harper, J.L. (1962). The comparative biology of closely related species living in the same area. V. Inter- and intraspecific interference within cultures of *Lemna* spp. and *Salvinia natans*. *Journal of Experimental Botany*, 13: 307–324.

Clausen, J., Keck, D.D., & Hiesey, W.M. (1940). Experimental studies on the nature of species. I. Effect of varied environments on Western North American plants. *Carnegie Institute of Washington Publication*, 520.

Clayton, G., & Robertson, A. (1955). Mutation and quantitative variation. *American Naturalist*, 89: 151–158.

Clayton, G.A., Knight, G.R., Morris, J.A., & Robertson, A. (1957). An experimental check on quantitative genetic theory. III. Correlated responses. *Journal of Genetics*, 55: 171–180.

Clayton, G.A., Morris, J.A., & Robertson, A. (1957). An experimental check on quantitative genetical theory. I. Short-term responses to selection. *Journal of Genetics*, 55: 131–151.

Clayton, G.A., & Robertson, A. (1957). An experimental check on quantitative genetical theory. II. The long-term effects of selection. *Journal of Genetics*, 55: 152–170.

Cohan, F.M. (1984). Genetic divergence under uniform selection. I. Similarity among populations of *Drosophila melanogaster* in their responses to artificial selection for modifiers of $ci^D$. *Evolution*, 38: 55–71.

Cohan, F.M., Hoffman, A.A., & Gayley, T.W. (1989). A test of the role of epistasis in divergence under uniform selection. *Evolution*, 43: 766–774.

Compton, W.A., Mumm, R.F., & Mathema, B. (1979). Progress from adaptive mass selection in incompletely adapted maize populations. *Crop Science*, 19: 531–533.

Condit, R. (1990). The evolution of transposable elements: conditions for establishment in bacterial populations. *Evolution*, 44: 347–359.

Conover, D.O., & Van Voorhees, D.A. (1990). Evolution of a balanced sex-ratio by frequency-dependent selection in a fish. *Science*, 250: 1556–1558.

Conover, D.O., Van Voorhees, D.A., & Ehtisham, A. (1992). Sex ratio selection and the evolution of environmental sex determination in laboratory populations of *Menidia menidia*. *Evolution*, 46: 1722–1730.

Cook, L.M., & Miller, P. (1977). Density-dependent selection on polymorphic prey—some data. *The American Naturalist*, 111: 594–598.

Cook, S.C.A., Lefébvre, C., & McNeilly, T. (1972). Competition between metal tolerant and normal plant populations on normal soil. *Evolution*, 26: 366–372.

Cox, E.C. (1976). Bacterial mutator genes and the control of spontaneous mutation. *Annual Reviews of Genetics*, 10: 135–156.

Cox, E.C., & Gibson, T.C. (1974). Selection for high mutation rates in chemostats. *Genetics*, 77: 169–184.

Craig, D.M. (1982). Group selection versus individual selection: an experimental analysis. *Evolution*, 36: 271–282.

Crampton, H.E. (1904). Experimental and statistical studies upon lepidoptera. I. Variation and elimination in *Philosamia cynthia*. *Biometrika*, 3: 113–130.

Crenshaw, J.W. (1965). Radiation-induced increases in fitness in the flour beetle *Tribolium confusum*. *Science*, 149: 426–427.

Crossley, S.A. (1974). Changes in mating behavior produced by selection for ethological isolation between *Ebony* and *Vestigial* mutants of *Drosophila melanogaster*. *Evolution*, 28: 631–647.

Daniels, S.B., Clark, S.H., Kidwell, M.G., & Chovnick, A. (1987). Genetic transformation of *Drosophila melanogaster* with an autonomous *P* element: phenotypic and molecular analyses of long-established transformed lines. *Genetics*, 115: 711–723.

Darlington, C.D. (1958). *Evolution of Genetic Systems*. Oliver and Boyd, Edinburgh.

David, J., Boucquet, C., Fouillet, P., & Arens, M. (1977). Tolérance génétique a l'alcool chez *Drosophila*: comparaison des effets de la séléction chez *Drosophila melanogaster* et *Drosophila simulans*. *Comptes rendus de l'Académie des Sciences de Paris*, 285: 405–408.

Davies, M.S., & Snaydon R.W. (1973a). Physiological differences among populations of *Anthoxanthum odoratum* L. collected from the Park Grass Experiment, Rothamsted. I. Response to calcium. *Journal of Applied Ecology*, 10: 33–45.

Davies, M.S., & Snaydon, R.W. (1973b). Physiological differences among populations of *Anthoxanthum odoratum* L. collected from the Park Grass Experiment, Rothamsted. II. Response to aluminium. *Journal of Applied Ecology*, 10: 47–55.

Davies, M.S., & Snaydon, R.W. (1973c). Physiological differences among populations of *Anthoxanthum odoratum* L. collected from the Park Grass Experiment, Rothamsted. III. Response to phosphorus. *Journal of Applied Ecology*, 10: 699–707.

Davies, M.S., & Snaydon, R.W. (1976). Rapid population differentiation in a mosaic environment. III. Measures of selection pressures. *Heredity*, 36: 59–66.

Davis, B.K. (1991). Kinetics of rapid RNA evolution *in vitro*. *Journal of Molecular Evolution*, 33: 343–356.

Dawood, M.M., & Strickberger, M.W. (1969). The effect of larval interaction on viability in *Drosophila melanogaster*. III. Effects of biotic residues. *Genetics*, 63: 213–220.

Dawson, P.S. (1965). Genetic homeostasis and developmental rate in *Tribolium*. *Genetics*, 51: 873–885.

Dawson, P.S. (1972). Evolution in mixed populations of *Tribolium*. *Evolution*, 26: 357–365.

Dean, A.M. (1989). Selection and neutrality in lactose operons of *Escherichia coli*. *Genetics*, 123: 441–454.

Dean, A.M., Dykhuizen, D.E., & Hartl, D.L. (1986). Fitness as a function of $\beta$-galactosidase activity in *Escherichia coli*. *Genetical Research*, 48: 1–8.

van Delden, W. (1970). Selection for competitive ability. *Drosophila Information Service*, 45: 169.

van Delden W. (1982). The alcohol dehydrogenase polymorphism in *Drosophila melanogaster*: selection at an enzyme locus. *Evolutionary Biology*, 15: 187–222.

Detlefson, J.A., & Roberts, E. (1921). Studies on crossing-over. I. The effect of selection on crossover values. *Journal of Experimental Zoology*, 32: 333–354.

Dewees, A.A. (1970). Two-way selection for recombination rates in *Tribolium castaneum*. *Genetics*, 64 (Suppl.): s16–s17.

Dickerson, G.E. (1955). Genetic slippage in response to selection for multiple objectives. *Cold Spring Harbor Symposia in Quantitative Biology*, 20: 213–224.

van Dijken, F.R., & Scharloo, W. (1979a). Divergent selection on locomotor activity in *Drosophila melanogaster*. I. Selection response. *Behavior Genetics*, 9: 543–553.

van Dijken, F.R., & Scharloo, W. (1979b). Divergent selection on locomotor activity in *Drosophila melanogaster*. II. Test for reproductive isolation between selected lines. *Behavior Genetics*, 9: 555–561.

Dingle, H., Evans, K.E., & Palmer, J.O. (1988). Responses to selection among life-history traits in a non-migratory population of milkweed bugs (*Oncopeltus fasciatus*). *Evolution*, 42: 79–92.

Dobzhansky, T., & Pavlovsky, O. (1971). Experimentally created incipient species of *Drosophila*. *Nature*, 230: 289–292.

Dobzhansky, T., Hunter, A.S., Pavlovsky, O., Spassky, B., and Wallace, B. (1963). Genetics of natural populations. XXXI. Genetics of an isolated marginal population of *D. pseudoobscura*. *Genetics*, 48: 91–103.

Dobzhansky, T., & Spassky, B. (1947). Evolutionary changes in laboratory cultures of *Drosophila pseudoobscura*. *Evolution*, 1: 191–216.

Dodd, D.M.B. (1989). Reproductive isolation as a consequence of adaptive divergence in *Drosophila pseudoobscura*. *Evolution*, 43: 1308–1311.

Dodd, D.M.B., & Powell, J.R. (1985). Founder-flush speciation: an update of experimental results with *Drosophila*. *Evolution*, 39: 1388–1392.

Dolan, R., & Robertson, A. (1975). The effect of conditioning the medium in *Drosophila*, in relation to frequency-dependent selection. *Heredity*, 35: 311–316.

Donald, C.M., & Hamblin, J. (1976). The convergent evolution of annual seed crops in agriculture. *Advances in Agronomy*, 36: 97–143.

Doney, D., Plaisted, R.L., & Peterson, L.C., (1965). Genotypic competition in progeny performance evaluation of potatoes. *Crop Science*, 5: 433–435.

Doyle, R.W., & Hunte, W. (1981a). Demography of an estuarine amphipod (*Gammarus lawrencianus*) experimentally selected for high "r": a model of the genetic effects of environmental change. *Canadian Journal of Fisheries and Aquatic Sciences*, 38: 1120–1127.

Doyle, R.W., & Hunte, W. (1981b). Genetic changes in "fitness" and yield of a crustacean population in a controlled environment. *Journal of Experimental Marine Biology and Ecology*, 52: 147–156.

Drickamer, L.C. (1981). Selection for age of sexual maturity in mice and the consequences for population regulation. *Behavioral and Neural Biology*, 31: 82–89.

Druger, M. (1967). Selection and the effect of temperature on scutellar bristle number in *Drosophila*. *Genetics*, 56: 39–47.

Dudley, J.W. (1977). 76 generations of selection for oil and protein percentage in maize. In E. Pollak, O. Kempthorne, & T.B. Bailey (eds.), *Proceedings of the International Conference on Quantitative Genetics* (pp 459–474). Iowa State University Press, Ames, IA.

Dudley, J.W., & Lambert, R.J. (1992). Ninety generations of selection for oil and protein in maize. *Maydica*, 37: 81–87.

Dudley, J.W., Lambert, R.J., & Alexander, D.E. (1974). Seventy generations of selection for oil and protein concentration in the maize kernel. In J.W. Dudley (ed.), *Seventy Generations of Selection for Oil and Protein in Maize* (pp. 181–212). Crop Science Society of America, Madison, WI.

Dunbar, M.J. (1960). The evolution of stability in marine environments: natural selection at the level of the ecosystem. *The American Naturalist*, 94: 129–136.

Dyke, G., George, B.J., Johnsyon, A.E., Poulton, P.R., & Todd, A.D. (1983). The Broadbalk Wheat experiment 1968–1978: Yields and plant nutrients in crops grown continuously and in rotation. *Rothamsted Experimental Station Report for 1982*, Pt. 2: 5–44.

Dykhuizen, D.E. (1978). Selection for tryptophan auxotrophs of *Escherichia coli* in glucose-limited chemostats as a test of the energy conservation hypothesis of evolution. *Evolution*, 32: 125–150.

Dykhuizen, D.E. (1990). Experimental studies of natural selection in bacteria. *Annual Reviews of Ecology and Systematics*, 21: 373–398.

Dykhuizen, D.E., & Davies, M. (1980). An experimental model: Bacterial specialists and generalists competing in chemostats. *Ecology*, 61: 1213–1227.

Dykhuizen, D.E., & Dean, A.M. (1990). Enzyme activity and fitness: evolution in solution. *Trends in Ecology and Evolution*, 5: 257–262.

Dykhuizen, D.E., Dean, A.M., & Hartl, D.L. (1987). Metabolic flux and fitness. *Genetics*, 114: 25–31.

Dykhuizen, D.E., & Hartl, D.L. (1980). Selective neutrality of 6PGD allozymes in *E. coli* and the effects of genetic background. *Genetics*, 96: 801–817.

Dykhuizen, D.E., & Hartl, D.L. (1983). Functional effects of PGI allozymes in *Escherichia coli*. *Genetics*, 105: 1–18.

Eagles, C.F. (1983). Relationship between competitive ability and yielding ability in mixtures and monocultures of populations of *Dactylis glomerata* L. *Grass and Forage Science*, 38: 21–24.

Ebinuma, H. (1987). Selective recombination system in *Bombyx mori*. I. Chromosome specificity of the modification effect. *Genetics*, 117: 521–531.

Ebinuma, H., & Yoshitake, N. (1981). The genetic system controlling recombination in the silkworm. *Genetics*, 99: 231–245.

Edley, M.T., & Law, R. (1988). Evolution of life histories and yields in experimental populations of *Daphnia magna*. *Biological Journal of the Linnean Society*, 34: 309–326.

Edlin, G., Lin, L., & Bitner, R. (1977). Reproductive fitness of P1, P2 and mu lysogens. *Journal of Virology*, 21: 560–564.

Ehrman, L., White, M.M., & Wallace, B. (1991). A long-term study involving *Drosophila melanogaster* and toxic media. *Evolutionary Biology*, 25: 175–209.

Eigen, M. (1983). Self-replication and molecular evolution. In D.S. Bendall (ed.), *Evolution from Molecules to Men* (pp. 105–130). Cambridge University Press, Cambridge, England.

Eisen, E.J. (1972). Long-term response for 12-day litter weight in mice. *Genetics*, 72: 129–142.

Eisen, E.J. (1978). Single trait and antagonistic index selection for litter size and body weight in mice. *Genetics*, 88: 781–811.

Eisen, E.J. (1980). Conclusions from long-term selection experiments with mice. *Journal of Animal Breeding and Genetics*, 97: 305–319.

Eisen, E.J. (1992). Restricted index selection in mice designed to change body fat without changing body weight: direct responses. *Theoretical and Applied Genetics*, 83: 973–980.

Elgin, J.H., & Ostazewski, S.A. (1982). Evaluation of selected alfalfa cultivars and related *Medicago* species for resistance to race 1 and race 2 of anthracnose. *Crop Science*, 22: 39–42.

El-Lakany, M.A., & Russell, W.A. (1971). Relationship of maize characters with yield in testcrosses of inbreds at different plant densities. *Crop Science*, 11: 698–701.

Ellis, R.E., Yuan, J., & Horvitz, H.R. (1991). Mechanisms and functions of cell death. *Annual Reviews of Cell Biology*, 7: 663–698.

Endler, J.A. (1986). *Natural Selection in the Wild*. Princeton University Press, Princeton, NJ.

Enfield, F.D. (1977). Selection experiments in *Tribolium* designed to look at gene action issues. In E. Pollak, O. Kempthorne, & T.B. Bailey (eds.), *Proceedings of the International Conference on Quantitative Genetics* (pp. 177–190). Iowa State University Press, Ames, IA.

Enfield, F.D. (1982). Long-term effects of selection; the limits to response. In A. Robertson (ed.), *Selection Experiments in Domestic Animals* (pp. 69–81). Commonwealth Agricultural Bureaux, Wallingford, U.K.

Enfield, F.D., Comstock, R.E., & Braskerud, O. (1966). Selection for pupa weight in *Tribolium castaneum*. I Parameters in base populations. *Genetics*, 54: 523–533.

Englert, D.C., & Bell, A.E. (1970). Selection for time of pupation in *Tribolium castaneum*. *Genetics*, 64: 541–552.

Engström, G., Liljedahl, L.-E., & Björklund, T. (1992). Expression of genetic and environmental variation during ageing. 2. Selection for increased lifespan in *Drosophila melanogaster*. *Theoretical and Applied Genetics*, 85: 26–32.

Ewing, H.E. (1914). Notes on regression in a pure line of plant lice. *Biological Bulletin*, 27: 164–168.

Falconer, D.S. (1953). Selection for large and small size in mice. *Journal of Genetics*, 51: 470–501.

Falconer, D.S. (1955). Patterns of response in selection experiments with mice. *Cold Spring Harbor Symposia on Quantitative Biology*, 20: 178–196.

Falconer, D.S. (1957). Selection for phenotypic intermediates in *Drosophila*. *Journal of Genetics*, 55: 551–561.

Falconer, D.S. (1960). Selection of mice for growth on high and low planes of nutrition. *Genetical Research*, 1: 91–113.

Falconer, D.S. (1971). Improvement of litter size in a strain of mice at a selection limit. *Genetical Research*, 17: 215–235.

Falconer, D.S. (1973). Replicated selection for body weight in mice. *Genetical Research*, 22: 291–321.

Falconer, D.S. (1990). Selection in different environments: effects on environmental sensitivity (reaction norm) and on mean performance. *Genetical Research*, 56: 57–70.

Falconer, D.S., & King, J.W.B. (1953). A study of selection limits in the mouse. *Journal of Genetics*, 51: 561–581.

Falconer, D.S., & Latyszewski, M. (1952). The environment in relation to selection for size in mice. *Journal of Genetics*, 51: 67–80.

Falk, R. (1961). Are induced mutations in *Drosophila* overdominant? II. Experimental results. *Genetics*, 46: 737–757.

Fatunla, T., & Frey, K.J. (1974). Stability indexes of radiated and nonradiated oat genotypes propagated in bulk populations. *Crop Science*, 14: 719–724.

Favro, L.D., Kuo, P.K., & McDonald, J.F. (1979). Population-genetic study of the effects of selective fishing on the growth-rate of trout. *Journal of the Fisheries Research Board of Canada*, 36: 552–561.

Feaster, C.V., Young, E.F., & Turcotte, E.L. (1980). Comparison of artificial and natural selection in American Pima cotton under different environments. *Crop Science*, 20: 555–558.

Finckh, M.R., & Mundt, C.C. (1993). Effects of stripe rust on the evolution of genetically diverse wheat populations. *Theoretical and Applied Genetics*, 85: 809–821.

Fisher, R.A. (1930). *The Genetical Theory of Natural Selection*. Oxford University Press, Oxford, England.

Flexon, P.B., & Rodell, C.F. (1982). Genetic recombination and directional selection for DDT resistance in *Drosophila melanogaster*. *Nature*, 298: 672–674.

Ford, E.B. (1940). Genetic research in the Lepidoptera. *Annals of Eugenics*, 10: 227–252.

Foster, P.L., & Trimarchi, J.M. (1994). Adaptive reversion of a frameshift mutation in *Escherichia coli* by simple base deletions in homopolymeric runs. *Science*, 265: 407–409.

Fowler, R.E., & Edwards, R.G. (1960). The fertility of mice selected for large or small body size. *Genetical Research*, 1: 393–407.

Fox, S.F. (1975). Natural selection on morphological phenotypes of the lizard *Uta stanburiana*. *Evolution*, 29: 95–107.

Frahm, R.R., & Brown, M.M. (1975). Selection for increased pre-weaning and post-weaning weight gain in mice. *Journal of Animal Science*, 41: 33–42.

Frahm, R.R., & Kojima, K. (1966). Comparison of selection responses on body weight under divergent larval density conditions in *Drosophila pseudoobscura*. *Genetics*, 54: 625–637.

Francis, J.C., & Hansche, P.E. (1972). Directed evolution of metabolic pathways in microbial populations. I. Modification of the acid phosphatase pH optimum in *Saccharomyces cerevisiae*. *Genetics*, 70: 59–73.

Francis, J.C., & Hansche, P.E. (1973). Directed evolution of metabolic pathways in microbial populations. II. A repeatable adaptation in *Saccharomyces cerevisiae*. *Genetics*, 74: 259–265.

Frankham, R. (1977). Optimum selection intensities in artificial selection programmes: an experimental evaluation. *Genetical Research*, 30: 115–119.

Frankham, R. (1982). Origin of genetic variation in selection lines. In A. Robertson (ed.), *Selection Experiments in Laboratory and Domestic Animals* (pp. 56–68). Commonwealth Agricultural Bureaux, Wallingford, U.K.

Frankham, R. (1990). Are responses to artificial selection for reproductive fitness characters consistently asymmetrical? *Genetical Research*, 50: 35–42.

Frankham, R., Jones, L.P., & Barker, J.S.F. (1968a). The effects of population size and selection intensity for a quantitative characters in *Drosophila*. I. Short-term response to selection. *Genetical Research*, 12: 237–248.

Frankham, R., Jones, L.P., & Barker, J.S.F. (1968b). The effects of population size and selection intensity for a quantitative characters in *Drosophila*. III. Analyses of the lines. *Genetical Research*, 12: 267–283.

Frankham, R., Yoo, B.H., & Sheldon, B.L. (1988). Reproductive fitness and artificial selection in animal breeding: culling on fitness prevents a decline in reproductive fitness in lines of *Drosophila melanogaster* selected for increased inebriation time. *Theoretical and Applied Genetics*, 76: 909–914.

Fredeen, H.T., & Mikami, H. (1986a). Mass selection in a pig population: Experimental design and responses to direct selection for rapid growth and minimum fat. *Journal of Animal Science*, 62: 1492–1508.

Fredeen, H.T., & Mikami, H. (1986b). Mass selection in a pig population: Realized heritabilities. *Journal of Animal Science*, 62: 1509–1522.

Frey, K.J., & Maldonado, U. (1967). Relative productivity of homogeneous and heterogeneous oat cultivars in optimum and sub-optimum environments. *Crop Science*, 7: 532–535.

de la Fuente, L.F., & San Primitivo, F. (1985). Selection for large and small litter size of the first three litters in mice. *Génétique, Sélection, Évolution*, 17: 251–264.

Futcher, B., Reid, E., & Hickey, D.A. (1988). Maintenance of the 2 $\mu$m circle plasmid of *Saccharomyces cerevisiae* by sexual transmission: an example of a selfish DNA. *Genetics*, 118: 411–415.

Futuyma, D.J. (1970). Variation in genetic response to intraspecific competition in laboratory populations of *Drosophila*. *The American Naturalist*, 104: 239–252.

Galiana, A., Moya, A., & Ayala, F.J. (1993). Founder-flush speciation in *Drosophila pseudoobscura*: A large-scale experiment. *Evolution*, 47: 432–444.

Gall, G.A.E. (1971). Replicated selection for 21-day pupal weight of *Tribolium castaneum*. *Theoretical and Applied Genetics*, 41: 164–173.

Gall, G.A.E., & Kyle, W.H. (1968). Growth of the laboratory mouse. *Theoretical and Applied Genetics*, 38: 304–308.

Garcia-Dorado, A., & López-Fanjul, C. (1983). Accumulation of lethals in highly selected lines of *Drosophila melanogaster*. *Theoretical and Applied Genetics*, 66: 221–223.

Gardner, C.O. (1961). An evaluation of effects of mass selection and seed irradiation with thermal neutrons on yield of corn. *Crop Science*, 1: 241–245.

Garwood, V.A., Lowe, P.C., & Bohren, B.B. (1980). An experimental test of the efficiency of family selection in chickens. *Theoretical and Applied Genetics*, 56: 5–9.

Gause, G.F. (1934). *The Struggle for Existence*. Williams & Wilkins, New York. (Reprinted 1971 by Dover, New York.)

Gengenbach, B.G., Green, C.E., & Donovan, C.M. (1977). Inheritance of selected pathotoxin-resistance in maize plants regenerated from cell cultures. *Proceedings of the National Academy of Sciences, USA*, 74: 5113–5117.

Genter, C.F. (1976). Mass selection in a composite of intercrosses of Mexican races of maize. *Crop Science*, 16: 556–558.

Gibson, J.B., & Bradley, B.P. (1974). Stabilizing selection in constant and fluctuating environments. *Heredity*, 33: 293–302.

Gibson, T.C., Scheppe, M.L., & Cox, E.C. (1970). Fitness of an *Escherichia coli* mutator gene. *Science*, 169: 686–688.

Gilinsky, N.L. (1986). Species selection as a causal process. *Evolutionary Biology*, 20: 249–273.

Gill, K.S., Nanda, G.S., & Singh, G. (1984). Stability analysis over seasons and locations of multilines of wheat (*Triticum aestivum* L.). *Euphytica*, 33: 489–495.

Gillespie, J.H. (1977). Natural selection for variance in offspring numbers: a new evolutionary principle. *The American Naturalist*, 111: 1010–1014.

Godfrey, M. (1968). Ten generations of selection for lysine utilization in Japanese quail. *Poultry Science*, 47: 1559–1560.

Goldstein, A.H. (1991). Plant cells selected for resistance to phosphate starvation show enhanced P use efficiency. *Theoretical and Applied Genetics*, 82: 191–194.

Good, A.G., Meister, G.A., Brock, H.W., Grigliatti, T.A., & Hickey, D.A. (1989). Rapid spread of transposable *P* elements in experimental populations of *Drosophila melanogaster*. *Genetics*, 122: 387–396.

Goodman, D. (1979). Competitive hierarchies in laboratory *Drosophila*. *Evolution*, 33: 207–219.

Goodnight, C.J. (1990a). Experimental studies of community evolution. I. The response to selection at the community level. *Evolution*, 44: 1614–1624.

Goodnight, C.J. (1990b). Experimental studies of community evolution. II. The ecological basis of the response to community selection. *Evolution*, 44: 1625–1636.

Goodwill, R. (1974). Comparison of three selection programs using *Tribolium castaneum*. *Journal of Heredity*, 65: 8–14.

Gorodetskii, V.P., Zhuchenko, A.A., & Korol, A.B. (1990). Efficiency of feedback selection for recombination in *Drosophila*. *Genetika*, 26: 1942–1952.

Gottschalk, W., & Wolff, G. (1983). *Induced Mutations in Plant Breeding*. Springer, Berlin.

Gould, F. (1979). Rapid host range evolution in a population of the phytophagous mite *Tetranychus urticae* Koch. *Evolution*, 33: 791–802.

Gould, S.J., and Vrba, E.S. (1982). Exaptation—a missing term in the science of form. *Paleobiology* 8: 4–15.

Grant, B., & Mettler, L.E. (1969). Disruptive and stabilizing selection on the "escape" behavior of *Drosophila melanogaster*. *Genetics*, 62: 625–637.

Grant, B.R. (1985). Selection on bill characters in a population of Darwin's finches, *Geospiza conirostris*, on Isla Genovesa, Galapagos. *Evolution*, 39: 523–532.

Grant, P.R., Grant, B.R., Smith, J.M.N., Abbott, I.J., & Abbott, I.K. (1976). Darwin's finches: population variation and natural selection. *Proceedings of the National Academy of Sciences, USA*, 13: 257–261.

Grant, P.R., & Grant, B.R. (1986). *Ecology and Evolution of Darwin's Finches*. Princeton University Press, Princeton, NJ.

Graves, J.L., Luckinbill, L.S., & Nichols A. (1988). Flight duration and wing beat frequency in long- and short-lived *Drosophila melanogaster*. *Journal of Insect Physiology*, 34: 1021–1026.

Graves, J.L., Toolson, E.C., Jeong, C., Vu, L.N., & Rose, M.R. (1992). Desiccation, glycogen and postponed senescence in *Drosophila melanogaster*. *Physiological Zoology*, 65: 268–286.

Green, D.M. (1991). Chaos, fractals and nonlinear dynamics in evolution and phylogeny. *Trends in Ecology and Evolution*, 6: 333–337.

Gregory, W.C. (1965). Mutation frequency, magnitude of change and the probability of improvement in adaptation. *Radiation Botany*, 5(Suppl.): 429–441.

Gromko, M.H., Briot, A., Jensen, S.C., & Fukui, H.H. (1991). Selection on copulation duration in *Drosophila melanogaster*: predictability of direct response versus unpredictability of correlated response. *Evolution*, 45: 69–81.

Gross, H.P. (1978). Natural selection by predators on the defensive apparatus of the three-spined stickleback, *Gasterosteus aculeatus* L. *Canadian Journal of Zoology*, 56: 398–413.

Gründl, E., & Dempfle, L. (1990). Effects of spontaneous and induced mutations on selection response. *Proceedings of the Fourth World Congress on Genetics Applied to Livestock Production*, 13: 177–184.

Hagen, D.W., & Gilbertson, L.G. (1973). Selective predation and the intensity of selection acting on the lateral plates of three-spine sticklebacks. *Heredity*, 30: 273–287.

Haldane, J.B.S. (1957). The cost of natural selection. *Journal of Genetics*, 55: 511–524.

Haley, C.S., & Birley, A.J. (1983). The genetical response to natural selection by varied environments. II. Observations on replicate populations in spatially varied environments. *Heredity*, 51: 581–606.

Halkka, O., Halkka, L., & Raatikainen, M. (1975). Transfer of individuals as a means of investigating natural selection in operation. *Hereditas*, 80: 27–34.

Hall, B.G. (1984). The evolved $\beta$-galactosidase system of *Escherichia coli*. In R.P. Mortlock (ed.), *Microorganisms as Model Systems for Studying Evolution* (pp. 165–186). Plenum, New York.

Hall, B.G. (1988). Adaptive evolution that requires multiple spontaneous mutations. I. Mutations involving an insertion sequence. *Genetics*, 120: 887–897.

Hall, B.G. (1990). Spontaneous point mutations that occur more often when advantageous than when neutral. *Genetics*, 126: 5–16.

Hallauer, A.R., & Sears, J.H. (1969). Mass selection for yield in two varieties of maize. *Crop Science*, 9: 47–50.

Halvorson, H.O., & Monroy, A. (eds.). (1985). *The Origin and Evolution of Sex*. Alan R. Liss, New York.

Hamilton, W.D. (1964a). The genetical evolution of social behaviour. I. *Journal of Theoretical Biology*, 7: 1–16.

Hamilton, W.D. (1964b). The genetical evolution of social behaviour. II. *Journal of Theoretical Biology*, 7: 17–52.

Hammond, J. (1947). Animal breeding in relation to nutrition and environmental conditions. *Biological Reviews*, 22: 195–213.

Hanel, E. (1908). Vererbung bei ungeschlechtlicher Fortplanzung von *Hydra grisea*. *Jenaische Zeitschrift*, 43: 321–372. (Not seen.)

Handford, P., Bell, G., & Reimchen, T. (1977). A gillnet fishery considered as an experiment in artificial selection. *Journal of the Fisheries Research Board of Canada*, 34: 954–961.

Hansche, P.E. (1975). Gene duplication as a mechanism of genetic adaptation in *Saccharomyces cerevisiae*. *Genetics*, 79: 661–674.

Hansen, S.R., & Hubbell, S.P. (1980). Single-nutrient microbial competition: qualitative agreement between experimental and theoretically forecast outcomes. *Science*, 207: 1491–1493.

Harding, J., Allard, R.W., & Smeltzer, D.G. (1966). Population studies in predominantly self-pollinated species. IX. Frequency-dependent selection in *Phaseolus lunatus*. *Proceedings of the National Academy of Sciences, USA*, 56: 99–104.

Harlan, H.V., & Martini, M.I. (1938). The effect of natural selection in a mixture of barley varieties. *Journal of Agricultural Research*, 57: 189–199.

Harris, R., Longerich, S., & Rosenberg, S.M. (1994). Recombination in adaptive mutation. *Science*, 264: 258–260.

Harrison, B.J. (1954). X-irradiation and selection. *Drosophila Information Service*, 28: 123–124.

Hartl, D.L., & Dykhuizen, D.E. (1981). Potential for selection among nearly neutral allozymes of 6-phosphogluconate isomerase in *Escherichia coli*. *Proceedings of the National Academy of Sciences, USA*, 78: 6344–6348.

Hartl, D.L., Dykhuizen, D.E., Miller, R.D., Green, L., & de Framond, J. (1983). Transposable element *IS*50 improves growth rate of *E. coli* cells without transposition. *Cell*, 35: 503–510.

Hartl, D.L., & Jungen, H. (1979). Estimation of average fitness of populations of *Drosophila melanogaster* and the evolution of fitness in experimental populations. *Evolution*, 33: 371–380.

Hartley, B.S. (1984). Experimental evolution of ribitol dehydrogenase. In R.P. Mortlock (ed.), *Microorganisms as Model Systems for Studying Evolution* (pp. 23–54). Plenum, New York.

Hartley, B.S., Altosaar, I., Dothie, J.M., & Neuberger, M.S. (1976). Experimental evolution of a xylitol dehydrogenase. In R. Markham & R.W. Horne (eds.), *Structure—Function Relationship of Proteins* (pp. 191–200). North-Holland, Amsterdam.

Harvey, P.H., Partridge, L., & Southwood, T.R.E. (eds.) (1991). *The Evolution of Reproductive Strategies*. The Royal Society, London.

Hastings, I.M. (1991). Germline selection: population genetic aspects of the sexual/asexual life cycle. *Genetics*, 129: 1167–1176.

Hedrick, P.W. (1976). Simulation of X-linked selection in *Drosophila*. *Genetics*, 83: 551–571.

Hedrick, P.W. (1980). Selection in finite populations. III. An experimental examination. *Genetics*, 87: 297–313.

Hedrick, P.W. (1986). Genetic polymorphism in heterogeneous environments: A decade later. *Annual Review of Ecology and Systematics*, 17: 535–566.

Hedrick, P.W., & Murray, E. (1983). Selection and measures of fitness. In M. Ashburner, H.L. Carson, & J.N. Thompson (eds.), *The Genetics and Biology of Drosophila* (Vol. 3d, pp. 61–104). Academic Press, London.

Helling, R.B., Vargas, C.N., & Adams, J. (1987). Evolution of *Esherichia coli* during growth in a constant environment. *Genetics*, 116: 349–358.

Hetzer, H.O. (1954). Effectiveness of selection for extension of black spotting in Beltsville no. 1 swine. *Journal of Heredity*, 45: 215–223.

Hetzer, H.O., & Harvey, W.R. (1967). Selection for high and low fatness in swine. *Journal of Animal Science*, 26: 1244–1251.

Hiesey, W.M. (1940). Environmental influences and transplant experiments. *Botanical Review*, 6: 181–203.

Hill, D., & Blumenthal, T. (1983). Does Q$\beta$ replicase synthesize RNA in the absence of template? *Nature*, 301: 350–352.

Hill, W.G., & Mackay, T.F.C. (eds.) (1988). *Evolution and Animal Breeding*. Commonwealth Agricultural Bureaux International.

Hillesheim, E., & Stearns, S.C. (1991). The responses of *Drosophila melanogaster* to artificial selection on body weight and its phenotypic plasticity in two larval food environments. *Evolution*, 45: 1909–1923.

Hinson, K., & Hanson, W.D. (1962). Competition studies in soybeans. *Crop Science*, 2: 117–123.

Hoard, K.G., & Crosbie, T.M. (1986). Effects of recurrent selection for cold tolerance on genotype-environment interactions for cold tolerance and agronomic traits in two maize populations. *Crop Science*, 26: 238–242.

Hoffman, A. (1984). Species selection. *Evolutionary Biology*, 18: 1–20.

Hoffmann, A.A., & Parsons, P.A. (1989). Selection for increased desiccation resistance in *Drosophila melanogaster*: Additive genetic control and correlated responses for other stresses. *Genetics*, 122: 837–845.

Hollingdale, B., & Barker, J.S.F. (1971). Selection for increased abdominal bristle number in *Drosophila melanogaster* with concurrent irradiation. I. Populations derived from an inbred line. *Theoretical and Applied Genetics*, 41: 208–215.

Hopkins, C.G. 1899. Improvement in the chemical composition of the corn kernel. *Illinois Agricultural Experimental Station Bulletin*, 55: 205–240.

Hormaza, J.I., & Herrero, M. (1992). Pollen selection. *Theoretical & Applied Genetics*, 83: 663–672.

Horowitz, N.H. (1945). On the evolution of biochemical synthesis. *Proceedings of the National Academy of Sciences, USA*, 31: 153–157.

Horowitz, N.H. (1965). The evolution of biochemical synthesis—retrospect and prospect. In V. Bryson & H.J. Vogel (eds.), *Evolving Genes and Proteins* (pp. 15–23). Academic, New York.

Huang, S.L., Singh, M., & Kojima, K. (1971). A study of frequency-dependent selection observed in the Esterase-6 locus of *Drosophila melanogaster* using a conditioned media method. *Genetics*, 68: 97–104.

Hudak, M.J., & Gromko M.H. (1989). Responses to selection for early and late development of sexual maturity in *Drosophila melanogaster*. *Animal Behaviour*, 38: 344–351.

Hunter, P.E. (1959). Selection of *Drosophila melanogaster* for length of larval period. *Zeitschrift für Vererbungslehre*, 90: 7–28.

Hurd, L.E., & Eisenberg, R.M. (1975). Divergent selection for geotactic response and evolution of reproductive isolation in sympatric and allopatric populations of houseflies. *The American Naturalist*, 109: 353–358.

Hurnik, J.F., Bailey, E.D., & Jerome, F.N. (1973). Selection for divergent lines of mice based on their performance in a T-maze. *Behavior Genetics*, 3: 45–55.

Hurst, L.D. (1993). The incidences, mechanisms and evolution of cytoplasmic sex ratio distorters in animals. *Biological Reviews*, 68: 121–193.

Hutchinson, E.W., & Rose, M.R. (1991). Quantitative genetics of postponed aging in *Drosophila melanogaster*. I. Analysis of outbred populations. *Genetics*, 127: 719–727.

Hutchinson, E.W., Shaw, A.J., & Rose, M.R. (1991). Quantitative genetics of postponed aging in *Drosophila melanogaster*. II. Analysis of selected lines. *Genetics*, 127: 729–737.

Ikeda, H. (1970). The cytoplasmically inherited sex ratio condition in natural and experimental populations of *Drosophila bifasciata*. *Genetics*, 65: 311–333.

Inderlied, C.B., & Mortlock, R.P. (1977). Growth of *Klebsiella aerogenes* on xylitol: Implications for bacterial enzyme evolution. *Journal of Molecular Evolution*, 9: 181–190.

Inger, R.F. (1942). Differential selection of variant juvenile snakes. *The American Naturalist*, 76: 104–109.

Inwang, E.E., Khan, M.A.Q., & Brown, A.W.A. (1967). DDT-resistance in West African and Asian strains of *Aedes aegypti* (L.). *Bulletin of the World Health Organization*, 36: 409–421.

Jackson, L.F., Kahler, A.L., Webster, R.K., & Allard, R.W. (1978). Conservation of scald resistance in barley composite cross populations. *Phytopathology*, 68: 645–650.

Jain, S.K. (1961). Studies on the breeding of self-pollinating cereals. The Composite Cross bulk population method. *Euphytica*, 10: 315–324.

Jain, S.K., & Bradshaw, A.D. (1966). Evolution in closely adjacent plant populations. I. The evidence and its theoretical analysis. *Heredity*, 21: 407–441.

Jennings, H.S. (1908). Heredity, variation and evolution in protozoa. II. Heredity and variation in size and form in *Paramecium*, with studies of growth, environmental action and selection. *Proceedings of the American Philosophical Society*, 47: 393–546.

Jennings, H.S. (1910). Experimental evidence on the effectiveness of selection. *The American Naturalist*, 44: 136–145.

Jennings, H.S. (1916). Heredity, variation and the results of selection in the uniparental reproduction of *Difflugia corona*. *Genetics*, 1: 407–534.

Jennings, P.R., & Herrera, R.M. (1968). Studies on competition in rice. II. Competition in segregating populations. *Evolution*, 22: 332–336.

Jennings, P.R., & de Jesus, J. (1968). Studies on competition in rice. I. Competition in mixtures of varieties. *Evolution*, 22: 119–124.

Jeon, K.W., & Jeon, M.S. (1976). Endosymbiosis in amoebae: recently established endosymbionts have become required cytoplasmic components. *Journal of Cellular Physiology*, 89: 337–344.

Jeon, K.W., & Lorch, I.J. (1967). Unusual intra-cellular bacterial infections in large, freeliving amoebae. *Experimental Cell Research*, 48: 236–240.

Jinks, J.L., & Connolly, V. (1973). Selection for specific and general response to environmental differences. *Heredity*, 30: 33–40.

Jinks, J.L., & Connolly, V. (1975). Determination of the environmental sensitivity of selection lines by the selection environment. *Heredity*, 34: 401–406.

Jinks, J.L., Jayasekara, N.E.M., & Boughey, H. (1977). Joint selection for both extremes of mean performance and of sensitivity to a macroenvironmental variable. II. Single seed descent. *Heredity*, 39: 345–355.

Jinks, J.L., & Pooni, H.S. (1982). Determination of the environmental sensitivity of selection lines of *Nicotiana rustica* by the selection environment. *Heredity*, 49: 291–294.

Johannsen, W. (1903). *Uber Erblichkeit in Populationen und in reinen Linien*. Gustav Fischer, Jena.

Johannsen, W. (1911). The genotype conception of heredity. *American Naturalist*, 45: 129–159.

Johnson, R. (1987). Selected examples of relationships between pathogenicity in cereal rusts and resistance in their hosts. In M.S. Wolfe & C.E. Caten (eds.), *Populations of Plant Pathogens* (pp. 181–192). Blackwell, Oxford, England.

Jones, L.P., Frankham, R., & Barker, J.S.F. (1968). The effects of population size and selection intensity for a quantitative characters in *Drosophila*. II. Long-term response to selection. *Genetical Research*, 12: 249–266.

Jones, J.S., Leith, B., & Rawlings, P. (1977). Polymorphism in *Cepaea*: A problem with too many solutions? *Annual Review of Ecology and Systematics*, 8: 109–143.

Jones, J.S., & Parkin, D.T. (1977). Attempts to measure selection by altering gene frequencies in natural populations. In F.B. Christiansen & T.M. Fenchel (eds.), *Measuring Selection in Natural Populations. Lecture Notes in Biomathematics* (Vol. 19, pp. 83–96).

de Jong, G. (1994). The fitness of fitness concepts and the description of natural selection. *Quarterly Review of Biology*, 69: 3–29.

Jordan, N. (1992). Path analysis of local adaptation in two ecotypes of the annual plant *Diodia teres* Walt. (Rubiaceae). *The American Naturalist*, 140: 149–165.

Joshi, A., & Mueller, L.D. (1988). Evolution of higher feeding rate in *Drosophila* due to density-dependent natural selection. *Evolution*, 42: 1090–1093.

Joyce, G.F. (1992). Directed molecular evolution. *Scientific American*, 267: 90–97.

Juchault, P., & Legrand, J.J. (1989). Sex determination and monogeny in terrestrial isopods

*Armadillidium vulgare* (Latreille, 1804) and *Armadillidium nasutum* Budde-Lund, 1885. *Monitore Zoologico Italiana Monographico*, 4: 359–375.

Kacser, H. (1988). Quantitative variation and the control analysis of enzyme systems. In W.G. Hill & T.F.C. Mackay (eds.), *Evolution and Animal Breeding* (pp. 219–226). Commonwealth Agricultural Bureaux International, London.

Kannenberg, L.W., & Hunter, R.B. (1972). Yielding ability and competitive influence in hybrid mixtures of maize. *Crop Science*, 12: 274–277.

Karn, M.N., & Penrose, L.S. (1951). Birth weight and gestation time in relation to maternal age, parity and infant survival. *Annals of Eugenics*, 16: 147–164.

Katz, A.J., & Enfield, F.D. (1977). Response to selection for increased pupa weight in *Tribolium castaneum* as related to population structure. *Genetical Research*, 30: 237–246.

Katz, A.J., & Young, S.Y.Y. (1975). Selection for high adult body weight in *Drosophila* populations with different structures. *Genetics*, 81: 163–175.

Kauffman, S.A. (1993). *The Origins of Order*. Oxford University Press, Oxford.

Kaufman, P.K., Enfield, F.D., & Comstock, R.E. (1977). Stabilizing selection for pupa weight in *Tribolium castaneum*. *Genetics*, 87: 327–341.

Kawano, K., Gonzalez, H., & Lucena, M. (1974). Intraspecific competition, competition with weeds, and spacing response in rice. *Crop Science*, 14: 841–845.

Kawano, K., & Thung, M.D. (1982). Intergenotypic competition and competition with associated crops in cassava. *Crop Science*, 22: 59–63.

Kayasthar, B.N., & Heyne, E.G. (1978). Interaction of near-isogenic populations of wheat in infested and uninfested environments of wheat soilborne mosaic virus. *Crop Science*, 18: 840–844.

Kelley, S.E. (1989). Experimental studies of the evolutionary significance of sexual reproduction. VI. A greenhouse test of the sib-competition hypotheses. *Evolution*, 43: 1068–1074.

Kerfoot, W.C. (1977). Competition in cladoceran communities: The cost of evolving defences against copepod predation. *Ecology*, 58: 303–313.

Kessler, S. (1966). Selection for and against ethological isolation between *Drosophila pseudoobscura* and *Drosophila persimilis*. *Evolution*, 20: 634–645.

Kessler, S. (1969). The genetics of *Drosophila* mating behaviour. II. The genetic architecture of mating speed in *Drosophila pseudoobscura*. *Genetics*, 62: 421–433.

Kettlewell, H.B.D. (1973). *The Evolution of Melanism: the Study of a Recurring Necessity*. Oxford University Press, Oxford, England.

Khalifa, M.A., & Qualset, C.O. (1974). Intergenotypic competition between tall and dwarf wheats. I. In mechanical mixtures. *Crop Science*, 14: 795–799.

Kidwell, M.G. (1972a). Genetic change of recombination value in *Drosophila melanogaster*. I. Artificial selection for high and low recombination and some properties of recombination-modifying genes. *Genetics*, 70: 419–432.

Kidwell, M.G. (1972b). Genetic change of recombination value in *Drosophila melanogaster*. II. Simulated natural selection. *Genetics*, 70: 433–443.

Kidwell, M.G., Kimura, K., & Black, D.M. (1988). Evolution of hybrid dysgenesis potential following *P* element contamination in *Drosophila melanogaster*. *Genetics*, 119: 815–828.

Kilan, T.C., & Keeling, B.L. (1990). Gene frequency changes in soybean bulk populations exposed to *Phytophthora* rot. *Crop Science*, 30: 575–578.

Kilias, G., Alahiotis, S.N., & Pelecanos, M. (1980). A multifactorial genetic investigation of speciation theory using *Drosophila melanogaster. Evolution*, 34: 730–737.

Kimura, M. (1983). *The Neutral Theory of Molecular Evolution*. Cambridge University Press, Cambridge.

Kindred, B. (1965). Selection for temperature sensitivity in scute *Drosophila. Genetics*, 52: 723–728.

King, J.C. (1955). Evidence of the integration of the gene pool from studies of DDT resistance in *Drosophila. Cold Spring Harbour Symposia in Quantitative Biology*, 20: 311–317.

Kinney, T.B., Bohren, B.B., Craig, J.V., & Lowe, P.C. (1970). Responses to individual, family or index selection for short term rate of egg production in chickens. *Poultry Science*, 49: 1052–1064.

Kirk, D.L. (1988). The ontogeny and phylogeny of cellular differentiation in *Volvox. Trends in Genetics*, 4: 32–36.

Kirkpatrick, M. (ed.) (1994). *The Evolution of Haploid-Diploid Life Cycles. Lectures on Mathematics in the Life Sciences*, American Mathematical Society, Providence, RI.

Kitagawa, O. (1967). The effects of X-ray irradiation on selection response in *Drosophila melanogaster. Japanese Journal of Genetics*, 42: 121–137.

Knowles, P.F., & Houston, B.R. (1953). Resistance of flax varieties to *Fusarium* wilt. *Agronomy Journal*, 45: 408–414.

Koepfer, H.R. (1987). Selection for sexual isolation between geographic forms of *Drosophila mojavensis*. I. Interactions between the selected forms. *Evolution*, 4: 135–148.

Kohane, M.J., & Parsons, P.A. (1989). Domestication: evolutionary changes under stress. *Evolutionary Biology*, 23: 31–48.

Kojima, K. (1971). Is there a constant fitness value for a given genotype? No! *Evolution*, 25: 281–285.

Kojima, K., & Kelleher, T.H. (1963). A comparison of purebred and crossbred selection schemes with two populations of *Drosophila pseudo-obscura. Genetics*, 48: 57–72.

Kojima, K., & Tobari, Y.N. (1969). The pattern of viability changes associated with genotype frequency at the alcohol dehydrogenase locus in a population of *Drosophila melanogaster. Genetics*, 61: 201–209.

Kojima, K., & Yarborough, K.M. (1967). Frequency-dependent selection at the Esterase-6 locus in *Drosophila melanogaster. Proceedings of the National Academy of Sciences of the USA*. 57: 645–649.

Koopman, K.F. (1950). Natural selection for reproductive isolation between *Drosophila pseudoobscura* and *Drosophila persimilis. Evolution*, 4: 135–148.

Korona, R., & Levin, B.R. (1993). Phage-mediated selection and the evolution and maintenance of restriction-modification. *Evolution*, 47: 556–575.

Koufopanou, V., & Bell, G. (1984). Measuring the cost of reproduction. IV. Predation experiments with *Daphnia pulex. Oecologia*, 64: 81–86.

Kramer, F.R., Mills, D.R., Cole, P.E., Nishihara, T., & Spiegelman, S. (1974). Evolution *in vitro*: Sequence and phenotype of a mutant RNA resistant to ethidium bromide. *Journal of Molecular Biology*, 89: 719–736.

Kurlandzka, A., Rosenzweig, R.F., & Adams, J. (1991). Identification of adaptive changes in an evolving population of *Escherichia coli*: the role of changes with regulatory and highly pleiotropic effects. *Molecular Biology and Evolution*, 8: 261–281.

Lack, D. (1947). *Darwin's Finches*. Cambridge University Press, Cambridge, England.

Lack, D. (1966). *Population Studies of Birds*. Clarendon Press, Oxford, England.

Lambio, A.L. (1981). Response to divergent selection for 4-week body weight, egg production and total plasma phosphorus in Japanese quail. *Dissertation Abstracts International* B, 42: 2694. (Not seen.)

Land, R.B., & Falconer, D.S. (1968). Genetic studies of ovulation rate in the mouse. *Genetical Research*, 13: 25–46.

Lande, R., & Arnold, S.J. (1983). The measurement of selection on correlated characters. *Evolution*, 37: 1210–1226.

Lang, A.L., Pendleton, J.W., & Dungan, G.H. (1956). Influence of population and nitrogen levels on yield and protein and oil contents of nine corn hybrids. *Agronomy Journal*, 48: 284–289.

Langer, I., Frey, K.J., & Bailey, T.B. (1978). Production response and stability characteristics of oat cultivars developed in different eras. *Crop Science*, 18: 938–942.

Lashley, K.S. (1916). Results of continued selection in *Hydra*. *Journal of Experimental Zoology*, 20: 19–26.

Latter, B.D.H. (1964). Selection for a threshold character in *Drosophila*. I. An analysis of the phenotypic variance on the underlying scale. *Genetical Research*, 5: 198–210.

Latter, B.D.H. (1966). Selection for a threshold character in *Drosophila*. II. Homeostatic behaviour on relaxation of selection. *Genetical Research*, 8: 205–218.

Latter, B.D.H., & Robertson, A. (1962). The effects of inbreeding and artificial selection on reproductive fitness. *Genetical Research*, 3: 110–138.

Lechowicz, M.J., & Bell, G. (1991). The ecology and genetics of fitness in forest plants. II. Microscale heterogeneity of the edaphic environment. *Journal of Ecology*, 79: 686–696.

Lee, R.D., & Hartgerink, A.P. (1986). Pollination intensity, fruit maturation pattern, and offspring quality in *Cassia fasciculata* (Leguminosae). In D.L. Mulcahy, G.B. Mulcahy, & E. Ottaviano (eds.), *Biotechnology and Ecology of Pollen* (pp. 417–422). Springer–Verlag, New York.

Lehman, N., & Joyce, G.F. (1993a). Evolution *in vitro*: analysis of a lineage of ribozymes. *Current Biology*, 3: 723–734.

Lehman, N., & Joyce, G.F. (1993b). Evolution *in vitro* of an RNA enzyme. *Nature*, 361: 182–185.

Lenski, R.E. (1988a). Experimental studies of pleiotropy and epistasis in *Escherichia coli*. I. Variation in competitive fitness among mutants resistant to virus T4. *Evolution*, 42: 425–432.

Lenski, R.E. (1988b). Experimental studies of pleiotropy and epistasis in *Escherichia coli*. II. Compensation for maladaptive effects associated with resistance to virus T4. *Evolution*, 42: 433–440.

Lenski, R.E., & Mittler, J.E. (1993). The directed mutation controversy and neo-Darwinism. *Science*, 259: 188–194.

Lenski, R.E., Simpson, S.C., & Nguyen, T.T. (1994). Genetic analysis of a plasmid-encoded host genotype-specific enhancement of bacterial fitness. *Journal of Bacteriology*, 176: 3140–3147.

Lenski, R.E., Slatkin, M., & Ayala, F.I. (1989). Mutation and selection in bacterial populations: alternatives to the hypothesis of directed mutation. *Proceedings of the National Academy of Sciences, USA*, 86: 2775–2778.

Lenski, R.E., Souza, V., Duong, L.P., Phan, Q.G., Nguyen, N.M., & Bertrand, K.P. (1994).

Epistatic effects of promotor and repressor functions of the *Tn*10 tetracycline-resistance operon on the fitness of *Escherichia coli*. *Molecular Ecology*, 3: 127–135.

Lenski, R.E., & Travisano, M. (1994). Dynamics of adaptation and diversification: a 10,000-generation experiment with bacterial populations. *Proceedings of the National Academy of Sciences, USA*, 91: 6808–6814.

Leonard, K.J. (1969). Selection in heterogeneous populations of *Puccinia graminis* f. sp. *avenae*. *Phytopathology*, 59: 1851–1857.

Leonard, K.J. (1987). The host population as a selective factor. In M.S. Wolfe & C.E. Caten (eds.), *Populations of Plant Pathogens* (pp. 163–179). Blackwell, Oxford, England.

Lerner, I.M., & Dempster, E.R. (1951). Attenuation of genetic progress under continued selection in poultry. *Heredity*, 5: 75–94.

Lerner, I.M., & Hazel, L.N. (1947). Population genetics of a poultry flock under artificial selection. *Genetics*, 32: 325–339.

Lerner, S.A., Wu, T.T., & Lin, E.C.C. (1964). Evolution of a catabolic pathway in bacteria. *Science*, 146: 1313–1314.

Levene, H., Pavlovsky, O., & Dobhansky, T. (1954). Interaction of the adaptive values in polymorphic experimental populations of *Drosophila pseudoobscura*. *Evolution*, 8: 335–349.

Levin, B.R. (1972). Coexistence of two asexual strains on a single resource. *Science*, 175: 1272–1274.

Levin, B.R. (1988). Frequency-dependent selection in bacterial populations. *Philosophical Transactions of the Royal Society of London*, 319: 459–472.

Lewontin, R.C. (1955). The effects of population density and composition on viability in *Drosophila melanogaster*. *Evolution*, 9: 27–41.

Lewontin, R.C. (1974). *The Genetic Basis of Evolutionary Change*. Columbia University Press, New York.

L'Héritier, P., & Teissier, G. (1934). Une expérience de sélection naturelle. Courbe d'élimination du géne "Bar" dans une population de Drosophiles en équilibre. *Compte Rendus de la Societé de Biologie de Paris*, 117: 1049–1051.

Lin, C.S., Binns, M.R., & Lefkovitch, L.P. (1986). Stability analysis: Where do we stand? *Crop Science*, 26: 894–900.

Lin, E.C.C., Hacking, A.J., & Aguilar, J. (1976). Experimental models of acquisitive evolution. *BioScience*, 26: 548–555.

Lin, E.C.C., & Wu, T.T. (1984). Functional divergence of the L-Fucose system in mutants of *Escherichia coli*. In R.P. Mortlock (ed.), *Microorganisms as Model Systems for Studying Evolution* (pp. 135–164). Plenum, New York.

Long, T. (1970). Genetic effects of fluctuating temperature in populations of *Drosophila melanogaster*. *Genetics*, 66: 401–416.

Lopez-Fanjul, C., & Hill, W.G. (1973). Genetic differences between populations of *Drosophila melanogaster* for a quantitative trait. I. Laboratory populations. *Genetical Research*, 22: 51–68.

Lovett Doust, L. (1981). Population dynamics and local specialization in a clonal perennial (*Ranunculus repens*). II. The dynamics of leaves, and a reciprocal transplant-replant experiment. *Journal of Ecology*, 69: 757–768.

Luckinbill, L.S. (1978). r and K selection in experimental populations of *Escherichia coli*. *Science*, 202: 1201–1203.

Luckinbill, L.S. (1979). Selection and the r/K continuum in experimental populations of protozoa. *The American Naturalist*, 113: 427–437.

Luckinbill, L.S. (1984). An experimental analysis of a life-history theory. *Ecology*, 65: 1170–1184.

Luckinbill, L.S., Arking, R., & Clare, M.J. (1984). Selection for delayed senescence in *Drosophila melanogaster. Evolution*, 38: 996–1003.

Luckinbill, L.S., Clare, M.J., Krell, W.L., Cirocco, W.C., & Richards, P.A. (1987). Estimating the number of genetic elements that defer senescence in *Drosophila. Evolutionary Ecology*, 1: 37–46.

Luckinbill, L.S., Graves, J.L., Reed, A.H., & Koetsawang, S. (1988). Localizing genes that defer senescence in *Drosophila melanogaster. Heredity*, 60: 367–374.

Luria, S.E., & Delbrück, M. (1943). Mutations of bacteria from virus sensitivity to virus resistance. *Genetics*, 28: 491–511.

Lush, J.L. (1947a). Family merit and individual merit as bases for selection. Part I. *The American Naturalist*, 81: 241–261.

Lush, J.L. (1947b). Family merit and individual merit as bases for selection. Part II. *The American Naturalist*, 81: 362–379.

Lyttle, T.W. (1977). Experimental population genetics of meiotic drive systems. I. Pseudo-Y chromosomal drive as a means of eliminating cage populations of *Drosophila melanogaster. Genetics*, 86: 413–445.

Lyttle, T.W. (1979). Experimental population genetics of meiotic drive systems. II. Accumulation of genetic modifiers of Segregation Distorter (*SD*) in laboratory populations. *Genetics*, 91: 339–357.

Lyttle, T.W. (1981). Experimental population genetics of meiotic drive systems. III. Neutralization of sex-ratio distortion in *Drosophila* through sex-chromosome aneuploidy. *Genetics*, 98: 317–334.

MacArthur, J.W. (1949). Selection for small and large body size in the house mouse. *Genetics*, 34: 194–209.

Macdowell, E.C. (1919). Bristle inheritance in *Drosophila*. II. Selection. *Journal of Experimental Zoology*, 23: 109–146.

Macha, A.M., & Becker, W.A. (1976). Comparison of predicted with actual body weight selection gains of *Coturnix coturnix japonica. Theoretical and Applied Genetics*, 47: 251–255.

Mackay, T.F.C. (1981). Genetic variation in varying environments. *Genetical Research*, 37: 79–83.

Mackay, T.F.C. (1985). Transposable element-induced response to artificial selection in *Drosophila melanogaster. Genetics*, 111: 351–374.

MacNeil, M.D., Kress, P.R., Flower, A.E., & Blackwell, R.L. (1984). Effects of mating system in Japanese Quail. 2. Genetic parameters, response and correlated response to selection. *Theoretical and Applied Genetics*, 67: 407–412.

MacRae, A.F., & Anderson, W.W. (1988). Evidence for non-neutrality of mitochondrial DNA haplotypes in *Drosophila pseudoobscura. Genetics*, 120: 485–494.

Madalena, F.E., & Robertson, A. (1975). Population structure in artificial selection: studies with *Drosophila melanogaster. Genetical Research*, 24: 113–126.

Maliga, P., Breznovits, S., Marton, L., & Job, F. (1975). Nonmendelian streptomycin-resistant tobacco mutant with altered chloroplasts and mitochondria. *Nature*, 255: 401–402.

Malmberg, R.L. (1977). The evolution of epistasis and the advantage of recombination in populations of bacteriophage. *Genetics*, 86: 607–621.

Manly, B.F.J., Miller, P., & Cook, L.M. (1972). Analysis of a selective predation experiment. *The American Naturalist*, 109: 719–736.

Manning, A. (1961). The effects of artificial selection for mating speed in *Drosophila melanogaster*. *Animal Behaviour*, 9: 82–92.

Manning, A. (1963). Selection for mating speed in *Drosophila melanogaster* based on the behaviour of one sex. *Animal Behaviour*, 11: 116–120.

Manning, A. (1968). The effects of artificial selection for slow mating in *Drosophila simulans*. *Animal Behaviour*, 16: 108–113.

Manning, A., & Hirsch, J. (1971). The effects of artificial selection for slow mating in *Drosophila simulans*. II. Genetic analysis of the slow mating line. *Animal Behaviour*, 19: 448–453.

Manson, J.M. (1973). Genetic change in the egg weight–body weight association in the fowl. *Proceedings of the Fourth European Poultry Conference* (pp. 247–256).

Mareck, J.H., & Gardner, C.O. (1979). Responses to mass selection in maize and stability of resulting populations. *Crop Science*, 19: 779–783.

Marien, D. (1958). Selection for developmental rate in *Drosophila pseudoobscura*. *Genetics*, 50: 3–15.

Marinkovic, D. (1968). Selection for higher fitness in populations of *Drosophila pseudoobscura*. *Science*, 160: 199–200.

Marks, H.L., & Lepore, P.D. (1968). Growth-rate inheritance in Japanese quail. 2. Early responses to selection under different nutritional environments. *Poultry Science*, 47: 1540–1546.

Marshall, D.L., & Ellstrand, N.C. (1986). Sexual selection in *Raphanus raphinastrum*: experimental data on non-random fertilization, maternal choice, and consequences of multiple paternity. *The American Naturalist*, 127: 446–461.

Marshall, D.R. (1977). The advantages and hazards of genetic homogeneity. *Annals of the New York Academy of Sciences*, 287: 1–20.

Martin, G.A., & Bell, A.E. (1960). An experimental check on the accuracy of prediction of response during selection. In O. Kempthorne (ed.), *Biometrical Genetics* (pp. 178–187). Pergamon Press, London.

Marton, L., & Maliga, P. (1975). Control of resistance in tobacco plants to 5-bromodeoxyuridine by a simple mendelian factor. *Plant Science Letters*, 5: 77–81.

Mason, L.G. (1964). Stabilizing selection for mating fitness in natural populations of *Tetraopes*. *Evolution*, 18: 492–497.

Mather, K. (1983). Response to selection. In M. Ashburner, H.L. Carson, & J.N. Thompson (eds.), *The Genetics and Biology of Drosophila* (Vol. 3c, pp. 155–221). Academic, New York.

Mather, K., & Caligari, P.D.S. (1983). Pressure and response in competitive interactions. *Heredity*, 51: 435–454.

Mather, K., & Cooke, P. (1962). Differences in competitive ability between genotypes of *Drosophila*. *Heredity*, 17: 381–407.

Mather, K., & Harrison, B.J. (1949). The manifold effect of selection. *Heredity*, 3: 1–52 and 131–162.

Mattern, U., Strobel, G., & Shepard, J. (1978). Reaction to phytotoxins in a potato population derived from mesophyll protoplasts. *Proceedings of the National Academy of Sciences, USA*, 75: 4935–4939.

May, H.G. (1917). Selection for higher and lower facet numbers in the bar-eyed race of *Drosophila* and the appearance of reverse mutations. *Biological Bulletin*, 33: 361–395.

Maynard Smith, J. (1964). Group selection and kin selection. *Nature*, 201: 1145–1147.

Maynard Smith, J. (1970). Time in the evolutionary process. *Studium Generale*, 23: 266–272.

Maynard Smith, J. (1976). Group selection. *Quarterly Review of Biology*, 51: 277–283.

Maynard Smith, J. (1982). *Evolution and the Theory of Games*. Cambridge University Press, Cambridge, England.

McCall, C., Mitchell-Olds, T., & Waller, D.M. (1989). Fitness consequences of outcrossing in *Impatiens capensis*: tests of the frequency-dependent and sib-competition models. *Evolution*, 43: 1075–1084.

McCall, C., Mitchell-Olds, T., & Waller, D.M. (1991). Distance between mates affects seedling characters in a population of *Impatiens capensis* (Balsaminaceae). *American Journal of Botany*, 78: 964–970.

McCann, J., Luedders, V.D., & Dropkin, V.H. (1982). Selection and reproduction of soybean cyst nematodes on resistant soybeans. *Crop Science*, 22: 78–80.

McCaskill, J.S., & Bauer, G.J. (1993). Images of evolution: the origin of spontaneous RNA replication waves. *Proceedings of the National Academy of Sciences, USA*, 90: 4191–4195.

McCauley, D.E., & Wade, M.J. (1980). Group selection: the genetic and demographic basis for the phenotypic differentiation of small populations of *Tribolium castaneum*. *Evolution*, 34: 813–821.

McDonald, B.A., McDermott, J.M., Allard, R.W., & Webster, R.K. (1989). Coevolution of host and pathogen populations in the *Hordeum vulgare–Rhynchosporium secalis* pathosystem. *Proceedings of the National Academy of Sciences, USA*, 86: 3924–3927.

McDonald, J.F., & Ayala, F.J. (1974). Genetic response to environmental heterogeneity. *Nature*, 250: 572–574.

McGill, A., & Mather, K. (1972). Competition in *Drosophila*. I. A case of stabilizing selection. *Heredity*, 27: 473–478.

McGraw, J.B., & Antonovics, J. (1983). Experimental ecology of *Dryas octopetala* ecotypes. I. Ecotypic differentiation and life-cycle stages of selection. *Journal of Ecology*, 71: 879–897.

McGuirk, B.J., Atkins, K.D., & Thompson, R. (1986). Long-term selection experiments with sheep. *Proceedings of the Third World Congress on Genetics Applied to Livestock Production*, 12: 181–198, Lincoln Nebrastra.

McNeal, F.H., Qualset, C.O., Baldridge, D.E., & Stewart, V.R. (1978). Selection for yield and yield components in wheat. *Crop Science*, 18: 795–801.

McNeilly, T. (1968). Evolution in closely adjacent plant populations. III. *Agrostis tenuis* on a small copper mine. *Heredity*, 23: 99–108.

McNeilly, T., & Antonovics, J. (1968). Evolution in closely adjacent plant populations. IV. Barriers to gene flow. *Heredity*, 23: 205–218.

McNeilly, T., & Bradshaw, A.D. (1968). Evolutionary processes in populations of copper tolerant *Agrostis tenuis* Sibth. *Evolution*, 22: 108–118.

McPhee, C.P., & Robertson, A. (1970). The effect of suppressing crossing-over on the response to selection in *Drosophila melanogaster. Genetical Research*, 16: 1–16.

Meffert, L.M., & Bryant, E.H. (1991). Mating propensity and courtship behaviour in serially bottlenecked lines of the housefly. *Evolution*, 45: 293–306.

Mendiola, N.B. (1919). Variation and selection within clonal lines of *Lemna minor. Genetics*, 4: 151–182.

Meredith, W.R., & Culp, T.W. (1979). Influence of longevity of use on cotton cultivars' performance. *Crop Science*, 19: 654–656.

Mertz, D.R. (1975). Senescent decline in flour beetle strains selected for early adult fitness. *Physiological Zoology*, 48: 1–23.

Meyer, H.H., & Enfield, F.D. (1975). Experimental evidence on limitations of the heritability parameter. *Theoretical and Applied Genetics*, 45: 268–273.

Michaelson, J. (1987). Cell selection in development. *Biological Reviews*, 62: 115–139.

Michod, R.E., & Levin, B.R. (eds.) (1988). *The Evolution of Sex*. Sinauer, Sunderland, MA.

Middleton, A.R. (1915). Heritable variations and the results of selection in the fission rate of *Stylonichia pustulata. Journal of Experimental Zoology*, 19: 451–503.

Mikkola, R., & Kurland, C.G. (1992). Selection of wild-type phenotype from natural isolates of *Escherichia coli* in chemostats. *Molecular Biology and Evolution*, 9: 394–402.

Milkman, R.D. (1964). The genetic basis of natural variation. V. Selection for crossveinless polygenes in new wild strains of *Drosophila melanogaster. Genetics*, 50: 625–632.

Milkman, R.D. (1979). The posterior crossvein in *Drosophila* as a model phenotype. In J.N. Thompson & J.M. Thoday (eds.), *Quantitative Genetic Variation* (pp. 157–176). Academic, New York.

Miller, P.A., & Rawlings, J.O. (1967). Selection for increased lint yield and correlated responses in upland cotton, *Gossypium hirsutum* L. *Crop Science*, 7: 637–640.

Mills, D.R., Kramer, F.R., Dobkin, C., Nishihara, T., & Spiegelman, S. (1975). Nucleotide sequence of microvariant RNA: Another small replicating molecule. *Proceedings of the National Academy of Sciences, USA*, 72: 4252–4256.

Mills, D.R., Peterson, R.L., & Spiegelman, S. (1967). An extracellular Darwinian experiment with a self-duplicating nucleic acid molecule. *Proceedings of the National Academy of Sciences, USA*, 58: 217–224.

Minawa, A., & Birley, A.J. (1978). The genetical response to natural selection by varied environments. I. Short-term observations. *Heredity*, 40: 31–50.

Mittler, J.E., & Lenski, R.E. (1990). New data on excisions of mu from *E. coli* MS2 cast doubt on the directed mutation hypothesis. *Nature*, 344: 173–175.

Modi, R.I., & Adams, J. (1991). Coevolution in bacterial-plasmid populations. *Evolution*, 45: 656–667.

Modi, R.I., Castilla, L.H., Puskas-Rozsa, S., Helling, R.B., & Adams, J. (1992). Genetic changes accompanying increased fitness in evolving populations of *Escherichia coli. Genetics*, 130: 241–249.

Moll, R.H., Cockerham, C.C., Stiber, C.W., & Williams, W.P. (1978). Selection responses, genetic-environmental interactions, and heterosis with recurrent selection for yield in maize. *Crop Science*, 18: 641–645.

Moore, J.A. (1952). Competition between *Drosophila melanogaster* and *Drosophila simulans*.

II. The improvement of competitive ability through selection. *Proceedings of the National Academy of Sciences, USA*, 38: 813–817.

Morgan, P. (1976). Frequency-dependent selection at two enzyme loci in *Drosophila melanogaster*. *Nature*, 263: 765–767.

Moriwaki, D., & Fuyama, Y. (1963). Responses to selection for rate of development in *Drosophila melanogaster*. *Drosophila Information Service*, 38: 74.

Moriwaki, D., & Kitigawa, O. (1957). Sex ratio J in *Drosophila bifasciata*. I. A preliminary note. *Japanese Journal of Genetics*, 32: 208–210.

Morris, J.A. (1963). Continuous selection for egg production using short-term records. *Australian Journal of Agricultural Research*, 14: 909–925.

Mortlock, R.P. (1980). Regulatory mutations and the development of new metabolic pathways by bacteria. *Evolutionary Biology*, 14: 205–268.

Mortlock, R.P. (1984). The utilization of pentitols in studies of the evolution of enzyme pathways. In R.P. Mortlock (ed.), *Microorganisms as Model Systems for Studying Evolution* (pp. 1–22). Plenum, New York.

Mortlock, R.P. (ed.) (1984). *Microorganisms as Model Systems for Studying Evolution* Plenum, New York.

Mortlock, R.P., Fossitt, D.D., Petering, D.H., & Wood, W.A. (1965). A basis for utilization of unnatural pentoses and pentitols by *Aerobacter aerogenes*. *Proceedings of the National Academy of Sciences, USA*, 54: 572–579.

Mortlock, R.P., & Wood, W.A. (1964). Metabolism of pentoses and pentitols by *Aerobacter aerogenes*. I. Demonstration of pentose isomerase. pentulokinase, and pentitol dehydrogenase enzyme families. *Journal of Bacteriology*, 88: 835–844.

Morton, R.A., & Hall, S.C. (1985). Response of dysgenic and non-dysgenic populations to mutation exposure. *Drosophila Information Service*, 61: 126–128.

Mueller, L.D. (1987). Evolution of accelerated senescence in laboratory populations of *Drosophila*. *Proceedings of the National Academy of Sciences, USA*, 84: 1974–1977.

Mueller, L.D. (1991). Ecological determinants of life-history evolution. *Philosophical Transactions of the Royal Society of London Series B*, 332: 25–30.

Mueller, L.D., & Ayala, F.J. (1981). Trade-off between r-selection and K-selection in *Drosophila* populations. *Proceedings of the National Academy of Sciences, USA*, 78: 1303–1305.

Mueller, L.D., Guo, P., & Ayala, F.J. (1991). Density-dependent natural selection and trade-offs in life history traits. *Science*, 253: 433–435.

Mueller, L.D., & Sweet, V.F. (1986). Density-dependent natural selection in *Drosophila*: evolution of pupation height. *Evolution*, 40: 1354–1356.

Mukherjee, A.S. (1961). Effect of selection on crossing-over in the males of *Drosophila ananassae*. *The American Naturalist*, 95: 57–59.

Mukai, T., Chigusa, S.T., Mettler, L.E., & Crow, J.F. (1972). Mutation rate and dominance of genes affecting viability in *Drosophila melanogaster*. *Genetics*, 72: 335–355.

Mulcahy, D.L., & Mulcahy, G.B. (1975). The influence of gametophytic competition on sporophytic quality in *Dianthus chinensis*. *Theoretical and Applied Genetics*, 46: 277–280.

Mulcahy, D.L., Mulcahy, G.B., & Ottaviano, E. (1978). Further evidences that gametophytic selection modifies the genetic quality of the sporophyte. *Société Botanique de France Actualités Botaniques*, 1: 57–60.

Mumaw, C.R., & Weber, C.R. (1957). Competition and natural selection in soybean varietal composites. *Agronomy Journal*, 49: 154–160.

Muona, O., Allard, R.W., & Webster, R.K. (1982). Evolution of resistance to *Rhynchosporium secalis* (Oud.) Davis in barley Composite Cross II. *Theoretical and Applied Genetics*, 61: 209–214.

Murphy, J.P., Hebsel, D.B., Elliott, A., Thro, A.M., & Frey, K.J. (1982). Compositional stability of an oat multiline. *Euphytica*, 31: 33–40.

Nagai, J., Eisen, E.J., Emsley, J.A.B., & McAllister, A.J. (1978). Selection for nursing ability and adult weight in mice. *Genetics*, 88: 761–780.

Narain, P., Joshi, C., & Prabhu, S.S. (1960). Response to selection for fecundity in *Drosophila melanogaster. Drosophila Information Service*, 36: 96–99.

Nassar, R., Muhs, H.J., & Cook, R.D. (1973). Frequency-dependent selection at the Payne inversion in *Drosophila melanogaster. Evolution*, 27: 558–564.

Nestmann, E.R., & Hill, R.F. (1973). Population changes in continuously growing mutator cultures of *Escherichia coli. Genetics* (Suppl.) 73: 41–44.

Neu, H. (1992). The crisis in antibiotic resistance. *Science*, 257: 1064–1072.

Newcombe, H.B. (1949). Origin of bacterial variants. *Nature*, 164: 150.

Newman, E.I., & Rovira, A.D. (1975). Allelopathy among some British grassland species. *Journal of Ecology*, 63: 727–737.

Newport, M.E.A. (1989). A test for proximity-dependent outcrossing in the alpine skypilot, *Polemonium viscosum. Evolution*, 43: 1110–1113.

Nguyen, T.N.M, Phan, Q.G., Duong, L.P., Bertrand, K.P., & Lenski, R.E. (1989). Effects of carriage and expression of the *Tn*10 tetracycline-resistance operon on the fitness of *Escherichia coli* K12. *Molecular Biology and Evolution*, 6: 213–225.

Nordskog, A.W., Tolman, H.S., Casey, D.W., & Lin, C.Y. (1974). Selection in small populations of chickens. *Poultry Science*, 53: 1188–1219.

Novick, A., & Szilard, L. (1950). Experiments with the chemostat on spontaneous mutations of bacteria. *Proceedings of the National Academy of Sciences, USA*, 36: 708–719.

Novick, A., & Szilard, L. (1951). Genetic mechanisms in bacteria and bacterial viruses. I. Experiments on spontaneous and chemically induced mutations of bacteria growing in the chemostat. *Cold Spring Harbor Symposia on Quantitative Biology*, 16: 337–343.

Oakeshott, J.G. (1979). Selection affecting enzyme polymorphisms in laboratory populations of *Drosophila melanogaster. Oecologia*, 43: 341–354.

O'Donald, P. (1973). A further analysis of Bumpus's data: the intensity of natural selection. *Evolution*, 27: 398–404.

de Oliveira, A.K., & Cordeiro, A.R. (1980). Adaptation of *Drosophila willistoni* populations to extreme pH medium. II. Development of incipient reproductive isolation. *Heredity*, 44: 123–130.

Orgel, L.E. (1979). Selection in *vitro. Proceedings of the Royal Society of London B*, 205: 435–442.

Ornston, L.N., & Yeh, W.K. (1979). Origins of metabolic diversity: evolutionary divergence by sequence repetition. *Proceedings of the National Academy of Sciences, USA*, 76: 3996–4000.

Osman, H., El, S., & Robertson, A. (1968). The introduction of genetic material from inferior into superior strains. *Genetical Research*, 12: 221–236.

Ottaviano, E., & Mulcahy, D.L. (1989). Genetics of angiosperm pollen. *Advances in Genetics*, 26: 1–64.

Ottaviano, E., Sari-Gorla, M., & Arenari, I. (1983). Male gametophyte competitive ability in maize: selection and implications with regard to the breeding system. In D.L. Mulcahy & E. Ottaviano (eds.), *Pollen: Biology and Implications for Plant Breeding*, (pp. 367–374). Elsevier, New York.

Ottaviano, E., Sari Gorla, M., & Pe, E. (1982). Male gametophytic selection in maize. *Theoretical and Applied Genetics*, 75: 252–258.

Ottaviano, E., Sari Gorla, M., & Villa, M. (1988). Pollen competitive ability in maize: within-population variability and response to selection. *Theoretical and Applied Genetics*, 76: 601–608.

Otte, D., & Endler, J.A. (eds.) (1989). *Speciation and its Consequences*. Sinauer, Sunderland, MA.

Palacco, J.C., & Palacco, M.L. (1977). Inducing and selecting a valuable mutant in plant cell culture: a tobacco mutant resistant to carboxin. *Annals of the New York Academy of Sciences*, 287: 385–400.

Paquin, C.E., & Adams, J. (1983). Relative fitness can decrease in evolving asexual populations of *S. cerevisiae*. *Nature*, 306: 368–370.

Palenzona, D.L., Fini, C., & Scossiroli, R.E. (1971). Comparative study of natural selection in *Cardium edule* L. and *Drosophila melanogaster* Meig. *Monitori Zoologica Italiana*, 5: 165–172.

Palmer, J.O., & Dingle, H. (1986). Direct and correlated responses to selection among life-history traits in milkweed bugs (*Oncopeltus fasciatus*). *Evolution*, 40: 767–777.

Paquin, C., & Adams, J. (1983). Frequency of fixation of adaptive mutations is higher in evolving diploid than haploid yeast populations. *Nature*, 302: 495–500.

Park, T., & Lloyd, M. (1955). Natural selection and the outcome of competition. *The American Naturalist*, 89: 235–240.

Park, Y.I., Hansen, C.T., Chung, C.S., & Chapman, A.B. (1966). Influence of feeding regime on the effects of selection for postweaning gain in the rat. *Genetics*, 54: 1315–1327.

Parsons, P.A. (1958). Selection for increased recombination in *Drosophila melanogaster*. *The American Naturalist*, 88: 255–256.

Pasyukova, E.G., Belyaeva, E.S., Kogan, G.L., Kaidanov, L.Z., & Gvozdev, V.A. (1986). Concerted transpositions of mobile genetic elements coupled with fitness changes in *Drosophila melanogaster*. *Molecular Biology and Evolution*, 34: 299–312.

Payne, F. (1912). *Drosophila ampelophila* Loew bred in the dark for sixty-nine generations. *Biological Bulletin*, 28: 297–301.

Pearl, R. (1917). The selection problem. *The American Naturalist*, 51: 65–91.

Pearl, R., & Surface, F.M. (1910). Selection in maize. *Maine Agriculture Experimental Station Annual Report*, 1910: 249–307.

Pignatelli, P.M., & Mackay, T.F.C. (1989). Hybrid dysgenesis-induced response to selection in *Drosophila melanogaster*. *Genetical Research*, 54: 183–195.

Pimentel, D., Levin, S.A., & Soans, A.B. (1975). On the evolution of energy balance in some exploiter-victim systems. *Ecology*, 56: 381–390.

Potts, D.C. (1984). Natural selection in experimental populations of reef-building corals (Scleractinia). *Evolution*, 38: 1059–1078.

Powell, J.R. (1971). Genetic polymorphism in varied environments. *Science*, 174: 1035–1036.

Powell, J.R. (1978). The founder-flush speciation theory: an experimental approach. *Evolution*, 32: 465–474.

Powell, J.R., & Andjelkovich, M. (1983). Population genetics of *Drosophila* amylase. IV. Selection in laboratory populations maintained on different carbohydrates. *Genetics*, 99: 675–689.

Powell, J.R., & Wistrand, H. (1978). The effect of heterogeneous environments and competition on genetic variation in *Drosophila*. *The American Naturalist*, 112: 935–947.

Powell, W., & Phillips, M.S. (1984). An investigation of genotype environment interactions in oat lines (*Avena sativa*) derived from composite populations. *Heredity*, 52: 171–178.

Prevosti, A. (1967). Inversion heterozygosity and selection for wing length in *Drosophila subobscura*. *Genetical Research*, 10: 81–93.

Price, M.V., & Waser, N.M. (1979). Pollen dispersal and optimal outcrossing in *Delphinium nelsoni*. *Nature*, 277: 294–297.

Primack, R.B. (1978). Regulation of seed yield in *Plantago*. *Journal of Ecology*, 66: 835–847.

Primack, R.B. (1979). Reproductive effort in annual and perennial species of *Plantago* (Plantinaginaceae). *The American Naturalist*, 114: 51–62.

Primack, R.B., & Antonovics, J. (1981). Experimental ecological genetics in *Plantago*. V. Components of seed yield in the ribwort plantain *Plantago lanceolata* L. *Evolution*, 35: 1069–1079.

Primack, R.B., & Antonovics, J. (1982). Experimental ecological genetics in *Plantago*. VII. Reproductive effort in populations of *P. lanceolata* L. *Evolution*, 36: 742–752.

Prout, T. (1962). The effects of stabilizing selection on the time of development in *Drosophila melanogaster*. *Genetical Research*, 3: 364–382.

Pruzan-Hotchkiss, A., Perelle, I.B., Hotchkiss, F.H.C., & Ehrman, L. (1980). Altered competition between two reproductively isolated strains of *Drosophila melanogaster*. *Evolution*, 34: 445–452.

Pym, R.A.E., & Nicholls, P.J. (1979). Selection for feed efficiency in broiler production. I. Direct and correlated responses to selection for body weight gain, food consumption and feed utilization efficiency. *British Poultry Science*, 20: 73–86.

Qualset, C.O. (1981). Barley mixtures: the continuing search for high-performing combinations. *Barley Genetics*, IV: 130–137.

Rahnefeld, G.W., & Garnett, I. (1976). Mass selection for post-weaning growth in swine. IV. Selection response and control population stability. *Canadian Journal of Animal Science*, 56: 783–790.

Rainey, P.B., Moxon, E.R., & Thompson, I.P. (1993). Intraclonal polymorphism in bacteria. *Advances in Microbial Ecology*, 13: 263–300.

Rasmuson, M. (1956). Reciprocal recurrent selection. Results of three model experiments on *Drosophila* for improvement of quantitative characters. *Hereditas*, 42: 397–414.

Rathie, K.A., & Nicholas, F.W. (1980). Artificial selection with differing population structures. *Genetical Research*, 36: 117–131.

Ray, T.S. (1991). An approach to the synthesis of life. In C. Langton, C. Taylor, D. Farmer, & S. Rasmussen, (eds.), Artificial Life II, pp. 371–408. Santa Fe Institute Studies in the Science of Complexity Vol. X. Addison-Wesley, Redwood City, California.

Rebound, X., & Bell, G. (1996). Experimental evolution in *Chlamydomonas*. III. Evolution of specialist and generalist types in environments that vary in space and time. Unpublished.

Redmond, A.M., Robbins, L.E., & Travis, J. (1989). The effects of pollination distance on seed production in three populations of *Amianthium muscaetoxicum*. (Liliaceae). *Oecologia*, 79: 260–264.

Reeve, E.C.R., & Robertson, F.W. (1953). Studies in quantitative inheritance. II. Analysis of a strain of *Drosophila melanogaster* selected for long wings. *Journal of Genetics*, 51: 276–316.

Rendel, J.M. (1943). Variation in the weights of hatched and unhatched duck's eggs. *Biometrika*, 33: 48–56.

Rendel, J.M., & Sheldon, B. (1960). Selection for canalization of the *scute* phenotype of *Drosophila*. *Australian Journal of Biological Science*, 13: 36–47.

Reznick, D. (1982). The impact of predation on life history evolution in Trinidadian guppies: genetic basis of observed life history patterns. *Evolution*, 36: 1236–1250.

Reznick, D., & Bryga, H. (1987). Life history evolution in guppies (*Poecilia reticulata*). I. Phenotypic and genetic changes in an introduction experiment. *Evolution*, 41: 1370–1385.

Reznick, D., & Endler, J.A. (1982). The impact of predation on life history evolution in Trinidadian guppies (*Poecilia reticulata*). *Evolution*, 36: 160–177.

Reznick, D., Bryga, H., & Endler, J.A. (1990). Experimentally induced life-history evolution in a natural population. *Nature*, 346: 357–359.

Rice, W.R., & Hostert, E.E. (1993). Laboratory experiments on speciation: What have we learned in forty years? *Evolution*, 47: 1637–1653.

Rice, W.R., & Salt, G.W. (1990). The evolution of reproductive isolation as a correlated character under sympatric conditions: experimental evidence. *Evolution*, 44: 1140–1152.

Rich, S.S., Bell, A.E., & Wilson, S.P. (1979). Genetic drift in small populations of *Tribolium*. *Evolution*, 33: 579–584.

Richardson, R.H., Kojima, K., & Lucas, H.L. (1968). An analysis of short-term selection experiments. *Heredity*, 23: 493–506.

Richmond, R.C., Gilpin, M.E., Perez Salas, S., & Ayala, F.J. (1975). A search for emergent competitive phenomena: the dynamics of multispecies *Drosophila* systems. *Ecology*, 56: 709–714.

Ricker, J.P., & Hirsch, J. (1985). Evolution of an instinct under long-term divergent selection for geotaxis in domesticated populations of *Drosophila melanogaster*. *Journal of Comparative Psychology*, 99: 380–390.

Ricker, J.P., & Hirsch, J. (1988). Genetic changes occurring over 500 generations in lines of *Drosophila melanogaster* selected divergently for geotaxis. *Behavior Genetics*, 18: 13–25.

Ricker, W.E. (1981). Changes in the average size and average age of Pacific salmon. *Canadian Journal of Fisheries and Aquatic Science*, 38: 1636–1656.

Rigby, P.W.J., Burleigh, B.D., & Hartley, B.S. (1974). Gene duplication in experimental enzyme evolution. *Nature*, 251: 200–204.

Ringo, J., Wood, D., Rockwell, R., & Dowse, H. (1985). An experiment testing two hypotheses of speciation. *The American Naturalist*, 126: 642–661.

Ripley, P.O. (1941). The influence of crops upon those which follow. *Scientific Agriculture*, 21: 522–583.

Roberts, R.C. (1966a). The limits to artificial selection for body weight in the mouse. I. The limits attained in earlier experiments. *Genetical Research*, 8: 347–360.

Roberts, R.C. (1966b). The limits to artificial selection for body weight in the mouse. II. The genetic nature of the limits. *Genetical Research*, 9: 73–85.

Roberts, R.C. (1967). The limits to artificial selection for body weight in the mouse. III. Selection from crosses between previously selected lines. *Genetical Research*, 9: 73–85.

Robertson, A. (1960). A theory of limits in artificial selection. *Proceedings of the Royal Society of London*, B. 153: 234–249.

Robertson, F.W. (1966). A test of sexual isolation in *Drosophila. Genetical Research*, 8: 181–187.

Robertson, F.W., & Reeve, E. (1952). Studies in quantitative inheritance. I. Effects of selection of wing and thorax length in *Drosophila melanogaster*. *Journal of Genetics*, 50: 414–448.

Roper, C., Pignatelli, P., & Partridge, L. (1993). Evolutionary effects of selection on age at reproduction in larval and adult *Drosophila melanogaster. Evolution*, 47: 445–455.

Rose, M.R. (1984). Laboratory evolution of postponed senescence in *Drosophila melanogaster. Evolution*, 38: 1004–1010.

Rose, M.R. (1991). *Evolutionary Biology of Aging*. Oxford University Press, Oxford, England.

Rose, M.R., & Charlesworth, B. (1980). A test of evolutionary theories of senescence. *Nature*, 287: 141–142.

Rose, M.R., & Charlesworth, B. (1981). Genetics of life history in *Drosophila melanogaster*. II. Exploratory selection experiments. *Genetics*, 97: 187–196.

Rose, M.R., Dorey, M.L., Coyle, A.M., & Service, P.M. (1984). The morphology of postponed senescence in *Drosophila melanogaster. Canadian Journal of Zoology*, 62: 1576–1580.

Rose, M.R., Vu, L.N., Park, S.U., & Graves, J.L. (1992). Selection on stress resistance increases longevity in *Drosophila melanogaster. Experimental Gerontology*, 27: 241–250.

Rosenberg, S.M., Longerich, S., Gee, P., & Harris, R.S. (1994). Adaptive mutation by deletions in small mononucleotide repeats. *Science*, 265: 405–407.

Roy, K.R. (1961). Interaction between rice varieties. *Journal of Genetics*, 57: 137–152.

Ruano, R.G., Orozco, F., & López-Fanjul, C. (1975). The effect of different selection intensities on selection response in egg-laying of *Tribolium castaneum. Genetical Research*, 25: 17–27.

Saffhill, R., Schneider-Bernloehr, H., Orgel, L.E., & Spiegelman, S. (1970). *In vitro* selection of bacteriophage Q$\beta$ RNA variants resistant to ethidium bromide. *Journal of Molecular Biology*, 51: 531–539.

Saadeh, H.K., Craig, J.V., Smith, L.T., & Wearden, S. (1968). Effectiveness of alternative breeding systems for increasing rate of egg production in chickens. *Poultry Science*, 47: 1057–1072.

Sacristan, M.D. (1982). Resistance responses to *Phoma lingam* of plants regenerated from selected cell and embryogenic cultures of haploid *Brassica napus*. *Theoretical and Applied Genetics*, 61: 193–200.

Saghai-Maroof, M.A., Webster, R.K., & Allard, R.W. (1983). Evolution of resistance to scald, powdery mildew, and net blotch in barley Composite Cross II populations. *Theoretical and Applied Genetics*, 66: 279–283.

Sakai, K.I., & Gotoh, K. (1955). Studies on competition in plants. IV. Competitive ability of $F_1$ hybrids in barley. *Journal of Heredity*, 46: 139–143.

Sanders, D.C. (1981). The Bethlem lines: genetic selection for high and low rearing activity in rats. *Behavior Genetics*, 11: 491–503.

Sang, J.H., & Clayton, G.A. (1957). Selection for larval development time in *Drosophila. Journal of Heredity*, 48: 265–270.

Schaffner, W., Ruegg, K.J., & Weissmann, C. (1977). Nanovariant RNAs: nucleotide sequence and interaction with bacteriophage Q$\beta$ replicase. *Journal of Molecular Biology*, 117: 877–907.

Scharloo, W. (1964). The effect of stabilizing and disruptive selection on a *cubitus interruptus* mutant in *Drosophila*. *Genetics*, 50: 553–562.

Scharloo, W., den Boer, M., & Hoogmoed, M.S. (1967). Disruptive selection on sternopleural chaeta number. *Genetical Research*, 9: 115–118.

Scharloo, W., Hoogmoed, M.S., & Ter Kuile, A. (1967). Stabilizing and disruptive selection on a mutant character in *Drosophila*. I. The phenotypic variance and its components. *Genetics*, 56: 709–726.

Scheiner, S.M. (1993). Genetics and evolution of phenotypic plasticity. *Annual Reviews of Ecology and Systematics*, 24: 35–68.

Scheiner, S.M., & Lyman, R.F. (1991). The genetics of phenotypic plasticity. II. Response to selection. *Journal of Evolutionary Biology*, 4: 23–50.

Scheiring, J.F. (1977). Stabilizing selection for size as related to mating fitness in *Tetraopes*. *Evolution*, 31: 447–449.

Schlichting, C.D., Stephenson, A.G., Davis, L.E., & Winsor, J.A. (1987). Pollen competition and offspring variance. *Evolutionary Trends in Plants*, 1: 35–39.

Schluter, D. (1994). Experimental evidence that competition promotes divergence in adaptive radiation. *Science*, 266: 798–801.

Schmidt, K.P., & Levin, D.A. (1985). The comparative demography of reciprocally sown populations of *Phlox drummondi* Hood. I. Survivorships, fecundities, and finite rates of increase. *Evolution*, 39: 396–404.

Schmitt, J., & Antonovics, J. (1986). Experimental studies of the evolutionary significance of sexual reproduction. IV. Effect of neighbor relatedness and aphid infestation on seedling performance. *Evolution*, 40: 830–836.

Schmitt, J., & Ehrhardt, D.W. (1987). A test of the sib-competition hypothesis for outcrossing advantage in *Impatiens capensis*. *Evolution*, 41: 579–590.

Schmitt, J., & Gamble, S.E. (1990). The effect of distance from the parental site on offspring performance and inbreeding depression in *Impatiens capensis*: a test of the local adaptation hypothesis. *Evolution*, 44: 2022–2030.

Scossiroli, R.E. (1954). Effectiveness of artificial selection under irradiation of plateaued populations of *Drosophila melanogaster*. *Symposium on the Genetics of Population Structure, International Union of Biological Sciences Publications Series B*, 15: 42–66.

Scossiroli, R.E., & Scossiroli, S. (1959). On the relative role of selection and recombination in responses to selection for polygenic traits in irradiated populations of *Drosophila melanogaster*. *International Journal of Radiation Biology*, 1: 61–69.

Scowcroft, W.R. (1968). Variation of scutellar bristles in *Drosophila*. XI. Selection for scutellar microchaetae and the correlated response of scutellar bristles. *Genetical Research*, 11: 125–134.

Seaton, A.P.C., & Antonovics, J. (1967). Population inter-relationships. I. Evolution in mixtures of *Drosophila* mutants. *Heredity*, 22: 19–33.

Semlitsch, R.D., & Wilbur, H.M. (1989). Artificial selection for paedomorphosis in the salamander *Ambystoma talpoideum*. *Evolution*, 43: 105–112.

Sen, B.K., & Robertson, A. (1964). An experimental examination of methods for the simultaneous selection of two characters using *Drosophila melanogaster. Genetics*, 50: 199–209.

Senior, E., Bull, A.T., & Slater, J.H. (1976). Enzyme evolution in a microbial community growing on the herbicide Dalapon. *Nature*, 263: 476–479.

Service, P.M. (1987). Physiological mechanisms of increased stress resistance in *Drosophila melanogaster* selected for postponed senescence. *Physiological Zoology*, 60: 321–326.

Service, P.M. (1993). Laboratory evolution of longevity and reproductive fitness components in male fruit flies: Mating ability. *Evolution*, 47: 387–399.

Service, P.M., Hutchinson, E.W., MacKinley, M.D., & Rose, M.R. (1985). Resistance to environmental stress in *Drosophila melanogaster* selected for postponed senescence. *Physiological Zoology*, 58: 380–389.

Service, P.M., Hutchinson, E.W., & Rose, M.R. (1988). Multiple genetic mechanisms for the evolution of senescence in *Drosophila melanogaster. Evolution*, 42: 708–716.

Shapiro, J.A. (1984). Observations on the formation of clones containing *araB-lacZ* cistron fusions. *Molecular and General Genetics*, 194: 79–90.

Sheldon, B.L. (1963a). Studies in artificial selection on quantitative characters. I. Selection for abdominal bristles in *Drosophila melanogaster. Australian Journal of Biological Science*, 16: 490–515.

Sheldon, B.L. (1963b). Studies in artificial selection on quantitative characters. II. Selection for body weight in *Drosophila melanogaster. Australian Journal of Biological Science*, 16: 516–541.

Sheridan, A.K. (1988). Agreement between estimated and realized genetic parameters. *Animal Breeding Abstracts*, 56: 877–889.

Sheridan, A.K., & Barker, J.S.F. (1974a). Two-trait selection and the genetic correlation. I. Prediction of responses in single-trait and two-trait selection. *Australian Journal of Biological Sciences*, 27: 75–88.

Sheridan, A.K., & Barker, J.S.F. (1974b). Two-trait selection and the genetic correlation. II. Changes in the genetic correlation during two-trait selection. *Australian Journal of Biological Sciences*, 27: 89–101.

Sherwin, R.N. (1975). Selection for mating ability in two chromosomal arrangements of *Drosophila pseudoobscura. Evolution*, 29: 519–530.

Shorter, R., & Frey, K.J. (1979). Relative yields of mixtures and monocultures of oat genotypes. *Crop Science*, 19: 548–553.

Siegel, P.B. (1965). Genetics of behavior: selection for mating ability in chickens. *Genetics*, 52: 1269–1277.

da Silva, J., & Bell, G. (1992). The ecology and genetics of fitness in *Chlamydomonas*. VI. Antagonism between natural selection and sexual selection. *Proceedings of the Royal Society of London Series B*, 249: 227–233.

Simmonds, N.W. (1991). Selection for local adaptation in a plant breeding programme. *Theoretical and Applied Genetics*, 82: 363–367.

Simpson, G.G. (1945). *Tempo and Mode in Evolution*. Columbia University Press, New York.

Slatkin, M. (1985). Somatic mutations as an evolutionary force. In P.J. Greenwood, P.H. Harrey, & M. Slatkin, (eds.), *Evolution: Essays in honour of John Maynard Smith*, pp. 19–30. Cambridge University Press, Cambridge.

Smith, K.P., & Bohren, B.B. (1974). Direct and correlated responses to selection for hatching time in the fowl. *British Poultry Science*, 15: 597–604.

Smith, L.H. (1908). Ten generations of corn breeding. *Illinois Agricultural Experimental Station Bulletin*, 55: 127–180.

Smith, O.D., Kleese, R.A., & Stuthman, D.D. (1970). Competition among oat varieties grown in hill plots. *Crop Science*, 10: 381–384.

Smouse, P., & Kosuda, K. (1977). The effects of genotypic frequency and population density on fitness differentials in *Escherichia coli*. *Genetics*, 64: 399–411.

Snaydon, R.W. (1970). Rapid population differentiation in a mosaic environment. I. The response of *Anthoxanthum odoratum* populations to soils. *Evolution*, 24: 257–269.

Snaydon, R.W., & Davies, M.S. (1972). Rapid population differentiation in a mosaic environment. II. Morphological variation in *Anthoxanthum odoratum*. *Evolution*, 26: 390–405.

Snaydon, R.W., & Davies, M.S. (1976). Rapid population differentiation in a mosaic environment. IV. Populations of *Anthoxanthum odoratum* at sharp boundaries. Heredity, 37: 9–25.

Snaydon, R.W., & Davies, M.S. (1982). Rapid divergence of plant populations in response to recent changes in soil conditions. *Evolution*, 36: 289–297.

Snow, A.A., & Mazer, S.J. (1988). Gametophytic selection in *Raphanus raphinastrum*: A test for heritable variation in pollen competitive ability. *Evolution*, 42: 1065–1075.

Soane, I.D., & Clarke, B. (1973). Evidence for apostatic selection by predators using olfactory cues. *Nature*, 241: 62–63.

Soans, A.B., Pimentel, D., & Soans, J.S. (1974). Evolution of reproductive isolation in allopatric and sympatric populations. *The American Naturalist*, 101: 493–504.

Sobrevila, C. (1988). Effects of distance between pollen donor and pollen recipient on fitness components in *Espeletia schultzii*. *American Journal of Botany*, 75: 701–724.

Sokal, R.R., Bryant, E.H., & Wool, D. (1970). Selection for changes in genetic facilitation: Negative results in *Tribolium* and *Musca*. *Heredity*, 25: 299–306.

Sokal, R.R., & Huber, I. (1963). Competition among genotypes in *Tribolium castaneum* at varying densities and gene frequencies (the *sooty* locus). *The American Naturalist*, 97: 169–184.

Sokal, R.R., & Sullivan, R.L. (1963). Competition between mutant and wild-type house-fly strains at varying densities. *Ecology*, 44: 314–322.

del Solar, E. (1966). Sexual isolation caused by selection for positive and negative phototaxis and geotaxis in *Drosophila pseudoobscura*. *Proceedings of the National Academy of Sciences*, USA, 56: 484–487.

Soliman, M.H. (1982). Directional and stabilizing selection for developmental time and correlated response in reproductive fitness in *Tribolium castaneum*. *Theoretical and Applied Genetics*, 63: 111–116.

Spiegelman, S. (1971). An approach to the experimental analysis of precellular evolution. *Quarterly Reviews of Biophysics*, 4: 213–253.

Spiess, E.B., & Stankevych, A.J. (1973). Mating speed selection and egg chamber correlation in *Drosophila persimilis*. *Egyptian Journal of Genetics and Cytology*, 2: 177–194.

Spiess, E.B., & Wilke, C.M. (1984). Still another attempt to achieve assortative mating by disruptive selection in *Drosophila*. *Evolution*, 38: 505–515.

Spuhler, K.P., Crumpacker, D.W., Williams, J.S., & Bradley, B.P. (1978). Response to selection for mating speed and changes in gene arrangement frequencies in descendants from a single population of *Drosophila pseudoobscura*. *Genetics*, 89: 729–749.

Stanton, M., & Young, H.J. (1994). Selecting for floral character associations in wild radish, *Raphanus sativus* L. *Journal of Evolutionary Biology*, 7: 271–285.

Stearns, S.C. (ed.) (1987). *The Evolution of Sex and its Consequences*. Birkhäuser Verlag, Basel, Switzerland.

Stephenson, A.G., Winsor, J.A., & Davis, L.E. (1986). Effects of pollen load size on fruit maturation and sporophyte quality in zucchini. In D.L. Mulcahy, G.B. Mulcahy, & E. Ottaviano (eds.), *Biotechnology and Ecology of Pollen* (pp. 429–434). Springer–Verlag, New York.

Stevens, L., & Wade, M.J. (1990). Cytoplasmically inherited reproductive incompatibility in *Tribolium* flour beetles: The rate of spread and effect on population size. *Genetics*, 124: 367–372.

Stewart, S.C., & Schoen, D.J. (1987). Pattern of phenotypic viability and fecundity selection in a natural population of *Impatiens pallida*. *Evolution*, 41: 1290–1301.

Stouthamer, R., Luck, R.F., & Hamilton, W.D. (1990). Antibiotics cause parthenogenetic *Trichogramma* (Hymenoptera/Trichogrammatidae) to revert to sex. *Proceedings of the National Academy of Sciences, USA*, 87: 2424–2427.

Sulzbach, D.S. (1980). Selection for competitive ability: Negative results in *Drosophila*. *Evolution*, 34: 431–436.

Sulzbach, D.S., & Emlen, J.M. (1979). Evolution of competitive ability in mixtures of *Drosophila melanogaster*: Populations with an initial asymmetry. *Evolution*, 33: 1138–1149.

Sumper, M., & Luce, R. (1975). Evidence for *de novo* production of self-replicating and environmentally adapted RNA structures by bacteriophage Q$\beta$ replicase. *Proceedings of the National Academy of Sciences, USA*, 72: 162–166.

Suneson, C.A. (1949). Survival of four barley varieties in a mixture. *Agronomy Journal*, 41: 459–461.

Suneson, C.A. (1956). An evolutionary plant breeding method. *Agronomy Journal*, 48: 188–191.

Suneson, C.A., & Stevens, H. (1953). Studies with bulked hybrid populations of barley. *United States Department of Agriculture Technical Bulletin*, 1067: 1–14.

Surface, F.M., & Pearl, R. (1915). Selection in oats. *Maine Agricultural Experimental Station Annual Report*, 1915: 1–40.

Sutherland, T.M., Biondini, P.E., Haverland, L.H., Pettus, D., & Owen, W.B. (1970). Selection for rate of gain, appetite, and efficiency of feed utilization in mice. *Journal of Animal Science*, 31: 1049–1057.

Tantawy, A.O., & Tayel, A.A. (1970). Studies on natural populations of *Drosophila*. X. Effects of disruptive and stabilizing selection on wing length and the correlated response in *Drosophila melanogaster*. *Genetics*, 65: 121–132.

Taylor, C.E., & Condra, C. (1980). $r$- and $K$-selection in *Drosophila pseudoobscura*. *Evolution*, 34: 1183–1193.

Taylor, L.H., & Atkins, R.E. (1954). Effects of natural selection in segregating generations upon bulk populations of barley. *Iowa State College Journal of Science*, 29: 147–162.

Ter-Avanesian, D.V. (1978). The effect of varying the number of pollen grains used in fertilization. *Theoretical and Applied Genetics*, 52: 77–79.

Terzaghi, E., & O'Hara, M. (1991). Microbial plasticity: The relevance to microbial ecology. *Advances in Microbial Ecology*, 11: 431–460.

Thanutong, P., Furusawa, I., & Yamamoto, M. (1983). Resistant tobacco plants from protoplast-derived calluses selected for their resistance to *Pseudomonas* and *Alternaria* toxins. *Theoretical and Applied Genetics*, 66: 209–215.

Thoday, J.M. (1959). Effects of disruptive selection. I. Genetic flexibility. *Heredity*, 14: 406–409.

Thoday, J.M., & Gibson, J.B. (1962). Isolation by disruptive selection. *Nature*, 193: 1164–1166.

Thoday, J.M., & Gibson, J.B. (1970). The probability of isolation by disruptive selection. *The American Naturalist*, 104: 219–230.

Thomas, E., King, P.J., & Potrykus, I. (1979). Improvement of crop plants via single cells *in vitro*: An assessment. *Zeitschrift für Pflanzenzücht*, 82: 1–30.

Thompson, L.W., & Krawiec, S. (1983). Acquisitive evolution of ribitol dehydrogenase in *Klebsiella aerogenes*. *Journal of Bacteriology*, 154: 1027–1031.

Thompson, V. (1977). Recombination and response to selection in *Drosophila melanogaster*. *Genetics*, 85: 125–140.

Thornhill, N.W. (ed.) 1994. *The Natural History of Inbreeding and Outbreeding*. The University of Chicago Press, Chicago, IL.

van Tienderen, P.H. (1992). Variation in a population of *Plantago lanceolata* along a topographical gradient. *Oikos*, 64: 560–572.

van Tienderen, P.H., & van der Toorn, J. (1991). Genetic differentiation between populations of *Plantago lanceolata*. II. Phenotypic selection in a transplant experiment in three contrasting habitats. *Journal of Ecology*, 79: 43–59.

Tilman, D. (1977). Resource competition between planktonic algae: An experimental and theoretical approach. *Ecology*, 58: 338–348.

Tilman, D. (1981). Experimental tests of resource competition theory using four species of Lake Michigan algae. *Ecology*, 62: 802–815.

Tilman, D. (1982). *Resource Competition and Community Structure*. Princeton University Press, Princeton, NJ.

Tilman, D., & Kilham, S. (1976). Phosphate and silicate growth and uptake kinetics of the diatoms *Asterionella formosa* and *Cyclotella meneghiniana* in batch and semicontinuous cultures. *Journal of Phycology*, 12: 375–383.

Tilman, D., Mattson, M., & Langer, S. (1981). Competition and nutrient kinetics along a temperature gradient: An experimental test of resource-based competition theory. *Science*, 192: 463–465.

Tindell, D., & Arze, C.G. (1965). Sexual maturity of male chickens selected for mating ability. *Poultry Science*, 44: 70–72.

Tonsor, S.J. (1989). Relatedness and intraspecific competition in *Plantago lanceolata*. *The American Naturalist*, 134: 897–906.

Toolson, E.C., & Kuper-Simbrón, R. (1989). Laboratory evolution of epicuticular hydrocarbon composition and cuticular permeability in *Drosophila pseudoobscura*: Effects on sexual dimorphism and thermal-acclimation ability. *Evolution*, 43: 468–473.

Torkamenzahi, A., Moran, C., & Nicholas, F.W. (1988). *P*-element-induced mutation and quantitative variation in *Drosophila melanogaster*: Lack of enhanced response to selection in lines derived from dysgenic crosses. *Genetical Research*, 51: 231–238.

Torkamenzahi, A., Moran, C., & Nicholas, F.W. (1992). *P* element transposition contributes substantial new variation for a quantitative trait in *Drosophila melanogaster*. *Genetics*, 131: 73–78.

Travisano, M., Mongold, J.A., Bennett, A.F., & Lenski, R.E. (1995). Experimental tests of the roles of adaptation, chance and history in evolution. *Science*, 267: 87–90.

Trenbath, B.R. (1974). Biomass productivity of mixtures. *Advances in Agronomy*, 26: 177–210.

Trivers, R.L. (1974). Parent-offspring conflict. *American Zoologist*, 14: 249–264.

Turesson, G. (1922). The genotypical response of the plant species to the habitat. *Hereditas*, 3: 211–350.

Turkington, R., & Harper, J.L. (1979). The growth, distribution and neighbour relationships of *Trifolium repens* in a permanent pasture. IV. Fine-scale biotic differentiation. *Journal of Ecology*, 67: 245–254.

Turner, J.R.G. (1979). Genetic control of recombination in the silkworm. I. Multigenetic control of chromosome 2. *Heredity*, 43: 273–293.

van Valen, L.M. (1988). Is somatic selection an evolutionary force? *Evolutionary Theory*, 8: 163–167.

de Varigny, H. (1892). *Experimental Evolution*. Macmillan, London.

Vasi, F., Travisano, M., & Lenski, R.E. (1994). Long-term experimental evolution in *Escherichia coli*. II. Changes in life-history traits during adaptation in a seasonal environment. *The American Naturalist*, 144: 432–456.

Via, S. (1984). The quantitative genetics of polyphagy in an insect herbivore. I. Genotype-environment interaction in larval performance on different host plant species. *Evolution*, 38: 881–895.

de Vilmorin, L.-P. (1840). *Transactions of the Horticultural Society*, 2: 348. [Not seen; described by de Varigny 1892.]

Waddington, C.H. (1953). Genetic assimilation of an acquired character. *Evolution*, 7: 118–126.

Waddington, C.H. (1960). Experiments on canalizing selection. *Genetical Research*, 1: 140–150.

Wade, M.J. (1977). An experimental study of group selection. *Evolution*, 31: 134–153.

Wade, M.J. (1980). Group selection, population growth rate, and competitive ability in the flour beetles, *Tribolium* spp. *Ecology*, 61: 1056–1064.

Wade, M.J. (1982). Group selection: migration and the differentiation of small populations. *Evolution*, 36: 949–961.

Wade, M.J. (1984). Changes in group-related traits that occur when group selection is relaxed. *Evolution*, 38: 1039–1046.

Wade, M.J., & McCauley, D.E. (1984). Group selection: the interaction of local deme size and migration in the differentiation of small populations. *Evolution*, 38: 1047–1058.

Walker, A.K., & Fehr, W.R. (1978). Yield stability of soybean mixtures and multiple pure stands. *Crop Science*, 18: 719–723.

Wallace, B. (1957). The effect of heterozygosity for new mutations on viability in *Drosophila melanogaster*: a preliminary report. *Proceedings of the National Academy of Sciences, USA*, 43: 404–407.

Wallace, B. (1958). The average effect of radiation-induced mutations on viability in *Drosophila melanogaster. Evolution*, 12: 532–556.

Wallace, B. (1959). The role of heterozygosity in *Drosophila* populations. *Proceedings of the 10th International Congress of Genetics*, 1: 408–419.

Wallace, B. (1963a). The elimination of an autosomal lethal from an experimental population of *Drosophila melanogaster*. *The American Naturalist*, 97: 65–66.

Wallace, B. (1963b). Further data on the overdominance of induced mutations. *Genetics*, 48: 633–651.

Wallace, B. (1982). *Drosophila melanogaster* populations selected for resistance to NaCl and $CuSO_4$ in both allopatry and sympatry. *Journal of Heredity*, 73: 35–42.

Wallinga, J.H., & Bakker, H. (1978). Effect of long-term selection for litter size in mice on lifetime reproduction. *Journal of Animal Science*, 46: 1563–1571.

Waser, N.M., & Price, M.V. (1991). Outcrossing distance effects in *Delphinium nelsoni*: Pollen loads, pollen tubes and seed set. *Ecology*, 72: 171–179.

Watson, I.A., & Luig, N.H. (1968). The ecology and genetics of host-pathogen relationships in wheat rusts in Australia. *Proceedings of the Third International Wheat Genetics Symposium*, 227–238.

Weber, K.E. (1990). Increased selection response in larger populations. I. Selection for wing-tip height in *Drosophila melanogaster* at three population sizes. *Genetics*, 125: 579–584.

Weber, K.E., & Diggins, L.T. (1990). Increased selection response in larger populations. II. Selection for ethanol vapor resistance in *Drosophila melanogaster* at two population sizes. *Genetics*, 125: 585–597.

Webster, R.K., Saghai-Maroof, M.A., & Allard, R.W. (1986). Evolutionary response of barley Composite Cross II to *Rhynchosporium secalis* analyzed by pathogenic complexity and by gene-by-race relationships. *Phytopathology*, 76: 661–668.

Weisbrot, D.R. (1966). Genotypic interactions among competing strains and species of *Drosophila*. *Genetics*, 53: 427–435.

Weldon, W.F.R. (1901). A first study of natural selection in *Clausilia laminata* (Montagu). *Biometrika*, 1: 109–124.

Wells, R., & Meredith, W.R. (1984a). Comparative growth of obsolete and modern cotton cultivars. I. Vegetative dry matter partitioning. *Crop Science*, 24: 858–862.

Wells, R., & Meredith, W.R. (1984b). Comparative growth of obsolete and modern cotton cultivars. II. Reproductive dry matter partitioning. *Crop Science*, 24: 863–868.

Wells, R., & Meredith, W.R. (1984c). Comparative growth of obsolete and modern cotton cultivars. III. Relationship of yield to observed growth characteristics. *Crop Science*, 24: 868–872.

Werren, J.H., Nur, U., & Eickbush, D. (1987). An extrachromosomal factor causing loss of paternal chromosomes. *Nature*, 327: 75–76.

Westermann, D.T., & Crothers, S.E. (1977). Plant population effects on the seed yield components of beans. *Crop Science*, 17: 493–496.

Widholm, J.M. (1977). Selection and characterization of amino acid analog resistant plant cell cultures. *Crop Science*, 17: 597–600.

Wilke, C.M., & Adams, J. (1992). Fitness effects of *Ty* transposition in *Saccharomyces cerevisiae*. *Genetics*, 131: 31–42.

Wilke, C.M., Maimer, E., & Adams, J. (1992). The population biology and evolutionary significance of *Ty* elements in *Saccharomyces cerevisiae*. *Genetica*, 86: 155–173.

Wilkinson, G.S., Fowler, K., & Partridge, L. (1990). Resistance of genetic correlation structure to directional selection in *Drosophila melanogaster*. *Evolution*, 44: 1990–2003.

Williams, G.C. (1966). *Adaptation and Natural Selection: A Critique of Some Current Evolutionary Thought*. Princeton University Press, Princeton, New Jersey.

Williams, G.C. (1992). *Natural Selection. Domains, Levels, and Challenges. Oxford Series in Ecology and Evolution*, Vol. 4. Oxford University Press, Oxford, England.

Wills, C. (1984). Structural evolution of yeast alcohol dehydrogenase in the laboratory. In R.P. Mortlock (ed.) *Microorganisms as Model Systems for Studying Evolution*, pp. 233–254. Plenum, New York and London.

Willson, M.F., Thomas, P.A., Hoppes, W.G., Katusic-Malmborg, P.L., Goldman, D.A., & Bothwell, J.L. (1987). Sibling competition in plants: an experimental study. *The American Naturalist*, 129: 304–311.

Wilson, D.S. (1980). *The Natural Selection of Populations and Communities*. Benjamin/Cummings, Menlo Park, CA.

Wilson D.S. (1983). The group selection controversy: history and current status. *Annual Reviews of Ecology and Systematics*, 14: 159–187.

Wilson, S.P. (1974). An experimental comparison of individual, family and combination selection. *Genetics*, 76: 823–836.

Wilson, S.P., Goodale, H.P., Kyle, W.H., & Godfrey, E.F. (1971). Long-term selection for body weight in mice. *Journal of Heredity*, 62: 228–234.

Winter, F.L. (1929). The mean and variability as affected by continuous selection for composition in corn. *Journal of Agronomical Research*, 39: 451–476.

Wolfe, M.S. (1985). The current status and prospects of multiline cultivars and variety mixtures for disease resistance. *Annual Reviews of Phytopathology*, 23: 251–273.

Wolfe, M.S. (1987). Trying to understand and control powdery mildew. In M.S. Wolfe & C.E. Caten (eds.), *Populations of Plant Pathogens*, pp. 253–273. Blackwell, Oxford, England.

Wolfe, M.S., & Barrett, J.A. (1977). Population genetics of powdery mildew epidemics. *Annals of the New York Academy of Sciences*, 287: 151–163.

Wolfe, M.S., & Barrett, J.A. (1980). Can we lead the pathogen astray? *Plant Disease*, 64: 148–155.

Wolfe, M.S., & Barrett, J.A. (1981). The agricultural value of variety mixtures. *Barley Genetics*, IV: 435–440.

Wooding, F.J. (1981). Performance of three barley cultivars grown in different cropping sequences in central Alaska. *Barley Genetics*, IV: 147–152.

Woodruff, L.L. (1911). Two thousand generations of *Paramecium. Archiv Für Protistenkunde*, 21: 263–266.

Woodruff, L.L. (1926). Eleven thousand generations of *Paramecium. Quarterly Review of Biology*, 1: 436–438.

Woodworth, C.M., & Jugenheimer, R.W. (1948). Breeding and genetics of high-protein corn. *Annual Report of the American Seed Trade Association (Hybrid Corn Division)*, 3: 75–83.

Woodworth, C.M., Leng, E.R., & Jugenheimer, R.W. (1952). Fifty generations of selection for protein and oil in corn. *Agronomy Journal*, 44: 60–65.

Wright, S. (1945). Tempo and mode in evolution: A critical review. *Ecology*, 26: 415–419.

Wu, T.T. (1976). Growth of a mutant of *Escherichia coli* K-12 on xylitol by recruiting enzymes for D-xylose and, L-1,2- propanediol metabolism. *Biochimica et Biophysica Acta*, 428: 656–663.

Wu, T.T., Lin, E.C.C., & Tanaka, S. (1968). Mutants of *Aerobacter aerogenes* capable of utilizing xylitol as a novel carbon source. *Journal of Bacteriology*, 96: 447–456.

Wynne-Edwards, V.C. (1962). *Animal Dispersion in Relation to Social Behaviour*. Oliver & Boyd, Edinburgh, Scotland.

Yamada, Y., Bohren, B.B., & Crittenden, L.B. (1958). Genetic analysis of a White Leghorn closed flock apparently plateaued for egg production. *Poultry Science*, 37: 565–580.

Yamazaki, T. (1971). Measurement of fitness at the Esterase-5 locus in *Drosophila pseudoobscura*. *Genetics*, 67: 579–603.

Yamazaki, T., Kasekabe, S., Tachida, H., Matsuda, M., & Mukai, T. (1980). Re-examination of diversifying selection of polymorphic allozyme genes by using population cages in *Drosophila melanogaster*. *Proceedings of the National Academy of Sciences, USA*, 80: 5789–5792.

Yoo, B.H. (1980a). Long-term selection for a quantitative character in large replicate populations of *Drosophila melanogaster*. I. Response to selection. *Genetical Research*, 35: 1–17.

Yoo, B.H. (1980b). Long-term selection for a quantitative character in large replicate populations of *Drosophila melanogaster*. II. Lethals and visible mutants with large effects. *Genetical Research*, 35: 19–31.

Yoo, B.H. (1980c). Long-term selection for a quantitative character in large replicate populations of *Drosophila melanogaster*. Part 3. The nature of residual genetic variability. *Theoretical and Applied Genetics*, 57: 25–32.

Zambrano, M.M., Siegele, D.A., Almirón, M., Tormo, A., & Kolter, R. (1993). Microbial competition: *Escherichia coli* mutants that take over stationary phase cultures. *Science*, 259: 1757–1760.

Zamerhof, S., & Eichhorn, H. (1967). Study of microbial evolution through the loss of biosynthetic functions. Establishment of 'defective' mutants. *Nature*, 216: 455–458.

Zeyl, C., Bell, G., & da Silva, J. (1994). Transposon abundance in sexual and asexual populations of *Chlamydomonas reinhardtii*. Evolution 48: 1406–1409.

Zeleny, C. (1921). Decrease in sexual dimorphism of bar-eye *Drosophila* during the course of selection for low and high facet number. *The American Naturalist*, 55: 404–411.

Zeleny, C. (1922). The effect of selection for eye facet number in the white bar-eye race of *Drosophila melanogaster*. *Genetics*, 7: 1–115.

Zeleny, C., & Mattoon, E.W. (1915). The effect of selection upon the 'bar-eye' mutant of *Drosophila*. *Journal of Experimental Zoology*, 19: 515–529.

Zevenboom, W., van der Does, J., Bruning, K., & Mur, L.B. (1981). A non-heterocystous mutant of *Aphanizomenon flos-aquae*, selected by competition in light-limited continuous culture. *FEBS Microbiology Letters*, 10: 11–16.

# *Index of Organisms**



*Taxonomy is carried to generic level only, except for *Drosophila*. Reference is to first page in text or box where organism is mentioned in a given section.

# General Index*



*Citations are to first page of text where topic occurs in given section; boxes are not indexed.